Biochemistry and Molecular Biology Compendium

Roger L. Lundblad

CRC Press
Taylor & Francis Group
Boca Raton   London   New York

CRC Press is an imprint of the
Taylor & Francis Group, an informa business

CRC Press
Taylor & Francis Group
6000 Broken Sound Parkway NW, Suite 300
Boca Raton, FL 33487-2742

© 2007 by Taylor & Francis Group, LLC
CRC Press is an imprint of Taylor & Francis Group, an Informa business

No claim to original U.S. Government works
Printed in the United States of America on acid-free paper
10 9 8 7 6 5 4 3 2 1

International Standard Book Number-10: 1-4200-4347-1 (Hardcover)
International Standard Book Number-13: 978-1-4200-4347-1 (Hardcover)

Library of Congress Cataloging-in-Publication Data

Lundblad, Roger L.
　Biochemistry and molecular biology compendium / Roger L. Lundblad.
　　　p. ; cm.
　Includes bibliographical references and index.
　ISBN-13: 978-1-4200-4347-1 (hardcover : alk. paper)
　ISBN-10: 1-4200-4347-1 (hardcover : alk. paper)
　1. Biochemistry--Handbooks, manuals, etc. 2. Molecular biology--Handbooks,
manuals, etc. I. Title.　[DNLM: 1. Biochemistry--Terminology--English. 2. Molecular
Biology--Terminology--English.　QU 15 L962b 2007] I. Title.

QP514.2.P73 1989 Suppl.
612í015--dc22　　　　　　　　　　　　　　　　　　　　　　　　　　　　　2006102979

Visit the Taylor & Francis Web site at
http://www.taylorandfrancis.com

and the CRC Press Web site at
http://www.crcpress.com

Contents

Preface

This book evolved out of the process of revising *The Practical Handbook of Biochemistry and Molecular Biology*, which was edited by the late Gerald Fasman. I had come to several conclusions: (1) I no longer understood the titles of articles in journals because such titles were increasingly written in tongues only understood by selected tribes of investigators; (2) I had forgotten most of the organic chemistry passed on to me by distinguished individuals including Chuck Anderson, Stan Moore, and Bill Stein; (3) many investigators now worked with kits of stuff and had little knowledge of the stuff; and (4) I was not alone with respect to (1) and (2). The enclosed material has been assembled to supplement *The Practical Handbook of Biochemistry and Molecular Biology*.

The content is biased toward my own particular interests and I would appreciate receiving comment regarding this specific issue. While I spent considerable time reading journals such as *The Journal of Biological Chemistry, Biochemistry, The Journal of Molecular Biology*, and *Nucleic Acids Research* from cover to cover and selecting terms and acronyms that I did not readily recognize for inclusion, I do recognize that the selected content will seem weak or incomplete in certain areas. In particular, I would appreciate guidance on acronyms: The list of accepted abbreviations appears to be an item of the past and authors are allowed to indulge their individual creativity in creating new and novel acronyms that, in turn, lead to interesting search results when using Internet search engines. The same holds for the invention of new terms to describe old phenomena; in general, biomedical investigators are not very good at brand naming.

Finally, I urge you to visit your local library. The amount of material that you can get from sitting in front of your computer is limited with respect to what you get with focused searches guided by an experienced reference librarian. It is not unlike the situation in the late Douglas Adams's *The Hitchhiker's Guide to the Galaxy*, where an answer is meaningless unless you thoroughly understand the question.

Acknowledgments

I want to first acknowledge my debt to those distinguished educators who have valiantly tried to provide me with insight into chemistry and biochemistry. They include Professor Charles D. Anderson of Pacific Lutheran University, Tacoma, Washington, Professor Earl W. Davie of the University of Washington, Seattle, and Professors Stanford Moore and William Stein of the Rockefeller Institute, New York.

Professor Charles Craik of the University of California at San Francisco and Professor Nicholas Price of the University of Glasgow, Scotland, UK, have provided invaluable support during the preparation of this book. I owe a special debt to Professor Bryce Plapp of the University of Iowa for demonstrating incredible patience in again working with the thermodynamically challenged. Danielle Jacobs of the Department of Chemistry at the University of North Carolina at Chapel Hill provided guidance in the preparation of the material on name reactions in organic chemistry.

Last, but not least, I would like to thank Dr. Judith Spiegel of Taylor & Francis for her continual support and to thank Helena Redshaw for her patience in working with the material provided in the preparation of the book.

1 Abbreviations and Acronyms

A	Absorbance
A23187	A calcium ionophore, Calcimycin
AAA	Abdominal aortic aneurysm; AAA+. ATPases associated with various cellular activities
AAAA	Association Against Acronym Abuse
AAG box	An upstream *cis*-element
AAS	Aminoalkylsilane; atomic absorption spectroscopy
AAT	Amino acid transporter; alpha-1-antitrypsin
AAV	Adenoassociated virus
ABA	Abscisic acid, a plant hormone
ABC	ATP-binding cassette; antigen-binding cell
ABC-Transporter Proteins	ATP-binding cassette transporter proteins
ABE	Acetone butanol ethanol
Abl	Retroviral oncogene derived from Abelson murine leukemia
ABRC	ABA response complex
ABRE	ABA response element
7-ACA	7-aminocephalosporanic acid
ACES	2-[(2-amino-2-oxyethyl)amino]-ethanesulfonic acid
Ach (AcCho)	Acetylcholine
AChR (AcChoR)	Acetylcholine receptor
ACME	Arginine catabolic mobile element
Acrylodan	6-acryloyl-2-(dimethylamino)-naphthalene
ACS	Active sequence collection
ACSF	Artificial cerebrospinal fluid
ACTH	Adrenocorticotropin
ADA	Adenosine deaminase; antidrug antibody
ADAM	A disintegrin and metalloproteinase
ADAMTS	A subfamily of disintegrin and metalloproteinase with thrombospondin motifs
ADCC	Antibody-dependent cell-mediated cytotoxicity as in NK cells attacking antibody-coated cells
ADH	Alcohol dehydrogenase; antidiuretic hormone
ADME	Adsorption, distribution, metabolism, excretion
ADME-Tox	ADME-Toxicology
AdoMet	*S*-adenosyl-L-methionine
AEC	Alveolar epithelial cell
AFLP	Amplified fragment-length polymorphism
AFM	Atomic force microscopy
AGE	Advanced glycation endproducts
AGO	Argonaute protein family
AGP	Acid glycoprotein
AID	Activation-induced cytodine deaminase

AKAP	A kinase anchoring proteins
Akt	A protein kinase
Akt	A retroviral oncogene derived from AKT8 murine T-cell lymphoma
Alk	Anaplastic lymphoma kinase; receptor member of insulin superfamily
ALL	Acute lymphocytic leukemia
ALP	Alkaline phosphatase
ALS	Anti-lymphocyte serum
ALT	Alanine aminotransferase
ALV	Avian leukosis virus
AML	Acute myeloid leukemia
AMPK	AMP-activated protein kinase
AMS	Accelerator mass spectrometry
AMT	Accurate mass tag
ANDA	Abbreviated new drug application
ANOVA	Analysis of variables (factorial analysis of variables)
ANS	1-anilino-8-napthlenesulfonate; autonomic nervous system
ANTH	AP180 N-terminal homology, as in ANTH-domain
2-AP	2-aminopyridine
6-APA	6-aminopenicillanic acid
APAF1	Apoptotic protease activating factor 1
Apg1	A serine/threonine protein kinase required for vesicle formation, which is essential for autophagy
APL	Acute promyelocytic leukemia
ApoB	Apolipoprotein B
AQP	Adenosine tetraphosphate
ARAP3	A dual Arf and Rho GTPase-activating protein
ARD	Acute respiratory disease; acireductone dioxygenase; automatic relevance determination; acid rock drainage
ARE	AU-rich elements
ARF	ADP-ribosylation factor
ARL	Arflike
ARM	Arginine-rich motif
ARS	Automatic replicating sequence or autonomously replicating sequence
ART	Mono-ADP-ribosyltransferase; family of proteins, large group of A-B toxins
AS	Antisense
ASD	Alternative splicing database: http://www.ebi.ac.uk/asd
ASPP	Ankyrin-repeat, SH3-domain, and proline-rich region containing proteins
AST	Aspartate aminotransferase
ATC	Aspartate transcarbamylase domain
ATCase	Aspartate transcarbamylase
ATP	Adenosine-5′-triphosphate
ATPγS	Adenosine-5′-3-O-(thiotriphosphate)
ATR-FTIR	Attenuated total reflectance-Fourier transform infrared
ATR-IR	Attenuated total reflection-infrared
AVT	Arginine vasotocin
Axl	Anexceleko; used in reference to a receptor kinase related to the Tyro 3 family
BA	Betaine aldehyde
BAC	Bacterial artificial chromosome; blood alcohol concentration
BAD	Member of the Bc102 protein family — considered to be a proapoptotic factor
BADH	Betaine aldehyde dehydrogenase
BAEC	Bovine aortic endothelial cells

BAEE	Benzoyl-arginine ethyl ester
BALT	Bronchial-associated lymph tissue
BBB	Blood–brain barrier
B-CAM	Basal cell adhesion molecule
BCG	Bacille–Calmette–Guérin
BCIP	5-bromo, 4-chloro, 3-indoyl phosphate
Bcl-2	Protein family regulating apoptosis
BCR	Breakpoint cluster region; B-cell receptor
BCRA-1	Breast cancer 1; a tumor suppressor gene associated with breast cancer
BCR-ABL	Fused gene that results from the *Philadelphia chromosome*; the BCR-ABL gene produces Bcr-Abl tyrosine kinase
BCS	Biopharmaceutical classification system for describing the gastrointestinal absorption of drugs; also Budd–Chiari syndrome
BDH	D-β-butyrate dehydrogenase
BDNF	Brain-derived growth factor
BEBO	An unsymmetrical cyanine dye for binding to the minor groove of DNA; 4-[(3-methyl-6-(6-methyl-benzothiazol-2-yl)-2,3,-dihydro(benzo-1,3-thiazole)-2-methylidene)]-1-methyl-pyridinium iodide
BET	An isotherm for adsorption phenomena in chromatography; acronym derived from Stephen Brunauer, Paul Emmet, and Edward Teller
B/F	Bound/free
bFGF	Basic fibroblast growth factor
BFP	Blue fluorescent protein
BGE	Background electrolyte
Bicine	*N,N*-bis(2-hydroxyethyl)glycine
BiFC	Bimolecular fluorescence complementation
BIND	Biomolecular interaction network database
BiP	Immunoglobulin heavy chain-binding protein
Bis-Tris	2,2-bis-(hydroxymethyl)-2,2′,2″ nitriloethanol
BLA	Biologic license application
BLAST	Basic local alignment search tool
BME	2-mercaptoethanol; β-mercaptoethanol
BMP	Bone morphogenic protein
BopA	Secreted protein required for biofilm formation
BPTI	Bovine pancreatic trypsin inhibitor
BrdU	Bromodeoxyuridine
BRE-luc	A mouse embryonic stem cell line used to study bone morphogenetic protein
BRET	Bioluminescence resonance energy transfer; see FRET
Brig	Polyoxyethylene lauryl ether
BSA	Bovine serum albumin
bZIP	Basic leucine zipper transcription factor
C1INH	C1 inhibitor; inhibitor of activated complement component 1, missing in hereditary angioneurotic edema
CA125	Cancer antigen 125; a glycoprotein marker used for prognosis in ovarian cancer; also referred to as MUC16
CAD	Multifunctional protein which initiates and regulates *de novo* pyrimidine biosynthesis; caspases-activated DNAse
CAK	Cdk-activating kinase
CALM	Clathrin assembly lymphoid myeloid leukemia, as in CALM gene
CAM (CaM)	Calmodulin; cell adhesion molecule
CAMK	Ca^{2+}/calmodulin-dependent protein kinase

CaMK	Calmodulin kinase, isoforms I, II, III
Can	Acetonitrile
CAN	Bacterial cell wall collagen-binding protein
CAPS	Cleavable amplified polymorphic sequences; cationic antimicrobial peptide
CArG	A promoter element [CC(A/T)$_6$G] gene for smooth muscle α-actin
CASP	Critical assessment of structural prediction
CASPASE	Cysteine-dependent aspartate-specific protease
CAT	Catalase; chloramphenicol acetyl transferase
CATH	Class, architecture, topology, homologous superfamily; hierarchical classification of protein domain structure
cATP	Chloramphenicol resistance gene; caged ATP; cation transporting P-type
Cbl	A signal transducing protein downstream of a number of receptors coupled to tyrosine kinases; a product of the *c-cbl* proto-oncogene
Cbs	Chromosomal breakage sequence
CBz	Carbobenzoxy
CCC	Concordance correlation coefficient
CCD	Charge couple device
CCK	Choleocystokinin
CCV	Clathrin-coated vesicles
CD	Clusters of differentiation; circular dichroism; cyclodextrin
CDC	Complement-dependent cytotoxicity; complement-mediated cell death
CDK (cDK)	Cyclin-dependent kinase
cDNA	Complementary DNA
Cdpk4	Ca^{2-} protein kinase
CDR	Complementary determining region
CDTA	1,2-cyclohexylenedinitriloacetic acid
CE	Capillary electrophoresis
CEC	Capillary electrochromatography
CELISA	Cellular enzyme-linked immunosorbent assay; enzyme-linked immunosorbent assay on live cells
CEPH	Centre d'Etude du Polymorphisme Humain
CERT	Ceramide transport protein
CE-SDS	Capillary electrophoresis in the presence of sodium dodecyl sulfate
CEX	Cation exchange
CFA	Complete Freund's adjuvant
CFP	Cyan fluorescent protein
CFTR	Cystic fibrosis transmembrane conductance region
Cfu	Colony-forming unit
CGE	Capillary gel electrophoresis
CGH	Comparative genome hybridization
CGN	*cis*-Golgi network
CH	Calponin homology
CHAPS	3-[(3-cholamidopropyl)dimethylammonio]-1-propanesulfonic acid
CHCA	α-cyano-4-hydroxycinnamic acid
CHEF	Chelation-enhanced fluorescence
CHES	2-(*N*-cyclohexylamino)ethanesulfonic acid
ChiP	Chromatin immunoprecipitation
CHO	Chinese hamster ovary; carbohydrate
CID	Collision-induced dissociation; collision-induced dimerization
CIDEP	Chemically induced dynamic electron polarization
CIDNP	Chemically induced dynamic nuclear polarization

CIEEL	Chemically initiated electron exchange luminescence
CLIP	Class II-associated invariant chain (Ii) peptide
CLT	Clotvinazole [1-(α2-chlorotrityl)imidazole]
CLUSTALW	A general purpose program for structural alignment of proteins and nucleic acids: http://www.ebi.ac.uk/clustalw/
cM	Centimorgan
CM	Carboxymethyl
CMCA	Competitive metal capture analysis
CML	Chronic myelogenous leukemia; carboxymethyl lysine
Cn	Calcineurin
CNC	Cap'n'Collar family of basic leucine zipper proteins
CNE	Conserved noncoding elements
CoA	Coenzyme A
COACH	Comparison of alignments by constructing hidden Markov models
COFFEE	Consistency-based objective function for alignment evaluation
COFRADIC	Combined fractional diagonal chromatography
COG	Conserved oligomeric Golgi; cluster of orthologous groups
COPD	Chronic obstructive pulmonary disease
COX	Cytochrome C oxidase
Cp	Ceruloplasmin
CPA	Carboxypeptidase A
CPB	Carboxypeptidase B
CPD	Cyclobutane pyrimidine dimer
CPDK	Calcium-dependent protein kinase
CpG	Cytosine-phosphate-guanine
CpG-C	Cytosine-phosphate-guanine class C
CPP	Cell-penetrating peptide; combinatorial protein pattern
CPSase	Carbamoyl-phosphate synthetase
CPY	Carboxypeptidase Y
CRAC	Calcium release-activated calcium (channels)
CRE	Cyclic AMP response element
CREA	Creatinine
CREB	cAMP response element-binding protein
Cre1	Cytokine response 1; a membrane kinase
CRM	Certified reference material
CRP	C-reactive protein; also cAMP receptor protein
CRY	Chaperone
CS	Chondroitin sulfate
CSF	Colony-stimulating factor
CSP	Cold-shock protein
CSR	Cluster-situated regulator; class-switch recombination
CSSL	Chromosome segment substitution lines
Cst3	Cystatin 3
Ct	Chloroplast
CT	Charge transfer
CTB	Cholera toxin B subunit
CTD	C-terminal domain
CTL	Cytotoxic T lymphocytes
CTLA	Cytotoxic T lymphocyte-associated antigen
CTLL	Cytotoxic T-cell lines
CTPSase	CTP synthetase

CtrA	A master regulator of cell cycle progression
CV	Coefficient of variation
Cvt	Cytosome to vacuole targeting
CW	Continuous wave (nonpulsed source of electromagnetic radiation)
CYP	Cytochrome P450 enzyme
CZE	Capillary zone electrophoresis
2D-DIGE	Two-dimensional difference gel electrophoresis
2DE	Two-dimensional electrophoresis
D	Diffusion
D$_{ax}$	Axial dispersion coefficient
DAB(p-dab)	p-dimethyl amino azo benzene
dABs	Domain antibodies
DABSYL	N,N-dimethylaminoazobenzene-4′-sulfonyl — usually as the chloride, DABSYL chloride
DAD	Diaphanous-autoregulatory domain
DAF	Decay accelerating factor
DAG	Diacyl glycerol
DALI	Distance matrix alignment: http://www.ebi.ac.uk/dali/
DANSYL	5-dimethylaminonapthalene-1-sulfonyl; usually as the chloride, DANSYL chloride
DAP	DNAX-activation protein; diaminopimelic acid
DAP12	DNAX-activating protein of 12kDa mass
DAS	Distributed annotated system; downstream activation site
DBD-PyNCS	4-(3-isocyanatopyrrolidin-1-yl)-7-(N,N-dimethylaminosulfonyl)-2-benzoxadiazole
DBTC	"Stains All"; 4,5,4′,5′-dibenzo-3,3′-diethyl-9-methylthiacarbocyanine bromide
DC	Dendritic cell
DCC	Dicyclohexylcarbodimide
DCCD	$N,N′$-dicyclohexylcarbodimide
dCNE	Duplicated CNE
DDBJ	DNA Data Bank of Japan: http://www.ddbj.nig.ac.jp
DDR1	Discoidin domain receptor1, CAK, CD167a, PTK3, Mck10
DDR2	Discoidin domain receptor2, NTRK3, TKT, Tyro10
DDRs	Discoidin domain receptors (DDR1, DDR2)
DEAE	Diethylaminoethyl
DEG	Differentially expressed gene(s)
DEX	Dendritic cell-derived exosomes
DFF	DNA fragmentation factor
DFP	Diisopropylfluorophosphate; diisopropylphosphorofluoridate
DHFR	Dihydrofolate reductase
DHO	Dihydroorotase domain
DHOase	Dihydroorotase
DHPLC (dHPLC)	Denaturing HPLC
DHS	DNase I hypersensitivity site
DIP	Database of interacting proteins: http://dip.doe-mbi.ucla.edu; also dictionary of interfaces in proteins: http://drug-redesign.de/superposition.html
Dipso	3-[N,N-bis(2-hydroxyethyl)amino]-2-hydroxypropanesulfonic acid
DLS	Dynamic light scattering
DM	An accessory protein located in the lysosome associated with MHC class II antigen presentation; located in the endosomal/lysosomal system of APC

DMBA	7,12-dimethylbenz[α]anthracene
DMD	Duchenne muscular dystrophy; also Doctor of Dental Medicine
DMEM	Dulbecco's Modified Eagle's Medium
DMF	Dimethylformamide; decayed, missing, filled (in dentistry)
DMS	Dimethyl sulfate
DMSO	Dimethyl sulfoxide
DMT1	Divalent metal transporter 1
ssDNA	Single-stranded DNA
DNAa	A bacterial replication initiation factor
DNAX	DNAase III, tau and gamma subunits
dNPT	Deoxynucleoside triphosphate
DO	An accessory protein located in the lysosome associated with MHC class II antigen presentation; DO has an accessory role to DM
DOTA	Tetraazacyclodecanetetraacetic acid
DPE	Downstream promoter element
DPI	Dual polarization interferometry
DPM	Disintegrations per minute
DPN	Diphosphopyridine dinucleotide (currently NAD)
DPPC	Dipalmitoylphosphatidylcholine
DPPE	1,2-dipalmitoyl-*sn*-glycerol-3-phosphoethanolamine
DPTA	Diethylenetriaminepentaacetic acid
DRE	Dehydration response element; dioxin response element
DRT	Dimensionless retention time (a value for chromatography)
DSC	Differential scanning calorimetry
dsDNA	Double-stranded DNA
DSP	Downstream processing
dsRBD	Double-stranded RNA binding domain
dsRNA	Double-stranded RNA
DTAF	Dichlorotriazinyl aminofluorescein
DTE	Dithioerythritol
DTNB	5,5′-dithio-bis(2-nitrobenzoic acid) Ellman's Reagent
DTT	Dithiothreitol
DUP	A duplicated yeast gene family
DVDF	Polyvinyl difluoride
E1	Ubiquitin-activating enzyme
E2	Ubiquitin carrier protein
E3	Ubiquitin-protein isopeptide ligase
E-64	*Trans*-epoxysuccinyl-L-leucylamino-(4-guanidino)-butane, proteolytic enzyme inhibitor
EAA	Excitatory amino acid
EBA	Expanded bead adsorption
EBV	Epstein-Barr virus
ECF	Extracytoplasmic factor; extracellular fluid
ECM	Extracellular matrix
EDC	1-ethyl-(3-dimethylaminopropyl)-carbodiimide
EDC (EADC)	1-ethyl-3-(3-dimethylaminopropyl) carbodiimide; *N*-ethyl-*N*′-(3-dimethyl-aminopropyl) carbodiimide
EDI	Electrodeionization
EDTA	Ethylenediaminetetraacetic acid, Versene, (ethylenedinitrilo)tetraacetic acid
EEO	Electroendoosmosis
EEOF	Electroendoosmotic flow

EF	Electrofiltration
EGF	Epidermal growth factor
EGFR	Epidermal growth factor receptor; Erb-1; HER1
EGTA	Ethyleneglycol-bis(β-aminoethylether)-N,N,N',N'-tetraacetic acid
eIF	Eukaryotic initiation factor
EK	Electrokinetic
EKLF	Erythroid Krüppel-like factor
ELISA	Enzyme-linked immunosorbent assay
EMBL	European Molecular Biology Laboratory
EMCV	Encephalomyocarditis virus
EMF	Electromotive force
EMMA	Enhanced mismatch mutation analysis
EMSA	Electrophoretic mobility shift assay
ENaC	Epithelial Na channel
EndoG	Endonuclease G
ENTH	Epsin N-terminal homology as ENTH-domain
ENU	N-ethyl-N-nitrosourea
EO	Ethylene oxide
EOF	Electroosmotic flow
Eph	A family of receptor tyrosine kinases; function as receptors/ligands for ephrins
EPL	Expressed protein ligation
Epps	4-(2-hydroxyethyl)-1-piperazinepropanesulfonic acid
EPR	Electron paramagnetic resonance
ER	Endoplasmic reticulum
ERAD	Endoplasmic reticulum-associated protein degradation
ErbB2	Epidermal growth factor receptor, HER2
ErbB3	Epidermal growth factor receptor, HER3
ErbB4	Epidermal growth factor receptor, HER4
ERK	Extracellular-regulated kinase
Erk ∫	P 42/44 extracellular signal-regulated kinase
Ero1p	A thiol oxidase that generates disulfide bonds inside the endoplasmic reticulum
ERSE	Endoplasmic reticulum (ER) stress-response element
ES	Embryonic stem, as in embryonic stem cell
ESI	Electrospray ionization
ESR	Electron spin resonance; erthyrocyte sedimentation rate
ESS	Exonic splicing silencer
EST	Expressed sequence tag
ETAAS	Electrothermal atomic absorption
5,6-ETE	5,6-epoxyeicosatrienoic acid
ETS	Family of transcription factors
EUROFAN	European Functional Analysis Network: http://mips.gsf.de/proj/eurofan/; European Programme for the Study and Prevention of Violence in Sport
Exo1	Exonuclease 1
EXP1	Expansion gene
FAAH	Fatty acid amide hydrolase
Fab	Antigen-binding fragment from immunoglobulin
FAB	Fast atom bombardment
FAB-MS/MS	Fast atom bombardment-mass spectrometry/mass spectrometry
FACE	Fluorophore-assisted carbohydrate electrophoresis
FACS	Fluorescence-activated cell sorting

FAD	Flavin adeninine dinucleotide
FADD	Fas-association death domain
FAK	Focal adhesion kinase
FBS	Fetal bovine serum
Fc	Fc region of an immunoglobulin representing the C-terminal region
FCCP	Carbonyl cyanide *p*-trifluoromethoxyphenyl-hydrazine
FcγR	Cell surface receptor for the Fc domain of IgG
FDA	Fluorescein diacetate
FDC	Follicular dendritic cells
FEAU	2′-fluoro-2′-deoxy-β-D-arabinofuranosyl-5-ethyluracil
FecA	Ferric citrate transporter
FEN	Flap endonuclease
FERM	FERM-domain (four-point-one; ezrin, radixin, moesin)
Fes	Retroviral oncogene derived from ST and GA feline sarcoma
FFAT	Two phenylalanyl residues in an acidic tract
FFPE	Formalin-fixed, paraffin-embedded
FGF	Fibroblast growth factor
FGFR	Fibroblast growth factor receptor
Fgr	Retroviral oncogene derived from GR feline sarcoma
FIAU	2′-fluoro-2′-deoxy-β-D-arabinofuranosyl-5-iodouracil
FIGE	Field-inversion gel electrophoresis
FITC	Fluoroscein isothiocyanate
FLAG™	An epitope "tag" that can be used as a fusion partner for recombinant protein expression and purification
FlhB	A component of the flagellum-specific export apparatus in bacteria
FLIP	Fluorescence loss in photobleaching
FLK-1	Vascular endothelial growth factor receptor (VEGFR)
FLT-1	Vascular endothelial growth factor receptor (VEGFR)
fMLP(FMLP)	*N*-formyl methionine leucine phenylalanine
fMOC	9-fluorenzylmethyloxycarbonyl
Fms	Retroviral oncogene derived from SM feline sarcoma
Fok1	A type IIS restriction endonuclease derived from *Flavobacterium okeanokoites*
Fos	Retroviral oncogene derived from FBJ murine osteosarcoma
FOX	Forkhead box
FpA	Fibrinopeptide A
FPC	Fingerprinted contigs
Fps	Retroviral oncogene from Fujiami avian sarcoma
FRAP	Fluorescence recovery after photobleaching
FRET	Fluorescence resonance energy transfer; Förster resonance energy transfer
FSSP	Fold classification based on structure alignment of proteins: http://www.ebi. ac.uk/dali/fssp/fssp.html
FT	Fourier transform
FTIR	Fourier transform infrared reflection
FTIR-ATR	Fourier transform infrared reflection-attenuated total reflection
FU	Fluorescence unit
5-Fu	5-fluorouracil
Fur	Ferric uptake receptor
Fur	Gene for fur
FYVE	Zinc-binding motif; acronym derived from four proteins containing this domain
G	Guanine

Gα	Heterotrimeric G protein, α-subunit
Gβ	Heterotrimeric G protein, β-subunit
Gγ	Heterotrimeric G protein, γ-subunit
G-6-PD	Glucose-6-phosphate dehydrogenase
GABA	Gamma (γ)-aminobutyric acid
GAG	Glycosaminoglycan
GalNac	*N*-acetylgalactosamine
GALT	Gut-associated lymphoid tissues
GAPDH	Glyceraldehyde 3-phosphate dehydrogenase
GAPS	GTPase-activating proteins
GAS6	A protein, member of the vitamin K-dependent protein family
GASP	Genome Annotation Assessment Project: http://www.fruitfly.org/GASP1/; also growth advantage in stationary phase
GBD	GTPase-binding domain
GC	Gas chromatography; granular compartment
GC-MS	Gas chromatography-mass spectroscopy
GC-MSD	Gas chromatography-mass selective detector
GCP	Good clinical practice
GcrA	A master regulator of cell cycle progression
G-CSF	Granulocyte colony-stimulating factor
GDH	Glutamate dehydrogenase
GDNF	Glial-derived neurotrophic factor
GdnHCl	Guanidine hydrochloride
GEFs	Guanine nucleotide exchange factors
GF-AAS	Graphite furnace atomic absorption spectroscopy
GFP	Green fluorescent protein
GGDEF	A protein family
GGT	Gamma-glutamyl transferase
GGTC	German Gene Trap Consortium: a reference library of gene trap sequence tags (GTST), http://www.genetrap.de/
GHG	Greenhouse gas
GI	Gastrointestinal; genomic islands
cGK	Cyclic GMP (cGMP)-dependent protein kinase
GlcNac	*N*-acetylglucosamine
GLD	Gelsolinlike domain
GLP	Good laboratory practice(s)
GlpD	Glyceraldehyde-3-phosphate dehydrogenase
GLUT	A protein family involved in transporting hexoses into mammalian tissues
Glut4	Facilitative glucose transporter, which is insulin-sensitive
Glut5	A fructose transporter, catalyzes the uptake of fructose
GM	Genetically modified
GM-CSF	Granulocyte-macrophage colony-stimulating factor
cGMP	Current good manufacturing practice
GMP-PDE (cyclic GMP-PDE)	Cyclic GMP-phosphodiesterase
GNSO	5-nitrosoglutathione
GPC	Gel permeation chromatography
GPCR	G-protein-coupled receptor
GPI	Glycosyl phosphatidylinositol
GRIP	A Golgi-targeting protein domain

GRP	Glucose-regulated protein
Grp78	A glucose-regulated protein; identical with BiP
GSH	Glutathione
GST	Glutathione-*S*-transferase; gene trap sequencing tag
GTF	General transcription factor
GTST (GST)	Gene trap sequence tags
GUS	Beta-glucuronidase
GXP(s)	A generic acronym for good practices including but not limited to good clinical practice, good laboratory practice, and good manufacturing processes
HA	Hemaglutin-A; hyaluronic acid; hydroxyapatite, $Ca_{10}(PO_4)_6(OH)_2$
HABA	[2-(4′-hydroxyazobenzene)]benzoic acid
HAS	Human serum albumin; hyaluron synthase
HAT	Histone acetyltransferase; hypoxanthine, aminopterin, and thymidine
HBSS	Hanks' balanced salt solution
H/D	Hydrogen/deuterium exchange
HDA	Heteroduplex analysis
HDAC	Histone deacetylase
HDL	High-density lipoprotein
HDLA	Human leukocyte differentiation antigen
HD-ZIP	Homeodomain-leucine zipper proteins
HEPT	Height equivalent to plate number
HERV	Human endogenous retrovirus
20-HETE	20-hydroxyeicosatetranenoic acid
HETP	Plate height (chromatography)
HexNac	*N*-acetylhexosamine
HGP	Human genome project
HH	Hereditary hemochromatosis
His-Tag **(His$_6$; H$_6$)**	Histidine tag; a hexahistidine sequence
HLA	Human leukocyte-associated antigen
HLA-DM	Enzyme responsible for loading peptides onto MHC class II molecules
HLA-DO	Protein factor that modulates the action of HLA-DM
HMGR	3-hydroxy-3-methylglutamyl-coenzyme A reductase
HMM	Hidden Markov models
HMP	Herbal medicinal product(s)
HMT	Histone
hnRNA	Heterologous nuclear RNA
HOG	High-osmolarity glycerol
HOPE	HEPES-glutaminic acid buffer-mediated organic solvent protein effect
HOX **(*HOX, hox*)**	Describing a family of transcription factors
HPAEC-PAD	High-performance anion-exchange chromatography-pulsed amperometric detection
5-HPETE	5-hydroperoxyeicosatetranenoic acid
HPRD	Human protein reference database
HPRT	Hypoxanthine phosphoribosyl transferase
HRP	Horseradish peroxidase
HS	Heparan sulfate
HSB	Homologous synteny blocks
HSC	Hematopoietic stem cell
HSCQ	Heteronuclear single quantum correlation

HSE	Heat-shock element
Hsp	Heat-shock protein
Hsp70	Heat-shock protein 70
5-HT	5-hydroxytryptamine
HTF	*HpaII* tiny fragments; distinct fragments from the *HpaII* digestion of DNA; *HpaII* is a restriction endonuclease
HTH	Helix-turn-helix
HTS	High-throughput screening
htSNP	Haplotype single-nucleotide polymorphism
HUGO	Human genome organization
HUVEC	Human umbilical vein endothelial cells
IAA	Iodoacetic acid
IAEDANS	*N*-iodoacetyl-*N'*-(5-sulfo-1-napthyl) ethylenediamine
IBD	Identical-by-descent; inflammatory bowel disease
IC	Ion chromatography
ICAM	Intercellular adhesion molecule
ICAT	Isotope-coded affinity tag
ICH	Intracerebral hemorrhage; a gene related to *Ice* involved in programmed cell death; historically, international chick unit; International Conference for Harmonisation
ICPMS	Inductively coupled plasma mass spectrometry
ID	Internal diameter
IDA	Interaction defective allele
IDMS	Isotope dilution mass spectrometry
IEC	Ion-exchange chromatography
IEF	Isoelectric focusing
IES	Internal eliminated sequences
IFE	Immunofixation electrophoresis
IFN	Interferon
Ig	Immunoglobulin
IGF	Insulinlike growth factor
IGFR	Insulinlike growth factor receptor
Ihh	Indian hedgehog
IκB	NF-κB inhibitor
IκK	IκB kinase
IL	Interleukin
iLAP	Integrated lysis and purification
ILGF	Insulinlike growth factor
ILGFR	Insulinlike growth factor receptor
ILK	Integrin-linked kinase
IMAC	Immobilized metal-affinity chromatography
IMINO	Na^+-dependent alanine-insensitive proline uptake system (SLC6A20)
IMP	Integrin-mobilferrin pathway: membrane protein system involved in the transport of ferric iron; also inosine-5'- monophosphate
iNOS	Inducible oxide synthetase
Inr	Initiator element
IP$_3$	Inositol 1,4,5-triphosphate
IPG	Immobilized pH gradient
IPTG	Isopropylthio-β-D-galactosidase
IPTH	Isopropylthio-β-D-galactopyranoside
IR	Inverted repeat; insulin receptor

IRES	Internal ribosome entry site
IRS	Insulin receptor substrate
ISE	Ion-specific electrode
ISO	International Standards Organization
ISS	Immunostimulatory sequence; intronic splicing silencer
ISS-ODN	Immunostimulatory sequence-oligodeoxynucleotide
ISSR	Inter-simple sequence repeats
IT	Isotocin
ITAF	IRES trans-acting factor
ITAM	Immunoreceptor tyrosine-based activation motif
ITC	Isothermal titration calorimetry
iTRAQ	Isobaric tags for relative and absolute quantitation of proteins in proteomic research
JAK	Janus kinase
JNK	*c*-Jun *N*-terminal kinase
KARAP	Killer cell-activating receptor-associated protein
Kb, kb	Kilobase
KDR	Kinase insert domain-containing receptor; KDR is the human homolog of the mouse FLK-1 receptor; the KDR and FLK-1 receptors are also known as VEGFR2: see VEGFR
Kit	Mast/stem cell growth factor receptor, CD 117
Kit	Retroviral oncogene derived from HZ4 feline sarcoma
KLF5	Krüppel-like factor 5, a transcription factor
LAK	Lymphokine-activated killer cells
LATE-PCR	Linear-after-the-exponential-PCR
LB	Luria–Bertani
LC$_{50}$	Median lethan concentration in air
Lck	Member of the Src family of protein kinases
LC-MS	Liquid chromatography-mass spectrometry
LCR	Low-copy repeat; locus control region; low-complexity region
LCST	Lower critical solution temperature
LD	As in LD motif, a leucine/aspartic acid-rich protein-binding domain; also used to refer to peptidases without stereospecificity; also long in domain, linkage disequilibrium, lactate dehydrogenase
LD$_{50}$	Median lethal dose
LDL	Low-density lipoprotein
LECE	Ligand exchange capillary electrophoresis
LED	Light-emitting diode
Lek	Lymphocyte-specific protein tyrosine kinase
LFA	Lymphocyte function-associated antigen
LGIC	Ligand-gated ion channel
LH	Luteinizing hormone
LIF	Laser-induced fluorescence
LIM	Domain involved in protein–protein interaction, originally described in transcription factors LIN1, ISL1, and MED3
LINE	Long interspersed nuclear element
LLE	Liquid–liquid extraction
LLOD	Lower limit of detection
LLOQ	Lower limit of quantification
lnRNP	Large nuclear ribonucleoprotein
LOD	Limit of detection; $\log_{10}$ of odds

LOLA	List of lists annotated
LOQ	Limit of quantitation
LP	Lysophospholipid
LPA	Lysophosphatidic acid
LPH	Lipotropic hormone
LPS	Lipopolysaccharide
LTB$_4$	Leukotriene B$_4$
LTH	Luteotropic hormone
Ltk	Leukocyte tyrosine kinase
LRP	Low-density lipoprotein receptor-related protein
LSPR	Localized surface plasmon resonance
LTR	Long terminal repeat
LUCA	Last universal cellular ancestor
M13	A bacteriophage used in phage display
M	Macrophage
Mab, MABq	Monoclonal antibody
MAC	Membrane attack complex
MAD	Multiwavelength anomalous diffraction
Maf	Retroviral oncogene derived from AS42 avian sarcoma
MAGE	Microarray and gene expression
MALDI-TOF	Matrix-assisted laser desorption ionization time-of-flight
MAP	Mitogen-activated protein, usually referring to a protein kinase such as MAP-kinase
MAPK	MAP-kinase
MAPKK	MAP-kinase kinase
MAPKKK	MAP-kinase kinase kinase
MAR	Matrix attachment region
Mb, mb	Megabase (10^6)
MB	Molecular beacon
MBL	Mannose-binding lectin
MBP	Myelin basic protein; maltose-binding protein
MCA	4-methylcoumaryl-7-acetyl
MCAT	Mass coded abundance tag
MCD	Magnetic circular dichroism
MCM	Mini-chromosome maintenance
MCS	Multiple cloning site
M-CSF	M-colony stimulating factor; macrophage-colony stimulating factor
MDA	Malondialdehyde
MDCK	Madin–Darby canine kidney
MDMA	3,4-methylenedioxymethamphetamine
MEF	Mouse embryonic fibroblasts
MEF-2	Myocyte enhancer factor 2
MEGA-8	Octanoyl-*N*-methylglucamide
MEGA-10	Decanoyl-*N*-methylglucamide
MEK	Mitogen-activated protein kinase/extracellular signal-regulated kinase kinase; also methylethyl ketone
MELC	Microemulsion liquid chromatography
MELK	Multi-epitope-ligand-kartographie
MEM	Minimal essential medium
Mer	A receptor protein kinase; also Mertk, Mer tyrosine kinase
MES	2-(*N*-morpholinoethanesulfonic acid)

Met	Receptor for hepatocyte growth factor
MFB	Membrane fusion protein
MGO	Methylglyoxal
MGUS	Monoclonal gammopathy of undetermined significance
MHC	Major histocompatibility complex
MIAME	Minimum information about a microarray experiment
Mil	Retroviral oncogene derived from Mill Hill-2 chicken carcinoma
MIP	Molecularly imprinted polymer; macrophage inflammatory protein; methylation induced premeiotically
MIPS	Munich Information Center for Protein Sequences
MIS	Mullerian inhibiting substance
MLCK	Myosin light chain kinase
MLCP	Myosin light chain phosphatase
MMP	Matrix metalloproteinase
MMR	Mismatch repair
MMTV	Mouse mammary tumor virus
MOPS	3-(*N*-morpholino)propanesulfonic acid; 4-morpholinopropanesulfonic acid
MOPSo	3-(N-morpholino)-2-hydroxypropanesulfonic acid
Mos	Retroviral oncogene derived from Moloney murine sarcoma
MPD	2-methyl-2,4-pentanediol
MPSS	Massively parallel signature sequencing
MR	Magnetic resonance
MRI	Magnetic resonance imaging
mRNA	Messenger RNA
MRP	Migratory inhibitory factor-related protein
MRTF	Myocardin-related transcription factor
MS	Mass spectrometry, also mechanosensitive (receptors), multiple sclerosis
MS/MS	Mass spectrometry/mass spectrometry
MS^3	Tandem mass spectrometry/mass spectrometry/mass spectrometry
MSP	Macrophage-stimulating protein
Mt	Mitochondrial
MTBE	Methyl-*t*-butyl ether
mt-DNA	Mitochondrial DNA
MTOC	Microtubule organizing center
mTOR	A eukaryotic regulator of cell growth and proliferation; see TOR
MTSP	Membrane-type serine proteases
MTT	Methylthiazoletetrazolium
MTX	Methotrexate
Mu	Mutator
MU	Miller units
MuDPiT	Multidimensional protein identification technology
MuLV	Muloney leukemia virus
MUSK	Muscle skeletal receptor tyrosine kinase
MWCO	Molecular weight cutoff
My	Million years
Myb	Retroviral oncogene derived from avian myeloblastosis
Myc	Retroviral oncogene derived from MC29 avian myelocytomatosis
MYPT	Myosin phosphatase targeting
Mys	Myristoylation site
NAA	Neutron activation analysis
Nabs	Neutralizing antibodies

nAChR	Nicotinic acetylcholine receptor
(nAcChoR)	
NAD	Nicotinamideadenine dinucleotide (DPN)
NADP	Nicotinamideadenine dinucleotide phosphate (TPN)
NAO	Nonanimal origin
NAT	Nucleic acid amplification testing; nucleic acid testing
NBD	Nucleotide-binding domain
NBD-PyNCS	4-(3-iosthiocyanatopyrrolidin-1-yl)-7-nitro-2,1,3-benzoxadiazole
Nbs$_2$	Ellman's reagent; 5,5′-dithiobis(2-nitrobenzene acid)
NBS	N-bromosuccinimide
NBT	Nitroblue tetrazolium
NCBI	National Center for Biotechnology Information
NCED	9-*cis*-epoxycarotenoid dioxygenase
NDA	New drug application
NDB	Nucleic acid databank
NDMA	N-methyl-D-aspartate
NDSB	3-(1-pyridinio)-1-propanesulfonate (nondetergent sulfobetaine)
NEM	N-ethylmaleimide
NEO	Neopterin
NEP	Nucleus-encoded polymerase (RNA polymerase)
NeuAc	N-acetylneuraminic acid
NeuGc	N-glycolylneuraminic acid
NF	National formulary
NFAT	Nuclear factor of activated T-cells, a transcription factor
NF-κB	Nuclear factor kappa B, a nuclear transcription factor
NGF	Nerve growth factor
NGFR	Nerve growth factor receptor
NHS	N-hydroxysuccinimide
Ni-NTA	Ni^{2+}-nitriloacetate
NIR	Near infrared
NIRF	Near-infrared fluorescence
NIST	National Institute of Standards and Technology
NK	Natural killer (as in cytotoxic T-cell)
NKCF	Natural killer cytotoxic factor
NKF	N-formylkynurenine
NMDA	N-methyl-D-aspartate
NMM	Nicotinamide mononucleotide
NMR	Nuclear magnetic resonance
NO	Nitric oxide
NOE	Nuclear Overhauser effect
NOESY	Nuclear Overhauser effect spectroscopy
NOHA	N^w-hydroxy-L-arginine
NORs	Specific chromosomal sites of nuclear reformulation
NOS	Nitric oxide synthetase
NPC	Nuclear pore complex
***p*NPP**	*p*-nitrophenyl phosphate
NSAID	Nonsteroid anti-inflammatory drug(s)
NSF	N-ethylmaleimide sensitive factor; National Science Foundation; N-ethylmaleimide-sensitive fusion
Nt, nt	Nucleotide
NTA	Nitriloacetic acid

NTPDases	Nucleoside triphosphate diphosphohydrolases; also known as apyrases, E-ATPases
NuSAP	Nucleolar spindle-associated protein
ODMR	Optically detected magnetic resonance
ODN	Oligodeoxynucleotide
OECD	Organization for Economic Cooperation and Development
OFAGE	Orthogonal-field-alternation gel electrophoresis
OHQ	8-hydroxyquinoline
OMG	Object management group
OMIM	Online Mendelian Inheritance in Man (database), OMIM220100: http://www.ncbi.nlm.nih.gov
OMP	Outer membrane protein; a protein family associated with membranes
OMT	Outer membrane transport
OPG	Osteoprotegerin
ORC	Origin recognition complex
ORD	Optical rotatory dispersion
ORF	Open reading frame
ORFan	Orphan open reading frame
ORFeome	The protein-coding ORFs of an organism
OSBP	Oxysterol-binding proteins
OVA	Ovalbumin
OXPHOS	Oxidative phosphorylation
OYE	Old yellow enzyme
p53	A nuclear phosphoprotein that functions as a tumor suppressor
PA	Peptide amphiphile
PAC	P1-derived artificial chromosome
PACAP	Pituitary adenylyl cyclase-activating polypeptide
PAD	Peptidylarginine deiminase; protein arginine deiminase (EC 3.5.5.15)
PADGEM	Platelet activator-dependent granule external membrane protein; GMP-140
PAGE	Polyacrylamide gel electrophoresis
PAH	Polycyclic aromatic hydrocarbon
PAK	P21-activated kinase
PAO	A redundant gene family (ser*ipao*parin)
PAR	Protease-activated receptor
PAS	Preautophagosomal structure
PAT1	H$^+$-coupled amino acid transporter (slc36a1)
PAZ	A protein interaction domain; PIWI-argonaute-zwille
PBS	Phosphate-buffered saline
PBST	Phosphate-buffered saline with Tween-20
PBP	Periplasmic-binding protein
PC	Polycystin; phosphatidyl choline
PCAF	p300/CBP-associated factor, a histone acetyltransferase
PCNA	Proliferating cell nuclear antigen; processing factor
PDB	Protein databank
PDE	Phosphodiesterase
PDGF	Platelet-derived growth factor
PDGFR	Platelet-derived growth factor receptor
PDI	Protein disulfide isomerase
PDMA	Polydimethylacrylamide
PDMS	Polydimethylsiloxane
pDNA	Plasmid DNA

PE	Phycoerythrin; polyethylene
PEC	Photoelectrochemistry
PECAM-1	Platelet/endothelial cell adhesion molecule-1
PEI	Polyethyleneimine
PEND protein	DNA-binding protein in the inner envelope membrane of the developing chloroplast
PEP	Phosphoenol pyruvate
PEP	Plastid-encoded polymerase (RNA polymerase)
PEPCK-C	Phosphoenolpyruvate carboxykinase, cytosolic form
PERK	Double-stranded RNA-activated protein kinaselike ER kinase
PES	Photoelectron spectroscopy
PET	Positron emission tomography
Pfam	Protein family database; protein families database of alignments
PFGE	Pulsed-field gel electrophoresis
PFK	Phosphofructokinase
PFU	Plaque-forming unit
PG	Phosphatidyl glycerol; prostaglandin
3-PGA	3-phospho-D-glycerate
PGO	Phenylglyoxal
PGP-Me	Archaetidylglycerol methyl phosphate
PGT box	An upstream *cis*-element
PGx (PGX)	Pharmacogenetics (PGx) is the use of genetic information to guide drug choice; prostaglandins (PGX) include thromboxanes and prostacyclins
PH	Pleckstrin homology
pHB (*p*-HB)	4-hydroxybenzoic acid; (*p*-hydroxybenzoate)
PHD	Plant homeodomain
PI	Propidium iodide
PIC	Preinitiation complex — complex of GTFs
PINCH	PINCH-protein; particularly interesting *cis*-his-rich protein
PIP$_3$	Phosphatidylinositol-3,4,5-triphosphate
PIP$_n$	Polyinositol polyphosphate
PIP$_n$S	Polyinositol polyphosphates
Pipes	1,4-piperazinediethanesulfonic acid
PIRLβ	Paired immunoglobulinlike type-2 receptor β
PKA	Protein kinase A; cAMP-dependent kinase; pKa, acid dissociation constant
PKC	Protein kinase C
Pkl	Paxillin kinase linker
PLL	Poly-L-Lysine
PLP	Pyridoxal-5-phosphate
PMA	Phenyl mercuric acetate; phorbol-12-myristate-13 acetate
PMCA	Plasma membrane Ca^{2+} as PMCA-ATPase, a PMCA pump
PMSF	Phenylmethylsulfonyl fluoride
PNA	Peptide nucleic acid; *p*-nitroanilide
PNGase	Endoglycosidase
PNP	*p*-nitrophenol (4-nitrophenol)
POD	Peroxidase
POET	Pooled ORF expression technology
POINT	Prediction of interactome database
Pol II	RNA polymerase II
POTRA	Polypeptide translocation associated
PP	Polypropylene

PPAR	Peroxisome proliferator-activated receptor
PPase	Phosphoprotein phosphatase
PQL	Protein quantity loci
PS	Position shift polymorphism
PS-1	Presenilin-1
PSG	Pregnancy-specific glycoprotein(s)
PSI	Photosystem I
PSI-BLAST	Position-specific interactive BLAST; position-shift iterated BLAST (software program)
PSII	Photosystem II
PTB	Polypyrimidine tract-binding protein, a repressive regulator of protein splicing; also pulmonary tuberculosis
PTD	Protein transduction domain
PTEN	Phosphatase and tensin homolog deleted on chromosome 10
PTFE	Polytetrafluoroethylene
PTGS	Posttranscriptional gene silencing
PTH	Phenylthiohydantoin
PTK	Protein-tyrosine kinase
PTPase	Protein-tyrosine phosphatase
PVA	Polyvinyl alcohol
PVDF	Polyvinylidine difluoride
QA	Quality assurance
QC	Quality control
QSAR	Quantitative structure–activity relationship(s)
QTL	Quantitative trait loci
Q-TOF	Quadruple time-of-flight
R_f	Retardation factor
RA	Rheumatoid arthritis; radiographic absorptiometry (bone density)
RAB-GAP	Rab-GTPase-activating protein
RACE	Rapid amplification of cDNA ends
Raf	Retroviral oncogene derived from 3611 murine sarcoma
RAGE	Receptors for advanced glycation endproducts; receptors for AGE; recombinase-activated gene expression
RAMP	Receptor activity-modified protein
RANK	Receptor activator of NF-κB
RANK-L	Receptor activator of NF-κB ligand
Rap	A family of GTPase-coupled signal transduction factors, which are part of the RAS superfamily
Rap1	A small GTPase involved in integrin activation and cell adhesion
RAPD	Randomly amplified polymorphic DNA
RARE	RecA-assisted restriction endonuclease
RAS	GTP-binding signal transducers
H-*ras*	Retroviral oncogene derived from Harvey murine sarcoma
K-*ras*	Retroviral oncogene derived from Kirsten murine sarcoma
RC	Recombinant cogenic
RCA	Rolling circle amplification
RCCX	RP-C4-CYP21-TNX module
RCFP	Reef coral fluorescent protein
RCP	Receptor component protein
RCR	Rolling circle replication
rDNA	Ribosomal DNA

REA	Restriction enzyme analysis
Rel	Avian reticuloendotheliosis
REMI	Restriction enzyme-mediated integration
RET	Receptor for the GDNF family
RF	A transcription factor, RFX family
Rfactor	Final crystallographic residual
RFID	Radio frequency identification device
RFLP	Restriction fragment-length polymorphism
RGD	A signature peptide sequence: arginine-glycine-aspartic acid found in protein, which binds integrins
RGS	Regulator of G-protein signaling
RHD	*Rel* homology domain
Rheb	*Ras* homolog enriched in brain
RhoA	*Ras* homologous; signaling pathway
RI	Random integration
RIP	Repeat-induced point mutation
RIS	Radioimmunoscintigraphy
RISC	RNA-induced silencing complex
RIT	Radioimmunotherapy
RM	Reference material
RNAi	RNA interference
dsRNA	Double-stranded RNA
hpRNAi	Hairpin RNA interference
ncRNA	Noncoding RNA
rRNA	Ribosomal RNA
shRNA	Small hairpin RNA
siRNA	Small interfering RNA
snRNA	Small nuclear RNA
snoRNA	Small nucleolar RNA
stRNA	Small temporal RNA
RNAse/RNAase	Ribonuclease
RNAse III	A family of ribonucleases (RNAses)
RNC	Ribosome-nascent chain complex
snRNP	Small nuclear ribonucleoprotein particle
RNS	Reactive nitrogen species
RO	Reverse osmosis
ROCK (ROK)	Rho kinase
ROESY	Rotating frame Overhauser effect spectroscopy
Ron	Receptor for macrophage-stimulating protein
Ros	Retroviral oncogenes derived from UR2 avian sarcoma
ROS	Reactive oxygen species
RP	Reverse-phase; also a nuclear serine/threonine protein kinase
RPA	Replication protein A
RPC	Reverse-phase chromatography
RP-CEC	Reverse-phase capillary electrochromatography
RPEL	A protein motif involved in the cytoskeleton
RP-HPLC	Reverse-phase high-performance liquid chromatography
RPMC	Reverse-phase microcapillary liquid chromatography
RPMI 1640	Growth media for eukaryotic cells
RPTP	Receptor protein-tyrosine kinase
RRM	RNA-recognition motif

RRS	*Ras* recruitment system; resonance Raleigh scattering
R,S	Designating optical activity of chiral compounds where R is rectus (right) and S is sinister (left)
RSD	Root square deviation
RT	Reverse transcriptase; room temperature
RTD	Residence time distribution
RTK	Receptor tyrosine kinase
RT-PCR	Reverse transcriptase-polymerase chain reaction
RTX	Repeat in toxins; pore-forming toxin of *E. coli* type (RTX toxin); also rituximab, resiniteratoxin, renal transplantation
Rub1	A ubiquitinlike protein, Nedd8
S1P	Sphingosine-1-phosphate
S100	S100 protein family
SA	Salicylic acid
SAGE	Serial analysis of gene expression
SALIP	Saposinlike proteins
SAM	Self-assembling monolayers
SAMK	A plant MAP kinase
SAMPL	Selective amplification of microsatellite polymorphic loci
Sap	Saposin
SAP	Sphingolipid activator protein; also serum amyloid P, shrimp alkaline phosphatase
SAR	Scaffold-associated region; structure–activity relationship
SATP	Heterobifunctional crosslinker; *N*-succinimidyl-*S*-acetylthiopropionate
SAXS	Small angle x-ray scattering
scFv	Single-chain Fv fragment of an antibody
SCID	Severe combined immunodeficiency
SCOP	Structural classification of proteins: http://scop.mrc-lmb.cam.ac.uk/scop
SCOPE	Structure-based combinatorial protein engineering
SDS	Sodium dodecyl sulfate
Sec	Secretory, usually related to protein translocation
SEC	Secondary emission chamber for pulse radiolysis; size exclusion chromatography
SELDI	Surface-enhanced laser desorption/ionization
SELEX	Systematic evolution of ligands by exponential enrichment
SERCA	Sarco/endoplasmic reticulum Ca^{2+} as in SERCA-ATPase, a calcium pump
SFC	Supercritical fluid
SH2	*Src* homology domain 2
SH3	*Src* homology domain 3
SHAP	Serum-derived hyaluron-associated protein
Shh	Sonic hedgehog
SHO	Yeast osmosensor
shRNA	Small hairpin RNA
SILAC	Stable-isotope labeling with amino acids in cell culture
SIMK	A plant MAP kinase
SINE	Short interspersed nuclear element
SINS	Sequenced insertion sites
SIPK	Salicylic acid–induced protein kinase
Sis	Retroviral oncogene derived from simian sarcoma
SISDC	Sequence-independent site-directed chimeragenesis
Ski	Retroviral oncogene derived from avian SK77
Skp	A chaperone protein

SLAC	Serial lectin affinity chromatography
SLE	Systemic lupus erythematoses
SLN1	Yeast osmosensor
S/MAR	Scaffold and matrix attachment region
SMC	Smooth muscle cell
SNAREs	Soluble *N*-ethylmaleimide-sensitive fusion (NSF; *N*-ethylmaleimide-sensitive factor) protein attachment protein receptors: can be either R-SNAREs or Q-SNARES depending on sequence homologies
SNM	SNARE motif
snoRNA	Small nucleolar RNA
SNP	Single nucleotide polymorphism
snRNA	Small nuclear RNA
snRNP	Small nuclear ribonucleoprotein particle
SOC	Soil organic carbon; store-operated channel
SOCS	Suppressors of cytokine signaling
SOD	Superoxide dismutase
SOD1s	CuZn-SOD enzyme (intracellular)
SOP	Standard operating procedure
SOS	Response of a cell to DNA damage; salt overly sensitive (usually plants); Son of Sevenless (signaling cascade protein)
SPA	Scintillation proximity assay
SPC	Statistical process control
SPECT	Sporozoite microneme protein essential for cell transversal; also single-photon emission-computed tomography
SPIN	Surface properties of protein-protein interfaces (database)
SPR	Surface plasmon resonance
SQL	Structured query language
SR	As in the SR protein family (serine- and arginine-rich proteins); also sarcoplasmic reticulum, scavenger receptor
SRCD	Synchrotron radiation circular dichroism
SRF	Serum response factor, a ubiquitous transcription factor
SRP	Signal recognition particle
SRPK	SR protein kinase
SRS	Sequence retrieval system; SOS recruitment system
SRWC	Short rotation woody crop
SSC	Saline sodium citrate
ssDNA	Single-stranded DNA
SSLP	Simple sequence length polymorphism
SSR	Simple sequence repeats
STAT	Signal transducers and activators of transcription
STC	Sequence-tagged connector
STM	Sequence-tagged mutagenesis
STORM	Systematic tailored ORF-data retrieval and management
STR	Short tandem repeats
STREX	Stress axis-related exon
stRNA	Small temporal RNA
SUMO	Small ubiquitinlike (UBL) modifier; small ubiquitin-related modifier; sentrin
SurA	A chaperone protein
SV40	Simian virus 40
SVS	Seminal vesicle secretion
S$_{w,20}$	Sedimentation coefficient corrected to water at 20°C

SWI/SNF	Switch/sucrose nonfermenting
TAC	Transcription-competent artificial chromosome
TACE	Tumor necrosis factor α-converting enzyme; also transcatheter arterial chemoembolization
TAFE	Transversely alternating-field electrophoresis
TAFs	TBP-associated factors
TAG	Triacyl glycerol
TAME	Tosyl-arginine methyl ester
TAP	Tandem affinity purification; also transporter associated with antigen processing
TAR	Transformation-associated recombination; *trans*-activation response region
TAT	*Trans*-activator of transcription
TATA	As in the TATA box, which is a TATA-rich region located upstream from the RNA-synthesis initiation site in eukaryotes and within the promoter region for the gene in question; analogous to the Pribnow box in prokaryotes
TBA-Cl	Tetrabutylammonium chloride
TBP	TATA-binding protein; telomere-binding protein
TCA	Trichloroacetic acid; tricarboxylic acid
TCR	T-cell receptor
TE	Therapeutic equivalence; transposable elements
TEA	Triethylamine
TEAA	Triethylammonium acetate
TEF	Toxic equivalency factor
TEM	Transmission electron microscopy
TEMED (TMPD)	*N,N,N′,N′*-tetramethylethylenediamine
TEP	Tobacco etch protease
TF	Tissue factor; transcription factor
TFA	Trifluoroacetic acid
TFIIIA	Transcription factor IIIA
TGN	*Trans*-Golgi network
TGS	Transcriptional gene silencing
TH	Thyroid hormone
THF	Tetrahydrofuran
TIGR	The Institute for Genomic Research
TIM	Translocase of inner mitochondrial membrane
TIP	Tonoplast intrinsic protein(s)
TIR	Toll/IL-1 receptor
TI-VAMP	Tetanus neurotoxin-insensitive VAMP
TLCK	Tosyl-lysyl chloromethyl ketone
TLR	Toll-like receptor
T_m	Tubular membrane; tension/mucosal; DNA thermal melting point; midpoint of thermal denaturation curve
TM	Transmembrane
TMAO	Trimethylamine oxide
TMD	Transmembrane domain
TMS	Trimethylsilyl; thimersol
TMV	Tobacco mosaic virus
TNA	Treose nucleic acid
TNB	5-thio-2-nitrobenzoate

TNBS	Trinitrobenzenesulfonic acid
TnI	Troponin I
TnC	Troponin C
TNF	Tumor necrosis factor
TNF-α (TNFα)	Tumor necrosis factor-α
TNR	Transferrin receptor
TnT	Troponin T
TNX	Tenascin-X
TOC	Total organic carbon
TOCSY	Total correlated spectroscopy
TOF	Time-of-flight
TOP	5′ tandem oligopyrimidine (terminal oligopyrimidine) tract
TOPRIN	Topoisomerase and primase in reference to a domain
TOR	Target of rapamycin; mTOR, mammalian target of rapamycin; dTOR, *Drosophila* target of rapamycin
TOX	Toxicology
TPCK	Tosylphenylalanylchloromethyl ketone
TPD	Temperature-programmed desorption
TPEN	$N′,N′$-tetrakis-(2-pyridyl-methyl)ethylenediamine
TPN	Triphosphopyridine dinucleotide (now NADP)
TRADD	A scaffold protein
TRAP	Tagging and recovery of associated proteins, as in RNA-TRAP; also thrombin receptor activation peptide
TRE	Thyroid hormone response elements
TRH	Thyrotropin-releasing hormone
TRI	As in TRI reagents, such as TRIZOL™ reagents used for RNA purification from cells and tissues
Tricine	N-(2-hydroxy-1,1-bis(hydroxymethyl)ethyl) glycine
TRIF	TIR domain-containing adaptor-inducing interferon-β
Tris	Tris-(hydroxymethyl)aminomethyl methane; 2-amino-2-hydroxymethyl-1, 3-propanediol
Bis-Tris	2-[bis(2-hydroxyethyl)amino]-2-(hydroxymethyl) propane-1,3-diol
Trk	Neurotrophic tyrosine kinase receptor
TRL	Time-resolved luminescence
TRP	Transient receptor potential, as in TRP-protein
TRs	Thyroid receptors
TSP	Thrombospondin; traveling salesman problem
TTSP	Transmembrane-type serine proteases
TUSC	Trait utility system for corn
Tween	Polyoxyethylsorbitan monolaurate
TX	Thromboxane; treatment
TyroBP	Tyro protein tyrosine kinase-binding protein, DNAX-activation protein 12, DAP12, KARAP
UAS	Upstream activation site
UBL	Ubiquitinlike modifiers
UCDS	Universal conditions direct sequencing
UDP	Ubiquitin-domain proteins; uridine diphosphate
UDP-GlcNAc	Uridine-5′-diphospho-N-acetylglucosamine
UNG	Uracil DNA glycosylase
uORF	Upstream open reading frame
UPA	Universal protein array; urokinaselike plasminogen activator

UPR	Unfolded protein response
URL	Uniform resource locator
URS	Upstream repression site
USP	United States pharmacopeia
USPS	Ubiquitin-based split protein sensor
UTR	Untranslated region
VAMP	Vesicle-associated membrane protein
VAP	VAMP-associated protein
VCAM	Vascular cellular adhesion molecule
VDAC	Voltage-dependent anion-selective channel
VDJ	Variable diversity joining; regions of DNA joined in recombination during lymphocyte development; see VDJ recombination
VDR	Vitamin D receptor
VEGF	Vascular endothelial growth factor
VEGFR	Vascular endothelial growth factor receptor
VGH	Nonacronymical use; a neuronal peptide
V_H	Variable heavy chain domain
VICKZ	A family of RNA-binding proteins recognizing specific *cis*-acting elements
VIGS	Virus-induced gene silencing
VIP	Vasoactive intestinal peptide
VLDL	Very low-density lipoprotein
VLP	Viruslike particle
VNC (VNBC)	Viable, but not cultivatable (bacteria)
VNTR	Variable number of tandem repeats
VOC	Volatile organic carbon
VPAC	VIP PACAP receptors
VSG	Variable surface glycoproteins
VSP	Vesicular sorting pathway
vsp10	Gene for Vsp10
Vsp10	A type I transmembrane receptor responsible for delivery of protein to lysozyme/vacuole
WGA	Whole-genome amplification
WT, Wt	Wild type
XBP	X-box binding protein
XO	Xanthine oxidase
Y2H	Yeast two-hybrid
YAC	Yeast artificial chromosome
YCp	Yeast centromere plasmid
YEp	Yeast episomal plasmid
YFP	Yellow fluorescent protein
Z	Benzyloxycarbonyl
ZDF	Zucker diabetic factor
Zif	Zinc finger domain peptides (i.e., Zif-1, Zif-3)
ZIP	Leucine zipper
ZZ domain	A tandem repeat dimer of the immunoglobulin-binding protein A from *Staphylococcus aureus*

2 Glossary of Terms Useful in Biochemistry and Molecular Biology and Related Disciplines

Abbreviated New Drug Application (ANDA)
This document contains data that, when submitted to FDA's Center for Drug Evaluation and Research (CDER), Office of Generic Drugs, provide for the review and ultimate approval of a generic drug product. This document does not contain preclinical or clinical data but must demonstrate that the drug in question is a bioequivalent to the currently licensed drug, which is also referred to as the innovator drug. See http://www.fda.gov/cder/drugsat-fda/glossary.htm.

ABC Transporter
The ATP-binding cassette transporter family consists of a large number of membrane proteins involved in the transport of a variety of substances including ions, steroids, metabolites, and drugs across extracellular and intracellular membranes. A defect in an ABC transporter is important in cystic fibrosis. See Schwiebert, E.M., ABC transporter-facilitated ATP conductive transport, *Am. J. Physiol.* 276, C1–C8, 1999; Dean, M., Rzhetsky, A., and Allikmets, R., The human ATP-binding cassette (ABC) transporter superfamily, *Genome Res.* 11, 1156–1166, 2001; Dean, M., Hamon, Y., and Chimini, G., The human ATP-binding cassette (ABC) transporter superfamily, *J. Lipid Res.* 42, 1007–1017, 2001; Georujon, C., Orelle, C., Steinfels, E. et al., A common mechanism for ATP hydrolysis in ABC transporter and helicase superfamilies, *Trends Biochem. Sci.* 26, 539–544, 2001; Schmitt, L., The first view of an ABC transporter: the X-ray crystal structure of MsbA from *E. coli*, *Chembiochem* 3, 161–165, 2002; Holland, I.B., Schmitt, L., and Young, J., Type 1 protein secretion in bacteria, the ABC-transporter dependent pathway, *Mol. Membr. Biol.* 22, 29–39, 2005; Blemans-Oldehinkel, E., Doeven, M.K., and Poolman, B., ABC transporter architecture and regulatory roles of accessory domains, *FEBS Lett.* 580, 1023–1035, 2006; Frelet, A. and Klein, M., Insight in eukaryotic ABC transporter function by mutation analysis, *FEBS Lett.* 580, 1064–1084, 2006; Crouzet, J., Trombik, T., Fraysse, A.S., and Boutry, M., Organization and function of the plant pleiotropic drug resistance ABC transporter family, *FEBS Lett.* 580, 1123–1130, 2006.

Ablation
A multifunctional word derived from the Latin *ablatus* (to carry away). In medicine, refers to the surgical removal of tissue or the elimination of cells by irradiation or immunological approaches. The surgery approach is used extensively in cardiology (Gillinov, A.M. and Wolf, R.K., Surgical ablation of atrial fibrillation, *Prog. Cardiovasc. Dis.* 48, 169–177, 2005) while irradiation or immunological approaches are used in oncology (Appelbaum, F.R., Badger, C.C., Bernstein, I.D. et al., Is there a better way to deliver total body irradiation? *Bone Marrow Transplantation* 10, (Suppl. 1), 77–81, 1992; van Bekkum, D.W., Immune ablation and stem-cell

therapy in autoimmune disease. Experimental basis for autologous stem-cell transplantation, *Arthritis Res.* 2, 281–284, 2000). It also refers to the reduction of particles into smaller sizes during erosion by other particles or the surrounding fluid (see Lindner, H., Koch, J., and Niema, K., Production of ultrafine particles by nanosecond laser sampling using orthogonal prepulse laser breakdown, *Anal. Chem.* 77, 7528–7533, 2005). It also has a definition in aerospace technology for the dissipation of heat generated by atmospheric friction upon reentry of a space vehicle.

Abscisic Acid A plant hormone. See Leung, J. and Giraudet, J., Abscisic acid signal transduction, *Annu. Rev. Plant Physiol. Plant Mol. Biol.* 25, 199–221, 1998; Finkelstein, R.R., Gampala, S.S., and Rock, C.D, Abscisic acid signaling in seeds and seedlings, *Plant Cell* 14 (Suppl.), S15–S45, 2002.

Absolute Oils See *Essential Oils*.

Absorption Generally refers to the ability of a material to absorb another substance (hydration) or energy (the ability of a substance to absorb light). See *Adsorption*.

Abzymes See *Catalytic Antibodies*.

Accuracy The difference between the measured value for an analyte and the true value. Absolute error is the difference between the measured value and the true value while the relative error is that fraction that the absolute error is of the measured amount and is usually expressed as a percentage or at ppt/ppm. See Meites, L., Ed., *Handbook of Analytical Chemistry*, McGraw-Hill, New York, 1963; Dean, J.A., *Analytical Chemistry Handbook*, McGraw-Hill, New York, 1995; Dean, J.A., *Dean's Analytical Chemistry Handbook,* McGraw-Hill, New York, 2005.

Accurate Mass Tag (AMT) A peptide of sufficiently distinctive and accurate mass and elution time from liquid chromatography that can be used as a single identifier of a protein. See Conrads, T.P., Anderson, G.A., Veenstra, T.D. et al., *Anal. Chem.* 72, 3349–3354, 2000; Smith, R.D., Anderson, G.A., Lipton, M.S. et al., An accurate mass tag strategy for quantitative and high-throughput proteome measurements, *Proteomics* 2, 513–523, 2002; Strittmatter, E.F., Ferguson, P.L., Tang, K., and Smith, R.D., Proteome analyses using accurate mass and elution time peptide tags with capillary LC time-of-flight mass spectrometry, *J. Am. Soc. Mass Spectrom.* 14, 980–991, 2003; Shen, Y., Tolic, N., Masselon, C. et al., Nanoscale proteomics, *Anal. Bioanal. Chem.* 378, 1037–1045, 2004; Zimmer, J.S., Monroe, M.E., Qian, W.J., and Smith, R.D., Advances in proteomics data analysis and display using an accurate mass and time tag approach, *Mass Spectrom. Rev.* 25, 450–482, 2006.

Active Ingredient Any component of a final drug product that provides pharmacological activity or another direct effect in the diagnosis, cure, mitigation, treatment, or prevention of disease or to affect the structure on any function of the body. Sometimes referred to as the active pharmaceutical ingredient (API). See http://www.fda.gov/cber; http://www.ich.org (see Q7, Good Manufacturing Guide for Active Pharmaceutical Ingredients).

Active Sequence Collection (ACS) A collection of active protein sequences or protein fragments or subsequences, collected in the form of function-oriented databases, http://bioinformatica. isa.cnr.it/ACS/. AIRS — Autoimmune Related Sequences; BAC — Bioactive Peptides; CHAMSE — Chameleon Sequences (sequences that can adopt both an alpha helix and beta sheet conformation; DORRS — Database of RGD-Related Sequences; DVP — Delivery Vector Peptides; SSP — Structure-Solved Peptides; TRANSIT — Transglutamation Sites.

Activity-Based Proteomics   Identification of proteins in the proteome by the use of reagents, which measure biological activity. Frequently the activity is measured by the incorporation of a "tag" into the active site of the enzyme. The earliest probes were derivatives of alkyl-fluorophosphonates, which were well-understood inhibitors of serine proteases. The technical approach is related to enzyme histochemistry/histocytochemistry. Most often used for enzymes where functional families of proteins can be identified. See Liu, Y., Patricelli, M.P., and Cravatt, B.F., Activity-based protein profiling: the serine hydrolases, *Proc. Natl. Acad. Sci. USA* 96, 14694–14699, 1999; Adam, G.C., Sorensen, E.J., and Carvatt, B.F., Chemical strategies for functional proteomics, *Mol. Cell. Proteomics* 1, 781–790, 2002; Speers, A.E. and Cravatt, B.F., Chemical strategies for activity-based proteomics, *ChemBioChem* 5, 41–47, 2004; Kumar, S., Zhou, B., and Liang, F., Activity- based probes for protein tyrosine phosphatases, *Proc. Nat. Acad. Sci. USA* 101, 7943–7948, 2004; Berger, A.B., Vitorino, P.M., and Bogyo, M., Activity-based protein profiling: applications to biomarker discovery, *in vivo* imaging, and drug discovery, *Am. J. Pharmacogenomics* 4, 371–381, 2004; Williams, S.J., Hekmat, O., and Withers, S.G., Synthesis and testing of mechanism-based protein-profiling probes for retaining endo-glycosidases, *ChemBioChem* 7, 116–124, 2006; Sieber, S.A. and Cravatt, B.F., Analytical platforms for activity-based protein profiling — exploiting the versatility of chemistry for functional proteomics, *Chem. Commun.* 22, 2311–2318, 2006; Schmidinger, H., Hermetter, A., and Birner-Gruenberger, R., Activity-based proteomics: enzymatic activity profiling in complex proteomes, *Amino Acids* 30, 333–350, 2006.

Acute Phase Proteins   Proteins that are either *de novo* or markedly elevated after challenge by infectious disease, inflammation, or other challenge to homeostasis. Another definition is any protein whose blood concentration increases (or decreases) by 25% or more during certain inflammatory disorders. Acute phase proteins include C-reactive protein, fibrinogen, and α-1-acid glycoprotein. Acute phase proteins are part of the acute phase response. Some acute phase proteins have been used for diagnosis of specific disorders such as C-reactive protein and cardiovascular disease. See Sutton, H.E., The haptoglobins, *Prog. Med. Genet.* 7, 163–216, 1970; Gordon, A.H., Acute-phase proteins in wound healing, *Ciba Found. Symp.* 9, 73–90, 1972; Bowman, B.H., *Hepatic Plasma Proteins: Mechanisms of Function and Regulation*, Academic Press, San Diego, CA, 1993; Mackiewicz, A. and Kushner, I., *Acute Phase Proteins: Molecular Biology, Biochemistry, and Clinical Applications,* CRC Press, Boca Raton, FL, 1993; Kerr, M.A. and Thorpe, R., *Immunochemistry Labfax*, Bios Scientific Publishers, Oxford, UK, 1994; Black, S., Kushner, I., and Samols, D., C-reactive protein, *J. Biol. Chem.* 279, 48487–48490, 2004; Du Clos, T.W. and Mold, C., C-reactive protein: an activator of innate immunity and a modulator of adaptive immunity, *Immunol. Res.* 30, 261–277, 2004; Garlanda, C., Bottazzi, B., Bastone, A., and Mantovani, A., Pentraxins at the crossroads between innate immunity, inflammation, matrix deposition, and female fertility, *Annu. Rev. Immunol.* 23, 337–366, 2005; Ceron, J.J., Eckersall, P.D., and Martynez-Subiela, S., Acute phase proteins in dogs and cats: current knowledge and future perspectives, *Vet. Clin. Pathol.* 34, 85–99, 2005; Sargent, P.J., Farnaud, S., and Evans, R.W., Structure/function overview of proteins involved in iron storage and transport, *Curr. Med. Chem.* 12, 2683–2693, 2005; Bottazzi, B., Garlanda, C., Salvatori, G. et al.,

Pentraxins as a key component on innate immunity, *Curr. Opin. Immunol.* 18, 10–15, 2006; Vidt, D.G., Inflammation in renal disease, *Am. J. Cardiol.* 97, 20A–27A, 2006; Armstrong, E.J., Morrow, D.A., and Sabatine, M.S., Inflammatory biomarkers in acute coronary syndromes. Part II: acute-phase reactants and biomarkers of endothelial cell activation, *Circulation* 113, e152–e155, 2006. See also *Heat-Shock Proteins*.

ADAM-TS   A disintegrin and metalloproteinase with thrombospondin motifs. A family of multidomain metalloproteinases with a variety of biological activities. ADMETS are part of the reprolysin family. ADAM-TS13, which is involved in the processing of the von Willebrand Factor, is the best-known member of this family. See Hooper, N.M., Families of zinc metalloproteases, *FEBS Lett.* 354, 1–6, 1994; Hurskainen, T.L., Hirohata, S., Seldin, M.F., and Apte, S.S., ADAM-TS5, ADAM-TS6, and ADAM-TS7, novel members of a new family of zinc metalloproteases. General features and genomic distribution of the ADAM-TS family, *J. Biol. Chem.* 274, 2555–2563, 1999; Sandy, J.D. and Verscharen, C., Analysis of aggrecan in human knee cartilage and synovial fluid indicates that aggrecanase (ADAMTS) activity is responsible for the catabolic turnover and loss of aggrecan whereas other protease activity is required for C-terminal processing *in vivo*, *Biochem. J.* 358, 615–626, 2001; Fox, J.W. and Serrano S.M., Structural considerations of the snake venom metalloproteinases, key members of the M12 reprolysin family of metalloproteinases, *Toxicon* 45, 969–985, 2005.

Adjuvant   A substance that increases an immune response. Frequently a component of the excipients in the formulation of vaccines. See Spriggs, D.R. and Koff, W.C., Eds., *Topics in Vaccine Adjuvant Research*, CRC Press, Boca Raton, FL, 1991; Powell, M.F., Ed., *Vaccine Design: The Subunit and Adjuvant Approach*, Plenum Press, New York, 1995; Brown, L.E. and Jackson, D.C., Lipid-based self-adjuvanting vaccines, *Curr. Drug Deliv.* 2, 283–393, 2005; Gluck, R., Burri, K.G., and Metcalfe, I., Adjuvant and antigen delivery properties of virosomes, *Curr. Drug Deliv.* 2, 395–400, 2005; Smales, M.C. and James, D.C., Eds., *Therapeutic Proteins: Methods and Protocols*, Humana Press, Totowa, NJ, 2005; Schijns, V.E.J.C. and O'Hagan, D.T., Eds., *Immunopotentiation in Modern Vaccines*, Elsevier, Amsterdam, 2006.

Adrenomedullin Adrenomedullin is a peptide originally isolated from a phenochromocytoma (Kitamura, K., Kangawa, K., Kawamoto, M. et al., Adenomedullin: a novel hypotensive peptide isolated from human phenochromocytoma, *Biochem. Biophys. Res. Commun.* 192, 553–560, 1993). Adrenomedullin elevated intracellular cAMP in platelets and caused hypotension. Since its discovery, adrenomedullin has been found in a variety of cells and tissues (Hinson, J.P., Kapas, S., and Smith, D.M., Adrenomedullin, a multifunctional regulatory peptide, *Endocrine Rev.* 21, 138–167, 2000). Adrenomedullin has been suggested to have a variety of physiological activities. See Poyner, D., Pharmacology of receptors for calcitonin gene-related peptide and amylin, *Trends Pharmacol. Sci.* 16, 424–428, 1995; Muff, R., Born, W., and Fischer, J.A., Calcitonin, calcitonin gene-related peptide, adrenomedullin, and amylin: homologous peptides, separate receptors, and overlapping biological actions, *Eur. J. Endocrinol.* 133, 17–20, 1995; Richards, A.M., Nicholls, M.G., Lewis, L., and Lainchbury, J.G., Adrenomedullin, *Clin. Sci.* 91, 3–16, 1996; Massart, P.E., Hodeige, D., and Donckier, J., Adrenomedullin: view on a novel vasodilatory peptide with naturetic properties, *Acta Cardiol.* 51,

259–269, 1996; Hay, D.L. and Smith, D.M., Adrenomedullin receptors: molecular identity and function, *Peptides* 22, 1753–1763, 2001; Julian, M., Cacho, M., Garcia, M.A. et al., Adrenomedullin: a new target for the design of small molecule modulators with promising pharmacological activities, *Eur. J. Med. Chem.* 40, 737–750, 2005; Shimosawa, T. and Fujita, T., Adrenomedullin and its related peptides, *Endocr. J.* 52, 1–10, 2005; Zudaire, E., Portal-Núñez, S., and Cuttitta, F., The central role of adrenomedullin in host defense, *J. Leuk. Biol.* 80, 237–244, 2006; Hamid, S.A. and Baxter, G.F., A critical cytoprotective role of endogenous adrenomedullin in acute myocardial infarction, *J. Mol. Cell Cardiol.* 41, 360–363, 2006.

Adsorption   The transfer of a substance from one medium to another such as the adsorption of a substance from a fluid onto a surface. The *adsorbent* is the substrate onto which material is adsorbed. The *adsorbate* is the material adsorbed onto a matrix.

Advanced Glycation Endproducts (AGE)   A heterogeneous group of products resulting from a series of chemical reactions starting with the formation of adducts between reducing sugars and protein nucleophiles such as nitrogen bases. Reaction with nucleic acid is also possible but has not been extensively described. The reactions involved are complex involving the Amadori reaction and the Maillard reaction. Some products include triosidines, *N*-carboxymethyl-lysine, and pentosidine-adducts. These products can undergo further reactions to form crosslinked products; advanced glycation endproducts are involved in the generation of reactive oxygen species (ROS). See Deyl, Z. and Mikšík, I., Post-translational non-enzymatic modification of proteins I. Chromatography of marker adducts with special emphasis to glycation reactions, *J. Chromatog.* 699, 287–309, 1997; Bonnefont-Rousselot, D., Glucose and reactive oxygen species, *Curr. Opin. Clin. Nutr.* 5, 561–568, 2002; Tessier, F.J., Monnier, V.M., Sayre, L.M., and Kornfield, J.A., Triosidines: novel Maillard reaction products and crosslinks from the reaction of triose sugars with lysine and arginine residues, *Biochem. J.*, 369, 705–710, 2003: Thornally, P.J., Battah, S., Ahmed, N., Karachalias, N., Agalou, S., Babaei-Jadidi, R., and Dawnay, A., Quantitative screening of advanced glycation endproducts in cellular and extracellular proteins by tandem mass spectrometry, *Biochem. J.* 375, 581–592, 2003; Ahmed, N., Advanced glycation endproducts — role in pathology of diabetic complications, *Diabetes Res. Clin. Pract.* 67, 3–21, 2005.

Aeration   The dispersion and/or dissolution of a gas into a liquid; generally refers to the process of dispersing air or an oxygen–gas mixture into a liquid such as culture media (Wang, D.I. and Humphrey, A.E., Developments in agitation and aeration of fermentation systems, *Prog. Ind. Microbiol.* 8, 1–34, 1968; Papoutsakis, E.T., Media additives for protecting freely suspended animal cells against agitation and aeration damage, *Trends Biotechnol.* 9, 316–324, 1991; Barberel, S.I. and Walker, J.R., The effect of aeration upon the secondary metabolism of microorganisms, *Biotechnol. Genet. Eng. Rev.* 17, 281–323, 2000). Also refers to the process of air dispersion in the pulmonary system, which can include both the inspiratory process and the exchange between the pulmonary system and the vascular bed, most frequently the latter (Newman, B. and Oh, K.S., Abnormal pulmonary aeration in infants and children, *Radiol. Clin. North Am.* 26, 323–339, 1988; Kothari, N.A. and Kramer, S.S., Bronchial diseases and lung aeration in children, *J. Thorac. Imaging* 16, 207–223, 2001).

Aerosol   A colloidlike dispersion of a liquid or solid material into a gas. There is
                       considerable interest in the use of aerosols as drug delivery vehicles. See
                       Sanders, P.A., *Aerosol Science*, Van Nostrand Reinhold, New York, 1970;
                       Sanders, P.A., *Handbook of Aerosol Technology*, Van Nostrand Reinhold,
                       New York, 1979; Davies, C.N., Ed., *Aerosol Science*, Academic Press,
                       London, 1996; Adjei, A.L. and Gupta, P.K., *Inhalation Delivery of Ther-
                       apeutic Peptides and Proteins*, Marcel Dekker, New York, 1997; Macalady,
                       D.L., *Perspectives in Environmental Chemistry*, Oxford University Press,
                       New York, 1998; Hinds, W.C., *Aerosol Technology: Properties, Behavior,
                       and Measurement of Airborne Particles*, John Wiley & Sons, New York,
                       1999; Roche, N. and Huchon, G.J., Rationale for the choice of an aerosol
                       delivery system, *J. Aerosol. Med.* 13, 393–404, 2000; Gautam, A., Wal-
                       drep, J.C., and Densmore, C.L., Aerosol gene therapy, *Mol. Biotechnol.*
                       23, 51–60, 2003; Densmore, C.L., The re-emergence of aerosol gene
                       delivery: a viable approach to lung cancer therapy, *Curr. Cancer Drug
                       Targets* 3, 275–286, 2003. See also *Colloid*.

Affibody   A phage-selected protein developed using a scaffold domain from Protein
                       A. Such a protein can be selected for specific binding characteristics. See
                       Ronnmark, J., Hansson, M., Nguyen, T. et al., Construction and charac-
                       terization of affibody-Fc chimeras produced in *Escherichia coli*, *J. Immunol.
                       Meth.* 261, 199–211, 2002; Eklund, M., Axelsson, L., Uhlen, M., and
                       Nygren, P.A., Anti-idiotypic protein domains selected from protein A-
                       based affibody libraries, *Proteins* 48, 454–462, 2002; Renberg, B.,
                       Shiroyama, I., Engfeldt, T. et al., Affibody protein capture microarrays:
                       synthesis and evaluation of random and directed immobilization of affibody
                       molecules, *Analyt. Biochem.* 341, 334–343, 2005; Orlova, A., Nilsson, F.Y.,
                       Wikman, M. et al., Comparative *in vivo* evaluation of technetium and iodine
                       labels on an anti-HER2 affibody for single-photon imaging of HER2
                       expression in tumors, *J. Nucl. Med.* 47, 512–519, 2006; Wahlberg, E. and
                       Hard, T., Conformational stabilization of an engineered binding protein,
                       J. Am. Chem. Soc. 128, 7651–7660, 2006; Lendel, C., Dogan, J., and Hard,
                       T., Structural basis of molecular recognition in an affibody: affibody complex,
                       J. Mol. Biol., 359, 1293–1304, 2006.

Affinity   The use of affinity reagents for the study of the proteome. The concept of
Proteomics   the design and use of affinity labels for the study of proteins is well
                       understood (see Plapp, B.V. and Chen, W.S., Affinity labeling with omega-
                       bromoacetamide fatty acids and analogs, *Methods Enzymol.* 72, 587–591,
                       1981; Plapp, B.V., Application of affinity labeling for studying structure
                       and function of enzymes, *Methods Enzymol.* 87, 469–499, 1982; Fan, F.
                       and Plapp, B.V., Probing the affinity and specificity of yeast alcohol
                       dehydrogenase I for coenzymes, *Arch. Biochem. Biophys.* 367, 240–249,
                       1999). For application of affinity technology to proteomics, see Larsson,
                       T., Bergstrom, J., Nilsson, C., and Karlsson, K.A., Use of an affinity
                       proteomics approach for the identification of low-abundant bacterial
                       adhesins as applied on the Lewis(b)-binding adhesin of *Helicobacter
                       pylori*, *FEBS Lett.* 469, 155–158, 2000; Agaton, C., Falk, R., Hoiden
                       Guthenberg, I. et al., Selective enrichment of monospecific polyclonal
                       antibodies for antibody-based proteomics efforts, *J. Chromatog. A*, 1043,
                       33–40, 2004; Strege, M.A. and Lagu, A.L., Eds., *Capillary Electrophoresis
                       of Proteins and Peptides*, Humana Press, Totowa, NJ, 2004; Stults, J.T.
                       and Arnott, D., Proteomics, *Methods Enzymol.* 402, 245–289, 2005; Monti, M.,

Orru, S., Pagnozzi, D., and Pucci, P., Interaction proteomics, *Biosci. Rep.* 25, 45–56, 2005; Zanders, E.D., Ed., *Chemical Genomics: Reviews and Protocols*, Humana Press, Totowa, NJ, 2005; Schou, C. and Heegaard, N.H., Recent applications of affinity interactions in capillary electrophoresis, *Electrophoresis* 27, 44–59, 2006; Niwayama, S., Proteomics in medicinal chemistry, *Mini Rev. Med. Chem.* 6, 241–246, 2006; Nedelkov, D. and Nelson, R.W., Eds., *New and Emerging Proteomics Techniques,* Humana Press, Totowa, NJ, 2006. See also *Activity-Based Proteomics.*

Agar/Agarose   Agar is a heterogeneous natural product derived from algae/seaweed. It is used as a gelatin-like "thickening" agent in cooking. Agar is also used as a matrix for growing microorganisms. See Turner, H.A., Theory of assays performed by diffusion in agar gel. I: General considerations, *J. New Drugs* 41, 221–226, 1963; Rees, D.A., Structure, conformation, and mechanism in the formation of polysaccharide gels and networks, *Adv. Carbohydr. Chem. Biochem.* 24, 267–332, 1969; Metcalf, D., Clinical applications of the agar culture technique for haematopoietic cells, *Rev. Eur. Etud. Clin. Biol.* 16, 855–859, 1971; Johnstone, K.I., *Micromanipulation of Bacteria: The Cultivation of Single Bacteria and Their Spores by the Agar Gel Dissection Techniques,* Churchill-Livingston, Edinburgh, UK, 1973; Watanabe, T., *Pictorial Atlas of Soil and Seed Fungi: Morphologies of Cultured Fungi and Key to Species,* CRC Press, Boca Raton, FL, 1973; Wilkinson, M.H.F., *Digital Image Analysis of Microbes: Imaging, Morphometry, Fluorometry, and Motility Techniques and Applications*, Wiley, Chichester, UK, 1988; Holt, H.M., Gahrn-Hansen, B., and Bruun, B., *Shewanella algae* and *Shewanella putrefaciens*: clinical and microbiological characteristics, *Clin. Microbiol. Infect.* 11, 347–352, 2005; Discher, D.E., Janmey, P., and Wang, Y.L., Tissue cells feel and respond to the stiffness of their substrate, *Science* 310, 1139–1143, 2005. Agar is composed of two primary components: agarose, which is a gelling component, and agaropectin, which is a sulfated, nongelling component. Agarose is used as a matrix for the separation of large molecules such as DNA. See Lai, E.H.C. and Birren, B.W., Eds., *Electrophoresis of Large DNA Molecules: Theory and Applications*, Cold Spring Harbor Laboratory Press, Cold Spring Harbor, NY, 1990; Birren, B.W. and Lai, E.H.C., *Pulsed Field Gel Electrophoresis: A Practical Guide,* Academic Press, San Diego, CA, 1993; Bickerstaff, G.F., *Immobilization of Enzymes and Cells,* Humana Press, Totowa, NJ, 1997; Westermeier, R., *Electrophoresis in Practice: A Guide to Methods and Applications of DNA and Protein Separations*, 3rd ed., Wiley-VCH, Weinheim, Germany, 2001.

Aggregation   The process of forming an ordered or disordered group of particles, molecules, bubbles, drops, or other physical components that bind together in an undefined fashion; a common physical analogy is concrete or brick. Aggregation is used to measure macromolecular interactions and the interactions of cells such as platelets and frequently involves nephelometry. Agglutination is a term used to describe the aggregation or clumping of blood cells or bacteria caused by antibodies or other biological or chemical factors. Aggregation of proteins is thought to be involved in the pathogenesis of diseases such as Parkinson's disease and Alzheimer's disease; these diseases are thought to be conformation diseases of proteins resulting in disorder structure and aggregation. Aggregation of blood platelets is an initial step in the hemostatic response. See Born, G.V., Inhibition of thrombogenesis by inhibition of platelet aggregation, *Thromb. Diath. Haemorrh. Suppl.* 21, 159–166, 1966; Zucker, M.B., ADP- and collagen-induced platelet aggregation *in vivo* and *in vitro, Thromb. Diath. Haemorrh. Suppl.* 26, 175–184, 1967; Luscher, E.F., Pfueller, S.L., and Massini, P., Platelet aggregation by large molecules, *Ser. Haematol.* 6, 382–391, 1973; Harris, R.H. and Mitchell, R., The role of polymers in microbial aggregation, *Ann. Rev. Microbiol.* 27, 27–50, 1973; Harrington, R.A., Kleimna, N.S., Granger, C.B. et al., Relation between inhibition of platelet aggregation and clinical outcomes, *Am. Heart J.* 136, S43–S50, 1998; Hoylaerts, M.F., Oury, C., Toth-Zamboki, E., and Vermylen, J., ADP receptors in platelet activation and aggregation, *Platelets* 11, 307–309, 2000; Kopito, R.R., Aggresomes, inclusion bodies, and protein aggregation, *Trends Cell Biol.* 10, 524–530, 2000; Savage, B., Cattaneo, M., and Ruggeri, Z.M., Mechanisms of platelet aggregation, *Curr. Opin. Hematol.* 8, 270–276, 2001; Valente, J.J., Payne, R.W., Manning, M.C. et al., Colloidal behavior of proteins: effects of the second virial coefficient on solubility, crystallization, and aggregation of proteins in aqueous solution, *Curr. Pharm. Biotechnol.* 6, 427–436, 2005; Schwarzinger, S., Horn, A.H., Ziegler, J., and Sticht, H., Rare large-scale subdomain motions in prion protein can initiate aggregation, *J. Biomol. Struct. Dyn.* 23, 581–590, 2006; Ellis, R.J. and Minton, A.P., Protein aggregation in crowded environments, *Biol. Chem.* 387, 485–497, 2006; Estada, L.D. and Soto, C., Inhibition of protein misfolding and aggregation by small rationally designed peptides, *Curr. Pharm. Des.* 12, 2557–2567, 2006.

Agonist   Generally a compound or substance that binds to a receptor site, which could be on a cell membrane or a protein and elicits a positive physiological response. See Gowing, L., Ali, R., and White, J., Opioid antagonists with minimal sedation for opioid withdrawal, *Cochrane Database Syst. Rev.* 2, no. CD002021, 2002; Bernardo, A. and Minghetti, L., PPAR-gamma agonists as regulators of microglial activation and brain inflammation, *Curr. Pharm. Des.* 12, 93–109, 2006; Bonuccelli, U. and Pavese, N., Dopamine agonists in the treatment of Parkinson's disease, *Expert Rev. Neurother.* 6, 81–89, 2006; Thobois, S., Proposed dose equivalence for rapid switch between dopamine receptor agonists in Parkinson's disease: a review of the literature, *Clin. Ther.* 28, 1–12, 2006; Schwartz, T.W. and Holst, B., Ago-allosteric modulation and other types of allostery in dimeric 7TM receptors, *J. Recept. Signal Transduct. Res.* 26, 107–128, 2006.

Albumin   A protein, most notably derived from plasma or serum and secondarily from egg (ovalbumin). It is the most abundant protein in blood/plasma, constituting approximately half of the total plasma protein. It functions in establishing plasma colloid strength, which preserves the fluid balance between the intravascular and extravascular space (Starling, E.H., On the absorption of fluids from the connective tissue spaces, *J. Physiol.* 19, 312–326, 1896). Albumin, particularly bovine serum albumin (BSA), is used as a model protein and as a standard for the measurement of protein concentration. See Foster, J.F., Plasma albumin, in *The Plasma Proteins*, Vol. 1, pp. 179–239, F.W. Putnam, Ed. Academic Press, New York, 1960; Tanford, C., Protein denaturation, *Adv. Protein Chem.* 23, 121–282, 1968; Peters, T., Jr., Serum albumin, *Adv. Clin. Chem.* 13, 37–111, 1970; Gillette, J.R., Overview of drug-protein binding, *Ann. N.Y. Acad. Sci.* 226, 6–17, 1973; Peters, T., *All about Albumin: Biochemistry, Genetics, and Medical Applications,*

Academic Press, San Diego, CA, 1996; Vo-Dinh, T., Protein nanotechnology: the new frontier in biosciences, *Methods Mol. Biol.* 300, 1–13, 2005; Quinlan, G.J., Martin, G.S., and Evans, T.W., Albumin: biochemical properties and therapeutic potential, *Hepatology* 41, 1211–1219, 2005; Rasnik, I., McKenney, S.A., and Ha, T., Surfaces and orientation: much to FRET about? *Acc. Chem. Res.* 38, 542–548, 2005; Smales, C.M. and James, D.C., Eds., *Therapeutic Proteins: Methods and Protocols*, Humana Press, Totowa, NJ, 2005; Yamakura, F. and Ikeda, K., Modification of tryptophan and tryptophan residues in proteins by reactive nitrogen species, *Nitric Oxide* 14, 152–161, 2006; Chuang, V.T. and Otagiri, M., Stereoselective binding of human serum albumin, *Chirality* 18, 159–166, 2006; Ascenzi, P., Bocedi, A., Notari, S. et al., Allosteric modulation of drug binding to human serum albumin, *Mini Rev. Med. Chem.* 6, 483–489, 2006. Albumin was the first protein biopharmaceutical (Newhauser, L.R. and Loznen, E.L., Studies on human albumin in military medicine: the standard Army-Navy package of serum albumin [concentrated], *U.S. Navy Med. Bull.* 40, 796–799, 1942; Heyl, J.T., Gibson, J.G., II, and Janeway, C.W., Studies on the plasma proteins. V. The effect of concentrated solutions of human and bovine serum albumin in man, *J. Clin. Invest.* 22, 763–773, 1943) and is used for a variety of clinical indications (Blauhut, B. and Lundsgaard-Hansen, P., Eds., *Albumin and the Systemic Circulation*, Karger, Berlin, 1986) including use in extracorporeal circulation as a "bridge-to-transplant" (Sen, S. and Williams, R., New liver support devices in acute liver failure: a critical evaluation, *Semin. Liver Dis.* 23, 283–294, 2003; Tan, H.K., Molecular absorbent recirculating system [MARS], *Ann. Acad. Med. Singapore* 33, 329–335, 2004; George, J., Artificial liver support systems, *J. Assoc. Physicians India* 52, 719–722, 2004; Barshes, N.R., Gay, A.N., Williams, B. et al., Support for the acutely failing liver: a comprehensive review of historic and contemporary strategies, *J. Am. Coll. Surg.* 201, 458–476, 2005). Albumin is also noted for its ability to interact with various dyes and the binding of bromocresol green is an example of a clinical assay method for albumin (Rodkey, F.L., Direct spectrophotometric determination of albumin in human serum, *Clin. Chem.* 11, 478–487, 1965; Hill, P.G., The measurement of albumin in serum and plasma, *Ann. Clin. Biochem.* 22, 565–578, 1985; Doumas, B.T. and Peter, T., Jr., Serum and urine albumin: a progress report on their measurement and clinical significance, *Clin. Chim. Acta* 258, 3–20, 1997; Duly, E.B., Grimason, S., Grimaon, P. et al., Measurement of serum albumin by capillary zone electrophoresis, bromocresol green, bromocresol purple, and immunoassay methods, *J. Clin. Pathol.* 56, 780–781, 2003). Albumin is a general designation to describe a fraction of simple proteins that are soluble in water and dilute salt solutions as opposed to the globulin fraction, which is insoluble in water but soluble in dilute salt solutions. This is an old classification and has many exceptions (Taylor, J.F., The isolation of proteins, in Neurath, H. and Bailey, K., Eds., *The Proteins: Chemistry, Biological Activity, and Methods*, Vol. 1, pp. 1–85, Academic Press, New York, 1953). Albumins also migrate faster than globulins on electrophoresis, which resulted in the development of the classification of plasma proteins as albumins and globulins (Cooper, G.R., Electrophoretic and ultracentrifugal analysis of normal human serum, in *The Plasma Proteins*, Putnam, F.W., Ed., Academic Press, New York, 1960, pp. 51–103).

Algorithm   The underlying iterative method or mathematic theory for any particular
                       computer programming technique; a precisely described routine process
                       that can be applied and systematically followed through to a conclusion;
                       a step-by-step procedure for solving a problem or accomplishing some
                       end. There are a variety of algorithms ranging from defining clinical
                       treatment protocols to aligning and predicting sequences of biopolymers.
                       See Rose, G.D. and Seltzer, J.P., A new algorithm for finding the peptide
                       chain turns in a globular protein, *J. Mol. Biol.* 113, 153–164, 1977; Gotoh,
                       O., An improved algorithm for matching biological sequences, *J. Mol.
                       Biol.* 162, 705–708, 1982; Dandekar, T. and Argos, P., Folding the main
                       chain of small proteins with the genetic algorithm, *J. Mol. Biol.* 236,
                       844–861, 1994; Rarey, M., Kramer, B., Langauer, T., and Klebe, G., A
                       fast, flexible docking method using an incremental construction algorithm,
                       J. Mol. Biol. 261, 470–489, 1996; Jones, G., Willett, P., Glen, R.C. et al.,
                       Development and validation of a genetic algorithm for flexible docking,
                       J. Mol. Biol. 267, 427–448, 1997; Samudrala, R. and Moult, J., A graph-
                       theoretic algorithm for comparative modeling of protein structure, *J. Mol. Biol.*
                       279, 287–302, 1998; Chacon, P., Diaz, J.F., Moran, F., and Andreu, J.M.,
                       Reconstruction of protein form with X-ray solution scattering and a genetic
                       algorithm, *J. Mol. Biol.* 299, 1289–1302, 2000; Mathews, D.H. and Turner,
                       D.H., Dyalign: an algorithm for finding the secondary structure common
                       to two RNA sequences, *J. Mol. Biol.* 317, 191–203, 2002; Herrmann, T.,
                       Guntert, P., and Wuthrich, K., Protein NMR structure determination with
                       automated NOE assignment using the new software CANDID and the
                       torsion angle dynamics algorithm DYANA, *J. Mol. Biol.* 319, 209–227, 2002;
                       Andronescu, M., Fejes, A.P., Hutter, F. et al., A new algorithm for RNA
                       secondary structure design, *J. Mol. Biol.* 336, 607–624, 2004; Fang, Q.
                       and Shortle, D., Protein refolding in silico with atom-based statistical
                       potentials and conformational search using a simple genetic algorithm, *J.
                       Mol. Biol.* 359, 1456–1467, 2006.

Alloantibody   Also an isoantibody. An antibody directed against a cell or tissue from an
                       individual of the same species. Transplantation antibodies, transfusion
                       antibodies, and antibodies against blood coagulation factors such as factor
                       VIII inhibitors are examples of alloantibodies. See Glotz, D., Antoine, C., and
                       Duboust, A., Antidonor antibodies and transplantation: how to deal with them
                       before and after transplantation, *Transplantation* 79 (Suppl. 3), S30–S32,
                       2005; Colvin, R.B. and Smith, R.N., Antibody-mediated organ-allograft
                       rejection, *Nat. Rev. Immunol.* 5, 807–817, 2005; Moll, S. and Pascual, M.,
                       Humoral rejection of organ allografts, *Am. J. Transplant.* 5, 2611–2618,
                       2005; Waanders, M.M., Roelen, D.L, Brand, A., and Class, F.H., The
                       putative mechanism for the immunomodulating effect of HLA-DR shared
                       allogeneic blood transfusion on the alloimmune response, *Transfus. Med.
                       Rev.* 19, 281–287, 2005.

Alloantigen   An antigen present in some, but not all, members of a species or strain. The
                       histocompatibility locus antigen (HLA) is an example. See Schiffman, G.
                       and Marcus, D.M., Chemistry of the ABH blood group substances, *Prog.
                       Hematol.* 27, 97–116, 1964; Race, R.R., Contributions of blood groups to
                       human genetics, *Proc. R. Soc. Lond. B. Biol. Sci.* 163, 151–168, 1965;
                       Dausset, J., Leucocyte and tissue groups, *Vox Sang.* 11, 263–275, 1966;
                       Amos, B., Immunologic factors in organ transplantation, *Am. J. Med.* 55,
                       767–775, 1968; Marcus, D.M., The ABO and Lewis blood-group system.

Immunochemistry, genetics, and relation to human disease, *N. Engl. J. Med.* 280, 994–1006, 1969; Bach, F.H., Histocompatibility in man — genetic and practical considerations, *Prog. Med. Genet.* 6, 201–240, 1969; Drozina, G., Kohoutek, J., Janrane-Ferrat, N., and Peterlin, B.M., Expression of MHC II genes, *Curr. Top. Microbiol. Immunol.* 290, 147–170, 2005; Serrano, N.C., Millan, P., and Paez, M.C., Non-HLA associations with autoimmune diseases, *Autoimmun. Rev.* 5, 209–214, 2006; Koehn, B., Gangappa, S., Miller, J.D., Ahmed, R., and Larsen, C.P., Patients, pathogens, and protective immunity: the relevance of virus-induced alloreactivity in transplantation, *J. Immunol.* 176, 2691–2696, 2006; Turesson, C. and Matteson, E.L., Genetics of rheumatoid arthritis, *Mayo Clin. Proc.* 81, 94–101, 2006.

Allosteric   Originally a term that described the interaction of small molecules with an enzyme at a site physically distant from the active site where such interaction influenced enzyme activity. These small molecules were generally related to the substrate or product of the enzyme action. More recently, it has been used to describe the modulation of enzyme activity by the binding of a large or small molecule to a site distant from the active site. See Changeux, J.-P., Allosteric interactions interpreted in terms of quaternary structure, *Brookhaven Symp. Biol.* 17, 232–249, 1964; Monod, J., From enzymatic adaptation to allosteric transitions, *Science* 154, 475–483, 1966; Stadtman, E.R., Allosteric regulation of protein activity, *Adv. Enzymol. Relat. Areas Mol. Biol.* 28, 41–154, 1966; Changeux, J.-P. and Kvamme, E., *Regulation of Enyzme Activity and Allosteric Interactions,* Academic Press, New York, 1968; Frieden, C., Protein–protein interaction and enzymatic activity, *Annu. Rev. Biochem.* 40, 653–696, 1971; Matthews, B.W. and Bernhard, S.A., Structure and symmetry of oligomeric enzymes, *Annu. Rev. Biophys. Bioeng.* 2, 257–317, 1973; Hammes, G.G. and Wu, C.W., Kinetics of allosteric enzymes, *Annu. Rev. Biophys. Bioeng.* 3, 1–33, 1974; Kurganov, B.I., *Allosteric Enzymes: Kinetic Behavior*, Wiley, Chichester, UK, 1982; Perutz, M.F., *Mechanisms of Cooperativity and Allosteric Regulation in Proteins*, Cambridge University Press, Cambridge, UK, 1990; Segal, L.A., *Biological Kinetics*, Cambridge University Press, Cambridge, UK, 1991; Ostermeier, M., Engineering allosteric protein switches by domain insertion, *Protein Eng. Des. Sel.* 18, 359–364, 2005; Horovitz, A. and Willison, K.R., Allosteric regulation of chaperonins, *Curr. Opin. Struct. Biol.* 15, 646–651, 2005; Ascenzi, P., Bocedi, A., Notari, S. et al., Allosteric modulation of drug binding to human serum albumin, *Mini Rev. Med. Chem.* 6, 483–489, 2006.

Alternative Splicing   Alternative splicing is a process by which biological diversity can be increased without change in DNA content. Alternative splicing is a mechanism by a single pre-mRNA and is processed in different ways (different splicing sites) to yield a diverse group of messenger RNA molecules. See Choi, E., Kuehl, M., and Wall, R., RNA splicing generates a variant light chain from an aberrantly rearranged kappa gene, *Nature* 286, 776–779, 1980; Mariman, E.C., van Beek-Reinders, R.J., and van Venrooij, W.J., Alternative splicing pathways exist in the formation of adenoviral late messenger RNAs, *J. Mol. Biol.* 163, 239–256, 1983; Lerivray, R., Mereau, A., and Osborne, H.B., Our favorite alternative splice site, *Biol. Cell.* 98, 317–321, 2006; Florea, L., Bioinformatics of alternative splicing and its regulation, *Brief Bioinform.* 7, 55–69, 2006; Xing, Y. and Lee, C., Alternative splicing and RNA selection pressure — evolutionary consequences for eukaryotic genomes, *Nat. Rev. Genet.* 7, 499–509, 2006. Alternative *trans*-splicing has also been demonstrated. See Maniatis, T. and Tasic, B., Alternative pre-mRNA splicing and proteome expansion in metazoans, *Nature* 418, 236–243, 2002; Garcia-Blanco, M.A., Messenger RNA reprogramming by spliceosome-mediated RNA *trans*-splicing, *J. Clin. Invest.* 112, 474–480, 2003; Kornblitt, A.R., de la Mata, M., Fededa, J.P. et al., Multiple links between transcription and splicing, *RNA* 10, 1489–1498, 2004; Horiuchi, T. and Aigaki, T., Alternative *trans*-splicing: a novel mode of pre-mRNA processing, *Biol. Chem.* 98, 135–140, 2006. The production of variants of fibronectin is one of the better-known examples of alternative splicing (see Schwarzbauer, J.E., Paul, J.T., and Hynes, R.O., On the origin of species of fibronectin, *Proc. Natl. Acad. Sci. USA* 82, 1424–1428, 1985).

Ambisense A genome or genome segment that contains regions that are positive-sense for some genes and negative-sense (antisense) for other genes as in an ambisense RNA as viral ssRNA genome or genome segment. See Bishop, D.H., Ambisense RNA viruses: positive and negative polarities combined in RNA virus genomes, *Microbiol. Sci.* 3, 183–187, 1986; Ngugen, M. and Naenni, A.L., Expression strategies of ambisense viruses, *Virus Res.* 93, 141–150, 2003; van Knippenberg, I., Goldbach, R., and Kormelink, R., Tomato spotted wilt virus S-segment mRNAs have overlapping 3′-ends containing a predicted stem-loop structure and conserved sequence motif, *Virus Res.* 110, 125–131, 2005; Barr, J.N., Rodgers, J.W., and Wertz, G.W., The Bunyamwera virus mRNA transcription signal resides within both the 3′ and the 5′ terminal regions and allows ambisense transcription from a model RNA segment, *J. Virol.* 79, 12602–12607, 2005.

Aminophos- Amino-containing phopholipids such as phosphatidyl ethanolamine and
pholipids phosphatidyl serine. Phosphatidyl serine is involved in specific membrane functions and changes in membrane distribution producing asymmetry are considered important for function. There are enzymes described as flippases, floppases, transporters, scramblease, and aminophospholipid translocase, which are responsible for this asymmetry, which results in aminophospholipids on the cytoplasmic side of the membrane and cholines and sphingolipids on the outer surface. See Devaux, P.F., Protein involvement in transmembrane lipid asymmetry, *Annu. Rev. Biophys. Biomol. Struct.* 21, 417–439, 1992; Schlegel, R.A., Callahan, M.K., and Williamson, P., *Ann. N.Y. Acad. Sci.* 926, 271–225, 2000; Daleke, D.L. and Lyles, J.V., Identification and purification of aminophospholipid flippases, *Biochem. Biophys. Acta* 1486, 108–127, 2000; Balasubramanian, K. and Schroit, A.J., Aminophospholipid asymmetry: a matter of life and death, *Annu. Rev. Physiol.* 65, 701–734, 2003; Daleke, D.L., Regulation of transbilayer plasma membrane phospholipid asymmetry, *J. Lipid. Res.* 44, 233–242, 2003.

Amorphous A solid form of a material that does not have a definite form such as a crystal
Powder structure. Differing from a crystal form, an amorphous form is thermodynamically unstable and does not have a defined melting point. The physical characteristics of an amorphous powder make it the desired physical state for drugs after lyophilization. See Izutsu, K., Yoshioka, S., and Kojima, S., Increased stabilizing effects of amphiphilic excipients on freeze-drying of lactate dehydrogenase (LDH) by dispersion into sugar matrices, *Pharm. Res.* 12, 838–843, 1995; Jennings, T.A., *Lyophilization Introduction and Basic Principles,* Interpharm Press, Denver, CO, 1999; Royall, P.G., Huang, C.Y., Tang, S.W. et al., The development of DMA for the detection

of amorphous content in pharmaceutical powdered material, *Int. J. Pharm.* 301, 181–191, 2005; Stevenson, C.L., Bennett, D.B., and Lechuga-Ballesteros, D., Pharmaceutical liquid crystals: the relevance of partially ordered systems, *J. Pharm. Sci.* 94, 1861–1880, 2005; Skakle, J., Applications of X-ray power diffraction in materials chemistry, *Chem. Rec.* 5, 252–262, 2005; Farber, L., Tardos, G.I., and Michaels, J.N., Micro-mechanical properties of drying material bridges of pharmaceutical excipients, *Int. J. Pharm.* 306, 41–55, 2005; Jovanovic, N., Bouchard, A., Hofland, G.W. et al., Distinct effects of sucrose and trehalose on protein stability during supercritical fluid drying and freeze-drying, *Eur. J. Pharm. Sci.* 27, 336–345, 2006; Jorgensen, A.C., Miroshnyk, I., Karjalainen, M. et al., Multivariate data analysis as a fast tool in evaluation of solid state phenomena, *J. Pharm. Sci.* 95, 906–916, 2006; Shah, S., Sharma, A., and Gupta, M.N., Preparation of crosslinked enzyme aggregates by using bovine serum albumin as a proteic feeder, *Anal. Biochem.* 351, 207–213, 2006; Reverchon, E. and Atanacci, A., Cyclodextrins micrometric powders obtained by supercritical fluid processing, *Biotechnol. Bioeng.*, 94, 753–761, 2006.

Amphipathic (Amphiphilic) A compound that has both hydrophilic (lyophilic) and hydrophobic (lyophobic) properties. This is important for the interaction of proteins with lipids and for the properties of cell-penetrating peptides. Detergents are amphipathic molecules. See Scow, R.O., Blanchette-Mackie, E.J., and Smith, L.C., Transport of lipids across capillary endothelium, *Fed. Proc.* 39, 2610–2617, 1980; Corr, P.B., Gross, R.W., and Sobel, B.E., Amphipathic metabolites and membrane dysfunction in ischemic myocardium, *Circ. Res.* 55, 135–154, 1984; Fasman, G.D., *Prediction of Protein Structures and the Principles of Protein Conformation,* Plenum Press, New York, 1989; Anantharamaiah, G.M., Brouillette, C.G., Engler, J.A. et al., Role of amphipathic helixes in HDL structure/function, *Adv. Exp. Med. Biol.* 285, 131–140, 1991; Epand, R.M., *The Amphipathic Helix*, CRC Press, Boca Raton, FL, 1993; Segrest, J.P., Garber, D.W., Brouillette, C.G. et al., The amphipathic alpha helix: a multifunctional structural motif in plasma apolipoproteins, *Adv. Protein Chem.* 45, 303–369, 1994; Lester, J.B. and Scott, J.D., Anchoring and scaffold proteins for kinases and phosphatases, *Recent Prog. Horm. Res.* 52, 409–429, 1997; Lesieur, C., Vecsey-Semjen, B., Abrami, L. et al., Membrane insertion: the strategies of toxins, *Mol. Membr. Biol.* 14, 45–64, 1997; Johnson, J.E. and Cornell, R.B., Amphitropic proteins: regulation by reversible membrane interactions, *Mol. Membr. Biol.* 16, 217–235, 1999; Tossi, A., Sandri, L., and Giangaspero, A., Amphipathic, alpha-helical antimicrobial peptides, *Biopolymers* 55, 4–30, 2000; Garavito, R.M. and Ferguson-Miller, S., Detergents as tools in membrane biochemistry, *J. Biol. Chem.* 276, 32403–32406, 2001; Langel, U., *Cell-Penetrating Peptides: Processes and Applications*, CRC Press, Boca Raton, FL, 2002; Simon, S.A. and McIntosh, T.J., Eds., *Peptide-Lipid Interactions*, Academic Press, San Diego, CA, 2002; El-Andaloussi, S., Holm, T., and Langel, U., Cell-penetrating peptides: mechanisms and applications, *Curr. Pharm. Des.* 11, 3597–3611, 2005; Deshayes, S., Morris, M.C., Divita, G., and Heitz, F., Interactions of primary amphipathic cell-penetrating peptides with model membranes: consequences on the mechanism of intracellular delivery of therapeutics, *Curr. Pharm. Des.* 11, 3629–3638, 2005.

Ampholyte An amphoteric electrolyte. In proteomics, this term is used to describe small multicharged organic buffers used to establish pH gradients in isoelectric focusing. See Righetti, P.G., Isoelectric focusing as the crow flies, *J. Biochem. Biophys. Methods* 16, 99–108, 1988; Patton, W.F., Pluskal, M.G., Skea, W.M. et al., Development of a dedicated two-dimensional gel electrophoresis system that provides optimal pattern reproducibility and polypeptide resolution, *Biotechniques* 8, 518–527, 1990; Hanash, S.M., Strahler, J.R., Neel, J.V. et al., Highly resolving two-dimensional gels for protein sequencing, *Proc. Natl. Acad. Sci. USA* 88, 5709–5713, 1991; Cade-Treyer, D., Cade, A., Darjo, A., and Jouvion-Moreno, M., Isoelectric focusing and titration curves in biomedicine and in agrofood industries: a multimedia teaching program, *Electrophoresis* 17, 479–482, 1996; Stoyanov, A.V. and Pawliszyn, J., Buffer composition changes in background electrolyte during electrophoretic run in capillary zone electrophoresis, *Analyst* 129, 979–982, 2004; Gorg, A., Weiss, W., and Dunn, M.J., Current two-dimensional technology for proteomics, *Proteomics* 4, 3665–3685, 2004; Kim, S.H., Miyatake, H., Ueno, T. et al., Development of a novel ampholyte buffer for isoelectric focusing: electric charge-separation of protein samples for X-ray crystallography using free-flow isoelectric focusing, *Acta Crystallogr. D Biol. Crystallogr.* 61, 799–802, 2005; Righetti, P.G., The Alpher, Bethe, Gamow of isoelectric focusing, the alpha-Centaury of electrokinetic methods, *Electrophoresis* 27, 923–938, 2006.

Amphoteric   Referring to a molecule such as a protein, peptide, or amino acid capable of having a positive charge, negative charge, or zero net charge. When at a zero net charge, it is also referred to as a zwitterion. See Haynes, D., The action of salts and non-electrolytes upon buffer solutions and amphoteric electrolytes and the relation of these effects to the permeability of the cell, *Biochem. J.* 15, 440–461, 1921; Akabori, S., Tani, H., and Noguchi, J., A synthetic amphoteric polypeptide, *Nature* 167, 1591–160, 1951; Coway-Jacobs, A. and Lewin, L.M., Isoelectric focusing in acrylamide gels: use of amphoteric dyes as internal markers for determination of isoelectric points, *Anal. Biochem.* 43, 294–400, 1971; Chiari, M., Pagani, L., and Righetti, P.G., Physico-chemical properties of amphoteric, isoelectric, macroreticulate buffers, *J. Biochem. Biophys. Methods* 23, 115–130, 1991; Blanco, S., Clifton, M.J., Joly, J.L., and Peltre, G., Protein separation by electrophoresis in a nonsieving amphoteric medium, *Electrophoresis* 17, 1126–1133, 1996; Tulp, A., Verwoerd, D., and Hart, A.A., Density-gradient isoelectric focusing of proteins in artificial pH gradients made up of binary mixtures of amphoteric buffers, *Electrophoresis* 18, 767–773, 1997; Akahoshi, A., Sato, K., Nawa, Y. et al., Novel approach for large-scale, biocompatible, and low-cost fractionation of peptides in proteolytic digest of food protein based on the amphoteric nature of peptides, *J. Agric. Food Chem.* 48, 1955–1959, 2000; Matsumoto, H., Koyama, Y., and Tanioka, A., Interaction of proteins with weak amphoteric-charged membrane surfaces: effect of pH, *J. Colloid Interface Sci.* 264, 82–88, 2003; Fortis, F., Girot, P., Brieau, O. et al., Amphoteric, buffering chromatographic beads for proteome prefractionation. I: theoretical model, *Proteomics* 5, 620–628, 2005; Kitano, H., Takaha, K., and Gemmei-Ide, M., Raman spectroscopic study of the structure of water in aqueous solutions of amphoteric polymers, *Phys. Chem. Chem. Phys.* 8, 1178–1185, 2006.

Amplicon   (Usually) the DNA product of a PCR reaction, usually an amplified segment of a gene or DNA. An RNA amplicon would be an RNA sequence and can be obtained by transcription-mediated amplification (See Bustin, S.A., Benes, V., Nolan, T., and Pfaffl, M.W., Quantitative real-time RT-PCR — a perspective, *J. Mol. Endocrinol.* 34, 597–601, 2005; Sarrazin, C., Highly sensitive hepatitis C virus RNA detection methods: molecular backgrounds and clinical significance, *J. Clin. Virol.* 25, S23–S29, 2002). This also refers to herpesvirus vectors for gene therapy (Oehmig, A., Fraefel, C., and Breakfield, X.O., Update on herpesvirus amplicon vectors, *Molecular Therapy* 10, 630–643, 2004).

Amyloid   A waxlike translucent insoluble material consisting largely of proteins that may or may not contain carbohydrates and is associated with tissue degeneration. Amyloid peptides/proteins are thought to be associated with Alzheimer's disease. Glenner, G.G., The pathogenetic and therapeutic implications of the discovery of the immunoglobulin origin of amyloid fibrils, *Hum. Pathol.* 3, 157–162, 1972; Franklin, E.C. and Zucker-Franklin, D., Current concepts of amyloid, *Adv. Immunol.* 15, 249–304, 1972; Glenner, G.G. and Terry, W.D., Characterization of amyloid, *Annu. Rev. Med.* 25, 131–135, 1974; Glenner, G.G. and Page, D.L., Amyloid, amyloidosis, and amyloidogenesis, *Int. Rev. Exp. Pathol.* 15, 1–92, 1976; Gorevic, P.D., Cleveland, A.B., and Franklin, E.C., The biologic significance of amyloid, *Ann. N.Y. Acad. Sci.* 389, 380–394, 1982; Reinhard, C., Herbert, S.S., and De Strooper, B., The amyloid-beta precursor protein: integrating structure with biological function, *EMBO J.* 24, 3996–4006, 2005; Meersman, F. and Dobson, C.M., Probing the pressure-temperature stability of amyloid fibrils provides new insights into their molecular properties, *Biochem. Biophys. Acta* 1764, 452–460, 2006; Tycko, R., Solid-state NMR as a probe of amyloid structure, *Protein Pept. Lett.* 13, 229–234, 2006; Torrent, J., Balny, C., and Lange, R., High pressure modulates amyloid formation, *Protein Pept. Lett.* 13, 271–277, 2006; Gorbenko, G.P. and Kinnuen, P.K., The role of lipid–protein interactions in amyloid-type protein fibril formation, *Chem. Phys. Lipids* 141, 72–82, 2006; Catalano, S.M., Dodson, E.C., Henze, D.A. et al., The role of amyloid-beta derived diffusible ligands (ADDLs) in Alzheimer's disease, *Curr. Top. Med. Chem.* 6, 597–608, 2006.

Anaphylatoxin(s) Fragment(s) of complement proteins released during complement activation. See Corbeil, L.B., Role of the complement system in immunity and immunopathology, *Vet. Clin. North Am.* 8, 585–611, 1978; Hugli, T.E. and Muller-Eberhard, H.J., Anaphylatoxins: C3a and C5a, *Adv. Immunol.* 26, 1–53, 1978; Hugli, T.E., The structural basis for anaphylatoxin and chemotactic functions of C3a, C4a, and C5a, *Crit. Rev. Immunol.* 1, 321–366, 1981; Hawlisch, H., Wills-Karp, M., Karp, C.L., and Kohl, J., The anaphylatoxins bridge innate and adaptive immune responses in allergic asthma, *Mol. Immunol.* 41, 123–131, 2004; Ali, H. and Panettieri, R.A., Jr., Anaphylatoxin C3a receptors in asthma, *Respir. Res.* 6, 19, 2005; Sunyer, J.O., Boshra, H., and Li, J., Evolution of anaphylatoxins, their diversity and novel roles in innate immunity: insights from the study of fish complement, *Vet. Immunol. Immunopathol.* 108, 77–89, 2005; Schmidt, R.E. and Gessner, J.E., Fc receptors and their interactions with complement in autoimmunity, *Immunol. Lett.* 100, 56–67, 2005; Chaplin, H., Jr., Review: the burgeoning history of the complement system, 1888–2005, *Immunohematol.* 21, 85–93, 2005; Lambrecht, B.N., An unexpected role for the anaphylatoxin C5a receptor in allergic sensitization, *J. Clin. Invest.* 116, 626–632, 2006.

Anergy   Lack of an immune response to an allergen (antigen); can refer to an indi-
                     vidual cell such as a B-cell or a T-cell, tissue, or intact organism; however,
                     it is used most frequently with respect to B-cells or T-cells and immuno-
                     logical tolerance. See Kantor, F.S., Infection, anergy, and cell-mediated
                     immunity, *N. Engl. J. Med.* 292, 629–634, 1975; Bullock, W.E., Anergy
                     and infection, *Adv. Intern. Med.* 21, 149–173, 1976; Dwyer, J.M., Anergy.
                     The mysterious loss of immunological energy, *Prog. Allergy* 35, 15–92,
                     1984; Brennan, P.J., Saouaf, S.J., Greene, M.I., and Shen. Y., Anergy and
                     suppression as coexistent mechanisms for the maintenance of peripheral
                     T-cell tolerance, *Immunol. Res.* 27, 295–302, 2003; Macian, F., Im, S.H.,
                     Garcia-Cozar, F.J., and Rao, A., T-cell anergy, *Curr. Opin. Immunol.* 16,
                     209–216, 2004; Mueller, D.L., E3 ubiquitin ligases as T-cell anergy factors,
                     Nat. Immunol. 5, 883–890, 2004; Faria, A.M. and Weiner, H.L., *Immunol.
                     Rev.* 206, 232–259, 2005; Akdis, M., Blaser, K., and Akdis, C.A., T regulatory
                     cells in allergy, *Chem. Immunol. Allergy* 91, 159–173, 2006; Ferry, H.,
                     Leung, J.C., Lewis , G. et al., B-cell tolerance, *Transplantation* 81, 308–315,
                     2006.

Angiopoietin   A protein family that binds to endothelial cells; specific for Tie2 receptor
                     kinase. See Plank, M.J., Sleeman, B.D., and Jones, P.F., The role of the
                     angiopoietins in tumor angiogenesis, *Growth Factors* 22, 1–11, 2004;
                     Oike, Y., Yasunaga, K., and Suda, T., Angiopoietin-related/angiopoietin-
                     like proteins regulate angiogenesis, *Int. J. Hematol.* 80, 21–28, 2004;
                     Giuliani, N., Colla, S., Morandi, F., and Rizzoli, V., Angiopoietin-1 and
                     myeloma-induced angiogenesis, *Leuk. Lymphoma* 46, 29033, 2005; Dhanabal,
                     M., Jeffers, M., LaRochelle, W.J., and Lichenstein, R.S., Angioarrestin: a
                     unique angiopoietin-related protein with anti-angiogenic properties, *Biochem.
                     Biophys. Res. Commun.* 333, 308–315, 2005; Armulik, A., Abramsson A., and
                     Betsholtz, C., Endothelial/pericyte interactions, *Circ. Res.* 97, 512–523, 2005.

Anisotropy   A difference in a physical property such as the melting point when measured
                     in different principal directions; antonym, isotropy. Anisotropy is also
                     defined as the property of being anisotropic as in the case of light trans-
                     mission, where different values are obtained when along axes in different
                     directions. Time-resolved fluorescence anisotropy decay measures the time
                     dependence of the depolarization of light emitted from a fluorophore
                     experiencing angular motions. In botany, anisotropy is defined as assuming
                     different positions in response to the action of external stimuli. See Kinosita,
                     K., Jr., Kawato, S., and Ikegami, A., Dynamic structure of biological and
                     model membranes: analysis by optical anisotropy decay measurement,
                     Adv. Biophys. 17, 147–203, 1984; Kinosita, K., Jr., and Ikegami, A.,
                     Dynamic structure of membranes and subcellular components revealed by
                     optical anisotropy decay methods, *Subcell. Biochem.* 13, 55–88, 1988;
                     Bucci, E. and Steiner, R.F., Anisotropy decay of fluorescence as an exper-
                     imental approach to proteins, *Biophys. Chem.* 30, 199–224, 1988; Matko,
                     J., Jenei, A., Matyus, L., Ameloot, M., and Damjanovich, S., Mapping of
                     cell surface protein-patterns by combined fluorescence anisotropy and
                     energy transfer measurements, *J. Photochem. Photobiol. B* 19, 69–73,
                     1993; Rachofsky, E.L. and Laws, W.R., Kinetic methods and data analysis
                     methods for fluorescence anisotropy decay, *Methods Enzymol.* 321,
                     216–238, 2000; Santos, N.C., Prieto, M., and Castanho, M.A., Quantifying
                     molecular partition into model systems of biomembranes: an emphasis on
                     optical spectroscopic methods, *Biochim. Biophys. Acta* 1612, 123–135,

2003; Vrielink, A. and Sampson, N., Sub-angstrom resolution x-ray structures: is seeing believing? *Curr. Opin. Struct. Biol.* 13, 709–715, 2003; Wang, J., Cao, Z., Jiang, Y. et al., Molecular signaling aptamers for real-time fluorescence analysis of proteins, *IUBMB Life* 57, 123–128, 2005; Dmitrienko, V.E., Ishida, K., Kirfel, A., and Ovchinnikova, E.N., Polarization anisotropy of X-ray atomic factors and "forbidden" resonant reflections, *Acta Crystallogr. A* 61, 481–493, 2005; Baskin, T.I., Anisotropic expansion of the plant cell wall, *Annu. Rev. Cell Dev. Biol.* 21, 203–222, 2005; Heilker, R., Zemanova, L., Valler, M.J., and Nienhaus, G.U., Confocal fluorescence microscopy for high-throughput screening of G-protein-coupled receptors, *Curr. Med. Chem.* 12, 2551–2559, 2005; Guthrie, J.W., Hamula, C.L., Zhang, H., and Le, X.C., Assays for cytokines using aptamers, *Methods* 38, 324–330, 2006.

Ankyrin-Repeat Domains/Proteins   A domain or motif, named after ankydrin, a cytoskeletal protein, is found in a large number of proteins. This domain, which was first described in a yeast cell cycle regulator (Swi6/cdc10) and *Drosphilia* (notch protein), consists of approximately 30 amino acids and is involved in protein–protein interactions. See Liou, H.C. and Baltimore, D., Regulation of the NF-kappa B/rel transcription factor and I kappa B inhibitor system, *Curr. Opin. Cell Biol.* 5, 477–487, 1993; Dedhar, S. and Hannigan, G.E., Integrin cytoplasmic interactions and bidirectional transmembrane signalling, *Curr. Opin. Cell Biol.* 8, 657–669, 1996; Sedgwick, S.G. and Smerdon, S.J., The ankyrin repeat: a diversity of interactions on a common structural framework, *Trends Biochem. Sci.* 24, 311–316, 1999; Yoganathan, T.N., Costello, P., Chen, X. et al., Integrin-linked kinase (ILK): a "hot" therapeutic target, *Biochem. Pharmacol.* 60, 1115–1119, 2000; Hryniewicz-Jankowska, A., Czogalla, A., Bok, E., and Sikorsk, A.F., Ankyrins, multifunctional proteins involved in many cellular pathways, *Folia Histochem. Cytobiol.* 40, 239–249, 2002; Lubman, O.Y., Korolev, S.V., and Kopan, R., Anchoring notch genetics and biochemistry; structural analysis of the ankyrin domain sheds light on existing data, *Mol. Cell.* 13, 619–626, 2004; Mosavi, L.K., Cammett, T.J., Desosiers, D.C., and Peng, Z.Y., The ankyrin repeat as molecular architecture for protein recognition, *Protein Sci.* 13, 1435–1448, 2004; Tanke, H.J., Dirks, R.W., and Raap, T., FISH and immunocytochemistry: toward visualizing single target molecules in living cells, *Curr. Opin. Biotechnol.* 16, 49–54, 2005; Trigiante, G. and Lu, X., ASPPs and cancer, *Nat. Rev. Cancer* 6, 217–226, 2006; Legate, K.R., Montañez, E., Kudlacek, O., and Fässler, R., ILK, PINCH, and parvin: the tIPP of integrin signaling, *Nat. Rev. Mol. Cell Biol.* 7, 20–31, 2006.

Annotation   Information added to a subject after the initial overall definition. Most frequently used in molecular biology for the addition of information regarding function to the initial description of a gene/gene sequence in a genome. See Brent, M.R., Genome annotation past, present, and future: how to define an ORF at each locus, *Genome Res.* 15, 1776–1786, 2005; Boutros, P.C. and Okey, A.B., Unsupervised pattern recognition: an introduction to the whys and wherefores of clustering microarray data, *Brief Bioinform.* 6, 331–343, 2005; Boeckman, B., Blatter, M.C., Famiglietti, L. et al., Protein variety and functional diversity: Swiss-Prot annotation in its biological context, *C. R. Biol.* 328, 882–899, 2005; Koonin, E.V., Orthologs, paralogs, and evolutionary genomics, *Annu. Rev. Genet.* 39, 309–338, 2005; Cahan, P., Ahmed, A.M., Burke, H. et al., List of list-annotated (LOLA): a database for annotation and comparison of published microarray gene lists, *Gene* 360, 78–82, 2005; Dong, Q., Kroiss, L., Oakley, F.D., Wang, B.B., and Brendel, V., Comparative EST analyses in plant systems, *Methods Enzymol.* 395, 400–418, 2005; Crockett, D.K., Seiler, C.E., III, Elenitoba-Johnson, K.S., and Kim, M.S., *J. Biomed. Tech.* 16, 341–346, 2005; Hermida, L., Schaad, O., Demougin, P., Descombes, P., and Primig, M., MIMAS: an innovative tool for network-based high-density oligonucleotide microarray data management and annotation, *BMC Bioinformatics* 7, 190, 2006; Huang, D., Wei, P., and Pan, W., Combining gene annotation and gene expression data in model-based clustering weighted method, *OMICS* 10, 28–39, 2006; Snyder, K.A., Feldman, H.J., Dumontier, M., Salama, J.J., and Hogue, C.W., Domain-based small molecule binding site annotation, *BMC Bioinformatics* 7, 152, 2006.

Anoikis   Apoptosis following loss of attachment to a matrix or specific anchorage site. See Grossman, J., Molecular mechanisms of "detachment-induced apoptosis—anoikis," *Apoptosis* 7, 247–260, 2002; Zvibel, I., Smets, F., and Soriano, H., Anoikis: roadblock to cell transplantation? *Cell Transplant.* 11, 621–630, 2002; Valentijn, A.J., Zouq, N., and Gilmore, A.P., Anoikis, *Biochem. Soc. Trans.* 32, 421–425, 2004; Zhan, M., Zhao, H., and Han, Z.C., Signalling mechanisms of anoikis, *Histol. Histopathol.* 19, 973–983, 2004; Reddig, P.J. and Juliano, R.L., Clinging to life: cell to matrix adhesion and cell survival, *Cancer Metastasis Rev.* 24, 425–439, 2005; Rennebeck, G., Martelli, M., and Kyprianou, N., Anoikis and survival connections in the tumor microenvironment: is there a role in prostate cancer metastasis? *Cancer Res.* 65, 11230–11235, 2005.

ANTH-Domain   A protein domain similar to the ENTH-domain and contained in proteins involved in endocytotic processes. See Stahelin, R.V., Long, F., Petter, B.J. et al., Contrasting membrane interaction mechanisms of AP180 N-terminal homology (ANTH) and epsin N-terminal homology (ENTH) domains, *J. Biol. Chem.* 278, 28993–28999, 2003; Sun, Y., Kaksonen, M., Madden, D.T. et al., Interaction of Slap2p's ANTH domain with PtdIns(4,5)P2 is important for actin-dependent endocytotic internalization, *Mol. Biol. Cell* 16, 717–730, 2005; Yao, P.J., Bushlin, I., and Petralia, R.S., Partially overlapping distribution of epsin1 and HIP1 at the synapse: analysis by immunoelectron microscopy, *J. Comp. Neurol.* 494, 368–379, 2006.

Antibody   A protein synthesized and secreted by a plasma cell. A plasma cell or antibody-secreting cell is derived from an undifferentiated B-cell. Antibodies are designated as the humoral immune response as opposed to the cellular immune response. Antibodies are usually synthesized and secreted in response to a foreign protein or bacteria. Natural antibody preparations are polyclonal in that such preparations are derived from a population of plasma cells. A monoclonal antibody is derived from a signal plasma cell clone. Antibodies can be formed against the self; such antibodies are referred to as autoantibodies. Disease resulting from the formation of antibodies are called autoimmune diseases and can result from disorders of the humoral immune system or the cellular immune system Antibodies are classified as IgG, IgM, IgE, IgA, and IgD. There are unusual naturally occurring antibodies such as camelid antibodies and artificial derivatives such as Fab fragments and scFv fragments.

Antibody-Dependent Cellular Cytotoxicity (ADCC)	This is usually the process by which an organism destroys bacterial and viral pathogens but also is the mechanism by which tumor cells are lysed secondary to treatment with antibodies. The process involves the recognition of epitopes by the Fab region of the IgG on the target cell surface, resulting in the binding of the antibody. The Fc domain is then recognized by a phagocytic cell such as a natural killer (NK) cell. The Fc region is critical for this process. See Santonine, A., Herberman, R.B., and Holden, H.T., Correlation between natural and antibody-dependent cell-mediated cytotoxicity against tumor targets in the mouse. II. Characterization of the effector cells, *J. Natl. Cancer Inst.* 63, 995–1003, 1979; Muller-Eberhard, H.J., The molecular basis of target cell killing by human lymphocytes and of killer cell self-protection, *Immunol. Rev.* 103, 87–98, 1981; Moretta, L., Moretta, A., Canonica, G.W. et al., Receptors for immunoglobulins on resting and activated human T-cells, *Immunol. Rev.* 56, 141–162, 1981; Dallegri, F. and Ottonello, D., Neutrophil-mediated cytotoxicity against tumor cells: state of the art, *Arch. Immunol. Ther. Exp.* 40, 39–42, 1992; Sissons, J.G. and Oldstone, M.B., Antibody-mediated destruction of virus-infected cells, *Adv. Immunol.* 29, 209–260, 2000; Perussia, B. and Loza, M.J., Assays for antibody-directed cell-mediated cytotoxicity (ADCC) and reverse ADCC (redirected cytotoxicity) in human natural killer cells, *Methods Mol. Biol.* 121, 179–192, 2000; Villamor, N., Montserrat, E., and Colomer, D., Mechanism of action and resistance to monoclonal antibody therapy, *Semin. Oncol.* 30, 424–433, 2003; Casadevall, A. and Pirofski, L.A., Antibody-mediated regulation of cellular immunity and the inflammatory response, *Trends Immunol.* 24, 474–478, 2003; Mellstedt, H., Monoclonal antibodies in human cancer, *Drugs Today* 39 (Suppl. C), 1–16, 2003; Gelderman, K.A., Tomlinson, S., Ross, G.D., and Gorter, A., Complement function in mAb-mediated cancer immunotherapy, *Trends Immunol.* 25, 158–164, 2004; Schmidt, R.E. and Gessner, J.E., Fc receptors and their interaction with complement in autoimmunity, *Immunol. Lett.* 100, 56–67, 2005; Iannello, A. and Ahmad, A. Role of antibody-dependent cell-mediated cytotoxicity in the efficacy of therapeutic anti-cancer monoclonal antibodies, *Cancer Metastasis Rev.* 24, 487–499, 2005.
Antibody Proteomics/ Antibody-Based Proteomics	The systematic generation and use of antibodies for the analysis of the proteome. An example would be the use of an antibody-based protein microarray. See Agaton, C., Falk, R., Hoiden Guthenberg, I. et al., Selective enrichment of monospecific polyclonal antibodies for antibody-based proteomics efforts, *J. Chromatog. A* 1043, 33–40, 2004; Nielsen, A.B. and Geierstanger, B.H., Multiplexed sandwich assays in a microwave format, *J. Immunol. Methods* 290, 107–120, 2004; Uhlen, M. and Ponten, F., Antibody-based proteomics for human tissue profiling, *Mol. Cell. Proteom.* 4, 384–393, 2005; Stenvall, M., Steen, J., Uhlen, M. et al., High-throughput solubility assay for purified recombinant protein immunogens, *Biochim. Biophys. Acta* 1752, 6–10, 2005; Uhlen, M., Bjorling, E., Agaton, C. et al., A human protein atlas for normal and cancer tissues based on antibody proteomics, *Mol. Cell Proteomics* 4, 1920–1932, 2005. See also *Immunoproteomics*.
Antibody Valency	Antibody valency refers to the number of antigen binding sites there are on a single antibody molecule. An IgG molecule, which consists of two heavy chains and two light chains (a dimer of heterodimers), has two antibody-binding sites and hence is bivalent. IgM, which is a pentamer of IgG, has a valency of 10. An scFv fragment is monovalent. An antibody with

increased valence is considered to have greater avidity. See Marrack, J.R., Hoch, H., and Johns, R.G., The valency of antibodies, *Biochem. J.* 48, xxi–xxii, 1951; Sela, M., Antibodies: shapes, homogeneity, and valency, *FEBS Lett.* 1, 83–85, 1968; van Regenmortel, M.H., Which value of antigenic valency should be used in antibody avidity calculations with multivalent antigens? *Mol. Immunol.* 25, 565–567, 1988; Gerdes, M., Meusel, M., and Spener, F., Influence of antibody valency in a displacement immunoassay for the quantitation of 2,4-dichlorophenoxyacetic acid, *J. Immunol. Methods* 223, 217–226, 1999; Hudson, P.J. and Kortt, A.A., High avidity scFv multimers: diabodies and triabodies, *J. Immunol. Methods* 231, 177–189, 1999; Hard, S.A. and Dimmock, N.J., Valency of antibody binding to virions and its determination by surface plasmon resonance, *Rev. Med. Virol.* 14, 123–135, 2004; Scallon, B., Cai, A., Radewonuk, J., and Naso, M., Addition of an extra immunoglobulin domain to two anti-rodent TNF monoclonal antibodies substantially increased their potency, *Mol. Immunol.* 41, 73–80, 2004; Adams, G.P., Tai, M.S., McCartney, J.E. et al., Avidity-mediated enhancement of *in vivo* tumor targeting by single-chain Fv dimers, *Clin. Cancer Res.* 12, 1599–1605, 2006.

Antigen   A material that can be of diverse substance and origin such as protein or microorganism and which elicits an immune response. An immune response can be the formation of an antibody directed against the antigen (humoral response; B-cell response) as well as a cellular response (T-cell response). Antigens can be separated into immunogens (complete antigens) that can elicit an immune response and haptens or incomplete immunogens, which do not by themselves elicit an immune response but can react with antibodies. Haptens require a combination with a larger molecule such as a protein to elicit antibody formation. See Nossal, G.J.V., *Antigens, Lymphoid Cells, and the Immune Response,* Academic Press, New York, 1971; Langone, J.J., *Antibodies, Antigens, and Molecular Mimicry,* Academic Press, San Diego, CA, 1989; Paul, W., Ed., *Fundamental Immunology,* Raven Press, New York, 1993; Cruse, J.M., Lewis, R.E., and Wang, H., *Immunology Guidebook,* Elsevier, Amsterdam, 2004.

Antigenic   An antigenic determinant is also an epitope; this is the region of an antigen
determinant   that binds to the reactive site of an antibody referred to as a paratope. The antigenic determinant elicits the antibody response. There are linear or continuous determinants, which would be a continuous amino acid sequence in a protein antigen, and conformation or discontinuous determinants where, for example with a protein, the epitope is formed by protein folding. A linear determinant is recognized by T-cells as well as B-cells and antibodies, while a discontinuous determinant is recognized only by B-cells and antibodies. See Eisen, H.N., The immune response to a simple antigenic determinant, *Harvey Lect.* 60, 1–34, 1966; Kabat, E.A., The nature of an antigenic determinant, *J. Immunol.* 97, 1–11, 1966; Franks, D., Antigens as markers on cultured mammalian cells, *Biol. Rev. Camb. Philos. Soc.* 43, 17–50, 1968; Stevanovic, S., Antigen processing is predictable: from genes to T-cell epitopes, *Transpl. Immunol.* 14, 171–174, 2005; McRobert, E.A., Tikoo, A., Gallicchio, M.A., Cooper, M.E., and Bach, L.A., Localization of the ezrin binding epitope for glycated proteins, *Ann. N.Y. Acad. Sci.* 1043, 617–624, 2005; Lovtich, S.B. and Unanue, E.R., Conformational isomers of a peptide-class II major histocompatibility complex, *Immunol. Rev.* 207, 293–313, 2005; Phillips, W.J., Smith, D.J.,

Bona, C.A., Bot, A., and Zaghouani, H., Recombinant immunoglobulin-based epitope delivery: a novel class of autoimmune regulators, *Int. Rev. Immunol.* 24, 501–517, 2005; De Groot, A.S., Knopp, P.M., and Martin, W., De-immunization of therapeutic proteins by T-cell epitope modification, *Dev. Biol. (Basel)* 122, 171–194, 2005; Burlet-Schlitz, O., Claverol, S., Gairin, J.E., and Monsarrat, B., The use of mass spectrometry to identify antigens from proteosome processing, *Methods Enzymol.* 405, 264–300, 2005.

Anti-Idiotypic   Usually in reference to antibodies whose specificity is directed against the idiotypic region of an antibody, most frequently with naturally occurring antibodies. Because receptors and antibodies share common binding characteristics, this term is sometimes used to describe antibodies directed against receptors. See Couraud, P.O. and Strosberg, A.D., Anti-idiotypic antibodies against hormone and neurotransmitter receptors, *Biochem. Soc. Trans.* 19, 147–151, 1991; Erlanger, B.F., Antibodies to receptors by an auto-anti-idiotypic strategy, *Biochem. Soc. Trans.* 19, 138–143, 1991; Greally, J.M., Physiology of anti-idiotypic interactions: from clonal to paratopic selection, *Clin. Immunol. Immunopathol.* 60, 1–12, 1991; Friboulet, A., Izadyar, L., and Avalle, B., Abzyme generation using an anti-idiotypic antibody as the "internal image" of an enzyme active site, *Appl. Biochem. Biotechnol.* 47, 229–237, 1994; Hebert, J. and Boutin, Y., Anti-idiotypic antibodies in the treatment of allergies, *Adv. Exp. Med. Biol.* 409, 431–437, 1996.

Antisense   Generally refers to a nucleotide sequence that is complementary to a sequence of messenger RNA, which is the product of the noncoding sequence of DNA. It also refers to the peptide products from the antisense sequence referred to as antisense peptides. Antisense peptides have been investigated for biological activity. siRNA are based on the processing of antisense RNA. See Korneev, S. and O'Shea, M., Natural antisense RNAs in the nervous system, *Rev. Neurosci.* 16, 213–222, 2005. See also *MicroRNA, siRNA, Antisense Peptides, Aptamers.*

Antisense Peptides   The products from the translation of antisense RNA. Some antisense peptides have been demonstrated to show affinity properties that appear to be unique to that sequence and not seen in scrambled sequences. See Schwabe, C., New thoughts on the evolution of hormone-receptor systems, *Comp. Biochem. Physiol. A* 97, 101–106, 1990; Chaiken, I., Interactions and uses of antisense peptides in affinity technology, *J. Chromatog.* 597, 29–36, 1992; Labrou, N. and Clonis, Y.D., The affinity technology in downstream processing, *J. Biotechnol.* 36, 95–119, 1994; Root-Bernstein, R.S. and Holsworth, D.D., Antisense peptides: critical mini-review, *J. Theoret. Biol.* 190, 107–119, 1998; Siemion, I.Z., Cebrat, M., and Kluczyk, A., The problem of amino acid complementarity and antisense peptides, *Curr. Protein Pept. Sci.* 5, 507–527, 2004.

Apical   In reference to a differentiated cell, that portion or apex of the cell that is pointed toward the lumen: For example, in endothelial cells the membrane protein distribution is frequently different between the apical domain and the basolateral domain. See *Basolateral*. See Alfalah, M., Wetzel, G., Fischer, I. et al., A novel type of detergent-resistant may contribute to an early protein sorting event in epithelial cells, *J. Biol. Chem.* 280, 42636–42643, 2005; Kellett, G.L., and Brot-Laroche, E., Apical GLUT2: a major pathway of intestinal sugar absorption, *Diabetes* 54, 3056–3062, 2005; Ito, K., Suzuki, H., Horie, T., and Sugiyama, Y., Apical/basolater surface expression of drug transporters and its role in vectoral drug transport,

Pharm. Res. 22, 1559–1577, 2005; Anderson, J.M., Van Itallie, C.M., and Fanning, A.S., Setting up a selective barrier at the apical junction complex, *Curr. Opin. Cell Biol.* 16, 140–145, 2004.

Apoptosis   Programmed cell death; an organized process by which cells undergo degradation and elimination. See Tomei, L.D. and Cope, F.O., *Apoptosis: The Molecular Basis of Cell Death*, Cold Spring Harbor Laboratory Press, Plainview, NY, 1991; Studzinski, G.P., *Cell Growth and Apoptosis: A Practial Approach*, Oxford University Press, Oxford, UK, 1995; Christopher, G.D., *Apoptosis and the Immune Response*, Wiley-Liss, New York, 1995; Kumar, S., *Apoptosis: Mechanisms and Role in Disease*, Springer, Berlin, 1998; Lockshin, R.A. and Zakeri, Z., *When Cells Die: A Comprehensive Evaluation of Apoptosis and Programmed Cell Death,* Wiley-Liss, New York, 1998; Jacobson, M.D. and McCarthy, N.J., *Apoptosis*, Oxford University Press, Oxford, UK, 2002; LeBlanc, A.C., Ed., *Apoptosis Techniques and Protocols*, 2nd ed., Humana Press, Totowa, NJ, 2002; Hughes, D. and Mehmet, H., Eds., *Cell Proliferation and Apoptosis*, Bios, Oxford, UK, 2003; Potten, C.S. and Wilson, J.W., *Apoptosis: The Life and Death of Cells*, Cambridge University Press, Cambridge, UK, 2004.

Apoptosome   A multiprotein complex that contains caspase-9 and is thought to represent a holoenzyme involved in apoptosis. See Tsujimoto, Y., Role of Bcl-2 family proteins in apoptosis: apoptosomes or mitochondria, *Genes Cells,* 3, 697–707, 1998; Adrain, C. and Martin, S.J., The mitochondrial apoptosome: a killer unleashed by the cytochrome seas, *Trends Biochem. Sci.* 26, 390–397, 2001; Salvesen, G.S. and Renatus, M., Apoptosome: the seven-spoked death machine, *Dev. Cell.* 2, 256–257, 2002; Cain, K., Bratton, S.B., and Cohen, G.M., The Apaf-1 apoptosome: a large caspase-activating complex, *Biochemie* 84, 203–214, 2002; Shi, Y., Apoptosome: the cellular engine for the activation of caspase-9, *Structure* 10, 285–288, 2002; Reed, J.C., Apoptosis-based therapies, *Nat. Rev. Drug Disc.* 1, 111–121, 2002; Adams, J.M. and Cory, S., Apoptosomes: engines for caspase activation, *Curr. Opin. Cell Biol.* 14, 715–720, 2002; Hajra, K.M. and Liu, J.R., Apoptosome dysfunction in human cancer, *Apoptosis* 9, 691–704, 2004.

Aprotinin   A small protein (single-chain protein, MW 6.5 kDa; 58 amino acids) also known as basic pancreatic trypsin inhibitor (BPTI) or the Kunitz pancreatic trypsin inhibitor. It is best known as an inhibitor of trypticlike serine proteases such as plasma kallikrein and plasmin. Aprotinin is also used as the model for protein folding. See Kellermeyer, R.W. and Graham, J.C., Jr., Kinins-possible physiologic and pathologic roles in man, *N. Eng. J. Med.* 279, 754–759, 1968; Schachter, M., Kallikreins (kininogenases) — a group of serine proteases with bioregulatory actions, *Pharmacol. Rev.* 31, 1–17, 1979; Creighton, T.E., Experimental studies of protein folding and unfolding, *Prog. Biophys. Mol. Biol.* 33, 231–297, 1978; Fritz, H. and Wunderer, G., Biochemistry and applications of aprotinin, the kallikrein inhibitor from bovine organs, *Arzneimittelforschung* 33, 479–494, 1983; Sharpe, S., De Meester, I., Hendriks, D. et al., Proteases and their inhibitors: today and tomorrows, *Biochimie* 73, 121–126, 1991; Creighton, T.E., Protein-folding pathways determined using disulphide bonds, *Bioessays* 14, 195–199, 1992; Day, R. and Daggett, V., All-atom simulations of protein folding and unfolding, *Adv. Protein Chem.* 66, 373–403, 2003.

Aptamers   Aptamers are relatively short oligonucleotides (generally 100 bp or less) that act as relatively specific ligands to a broad range of targets. Aptamers are generally selected by combinatorial chemistry techniques. See Ellington, A.D. and Szostak, J.W., *In vitro* selection of RNA molecules that bind specific ligands, *Nature* 346, 818–822, 1990; Burke, J.M. and Berzal-Herranz, A., *In vitro* selection and evolution of RNA: applications for catalytic RNA, molecular recognition, and drug discovery, *FASEB J.* 7, 106–112, 1993; Stull, R.A. and Szoka, F.C., Jr., Antigene, ribozyme, and aptamer nucleic acid drugs: progress and prospects, *Pharm. Res.* 12, 465–483, 1995; Uphoff, K.W., Bell, S.D., and Ellington, A.D., *In vitro* selection of aptamers: the dearth of pure reason, *Curr. Opin. Struct. Biol.* 6, 281–288, 1996; Collett, J.R., Cho, E.J., and Ellington, A.D., Production and processing aptamers microarrays, *Methods* 37, 4–15, 2005; Nutiu, R. and Li, Y., Aptamers with fluorescence-signaling properties, *Methods* 37, 16–25, 2005; Proske, D., Blank, M., Buhmann, R., and Resch, A., Aptamers — basic research, drug development, and clinical applications, *Appl. Microbiol. Biotechnol.* 69, 367–374, 2005; Pestourie, C., Tavitian, B., and Duconge, F., Aptamers against extracellular targets for *in vivo* applications, *Biochimie* 87, 921–930, 2005. It is noted that the term intramer is used to describe intracellular aptamers (see Famulok, M., Blind, M., and Mayer, G., Intramers as promising new tools in functional proteomics, *Chem. Biol.* 8, 931–939, 2001; Famulok, M. and Mayer, G., Intramers and aptamers: applications in protein-function analyses and protential for drug screening, *ChemBioChem* 6, 19–26, 2005).

Aquaporin   Water-specific membrane pores that facilitate osmosis. See Sabolic, I. and Brown, D., Water transport in renal tubules is mediated by aquaporins, *Clin. Investig.* 72, 689–700, 1994; van Lieburg, A.F., Knoers, N.V., and Deen, P.M., Discovery of aquaporins: a breakthrough in research on renal water transport, *Ped. Nephrol.* 9, 228–234, 1995; King, L.S. and Agre, P. Pathophysiology of the aquaporin water channels, *Annu. Rev. Physiol.* 58, 619–648, 1996; Chaumont, F., Moshelion, M., and Daniels, M.J., Regulation of plant aquaporin activity, *Biol. Cell.* 97, 749–764, 2005; Castle, N.A., Aquaporins as targets for drug discovery, *Drug. Discov. Today* 10, 485–493, 2005.

Arabidopsis thaliana   A small plant in the mustard family that is the model for studies of the plant genome. Meinke, D.W., Cheng, D.M., Dean, C., Rounsley, S.D., and Koorneeft, M., *Arabidopsis thaliana:* a model plant for genome analysis, *Science* 282, 662–682, 1998; Glick, B.R. and Thompson, J.E., Eds., *Methods in Plant Molecular Biology and Biotechnology*, CRC Press, Boca Raton, FL, 1993; Meyerowitz, E.R. and Somerville, C.R., Eds., *Arabidopsis*, Cold Spring Harbor Laboratory Press, Cold Spring Harbor, NY, 1994; Anderson, M. and Roberts, J.A., Eds., *Arabidopsis*, Sheffield Academic Press, Sheffield, UK, 1998; Salinas, J. and Sánchez-Serrano, J.J., Eds., *Arabidopsis Protocols*, Humana Press, Totowa, NJ, 2006.

ARF Family GTPases   The ADP-ribosylation family of GTPases. The ADP-ribosylation factor of small GTPases have a role in the regulation of vesicular function via the recruitment of coat proteins and regulation of phospholipid metabolism. See Goud, B., Small GTP-binding proteins as compartmental markers, *Semin. Cell Biol.* 3, 301–307, 1992; Kjeldgaard, M., Nyborg, J., and Clark, B.F., The GTP-binding motif: variations on a theme, *FASEB J.* 10, 1347–1368, 1996; Donaldson, J.G. and Jackson, C.L., Regulators and effectors of the ARF GTPases, *Curr. Opin. Cell Biol.* 12, 475–482, 2000; Takai, Y., Sasaki, T., and Matozaki, T., Small GTP-binding proteins, *Physiol. Rev.* 81,

153–208, 2001; Kahn, R.A., Ed., *ARF Family GTPases*, Kluwer, Dordrecht, Netherlands, 2003; Munro, S., The ARF-like GTPase Arl1 and its role in membrane traffic, *Biochem. Soc. Trans.* 33, 601–605, 2005; Kahn, R.A., Cherfils, J., Elias, M. et al., Nomenclature for the human ARF family of GTP-binding proteins: ARF, ARL, and SAR proteins, *J. Cell Biol.* 172, 645–650, 2006; Nie, Z. and Randazzo, P.A., ARF GAPs and membrane traffic, *J. Cell Sci.* 119, 1203–1211, 2006; D'Souza-Schorey, C. and Chavrier, P., ARF proteins: roles of membrane traffic and beyond, *Nat. Rev. Mol. Cell Biol.* 7, 347–358, 2006. Also includes ARL and SAR proteins.

Arrhenius Energy of Activation An operationally defined quantity that relates rate constants to temperature by the following equation: $k = Ae^{-Ea/RT}$ where k is a rate constant; A and Ea are constants, R is the gas constant, and T the absolute temperature. A plot of ln k vs. 1/T (Arrhenius plot) yields the Arrhenius energy of activation. See Van Tol, A., On the occurrence of a temperature coefficient (Q10) of 18 and a discontinuous Arrhenius plot for homogeneous rabbit muscle fructosediphosphatase, *Biochem. Biophys. Res. Commun.* 62, 750–756, 1975; Ceuterick, F., Peeters, J., Heremans, K. et al., Involvement of lipids in the break of the Arrhenius plot of *Azobacter nitrogenase*, *Arch. Int. Physiol. Biochim.* 84, 587–588, 1976; Ceuterick, F., Peeters, J., Heremans, K. et al., Effect of high pressure, detergents, and phospholipase on the break in the Arrhenius plot of *Azobacter nitrogenase*, *Eur. J. Biochem.* 87, 401–407, 1978; Stanley, K.K. and Luzio, J.P., The Arrhenius plot behaviour of rat liver 5′-nucleotidase in different lipid environments, *Biochim. Biophys. Acta* 514, 198–205, 1978; De Smedt, H., Borghgraef, R., Ceuterick, F., and Heremans, K., The role of lipid–protein interactions in the occurrence of a nonlinear Arrhenius plot for (sodium–potassium)-activated ATPase, *Arch. Int. Physiol. Biochim.* 87, 169–170, 1979; Biosca, J.A., Travers, F., and Barman, T.E., A jump in an Arrhenius plot can be the consequence of a phase transition. The binding of ATP to myosin subfragment 1, *FEBS Lett.* 153, 217–220, 1983; Haeffner, E.W. and Friedel, R., Induction of an endothermic transition in the Arrhenius plot of fatty acid uptake by lipid-depleted ascites tumor cells, *Biochim. Biophys. Acta* 1005, 27–33, 1989; Muench, J.L., Kruuv, J., and Lepock, J.R., A two-step reversible-irrreversible model can account for a negative activation energy in an Arrhenius plot, *Cryobiology* 33, 253–259, 1996; Rudzinski, W., Borowieki, T., Panczyk, T., and Dominko, A., On the applicability of Arrhenius plot methods to determine surface energetic heterogeneity of adsorbents and catalysts surfaces from experimental TPD spectra, *Adv. Colloid Interface Sci.* 84, 1–26, 2000.

Atomic Force Microscopy A high-resolution form of microscopy that involves a probe or tip moving over a surface (alternatively, the sample can move with a static tip; the detection method is the same). As the probe changes position in response to sample topography, the movement is tracked by deflection of a laser beam, which is recorded by a detector (Gadegaard, N., Atomic force microscopy in biology: technology and techniques, *Biotechnic and Histochem.* 81, 87–97, 2006). See Hansma, P.K., Elings, V.B., Marti, O., and Bracker, C.E., Scanning tunneling microscopy: application to biology and technology, *Science* 242, 209–216, 1988; Yang, J., Tamm, L.K., Somlyo, A.P., and Shao, Z., Promises and problems of biological atomic force microscopy, *J. Microscop.* 171, 183–198, 1993; Hansma, H.G. and Hoh, J.H., Biomolecular imaging with the atomic force microscope, *Annu. Rev. Biophys. Biomol.*

Struct. 23, 115–139, 1994; Tendler, S.J., Davies, M.C., and Roberts, C.J., Molecules under the microscope, *J. Pharm. Pharmacol.* 48, 2–8, 1996; Lekka, M., Lekki, J., Shoulyarenko, A.P. et al., Scanning force microscopy of biological samples, *Pol. J. Pathol.* 47, 51–55, 1996; Ivanov, Y.D., Govorum, V.M., Bykov, V.A., and Archakov, A.I., Nanotechnologies in proteomics, *Proteomics* 6, 1399–1414, 2006; Connell, S.D. and Smith, D.A., The atomic force microscope as a tool for studying phase separation in lipid membranes, *Mol. Membr. Biol.* 23, 17–28, 2006; Bai, L., Santangelo, T.J., and Wang, M.D., Single-molecule analysis of RNA polymerase transcription, *Annu. Rev. Biophys. Biomol. Struct.* 35, 343–360, 2006; Guzman, C., Jeney, S., Kreplak, L. et al., Exploring the mechanical properties of single vimentin intermediate filaments by atomic force microscopy, *J. Mol. Biol.* 360, 623–630, 2006; De Jong, K.L., Incledon, B., Yip, C.M., and Defelippis, M.R., Amyloid fibrils of glucagon characterized by high-resolution atomic force microscopy, *Biophys. J.* 91, 1905–1914, 2006; Xu, H., Zhao, X., Grant, C. et al., Orientation of a monoclonal antibody adsorbed at the solid/solution interface: a combined study using atomic force microscopy and neutron reflectivity, *Langmuir* 22, 6313–6320, 2006.

Atomic Radius   A measurement of an atom that is not considered precise; generally half the distance between adjacent atoms of the same type in a crystal or molecule. It may be further described as a covalent radius, an ionic radius, or a metallic radius. The inability of cysteine to effectively substitute for serine in serine proteases is due, in part, to the increased atomic radius of sulfur compared to oxygen. See Alterman, M.A., Chaurasia, C.S., Lu, P., and Hanzlik, R.P., Heteroatom substitution shifts regioselectivity of lauric acid metabolism from omega-hydroxylation to (omega-1)-oxidation, *Biochem. Biophys. Res. Commun.* 214, 1089–1094, 1995; Zhang, R., Villeret, V., Lipscomb, W.N., and Fromm, H.J., Kinetics and mechanisms of activation and inhibition of porcine liver fructose-1,6-bisphosphatase by monovalent cations, *Biochemistry* 35, 3038–3043, 1996; Wachter, R.M. and Brachaud, B.P., Thiols as mechanistic probes for catalysis by the free radical enzyme galactose oxidase, *Biochemistry* 35, 14425–14435, 1996; Wagner, M.A., Trickey, P., Chen, Z.W. et al., Monomeric sarcosine oxidase: 1. Flavin reactivity and active site-binding determinants, *Biochemistry* 39, 8813–8824, 2000; Lack, J.G., Chaudhuri, S.K., Kelly, S.D. et al., Immobilization of radionucleotides and heavy metals through anaerobic bio-oxidation of Fe(II), *Appl. Environ. Microbiol.* 68, 2704–2710, 2002; Hamm, M.L., Rajguru, S., Downs, A.M., and Cholera, R., Base pair stability of 8-chloro- and 8-iodo-2′-deoxyguanosine opposite 2′-deoxycytidine: implications regarding the bioactivity of 8-oxo-2′-deoxyguanosine, *J. Am. Chem. Soc.* 127, 12220–12221, 2005.

Autoantigen   A component of self that can elicit an immune response, an autoimmune reaction; frequently with pathological complications such as the destruction of pancreatic beta cells (Islets of Langerhans), resulting in Type 1 diabetes. See Sigurdsson, E. and Baekkeskov, S., The 64-kDa beta cell membrane autoantigen and other target molecules of humoral autoimmunity in insulin-dependent diabetes mellitus, *Curr. Top. Microbiol. Immunol.* 164, 143–168, 1990; Werdelin, O., Autoantigen processing and the mechanisms of tolerance to self, *Immunol. Ser.* 52, 1–9, 1990; Manfredi, A.A., Protti, M.P., Bellone, M. et al., Molecular anatomy of an autoantigen: T and B epitopes on the nicotinic acetylcholine receptor in myasthenia gravis,

J. Lab. Clin. Med. 120, 13–21, 1992; Sedgwick, J.D., Immune surveillance and autoantigen recognition in the central nervous system, *Aust. N.E. J. Med.* 25, 784–792, 1995; Utz, P.J., Gensler, T.J., and Anderson, P., Death, autoantigen modifications, and tolerance, *Arthritis Res.* 2, 101–114, 2000; Narendran, P., Mannering, S.I., and Harrison, L.C., Proinsulin — a pathogenic autoantigen in type 1 diabetes, *Autoimmun. Rev.* 2, 204–210, 2003; Gentile, F., Conte, M., and Formisano, S., Thyroglobulin as an autoantigen: what can we learn about immunopathogenicity from the correlation of antigenic properties with protein structure? *Immunology* 112, 13–25, 2004; Pendergraft, W.F., III, Pressler, E.M., Jennette, J.C. et al., Autoantigen complementarity: a new theory implicating complementary proteins as initiators of autoimmune disease, *J. Mol. Med.* 83, 12–25, 2005; Wu, C.T., Gershwin, M.E., and Davis, P.A., What makes an autoantigen an autoantigen? *Ann. N.Y. Acad. Sci.* 1050, 134–1045, 2005; Wong, F.S., Insulin—a primary autoantigen in type 1 diabetes? *Trends Mol. Med.* 11, 445–448, 2005; Jasinski, J.M. and Eisenbarth, G.S., Insulin as a primary autoantigen for type 1A diabetes, *Clin. Dev. Immuol.* 12, 181–186, 2005.

Autocoid   An internal physiological secretion of uncertain or unknown classification. Adenosine is one of the better examples because, apart from its role as a purine base in RNA and DNA, it has diverse physiologic functions. See Boyan, B.D., Schwartz, Z., and Swain, L.D., Cell maturation-specific autocrine/paracrine regulation of matrix vesicles, *Bone Miner.* 17, 263–268, 1992; Polosa, R., Holgate, S.T., and Church, M,K., Adenosine as a pro-inflammatory mediator in asthma, *Pulm. Pharmacol.* 2, 21–26, 1989; Yan. L., Burbiel, J.C., Maass, A., and Muller, C.E., Adenosine receptor agonists: from basic medicinal chemical to clinical development, *Expert Opin. Emerg. Drugs* 8, 537–576, 2003; Tan, D.X., Manchester, L.C., Hadeland, R. et al., Melatonin: a hormone, a tissue factor, an autocoid, a paracoid, and an antioxidant vitamin, *J. Pineal Res.* 34, 75–78, 2003.

Autocrine   Usually in reference to a hormone or other biological effector such as a peptide growth factor or cytokine, which has an effect on the cell or tissue responsible for the synthesis of the given compound. Differentiated from endocrine or paracrine phenomena. See Sporn, M.B. and Roberts, A.B., Autocrine, paracrine, and endocrine mechanisms of growth control, *Cancer Surv.* 4, 627–632, 1985; Heldin, C.H. and Westermark, B., PDGF-like growth factors in autocrine stimulation of growth, *J. Cell Physiol.* (Suppl. 5), 31–34, 1987; Ortenzi, C., Miceli, C., Bradshaw, R.A., and Luporini, P., Identification and initial characterization of an autocrine pheromone receptor in the protozoan cilitate *Euplotes raikovi, J. Cell. Biol.* 111, 607–614, 1990; Vallesi, A., Giuli, G., Bradshaw, R.A., and Luporini, P., Autocrine mitogenic activity of pheromones produced by the protozoan ciliate *Euplotes raikovi*, *Nature* 376, 522–524, 1995; Bischof, P., Meissner, A., and Campana, A., Paracrine and autocrine regulators of trophoblast invasion — a review, *Placenta* 21 (Suppl. A), S55–S60, 2000; Bilezikjian, L.M., Blount, A.L., Leal, A.M. et al., Autocrine/paracrine regulation of pituitary function of activin, inhibin, and follistatin, *Mol. Cell. Endocrinol.* 225, 29–36, 2004; Singh, A.B. and Harris, R.C., Autocrine, paracrine, and juxtacrine signaling by EGFR ligands, *Cell. Signal.* 17, 1183–1193, 2005; Ventura, C. and Branzi, A., Autocrine and intracrine signaling for cardiogenesis in embryonic stem cells: a clue for the development of novel differentiating agents, *Handb. Exp. Pharmacol.* 174, 123–146, 2006.

Autophagy	A pathway for the physiological degradation of cellular macromolecules and subcellular structures mediated by intracellular organelles such as lysosomes. It can be considered to be a process by which there is a membrane reorganization to separate or sequester a portion of the cytoplasm or cytoplasmic contents for subsequent delivery to an intracellular organelle such as a lysosome for degradation. This pathway of "self-destruction" is separate from proteosome-mediated degradation of macromolecules internalized from outside the cell. See Wang, C-W. and Klianksy, D.J., The molecular mechanism of autophagy, *Molec. Med.* 9, 65–76, 2003; Kroemer, G. and Jaattela, M., Lysosomes and autophagy in cell death control, *Nat. Rev. Cancer* 5, 886–897, 2005; Deretic, V., Autophagy in innage and adaptive immunity, *Trends Immunol.* 26, 523–528, 2005; Baehrecke, E.H., Autophagy: dual roles in life and death? *Nat. Rev. Mol. Cell Biol.* 6, 505–510, 2005; Klinosky, D.J., Autophagy, *Curr. Biol.* 15, R282–F283, 2005.
Auto-phosphorylation	A process by which a substrate protein, usually a receptor, catalyzes self-phosphorylation, usually at a tyrosine residue. The mechanism can be either intramolecular (*cis*) or intermolecular (*trans*) although at least one system has been described with both *cis* and *trans* processes. See Cobb, M.H., Sang, B.-C., Gonzalez, R., Goldsmith, E., and Ellis, L., Autophosphorylation activates the soluble cytoplasmic domain of the insulin receptor in an intermolecular reaction, *J. Biol. Chem.* 264, 18701–18706, 1989; Frattali, A.L., Treadway, J.L., and Pessin, J.E., Transmembrane signaling by the human insulin receptor kinase. Relationship between intramolecular subunit *trans-* and *cis*-autophosphorylation and substrate kinase activation, *J. Biol. Chem.* 267, 19521–19528, 1992; Rim, J., Faurobert, E., Hurley, J.B., and Oprian, D.D., *In vitro* assay for *trans*-phosphorylation of rhodopsin by rhodopsin kinase, *Biochemistry* 36, 7064–7070, 1997; Cann, A.D., Bishop, S.M., Ablooglu, A.J., and Kobanski, R.A., Partial activation of the insulin receptor kinase domain by juxtamembrane autophosphorylation, *Biochemistry* 37, 11289–11300, 1998; Iwasaki, Y., Nishiyama, H., Suzuki, K., and Koizumi, S., Sequential *cis/trans* autophosphorylation in TrkB tyrosine kinase, *Biochemistry* 36, 2694–2700, 1997; Cohen, P., The regulation of protein function by multisite phosphorylation — a 25 year update, *Trends in Biochem. Sci.* 25, 596–601, 2000; Wick, M.J., Ramos, F.J., Chen, H. et al., Mouse 3-phosphoinositide–dependent protein kinase-1 undergoes dimerization and *trans*-phosphorylation in the activation loop, *J. Biol. Chem.* 278, 42913–42919, 2003; Wu, S. and Kaufman, R.J., *trans*-autophosphorylation by the isolated kinase domain is not sufficient for dimerization or activation of the dsRNA-activated protein kinase PKR, *Biochemistry* 43, 11027–11034, 2004.
B-Lymphocytes	Also called B-cells; the name derives from original studies involving the cells from the bursa of chickens. B-cells are best known for the production of antibodies but recent studies show increased complexity of function. See Greaves, M.F., Owen, J.J.T., and Raff, M.C., Eds., *T and B Lymphocytes: Origins, Properties and Roles in Immune Responses,* Excerpta Medica, New York, 1973; Loor, F. and Roelants, G.E., Eds., *B and T Cells in Immune Recognition*, John Wiley & Sons, New York, 1977; Pernis, B. and Vogel, H.J., Eds., *Cells of Immunoglobulin Synthesis*, Academic Press, New York, 1979; Bach, F.H., Ed., *T and B Lymphocytes: Recognition and Function*, Academic Press, New York, 1979; Cambier, J.C., *B-Lymphocyte Differentiation*, CRC Press, Boca Raton, FL, 1986; Chiorazzi, N., Ed., *B*

Lymphocytes and Autoimmunity, New York Academy of Sciences, New York, 1997; Cruse, J.M. and Lewis, R.E., *Atlas of Immunology,* CRC Press, Boca Raton, FL, 1999; Paul, W.E., Ed., *Fundamental Immunology,* Lippincott, Williams, and Wilkins, Philadelphia, 2003.

Backflushing (or Back Flushing)   A method for cleaning filters involving reverse flow through the membrane; occasionally used for cleaning large-scale chromatographic columns. See Tamai, G., Yoshida, H., and Imai, H., High-performance liquid chromatographic drug analysis by direct injection of whole blood samples. III. Determination of hydrophobic drugs adsorbed on blood cell membranes, *J. Chromatog.* 423, 163–168, 1987; Kim, B.S. and Chang, H.N., Effects of periodic backflushing on ultrafiltration performance, *Bioseparation* 2, 23–29, 1991; Dai, X.P., Luo, R.G., and Sirkar, K.K., Pressure and flux profiles in bead-filled ultrafiltration/microfiltration hollow fiber membrane modules, *Biotechnol. Prog.* 16, 1044–1054, 2000; Seghatchian, J. and Krailadsiri, P., Validation of different enrichment strategies for analysis of leucocyte subpopulations: development and application of a new approach, based on leucofiltration, *Transfus. Apher. Sci.* 26, 61–72, 2002; Kang, I.J., Yoon, S.H., and Lee, C.H., Comparison of the filtration characteristics or organic and inorganic membranes in a membrane-coupled anaerobic bioreactor, *Water Res.* 36, 1803–1813, 2002; Nemade, P.R. and Davis, R.H., Secondary membranes for flux optimization in membrane filtration of biologic suspensions, *Appl. Biochem. Biotechnol.* 113–116, 417–432, 2004.

Bacterial Artificial Chromosome   A DNA construct based on a fertility plasmid; used for transforming and cloning in bacteria. It has an average insert size of 150 kbp with a range of approximately 100 kbp to 300 kbp. Bacterial artificial chromosomes are frequently used to sequence genomes where the PCR reaction is used to prepare a region of genomic DNA and then sequenced; in other words, a bacterial artificial chromosome (BAC) is a vehicle based on the bacteria *Escherichia coli* that is used to copy, or clone, fragments of DNA that are 150,000 to 180,000 base pairs (bp) long. These DNA fragments are used as starting material for DNA sequencing. See Schalkwyk, L.C., Francis, F., and Lehrach, H., Techniques in mammalian genome mapping, *Curr. Opin. Biotechnol.* 6, 37–43, 1995; Zhang, M.B. and Wing, R.A., Physical mapping of the rice genome with BACs, *Plant Mol. Biol.* 35, 115–127, 1997; Zhu, J., Use of PCR in library screening. An overview, *Methods Mol. Biol.* 192, 353–358, 2002; Ball, K.D. and Trevors, J.T., Bacterial genomics: the use of DNA microarrays and bacterial artificial chromosomes, *J. Microbiol. Methods* 49, 275–284, 2002; Miyake, T. and Amemiya, C.T., BAC libraries and comparative genomics of aquatic chordate species, *Comp. Biochem. Physiol. C Toxicol. Pharmacol.* 138, 233–244, 2004; Ylstra, B., van den IJssel, P., Carvalho, B., Brakenhoff, R.H., and Maijer, G.A., BAC to the future! or oligonucleotides: a perspective for micro array comparative genomic hybridization (array CGH), *Nucleic Acids Res.* 34, 445–450, 2006.

Balanced Translocation   A chromosomal relocation that does not involve the net gain or loss of DNA; also referred to as reciprocal translocation. See Fraccaro, M., Chromosome abnormalities and gamete production in man, *Differentiation* 23 (Suppl.), S40–S43, 1983; Davis, J.R., Rogers, B.B., Hagaman, R.M., Thies, C.A., and Veomett, I.C., Balanced reciprocal translocations: risk factors for aneuploid segregant viability, *Clin. Genet.* 27, 1–19, 1985; Greaves, M.F., Biological models for leukemia and lymphoma, *IARC Sci. Publ.* 157, 351–372, 2004;

Benet, J., Oliver-Bonet, M., Cifuentes, P., Templado, C., and Navarro, J., Segregation of chromosomes in sperm of reciprocal translocation carriers: a review, *Cytogenet. Genome Res.* 111, 281–290, 2005; Aplan, P.D., Causes of oncogenic chromosomal translocation, *Trends in Genetics* 22, 46–55, 2006.

Basolateral   Located on the bottom opposite from the apical end of a differentiated cell. See *Apical*. See Terada, T. and Inui, K., Peptide transporters: structure, function, regulation, and application for drug delivery, *Curr. Drug Metab.* 5, 85–94, 2004; Brone, B. and Eggermont, J., PDZ proteins retain and regulate membrane transporters in polarized epithelial cell membranes, *Am. J. Physiol. Cell Physiol.* 288, C20–C29, 2005; Rodriquez-Boulan, E. and Musch, A., Protein sorting in the Golgi complex: shifting paradigms, *Biochim. Biophys. Acta* 1744, 455–464, 2005; Vinciguerra, M., Mordasini, D., Vandewalle, A., and Feraille, E., Hormonal and nonhormonal mechanisms of regulation of the NA,K-pump in collecting duct principal cells, *Semin. Nephrol.* 25, 312–321, 2005.

Bathochromic Shift   A shift in the absorption/emission of light to a longer wavelength ($\lambda > \lambda_o$); a "red" shift. See Waleh, A. and Ingraham, L.L., A molecular orbital study of the protein-controlled bathochromic shift in a model of rhodopsin, *Arch. Biochem. Biophys.* 156, 261–266, 1973; Heathcote, P., Vermeglio, A., and Clayton, R.K., The carotenoid band shift in reaction centers from the *Rhodopseudomonas sphaeroides*, *Biochim. Biophys. Acta* 461, 358–364, 1977; Kliger, D.S., Milder, S.J., and Dratz, E.A., Solvent effects on the spectra of retinal Schiff bases — I. Models for the bathochromic shift of the chromophore spectrum in visual pigments, *Photochem. Photobiol.* 25, 277–286, 1977; Cannella, C., Berni, R., Rosato, N., and Finazzi-Agro, A., Active site modifications quench intrinsic fluorescence of rhodanese by different mechanisms, *Biochemistry* 25, 7319–7323, 1986; Hermel, H., Holtje, H.D., Bergemann, S. et al., Band-shifting through polypeptide beta-sheet structures in the cyanine UV-Vis spectrum, *Biochim. Biophys. Acta* 1252, 79–86, 1995; Zagalsky, P.F., β-crustacyanin, the blue-purple carotenoprotein of lobster carapace: consideration of the bathochromic shift of the protein-bound astaxanthin, *Acta Chrystallogr. D Biol. Chrystallogr.* 59, 1529–1531, 2003.

Betaine   Glycine betaine, (carboxymethyl)trimethylamonnium inner salt, Cystadane®. Derived from choline; serves as methyl donor in the synthesis of methionine from homocysteine. Also functions as an osmoprotectant and this function is similar to trehalose in plants. See Chambers, S.T., Betaines: their significance for bacteria and the renal tract, *Clin. Sci.* 88, 25–27, 1995; Nuccio, M.L., Rhodes, D., McNeil, S.D., and Hanson, A.D., Metabolic engineering of plants for osmotic stress resistance, *Curr. Opin. Plant Biol.* 2, 129–134, 1999; Craig, S.A., Betaine in human nutrition, *Am. J. Clin. Nutr.* 80, 539–549, 2004; Zou, C.G. and Banerjee, R., Homocysteine and redox signaling, *Antioxid. Redox. Signal.* 7, 547–559, 2005; Fowler, B., Homocysteine: overview of biochemistry, molecular biology, and role in disease processes, *Semin. Vasc. Med.* 5, 77–86, 2005; Ueland, P.M., Holm, P.I., and Hustad, S., Betaine: a key modulator of one-carbon metabolism and homocysteine status, *Clin. Chem. Lab. Med.* 43, 1069–1075, 2005.

Bibody   One scFv fragment coupled to the C-terminus of the C_{H1} domain of a Fab fragment. See Schoonjans, R., Willems, A., Schoonooghe, S. et al., Fab chains as an efficient heterodimerization scaffold for the production of recombinant bispecific and trispecific antibody derivatives, *J. Immunol.* 165, 7050–7057, 2000. See also *Diabody*; *Tribody*.

Bicoid Protein   A transcription-factor protein produced in *Drosophila*. See Lawrence, P.A., Background to bicoid, *Cell* 54, 1–2, 1988; Stephenson, E.C. and Pokrywka, N.J., Localization of bicoid message during *Drosophila* oogenesis, *Curr. Top. Dev. Biol.* 26, 23–34, 1992; Johnstone, O. and Lasko, P., Translational regulation and RNA localization in *Drosophila* oocytes and embryos, *Annu. Rev. Genet.* 35, 365–406, 2001; Lynch, J. and Desplan, C., Evolution of development: beyond bicoid, *Curr. Biol.* 12, R557–R559, 2003.

BIND   Biomolecular Interaction Data Base, which is designed to store full descriptions of interactions, molecular complexes, and metabolic pathways. See Bader, G.D., Donaldson, I., Wolting, C. et al., BIND—the biomolecular interaction network database, *Nucleic Acids Res.* 29, 242–245, 2001; Alfarano, C. Andrade, C.E., Anthony, K. et al., The biomolecular interaction network database and related tools: 2005 update, *Nucleic Acids Res.* 33, D418–D424, 2005; Shah, S.P., Huang, Y., Xu, T. et al., Atlas—a data warehouse for integrative bioinformatics, *BMC Bioinformatics* 6, 34, 2005; Aytuna, A.S., Gursoy, A., and Keskin, O., Prediction of protein–protein interactions by combining structure and sequence conservation in protein interfaces, *Bioinformatics* 21, 2850–2855, 2005; Gilbert, D., Biomolecular interaction network database, *Brief Bioinform.* 6, 194–198, 2005.

Bioassay   Generally used to describe an assay for a drug/biologic after administration to subject. As such, a bioassay usually involves the sampling of a biological fluid such as blood. Bioassay can also describe an assay that uses a biological substrate such as a cell or an organism. The term bioassay does not define a technology. See Yamamoto, S., Urano, K., and Nomura, T., Validation of transgenic mice harboring the human prototype c-Ha-ras gene as a bioassay model for rapid carcinogenicity testing, *Toxicol. Lett.* 28, 102–103, 1998; Colburn, W.A. and Lee, J.W., Biomarkers, validation, and pharmacokinetic-pharmacodynamic modeling, *Clin. Pharmacokinet.* 42, 997–1022, 2003; Tuomela, M., Stanescu, I., and Krohn, K., Validation overview of bio-analytical methods, *Gene Ther.* 22 (Suppl. 1), S131–S138, 2005; Indelicato, S.R., Bradshaw, S.L., Chapman, J.W., and Weiner, S.H., Evaluation of standard and state of the art analytical technology-bioassays, *Dev. Biol. (Basal)* 122, 102–114, 2005.

Bioequivalence   Similarity of biological properties; used in the characterization of pharmaceuticals to demonstrate therapeutic equivalence. See Levy, R.A., Therapeutic inequivalence of pharmaceutical alternatives, *Am. Pharm.* NS23, 28–39, 1985; Durrleman, S. and Simon, R., Planning and monitoring of equivalence studies, *Biometrics* 46, 329–336, 1990; Schellekens, H., Bioequivalence and the immunogenicity of biopharmaceuticals, *Nat. Rev. Drug. Disc.* 1, 457–462, 2002; Lennernas, H. and Abrahamsson, B., The use of biopharmaceutical classification of drugs in drug discovery and development: current status and future extension, *J. Pharm. Pharmacol.* 57, 273–285, 2005; Bolton, S., Bioequivalence studies for levothyroxine, *AAPS J.* 7, E47–E53, 2005.

Bioinformatics   The use of information technology to analyze data obtained from proteomic analysis. An example is the use of databases such as SWISSPROT to identify proteins from sequence information determined by the mass spectrometric analysis of peptides. See Wang, J.T.L., *Data Mining in Bioinformatics*, Springer, London, 2005; Lesk, A.M., *Introduction to Bioinformatics*, Oxford University Press, New York, 2005; Englbrecht, C.C. and Facius, A., Bioinformatics challenges in proteomics, *Comb. Chem. High*

Throughput Screen. 8, 705–715, 2005; Brent, M.R., Genome annotation past, present, and future: how to define an ORF at each locus, *Genome Res.* 15, 1777–1786, 2005; Chalkley, R.J., Hansen, K.C., and Baldwin, M.A., Bioinformatic methods to exploit mass spectrometric data for proteomic applications, *Methods Enzymol.* 402, 289–312, 2005; Evans, W.J., *Statistical Methods in Bioinformatics: An Introduction,* Springer, New York, 2005; Buehler, L.K. and Rashidi, H.H., *Bioinformatics Basics: Applications in Biological Sciences and Medicine,* Taylor & Francis, Boca Raton, FL, 2005; Baxevanis, A.D. and Ouellette, B.F.F., *Bioinformatics: A Practical Guide to the Analysis of Genes and Proteins*, Wiley, Hoboken, NJ, 2005; Chandonia, J.M. and Brenner, S.E., The impact of structural genomics: expectations and outcomes, *Science* 311, 347–351, 2006; Allison, D.B, Cui, X., Page, G.P., and Sabripour, M., Microarray data analysis: from disarray to consolidation and consensus, *Nat. Rev. Genet.* 7, 55–65, 2006.

Biologicals   A biological product is any virus, serum, toxin, antitoxin, blood, blood component, or derivative, allergenic product, or analogous product applicable to the prevention, treatment, or cure of diseases or injury. Biologic products are a subset of "drug products" distinguished by their manufacturing processes (biological process vs. a chemical process). In general, "drugs" include biological products. Within the United States, the regulation of biologicals is the purview of the FDA Center for Biologicals Evaluation and Research (CBER) and drugs within the FDA Center for Drug Evaluation and Research (CDER). There has been a recent shift of some drug products, which were traditionally in CBER, such as monoclonal antibodies and peptide growth factors, to CDER. See Steinberg, F.M. and Raso, J., Biotech pharmaceuticals and biotherapy: an overview, *J. Pharm. Pharm. Sci.* 1, 48–59, 1998; Vincent-Gattis, M., Webb, C., and Foote, M., Clinical research strategies in biotechnology, *Biotechnol. Annu. Rev.* 5, 229–267, 2000; Stein, K.E. and Webber, K.O., The regulation of biologic products derived from bioengineered plants, *Curr. Opin. Biotechnol.* 12, 308–311, 2001; Morrow, K.S. and Slater, J.E., Regulatory aspects of allergen vaccines in the United States, *Clin. Rev. Allergy Immunol.* 21, 141–152, 2001; Hudson, P.J. and Souriau, C., Recombinant antibodies for cancer diagnosis and therapy, *Expert Opin. Biol. Ther.* 1, 845–855, 2001; Monahan, T.R., Vaccine industry perspective of current issues of good manufacturing practices regarding product inspections and stability testing, *Clin. Infect. Dis.* 33 (Suppl. 4), S356–S361, 2001; Hsueh, E.C. and Morton, D.L., Angiten-based immunotherapy of melanoma: canvaxin therapeutic polyvalent cancer vaccine, *Semin. Cancer Biol.* 13, 401–407, 2003; Miller, D.L. and Ross, J.J., Vaccine INDs: review of clinical holds, *Vaccine* 23, 1099–1101, 2005; Sobell, J.M., Overview of biologic agents in medicine and dermatology, *Semin. Cutan. Med. Surg.* 23, 2–9, 2005; Morenweiser, R., Downstream processing of viral vectors and vaccines, *Gene Ther.* 12 (Suppl. 1), S103–S110, 2005.

Biomarker   A change in response to an underlying pathology; current examples of molecular changes include C-reactive protein, fibrin D-dimer, and troponin; the term biomarker is also used to include higher-level responses such as behavior changes or anatomic changes. See Tronick, E.Z., The neonatal behavioral assessment scale as a biomarker of the effects of environmental agents on the newborn, *Environ. Health Perspect.* 74, 185–189, 1987; Salvaggio, J.E., Use and misuse of biomarker tests in "environmental conditions,"

J. Allergy Clin. Immunol. 94, 380–384, 1994; Den Besten, P.K., Dental fluorosis: its use as a biomarker, *Adv. Dent. Res.* 8, 105–110, 1994; Lohmander, L.S. and Eyre, D.R., From biomarker to surrogate outcome to osteoarthritis—what are the challenges? *J. Rheumatol.* 32, 1142–1143, 2005; Vineis, P. and Husgafvel-Pursiainen, K., Air pollution and cancer: biomarker studies in human populations, *Carcinogenesis* 26, 1846–1855, 2005; Seligson, D.B., The tissue micro-array as a translational research tool for biomarker profiling and validation, *Biomarkers* 10 (Suppl. 1), S77–S82, 2005; Danna, E.A. and Nolan, G.P., Transcending the biomarker mindset: deciphering disease mechanisms at the single cell level, *Curr. Opin. Chem. Biol.* 10, 20–27, 2006; Felker, G.M., Cuculich, P.S., and Gheorghiade, M., The Valsalva maneuver: a bedside "biomarker" test for heart failure, *Am. J. Med.* 119, 117–122, 2006; Allam, A. and Kabelitz, D., TCR *trans*-rearrangements: biological significance in antigen recognition vs. the role as lymphoma biomarker, *J. Immunol.* 176, 5707–5712, 2006.

Biopharma-ceutical Classification System   The biopharmaceutical classification system (BCS) provides a classification of gastrointestinal absorption. See Amidon, G., Lennernas, H., Shah, V.P., and Crison, J.A., A theoretical basis for a biopharmaceutic drug classification: the correlation of *in vitro* drug product dissolution and *in vivo* bioavailability, *Pharm. Res.* 12, 413–420, 1995; Wilding, I.R., Evolution of the biopharmaceutics classification system (BCS) to oral modified release (MR) formulations: what do we need to consider? *Eur. J. Pharm. Sci.* 8, 157–159, 1999; Dressman, J.B. and Reppas, C., *In vitro–in vivo* correlations for lipophilic, poorly water-soluble drugs, *Eur. J. Pharm. Sci.* 11 (Suppl. 2), S73–S80, 2000; Taub, M.E., Kristensen, L., and Frokjaer, S., Optimized conditions for MDCK permeability and turbidimetric solubility studies using compounds representative of BCS class I–IV, *Eur. J. Pharmaceut. Sci.* 15, 311–340, 2002; Huebert, N.D., Dasgupta, M., and Chen, Y.M., Using *in vitro* human tissues to predict pharmacokinetic properties, *Cur. Opin. Drug. Disc. Dev.* 7, 69–74, 2004; Lennernas, H. and Abrahamsson, B., The use of biopharmaceutic classification of drugs in drug discovery and development: current status and future extension, *J. Pharm. Pharmcol.* 57, 273–285, 2005.

Bone Morphogenetic Protein(s) (BMP)   A group of peptide/proteins that are multifunctional growth factors and members of the TGF β superfamily. There are multiple forms of bone morphogenetic proteins, which all function as differentiation factors for the maturation of mesenchymal cells into chondrocytes and osteoblasts. See Hauschka, P.V., Chen, T.L., and Mavrakos, A.E., Polypeptide growth factors in bone matrix, *Ciba Found. Symp.* 136, 207–225, 1988; Wozney, J.M., Bone morphogenetic proteins, *Prog. Growth Factor Res.* 1, 267–280, 1989; Rosen, V. and Thies, R.S., The BMP proteins in bone formation and repair, *Trends Genet.* 8, 97–102, 1992; Wang, E.A., Bone morphogenetic proteins (BMPs): therapeutic potential in healing bony defects, *Trends Biotechnol.* 11, 379–383, 1993; Kirker-Head, C.A., Recombinant bone morphogenetic proteins: novel substances for enhancing bone healing, *Vet. Surg.* 24, 408–419, 1995; Ramoshibi, L.N., Matsaba, J., Teare, L. et al., Tissue engineering: TGF- superfamily members and delivery systems in bone regeneration, *Expert Rev. Mol. Med.* 2002, 1–11, 2002; Monteiro, R.M., de Sousa Lopez, S.M., Korchynskyi, O. et al., Spatio-temporal activation of Smad1 and Smad5 *in vivo*: monitoring transcriptional activity of Smad proteins, *J. Cell Sci.* 117, 4653–4663, 2004; Canalis, E., Deregowski,

V., Pereira, R.C., and Gazzerro, E., Signals that determine the fate of osteoblastic cells, *J. Endocrinol. Invest.* 28 (Suppl. 8), 3–7, 2005; Franceschi, R.T., Biological approaches to bone regeneration by gene therapy, *J. Dent. Res.* 84, 1093–1103, 2005; Ripamonti, U., Teare, J., and Petit, J.C., Pleiotropism of bone morphogenetic proteins: from bone induction to cementogenesis and periodontal ligament regeneration, *J. Int. Acad. Periodontol.* 8, 23–32, 2006; Logeart-Avramoglou, D., Bourguignon, M., Oudina, K., Ten Dijke, P., and Petite, H., An assay for the determination of biologically active bone morphogenetic proteins using cells transfected with an inhibitor of differentiation promoter-luciferase construct, *Anal. Biochem.* 349, 78–86, 2006.

Bottom-Up Proteomics Identification of unknown proteins by analysis of peptides obtained from unknown proteins by enzymatic (usually trypsin) hydrolysis. See Brock, A., Horn, D.M., Peters, E.C. et al., An automated matrix-assisted laser desorption/ionization quadrupole Fourier transform ion cyclotron resonance mass spectrometer for "bottom-up" proteomics, *Anal. Chem.* 75, 3419–3428, 2003; Wennder, B.R. and Lynn, B.C., Factors that affect ion trap data-dependent MS/MS in proteomics, *J. Am. Soc. Mass Spectrom.* 15, 150–157, 2004; Amoutzias, G.D., Robertson, D.L., Oliver, S.G., and Bornberg-Bauer, E., Convergent evolution of gene networks by single-gene duplications in higher eukaryotes, *EMBO Rep.* 5, 274–279, 2004; Ren, D., Julka, S., Inerowicz, H.D., and Regnier, F.E., Enrichment of cysteine-containing peptides from tryptic digests using a quaternary amine tag, *Anal. Chem.* 76, 4522–4530, 2004; Listgarten, J. and Emili, A., Statistical and computational methods for comparative proteomic profiling using liquid chromatography-tandem mass spectrometry, *Mol. Cell. Proteomics* 4, 419–434, 2005; Slysz, G.W. and Schriemer, D.C., Blending protein separation and peptide analysis through real-time proteolytic digestion, *Anal. Chem.* 77, 1572–1579, 2005; Zhong, H., Marcus, S.L., and Li, L., Microwave-assisted acid hydrolysis of proteins combined with liquid chromatography MALDI MS/MS for protein identification, *J. Am. Soc. Mass Spectrom.* 16, 471–481, 2005; Putz, S., Reinders, J., Reinders, Y., and Sickmann, A., Mass spectrometry-based peptide quantification: applications and limitations, *Expert Rev. Proteomics* 2, 381–392, 2005; Riter, L.S., Gooding, K.M., Hodge, B.D., and Julian, R.K., Jr., Comparison of the Paul ion trap to the linear ion trap for use in global proteomics, *Proteomics* 6, 1735–1740, 2006.

Brand-Name Drugs A brand-name drug is a drug marketed under a proprietary, trademark-protected name.

BRET Bioluminesence Resonance Energy Transfer. Similar to FRET in BRET is a technique that can be used to measure physical interactions between molecules. Intrinsic bioluminescence is used in this procedure, such as different fluorescent protein (e.g., green fluorescent protein and blue fluorescent protein). See De, A. and Gambhir, S.S., Noninvasive imaging of protein–protein interactions from live cells and living subjects using bioluminescence resonance energy transfer, *FASEB J.* 19, 2017–2019, 2005.

Brownian Movement The random movement of small particles in a suspension, where the force of collision between particles is not lost but retained in part by the particle. The practical effect is to set the lower limit of particle size for settling from a suspension. Brownian movements are usually restricted to particles of 1 μm in diameter and are not observed with particles of 5 μm.

Bulk Solution   Any macroscropic volume of a substance. In the case of an electrolyte, a
                      bulk solution is charge neutral; intracellular and extracellular solutions
                      possess a neutral charge, even the presence of a membrane potential. Also
                      used to describe the difference between water structure in the hydration
                      layer immediately around a macromolecule such as protein and the bulk
                      solvent space. See Nakasako, M., Large-scale networks of hydration water
                      molecules around proteins investigated by cryogenic X-ray crystallography,
                      Cell. Mol. Biol. 47, 767–790, 2001; Lever, M., Blunt, J.W., and Maclagan,
                      R.G., Some ways of looking at compensatory kosmotropes and different
                      water environments, *Comp. Biochem. Physiol. A Mol. Integr. Physiol.* 130,
                      471–486, 2001; Halle, B., Protein hydration dynamics in solution: a critical
                      survey, *Philos. Trans. R. Soc. Biol. Sci.* 359, 1207–1223, 2004; Levicky,
                      R. and Horgan, A., Physicochemical perspectives on DNA microarray and
                      biosensor technologies, *Trends Biotechnol.* 23, 143–149, 2005.

CAD   A multifunctional protein that initiates and regulates *de novo* pyrimidine
                      biosynthesis. See Carrey, E.A., Phosphorylation, allosteric effectors, and
                      inter-domain contacts in CAD: their role in regulation of early steps of
                      pyrimidine biosynthesis, *Biochem. Soc. Trans.* 21, 191–195, 1993; Davidson,
                      J.N., Chen, K.C., Jamison, R.S., Musmanno, L.A., and Kern, C.B., The
                      evolutionary history of the first three steps in pyrimidine biosynthesis,
                      Bioessays 15, 157–164, 1993; Evans, D.R. and Guy, H.I., Mammalian
                      pyrimidine biosynthesis: fresh insights into an ancient pathway, *J. Biol.
                      Chem.* 279, 33035–33038, 2005.

Cadherins   A group of cell adhesion proteins that enable cells to interact with other cells
                      and extracellular matrix components. See Obrink, B., Epithelial cell adhe-
                      sion molecules, *Exp. Cell Res.* 163, 1–21, 1986; Takeichi, M., Cadherins:
                      a molecular family important in selective cell–cell adhesion, *Annu. Rev.
                      Biochem.* 59, 237–252, 1990; Geiger, B. and Ayalon, O., Cadherins, *Annu.
                      Rev. Cell Biol.* 8, 307–332, 1992; Tanoue, T. and Takeichi, M., New insights
                      into Fat cadherins, *J. Cell Sci.* 118, 2347–2353, 2005; Lecuit, T., Cell
                      adhesion: sorting out cell mixing with echinoid? *Curr. Biol.* 15,
                      R505–R507, 2005; Gumbiner, B.M., Regulation of cadherin-mediated
                      adhesion in morphogenesis, *Nat. Rev. Mol. Cell Biol.* 6, 622–634, 2005;
                      Bamji, S.X., Cadherins: actin with the cytoskeleton to form synapses,
                      Neuron 47, 175–178, 2005; Junghans, D., Hass, I.G., and Kemler, R.,
                      Mammalian cadherins and protocadherins: about cell death, synapses, and
                      processing, *Curr. Opin. Cell Biol.* 17, 446–452, 2005; Cavallaro, U.,
                      Liebner, S., and Dejana, E., Endothelial cadherins and tumor angiogenesis,
                      Exp. Cell Res. 312, 659–667, 2006; Redies, C., Vanhalst, K., and Roy, F.,
                      delta-protocadherins: unique structures and functions, *Cell Mol. Life Sci.*
                      62, 2840–2852, 2005; Collona, M., Cytolytic responses: cadherins put out
                      the fire, *J. Exp. Med.* 203, 289–295, 2006; Chan, A.O., E-cadherin in
                      gastric cancer, *World J. Gastroenterol.* 12, 199–203, 2006.

Caenorhabditis   A free-living roundworm that has been used extensively for genomic studies.
elegans   It is notable for the discovery of RNA silencing/RNA interference (Fire,
                      A., Xu, S., Montgomery, M.K. et al., Potent and specific genetic interference
                      by double-stranded RNA in *Caenorhabditis elegans*, *Nature* 391, 806–811,
                      1998). For general aspects of *Caenorhabditis elegans*, see Zuckerman,
                      B.M. and Merton, B., *Nematodes as Biological Models,* Academic Press,
                      New York, 1980; Emmons, S.W., Mechanisms of *C. elegans* development,
                      Cell 51, 881–883, 1987; Blumenthal, T. and Thomas, J., *Cis* and *trans*

splicing in *C. elegans*, *Trends Genet.* 4, 305–308, 1988; Wood, W.B., *The Nematode Caenorhabditis elegans*, Cold Spring Harbor Labortory Press, Cold Spring Harbor, NY, 1988; Greenwald, I., Cell–cell interactions that specify certain cell fates in *C. elegans* development, *Trends Genet.* 5, 237–241, 1989; Coulson, A., Kozono, Y., Lutterbach, B. et al., YACs and the *C. elegans* genome, *Bioessays* 13, 413–417, 1991; Plasterk, R.H., Reverse genetics of *Caenorhabditis elegans*, *Bioessays* 14, 629–633, 1992; Burglin, T.R. and Ruvkun, G., The *Caenorhabditis elegans* homeobox gene cluster, *Curr. Opin. Genet. Dev.* 3, 615–620, 1993; Selfors, L.M. and Stern, M.J., MAP kinase function in *C. elegans*, *Bioessays* 16, 301–304, 1994; Stern, M.J. and DeVore, D.L., Extending and connecting signaling pathways in *C. elegans*, *Dev. Biol.* 166, 443–459, 1994; Kayne, P.S. and Sternberg, P.W., Ras pathways in *Caehorhabditis elegans*, *Curr. Opin. Genet. Dev.* 5, 38–43, 1995; Hope, I.A., *C. elegans; A Practical Approach*, Oxford University Press, Oxford, UK, 1999; Brown, A., *In the Beginning Was the Worm: Finding the Secrets of Life in a Tiny Hermaphrodite*, Columbia University Press, NY, 2003; Filipowicz, W., RNAi: the nuts and bolts of the RISC machine, *Cell* 122, 17–20, 2005; Grishok, A., RNAi mechanisms in *Caenorhabditis elegans*, *FEBS Lett.* 579, 5932–5939, 2005; Hillier, L.W., Coulson, A., Murray, J.I. et al., Genomics in *C. elegans*; so many genes, such a little worm, *Genome Res.* 15, 1651–1660, 2005; Hobert, O. and Loria, P., Uses of GFP in *Caenorhabditis elegans*, *Methods Biochem. Anal.* 47, 203–226, 2006; http://www.nematodes.org/ Caenorhabditis; http://www.wormbook.org; http://elegans.swmed.edu.

Calcineurin   A protein phosphatase that activates IL-2 transcription; IL-2 stimulates the T-cell response. Calcineurin is inhibited by immunosuppressive drugs such as cyclosporine and FK506 (tacrolimus). See Pallen, C.J. and Wang, J.H., A multifunctional calmodulin-stimulated phosphatase, *Arch. Biochem. Biophys.* 237, 281–291, 1985; Klee, C.B., Draetta, G.F., and Hubbard, M.J., Calcineurin, *Adv. Enzymol. Relat. Areas Mol. Biol.* 61, 149–200, 1988; Siekierka, J.J. and Sigal, N.H., FK-506 and cyclosporine A: immunosuppressive mechanism of action and beyond, *Curr. Opin. Immunol.* 4, 484–552, 1992; Groenendyk, J., Lynch, J., and Michalak, M., Calreticulin, Ca^{2+}, and calcineurin — signaling from the endoplasmic reticulum, *Mol. Cells* 30, 383–389, 2004; Michel, R.N., Dunn, S.E., and Chin, E.R., Calcineurin and skeletal muscle growth, *Proc. Nutr. Soc.* 63, 341–349, 2004; Im, S.H. and Rao, A., Activation and deactivation of gene expression by Ca^{2+}/calcineurin-NFAT-mediated signaling, *Mol. Cells* 16, 1–9, 2004; Bandyopadhyay, J., Lee, J., and Bandopadhyay, A., Regulation of calcineurin, a calcium/calmodulin-dependent protein phosphatase in *C. elegans*, *Mol. Cells* 18, 10–16, 2004; Taylor, A.L., Watson, C.J., and Bradley, J.A., Immunosuppresive agents in solid organ transplantation: mechanisms of action and therapeutic efficacy, *Crit. Rev. Oncol. Hematol.* 56, 23–46, 2005; Crespo-Leiro, M.G., Calcineurin inhibitors in heart transplantation, *Transplant. Proc.* 37, 4018–4020, 2005.

Calcium Transients   Physiology phenomena resulting from changes in calcium concentration across membranes such as the diversity in Ca^{++}-stimulated transcriptional phenomona. See Thorner, M.O., Holl, R.W., and Leong, D.A., The somatotrope: an endocrine cell with functional calcium transients, *J. Exp. Biol.* 139, 169–179, 1988; Morgan, K.G., Bradley, A., and DeFeo, T.T., Calcium transients in smooth muscle, *Ann. N.Y. Acad. Sci.* 522, 328–337, 1988; Fumagalli,

G., Zacchetti, D., Lorenzon, P., and Grohovaz, F., Fluorimetric approaches to the study of calcium transients in living cells, *Cytotechnology* 5 (Suppl. 1), 99–102, 1991; Fossier, P., Tauc, L., and Baux, G., Calcium transients and neurotransmitter release at an identified synapse, *Trends Neurosci.* 22, 161–166, 1999; Spitzer, N.C., Lautermilch, N.J., Smith, R.D., and Gomez, T.M., Coding of neuronal differentiation by calcium transients, *Bioessays* 22, 811–817, 2000; Afroze, T. and Husain, M., Cell cycle-dependent regulation of intracellular calcium concentration in vascular smooth muscle cells: a potential target for drug therapy, *Curr. Drug Targets Cardiovasc. Haematol. Disord.* 1, 23–40, 2001; Komura, H., and Kumada, T., Ca^{2+} transients control CNS neuronal migration, *Cell Calcium* 37, 387–393, 2005.

CALM-Domain/ CALM Protein
Clathrin assembly lymphoid myeloid-domain, related to ANTH-domain proteins and involved in endocytosis, formation of clathrin-coated pits; binds to lipids. See Kim, J.A., Kim, S.R., Jung, Y.K. et al., Properties of GST-CALM expressed in *E. coli. Exp. Mol. Med.* 32, 93–99, 2000; Kusner, L. and Carlin, C., Potential role for a novel AP180-related protein during endocytosis in MDCK cells, *Am. J. Physiol. Cell Physiol.* 285, C995–C1008, 2003; Archangelo, L.F., Glasner, J., Krause, A., and Bohlander, S.K., The novel CALM interactor CATS influences the subcellular localization of the leukemogenic fusion protein CALM/AF10, *Oncogene*, 25, 4099–4109, 2006.

Calnexin
A lectin protein associated with the endoplasmic reticulum and which functions as a chaperone. See Cresswell, P., Androlewicz, M.J., and Ortmann, B., Assembly and transport of class I MHC-peptide complexes, *Ciba Found. Symp.* 187, 150–162, 1994; Bergeron, J.J., Brenner, M.B., Thomas, D.Y., and Williams, D.B., Calnexin: a membrane-bound chaperone for the endoplasmic reticulum, *Trends Biochem. Sci.* 19, 124–128, 1995; Parham, P., Functions for MHC class I carbohydrates inside and outside the cell, *Trends Biochem. Sci.* 21, 427–433, 1996; Trombetta, E.S. and Helenius, A., Lectins as chaperones in glycoprotein folding, *Curr. Opin. Struct. Biol.* 8, 587–592, 1998; Huari, H., Appenzeller, C., Kuhn, F., and Nufer, O., Lectins and traffic in the secretory pathway, *FEBS Lett.* 476, 32–37, 2000; Ellgaard, L. and Frickel, E.M., Calnexin, calreticulin, and ERp57: teammates in glycoprotein folding, *Cell. Biochem. Biophys.* 39, 223–247, 2003; Spiro, R.G., Role of *N*-linked polymannose oligosaccharides in targeting glycoproteins for endoplasmic reticulum-associated degradation, *Cell. Mol. Life Sci.* 61, 1025–1041, 2004; Bedard, K., Szabo, E., Michalak, M., and Opas, M., Cellular functions of endoplasmic reticulum chaperones calreticulin, calnexin, and ERp57, *Int. Rev. Cytol.* 245, 91–121, 2005; Ito, Y., Hagihara, S., Matsuo, I., and Totani, K., Structural approaches to the study of oligosaccharides in glycoprotein quality control, *Curr. Opin. Struct. Biol.* 15, 481–489, 2005.

Calponin
A family of actin-binding proteins that exist in various isoforms. As with other protein isoforms or isoenzymes, the expression of the isoforms is tissue-specific. The interaction of calponin with actin inhibits the actomyosin Mg-ATPase activity. See Winder, S. and Walsh, M., Inhibition of the actinomyosin MgATPase by chicken gizzard calponin, *Prog. Clin. Biol. Res.* 327, 141–148, 1990; Winder, S.J., Sutherland, C., and Walsh, M.P., Biochemical and functional characterization of smooth muscle calponin, *Adv. Exp. Med. Biol.* 304, 37–51, 1991; Winder, S.J. and Walsh, M.P., Calponin: thin filament-linked regulation of smooth muscle contraction, *Cell Signal.* 5, 677–686, 1993; el-Mezgueldi, M., Calponin, *Int. J. Biochem.*

Cell Biol. 28, 1185–1189, 1996; Szymanski, P.T., Calponin (CaP) as a latch-bridge protein — a new concept in regulation of contractility in smooth muscle, *J. Muscle Res. Cell Motil.* 25, 7–19, 2004; Lehman, W., Craig, R., Kendrick-Jones, J., and Sutherland-Smith, A.J., An open or closed case for the conformation of calponin homology domains on F-actin? *J. Muscle Res. Cell Motil.* 25, 351–358, 2004; Ferjani, I., Fattoum, A., Maciver, S.K. et al., A direct interaction with calponin inhibits the actin-nucleating activity of gelsolin, *Biochem. J.* 396, 461–468, 2006.

Calreticulin   A 50–60 kDa protein found in the endoplasmic reticulum. Calreticulin binds calcium ions tightly and is thought to play a role in calcium homeostasis. Calreticulin also functions as a chaperone. See Koch, G.L. and Smith, M.J., The analysis of glycoproteins in cells and tissues by two-dimensional polyacrylamide gel electrophoresis, *Electrophoresis* 11, 213–219, 1990; Krause, K.H., Ca(2+)-storage organelles, *FEBS Lett.* 285, 225–229, 1991; Herbert, D.N., Simons, J.F., Peterson, J.R., and Helenius, A., Calnexin, calreticulin, and Bip/Kar2p in protein folding, *Cold Spring Harbor Symp. Quant. Biol.* 60, 405–415, 1995; Groenendyk, J., Lynch, J., and Michalak, M., Calreticulin, Ca^{2+}, and calcineurin — signaling from the endoplasmic reticulum, *Mol. Cells* 17, 383–389, 2004; Michalak, M., Guo, L., Robertson, M., Lozak, M., and Opas, M., Calreticulin in the heart, *Mol. Cell. Biochem.* 263, 137–142, 2004; Gelebart, P., Opas, M., and Michalak, M., Calreticulin, a Ca^{2+}-binding chaperone of the endoplasmic reticulum, *Int. J. Biochem. Cell Biol.* 37, 260–266, 2005; Bedard, K., Szabo, E., Michalak, M., and Opas, M., Cellular functions of endoplasmic reticulum chaperones calreticulin, calnexin, and ERp57, *Int. Rev. Cytol.* 245, 91–121, 2005; Ito, Y., Hagihara, S., Matsuo, I., and Totani, K., Structural approaches to the study of oligosaccharides in glycoprotein quality control, *Curr. Opin. Struct. Biol.* 15, 481–489, 2005; Garbi, N., Tanaka, S., van den Broek, M. et al., Accessory molecules in the assembly of major histocompatibility complex class I/peptide complexes: how essential are they for CD(+) T-cell immune responses? *Immunol. Rev.* 207, 77–88, 2005; Cribb, A.E., Peyrou, M., Muruganandan, S., and Schneider, L., The endoplasmic reticulum in xenobiotic toxicity, *Drug Metab. Rev.* 37, 405–442, 2005; Hansson, M., Olsson, I., and Nauseef, W.M., Biosynthesis, processing, and sorting of human myeloperoxidase, *Archs. Biochem. Biophys.* 445, 214–224, 2006.

Camelid   Unique antibodies from members of the *Camelidae* family. The antibody
Antibodies   structure consists of a heavy chain consisting of a variable region V_HH but no light chain. The structure also misses the first constant domain (C_H1) but retains the other constant region's C-terminal from the hinge region. Thus, while a classical IgG is a dimer of heterodimers, the camelid antibody described herein is a homodimer. See Ghahroudi, M.A., Desmyter, A., Wyns, L., Hamers, R., and Muyldermans, S., Selection and identification of single domain antibody fragments from camel heavy-chain antibodies, *FEBS Lett.* 414, 521–526, 1997; Nguyen, V.K., Desmyter, A., and Muyldermans, S., Functional heavy-chain antibodies in *Camelidae*, *Adv. Immunol.* 79, 261–296, 2001; Muyldermans, S., Single domain camel antibodies: current status, *J. Biotechnol.* 74, 277–302, 2001; Nguyen, V.K., Su., C., Muyldermans, S., and van der Loo, W., Heavy-chain antibodies in *Camelidae*, a case of evolutionary innovation, *Immunogenetics* 54, 39–47, 2002; Conrath, K.E., Wernery, U., Mulydermans, S., and Nguyen, V.K., Emergence and evolution of

functional heavy-chain antibodies in *Camelidae*, *Dev. Comp. Immunol.* 27, 87–103, 2003; Rahbarisadeh, F., Rasaee, M.J., Forouzandeh, M. et al., The production and characterization of novel heavy-chain antibodies against the tandem repeat region of MUC1 mucin, *Immunol. Invest.* 34, 431–452, 2005.

Cap   Structure at the 5′end of eukaryotic RNA, introduced after transcription by linking the terminal phosphate of 5′GTP to the terminal base of the mRNA. The guanine base can be nucleated ($^{7Me}G^{5'}$-ppp^{5}Np). CAP is also an acronym for catabolic activator protein. See Banerjee, A.K., 5′-terminal cap structure in eukaryotic messenger ribonucleic acids, *Microbiol. Rev.* 44, 175–205, 1980; Miura, K., The cap structure in eukaryotic RNA as a mark of a strand carrying protein information, *Adv. Biophys.* 14, 205–238, 1981; Lewin, B., *Genes IV*, Oxford University Press, Oxford, UK, 1990; Cougot, N., van Dijk, E., Babajko, S., and Seraphin, B., "Cap-tabolism," *Trends Biochem. Sci.* 29, 436–444, 2004; Gu, M. and Lima, C.D., Processing the message: structural insights into capping and decapping mRNA, *Curr. Opin. Struct. Biol.* 15, 99–106, 2005; Bentley, D.L., Rules of engagement: co-transcriptional recruitment of pre-mRNA processing factors, *Curr. Opin. Cell Biol.* 17, 251–256, 2005; Liu, H. and Kiledjian, M., Decapping the message: a beginning or an end, *Biochem. Soc. Trans.* 34, 35–38, 2006; Simon, E., Camier, S., and Seraphin, B., New insights into the control of mRNA decapping, *Trends in Biochem. Sci.* 31, 241–243, 2006.

CASP   Critical Assessment of Structure Prediction describes a process for the evaluation of protein model building. See Moult, J., Predicting protein three-dimensional structure, *Curr. Opin. Biotechnol.* 10, 583–588, 1999; Moult, J., Fidelis, K., Rost, B., Hubbard, T., and Tramontano, A., Critical assessment of methods of protein structure prediction (CASP) — round 6, *Proteins* 61 (Suppl. 7), 3–7, 2005; Giorgetti, A., Raimondo, D., Miele, A.E., and Tramontano, A., Evaluating the usefulness of protein structure models for molecular replacement, *Bioinformatics* 21 (Suppl. 2), ii72–ii76, 2005; Espejo, F. and Patarroyo, M.E., Determining the 3-D structure of human ASC2 protein involved in apoptosis and inflammation, *Biochem. Biophys. Res. Commun.* 340, 860–864, 2006; Moult, J., Rigorous performance evaluation in protein structure modeling and implications for computational biology, *Philos. Trans. R. Soc. Lond. B Biol. Sci.* 361, 453–458, 2006.

Caspases   A family of intracellular cysteine proteases that are involved in the process of apoptosis (programmed cell death). Caspases are synthesized as precursor or zymogen forms, which require activation prior to function. One such activation process involves granzymes. Caspases also function in other intracellular processes. See Jacobson, M.D. and Evan, G.I., Apoptosis. Breaking the ice, *Curr. Biol.* 4, 337–340, 1994; Patel, T., Gores, G.B.J., and Kaufmann, S.H., The role of proteases during apoptosis, *FASEB J.* 10, 587–597, 1996; Alnemri, E.S., Mammalian cell death proteases: a family of highly conserved aspartate specific cysteine proteases, *J. Cell Biochem.* 64, 33–42, 1997; Zhivotovsky, B., Caspases: the enzymes of death, *Essays Biochem.* 39, 25–40, 2003; Twomey, C. and McCarthy, J.V., Pathways of apoptosis and importance in development, *J. Cell Mol. Med.* 9, 345–359, 2005; Ashton-Rickardt, P.G., The granule pathway of programmed cell death, *Crit. Rev. Immunol.* 25, 161–182, 2005; Yan, N. and Shi, Y., Mechanisms of apoptosis through structural biology, *Annu. Rev. Cell Dev. Biol.* 21, 35–56, 2005; Harwood, S.M., Yaqoob, M.M., and Allen, D.A.,

Caspase and calpain function in cell death: bridging the gap between apoptosis and necrosis, *Ann. Clin. Biochem.* 42, 415–431, 2005; Ho, P.K. and Hawkins, C.J., Mammalian initiator apoptotic caspases, *FEBS J.* 272, 5436–5453, 2005; Fardeel, B. and Orrenius, S., Apoptosis: a basic biological phenomenon with wide-ranging implications in human disease, *J. Intern. Med.* 258, 479–517, 2005; Cathelin, S., Rébe, C., Haddaoui, L. et al., Identification of proteins cleaved downstream of caspase activation in monocytes undergoing macrophage differentiation, *J. Biol. Chem.* 281, 17779–17788, 2006.

Catabolic Activator Protein
A transcription-regulating protein that binds to DNA in the promoter loop. See Benoff, B., Yang, H., Lawson, C.L. et al., Structural basis of transcription activation: the CAP-alpha CTD-DNA complex, *Science* 297, 1562–1566, 2002; Balaeff, A., Madadevan, L., and Schulten, K., Structural basis for cooperative DNA binding by CAP and lac repressor, *Structure* 12, 123–132, 2004; Akaboshi, E., Dynamic profiles of DNA: analysis of CAP- and LexA protein-binding regions with endonucleases, *DNA Cell Biol.* 24, 161–172, 2005.

Catalomics
The study of the enzymes in a proteome; the study of catalysis in a proteome. See Hu, Y., Uttamchandani, M., and Yao, S.Q., Microarray: a versatile platform for high-throughput functional proteomics, *Comb. Chem. High Throughput Screen.* 9, 201–212, 2006.

Catalytic Antibodies
Antibodies that demonstrate catalytic activity. The early development of these antibodies was based on the use of haptens, which mirrored transition state intermediates for enzyme-catalyzed reactions. Catalytic antibodies can be referred to as abzymes. See Kraut, J., How do enzymes work? *Science* 242, 533–540, 1988; Lerner, R.A. and Tramontano, A., Catalytic antibodies, *Sci. Am.* 258, 65–70, 1988; Green, B.S., Catalytic antibodies and biomimetics, *Curr. Opin. Biotechnol.* 2, 395–400, 1991; Jacobs, J.W., New perspectives on catalytic antibodies, *Biotechnology* 9, 258–262, 1991; Blackburn, G.M., Kingsbury, G., Jayaweera, S., and Burton, D.R., Expanded transition state analogues, *Ciba Found. Symp.* 159, 211–222, 1991; O'Kennedy, R. and Roben, P., Antibody engineering: an overview, *Essays Biochem.* 26, 59–75, 1991; Stewart, J.D., Krebs, J.F., Siuzdak, G. et al., Dissection of an antibody-catalyzed reaction, *Proc. Natl. Acad. Sci. USA* 91, 7404–7409, 1994; Posner, B., Smiley, J., Lee, I., and Benkovic, S., Catalytic antibodies: perusing combinatorial libraries, *Trends Biochem. Sci.* 19, 145–150, 1994; Kikuchi, K. and Hilvert, D., Antibody catalysis via strategic use of hepatenic charge, *Acta Chem. Scand.* 50, 333–336, 1996; Wentworth, P., Jr. and Janda, K.D., Catalytic antibodies: structure and function, *Cell. Biochem. Biophys.* 35, 63–87, 2001; Ostler, E.L., Resmini, M., Brocklehurst, K., and Gallacher, G., Polyclonal catalytic antibodies, *J. Immunol. Methods* 269, 111–124, 2002; Hanson, C.V., Nishiyama, Y., and Paul, S., Catalytic antibodies and their applications, *Curr. Opin. Biotechnol.* 16, 631–666, 2005. There has also been considerable interest in catalytic antibodies in pathological processes and as potential therapeutic agents. See Lacroix-Demazes, S., Kazatchkine, M.D., and Kaveri, S.V., Catalytic antibodies to factor VIII in haemophilia A, *Blood Coag. Fibrinol.* 14 (Suppl. 1), S31–S34, 2003; Poloukhina, D.I., Kanyshkova, T.G., Doronin, B.M. et al., Hydrolysis of myelin basic protein by poly-clonal catalytic IgGs from the sera of patients with multiple sclerosis, *J. Cell. Mol. Med.* 8, 359–368, 2004; Paul, S., Nishiyama, Y., Planque, S.

et al., Antibodies as defensive enzymes, *Springer Semin. Immunopathol.* 26, 485–503, 2005; Ponomarenko, N.A., Vorobiev, I.I., Alexandrova, E.S. et al., Induction of a protein-targeted catalytic response in autoimmune prone mice: antibody-mediated cleavage of HIV-1 glycoprotein GP120, *Biochemistry* 45, 324–330, 2006; Lacroix-Demazes, S., Wootla, B., Delignat, S. et al., Pathophysiology of catalytic antibodies, *Immunol. Lett.* 103, 3–7, 2006.

CATH   A classification process for protein domain structures based on class (C), architecture (A), topology (T), and homology superfamily (H). See Orengo, C.A., Michie, A.D., Jones, S. et al., CATH — a hierarchic classification of protein domain structures, *Structure* 5, 1093–1108, 1997; Bray, J.E., Todd, A.E., Pearl, F.M., Thornton, J.M., and Orengo, C.A., The CATH dictionary of homologous superfamilies (DHS): a consensus approach for identifying distant structural homologues, *Protein Eng.* 13, 153–165, 2000; Ranea, J.A., Buchan, D.W., Thornton, J.M., and Orengo, C.A., Evolution of protein superfamilies and bacterial genome size, *J. Mol. Biol.* 336, 871–887, 2004; Velazquez-Muriel, J.A., Sorzano, C.O., Scheres, S.H., and Carazo, J.M., SPI-EM: towards a tool for predicting CATH superfamilies in 3D-EM maps, *J. Mol. Biol.* 345, 759–771, 2005; Sillitoe, I., Dibley, M., Bray, J., Addou, S., and Orengo, C., Assessing strategies for improved superfamily recognition, *Protein Sci.* 14, 1800–1810, 2005.

Cathepsins   A family of intracellular thiol proteases involved in lysosomal digestion of proteins. See Janoff, A., Mediators of tissue damage in human polymorphonuclear neutrophils, *Ser. Haematol.* 3, 96–130, 1970; Harris, E.D., Jr. and Krane, S.M., Collagenases, *N. Engl. J. Med.* 291, 605–609, 1974; Larzarus, G.S., Hatcher, V.B., and Levine, N., Lysosomes and the skin, *J. Invest. Dermatol.* 65, 259–271, 1975; Ballard, F.J., Intracellular protein degradation, *Essays Biochem.* 13, 1–37, 1977; Barrett, A.J., Cathepsin D: the lysosomal aspartic proteinase, *Ciba Found. Symp.* 75, 37–50, 1979; Barrett, A.J. and Kischeke, H., Cathepsin B, cathepsin H, and cathepsin L, *Methods Enzymol.* 80, 535–561, 1981; Groutas, W.C., Inhibitors of leukocyte elastase and leukocyte cathepsin G. Agents for the treatment of emphysema and related ailments, *Med. Res. Rev.* 7, 227–241, 1987; Stoka, V., Turk, B., and Turk, V., Lysosomal cathepsin proteases: structural features and their role in apoptosis, *IUBMB Life* 57, 347–353, 2005; Roberts, R., Lysosomal cysteine proteases: structure, function, and inhibition of cathepsins, *Drug News Perspect.* 18, 605–614, 2005; Chwieralski, C.E., Welte, T., and Buhling, F., Cathepsin-regulated apoptosis, *Apoptosis* 11, 143–149, 2006. There is particular interest in the role of cathepsins in antigen processing (Honey, K. and Rudensky, A.Y., Lysosomal cysteine proteases regulate antigen presentation, *Nat. Rev. Immunol.* 3, 472–482, 2003; Bryant, P. and Ploegh, H., Class II MHC peptide loading by the professionals, *Curr. Opin. Immunol.* 16, 96–102, 2004; Liu, W. and Spero, D.M., Cysteine protease cathepsin S as a key step in antigen presentation, *Drug News Perspect.* 17, 357–363, 2004; Hsing, L.C. and Rudensky, A.Y., The lysosomal cysteine proteases in MHC Class II antigen presentation, *Immunol. Rev.* 207, 229–241, 2005).

CELISA   Enzyme-linked immunoassay on live cells. See Geraghyty, R.J., Jogger, C.P., and Spear, P.B., Cellular expression of alphaherpesvirus gD interferes with entry of homologous and heterologous alphaviruses by blocking access to a shared gD receptor, *Virology* 68, 147–156, 2000; Lee, R.B., Hassone,

D.C., Cottle, D.L., and Picket, C., Interactions of *Campylobacter jejuni* cytolethal distending toxin subunits Cdta and Cdtc with HeLa cells, *Infect. Immun.* 71, 4883–4890, 2003.

Cell-Based Assays
This is a broad classification for assays where cells are used as the substrate or indictor for the action of a drug. Examples include platelet aggregation, cell-based ELISA (see below), gene expression assays, receptor-ligand interactions, etc. See Nuttall, M.E., Drug discovery and target validation, *Cells Tissues Organs* 169, 265–271, 2001; Bhadriraju, K. and Chen, C.S., Engineering cellular microenvironments to improve cell-based drug testing, *Drug Discov. Today* 7, 612–620, 2002; Indelicato, S.R., Bradshaw, S.L., Chapman, J.W., and Weiner, S.H., Evaluation of standard and state of the art analytical technology-bioassays, *Dev. Biol.* 122, 103–114, 2005; Stacey, G.N., Standardization of cell lines, *Dev. Biol.* 111, 259–272, 2002; Qureshi, S.A., Sanders, P., Zeh, K. et al., A one-arm homologous recombination approach for developing nuclear receptor assays in somatic cells, *Assay Drug Dev. Technol.* 1, 767–776, 2003; Wei, X., Swanson, S.J., and Gupta, S., Development and validation of a cell-based bioassay for the detection of neutralizing antibodies against recombinant human erythropoietin in clinical studies, *J. Immunol. Methods* 293, 115–126, 2004; Pietrak, B.L., Crouthamel, M.C., Tugusheva, K. et al., Biochemical and cell-based assays for characterization of BACE-1 inhibitors, *Anal. Biochem.* 342, 144–151, 2005; Chen. T., Hansen, G., Beske, O. et al., Analysis of cellular events using CellCard™ System in cell-based high-content multiplexed assays, *Expert Rev. Mol. Diagn.* 5, 817–829, 2005.

Cell-Based ELISA
Cell-based ELISA are indirect or direct ELISA systems that use intact cells as antigen samples. Cells may be dried onto the microplate surface or a microplate surface treated with polylysine, chemically fixed with glutaraldehyde or similar reagents, or pelleted onto the surface. See Hoffman, T. and Herberman, R.B., Enzyme-linked immunosorbent assay for screening monoclonal antibody production: use of intact cells as antigen, *J. Immunol. Methods* 39, 309–316, 1980; Krakauer, H., Hartman, R.J., and Johnson, A.H., Monoclonal antibodies specific for human polymorphic cell surface antigens. I. Evaluation of methodology. Report of a workshop, *Human Immunol.* 4, 167–181, 1982; Bishara, A., Brautbar, C., Marbach, A., Bonvida, B., and Nelken, D., Enzyme-linked immunosorbent assay for HLA determination on fresh and dried lymphocytes, *J. Immunol. Methods* 62, 265–271, 1983; Sharon, R., Duke-Cohan, J.S., and Galili, U., Determination of ABO blood group zygosity by an antiglobulin resetting technique and cell-based enzyme immunoassay, *Vox Sang.* 50, 245–249, 1986; Zhao, Q., Lu, H., Schols, D., de Clercq, E., and Jiang, S., Development of a cell-based enzyme-linked immunosorbent assay for high-throughput screening of HIV-type enzyme inhibitors targeting the coreceptor CXCR4, *AIDS Res. Human Retrovirus* 19, 947–955, 2003; Yang, X.Y., Chen, E., Jiang, H. et al., Development of a quantitative cell-based ELISA, for a humanized anti-IL-2/IL-15 receptor beta antibody (HuMikbeta(1)), and correlation with functional activity using an antigen-transferred murine cell line, *J. Immunol. Methods* 311, 71–80, 2006. In some cases, a cell homogenate could be used as the sample. (See Franciotta, D., Martino, G., Brambilla, E. et al., TE671 cell-based ELISA for anti-acetylcholine receptor antibody determination in myasthenia gravis, *Clin. Chem.* 45, 400–405, 1999). The cell-based ELISA is distinct from the ELISPOT assay where there is a capture antibody on

the membrane (Arvilommi, H., Elispot for detecting antibody-secreting cells in response to infections and vaccination, *APMIS* 104, 401–410, 1996).

Cell Culture   The maintenance of dispersed animal or plant cells in a specialized media (cell culture media). In biotechnology manufacturing, cell culture is used for the production of protein biopharmaceuticals using cells such as Chinese hamster ovary (CHO) cells or baby hamster kidney (BHK) cells. The use of the term "cell culture" differentiates such a process from fermentation. See Mantell, S.H. and Smith, H., *Plant Biotechnology*, Cambridge University Press, Cambridge, UK, 1983; *Applications of Plant Cell and Tissue Culture*, John Wiley & Sons, Chichester, UK, 1988; Freshney, R.I., *Animal Cell Culture: A Practical Approach*, IRL Press at Oxford University Press, Oxford, UK, 1992; Morgan, S.J. and Darling, D.C., *Animal Cell Culture*, Bios/Biochemical Society, London, UK, 1993; Davis, J.M., *Basic Cell Culture*: *A Practical Approach*, IRL Press at Oxford University Press, Oxford, UK, 1994; Dodds, J.H. and Roberts, L.W., *Experiments in Plant Tissue Culture*, Cambridge University Press, Cambridge, UK, 1995; Spier, R., *Encyclopedia of Cell Technology*, Wiley Interscience, New York, 2000; Hesse, F. and Wagner, R., Development and improvements in the manufacture of human therapeutics with mammalian cell culture, *Trends Biotechnol.* 18, 173–180, 2000; James, E. and Lee, J.M., The production of foreign proteins from genetically modified plant cells, *Adv. Biochem. Eng. Biotechnol.* 72, 127–156, 2001; Kaeffer, B., Mammalian intestinal epithelial cells in primary culture: a mini-review, In Vitro *Cell Dev. Biol. Animal* 38, 128–134, 2002; Ikonomou, L., Schneider, Y.J., and Agathos, S.N., Insect cell culture for industrial production of recombinant proteins, *Appl. Microbiol. Biotechnol.* 62, 1–20, 2003; Kallos, M.S., Sen, A., and Behie, L.A., Large-scale expansion of mammalian neural stem cells: a review, *Med. Biol. Eng. Comput.* 41, 271–282, 2003; Schiff, L.J., Review: production, characterization, and testing of banked mammalian cell substrates used to produce biological products *in vitro, Cell Dev. Biol. Animal* 41, 65–70, 2005; Evan, M.S., Sandusky, C.B., and Barnard, N.D., Serum-free hybridoma culture: ethical, scientific, and safety considerations, *Trends Biotechnol.* 24, 105–108, 2006.

Cell-Penetrating   Cell-penetrating peptides are relatively small peptides, usually less than 30
Peptide   amino acids in length, which have the ability to pass through or translocate the cellular membrane via mechanisms which appear to be both receptor-independent as well as distinct from an endocytotic process. Such peptides have been demonstrated to "transport" diverse cargo and are being evaluated for drug delivery. See Lundberg, P. and Langel, U., A brief introduction to cell-penetrating peptides, *J. Mol. Recognit.*, 16, 227–233, 2003; Temsamani, J. and Vidal, P., The use of cell-penetrating peptides for drug delivery, *Drug Discov. Today* 9, 1012–1019, 2004; Gupta, B., Levchenko, T.S., and Torchilin, V.P., Intracellular delivery of large molecules and small particles by cell-penetrating proteins and peptides, *Adv. Drug Deliv. Rev.* 57, 637–651, 2005; Deshayes, S., Morris, M.C., Divta, G., and Heitz, F., Cell-penetrating peptides: tools for intracellular delivery of therapeutics, *Cell. Mol. Life Sci.* 62, 1839–1849, 2005. See also *Amphipathic*.

Centimorgan   A measure of genetic distance that tells how far apart physically two genes are, based on the frequency of recombination or crossover between the two gene loci. A frequency of 1% recombination in meiosis is 1 centimorgan and equals about 1 million base pairs. See Southern, E.M., Prospects for a complete molecular map of the human genome, *Philos. Trans. R. Soc. Lond. B Biol. Sci.* 319, 299–307, 1988; White, R., Lalauel, J.M., Leppert, M. et al., Linkage maps of human chromosomes, *Genome* 31, 1066–1072, 1989; Smith, L.H., Jr., Overview of hemochromatosis, *West. J. Med.* 153, 296–308, 1990; Crabbe, J.C., Alcohol and genetics: new models, *Am. J. Med. Genet.* 114, 969–974, 2002.

CentiRay   A measure of the frequency of chromosome breakage between DNA markers in radiation-reduced somatic cell hybrids (radiation hybrids). One centi-Ray is equivalent to a 1% probability that a chromosome will break (centiRay distances are generally proportional to physical distance and are measured in centimorgans). See Hukriede, N.A., Joly, L., Tsang, M. et al., Radiation-hybrid mapping of the zebrafish genome, *Proc. Natl. Acad. Sci. USA* 96, 9745–9750, 1999; Hamasima, N., Suzuki, H., Mikawa, A. et al., Construction of a new porcine whole-genome framework map using a radiation-hybrid panel, *Anim. Genet.* 34, 216–220, 2003; Voigt, C., Moller, S., Ibrahim, S.M., and Serrano-Fernandez, P., Nonlinear conversion between genetic and physical chromosomal distances, *Bioinformatics* 20, 1966–1977, 2004.

Chameleon Sequences   Identical sequences in a protein, which can adopt either an alpha-helical conformation or a beta-sheet conformation. See Minor, D.L., Jr. and Kim, P.S., Context-dependent secondary structure formation of a designed peptide sequence, *Nature* 380, 730–734, 1996; Mezei, M., Chameleon sequences in the PDB, *Protein. Eng.* 11, 411–414, 1998; Tidow, H. et al., The solution structure of a chimeric LEKTI domain reveals a chameleon sequence, *Biochemistry* 43, 11238–11247, 2004.

Chaotropic   Describing a reagent that disrupts the structure of water and macromolecules such as proteins. Chaotropic is sometimes confined to uncharged molecules such as urea or thiourea but is usually extended to include reagents such as guanidine hydrochloride and sodium thiocyanate. See Dandliker, W.B., Alonso, R., de Saussure, V.A. et al., The effect of chaotropic ions on the dissociation of antigen–antibody complexes, *Biochemistry* 6, 1460–1467, 1967; Hanstein, W.G., Davis, K.A., and Hatefi, Y., Water structure and the chaotropic properties of haloacetates, *Arch. Biochem. Biophys.* 147, 534–544, 1971; Sawyer, W.H. and Puckridge, J., The dissociation of proteins by chaotropic salts, *J. Biol. Chem.* 248, 8429–8433, 1973; Hatefi, Y. and Hanstein, W.G., Destabilization of membranes with chaotropic ions, *Methods Enzymol.* 31, 770–790, 1974; McLaughlin, S., Bruder, A., Chen, S., and Moser, C., Chaotropic anions and the surface potential of bilayer membranes, *Biochim. Biophys. Acta* 394, 304–313, 1975; Stein, M., Lazaro, J.J., and Wolsiuk, R.A., Concerted action of cosolvents, chaotropic anions, and thioredoxin on chloroplast fructose-1,6-bisphosphatase. Reactivity to iodoacetate, *Eur. J. Biochem.* 185, 425–431, 1989; Lever, M., Blunt, J.W., and MacLagan, R.G., Some ways of looking at compensatory kosmotropes and different water environments, *Comp. Biochem. Physiol. A Integr. Physiol.* 130, 471–486, 2001; Pilorz, K. and Choma, I., Isocratic reversed-phase high-performance liquid chromatographic separation of tetracyclines and flumequine controlled by a chaotropic effect, *J. Chromatog. A.* 1031, 303–306, 2004; Moelbert, S., Normand, B., and De Los Rios, P., Kosmotropes and chaotropes: modeling preferential exclusion, binding, and aggregate stability, *Biophys. Chem.* 112, 45–57, 2004; Salvi, G., De Los Rios, P., and Vendruscolo, M., Effective interactions between chaotropic agents and proteins, *Proteins* 61, 492–499, 2005; LoBrutto, R. and Kazakevich, Y.V., Chaotropic effects in RP-HPLC, *Adv. Chromatog.* 44, 291–315, 2006.

Chaperone An intracellular factor, most frequently a protein, that guides the intracellular folding/assembly of another protein. Examples include heat-shock proteins, chaperoinins. See Gregerson, N., Bolund, L., and Bross, P., Protein misfolding, aggregation, and degradation in disease, *Mol. Biotechnol.* 31, 141–150, 2005; Anken, E., Braakman, I., and Craig, E., Versatility of the endoplasmic reticulum protein folding factory, *Critl. Rev. Biochem. Mol. Biol.* 40, 191–288, 2005; Macario, A.J. and Conway de Marcario, E., Sick chaperones, cellular stress, and disease, *New Eng. J. Med.* 353, 1489–1501, 2005; Weibezahn, J., Schlieker, C., Tessarz, P., Mogk, A., and Bukau, B., Novel insights into the mechanism of chaperone-assisted protein disaggregation, *Biol. Chem.* 386, 739–744, 2005.

Chemical Biology The application of chemical techniques to problems in biology — the emphasis is directed toward study of the interaction of small molecules with proteins and other macromolecules. See Li, C.H., Current concepts on the chemical biology of pituitary hormones, *Perspect. Biol. Med.* 11, 498–521, 1968; Malmstrom, B.G. and Leckner, J., The chemical biology of copper, *Curr. Opin. Chem. Biol.* 2, 286–292, 1998; Bertini, I. and Luchinat, C., New applications of paramagnetic NMR in chemical biology, *Curr. Opin. Chem. Biol.* 3, 145–151, 1999; Volkert, M., Wagner, M., Peters, C., and Waldmann, H., The chemical biology of Ras lipidation, *Biol. Chem.* 382, 1133–1145, 2001; Hahn, M.E. and Muir, T.W., Manipulating proteins with chemistry: a cross-section of chemical biology, *Trends Biochem. Sci.* 30, 26–34, 2005; Cambell-Valois, F.X. and Michnick, S., Chemical biology on PINs and NeeDLes, *Curr. Opin. Chem. Biol.* 9, 31–37, 2005; Doudna, J.A., Chemical biology at the crossroads of molecular structure and mechanism, *Nat. Chem. Biol.* 1, 300–303, 2005.

Chemical Proteomics Use of chemical modification to identify enzymes in the proteome and to identify signaling pathways. See Jeffery, D.A. and Bogyo, M., Chemical proteomics and its application to drug discovery, *Curr. Opin. Biotechnol.* 14, 87–95, 2003; Daub, H., Godl, K., Brehmer, D. et al., Evaluation of kinase inhibitor selectivity by chemical proteomics, *Assay Drug Dev. Technol.* 2, 215–224, 2004; Piggott, A.M. and Karuso, P., Quality, not quantity: the role of natural products and chemical proteomics in modern drug discovery, *Comb. Chem. High Throughput Screen.* 7, 607–630, 2004; Beillard, E. and Witte, O.N., Unraveling kinase signaling pathways with chemical genetic and chemical proteomic approaches, *Cell Cycle* 4, 434–437, 2005; Sem, D.S., Chemical proteomics from a nuclear magnetic resonance spectroscopy perspective, *Expert Rev. Proteomics* 1, 165–178, 2004; Daub, H., Characterization of kinase-selective inhibitors by chemical proteomics, *Biochim. Biophys. Acta* 1754, 183–190, 2005; Verdoes, M., Berkers, C.R., Florea, B.I. et al., Chemical proteomics profiling of proteosome activity, *Methods Mol. Biol.* 328, 51–69, 2006.

Chemokines A large family of cytokines having a wide variety of biological actions that are generally associated with inducing mobilization and activation of immune cells; a contraction of chemotactic cytokines. See Atkins, P.C. and Wasserman, S.I., Chemotactic mediators, *Clin. Rev. Allergy* 1, 385–395, 1983; Hayashi, H., Honda, M., Shimokawa, Y., and Hirashima, M., Chemotactic factors associated with leukocyte emigration in immune tissue injury: their separation, characterization, and functional specificity,

Int. Rev. Cytol. 89, 179–250, 1984; Bignold, L.P., Measurement of chemotaxis of polymorphonuclear leukocytes *in vitro*. The problems of the control of gradients of chemotactic factors, of the control of the cells and of the separation of chemotaxis from chemokinesis, *J. Immunol. Methods* 108, 1–18, 1988; Horuk, R., *Chemokine Receptors*, Academic Press, San Diego, CA, 1997; Vaddi, K., Keller, M., and Newton, R.C., *The Chemokine Factbook*, Academic Press, San Diego, CA, 1997; Hebert, C., *Chemokines in Disease: Biology and Clinical Research,* Humana Press, Totowa, NJ, 1999; Proudfoot, A.E.I. and Well, T.N.C., *Chemokine Protocols*, Humana Press, Totowa, NJ, 2000; Schwarz, M.K., and Wells, T.N.C., New therapeutics that modulate chemokine networks, *Nat. Rev. Drug Disc.* 1, 342–358, 2002; White, F.A., Bhangoo, S.K., and Miller, R.J., Chemokines: integrators of pain and inflammation, *Nat. Rev. Drug Disc.* 4, 834–844, 2005; Schwiebert, L.M., *Chemokines, Chemokine Receptors, and Disease,* Elsevier, Amsterdam, 2005; Steinke, J.W. and Borish, L., Cytokines and chemokines, *J. Allergy Clin. Immunol.* 117 (Suppl. 2), S441–S445, 2006; Charo, I.F. and Ranosohoff, R.M., The many roles of chemokines and chemokine receptors in inflammation, *N. Engl. J. Med.* 354, 610–621, 2006; Laudanna, C. and Alon, R., Right on the spot. Chemokine triggering of integrin-mediated arrest of rolling leukocytes, *Thromb. Haemostas.* 95, 5–11, 2006.

Chemoproteomics The use of small molecules as affinity materials for the discovery of specific binding proteins in the proteome; the application of chemogenomics for proteomic research. See Beroza, P., Villar, H.O., Wick, M.M., and Martin, G.R., Chemoproteomics as a basis for post-genomic drug discovery, *Drug Discov. Today* 7, 807–814, 2002; Gagna, C.E., Winokur, D., and Lambert, W.C., Cell biology, chemogenomics, and chemoproteomics, *Cell Biol. Int.* 28, 755–764, 2004; Shin, D., Heo, Y.S., Lee, K.J. et al., Structural chemoproteomics and drug discovery, *Biopolymers* 80, 258–263, 2005; Hall, S.E., Chemoproteomics-driven drug discovery: addressing high attrition rates, *Drug Discov. Today* 11, 495–502, 2006.

Chondrocyte A cartilage cell. See von der Mark, K. and Conrad, G., Cartilage cell differentiation: review, *Clin. Orthop. Relat. Res.* 139, 195–205, 1979; Serni, U. and Mannoni, A., Chondrocyte physiopathology and drug efficacy, *Drug Exp. Clin. Res.* 17, 75–79, 1991; Urban, J.P., The chondrocytes: a cell under pressure, *Br. J. Rheumatol.* 33, 901–908, 1994; Yates, K.E., Shortkroff, S., and Reish, R.G., Wnt influence on chondrocyte differentiation and cartilage function, *DNA Cell Biol.* 24, 446–457, 2005; Wendt, D., Jakob, M., and Martin, I., Bioreactor-based engineering of osteochondral grafts: from model systems to tissue manufacturing, *J. Biosci. Bioeng.* 100, 489–494, 2005; Goldring, M.B., Tsduchmochi, K., and Ijiri, K., The control of chondrogenesis, *J. Cell Biochem.* 97, 33–44, 2006; Ruano-Ravina, A. and Diaz, M.J., Autologous chondrocytes implantation: a systematic review, *Osteoarthritis Cartilage* 14, 47–51, 2006; Toh, W.S., Yang, Z., Heng, B.C., and Cao, T., New perspectives in chondrogenic differentiation of stem cells for cartilage repair, *Scientific World Journal* 6, 361–364, 2006.

Chromatin Chromatin consists of a repeating fundamental nucleoprotein complex, the nucleosome; DNA wrapped around histones where the histones mediate the folding of DNA into chromatin. See Wolfe, A., *Chromatin: Structure and Function*, 3rd ed., Academic Press, San Diego, CA, 1998; Woodcock, C.L., Chromatin architecture, *Curr. Opin. Struct. Biol.* 16, 213–220, 2006; Aligianni, S.

and Varga-Weisz, P., Chromatin-remodeling factors and the maintenance of transcriptional states through DNA replication, *Biochem. Soc. Symp.* (73), 97–108, 2006; de la Serna, I.L., Ohkawa, Y., and Imbalzano, A.N., Chromatin remodeling in mammalian differentiation: lessons from ATP-dependent remodelers, *Nat. Rev. Genet.* 7, 461–473, 2006; Mersfelder, E.L. and Parthun, M.R., The tail beyond the tail: histone core domain modifications and the regulation of chromatin structure, *Nucleic Acids Res.* 34, 2653–2662, 2006.

Chromatin Remodeling   The dynamic structural change in chromatin by nucleosome sliding or posttranslational modifications (acetylation, methylation) of the histones. See Becker, P.B., The chromatin accessibility complex: chromatin dynamics through nucleosome sliding, *Cold Spring Harb. Symp. Quant. Biol.* 69, 281–287, 2004; Henikoff, S. and Ahmed, K., Assembly of variant histones into chromatin, *Annu. Rev. Cell Dev. Biol.* 21, 133–153, 2005; Dhananjayan, S.C., Ismail, A., and Nawaz, Z., Ubiquitin and control of transcription, *Essays Biochem.* 41, 69–80, 2005; Lucchesi, J.C., Kelly, W.G., and Panning, B., Chromatin remodeling in dosage compensation, *Annu. Rev. Genet.* 39, 615–651, 2005; Saha, A., Wittmeyer, J., and Cairns, B.R., Chromatin remodeling: the industrial revolution of DNA around histones, *Nat. Rev. Mol. Cell Biol.* 7, 437–447, 2006.

Chromatography   The physical separation of two or more components of a solution mixture based on the distribution of said individual components between a stationary phase and a mobile phase. Chromatography can occur within an enclosed column or tube (column chromatography — gas chromatography being a variant of column chromatography with a gaseous mobile phase) or a planar surface as in paper chromatography or thin-layer chromatography. A *chromatogram* is (usually) a graphical representation of a specific solute concentration at a given moment either in time or elution volume. In the case of planar chromatography, the term chromatography can refer to the actual paper or layer on which separation has occurred. The stationary phase may be a solid, gel, or liquid adsorbed onto a solid matrix. The mobile phase may be liquid or gaseous in nature. See Lederer, E. and Lederer, M., *Chromatography: A Review of Principles and Applications,* Elsevier, Amsterdam, 1957; Bobbit, J.M., *Thin-Layer Chromatography,* Reinhold, New York, 1963; Zweig, G. and Sherma, J., *CRC Handbook of Chromatography,* CRC Press, Cleveland, OH, 1972; Ettre, L.S., Nomenclature for chromatography, *Pure Appl. Chem.* 65, 819–872, 1993; Snyder, L.R., Kirkland, J.J., and Glajch, J.L., *Practical HPLC- Method Development,* 2nd ed., John Wiley & Sons, New York, 1997; Miller, J.M., *Chromatography: Concepts and Contrasts,* John Wiley & Sons, New York, 2005; Wall, P.E., *Thin-Layer Chromatography: A Modern Practical Approach,* Royal Society of Chemistry, Cambridge, UK, 2005; Cazes, J., *Encyclopedia of Chromatography,* Taylor & Francis, Boca Raton, FL, 2005; Perssen, P., Gustavsson, P.-E., Zacchi, G., and Nilsson, B., Aspects of estimating parameter dependencies in a detailed chromatography model based on frontal experiments, *Process Biochem.* 41, 1812–1821, 2006; Alpert, A.J., Chromatography of difficult and water-soluble proteins with organic solvents, *Adv. Chromatog.* 44, 317–329, 2006; Lundanes, E. and Greibrokk, T., Temperature effects in liquid chromatography, *Adv. Chromatog.* 44, 45–77, 2006.

Circadian   Used to describe an approximate 24-hour period; a phenomenon has demonstrated a circadian variation if it occurs with a certain frequency within an approximate 24-hour period. See Mills, J.N., Human circadian rhythms, *Physiol. Rev.* 46, 128–171, 1966; Brady, J., How are insect circadian rhythms controlled? *Nature* 223, 781–784, 1969; Menaker, M., Takahashi, J.S., and Eskin, A., The physiology of circadian pacemakers, *Annu. Rev. Physiol.* 40, 501–526, 1978; Soriano, V., The circadian rhythm embraces the variability that occurs within 24 hours, *Int. J. Neurol.* 15, 7–16, 1981; Gardner, M.J., Hubbard, K.E., Hatta, C.T. et al., How plants tell the time, *Biochem. J.* 397, 15–24, 2006; McClung, C.R., Plant circadian rhythms, *Plant Cell* 18, 792–803, 2006; Brunner, M. and Schafmeier, T., Transcriptional and posttranscriptional regulation of the circadian clock of cyanobacteria and neurospora, *Genes Dev.* 20, 1061–1074, 2006; Hardin, P.E. and Yu, W., Circadian transcription: passing the HAT to CLOCK, *Cell* 125, 424–426, 2006; Lewy, A.J., Emens, J., Jackman, A., and Yuhas, K., Circadian uses of melatonin in humans, *Chronobiol. Int.* 23, 403–412, 2006; Rosato, E., Tauber, E., and Kyriacou, C.P., Molecular genetics of the fruit-fly circadian clock, *Eur. J. Hum. Genet.* 14, 729–738, 2006.

Circular Dichroism The differential absorption of plane-polarized light passing through a solution; expressed as molar ellipticity $[\theta]_m$. See Greenfield, N.J., Analysis of circular dichroism data, *Meth. Enzymol*, 383, 282–317, 2004; Bayer, T.M., Booth, L.N., Knudsen, S.M., and Ellington, A.D., Arginine-rich motifs present multiple interfaces for specific binding by RNA, *RNA* 11, 1848–1857, 2005; Miles, A.J. and Wallace, B.A., Synchrotron radiation circular dichroism spectroscopy of proteins and applications in structural and functional genomics, *Chem. Soc. Rev.* 35, 39–51, 2006; Paramonov, S.E., Jun, H.W., and Hartgerink, J.D., Modulation of peptide-amphiphile nanofibers via phospholipid inclusions, *Biomacromolecules* 7, 24–26, 2006; Harrington, A., Darboe, N., Kenjale, R. et al., Characterization of the interaction of single tryptophan containing mutants of IpaC from *Shingella flexneri* with phospholipid membranes, *Biochemistry* 45, 626–636, 2006.

***Cis*-Element; *Cis*-Locus; *Cis*-Factors** A region or regions on a DNA molecule that affect(s) activity of DNA sequences on DNA molecule(s); an intramolecular effect; usually, but does not always, code for the expression of protein. A *cis*-element or regulatory region can be complex and may contain several regulatory sequences. See Gluzman, Y., *Eukaryotic Transcription: The Role of Cis- and Trans-Acting Elements in Initiation*, Cold Spring Harbor Laboratory Press, Cold Spring Harbor, NY, 1985; Tanaka, N. and Taniguchi, T., Cytokine gene regulation: regulatory *cis*-elements and DNA binding factors involved in the interferon system, *Adv. Immunol.* 52, 263–281, 1992; Hames, B.D. and Higgins, S.J., *Gene Transcription: A Practical Approach*, IRL Press at Oxford, Oxford, UK, 1993; Galson, D.L., Blanchard, K.L., Fandrey, J., Goldberg, M.A., and Bunn, H.F., *Cis* elements that regulate the erythropoietin gene, *Ann. N.Y. Acad. Sci.* 718, 21–30, 1994; Hapgood, J.P., Riedemann, J., and Scherer, S.D., Regulation of gene expression by GC-rich DNA *cis*-elements, *Cell Biol. Int.* 25, 71–31, 2001; Tumpel, S., Maconochie, M., Wiedmann, L.M., and Krumlauf, R., Conservation and diversity in the *cis*-regulatory networks that integrate information controlling expression of Hoxa2 in hindbrain and cranial neural crest cells in vertebrates, *Dev. Biol.* 246, 45–56, 2002; Moolla, N., Kew, M., and Arbuthnot, P., Regulatory elements of hepatitis B virus transcription, *J. Viral Hepat.* 9, 323–331, 2002; Manna, P.R., Wang, X.J., and Stocco, D.M., Involvement of multiple transcription factors in the regulation of steroidogenic acute regulatory protein gene expression, *Steroids* 68, 1125–1134, 2003; Gambari, R., New trends in the development of transcription factor decoy (TFD) pharmacotherapy, *Curr. Drug Targets* 5, 419–430, 2004; McBride, D.J. and Kleinjan, D.A., Rounding up active *cis*-elements in the triple C corral: combining conservation, cleavage, and conformation capture for the analysis of regulatory gene domains, *Brief Funct. Genomic Proteomic* 3, 267–279, 2004.

Claisen Condensation
Base-catalyzed reaction of an ester with an α-carbon hydrogen with another ester (same or different) to yield a β-keto ester. A model for thiolase reactions. See Johnson, T.B. and Hill, A.J., Catalytic action of esters in the Claisen condensation, *J. Amer. Chem. Soc.*, 35, 1023–1034, 1913; Clark, J.D., O'Keefe, S.J., and Knowles, J.R., Malate synthase: proof of a stepwise Claisen condensation using the double-isotope fractionation test, *Biochemistry* 27, 5961–5971 1988; Modia, Y. and Wierenga, R.K., A biosynthetic thiolase in complex with a reaction intermediate: the crystal structure provides new insight into the catalytic mechanism, *Structure* 7, 1279–1290, 1999; Watanabe, A. and Ebizuka, Y., Unprecedented mechanism of chain length determination in fungal aromatic polyketide synthases, *Chem. Biol.* 11, 1101–1106, 2004; Veyron-Churlet, R., Bigot, S., Guerrini, O. et al., The biosynthesis of mycolic acids in *Mycobaceterium tuberculosis* relies on multiple specialized elongation complexes interconnected by specific protein–protein interactions, *J. Mol. Biol.* 353, 847–858, 2005; von Wettstein-Knowles, P., Olsen, J.G., McGuire, K.A., and Henriksen, A., Fatty acid synthesis. Role of active site histidines and lysine in Cys-His-His-type beta-ketoacyl-acyl carrier protein synthases, *FEBS J.* 273, 695–710, 2006; Ryu, Y., Kim, K.J., Rosennser, C.A., and Scott, A., Decarboxylative Claisen condensation catalyzed by *in vitro* selected ribozymes, *Chem. Commun.* 7, 1439–1441, 2006.

Class Switch Recombination
A process by which one constant region gene segment is switched with another gene segment during B-cell development when immunoglobulin production changes from IgM to IgA, IgE, or IgG. See Davis, M.M., Kim, S.K., and Hood, L.E., DNA sequences mediating class switching in alpha-immunoglobulin, *Science* 209, 1360–1365, 1980; Geha, R.S., Jabara, H.H., and Brodeur, S.R., The regulation of immunoglobulin E class-switch recombination, *Nat. Rev. Immunol.* 3, 721–732, 2003; Yu, K. and Lieber, M.R., Nucleic acid structures and enzymes in the immunoglobulin class switch recombination mechanism, *DNA Repair* 2, 1163–1174, 2003; Diamant, E. and Melamed, D., Class switch recombination in B lymphopoiesis: a potential pathway for B-cell autoimmunity, *Autoimmun. Rev.* 3, 464–469, 2004; Min, I.M. and Selsing, E., Antibody class switch recombination: roles for switch sequence and mismatch repair proteins, *Adv. Immunol.* 87, 297–328, 2005.

Classical Proteomics
Proteomic analysis based on the direct analysis of the expressed proteome such as an extract obtained from lysis of a cell; also referred to as forward proteomics as compared to reverse proteomics. More generally, classical proteomics is taken to mean protein separation followed by characterization. See Klade, C.S., Proteomics approaches toward antigen discovery and vaccine development, *Curr. Opin. Mol. Ther.* 4, 216–223, 2002; Vondriska, T.M. and Ping, P., Functional proteomics to study protection of the ischaemic myocardium, *Expert Opin. Ther. Targets* 6, 563–570, 2002; Thiede, B. and Rudel, T., Proteome analysis of apoptotic cells, *Mass Spectrom. Rev.* 23, 333–349, 2004; Gottlieb, D.M., Schultz, J., Bruun, S.W. et al., Multivariate approaches in plant science, *Phytochemistry* 65, 1531–1548, 2004.

Clinomics   Application of oncogenomics to cancer care. See Workman, P. and Clarke, P.A., Innovative cancer drug targets: genomics, transcriptomics, and clinomics, *Expert Opin. Pharmacother.* 2, 911–915, 2001.

Clonal   The selection of a clone. Most often used to describe the process by which
Selection   a B-cell is challenged by a specific antigen to produce a committed plasma cell or the differentiation of T-cells. More generally, the selection of a stem cell to become committed to a specific antigen. See Burnet, F.M., *The Clonal Selection Theory of Acquired Immunity*, Vanderbilt University Press, Nashville, TN, 1959; Williamson, A.R., The biological origin of antibody diversity, *Annu. Rev. Biochem.* 45, 467–500, 1976; D'Eustachio, P., Rutishauser, U.S., and Edelman, G.M., Clonal selection and the ontogeny of the immune response, *Int. Rev. Cytol. Suppl.* 5, 1–60, 1977; Mazumdar, P.M.H., *Immunology 1930–1980: Essays on the History of Immunology*, Wall & Thompson, Toronto, 1989; Coutinho, A., Beyond clonal selection and network, *Immunol. Rev.* 110, 63–87, 1989; Podolsky, S.H. and Tauber, A.I., *The Generation of Diversity: Clonal Selection Theory and the Rise of Molecular Immunology*, Harvard University Press, Cambridge, MA, 1997; Cohen, I.R., Antigenic mimicry, clonal selection, and autoimmunity, *J. Autoimmun.* 16, 337–340, 2001; Defrance, T., Casamayor-Palleja, M., and Krammer, P.H., The life and death of a B-cell, *Adv. Cancer Res.* 86, 195–225, 2002; van Boehmer, H., Aifantis, I., Gounari, F. et al., Thymic selection revisited: how essential is it? *Immunol. Rev.* 191, 62–78, 2003; McHeyzer-Williams, L.J. and McHeyzer-Williams, M.G., Antigen-specific memory B-cell development, *Annu. Rev. Immunol.* 23, 487–513, 2005; Bock, K.W. and Kohle, C., Ah receptor- and TCDD-mediated liver tumor promotion: clonal selection and expansion of cells evading growth arrest and apoptosis, *Biochem. Pharmacol.* 69, 1403–1408, 2005.

Clone   A cell or organism descended from and genetically identical to a single common ancestor. Clone also refers to a DNA sequence encoding a product or an entire gene sequence from an organism that is replicated by genetic engineering. Such material can be transferred to another organism for the expression of such cDNA or gene. See Cunningham, A.J., Antibody formation studied at the single-cell level, *Prog. Allergy* 17, 5–50, 1973; Hamer, D.H. and Thomas, C.A., Jr., Molecular cloning, *Adv. Pathobiol.* 6, 306–319, 1977; von Boehmer, H., Haas, W., Pohlit, H., Hengartner, H., and Nabholz, M., T-cell clones: their use for the study of specificity, induction, and effector-function of T-cells, *Springer Semin. Immunopathol.* 3, 23–37, 1980; Fung, J.J., Gleason, K., Ward. R., and Kohler, H., Maturation of B-cell clones, *Prog. Clin. Biol. Res.* 42, 203–214, 1980; Veitia, R.A., Stochasticity or the fatal "imperfection" of cloning, *J. Biosci.* 30, 21–30, 2005; Kettman, J.R., From clones of cells to cloned genes and their proteinpaedia, *Scand. J. Immunol.* 62 (Suppl. 1), 119–122, 2005; Vats, A., Bielby, R.C., Tolley, N.S., Nerem, R., and Polak, J.M., Stem cells, *Lancet* 366, 592–602, 2005; Wells, D.N., Animal cloning: problems and prospects, *Rev. Sci. Tech.* 24, 251–264, 2005; Diep, B.A., Gill, S.R., Chang, R.F. et al., Complete genome sequence of USA300, an epidemic clone of community-acquired methicillin-resistant *Staphylococcus aureus*, *Lancet* 367, 731–739, 2006.

Coefficient of   Ration of the change in length per degree C to length at O°C. The coefficient
Linear Thermal   of linear thermal expansion (CTLE) is used to describe the changes in the
Expansion (CLTE)   structure of proteins and other polymers as a function of temperature; the CTLE has also been used to describe thermal changes in micelles.

See Frauenfelder, H., Hartmann, H., Karplus, M. et al., Thermal expansion of a protein, *Biochemistry* 26, 254–261, 1987; Schulenberg, P.J., Rohr, M., Gartner, W., and Braslavsky, S.E., Photoinduced volume changes associated with the early transformations of bacteriorhodopsin: a laser-induced optoacoustic spectroscopy study, *Biophys. J.* 66, 838–843, 1994; Marsh, D., Intrinsic curvature in normal and inverted lipid structures and I membranes, *Biophys. J.* 70, 2248–2255, 1996; Daniels, B.V., Schoenborn, B.P., and Korszun, Z.R., A low-resolution low-temperature neutron diffraction study of myoglobin, *Acta Crystallogr. D. Biol. Crystallogr.* 53, 544–550, 1997; Cordier, F. and Grzesiek, S., Temperature-dependence of protein hydrogen bond properties as studied by high-resolution NMR, *J. Mol. Biol.* 317, 739–752, 2002; Pereira, F.R., Machado, J.C., and Foster, F.S., Ultrasound characterization of coronary artery wall *in vitro* using temperature-dependent wave speed, *IEEE Trans Ultrason. Ferroelectr. Freq. Control* 50, 1474–1485, 2003; Bhardwaj, R., Mohanty, A.K., Drzal, L.T. et al., Renewable resource-based composites from recycled cellulose fiber and poly(3-hydroxybutyrate-co-3-hydroxyvalerate) bioplastic, *Biomacromolecules* 7, 2044–2051, 2006.

Cold-Chain Product

A product or reagent that must be kept cold during transit and storage; most often between 4° and 8°C. See Elliott, M.A. and Halbert, G.W., Maintaining the cold chain shipping environment for phase I clinical trial distribution, *Int. J. Pharm.* 299, 49–54, 2005; Streatfield, S.J., Mucosal immunization using recombinant plant-based oral vaccines, *Methods* 38, 150–157, 2005.

Cold-Shock Protein

A group of proteins that are synthesized by plant cells, prokaryotic cells, and eukaryotic cells in response to cold stress. It has been suggested that cold-shock proteins (CSPs) function as "chaperones" for mRNA. See Graumann, P.L. and Marshiel, M.A., A superfamily of proteins that contain the cold-shock domain, *Trends Biochem. Sci.* 23, 286–290, 1998; Phadtare, S., Alsina, J., and Inouye, M., Cold-shock response and cold-shock proteins, *Curr. Opin. Microbiol.* 2, 175–180, 1999; Sommerville, J., Activities of cold-shock domain proteins in translational control, *Bioessays* 21, 319–325, 1999; Graumann, P.L. and Marahiel, M.A., Cold shock response in *Bacillus subtilis*, *J. Mol. Microbiol. Biotechnol.* 1, 203–209, 1999; Loa, D.A. and Murata, N., Responses to cold shock in cyanobacteria, *J. Mol. Microbiol. Biotechnol.* 1, 221–230, 1999; Ermolenko, D.N. and Makhatadze, G.I., Bacterial cold-shock proteins, *Cell. Mol. Life. Sci.* 59, 1902–1913, 2002; Alfageeh, M.B., Marchant, R.J., Carden, M.J., and Smales, C.M., The cold-shock response in cultured mammalian cells: harnessing the response for the improvement of recombinant protein production, *Biotechnol. Bioengineer.* 93, 829–835, 2006; Al-Fageeh, M.B. and Smales, C.M., Control and regulation of the cellular response to cold shock: the responses in yeast and mammalian systems, *Biochem. J.* 397, 247–259, 2006; Fraser, K.R., Tuite, N.L., Bhagwat, A., and O'Byrne, C.P., Global effects of homocysteine on transcription in *Escherichia coli*; induction of the gene for the major cold-shock protein, CspA, *Microbiology* 152, 2221–2231, 2006; Magg, C., Kubelka, J., Holtermann, G. et al., Specificity of the initial collapse in the folding of the cold-shock protein, *J. Mol. Biol.* 360, 1067–1080, 2006; Sauvageot, N., Beaufils, S., and Maze, A., Cloning and characterization of a gene encoding a cold-shock protein in *Lactobacillus casei*, *FEMS Microbiol. Lett.* 254, 55–62, 2006; Narberhaus, F., Waldminghous, T., and Chowdhury, S., RNA thermometers, *FEMS Microbiol. Lett.* 30, 3–16, 2006.

Colloid   A particle with dimensions between 1 nm and 1 μm, although it is not necessary for all three dimensions to be in this size range. For example, a thin fiber might only have two dimensions in this size range. A colloidal dispersion is a system where colloid particles are dispersed in a continuous phase of a different composition such as a suspension (particles in a liquid), an emulsion (colloids of one liquid are suspended in another liquid where the two liquids are immiscible such as oil and water), a foam (gas dispersed in a liquid or gel), or an aerosol (a colloid in a gas such as air; a fog is a liquid colloid dispersed in a gas). See Tolson, N.D., Boothroyd, B., and Hopkins, C.R., Cell surface labeling with gold colloid particulates: the use of aviden and staphylococcal protein A-coated gold in conjunction with biotin and fc-bearing ligands, *J. Microsc.* 123, 215–226, 1981; Rowe, A.J., Probing hydration and the stability of protein solution — a colloid science approach, *Biophys. Chem.* 93, 93–101, 2001; Bolhuis, P.G., Meijer, E.J., and Louis, A.A., Colloid-polymer mixtures in the protein limit, *Phys. Rev. Lett.* 90, 068304, 2003; Zhang, Z. and van Duijneveldt, J.S., Experimental phase diagram of a model colloid-polymer mixture in the protein limit., *Langmuir* 22, 63–66, 2006; Xu, L.C. and Logan, B.E., Adhesion forces between functionalized latex microspheres and protein-coated surfaces evaluated using colloid probe atomic force microscopy, *Colloids Surf. B. Biointerfaces* 48, 84–94, 2006.

Colloid Osmotic   The difference in osmotic pressure between two sides of a semipermeable
Strength/Colloid   membrane (permeable to solvent, such as water, but not to the colloids). See
Osmotic Pressure   Harry, S.B. and Steiner, R.F., Characterization of the self-association of a soybean proteinase inhibitor by membrane osmometry, *Biochemistry* 8, 5060–5064, 1969; de Bruijne, A.W. and van Steveninck, J., Apparent nonsolvent water and osmotic behavior of yeast cells, *Biochim. Biophys. Acta* 196, 45–52, 1970; Keshaviah, P.R., Constantini, E.G., Luehmann, D.A., and Shapiro, F.L., Dialyzer ultrafiltration coefficients: comparison between in vitro and in vivo values, *Artif. Organs* 6, 23–26, 1982; Boudinot, F.D. and Jusko, W.J., Fluid shifts and other factors affecting plasma protein binding of prednisolone by equilibrium dialysis, *J. Pharm. Sci.* 73, 774–780, 1984; McGrath, J.J., A microscopic diffusion chamber for the determination of the equilibrium and non-equilibrium osmotic response of individual cells, *J. Microsc.* 139, 249–263, 1985; Wiig, H., Reed, R.K., and Aukland, K., Measurement of interstitial fluid pressure: comparison of methods, *Ann. Biomed. Eng.* 14, 139–151, 1986; Cameron, I.L., Kanal, K.M., and Fullerton, G.D., Role of protein conformation and aggregation in pumping water in and out of a cell, *Cell Biol. Int.* 30, 78–85, 2006; Clarke, H.G., Hope, S.A., Byers, S., and Rodgers, R.J., Formation of ovarian follicular fluid may be due to the osmotic potential of large glycosaminoglycans and proteoglycans, *Reproduction* 132, 119–131, 2006. An example is the difference in osmotic strength between the intravascular bed and the extravascular bed which balance the flow pressures in the vascular system. Albumin and/or dextran are therapeutic agents to restore osmotic strength to the vascular system. See Bevan, D.R., Colloid osmotic pressure, *Anaesthesia* 35, 263–270, 1980; Webster, H.L., Colloid osmotic pressure: theoretical aspects and background, *Clin. Perinatol.* 9, 505–521, 1982; Lundsgaard-Hansen, P., Physiology and pathophysiology of colloid osmotic pressure and albumin metabolism, *Curr. Stud. Hematol. Blood Transfus.* (53), 1–17, 1986; Burton, R.F., The protein content of

extracellular fluids and its relevance to the study of ionic regulation: net charge and colloid osmotic pressure, *Comp. Biochem. Physiol. A* 90, 11–16, 1988; Kaminski, M.V., Jr. and Haase, T.J., Albumin and colloid osmotic pressure implications for fluid resuscitation, *Crit. Care Clin.* 8, 311–321, 1992; Blackwell, M.M., Riley, J., McCall, M., et al., An evaluation of three methods for determining colloid osmotic pressure, *J. Extra. Corpor. Technol.* 26, 18–22, 1994; Mbaba Mena, J., De Backer, D., and Vincent, J.L., Effects of a hydroxyethylstarch solution on plasma colloid osmotic pressure in acutely ill patients, *Acta Anaesthesiol. Belg.* 51, 39–42, 2000; Gupta, S. and Tasker, R.C., Does giving albumin infusion in hypoalbumaeimic children with oncological disease affect colloid osmotic pressure and outcome?, *Arch. Dis. Child.* 86, 380–381, 2002; Rosengren, B.I., Rippe, B., Tenstad, O., and Wiig, H. Acute peritoneal dialysis in rat results in a marked reduction of interstitial colloid osmotic pressure, *J. Am. Soc. Nephrol.* 15, 3111–3116, 2004. Colloid osmotic pressure can also play a role in dialysis. See Ehrlich, S., Wolff, N., Scheiderman, R., et al., The osmotic pressure of chondroitin sulphate solutions: experimental measurements and theoretical analysis, *Biorheology* 35, 383–397, 1998; Armando, G.R., de la Torre, L., Garcia-Serrano, L.A., and Aguilar-Eliguezabal, A., Effect of dialysis treatment on the aggregation state of montmorillonite clay, *J. Colloid Interface Sci.* 274, 550–554, 2004; Tessier, P.M., Verruto, V.J., Sandler, S.I., and Lenhoff, A.M., Correlation of dialfiltration sieving behavior of lysozyme-BSA mixtures with osmotic second virial cross-coefficients, *Biotechnol. Bioeng.* 87, 303–310, 2004; Rosenbloom, A.J., Sipe, D.M., and Weedn, V.W., Microdialysis of proteins: performance of the CMA/20 probe, *J. Neurosci. Methods* 148, 147–153, 2005.

Combination Electrode An ion-selective electrode and an external reference electrode combined into a single functional unit. A separate reference electrode is not required.

Combination Product A regulatory term used to describe a final drug product composed of, for example, two separate drugs: a drug and a biologic or a drug and a device. See Leyden, J.J., Hickman, J.G., Jarratt, M.T. et al., The efficacy and safety of a combination benzoyl peroxide/clindemycin topical gel compared with benzoyl peroxide alone and a benzoyl peroxide/erythromycin combination product, *J. Cutan. Med. Surg.* 5, 37–42, 2001; Bays, H.E., Extended-release niacin/lovastatin: the first combination product for dyslipidemia, *Expert Rev. Cardiovasc. Ther.* 2, 485–501, 2004; Anon., Definition of the primary mode of action of a combination product. Final rule, *Fed. Regist.* 70, 49848–49862, 2005.

Complement A combination or system of plasma/serum proteins that interact to form a membrane attack complex, which results in the lysis of bacterial pathogens and other cell targets such as tumor cells. There are three pathways of complement activation: the classical pathway, the alternative pathway, and the MBLectin (mannose-binding lectin; a plasma protein) pathway. The classical pathway is activated by an antigen-antibody complex (a free antibody does not activate complement) via the Fc domain of the antibody; there are other mechanisms for classical pathway activation that make minor contributions. The alternative pathway is activated by direct recognition of foreign materials in an antibody-independent manner and is driven by the autocatalytic action of C3b. The alternative pathway that is thought the oldest of the three pathways is phyllogenetic development.

The MBlectin pathway is initiated by the interaction of the MBlectin with a bacterial cell surface polysaccharide. The activation of complement component C3 is common to all three pathways. It is noted that there are similarities to the blood coagulation cascade. See Sim, R.B., Ed., *Activators and Inhibitors of Complement*, Kluwer Academic, Dordrecht, Netherlands, 1993; Whaley, K., Loos, M., and Weiler, J., Eds., *Complement in Health and Disease*, 2nd ed., Kluwer Academic, Dordrecht, Netherlands, 1993; Rother, K., Till, G.O., and Hansch, G.M., Eds., *The Complement System*, 2nd ed., Springer, Berlin, 1998; Volanakis, J.E. and Frank, M.M., Eds., *The Human Complement System in Health and Disease*, Marcel Dekker, New York, 1998; Prodinger, W.M., Würznen, R., Erdei, A., and Dierich, M.P., Complement, in *Fundamental Immunology*, Paul, W.E., Ed., Lippincott-Raven, Philadelphia, 1999, pp. 967–995; Lambis, J.D. and Holer, K.M., Eds., *Therapeutic Interventions in the Complement System*, Humana Press, Totowa, NJ, 2000; Szebeni, J., *The Complement System: Novel Roles in Health and Disease,* Kluwer Academic, Boston, 2004.

Complement Fixation   The binding of the first component of the complement pathway, C1, to an IgG- or IgM-antigen complex. The antigen is usually a cell surface protein. Free antibody does not fix complement. Productive binding of the antigen-antibody complex (binding involves the Fc portion of the antibody and a minimum of two Fc domains is required, thus two intact antibody molecules) results in complement activation. An antibody that activates complements is described as having fixed complement. Complement fixation has formed the basis for many serological tests, but most have been replaced by ELISA assays for the diagnosis of infectious disease. See Juji, T., Saji, H., Sataki, M., and Tukinaga, K., Typing for human platelet alloantigens, *Rev. Immunogenet.* 1, 239–254, 1999; Pappagianus, D., Serological studies in coccidiomycosis, *Semin. Respir. Infect.* 16, 242–250, 2001; Nielsen, K., Diagnosis of brucellosis by serology, *Vet. Microbiol.* 90, 447–459, 2002; Al-Dahouk, S., Tomaso, H., Nackler, E. et al., Laboratory-based diagnosis of brucellosis — review of the literature. Part I. Techniques for direct detection and identification of *Brucella sp., Clin. Lab.* 49, 387–404, 2003; Taggart, E.W., Hill, H.R., Martins, T.B., and Litwin, C.M., Comparison of complement fixation with two enzyme-linked immunosorbent assays for the detection of antibodies to respiratory viral antigens, *Amer. J. Clin. Path.* 125, 460–466, 2006. Complement fixation is usually measured by the lysis of sensitized cells (e.g., hemolysis of sensitized sheep red blood cells; CH_{50} assay. See Morgen, P.B., Complement, in *Immunochemistry*, van Oss, C.J. and van Regenmortel, M.C.H., Eds., Marcel Dekker, New York, 1994, pp. 903–923). The concept of complement fixation is still discussed with respect to *in vivo* antigen-antibody reactions such as those seen with transplantation antigens and alloantibodies. See Feucht, H.E., Felber, E., Gokel, M.J. et al., Vascular deposition of complement-split products in kidney allografts with cell-mediated rejection, *Clin. Exp. Immunol.* 86, 464–470, 1991; Feucht, H.E., Complement C4d in graft capillaries — the missing link in the recognition of humoral alloreactivity, *Am. J. Transplant.* 3, 646–652, 2003; Colvin, R.B. and Smith, R.N., Antibody-mediated organ-allograft rejection, *Nat. Rev. Immunol.* 5, 807–817, 2005; Rickert, R.C., Regulation of B lymphocyte activation by complement C_3 and the B-cell coreceptor complex, *Curr. Opin. Immunol.* 17, 237–243, 2005.

Confocal   A fluorescent microscopy technique that uses a highly focused beam of light
Microscopy   with suppression of fluorescence above and below the point of optimum
                      focus. An image is obtained by moving the excitation beam and measure-
                      ment aperture over the sample with point-by-point measurement. See
                      Cherry, R.J., *New Techniques of Optical Microscropy and Microspectros-*
                      copy, CRC Press, Boca Raton, FL, 1991; Stelzer, E.H., Wacker, I., and
                      De Mey, J.R., Confocal fluorescence microscopy in modern cell biology,
                      Sermin. Cell Biol. 2, 145–152, 1991; Stevens, J.K. and Mills, L.R., *Three-*
                      Dimensional Confocal Microscopy: Volume Investigation of Biological
                      Specimens, Academic Press, San Diego, CA, 1994; Smith, R.F., *Micros-*
                      copy and Photomicrography: A Working Manual, CRC Press, Boca Raton,
                      FL, 1994; Pawley, J.B., *Handbook of Biological Confocal Microscropy*,
                      Plenum Press, New York, 1995; Fay, F.S., Optical methods in cell physi-
                      ology, in *Handbook of Physiology, Section 14, Cell Physiology*, Hoffman,
                      J.F. and Jamieson, J.D., Eds., Oxford University Press, New York, 1997;
                      Paddock, S.W., *Confocal Microscopy Methods and Protocols*, Humana
                      Press, Totowa, NJ, 1999; Brelje, T.C., Wessendorf, M.W., and Sorenson,
                      R.L., Multicolor laser scanning confocal immunofluorescence micros-
                      copy: practical application and limitations, *Methods Cell Biol.* 70,
                      165–244, 2002; Bacia, K. and Schwille, P., A dynamic view of cellular
                      processes by *in vivo* fluorescence auto- and cross-correlation spectroscopy,
                      Methods 29, 74–85, 2003; Miyashita, T., Confocal microscopy for intra-
                      cellular co-localization of proteins, *Methods Mol. Biol.* 261, 399–410,
                      2004; Heilker, R., Zemanova, L., Valler, M.J., and Nienhaus, G.U., Con-
                      focal fluorescence microscopy for high-throughput screening of G-protein-
                      coupled receptors, *Curr. Med. Chem.* 12, 2551–2559, 2005; Becker, B.E.
                      and Gard, D.L., Visualization of the cytoskeleton in *Xenopus* oocytes and
                      eggs by confocal immunofluorescence microscopy, *Methods Mol. Biol.*
                      322, 69–86, 2006.

Conjugate   Coupling of a weak immunogen such as a polysaccharide to a protein to
Vaccine   improve/enhance immunogenicity. See Cryz, S.J., Jr., Furer, E., Sadoff,
                      J.C. et al., Use of *Pseudomonas aeruginosa* toxin A in the construction
                      of conjugate vaccines and immunotoxins, *Rev. Infect. Dis.* 9 (Suppl. 5),
                      S644–S649, 1987; Garner, C.V. and Pier, G.B., Immunologic consider-
                      ations for the development of conjugate vaccines, *Contrib. Microbiol.*
                      Immunol. 10, 11–17, 1989; Dintzis, R.Z., Rational design of conjugate
                      vaccines, *Pediatr. Res.* 32, 376–385, 1992; Ellis, R.W. and Douglas, R.G.,
                      Jr., New vaccine technologies, *JAMA* 272, 929–931, 1994; Lindberg, A.A.
                      and Pillai, S., Recent trends in the developments of bacterial vaccines,
                      Dev. Biol. Stand. 87, 59–71, 1996; Zimmer, S.M. and Stephens, D.S.,
                      Meningococcal conjugate vaccines, *Expert Opin. Pharmacother.* 5,
                      855–863, 2004; Finn, A., Bacterial polysaccharide-protein conjugate vac-
                      cines, *Br. Med. Bull.* 70, 1–14, 2004; Shape, M.D. and Pollard, A.J.,
                      Meningococcal polysaccharide-protein conjugate vaccines, *Lancet Infect.*
                      Dis. 5, 21–30, 2005; Finn, A. and Heath, P., Conjugate vaccines, *Arch. Dis.*
                      Child. 90, 667–669, 2005; Jones, C., NMR assays for carbohydrate-based
                      vaccines, *J. Pharm. Biomed. Anal.* 38, 840–850, 2005; Whitney, C.G, Impact
                      of conjugate pneumococcal vaccines, *Pediatr. Infect. Dis.* 24, 729–730,
                      2005; Lee, C.J. , Lee, L.H., and Gu, X.X., Mucosal immunity induced by
                      pneumococcal glycoconjugate, *Crit. Rev. Microbiol.* 31, 137–144, 2005.

Connexins   A protein subunit of connexon, which forms gap junctions critical for inter-cellular communication. Mutations in the connexins are responsible for a diversity of diseases, including deafness, skin disorders, and idiopathic atrial fibrillation. Connexins have been designated by their molecular mass while another system separates connexins on the basis of sequence homology. See Beyer, E.C., Paul, D.L., and Goodenough, D.A., Connexin family of gap junction proteins, *J. Membr. Biol.* 116, 187–194, 1990; Revel, J.P., Nicholson, B.J., and Yancey, S.B., Chemistry of gap junctions, *Annu. Rev. Physiol.* 47, 263–279, 1985; Revel, J.P., Yancey, S.B., Nicholson, B., and Hoh, J., Sequence diversity of gap junction proteins, *Ciba Found. Symp.* 125, 108–127, 1987; Stains, J.P. and Civitelli, R., Gap junctions in skeletal development and function, *Biochim. Biophys. Acta* 1719, 69–81, 2005; Anand, R.J. and Hackam, D.J., The role of gap junctions in health and disease, *Crit. Care Med.* 33 (Suppl. 12), S535–S538, 2005; Michon, L., Nlend Nlend, R., Bavamian, S. et al., Involvement of gap junctional communication in secretion, *Biochim. Biophys. Acta* 1719, 82–101, 2005; Vinken, M., Vanhaecke, T., Papeleu, P. et al., Connexins and their channels in cell growth and cell death, *Cell Signal.* 18, 592–600, 2006; Petit, C., From deafness genes to hearing mechanisms: harmony and counterpoint, *Trends Mol. Med.* 12, 57–64, 2006; Evans, W.H., De Vuyst, E., and Leybaert, L., The gap junction cellular internet: connexin hemichannels enter the signaling limelight, *Biochem. J.* 397, 1–14, 2006; Gollob, M.H., Cardiac connexins as candidate genes for idiopathic atrial fibrillation, *Curr. Opin. Cardiol.* 21, 155–158, 2006.

Contig   Originally defined as a set of overlapping DNA sequences; expanded to include a set of overlapping DNA clones. Specifically, it refers to a set of gel bands that can be related to each other by overlap sequences; see http://staden.sourceforge.net/contig.html. See Staden, R., A new computer method for the storage of any manipulation of DNA gel reading data, *Nucleic Acids Res.* 8, 3673–3694, 1980; Carrano, A.V., de Jong, P.J., Branscomb, E. et al., Constructing chromosome- and region-specific cosmid maps for the human genome, *Genome* 31, 1059–1065, 1989; Schalkwyk, L.C., Francis, F., and Lehrach, H., Techniques in mammalian genome mapping, *Curr. Opin. Biotechnol.* 6, 37–43, 1995; Presting, G.G., Budiman, M.A., Wood, T. et al., A framework for sequencing the rice genome, *Novartis Found. Symp.* 236, 13–24, 2001; Dodgson, J.B., Chicken genome sequence: a centennial gift to poultry genetics, *Cytogenet. Genome Res.* 102, 291–296, 2003.

Contour Length   End-to-end length of a stretched DNA molecule (see Wellauer, P., Weber, R., and Wyler, T., Electron microscopic study of the influence of the preparative conditions on contour length and structure of mitrochondrial DNA of mouse liver, *J. Ultrastruct. Res.* 42, 377–393, 1973; Geller, K. and Reinert, K.E., Evidence for an increase of DNA contour length at low ionic strength, *Nucleic Acids Res.* 8, 2807–2822, 1980; Motejlek, K., Schindler, D., Assum, G., and Krone, W., Increased amount and contour length distribution of small polydisperse circular DNA [spcDNA] in Fanconi anemia, *Mutat. Res.* 293, 205–214, 1993; Gast, F.U. and Sanger, H.L., Gel dependence of electrophoretic mobilities of double-stranded and viroid RNA and estimation of the contour length of a viroid by gel electrophoresis, *Electrophoresis* 15, 1493–1498, 1994; Sanchez-Sevilla, A.,

Thimonier, J., Marilley, M. et al., Accuracy of AFM measurements of the contour length of DNA fragments adsorbed on mica in air and in aqueous buffer, *Ultramicroscopy* 92, 151–158, 2002). The term has been used to describe very long proteins such as titin (Helmes, M., Trombitas, K., Centner, T. et al., Mechanically driven contour-length adjustment in rat cardiac titin's unique N2B sequence: titin is an adjustable spring, *Circ. Res.* 84, 1339–1352, 1999).

Core Promoter   A region immediately (+/− 30 bp) around the transcription start site that contains consensus sequence elements (TATA boxes, lnr, DPEs); *in vitro*, the core promoter is the minimal required sequence that is recognized by general transcription factors that activate correct transcription by RNA polymerase II. See Gill, G., Transcriptional initiation, *Curr. Biol.* 4, 374–376, 1994; Gill, G., Regulation of the initiation of eukaryotic transcription, *Essays Biochem.* 37, 33–43, 2001; Butler, J.E. and Kadonaga, J.T., The RNA polymerase II core promoter: a key component in the regulation of gene expression, *Genes Dev.* 16, 2583–2592, 2002; Kadonaga, J.T., The DPE, a core promoter element for transcription by RNA polymerase II, *Exp. Mol. Med.* 34, 259–264, 2002; Smale, C.T. and Kadonaga, J.T., The RNA polymerase II core promoter, *Annu. Rev. Biochem.* 72, 449–479, 2003; Lewis, B.A. and Reinberg, D., The mediator coactivator complex: functional and physical roles in transcriptional regulation, *J. Cell Sci.* 116, 3667–3675, 2003; Mulle, F. and Tora, L., The multicolored world of promoter recognition complexes, *EMBO J.* 23, 2–8, 2004; Chen, K., Organization of MAO A and MAO B promoters and regulation of gene expression, *Neurotoxicity* 25, 31–36, 2004; Hasselbach, L., Haase, S., Fischer, D., Kolberg, H.C., and Sturzbecher, H.W., Characterization of the promoter region of the human DNA-repair gene Rad51, *Eur. J. Gynecol. Oncol.* 26, 589–598, 2005.

Cosolvent   A miscible solvent added to a primary solvent to enhance salvation or stability of a specific solute. Such solvents have been used extensively in studies on enzymes where cosolvents were required to dissolve the substrate. Cosolvents are also used in the formulation of pharmaceuticals and in liquid chromatography. See Tan, K.H. and Lovrien, R., Enzymology in aqueous-organic cosolvent binary mixtures, *J. Biol. Chem.* 247, 3278–3285, 1972; Richardson, N.E. and Meaekin, B.J., The influence of cosolvents and substrate substituents on the sorption of benzoic acid derivatives by polyamides, *J. Pharm. Pharmcol.* 27, 145–151, 1975; Pescheck, P.S. and Lovrien, R.E., Cosolvent control of substrate inhibition I cosolvent stimulation of beta-glucuronidase activity, *Biochem. Biophys. Res. Commun.* 79, 417–421, 1977; Bulone, D., Cupane, A., and Cordone, L, Conformational and functional properties of hemoglobin in water-organic cosolvent mixtures: effect of ethylene glycol and glycerol on oxygen affinity, *Biopolymers* 22, 119–123, 1983; Rubino, J.T. and Berryhill, W.S., Effects of solvent polarity on the acid dissociation constants of benzoic acids, *J. Pharm. Sci.* 75, 182–186, 1986; Buck, M., Trifluoroethanol and colleagues: cosolvents come of age. Recent studies with peptides and proteins, *Q. Rev. Biophys.* 31, 297–355, 1998; Jouyban-Gharamaleki, A., Valaee, L., Barzegar-Jalali, M. et al., Comparison of various cosolvency models for calculating solute solubility in water-cosolvent mixtures, *Int. J. Pharm.* 177, 93–101, 1999; Lee, J.C., Biopharmaceutical formulation, *Curr. Opin. Biotechnol.* 11, 81–84, 2000; Moelbert, S., Normand, B., and de los Rios, P., Kosmotropes and chaotropes: modeling preferential exclusion, binding, and aggregate stability, *Biophys. Chem.* 112, 45–57, 2004; Scharnagl, C., Reif, M., and Friedrich, J., Stability of proteins: temperature, pressure, and the role of solvent, *Biochim. Biophys. Acta* 1749, 17–213, 2005.

Coupled Enzyme Systems   Most metabolic systems are composed of enzymes in a pathway where there is the sequential transformation of a substrate into a product through a series of separate enzyme-catalyzed reactions. One of the more simple coupled systems is the detoxification of ethyl alcohol (Plapp, B.V., Rate-limiting steps in ethanol metabolism and approaches to changing these rates biochemically, *Adv. Expt. Biol. Med.* 56, 77–109, 1975) or more complex (Brooks, S.P.J., Enzymes in the cell. What's really going on? in *Function and Metabolism*, Storey, K.B., Ed., Wiley-Liss, Hoboken, NJ, 2004, pp. 55–86). Coupled enzyme systems are also used extensively in clinical chemistry where they are also referred to as indicator enzyme systems (Russell, C.D. and Cotlove, E., Serum glutamic-oxaloacetic transaminase: evaluation of a coupled-reaction enzyme assay by means of kinetic theory, *Clin. Chem.* 17, 1114–1122, 1971; Bais, R. and Pateghini, M., Principles of clinical enzymology, in *Tietz Textbook of Clinical Chemistry and Molecular Diagnostics*, Burtis, C.A., Ashwood, E.R., and Bruns, D.E., Eds., Elsevier/Sanders, St. Louis, MO, 2006, pp. 191–218). The assay for creatine kinase is a coupled enzyme system as are some of the assays for glucose oxidase. An enzyme assay system is coupled to an immunological reaction in many solid-phase immunoassays such as ELISA assays (Kircks, L.J., Selected strategies for improving sensitivity and reliability of immunoassays, *Clin. Chem.* 40, 347–357, 1994). See Wimmer, M.C., Artiss, J.D., and Zak, B., Peroxidase-coupled method for kinetic colorimetry of total creatine kinase activity in serum, *Clin. Chem.* 31, 1616–1620, 1965; Shin, T., Murao, S., and Matsumura, E., A chromogenic oxidative coupling reaction of laccase: applications for laccase and angiotensin I converting enzyme assay, *Anal. Biochem.* 166, 380–388, 1987.

Creatine   A nitrogenous compound that is synthesized from arginine, glycine, and *S*-adenosylmethionine. See Van Pilsum, J.F., Stephens, G.C., and Taylor, D., Distribution of creatine, guanidinoacetate, and the enzymes for their biosynthesis in the animal kingdom, *Biochem. J.* 126, 325–345, 1972; Walker, J.B. and Hannan, J.K., Creatine biosynthesis during embryonic development. False feedback suppression of liver amidinotransferase by *N*-acetimdoylsarcosine and 1-carboxymethy-2-iminoimdazolidine (cyclo-creatine), *Biochemistry* 15, 2519–2522, 1976; Walker, J.B., Creatine: biosynthesis, regulation, and function, *The Enzymes* 50, 177–242, 1979; Wyss, M. and Wallimann, T., Creatine metabolism and the consequences of creatine depletion in muscle, *Mol. Cell. Biochem.* 133–134, 51–66, 1994; Wu, G. and Morris, S.M., Jr., Arginine metabolism: nitric oxide and beyond, *Biochem. J.* 336, 1–17, 1998; Brosnan, M.E. and Brosnan, J.T., Renal arginine metabolism, *J. Nutr.* 134 (Suppl. 10), 2791S–2795S, 1994; Morris, S.M., Jr., Enzymes of arginine metabolism, *J. Nutr.* (Suppl. 10), 2743S–2747S, 1994. Creatine is used as a biomarker for erthyrocytes (Beyer, C. and Alting, I.H., Enzymatic measurement of creatine in erythrocytes, *Clin. Chem.* 42, 313–318, 1996; Jiao, Y., Okumiya, T., Saibara, T. et al., An enzymatic assay for erythrocyte creatine as an index of the erythrocyte lifetime, *Clin. Biochem.* 31, 59–65, 1998; Takemoto, Y.,

Okumiya, T., Tsuchida, K. et al., Erythrocyte creatine as an index of the erythrocyte life span and erythropoiesis, *Nephron* 86, 513–514, 2000; Okumiya, T., Ishikawa-Nishi, M., Doi, T. et al., Evaluation of intravascular hemolysis with erythrocyte creatine in patients with cardiac valve prostheses, *Chest* 125, 2115–2120, 2004). There is increased use of creatine as a nutritional supplement (Korzun, W.J., Oral creatine supplements lower plasma homocysteine concentrations in humans, *Clin. Lab. Sci.* 17, 102–106, 2004; Pearlman, J.P. and Fielding, R.A., Creatine monohydrate as a therapeutic aid in muscular dysthrophy, *Nutr. Rev.* 64, 80–88, 2006; Hespel, P., Maughan, R.J., and Greenhaff, P.L., Dietary supplements for football, *J. Sports Sci.* 24, 749–761, 2006; Shao, A. and Hathcock, J.N., Risk assessment for creatine monohydrate, *Regul. Toxicol. Pharmacol.*, 45, 242–251, 2006).

Creatine Kinase Adenosine triphosphate: creatine *N*-phosphotransferase (EC 2.7.3.2), also creatine phosphokinase. Creatine kinase is found in muscle and is responsible for the formation of creatine phosphate from creatine and adenosine triphosphate; creatine phosphate is a higher energy source for muscle contraction. Creatine kinase is elevated in all forms of muscular dystrophy. Creatine kinase is dimer and is present as isozymes (CK-1, BB; CK-2, MB; CK-3, MM) and Ck-mt (mitochondrial). Creatine kinase is also used to measure cardiac muscle damage in myocardial infarction. See Bais, R. and Edwards, J.B., Creatine kinase, *CRC Crit. Rev. Clin. Lab. Sci.* 16, 291–355, 1982; McLeish, M.J. and Kenyon, G.L., Relating structure to mechanism in creatine kinase, *Crit. Rev. Biochem. Mol. Biol.* 40, 1–20, 2005.

Creatinine A catabolic product of creatine, which should be in blood as a constant quantity. An increase in creatinine is associated with a loss of kidney function. See Hodgkinson, A. and Edwards, N.A., Laboratory determinations of renal function, *Biochem. Clin.* 2, 77–86, 1963; Blainey, J.D., The renal excretion of higher molecular weight substances, *Curr. Probl. Clin. Biochem.* 2, 85–100, 1968; Cook, J.G., Factors influencing the assay of creatinine, *Ann. Clin. Biochem.* 12, 219–232, 1975; Greenberg, N., Smith, T.A., and VanBrunt, N., Interference in the Vitros CREA method when measuring urine creatinine on samples acidified with acetic acid, *Clin. Chem.* 50, 1273–1275, 2004; Price, C.P., Newall, R.G., and Boyd, J.C., Prediction of significant proteinuria: a systematic review, *Clin. Chem.* 51, 1577–1586, 2005; Verhoeven, N.M., Salmons, G.S., and Jakobs, C., Laboratory diagnosis of defects of creatine biosynthesis and transport, *Clin. Chim. Acta* 361, 1–9, 2005; Wishart, D.S., Metabolomics: the principles and potential applications to transplantation, *Am. J. Transplant.* 5, 2814–2820, 2005; Seron, D., Fulladosa, X., and Moreso, F., Risk factors associated with the deterioration of renal function after kidney transplantation, *Kidney Int. Suppl.* 99, S113–S117, 2005; Schrier, R.W., Role of diminished renal function in cardiovascular mortality: marker or pathogenic factor? *J. Am. Coll. Cardiol.* 47, 1–8, 2006.

Critical Pressure The minimum pressure required to condense gas to liquid at the critical temperature.

Critical Temperature The critical point (end of a vapor pressure curve in a phase diagram); above this temperature, a gas cannot be liquefied.

Crowding The general effect of polymers including proteins and polysaccharides on the solution properties of proteins. See Zimmerman, S.B., Macromolecular crowding effects on macromolecular interactions: some implications for genome structure and function, *Biochim. Biophys. Acta* 1216, 175–185, 1993; Minton, A.P., Molecular crowding: analysis of effects of high concentrations of inert cosolutes on biochemical equilibria and rates in terms of volume exclusion, *Methods Enzymol.* 295, 127–149, 1998; Johansson, H.O., Brooks, D.E., and Haynes, C.A., Macromolecular crowding and its consequences, *Int. Rev. Cytol.* 192, 155–170, 2000; Ellis, R.J., Macromolecular crowding: obvious but underappreciated, *Trends Biochem. Sci.* 26, 597–604, 2001; Bernardo, P., Garcia de la Torre, J., and Pons, M., Macromolecular crowding in biological systems: hydrodynamic and NMR methods, *J. Mol. Recognit.* 17, 397–407, 2004; Martin, J., Chaperon function — effects of crowding and confinement, *J. Mol. Recognit.* 17, 465–472, 2004; Minton, A.P., Influence of macromolecular crowding upon the stability and state of association of proteins: predictions and observations, *J. Pharm. Sci.* 94, 1668–1675, 2005; del Alamo, M., Rivas, G., and Mateu, M.G., Effect of macromolecular crowding agents on human immunodeficiency virus type 1 capsid protein assembly *in vitro*, *J. Virol.* 79, 14271–14281, 2005; Despa, F., Orgill, D.P., and Lee, R.C., Molecular crowding effects on protein stability, *Ann. N.Y. Acad. Sci.* 1066, 54–66, 2006; Szymanski, J., Patkowski, A., Gapinski, J. et al., Movement of proteins in an environment crowded by surfactant micelles: anomalous versus normal diffusion, *J. Phys. Chem. B. Condens. Matter Mater. Surf. Interfaces Biophys.* 110, 7367–7373, 2006; Derham, B.K. and Harding, J.J., The effect of the presence of globular proteins and elongated polymers on enzyme activity, *Biochim. Biophys. Acta* 1764, 1000–1006, 2006; Grailhe, R., Merola, F., Ridard, J. et al., Monitoring protein interactions in the living cell through the fluorescence decays of the cyan fluorescent protein, *Chemphyschem.* 7, 1442–1454, 2006; McPhie, P., Ni, Y.S., and Minton, A.P., Macromolecular crowding stabilizes the molten globule form of apomyoglobin with respect to both cold and heat unfolding, *J. Mol. Biol.* 361, 7–10, 2006.

Crown Gall Disease/Crown Gall Tumors   Crown gall is caused by a bacteria (*Agrobacterium tumefaciens*). These galls begin with tumorlike cell growth at or just below the soil's surface, near the base of the plant and commonly on bud unions. Galls usually begin as green, pliable tissue, then develop into dark, crusty growths. Crown gall disease has been used to study transformation with relevance to tumor formation. See Knoft, U.C., Crown-gall and *Agrobacterium tumefaciens*: survey of a plant-cell-transformation system of interest to medicine and agriculture, *Subcell. Biochem.* 6, 143–173, 1978; Zhu, J., Oger, P.M., Schrammeijer, B. et al., The bases of crown gall tumorigenesis, *J. Bacteriol.* 182, 3885–3895, 2000; Escobar, M.A., and Dadekar, A.M., *Agrobacterium tumefaciens* as an agent of disease, *Trends Plant. Sci.* 8, 380–386, 2003; Brencic, A. and Winans, S.C., Detection of and response to signals involved in host-microbe interactions by plant-associated bacteria. *Micobiol. Mol. Biol. Rev.* 69, 155–194, 2005.

Cryosection   A tissue section cut from a frozen specimen; in this situation, ice is the supporting matrix. See Yamada, E. and Watanabe, H., High voltage electron microscopy of critical-point dried cryosection, *J. Electron Microsc.* 26 (Suppl.), 339–342, 1977; Maddox, P.H., Tay, S.K., and Jenkins, D., A new fixed cryosection technique for the simultaneous immunocytochemical demonstration of T6 and S100 antigens, *Histochem. J.* 19, 35–38,

1987; Sod, E.W., Crooker, A.R., and Morrison, G.H., Biological cryosection preparation and practical ion yield evaluation for ion microscopic analysis, *J. Microsc.* 160, 55–65, 1990; Lewis Carl, S.A., Gillete-Ferguson, I., and Ferguson, D.G., An indirect immunofluorescence procedure for staining the same cryosection with two mouse monoclonal primary antibodies, *J. Histochem. Cytochem.* 41, 1273–1278, 1993; Jensen, H.L. and Norrild, B., Easy and reliable double-immunogold labeling of herpes simplex virus type-1 infected cells using primary antibodies and studied by cryosection electron microscopy, *Histochem. J.* 31, 525–533, 1999; Gou, D. and Catchpoole, D.R., Isolation of intact RNA following cryosection of archived frozen tissue, *Biotechniques* 34, 48–50, 2003; Rieppo, J., Hyttinen, M.M., Jurvelin, J.S., and Helminen, H.J., Reference sample method reduces the error caused by variable cryosection thickness in Fourier transform infrared imaging, *Appl. Spectrosc.* 58, 137–140, 2004; Takizawa, T. and Robinson, J.M., Thin is better! Ultrathin cryosection immunocytochemistry, *J. Nippon Med. Sch.* 71, 306–307, 2004.

Cyanine Dyes (CyDyes)
A family of fluorescent polymethine dyes containing a -CH = group-linking two nitrogen-containing heterocyclic rings; developed as a sensitizer for photographic emulsions. Used in biochemistry and molecular biology on nucleic acid probes for DNA microarrays and for labeling proteins for electrophoretic analysis. See Ernst, L.A., Gupta, R.K., Mujumdar, R.B., and Waggoner, A.S., Cyanine dye labeling reagents for sulfydryl groups, *Cytometry* 10, 3–10, 1989; Mujumdar, P.S., Ernst, L.A., Mujumdar, S.R., and Waggoner, A.S., Cyanine dye labeling reagents containing isothiocyanate groups, *Cytometry* 10, 11–19, 1989; Southwick, P.L., Ernst, L.A., Tauriello, E.W. et al., Cyanine dye labeling reagents — carboxymethylindocyanine succinimidyl esters, *Cytometry* 11, 418–430, 1990; Mujumdar, R.B., Ernst, L.A., Mujumdar, S.R., et al., Cyanine dye labeling reagents: sulfoindocyanine succinimidyl esters, *Bioconjug. Chem.* 4, 105–111, 1993; Benchaib, M., Delorme, R., Pluvinage, M. et al., Evaluation of five green fluorescence-emitting streptavidin-conjugated fluorochromes for use in immunofluorescence microscopy, *Histochem. Cell Biol.* 106, 253–256, 1996; Mujumdar, S.R., Mujumdar, R.B., Grant, C.M., and Waggoner, A.S., Cyanine-labeling reagents: sulfobenzindocyanine succinimidyl esters, *Bioconjug. Chem.* 7, 356–362, 1996; Karp, N.A. and Lilley, K.S., Maximizing sensitivity for detecting changes in protein expression: experimental design using minimal CyDyes, *Proteomics* 5, 3105–3115, 2005; Heilmann, M., Margeat, E., Kasper, R. et al., Carbocyanine dyes as efficient reversible single-molecule optical switch, *J. Am. Chem. Soc.* 127, 3801–3806, 2005; Wu, T.L., Two-dimensional difference gel electrophoresis, *Methods Mol. Biol.* 328, 71–95, 2006; Boisseau, S., Mabrouk, K., Ram, N. et al., Cell penetration properties of maurocalcine, a natural venom peptide active on the intracellular ryanodine receptor, *Biochim. Biophys. Acta* 1758, 308–319, 2006. There is also use of these dyes for the measurement of membrane potentials. See Miller, J.B. and Koshland, D.E., Effects of cyanine dye membrane probes on cellular properties, *Nature* 272, 83–84, 1978; Klausner, R.D. and Wolf, D.E., Selectivity of fluorescent lipid analogues for lipid domains, *Biochemistry* 19, 6199–6203, 1980; Kragh-Hansen, U., Jorgensen, K.E., and Sheikh, M.I., The use of potential-sensitive cyanine dye for studying ion-dependent electrogenic renal transport of organic solutes. Spectrophotometric

measurements, *Biochem. J.* 208, 359–368, 1982; Johnstone, R.M., Laris, P.C., and Eddy, A.A., The use of fluorescent dyes to measure membrane potentials: a critique, *J. Cell Physiol.* 112, 298–300, 1982; Toyomizu, M., Okamoto, K., Akiba, Y. et al., Anacardic acid–mediated changes in membrane potential and pH gradient across liposomal membranes, *Biochim. Biophys. Acta* 1558, 54–62, 2002.

Cyclitols   Term used to describe derivatives of hexhydroxyhexane (1,2,3,4,5,6-hexahydroxyhexane). An analogue to saccharides and serves as a matrix for the development of inhibitors and activators based on saccharide structure. See Tentative rules for cyclitol nomenclature, *Biochim. Biophys. Acta* 165, 1–21, 1968; Orthen, B. and Popp, M., Cyclitols as cryoprotectants for spinach and chickpea thylakoids, *Environ. Exp. Bot.* 44, 125–132, 2000; Pelyvas, I.F., Toth, Z.G., Vereb, G. et al., Synthesis of new cyclitol compounds that influence the activity of phosphatidylinositol 4-kinase isoforms, PI4K230, *J. Med. Chem.* 44, 627–632, 2001; Sureshan, K.M., Shashidhar, M.S., and Varma, A.J., Cyclitol-based metal-complexing agents. Effect of the relative orientation of oxygen atoms in the ionophoric ring on the cation-binding ability of myo-inositol-based crown ethers, *J. Org. Chem.* 67, 6884–6888, 2002; Freeman, C., Liu, L., Banwell, M.G. et al., Use of sulfated linked cyclitols as heparin sulfate mimetics to probe the heparin/heparin sulfate binding specificity of proteins, *J. Biol. Chem.* 280, 8842–8849, 2005; Cochran, S., Li, C.P., and Bytheway, I., An experimental and molecular-modeling study of the binding of linked sulfated tetracyclitols to FGF-1 and FGF-2, *ChemBioChem* 6, 1882–1890, 2005.

Cytochrome P-450 Enzymes (CPY)   A family of enzymes that have monooxygenase activity and are involved in the metabolism/catabolism of drugs. Cytochrome P450 proteins are found in a high concentration in the liver. See Grengerich, E.P., Cytochrome P450 enzymes in the generation of commercial products, *Nat. Rev. Drug Disc.* 1, 359–366, 2002; Jung, C., Schunemann, V., and Lendzian, F., Freeze-quenched iron-oxo intermediate in cytochrome P450, *Biochem. Biophys. Res. Commun.* 338, 355–364, 2005; Johnson, E.F. and Stout, C.D., Structural diversity of human xenobiotic-metabolizing cytochrome P450, *Biochem. Biophys. Res. Commun.* 338, 331–336, 2005; Tang, W., Wang, R.W., and Lu, A.Y., Utility of recombinant cytochrome P450 enzymes: a drug metabolism perspective, *Curr. Drug Metab.* 6, 503–517, 2005; Krishna, D.R. and Shekar, M.S., Cytochrome P450 3A: genetic polymorphisms and inter-ethnic differences, *Methods Find. Exp. Clin. Pharmacol.* 27, 559–567, 2005; Sarlis, N.J. and Gourgiotis, L., Hormonal effects on drug metabolism through the CYP system: perspectives on their potential significance in the era of pharmacogenomics, *Curr. Drug Targets Immune Endocr. Metabol. Disord.* 5, 439–448, 2005.

Cytokeratin   Intermediate filament keratins found in epithelial tissue. There are two types of cytokeratins: the acidic type I cytokeratins and the basic or neutral type II cytokeratins. Cytokeratins are thought to play a role in the activation of plasma prekallikrein and plasminogen. See Crewther, W.G., Fraser, R.D., Lennox, F.G., and Lindley, H., The chemistry of keratins, *Adv. Protein Chem.* 20, 191–346, 1965; Masri, M.S. and Friedman, M., Interactions of keratins with metal ions: uptake profiles, mode of binding, and effects on the properties of wool, *Adv. Exp. Med. Biol.* 48, 551–587, 1974; Fuchs, E. and Green, H., Multiple keratins of cultured human epidermal cells are translated from different mRNA molecules, *Cell* 17, 573–582, 1979;

Fraser, R.D. and Macrae, T.P., Molecular structure and mechanical properties of keratins, *Symp. Soc. Exp. Biol.* 34, 211–246, 1980; Moll, R., Franke, W.W., Schiller, D.L. et al., The catalog of human cytokeratins: patterns for expression in normal epithelia, tumors, and cultured cells, *Cell* 31, 11–24, 1982; Lazarides, E., Intermediate filaments: a chemically heterogeneous, developmentally regulated class of proteins, *Annu. Rev. Biochem.* 51, 219–250, 1982; Gonias, S.L., Hembrough, T.A., and Sankovic, M., Cytokeratin 8 functions as a major plasminogen receptor in select epithelial and carcinoma cells, *Front. Biosci.* 6, D1403–D1411, 2001; Kaplan, A.P., Joseph, K., and Silverberg, M., Pathways for bradykinin formation and inflammatory diseases, *J. Allergy Clin. Immunol.* 109, 195–209, 2002; Shariat-Madar, Z., Mahdi, F., and Schmaier, A.H., Assembly and activation of the plasma kallikrein/kinin system: a new interpretation, *Int. Immunopharmacol.* 2, 1841–1849, 2002; Langbein, L. and Schweizer, J., Keratins of the human hair follicle, *Int. Rev. Cytol.* 243, 1–78, 2005; Gusterson, B.A., Ross, D.T., Heath, V.J., and Stein, T., Basal cytokeratins and their relationship to the cellular origin and functional classification of breast cancer, *Breast Cancer Res.* 7, 143–148, 2005; Skakle, J., Applications of X-ray powder diffraction in materials chemistry, *Chem. Rec.* 5, 252–262, 2005. See also *Keratin.*

Cytokines   Nonantibody proteins secreted by immune system cells. This is a large category and includes the various interferons and interleukins as well as other protein substances. See Henle, W., Interference and interferon in persistent viral infections of cell cultures, *J. Immunol.* 91, 145–150, 1963; Isaacs, A., Interferon, *Adv. Virus Res.* 10, 1–38, 1963; Baron, S. and Levy, H.B., Interferon, *Annu. Rev. Microbiol.* 20, 291–318, 1966; Silverstein, S., Macrophages and viral immunity, *Semin. Hematol.* 7, 185–214, 1970; Bloom, B.R., *In vitro* approaches to the mechanism of cell-mediated immune reactions, *Adv. Immunol.* 13, 101–208, 1971; Granger, G.A., Lymphokines — the mediators of cellular immunity, *Ser. Hematol.* 5, 8–40, 1972; Valentine, F.T., Soluble factors produced by lymphocytes, *Ann. N.Y. Acad. Sci.* 221, 317–323, 1974; Ward, P.A., Leukotaxis and leukotactic disorders. A review, *Am. J. Pathol.* 77, 520–538, 1974; DeMaeyer, E.M. and Demaeyer-Guignard, J., *Interferons and Other Regulatory Cytokines,* John Wiley & Sons, New York, 1988; Plotnikoff, N.P., *Cytokines: Stress and Immunity,* CRC Press, Boca Raton, FL, 1999; Cruse, J.M. and Lewis, R.E., *Atlas of Immunology*, CRC Press, Boca Raton, FL, 1999; Rott, I.M. and Brostoff, J., *Immunology*, Mosby, Edinburgh, UK, 2001; Keisari, Y. and Ofek, I., *The Biology and Pathology of Innate Immunity Mechanisms*, Kluwer Academic, New York, 2002; Salazar-Mather, T.P. and Hokeness, K.L., Cytokine and chemokine networks: pathways to viral defense, *Curr. Top. Microbiol. Immunol.* 303, 29–46, 2006; Akira, S., Uematsu, S., and Takeuchi, O., Pathogen recognition and innate immunity, *Cell* 124, 783–801, 2006; Tedgui, A. and Mallat, Z., Cytokines in atherosclerosis: pathogenic and regulatory pathways, *Physiol. Rev.* 86, 515–581, 2006.

Cytokinesis   Cell division; the division of the cytoplasm of a cell following the division of the nucleus. See Robinson, D.N. and Spudich, J.A., Mechanics and regulation of cytokinesis, *Curr. Opin. Cell Biol.* 16, 181–188, 2004; Mayer, U. and Jurgens, G., Cytokinesis: lines of division taking shape, *Curr. Opin. Plant Biol.* 7, 599–604, 2004; Albertson, R., Riggs, B., and Sullivan, W.,

Membrane traffic: a driving force in cytokinesis, *Trends Cell Biol.* 15, 92–101, 2005; Glotzer, M., The molecular requirements for cytokinesis, *Science* 307, 1735–1739, 2005; Burgess, D.R., Cytokinesis: new roles for myosin, *Curr. Biol.* 15, R310–R311, 2005; Darenfeld, H. and Mandato, C.A., Wound-induced contractile ring: a model for cytokinesis, *Biochem. Cell Biol.* 83, 711–720, 2005; Konopka, C.A., Scheede, J.B., Skop, A.R., and Bednarek, S.Y., Dynamin and cytokinesis, *Traffic* 7, 239–247, 2006.

Cytomics   The molecular analysis of heterogeneous cellular systems. See Davies, E., Stankovic, B., Azama, K. et al., Novel components of the plant cytoskeleton: a beginning to plant "cytomices," *Plant Sci.* 160, 185–196, 2001; Bernas, T., Gregori, G., Asem, E.K., and Robinson, J.P., Integrating cytomics and proteomics, *Mol. Cell. Proteomics* 5, 2–13, 2006; Van Osta, P., Ver Donck, K., Bols, L., and Geysen, J., Cytomics and drug discovery, *Cytometry A* 69, 117–118, 2006; Tarnok, A., Slide-based cytometry for cytomics — a minireview, *Cytometry A* 69, 555–562, 2006; Herrera, G., Diaz, L., Martinez-Romero, A. et al., Cytomics: a multiparametric, dynamic approach to cell research, *Toxicol.* In Vitro, 21, 176–182, 2007; Valet, G., Cytomics as a new potential for drug discovery, *Drug Discov. Today* 11, 785–791, 2006.

Cytoskeleton   The internal framework of the cell; the cytoskeleton is composed largely of actin filaments and microtubules. See Wasteneys, G.O. and Yang, Z., New views on the plant cytoskeleton, *Plant Physiol.* 136, 3884–3891, 2004; Moller-Jensen, J. and Lowe, J., Increasing complexity of the bacterial cytoskeleton, *Curr. Opin. Cell Biol.* 17, 75–81, 2005; Smith, L.G. and Oppenheimer, D.G., Spatial control of cell expansion by the plant cytoskeleton, *Annu. Rev. Cell Dev. Biol.* 21, 271–295, 2005; Munro, E.M., PAR proteins and the cytoskeleton: a marriage of equals, *Curr. Opin. Cell Biol.* 18, 86–94, 2006; Boldogh, I.R. and Pon, L.A., Interactions of mitochondria with the actin cytoskeleton, *Biochim. Biophys. Acta* 1763, 405–462, 2006; Larsson, C., Protein kinase C and the regulation of the actin cytoskeleton, *Cell Signal.* 18, 276–284, 2006; Logan, M.R and Mandato, C.A., Regulation of the actin cytoskeleton by PIP_2 in cytokinesis, *Biol. Cell.* 98, 377–388, 2006; Sheetz, M.P., Sable, J.E., and Dobereiner, H.G., Continuous membrane-cytoskeleton adhesion requires continuous accommodation to lipid and cytoskeleton dynamics, *Annu. Rev. Biophys. Biomol. Struct.* 35, 417–434, 2006; Becker, B.E. and Gard, D.L., Visualization of the cytoskeleton in *Xenopus* oocytes and eggs by confocal immunofluorescence microscropy, *Methods Mol. Biol.* 322, 69–86, 2006; Popowicz, G.M., Scheicher, M., Noegel, A.A., and Holak, A.A., Filamins: promiscuous organizers of the cytoskeleton, *Trends Biochem. Sci.* 31, 411–419, 2006.

Cytotoxic T-Cells; Cytotoxic T-Lymphocytes   Also known as killer cells, killer T-cells, null cells. A differentiated T-cell (CD8 positive) that attacks and lyses target cells bearing specific antigens. Used in patient-specific immunotherapy with cells grown in culture. See Gillis, S., Baker, P.E., Ruscetti, F.W., and Smith, K.A., Long-term culture of human antigen-specific cytotoxic T-cell lines, *J. Exptl. Med.* 148, 1093–1098, 1978.

Database of Interacting Proteins (DIP)   The database of interacting proteins integrates the experimental evidence available on protein interactions into a single on-line resource: http://dip. doe-mbi.ucla.edu. See Xenarious, I., Fernandez, E., Salwinski, L., Duan, X.J. et al., DIP: the database of interacting proteins: 2001 update, *Nucleic Acids Res.* 29, 239–241, 2001; Deane, C.M., Salwinski, L., Xenarios, I., and Eisenberg, D., Protein interactions: two methods for assessment of the reliability of high throughput observations, *Mol. Cell Proteomics* 1, 349–356, 2002; Salwinski, L., Miller, C.S., Smith, A.J. et al., *Nucleic Acids Res.* 32, D449–D451, 2004; Han, D., Kim, H.S., Seo, J., and Jang, W., A domain combination based on probabilistic framework for protein–protein interaction prediction, *Genome Inform. Ser. Workshop Genome Inform.* 14, 250–259, 2003; Espadaler, J., Romero-Isart, O., Jackson, R.M., and Oliva, B., Prediction of protein–protein interactions using distant conservation of sequence patterns and structure relationships, *Bioinformatics* 21, 3360–3368, 2005.

Deconvolution
An algorithm used in electrospray mass spectrometry to translate the spectra of multiply charged ions into a spectrum of molecular species.

Dendrimers
A novel polymeric material containing a highly branched and well-defined structure. Dendrimers have been used for drug delivery, a biological matrix, and for model drug distribution studies. Dendrimers are similar to dendrites, which are branched crystals in which branches of crystallization proceed at different rates. See Meldal, M. and Hilaire, P.M., Synthetic methods of glycopeptide assembly, and biological analysis of glycopeptide products, *Curr. Opin. Chem. Biol.* 1, 552–563, 1997; Sadler, K. and Tam, J.P., Peptide dendrimers: applications and synthesis, *J. Biotechnol.* 90, 195–229, 2002; Turnbull, W.B. and Stoddart, J.F., Design and synthesis of glycodendrimers, *J. Biotechnol.* 90, 231–255, 2002; Kobayashi, H. and Brechbiel, M.W., Dendrimer-based macromolecular MRI contrast agents: characteristics and application, *Mol. Imaging* 2, 1–10, 2003; Lee, C.C., MacKay, J.A., Frechet, J.M., and Szoka, F.C., Designing dendrimers for biological applications, *Nat. Biotechnol.* 23, 1517–1526, 2005; Qiu, L.Y. and Bae, Y.H., Polymer architecture and drug delivery, *Pharm. Res.* 23, 1–30, 2006; Gupta, V., Agashe, H.B., Asthana, A., and Jain, N.K., Dendrimers: novel polymeric nanoarchitecture for solubility enhancement, *Biomacromolecules* 7, 649–658, 2006; Söntjens, S.H.M., Nettles, D.L., Carnahan, M.A. et al., Biodendrimer-based hydrogel scaffolds for cartilage tissue repair, *Biomacromolecules* 7, 310–316, 2006.

Desorption
Process by which molecules in solid or liquid form are transformed into a gas phase.

Detergent Perturbation
Treatment of total human plasma proteins with sodium cholate and subsequent removal; resulting in "remodeling" of the lipoproteins. See Pownall, H.J., Remodeling of human plasma lipoproteins by detergent perturbation, *Biochemistry* 44, 9714–9722, 2005.

Deterministic Series
A series or model that contains no random or probabilistic elements. See Everitt, B.S., Ed., *The Cambridge Dictionary of Statistics*, Cambridge University Press, Cambridge, UK, 1998.

Diabodies
An engineered noncovalent dimer of an scFv fragment, which has two antigen-binding sites that may either be homologous or heterologous. The normal linker engineered between the V_H and V_L domains is 15 residues (usually glycine and serine to promote maximum flexibility), which yields a monomer; if the linker is reduced to 10 residues, a dimer (diabody) is formed while with no linker there is a trimer or higher-order polymer. See Atwell, J.L., Breheney, K.A., Lawrence, L.J. et al., scFv multimers of the anti-neuraminidase antibody NC10: length of the linker between V_H and V_L domains dictates precisely the transition between diabodies and triabodies, *Protein Eng.* 12, 597–604, 1999; Todorovska, A., Roovers, R.C.,

Dolezal, O. et al., Design and application of diabodies, triabodies, and tetrabodies for cancer targeting, *J. Immunol. Methods* 248, 47–66, 2001; Holliger, P. and Hudson, P.J., Engineered antibody fragments and the rise of single domains, *Nature Biotechnol.* 23, 1126–1136, 2005. While diabodies are noncovalent complexes of engineered scFv constructs based on the association of the V_H domain with the most available V_L domain, a covalent diabody was observed with an engineered anticarcinoembryonic antigen (CEA) diabody with cysteine residues inserted for coupling with a radiolabel. The formation of a disulfide-linked diabody was an unexpected consequence (see Olafsen, T., Cheung, C.-W., Yazaki, P.J. et al., Covalent disulfide-linked anti-CEA diabody allows site-specific conjugation and radiolabeling for tumor targeting applications, *Prot. Eng. Des. Sel.* 17, 21–27, 2004). It has been observed that if the order of the variable regions are switched in scFv construct (V_L-V_H instead of V_H-V_L), the engineered scFv with a zero-length linker formed a dimer (diabody) instead of the expected trimer (see Arndt, M.A.E., Krauss, J., and Rybak, S.M., Antigen binding and stability properties of non-covalently linked anti-CD22 single-chain Fv dimers, *FEBS Lett.* 578, 257–261, 2004). See *Bibody*; *Single-Chain Fv Fragment*; *Triabody*.

Diapedesis   The migration of a leukocyte through the interendothelial junction space and the extracellular matrix/basement membrane to the site of tissue inflammation; a process driven by chemotaxis.

Dicer   Dicer is an RNAse III nuclease (class III), which is specific for double-stranded RNA and yields siRNAs. Structurally it consists of an amino terminal helicase domain, a PAZ domain, two RNAse III motifs, and a dsRNA binding motif. See Carmell, M.A., and Hannan, G.J., RNAse III enzymes and their initiation of gene silencing, *Nat. Struct. Mol. Biol.* 11, 214–218, 2004; Myers, J.W. and Ferrell, J.E., Jr., Silencing gene expression with Dicer-generated siRNA pools, in *RNA Silencing: Methods and Protocols*, Carmichael, G.G., Ed., Humana Press, Totowa, NJ, 2005; Hammond, S.M., Dicing and slicing. The core machinery of the RNA interference pathway, *FEBS Lett.* 579, 5822–5829, 2005.

Dictionary of Interfaces in Proteins (DIP)   A database that collects the 3-D structures of protein domains involved in interactions (patches). See Preissner, R., Goode, A., and Frommel, C., Dictionary of interfaces in proteins (DIP). Data bank of complementary molecules, *J. Mol. Biol.* 280, 535–550, 1998; Frommel, C., Gille, C., Goede, A. et al., Accelerating screening of 3-D protein data with a graph theoretical approach, *Bioinformatics* 19, 2442–2447, 2003.

Differential Scanning Calorimetry (DSC)   A physical technique for the study of conformation based on measuring changes in heat capacity of a molecule under various conditions. See Zecchinon, L., Oriol, A., Netzel, U. et al., Stability domains, substrate-induced conformational changes, and hinge-bending motions in a psychrophilic phosphoglycerate kinase. A microcalorimetric study, *J. Biol. Chem.* 280, 41307–41314, 2005.

Dipolar Couplings   Also residual dipolar couplings or ligand binding; measures the interaction between nuclei in an applied magnetic field; used for the determination of the solution structure of peptides, proteins, nucleic acids, and carbohydrates. See Post, C.B., Exchange-transferred NOE spectroscopy and bound ligand structure determination, *Curr. Opin. Struct. Biol.* 13, 581–588, 2003; Bush, C.A., Martin-Pastor, M., and Imberty, A., Structure and conformation of complex carbohydrates of glycoproteins, glycolipids, and bacterial polysaccharides, *Ann. Rev. Biophys. Biomol. Struct.* 28, 269–293, 1999; MacDonald, D., and Lu, P., Residual dipolar couplings in nucleic acid structure determination, *Curr. Opin. Struct. Biol.* 12, 337–343, 2002.

Directed Library   Also focused library. A screening library of chemical compounds that may be prepared by parallel synthesis, combinatorial chemistry, phage display, or similar multiplexed technologies. See Miller, J.L., Recent developments in focused library design: targeting gene-families, *Curr. Top. Med. Chem.* 6, 19–29, 2006; Xu, Y., Shi, J., Yamamoto, N. et al., A credit-card library approach for disrupting protein–protein interactions, *Bioorg. Med. Chem.*, 14, 2660–2673, 2006; Subramanian, T., Wang, Z., Troutman, J.M. et al., Directed library of anilinogeranyl analogues of farnesyl diphosphate via mixed solid- and solution-phase synthesis, *Org. Lett.* 7, 2109–2112, 2005; McGregor, M.J., and Muskal, S.M., Pharmacophore fingerprinting. 1. Application to QSAR and focused library design, *J. Chem. Inf. Comput. Sci.* 39, 569–574, 1999.

Distributed Annotation System (DAS)   The distributed annotation system is a communication protocol for the exchange of biological annotations. (In genetics, the process of identifying the locations and coding regions of genes in a genome and determining what those genes do. An annotation is a note added to comment on the function of the gene and/or coding region.) See Hubbard, T., Biological information: making it accessible and integrated (and trying to make sense of it), *Bioinformatics* 18 (Suppl. 2), S140, 2002; Olason, P.I., Integrating protein annotation resources through the Distributed Annotation System, *Nucleic Acids Res.* 33, W468–W470, 2005; Prlic, A., Down, T.A., and Hubbard, J.T., Adding some SPICE to DAS, *Bioinformatics* 21 (Suppl. 2), ii40–ii41, 2005; Stamm, S., Riethovan, J.J., Le Texier, V. et al., ASD: a bioinformatics resource on alternative splicing, *Nucleic Acids Res.* 32, D46–D55, 2006. See also http://www.cbs.dtu.dk/; http://www.cbs.dtu.dk/cgi-bin/das.

DNA Fingerprinting   This procedure is also referred to as chromosomal fingerprinting, restriction enzyme analysis (REA). This is a process whereby DNA is cleaved by a restriction endonuclease (restriction enzyme). The resulting DNA fragments are separated by gel electrophoresis and detected by specific and nonspecific probes. DNA fingerprinting is extensively used for forensic purposes. See Owen, R.J., Chromosomal DNA fingerprinting — a new method of species and strain identification applicable to microbial pathogens, *J. Med. Microbiol.* 30, 89–99, 1989; Cawood, A.H., DNA fingerprinting, *Clin. Chem.* 35, 1832–1837, 1989; Gazit, E. and Gazit, E., DNA fingerprinting, *Isr. J. Med. Sci.* 26, 158–162, 1990; de Gouyon, B., Julier, C., Avner, P., Georges, M., and Lathrop, M., Human variable number of tandem repeat probes as a source of polymorphic markers in experimental animals, *EXS* 58, 85–94, 1991; Webb, M.B. and Debenham, P.G., Cell line characterization by DNA fingerprinting: a review, *Dev. Biol. Stand.* 76, 39–42, 1992; Debenham, P.G., Probing identity: the changing face of DNA fingerprinting, *Trends Biotechnol.* 10, 96–102, 1992; McClelland, M. and Welsh, J., DNA fingerprinting by arbitrarily primed PCR, *PCR Methods Appl.* 4, S59–S65, 1994; Kuff, E.L. and Mietz, J.A., Analysis of DNA restriction enzyme digests by two-dimensional electrophoresis in agaraose gels, *Methods Mol. Biol.* 31, 177–186, 1994; Caetano-Anolles, G., Scanning of nucleic acids by *in vitro* amplification: new developments and applications, *Nat. Biotechnol.* 14, 1668–1674, 1996.

DNA Footprinting — DNA is incubated with a putative binding protein and then modified with dimethyl sulfate. Methylation of DNA bases occurs at regions not protected by the protein binding. The DNA can be cleaved at guanine residues and are then cleaved by piperidine. Footprinting can also be achieved by the use of DNAse I hydrolysis, reaction with hydroxyl radicals, or with metal ion-chelate complexes. With either enzymatic or chemical fragmentation, the DNA is end-labeled with ^{32}P-phosphate to permit identification by autoradiography. This has been used to identify the sites of transcription factor binding to *cis*-regions on DNA. See Guille, M.J. and Kneale, G.G., Methods of the analysis of DNA-protein interactions, *Mol. Biotechnol.* 8, 35–52, 1997; Cappabianca, L., Thomassin, H., Pictet, R., and Grange, T., Genomic footprinting using nucleases, *Methods Mol. Biol.* 199, 427–442, 1999; Angelov, D., Khochbin, S., and Dimitrov, S., U.S. laser footprinting and protein-DNA crosslinking. Application to chromatin, *Methods Mol. Biol.* 119, 481–495, 1999; Gao, B. and Kunos, G., DNase I footprinting analysis of transcription factors recognizing adrenergic receptor gene promoter sequences, *Methods Mol. Biol.* 126, 419–429, 2000; Brenowitz, M., Chance, M.R., Dhavan, G., and Takamoto, K., Probing the structural dynamics of nucleic acids by quantitative time-resolved and equilibrium hydroxyl radical "footprinting," *Curr. Opin. Struct. Biol.* 12, 648–653, 2002; Knight, J.C., Functional implications of genetic variation in noncoding DNA for disease susceptibility and gene regulation, *Clin. Sci.* 104, 493–501, 2003.

DNA Methylation — Modification (methylation) of DNA catalyzed by DNA methyltransferase enzymes. Modification occurs at cytosine and adenosine. In multicellular organisms, methylation appears to be confined to cytosine residues. See van Steensel, B. and Henikoff, S., Epigenomic profiling using microarrays, *Biotechniques* 35, 346–350, 2003; El-Maarri, O., Methods: DNA methylation, *Adv. Exp. Med. Biol.* 544, 197–204, 2003; Gut, I.G., DNA analysis by MALDI-TOF mass spectrometry, *Hum. Mutat.* 23, 437–441, 2004; Kapoor, A., Agius, F., and Zhu, J.K., Preventing transcriptional gene silencing by active DNA demethylation, *FEBS Lett.* 579, 5889–5898, 2005; Klose, R.J. and Bird, A.P., Genomic DNA methylation: the mark and its mediators, *Trends in Biochem. Sci.* 31, 81–97, 2006.

DNAse I Hypersenstivity Site — Preferred site(s) of DNA I cleavage; typically at regions where clusters of transcriptional activators bind to DNA and usually reflect a change in chromatin structure. See McGinnis, W., Shermoen, A.W., Heemskerk, J., and Beckendorf, S.K., DNA sequence changes in an upstream DNAse I-hypersensitive region are correlated with reduced gene expression, *Proc. Natl. Acad. Sci. USA* 80, 1063–1067, 1983; Cereghini, S., Saragosti, S., Yaniv, M., and Hamer, D.H., SV40-alpha-globulin hybrid minichromosomes. Differences in DNase I hypersensitivity of promoter and enhancer sequences, *Eur. J. Biochem.* 144, 545–553, 1984; Rothenberg, E.V. and Ward, S.B., A dynamic assembly of diverse transcription factors integrates activation and cell-type information for interleukin 2 gene regulation, *Proc. Natl. Acad. Sci. USA* 93, 9358–9365, 1996; Ishii, H., Sen, R., and Pazin, M.J., Combinatorial control of DNase I-hypersensitive site formation and erasure by immunoglobulin heavy chain enhancer-binding proteins, *J. Biol. Chem.* 279, 7331–7338, 2004; Hermann, B.P. and Heckert, L.L., Silencing of Fshr occurs through a conserved, hypersensitive site in the first intron, *Mol. Endocrinol.* 19, 2112–2131, 2005; Sun, D., Guo, K.,

Rusche, J.J., and Hurley, L.H., Facilitation of a structural transition in the polypurine/polypyrimidine tract within the proximal promoter region in the human VEGF gene by the presence of potassium and G-quadruplex-interactive agents, *Nucleic Acids Res.* 33, 6070–6080, 2005.

DNAzyme   A DNA molecule that contains a catalytic motif that cleaves bound RNA in a hydrolytic reaction; also known as deoxyribozymes or DNA enzymes. See Joyce, G.F., Directed evolution of nucleic acid enzymes, *Annu. Rev. Biochem.* 73, 791–836, 2004; Achenbach, J.C., Chiuman, W., Cruz, R.P., and Li., Y., DNAzymes: from creation *in vitro* to application *in vivo*, *Curr. Pharm. Biotechnol.* 5, 321–336, 2004; Sioud, M. and Iversen, P.O., Ribozymes, DNAzymes, and small interfering RNAs as therapeutics, *Curr. Drug Targets* 6, 647–653, 2005; Fiammengo, R. and Jaschke, A., Nucleic acid enzymes, *Curr. Opin. Biotechnol.* 16, 614–621, 2005.

Domain   A contiguous (usually) series of monomer units (amino acids in proteins; nucleic acid bases in nucleic acids; monosaccharide in oligosaccharides/polysaccharides). A domain can be continuous or discontinuous and is identified by a unique function such as catalysis or binding; domains are frequently identified by homology and used to group proteins into families.

Domain Antibodies Antibodies containing a single antigen-binding domain, most often the V_H region or the highly variable regions from the V_H and V_L regions. These antibodies are naturally occurring in camelids (members of the order *Camelidae*, which include llamas and camels). See Dick, H.M., Single domain antibodies, *BMJ* 300, 959, 1990; Riechman, L. and Muyldermans, S., Single domain antibodies: comparison of camel VH and camelized human VH domains, *J. Immunol. Methods* 231, 25–38, 1999; Stockwin, L.H. and Holmes, S., Antibodies as therapeutic agents: vive la renaissance! *Expert Opin. Biol. Ther.* 3, 1133–1152, 2003; Holt, L.J., Herring, C., Jespers, L.S., Woolven, B.P., and Tomlinson, I.M., Domain antibodies: proteins for therapy, *Trends Biotechnol.* 21, 484–490, 2003.

Drosha   A member of the RNAse III family of double-stranded specific endonucleases. Drosha is a member of Class II, in which each member contains tandem RNAse III catalytic motifs and one C-terminal dsRNA-binding domain. Class I proteins contain only one RNAse III catalytic domain and a dsRNA binding domain. Class III (see Dicer) contains a PAZ domain, a DUF283 domain, the tandem nuclease domains, and a dsRNA-binding domain. See Carmell, M.A., and Hannan, G.J., RNAse III enzymes and their initiation of gene silencing, *Nat. Struct. Mol. Biol.* 11, 214–218, 2004.

Drug   A drug is defined as (1) a substance recognized by an official pharmacopoeia or formulary; (2) a substance intended for use in the diagnosis, cure, mitigation, treatment, or prevention of disease; (3) a substance (other than food) intended to affect the structure or any function of the body; or (4) a substance intended for use as a component of a medicine but not of a device or a component, part, or accessory of a device. Biological products are included within this definition and are generally covered by the same laws and regulations, but differences exist regarding their manufacturing processes (chemical processes vs. biological processes).

Drug Master File Drug Master Files (DMF) contain information on the processes and facilities used in drug or drug component manufacture and storage and are submitted to the FDA for examination and approval.

Drug Product   The final dosage form, which contains a drug substance or drug substances as well as inactive materials that are also considered as excipients. The drug product is differentiated from the drug substance but may or may not be the same as the drug substance. See http://www.fda.gov/cder/drugsatfda/glossary.htm; http://www.fda.gov/cder/ondc/Presentations/2002/01-10- 19_DIA_JS.pps.

Drug Targeting   The ability to target a compound to a specific organ or cell type within an organism. The compound can be a drug/pharmaceutical or it can be a compound, such as a radioisotope, which can be used as a diagnostic. See Muzykantov, V.R., Biomedical aspects of targeted delivery of drugs to pulmonary endothelium, *Expert Opin. Drug Deliv.* 2, 909–926, 2005; Weissig, V., Targeted drug delivery to mammalian mitochondria in living cells, *Expert Opin. Drug Deliv.* 2, 89–102, 2005; Hilgenbrink, A.R. and Low, P.S., Folate-receptor–mediated drug targeting: from therapeutics to diagnostics, *J. Pharm. Sci.* 94, 2135–2146, 2005.

Dye(s)   A chemical compound with a structure that yields a color (a chromophore), which can be coupled either covalently or noncovalently to a substrate matrix. The ability of the compound to yield color is based on its ability to absorb light in the visible spectrum (400–700 nm). Dyes can be classified by various characteristics including mechanism/chemistry (e.g., basic dyes, acid dyes; acid/base indicators/redox dyes), structure (nitroso, acridine dyes, thiazole dyes), and process use (e.g., vat dyes). A dye is a colorant (a substance that yields color) as is a pigment. A dye is chemically different from a pigment, which is a particle suspended in a medium such as particles in paint. More recently, the term "dye" has expanded to include fluorescent compounds. See Conn, H.J., *Biological Stains: A Handbook on the Nature and Uses of the Dyes Employed in the Biological Handbook*, Williams & Wilkins, Baltimore, MD, 1961; Kasten, F.H., Cytochemical studies with acridine orange and the influence of dye contaminants in the staining of nucleic acids, *Int. Rev. Cytol.* 21, 141–202, 1967; Meyer, M.C. and Guttman, D.E., The binding of drugs by plasma proteins, *J. Pharm. Sci.* 57, 895–918, 1968; Adams, C.W., Lipid histochemistry, *Adv. Lipid Res.* 7, 1–62, 1969; Horobin, R.W., The impurities of biological dyes: their detection, removal, occurrence, and histochemical significance — a review, *Histochem. J.* 1, 231–265, 1969; Biswas, B.B., Basu, P.S., and Pai, M.K., Gram staining and its molecular mechanism, *Int. Rev. Cytol.* 29, 1–27, 1970; Gurr, E., *Synthetic Dyes in Biology, Medicine, and Chemistry*, Academic Press, London, 1971; Venkataraman, K., *The Chemistry of Synthetic Dyes*, Academic Press, New York, 1978; Egan, H. and Fishbein, L., *Some Aromatic Amines and Azo Dyes in the General and Industrial Environment*, International Agency for Research on Cancer, Lyon, France, 1981; Clark, G. and Koastan, F.H., *History of Staining*, 3rd ed., Williams & Wilkins, Baltimore, MD, 1983; Zollinger, H., *Color Chemistry: Syntheses, Properties, and Applications of Organic Dyes and Pigments*, 2nd ed., VCH, Weinheim, Germany, 1991; Peters, A.T. and Freeman, H.W., Eds., *Physico-Chemical Principles of Color Chemistry*, Blackie Academic and Professional, London, 1996; Mason, W.T., *Fluorescent and Luminescent Probes for Biological Activity: A Practical Guide to Technology for Quantitative Real-Time Analysis*, Academic Press, San Diego, CA, 1999; Horobin, R.W. and Kiernan, J.A., Eds., *Conn's Biological Stains: A Handbook of Dyes, Stains, and Fluorochromes for Use in Biology and Medicine*, 10th ed., Bios, Oxford, UK, 2002; Anumula, K.R., Advances in fluorescence derivatization methods for high-performance liquid chromatographic analysis of glycoprotein carbohydrates, *Anal. Biochem.* 350, 1–23, 2006; Waggoner, A., Fluorescent labels for proteomics and genomics, *Curr. Opin. Chem. Biol.* 10, 62–66, 2006; Chen, H., Recent advances in azo dye degrading enzyme research, *Curr. Protein Pept. Sci.* 7, 101–111, 2006; Mondal, K. and Gupta, M.N., The affinity concept in bio-separation: evolving paradigms and expanding range of applications, *Biomol. Eng.* 23, 59–76, 2006.

Ectodomain   The extracellular domain of a transmembrane protein. The proteolysis of the ectodomain regions of specific proteins is described as ectodomain shedding and is catalyzed by ADAM proteases. See Rapraeger, A. and Bernfield, M., Cell surface proteoglycan of mammary epithelial cells. Protease releases a heparan sulfate-rich ectodomain from a putative membrane-anchored domain, *J. Biol. Chem.* 260, 4103–4109, 1985; Johnson, J.D., Wong, M.L., and Rutter, W.J., Properties of the insulin receptor ectodomain, *Proc. Natl. Acad. Sci. USA* 85, 7516–7520, 1988; Schaefer, E.M., Erickson, H.P., Federwisch, M. et al., Structural organization of the human insulin receptor ectodomain, *J. Biol. Chem,* 267, 23393–23402, 1992; Attia, J., Hicks, L., Oikawa, K. et al., Structural properties of the myelin-associated glycoprotein ectodomain, *J. Neurochem.* 61, 718–726, 1993; Couet, J., Sar. S., and Jolviet, A., Shedding of human thyrotropin receptor ectodomain. Involvement of a matrix metalloproteinase, *J. Biol. Chem.* 271, 4545–4552, 1996; Petty, H.R., Kindzelskii, A.L., Adachi, Y. et al., Ectodomain interactions of leukocyte integrins and pro-inflammatory GPI-linked membrane proteins, *J. Pharm. Biomed. Anal.* 15, 1405–1416, 1997; Schlondorff, J. and Blobel, C.P., Metalloprotease-disintegrins: modular proteins capable of promoting cell–cell interactions and triggering signals by protein-ectodomain shedding, *J. Cell Sci.* 112, 3603–3617, 1999; Dello Sbarba, P. and Rovida, E., Transmodulation of cell surface regulatory molecules via ectodomain shedding, *Biol. Chem,* 383, 69–83, 2002; Arribas, J. and Borroto, A., Protein ectodomain shedding, *Chem. Rev.* 102, 4627–4638, 2002; Smalley, D.M. and Ley, K., L-Selectin: mechanisms and physiological significance of ectodomain cleavage, *J. Cell. Mol. Med.* 9, 255–266, 2005; Higashiyama, S. and Nanba, D., ADAM-mediated ectodomain shedding of HB-EGF in receptor cross-talk, *Biochim. Biophys. Acta* 1751, 110–117, 2005; Garton, K.J., Gough, P.J., and Raines, E.W., Emerging roles for ectodomain shedding in the regulation of inflammatory responses, *J. Leuk. Biol.* 79, 1105–1116, 2006.

Electrode Potential (E^o)   The potential measured with an electrode in contact with a solution of its ions. Electrode potential values will predict whether a substance will be reduced or oxidized. Values are usually expressed as a reduction potential ($M^{n+} \rightarrow M$). A positive electrode potential would indicate that reduction is spontaneous. A negative potential for this reaction would suggest that the oxidation reaction ($M \rightarrow M^{n+}$) would be spontaneous.

Electronegativity   The tendency of an atom to pull an electron toward it in a chemical bond; the difference in electronegativity between atoms in a molecule indicates polarity such that in bromoacetic acetamide, it permits an attack on a nucleophile such as cysteine in the protein.

Electrophoresis/ MS   Proteins are separated by one-dimensional, or more often, two-dimensional gel electrophoresis. The separated proteins are subjected to *in situ* tryptic digestion, and the peptides are separated by liquid chromatography and identified by mass spectrometry. See Nishihara, J.C. and Champion, K.M., Quantitative evaluation of proteins in one- and two-dimensional polyacrylamide gels using a fluorescent stain, *Electrophoresis* 23, 2203–2215, 2002.

ELISA	Enzyme-linked immunosorbent assay. An assay based on the reaction of antibody and antigen. There are direct, indirect, direct sandwich, and indirect sandwich assays. See Maggio, E.T., *Enzyme-Immunoassay*, CRC Press, Boca Raton, FL, 1980; Kemeny, D.M. and Challacombe, S.J., *ELISA and Other Solid Phase Immunoassays: Theoretical and Practical Aspects*, Wiley, Chichester, UK, 1988; Kemeny, D.M., *A Practical Guide to ELISA*, Pergamon Press, Oxford, UK, 1991; Kerr, M.A. and Thorpe, R., *Immunochemistry LabFax*, Bios, Oxford, UK, 1994; Law, B., *Immunoassay: A Practical Guide*, Taylor & Francis, London, 1996; Crowther, J.R., *The ELISA Guidebook*, Humana Press, Totowa, NJ, 2001; Burns, R., *Immunochemical Protocols*, Humana Press, Totowa, NJ, 2005.
Elispot	The use of membranes to measure cells secreting a specific product such as an antibody or a cytokine. A membrane (nitrocellulose or PDVF) containing an antibody or other specific binding protein is placed in a microtiter plate. Cells secreting a product, such as a cytokine, are grown in this plate and the secretion of the specific product evaluated in response to stimuli. As product is secreted from an individual cell, it is captured immediately by the antibody or other specific binding protein on the membrane and subsequently detected with a probe. An individual spot then corresponds to the secretion from a single cell. There are a number of instruments designed to measure such spots. See Stot, D.I., Immunoblotting, dot-blotting, and ELISPOT assay: methods and applications, in *Immunochemistry*, van Oss, C.J. and van Regenmortel, M.H.V., Eds., Marcel Dekker, New York, 1994, pp. 925–948; Arvilommi, H., ELISPOT for detecting antibody-secreting cells in response to infections and vaccination, *APMIS* 104, 401–410, 1996; Stott, D.I., Immunoblotting, dot-blotting, and ELISPOT assays: methods and applications, *J. Immunoassay* 21, 273–296, 2000; Anthony, D.D. and Lehmann, P.V., T-cell epitope mapping using the ELISPOT approach, *Methods* 29, 260–269, 2003; Ghanekar, S.A. and Maecker, H.T., Cytokine flow cytometry: multiparametric approach to immune function analysis, *Cytotherapy* 5, 1–6, 2003; Letsch, A. and Scheibenbogen, C., Quantification and characterization of specific T-cells by antigen-specific cytokine production using ELISPOT assay or intracellular cytokine staining, *Methods* 31, 143–149, 2003; Hernandez-Fuentes, M.P., Warrens, A.N., and Lechler, R.I., Immunologic monitoring, *Immunol. Rev.* 196, 247–264, 2003; Kalyuzhny, A.E., Chemistry and biology of the ELISPOT system, *Methods Mol. Biol.* 302, 15–31, 2005; Kalyuzhny, A., *Handbook of ELISPOT: Methods and Protocols*, Humana Press, Totowa, NJ, 2005; Periwal, S.B., Spagna, K., Shahabi, K. et al., Statistical evaluation for detection of peptide-specific interferon-gamma secreting T-cells induced by HIV vaccine determined by ELISPOT assay, *J. Immunol. Methods* 305, 128–134, 2005.
Embedding	Infiltration of a specimen with a liquid medium (paraffin) that can be solidified/ polymerized to form a matrix to support the tissue for subsequent manipulation.
Endocrine	Usually in reference to a hormone or other biological effector such as peptide growth factor or cytokine, which has a systemic effect.
Endoplasmic Reticulum-Associated Protein Degradation (ERAD)	A highly specific pathway for the degradation of misfolded proteins in the endoplasmic reticulum, which serves as a control mechanism for protein synthesis. See Werner, E.D., Brodsky, J.L., and McCracken, A.A., Proteasome-dependent endoplasmic reticulum–associated protein degradation: an unconventional route to a familiar fate, *Proc. Natl. Acad. Sci. USA* 93,

13797–13801, 1996; Yamaski, S., Yagishita, N., Tsuchimochi, K., Nishioka, K., and Nakajima, T., Rheumatoid arthritis as a hyper-endoplasmic reticulum-associated degradation disease, *Arthritis Res. Ther.* 7, 181–186, 2005; Meusser, B., Hirsch, C., Jarosch, E., and Sommer, T., ERAD: the long road to destruction, *Nature Cell Biol.* 7, 766–772, 2005.

Endosome   A physically distinct compartment resulting from the process of endocytosis and isolated from the rest of the cell with a permeable membrane. The endosome provides a pathway for transport of ingested materials to the lysosome. There is particular interest in this pathway for the process of antigen presentation. See Stahl, P. and Schwartz, A.L., Receptor-mediated endocytosis, *J. Clin. Invest.* 77, 657–662, 1986; Wagner, H., Heit, A., Schmitz, F., and Bauer, S., Targeting split vaccines to the endosome improves vaccination, *Curr. Opin. Biotechnol.* 15, 538–542, 2004; Boes, M., Cuvillier, A., and Ploegh, H., Membrane specializations and endosome maturation in dendritic cells and B-cells, *Trends Cell Biol.* 14, 175–183, 2004; Karlsson, L., DM and DO shape the repertoire of peptide-like-MHC-class-II complexes, *Curr. Opin. Immunol.* 17, 65–70, 2005; Li, P., Gregg, J.L., Wang, N. et al., Compartmentalization of class II antigen presentation: contribution of cytoplasmic and endosomal processing, *Immunol. Rev.* 207, 206–217, 2005.

Enhancer   DNA sequences that increase transcription from a linked promoter region
Elements   independent of operation and position (in contrast to proximal promoter elements). Enhancer elements are located at varying distances upstream and downstream of the linked gene. See Hankinson, O., Role of coactivators in transcriptional activation by the aryl hydrocarbon receptor, *Archs. Biochem. Biophys.* 433, 379–386, 2005; West, A.G. and Fraser, P., Remote control of gene transcription, *Hum. Mol. Genet.* 14 (Spec. No. 1), R101–R111, 2005; Sipos, L. and Gyurkovics, H., Long-distance interactions between enhancers and promoters, *FEBS J.* 272, 3253–3259, 2005; Zhao, H. and Dean, A., Organizing the genome: enhancers and insulators, *Biochem. Cell. Biol.* 83, 516–524, 2005.

Ensembl   A database (http://www.ensembl.org) maintained by the European Bioinformatics Institute (EBI). This database organizes large amounts of biological information around the sequences of large genomes. See Baxevanis, A.D., Using genomic databases for sequence-based biological discovery, *Mol. Med.* 9, 185–192, 2003; Birney, E., Andrews, T.D., Bevan, P. et al., An overview of Ensembl, *Genome Res.* 14, 925–928, 2004; Stabenau, A., McVicker, G., Melsopp, C. et al., The Ensembl core software libraries, *Genome Res.* 14, 929–933, 2004; Yanai, I., Korbel, J.O., Boue, S. et al., Similar gene expression profiles do not imply similar tissue functions, *Trends Genet.* 22, 132–138, 2006.

Ensemble Theory A proposition that several discrete compounds (proteins, nucleics, acids, carbohydrates) form a structural whole or functional whole. The term "ensemble" is frequently used to describe the population of discrete intermediates during the process of protein folding. See Dietrich, A., Buschmann, V., Muller, C., and Sauer, M., Fluorescence resonance energy transfer (FRET) and competing processes in donor-acceptor substituted DNA strands: a comparative study of ensemble and single-molecule data, *J. Biotechnol.* 82, 211–231, 2002; Sridevi, K., Lakshmikanth, G.S., Krishnamoorthy, G., and Udgaonkar, J.B., Increasing stability reduces conformational heterogeneity in a protein folding ensemble,

J. Mol. Biol. 337, 699–711, 2004; Thirumalai, D. and Hyeon, C., RNA and protein folding: common themes and variations, *Biochemisry* 44, 4957–4970, 2005.

Enthalpy (ΔHᵒ) This is the energy change or heat of reaction for either synthetic or degradative reaction in the standard state. See *Standard Free Energy.*

Entropy (S) A thermodynamic quantity that is a measure of the "disorder" or randomness in a system. For example, a crystal structure changing to a liquid is associated with an increase in entropy as, for example, the melting of ice crystals forming water under standard conditions. Entropy increases for a spontaneous process. "S" refers to entropy values in standard states of substances.

Eosinophil "Acid" staining leukocyte; associated with allergic inflammation. See Lee, J.J. and Lee, N.A, Eosinophil degranulation: an evolutionary vestige or a universally destructive effector function, *Clin. Exp. Allergy* 35, 986–994, 2005.

Eph Receptors/ Ephrin Eph receptors are the largest family of receptor tyrosine kinases. The structure of Eph receptors is comprised of an extracellular domain and an intracellular domain that are linked by a transmembrane segment. Ephrin ligands bind to Eph receptors, which are classified on the quality of the ephrin ligand; ephrin-A ligands bind to EphA receptors while ephrin-B ligands bind to EphB receptors. Eph receptors and ephrin ligands are integral components of cell surfaces and their interactions mediate growth and development. See Foo, S.S., Turner, C.J., Adams, S. et al., Ephrin-B2 controls cell motility and adhesion during blood-vessel-wall assembly, *Cell* 124, 161–173, 2006; Zhang, J. and Hughes, S., Role of the ephrin and Eph receptor tyrosine kinase families in angiogenesis and development of the cardiovascular system, *J. Pathol.* 208, 453–461, 2006; Haramis, A.P. and Perrakis, A., Selectivity and promiscuity in Eph receptors, *Structure* 14, 169–171, 2006; Chrencik, J.E., Brooun, A., Recht, M.I. et al., Structure and thermodynamic characterization of the EphB4/Ephrin-B2 antagonist peptide complex reveals the determinants for receptor stability, *Structure* 14, 321–330, 2006.

Epistasis Masking of a phenotype caused by mutation of one gene by a mutation in another gene; epistasis analysis can define order of gene expression in a genetic pathway.

Epitome All epitopes present in the antigenic universe; also defined as example, paradigm; a brief presentation or statement in most dictionaries.

Erk 1/2 p42/44 extracellular signal-regulated kinase, phosphorylated as a result of GPCR activation. A number of GPCR appear to converge at Erk 1/2. See Dhillon, A.S. and Kolch, W., Untying the regulation of the Raf-1 kinase, *Arch. Biochem. Biophys.* 404, 3–9, 2002; Chu, C.T., Levinthal, D.J., Kulich, S.M. et al., Oxidative neuronal injury. The dark side of ERK 1/2, *Eur. J. Biochem.* 271, 2060–2066, 2004; Clark, M.J. and Traynor, J.R., Assays for G-protein-coupled receptor signaling using RGS-insensitive Galpha subunits, *Methods Enzymol.* 389, 155–169, 2004; Clark, A. and Sugden, P.M., Signaling through the extracellular signal-regulated kinase 1/2 cascade in cardiac myocytes, *Biochem. Cell Biol.* 82, 603–609, 2004.

Essential Oils A heterogeneous mixture of lipophilic substances obtained from a plant. Also referred to as absolute oils. Originally referred to as the steam distillate of the rinds of certain citrus fruits but extends for more recently used materials such as tea tree oil, which is suggested to have some pharmacological use.

These products are also used in aromatherapy. See Ranganna, S., Govin-
darajan, V.S., and Ramana, K.V., Citrus fruits — varieties, chemistry,
technology, and quality evaluation. A chemistry, *Crit. Rev. Food Sci. Nutri.*
18, 313–386, 1983; Kalemba, D. and Kunicka, A., Antibacterial and antifungal properties of essential oils, *Curr. Med. Chem.* 10, 813–829, 2003;
Halcon, L. and Milkus, K., *Staphyloccus aureus* and wounds: a review of
tea tree oil as a promising antimicrobial, *Amer. J. Infect. Control* 32,
402–408, 2004.

EUROFAN   EUROFAN (European Functional Analysis Network) was established to
elucidate the physiological and biochemical functions of open reading
frames in yeast; http://mips.gsf.de/proj/eurofan/. See Sanchez, J.C., Golaz,
O., Frutiger, S. et al., The yeast SWISS-2DPAGE database, *Electrophoresis*
17, 556–565, 1996; Dujon, B., European Functional Analysis Network
(EUROFAN) and the functional analysis of the *Saccharomyces cerevisiae*
genome, *Electrophoresis* 19, 617–624, 1998; Bianchi, M.M., Ngo, S.,
Vandenbol, M. et al., Large-scale phenotypic analysis reveals identical
contributions to cell functions of known and unknown yeast genes, *Yeast*
18, 1397–1412, 2001; Avaro, S., Belgareh, N., Sibella-Arguelles, C. et al.,
Mutants defective in secretory/vacuolar pathways in the EUROFAN collection of yeast disruptants, *Yeast* 19, 351–371, 2002; Castrillo, J.I., Hayes,
A., Mohammed, S., Gaskell, S.J., and Oliver, S.G., An optimized protocol for metabolome analysis in yeast using direct infusion electrospray
mass spectrometry, *Phytochemistry* 62, 929–937, 2003; Davydenko,
S.G., Juselius, J.K., Munder, T. et al., Screening for novel essential genes
of *Saccharomyces cerevisiae* involved in protein secretion, *Yeast* 21,
463–471, 2004.

Eutectic   A mixture of components in such proportions that said mixture melts and
solidifies at a single temperature lower than the melting points of the constituents or any other mixture thereof; a minimum transformation temperature
between a solid solution and a mechanical mixture. This is an issue with
cryobiology and therapeutic protein processing processes such as
lyophilization. See Gutierrez-Merino, C., Quantitation of the Forster
energy transfer for two-dimensional systems. II. Protein distribution and
aggregation state in biological membranes, *Biophys. Chem.* 14, 259–266,
1981; Gatlin, L.A. and Nail, S.L., Protein purification process engineering.
Freeze drying: a practical overview, *Bioprocess Technol.* 18, 317–367, 1994;
Nail, S.L., Jiang, S., Chongprasert, S., and Knopp, S.A., Fundamentals of
freeze-drying, *Pharm. Biotechnol.* 14, 281–360, 2002; Han, B., and Bischof,
J.C., Thermodynamic nonequilibrium phase change behavior and thermal
properties of biological solutions for cryobiology applications, *J. Biomech.
Eng.* 126, 196–203, 2004.

Exosome   A precise definition is a work in process, but an exosome can be considered
to be an intracellular membrane vesicle derived from fusion of endosomes
with the plasma membrane. It is suggested that exosomes are involved in
the intracellular transfer of molecules. See Févier, B. and Raposo, G.,
Exosomes: endosomal-derived vesicles shipping extracellular messages,
Curr. Opin. Cell Biol. 16, 415–421, 2004; de Gassart, A., Géminard, C.,
Hoekstra, D., and Vidal, M., Exosome secretion: the art of reutilizing
nonrecycled proteins? *Traffic* 5, 896–903, 2004; Chaput, N., Taïeb, J.,
Schartz, N. et al., The potential of exosomes in immunotherapy of cancer,
Blood Cells, Mol. Dis. 35, 111–115, 2005; Seaman, M.N.J., Recycle your

receptors with retromer, *Trends Cell Biol.* 15, 68–75. 2005; Lencer, W.I. and Blumberg, R.S., A passionate kiss, then run: exocytosis and recycling of IgG by FcRn, *Trends Cell Biol.* 15, 5–9, 2005.

Exotoxico-genomics Study of the expression of genes important in adaptive responses to toxic exposures.

Expansins Family of plant proteins essential for acid-induced cell wall loosening. See Cosgrove, D.J., Relaxation in a high-stress environment: the molecular basis of extensible cell walls and cell enlargement, *Plant Cell* 9, 1031–1041, 1997.

Expressed Sequence Tag Usually an incomplete DNA sequence, which can be "read" from either end of a gene fragment and which is used as a "marker" or a "window" of gene presence in a genome; a short strand of DNA (approximately 200 base pairs long) that is usually unique to a specific cDNA and therefore can be used to identify genes and map their positions in a genome. See Wilcox, A.S., Khan, A., Hopkins, J.A., and Sikela J.M., Use of 3′ untranslated sequences of human cDNA for rapid chromosome assignment and conversion to STS's: implications for an expression map of the genome, *Nucl. Acid Res.* 19, 1837–1842, 1991; Hartl, D.L., EST!EST!!EST!!! *Bioessays* 18, 1021–1023, 1996; Gerhold, D. and Caskey, C.T., It's the genes! EST access to human genome content, *Bioessays* 18, 973–981, 1996; Robson, P., The maturing of the human embryonic stem cell transcriptome profile, *Trends Biotech.* 22, 609–612, 2004; Hoffman, M., Gene expression patterns in human and mouse B-cell development, *Curr. Top. Microbiol. Immunol.* 294, 19–29, 2005.

Expression Profiling The measurement or determination of DNA expression by the measurement of RNA (transcriptomics); also used to refer to protein expression as determined by proteomic technology.

Expressional Leakage A concept where the functionally important expression of one gene can result in the ectopic expression of a neighboring gene, resulting in apparent expression similarity between tissues. See de Marco, A. and de Marco, V., Bacteria co-transformed with recombinant proteins and chaperones cloned in independent plasmids are suitable for expression tuning, *J. Biotechnol.* 109, 45–52, 2004; Yanai, I., Korbel, J.O., Boue, S. et al., Similar gene expression profiles do not imply similar tissue functions, *Trends Genet.* 22, 132–138, 2006.

Families of Structurally Similar Proteins (FSSP) A database based on three-dimensional comparisons of protein structures: http://ekhidna.biocenter.helsinki.fi/dali/start. See Holm, L., Ouzounis, C., Sander, C., Tuparev, G., and Vriend, G., A database of protein structure families with common folding motifs, *Protein Sci.* 1, 1691–1698, 1992; Holm, L. and Sander, C., The FSSP database: fold classification based on structure alignment of proteins, *Nucleic Acids Res.* 24, 206–209, 1996; Notredame, C., Holm, L., and Higgins, D.G., COFFEE: an objective function for multiple sequence alignments, *Bioinformatics* 14, 407–422, 1998; Hadley, C. and Jones, D.T., A systematic comparison of protein structure classifications: SCOP, CATH, and FSSP, *Structure* 7, 1099–1112, 1999; Getz, G., Vendruscolo, M., Sachs, D., and Domany, E., Automated assignment of SCOP and CATH protein structure classifications from FSSP scores, *Proteins* 46, 405–415, 2002; Edgar, R.C. and Sjolander, K., A comparison of scoring functions for protein sequence profile alignment, *Bioinformatics* 20, 1301–1308, 2004; Edgar, R.C. and Sjolander, K., COACH: profile–profile alignment of protein families using hidden Markov models, *Bioinformatics* 20, 1309–1318, 2004.

Fenton Reaction Ferrous ion-dependent formation of hydroxyl radical from hydrogen peroxide; can be coupled with the oxidation of hydroxyl function to ketone/aldehydes. See Fenton, H.J.H., Oxidation of certain organic acids in the presence of ferrous salts, *Proc. Chem. Soc.* 15, 224–228, 1899; Goldstein, S., Meyerstein, D., and Czapski, G., The Fenton reagents, *Free Rad. Biol. Med.* 15, 435–445, 1993; Stadtman, E.R., Role of oxidized amino acids in protein breakdown and stability, *Meth. Enzymol.* 258, 379–393, 1995; Odyuo, M.M. and Sharan, R.N., Differential DNA strand breaking abilities of OH and ROS generating radiomimetic chemicals and γ-rays: study of plasmid dNA, pMTa4, *in vitro, Free Rad. Res.* 39, 499–505, 2005.

Fermentation The controlled aerobic or anaerobic process where a product is produced by yeast, molds, or bacteria from a substrate. Historically, fermentation was used to describe the action of a leavan (yeast) on a carbohydrate (saccharine) as in the production of beers and wines or a dough such as in making bread. In biotechnology manufacturing, fermentation is used to describe the product of a biopharmaceutical by yeast or bacteria while the term cell culture is used to describe the use of animal cells or plant cells. See Wiseman, A., *Principles of Biotechnology*, Chapman and Hall, New York, 1983; Sinclair, C.G., Kristiansen, B., and Bu'Lock, L.D., *Fermentation Kinetics and Modeling*, Open University Press, New York, 1987; Flickinger, M.C. and Drew, S.W., Eds., *The Encyclopedia of Bioprocess Technology*, John Wiley & Sons, New York, 1999; Walker, J.M. and Rapley, R., Eds., *Molecular Biology and Biotechnology*, Royal Society of Chemistry, Cambridge, UK, 2000; Badal, S.C., Ed., *Fermentation Biotechnology*, American Chemical Society, Washington, DC, 2003.

Ferredoxin A small protein that functions in the transport of electrons (reducing potential) in a variety of organisms. There are several classes of ferredoxins based on the nature of the chemistry of iron binding: Fe_2S_2, Fe_3S_4, Fe_4S_4. The iron is bound to cysteine residues in a cluster that also contains inorganic sulfur. See Mortenson, L.E., Nitrogen fixation: role of ferredoxin in anaerobic metabolism, *Annu. Rev. Microbiol.* 17, 115–138, 1963; Knaff, D.B., and Hirasawa, M., Ferredoxin-dependent chloroplast enzymes, *Biochim. Biophys. Acta* 1056, 93–125, 1991; Dai, S., Schwendtmayer, C., Johansson, K. et al., How does light regulate chloroplast enzymes? Structure-function studies of the ferredoxin/thioredoxin system, *Q. Rev. Biophys.* 33, 67–108, 2000; Schurmann, P., Redox signaling in the chloroplast: the ferredoxin/thioredoxin system, *Antioxid. Redox. Signal.* 5, 69–78, 2003; Carrillo, N., and Ceccarelli, E.A., Open questions in ferredoxin-NADP$^+$ reductase catalytic mechanism, *Eur. J. Biochem.* 270, 1900–1915, 2003; Karplus, P.A. and Faber, H.R., Structural aspects of plant ferredoxin: NADP$^+$ oxidoreductases, *Photosynth. Res.* 81, 303–315, 2004; Glastas, P., Pinotsis, N., Efthymiou, G. et al., The structure of the 2[4Fe-4S] ferredoxin from *Pseudomonas aeruginosa* at 1.32-A resolution: comparison with other high-resolution structures of ferredoxins and contributing structural features to reduction potential values, *J. Biol. Inorg. Chem.* 11, 445–458, 2006; Eckardt, N.A., Ferredoxin-thioredoxin system plays a key role in plant response to oxidative stress, *Plant Cell* 18, 1782, 2006.

Ferret Diameter The longest chord of the project of a regular or irregular object at specific angles. Maximum, minimum, and average Ferret diameters can be determined by successive measurements; a value used in particle characterization.

See M. Levin, Particle characterization — tools and methods, *Lab. Equip.,* Nov., 2005.

Fibrillation Forming fibers from small, soluble polymeric materials. This is observed with amyloid fibrils in Alzheimer's disease and with proteins during pharmaceutical processing. The term fibrillation was used in the nineteenth century to describe the physical changes in blood before the elucidation of fibrinogen clotting. Fibrillation is also used to describe physical changes in structural materials with ligaments and tendons. See Arvinte, T., Cudd, A., and Drake, A.F., The structure and mechanism of formation of calcitonin fibrils, *J. Biol. Chem.* 268, 6415–6422, 1993; Ghosh, P. and Smith, M., The role of cartilage-derived antigens, pro-coagulant activity, and fibrinolysis in the pathogenesis of osteoarthritis, *Med. Hypotheses* 41, 190–194, 1993; Bronfman, F.C., Garrido, J., Alvarez, A., Morgan, C., and Inestrosa, N.C., Laminin inhibits amyloid-beta-peptide fibrillation, *Neurosci. Lett.* 218, 201–203, 1996; Martin, J.A. and Buckwalter, J.A., Roles of articular cartilage aging and chondrocytes senescence in the pathogenesis of osteoarthritis, *Iowa Orthop. J.* 21, 1–7, 2001; Seyferth, S. and Lee, G., Structural studies of EDTA-induced fibrillation of salmon calcitonin, *Pharm. Res.* 20, 73–80, 2003; Librizzi, F. and Rischel, C., The kinetic behavior of insulin fibrillation is determined by heterogeneous nucleation pathways, *Protein Sci.* 14, 3129–3134, 2005; Westermark, P., Aspects on human amyloid forms and their fibril polypeptides, *FEBS J.* 272, 5942–5949, 2005; Pedersen, J.S., Dikov, D., Flink, J.L. et al., The changing face of glucagon fibrillation: structural polymorphism and conformational imprinting, *J. Mol. Biol.* 355, 501–523, 2006.

Fibroblast Growth Factor A group of peptide growth factors which regulate cell growth and proliferation and wound healing. There are more than 20 fibroblast growth factors including acidic and basic fibroblast growth factors. See Baird, A. and Klagsbrun, M., *The Fibroblast Growth Factor Family,* New York Academy of Sciences, New York, NY, USA, 1991; Barnes, D.W. and Mather, J.P., *Peptide Growth Factors Part C*, Academic Press, San Diego, CA, USA, 1991; Nielsen-Hamilton, M., *Growth Factors and Signal Transduction in Development*, Wiley-Liss, New York, NY, USA, 1994. See also Gospodarowicz, D. and Mescher, A.L., Fibroblast growth factor and the control of vertebrate regeneration and repair, *Ann. N.Y. Acad. Sci.* 339, 151–174, 1980; Schweigerer, L., Basic fibroblast growth factor as a wound healing hormone, *Trends Pharmacol. Sci.* 9, 427–428, 1988; Rifkin, D.B. and Moscatelli, D., Recent developments in the cell biology of basic fibroblast growth factor, *J. Cell Biol.* 109, 1–6, 1989; Burgess, W.H., Structure-function studies of acidic fibroblast growth factor, *Ann. N.Y. Acad. Sci.* 638, 89–97, 1991; Turnbull, J.E. and Gallagher, J.T., Heparan sulphate: functional role as a modulator of fibroblast growth factor activity, *Biochem. Soc. Trans.* 21, 477–482, 1993; Wang, Y.J., Shahrokh, E., Vemuri, S., et al., Characterization, stability, and formulation of basic fibroblast growth factor, *Pharm. Biotechnol.* 9, 141–180, 1996; Faham, S., Linhardt, R.J., and Rees, D.C., Diversity does make a difference: fibroblast growth factor-heparin interactions, *Curr. Opin. Struct. Biol.* 8, 578–586, 1998; Goldfarb, M., Fibroblast growth factor homologous factors: evolution, structure, and function, *Cytokine Growth Factor Rev.* 16, 215–220, 2005; Harmer, N.J., Insights into the role of heparan sulphate in fibroblast growth signalling, *Biochem. Soc. Trans.* 34, 442–445, 2006.

Fibroblast Growth Factor Receptor(s)   Receptor kinases (usually tyrosine kinases) which are activated by dimerization after ligand binding; Include FGFR1, FGFR2, FGFR3, FGFR4, FGFR5. Frequently functions in concert with cell-bound heparan sulfate although circulating heparan/heparin can bind synergistically with fibroblast growth factors. See Jaye, M. Schlessinger, J., and Dionne, C.A., Fibroblast growth factor receptor tyrosine kinases: molecular analysis and signal transduction, *Biochim. Biophys. Acta* 1135, 185–199, 1992; McKeehan, W.L. and Kan, M., Heparan sulfate fibroblast growth factor receptor complex: structure-function relationships, *Mol. Reprod. Dev.* 39, 69–81, 1994; De Moerlooze, L. and Dickson, C., Skeletal disorders associated with fibroblast growth factor receptor mutations, *Curr. Opin. Genet. Dev.* 7, 378–385, 1997; Friesel, R. and Maciag, T., Fibroblast growth factor prototype release and fibroblast growth factor receptor signaling, *Thromb. Haemost.* 82, 748–754, 1999; Manetti, F. and Botta, M., Small-molecule inhibitors of fibroblast growth factor receptor (FGFR) tyrosine kinases (TK), *Curr. Pharm. Des.* 9, 567–581, 2003; Mohammadi, M., Olsen, T.K., and Ibrahimi, O.A., Structural basis for fibroblast growth factor receptor activation, *Cytokine Growth Factor Rev.* 16, 107–137, 2005; Hung, K.W., Kumar, T.K., Kathir, K.M., et al., Solution structure of the ligand binding domain of the fibroblast growth factor receptor: role of heparin in the activation of the receptor, *Biochemistry* 44, 15787–15798, 2005; Kyu, E.K., Cho, K.J., Kim, J.K., et al., Expression and purification of recombinant human fibroblast growth factor receptor in *Escherichia coli*, *Protein Expr. Purif.* 49, 15–22, 2006; Duchesne, L., Tissot, B., Rudd, T.R., N-glycosylation o fibroblast growth factor receptor 1 regulates ligand and heparan sulfate co-receptor binding, *J. Biol. Chem.* 281, 27178–27189, 2006; Miliette, E., Rauch, B.R., Kenagy, R.D., et al., Platelet-derived growth factor-BB trans-activates the fibroblast growth factor receptor to induce proliferation in human smooth muscle cells, *Trends Cardiovasc. Med.* 16, 25–28. 2006.

FixJ-FixL   A two-component transcription regulatory system that is a global regulator of nitrogen fixation in *Rhizobium meliloti*. See Kahn, D. and Ditta, G., Molecular structure of Fix J: homology of the transcriptional activator domain with the −35 binding domain of sigma factors, *Mol. Microbiol.* 5, 987–997, 1991; Sousa, E.H.S., Gonzalez, G., and Gilles-Gonazalez, M.A., Oxygen blocks the reaction of FixL-FixJ complex with ATP but does not influence binding of FixJ or ATP to FixL, *Biochemistry* 44, 15359–15365, 2005.

FLAG™   FLAG™ has the sequence of AspTyrLysAspAspAsp-AspLys, which includes an enterokinase cleavage site. This epitope tag can be used as a fusion partner for the expression and purification of recombinant proteins. See Einhauer, A. and Jungbauer, A., The FLAG™ peptide, a versatile fusion tag for the purification of recombinant proteins, *J. Biochem. Biophys. Methods* 49, 455–465, 2001; Terpe, K., Overview of tag protein fusions: from molecular and biochemical fundamentals to commercial systems, *Appl. Microbiol. Biotechnol.* 60, 523–533, 2003; Lichty, J.J., Malecki, J.L., Agnew, H.D., Michelson-Horowitz, D.J., and Tan, S., Comparison of affinity tags for protein purification, *Protein Exp. Purif.* 41, 98–105, 2005.

Flap-Endonuclease (FEN; FEN-1)   An enzyme with endonuclease and exonuclease activity encoded by the *FEN-1* gene. Cleaves branched DNA structures including the 5′ end of Okazaki fragments. See Kunkel, T.A., Resnick, M.A., and Gordenin, D.A., Mutator specificity and disease: looking over the FENce, *Cell* 88, 155–158, 1997;

Shen, B., Qiu, J., Hosfield, D., and Tainer, J.A., Flap endonuclease homologs in archaebacteria exist as independent proteins, *Trends Biochem. Sci.* 23, 171–173, 1998; Henneke, G., Freidrich-Heineken, E., and Hubscher, U., Flap endonuclease 1: a novel tumor suppressor protein, *Trends Biochem. Sci.* 28, 384–390, 2003; Kao, H.I. and Bambara, R.A., The protein components and mechanism of eukaryotic Okazaki fragment maturation, *Crit. Rev. Biochem. Mol. Biol.* 38, 433–452, 2003; Garg, P. and Burgers, P.M., DNA polymerases that propagate the eukaryotic DNA replication fork, *Crit. Rev. Biochem. Mol. Biol.* 40, 115–128, 2005; Olivier, M., The invader assay for SNF genotyping, *Mutat. Res.* 573, 103–110, 2005; Shen, B., Singh, P., Liu, R. et al., Multiple but dissectible functions of FEN-1 nucleases in nucleic acid processing, genome stability, and diseases, *Bioessays* 27, 717–729, 2005.

Flux   Flux is the continuous flow of a substance. Flux can occur with electrons (Gutman, M., Electron flux through the mitochondrial ubiquinone, *Biochim. Biophys. Acta* 594, 53–84, 1980) and protons (Wang, J.H., Coupling of proton flux to the hydrolysis and synthesis of ATP, *Annu. Rev. Biophys. Bioeng.* 12, 21–34, 1983) as well as with ions and other substances. See Schwartz, A., Cell membrane Na^+, K^+-ATPase, and sarcoplasmic reticulum: possible regulators of intracellular ion activity, *Fed. Proc.* 35, 1279–1282, 1976; Mukohata, Y. and Packer, L., Eds., *Cation Flux across Biomembranes*, Academic Press, New York, 1979; Meissner, G., Monovalent ion and calcium ion fluxes in sarcoplasmic reticulum, *Mol. Cell. Biochem.* 55, 65–82, 1983; Jones, D.P., Intracellular diffusion gradients of O_2 and ATP, *Am. J. Physiol.* 250, C663–C675, 1986; Hunter, M., Kawahara, K., and Giebisch, G., Calcium-activated epithelial potassium channels, *Miner. Electrolyte Metab.* 14, 48–57, 1988; Weir, E.K. and Hume, J.R., *Ion Flux in Pulmonary Vascular Control*, Plenum Press, New York, 1993. Flux is defined in several ways: unidirectional influx is defined as the molar quantity of a solute passing across 1 cm^2 membrane in a unit period of time; unidirectional efflux is defined as the molar quantity of a solute crossing 1 cm^2 membrane outward from a cell in a unit period of time. Net flux is the difference between unidirectional influx and unidirectional efflux in a unit period of time. Understanding net flux is of importance in the design and interpretation of microdialysis studies (Schuck, V.J., Rinas, I., and Derendorf, H., *In vitro* microdialysis sample of docetaxel, *J. Pharm. Biomed. Anal.* 36, 807–813, 2004; Cano-Cebrian, M.J., Zornoza, T., Polache, A., and Granero, L., Quantitative *in vivo* microdialysis in pharmacokinetic studies: some reminders, *Curr. Drug. Metab.* 6, 83–90, 2005; Abrahamsson, P. and Winso, O., An assessment of calibration and performance of the microdialysis system, *J. Pharm. Biomed. Anal.* 39, 730–734, 2005).

Focal Adhesion   A membrane area for cellular adhesion via actin filaments to the extracellular matrix/fibronectin resulting from the clustering of integrins. The interaction with fibronectin results in the formation of fibrillar adhesions considered to be more mature structures. Other intracellular proteins such as vincullin and focal adhesion kinase (FAK) are recruited to the actin cytoskeleton structure. See Otey, C.A. and Burridge, K., Patterning of the membrane cytoskeleton by the extracellular matrix, *Semin. Cell Biol.* 1, 391–399, 1990; Arikama, S.K., Integrins in cell adhesion and signaling, *Hum. Cell* 9, 181–186, 1996; Bershadsky, A.D., Balaban, N.Q., and Geiger, B., Adhesion-dependent cell mechanosensitivity, *Annu. Rev. Cell Dev.*

Biol. 19, 677–695, 2003; Wozniak, M.A., Modzelekska, K., Kwong, L., and Keeley, P.J., Focal adhesion regulation of cell behavior, *Biochim. Biophys. Acta* 1692, 103–119, 2004; Small, J.V. and Resch, G.P., The comings and goings of actin: coupling protrusion and retraction in cell motility, *Curr. Opin. Cell Biol.* 17, 517–523, 2005; Wu, M.H., Endothelial focal adhesions and barrier function, *J. Physiol.* 569, 359–366, 2005; Cohen, L.A. and Guan, J.L., Mechanisms of focal adhesion kinase regulation, *Curr. Cancer Drug Targets* 5, 629–643, 2005; Romer, L.H., Birukov, K.G., and Garcia, J.G., Focal adhesions: paradigm for a signaling nexus, *Circ. Res.* 98, 606–616, 2006; Legate, K.R., Montañez, O., and Fässler, R., ILK, PINCH, and parvin: the tIPP of integrin signaling, *Nat. Rev. Mol. Cell. Biol.* 7, 20–31, 2006.

Fok1 Restriction Endonuclease   A type II restriction endonuclease isolated from *Flavobacterium okeanokoites*, which has been used to identify DNA polymorphisms. There has been extensive use in the study of the vitamin D receptor gene (VDR gene). See Sugisaki, H. and Kanazawa, S., New restriction endonucleases from *Flavobacterium okeanokoites* (FokI) and *Micrococcus luteus* (MluI), *Gene* 16, 73–78, 1981; Kato, A., Yakura, K., and Tanifuji, S., Sequence analysis of *Vicia faba* repeated DNA, the FokI repest element, *Nucleic Acids Res.* 24, 6415–6426, 1984; Posfai, G. and Szybalski, W., A simple method for locating methylated bases in DNA using class-IIS restriction enzymes, *Gene* 74, 179–181, 1988; Kita, K., Kotani, H., Sugisaki, H., and Tanami, M., The foci restriction-modification system. I. Organization and nucleotide sequences of the restriction and modification genes, *J. Biol. Chem.* 264, 5751–5756, 1989; Aggarwal, A.K. and Wah, D.A., Novel site-specific DNA endonucleases, *Curr. Opin. Struct. Biol.* 8, 19–25, 1998; Kovall, R.A. and Matthews, B.W., Type II restriction endonucleases: structural, functional, and evolutionary relationships, *Curr. Opin. Chem. Biol.* 3, 578–583, 1999; Akar, A., Orkunoglu, F.E., Ozata, M., Sengul, A., and Gur, A.R., Lack of association between vitamin D receptor FokI polymorphism and alopecia areata, *Eur. J. Dermatol.* 14, 156–158, 2004; Guy, M., Lowe, L.C., Bretherton-Watt, D. et al., Vitamin D receptor gene polymorphisms and breast cancer risk, *Clin. Cancer Res.* 10, 5472–5481, 2004; Claassen, M., Nouwen, J., Fang, Y. et al., *Staphylococcus aureus* nasal carriage is not associated with known polymorphism in the Vitamin D receptor gene, *FEMS Immunol. Med. Microbiol.* 43, 173–176, 2005; Bolu, S.E., Orkunoglu Suer, F.E., Deniz, F. et al., The vitamin D receptor foci start codon polymorphism and bone mineral density in male hypogonadotrophic hypogonadism, *J. Endocrinol. Invest.* 28, 810–814, 2005.

Foldamers   Single chain polymers that can adopt a secondary structure in solution and thus mimic proteins, nucleic acids, and polysaccharides; polymeric backbones have well-defined and predictable folding properties in the solvent of choice. See Appella, D.H., Christianson, L.A., Klein, D.A. et al., Residue-based control of helix shape in beta-peptide oligomers, *Nature* 387, 381–384, 1997; Tanatani, A., Mio, M.J., and Moore, J.S., Chain length-dependent affinity of helical foldamers for a rodlike guest, *J. Amer. Chem. Soc.* 123, 1792–1793, 2001; Cubberley, M.S. and Iverson, B.L., Models of higher-order structure: foldamers and beyond, *Curr. Opin. Chem. Biol.* 5, 650–653, 2001; Hill, D.J., Mio, M.J., Prince, R.B., Hughes, T.S., and Moore, J.S., A field guide to foldamers, *Chem. Rev.* 101, 393–4012, 2001; Martinek, T.A. and Fulop, F., Side-chain control of beta-peptide secondary structures, *Eur. J. Biochem.* 270, 3657–3666, 2003; Sanford, A.R., Yamato, K., Yang, X. et al., Well-defined secondary structures, *Eur. J. Biochem.* 271, 1416–1425, 2004; Cheng, R.P., Beyond *de novo* protein design — *de novo* design of nonnatural folded oligomers, *Curr. Opin. Struct. Biol.* 14, 512–520, 2004; Stone, M.T., Heemstra, J.M., and Moore, J.S., The chain-length dependence test, *Acc. Chem. Res.* 39, 11–20, 2006; Schmitt, M.A., Choi, S.H., Guzei, I.A., and Gellman, S.H., New helical foldamers: heterogeneous backbones with 1:2 and 2:1 alpha:beta-amino acid residue patterns, *J. Am. Chem. Soc.* 128, 4538–4539, 2006.

Fragnomics The use of smaller molecules in the drug discovery process. See Zartler, E.R. and Shapiro, M.J., Fragnomics: fragment-based drug discovery, *Curr. Opin. Chem. Biol.* 9, 366–370, 2005.

Frass Debris or excrement produced by insects. This material is thought to be involved with the role of cockroaches in the development of asthma. See Page, K., Hughes, V.S., Bennett, G.W., and Wong, H.R., German cockroach proteases regulate matrix metalloproteinase-9 in human bronchial epithelial cells, *Allergy* 61, 988–995, 2006.

Free Radical/ Radical A molecule containing an unpaired electron; can be electrically neutral. Free radicals may be created by the hemolytic cleavage of a precursor molecule. Free radicals can be formed by thermolytic cleavage, photolysis (ultraviolet light photolysis of hydrogen peroxide to form hydroxyl radicals), radiolysis (ionizing radiation of water to form hydroxyl radicals), or by homolytic cleavage with the participation of another molecule (i.e., Fenton reaction). Perkins, J., *Radical Chemistry: The Fundamentals*, Oxford University Press, Oxford, UK, 2000.

FRET (Fluorescence Resonance Energy Transfer) A technique for assaying the proximity of region by observed energy transfer between fluorophores. A concept similar to fluorescence quenching. With two-photon excitation, studies can be extended to the study of *in vivo* interactions with microscopy. See Zal, T. and Gascoigne, N.R., Using live FRET imaging to reveal early protein–protein interactions during T-cell activation, *Curr. Opin. Immunol.* 16, 674–683, 2004; Milligan, G. and Bouvier, M., Methods to monitor the quaternary structure of G-protein-coupled receptors, *FASEB J.* 272, 2914–2925, 2005; Rasnik, I., McKinney, S.A., and Ha, T., Surfaces and orientations: much to FRET about? *Acc. Chem. Res.* 38, 542–548, 2005; Gertler, A., Biener, E., Ramamujan, K.V., Dijiane, J., and Herman, B., Fluorescence resonance energy transfer (FRET) microscopy in living cells as a novel tool for the study of cytokine action, *J. Dairy Res.* 72 (Spec. No.), 14–19, 2005; Cudakov, D.M., Lukyanov, S., and Lukyanov, K.A., Fluorescent proteins as a toolkit for *in vivo* imaging, *Trends Biotechnol.* 23, 605–613, 2005. See also *BRET*.

Freund's Adjuvant A mixture of killed/lyophilized *Mycobacterium bovis* or *Mycobacterium tuberculosis* cells and oil resulting in an emulsion (referred to as complete Freund's adjuvant) used with an antigen to improve the immune response (antibody formation secondary to B-cell activation). Incomplete Freund's adjuvant does not contain the bacterial cells and is used to avoid an inflammatory response. See White, R.G., Factor affecting the antibody response, *Br. Med. Bull.* 19, 207–213, 1963; White, R.G., Antigen adjuvants, *Mod. Trends Immunol.* 2, 28–52, 1967; Myrvik, Q.N., Adjuvants, *Ann. N.Y. Acad. Sci.* 221, 324–330, 1974; Osebold, J.W., Mechanisms for action by immunologic adjuvants, *J. Am. Vet. Med. Assoc.* 181, 983–987, 1982; Warren, H.S., Vogel, F.R., and Chedid, L.A., Current status of immunological adjuvants, *Annu. Rev. Immunol.* 4, 369–388, 1986; Claassen, E., de Leeuw, W., de Greeve, P., Hendriksen, C., and Boersma, W., Freund's complete adjuvant: an effective but disagreeable formula, *Res. Immunol.* 143, 478–483, 1992; Billiau, A. and Matthys, P., Modes of action of Freund's adjuvants in experimental models of autoimmune diseases, *J. Leukoc. Biol.* 70, 849–860, 2001; Cachia, P.J., Kao, D.J., and Hodges, R.S., Synthetic peptide vaccine development: measurement of polyclonal antibody affinity and cross-reactivity using a new peptide capture and release system for surface plasmon resonance spectroscopy, *J. Mol. Recog.* 17, 540–557, 2004; Stills, H.F., Jr., Adjuvants and antibody production: dispelling the myths associated with Freund's complete and other adjuvants, *ILAR J.* 46, 280–293, 2005; Miller, L.H., Saul, A., and Mahanty, S., Revisiting Freund's incomplete adjuvant for vaccines in the developing world, *Trends Paristol.* 21, 412–414, 2005.

Functional Genomics   Functional genomics refers to establishing a verifiable link between gene expression and cell/organ/tissue function/dysfunction. See Evans, M.J., Carlton, M.B., and Russ, A.P., Gene trapping and functional genomics, *Trends Genet.* 13, 370–374, 1997; Schena, M., Heller, R.A., Theriault, T.P. et al., Microarrays: biotechnology's discovery platform for functional genomics, *Trends Biotechnol.* 16, 301–306, 1998; Hunt, S.P. and Livesey, R., *Functional Genomics: A Practical Approach,* Oxford University Press, Oxford, UK, 2000; Holtorf, H., Guitton, M.C., and Reski, R., Plant functional genomics, *Naturewissenschaften* 89, 235–249, 2002; Bader, G.D., Heilbut, A., Andrews, B. et al., Functional genomics and proteomics: charting a multidimensional map of the yeast cell, *Trends Cell Sci. Biol.* 13, 344–356, 2003; Kemmeren, P. and Holstege, F.C., Integrating functional genomics data, *Biochem. Soc. Trans.* 31, 1484–1487, 2003; Brownstein, M.J. and Khodursky, A.B., Eds., *Functional Genomics: Methods and Protocols*, Humana Press, Totowa, NJ, 2003; Grotewold, E., *Plant Functional Genomics,* Humana Press, Totowa, NJ, 2003; Zhou, J., *Microbial Functional Genomics*, Wiley-Liss, Hoboken, NJ, 2004; Werner, T., Proteomics and regulomics: the yin and yang of functional genomics, *Mass Spectrom. Rev.* 23, 25–33, 2004; Brunner, A.M., Busov, V.B., and Strauss, S.H., Poplar genome sequence: functional genomics in an ecologically dominant plant species, *Trends Plant Sci.* 9, 49–56, 2004; Hughes, T.R., Robinson, M.D., Mitsakakis, N., and Johnston, M., The promise of functional genomics: completing the encyclopedia of a cell, *Curr. Opin. Microbiol.* 7, 546–554, 2004; Kramer, R. and Cohen, D., Functional genomics to new drug targets, *Nat. Rev. Drug Discov.* 3, 965–972, 2004; Vanhecke, D. and Janitz, M., Functional genomics using high-throughput RNA interference, *Drug Discov. Today* 10, 205–212, 2005; Sauer, S., Lange, B.M., Gobom, J. et al., Miniaturization in functional genomics and proteomics, *Nat. Rev. Genet.* 6, 465–476, 2005; Stoeckert, C.J., Jr., Functional genomic databases on the web, *Cell Microbiol.* 7, 1053–1059, 2005; Foti, M., Grannuci, F., Pelizzola, M. et al., Dendritic cells in pathogen recognition and induction of immune response: a functional genomics approach, *J. Leukoc. Biol.* 79, 913–916, 2006.

Functional Proteomics   A broad area of inquiry encompassing the study of the function of proteins in the proteome, the study of changes in protein expression within the proteome, and the use of reactive chemical probes to identify enzymes in the proteome. This short list is not meant to be wholly inclusive. See

Lawrence, D.S., Functional proteomics: large-scale analysis of protein kinase activity, *Genome Biol.* 2, 1007, 2001; Famulok, M., Blind, M., and Mayer, G., Intramers as promising new tools in functional proteomics, *Chem. Biol.* 8, 931–939, 2001; Guengerich, F.P., Functional genomics and proteomics applied to the study of nutritional metabolism, *Nutr. Rev.* 59, 259–263, 2001; Strosberg, A.D., Functional proteomics to exploit genome sequences, *Cell. Mol. Biol.* 47, 1295–1299, 2001; Yanagida, M., Functional proteomics: current achievements, *J. Chromatog. B. Analyt. Technol. Biomed. Life Sci.* 771, 89–106, 2002; Hunter, T.C., Andon, N.L., Koller, A., Yates, J.R., and Haynes, P.A., The functional proteomics toolbox: methods and applications, *J. Chromatog. B. Analyt. Technol. Biomed. Life Sci.* 782, 165–181, 2002; Graves, P.R. and Haystead, T.A., A functional proteomics approach to signal transduction, *Recent Prog. Horm. Res.* 58, 1–14, 2003; Ilag, L.L., Functional proteomic screens in therapeutic protein drug discovery, *Curr. Opin. Mol. Ther.* 7, 538–542, 2005; Wagner, V., Gessner, G., and Mittag, M., Functional proteomics: a promising approach to find novel components of the circadian system, *Chronobiol. Int.* 22, 403–415, 2005; Monti, M., Orru, S., Pagnozzi, D., and Pucci, P., Functional proteomics, *Clin. Chim. Acta* 357, 140–150, 2005.

Furin   Furin is a subtilisinlike regulatory protease (subtilisinlike pro-protein convertases located in the *trans*-Golgi network), which functions in processing precursor proteins in the secretory pathway. See Molloy, S.S., Bresnahan, P.A., Leppla, S.H. et al., Human furin is a calcium-dependent serine endoprotease that recognizes the sequence Arg-X-X-Arg and efficiently cleaves anthrax protective antigen, *J. Biol. Chem.* 267, 16396–16402, 1992; Yanagita, M., Hoshimo, H., Nakayama, K., and Takeuchi, T., Processing of mutated proteins with tetrabasic cleavage sites to mature insulin reflects the expression of furin in nonendocrine cell lines, *Endocrinology* 133, 639–644, 1993; Brennan, S.O. and Nakayama, K., Furin has the proalbumin substrate specificity and serpin inhibitor properties of an *in situ* convertase, *FEBS Lett.* 338, 147–151, 1994; Roebroek, A.J., Creemers, J.W., Ayoubi, T.A., and Van de Ven, W.J., Furin-mediated pro-protein processing activity: involvement of negatively charged amino acid residues in the substrate binding site, *Biochemie* 76, 210–216, 1994; Denault, J.B. and Leduc, R., Furin/PACE/SPC1: a convertase involved in exocytic and endocytic processing of precursor proteins, *FEBS Lett.* 379, 113–116, 1996; Nakayama, K., Furin: a mammalian subtilisin/Kex2p-like endoprotease involved in processing of a wide variety of precursor proteins, *Biochem. J.* 327, 625–635, 1997; Rockwell, N.C., Krysan, D.J., Komiyama, T., and Fuller, B.S., Precursor processing by Kex2/furin proteases, *Chem. Rev.* 102, 4525–4548, 2002; Fugere, M., Limperis, P.C., Beaulieu-Audy, V. et al., Inhibitory potency and specificity of subtilisin-like pro-protein convertase (SPC) prodomains, *J. Biol. Chem.* 277, 7648–7656, 2002; Rockwell, N.C. and Thorner, J.W., The kindest cuts of all: crystal structures of Kex2 and furin reveal secrets of precursor processing, *Trends Biochem. Sci.* 29, 80–87, 2004. The first pro-protein processing proteins were described as Kex2 protease (kexin) in *Saccharomyces cerevesiae* (Leibowitz, M.J. and Wickner, R.B., A chromosomal gene required for killer plasmid expression, mating, and spore maturation in *Saccharomyces cerevisiae*, *Proc. Natl. Acad. Sci. USA* 73, 2061–2065, 1976; Rogers, D.T., Saville, D., and Bussey, H., *Saccharomyces cerevisiae* expression mutant

Kex2 has altered secretory proteins and glycoproteins, *Biochem. Biophys. Res. Commun.* 90, 187–193, 1979; Julius, D., Brake, A., Blair, L. et al., Isolation of the putative structural gene for the lysine-arginine-cleaving endopeptidase required for processing of yeast prepro-alpha-factor, *Cell* 37, 1075–1089, 1984). Furin is important for the secretion of recombinant proteins in mammalian cell lines (Mark, M.R., Lokker, N.A., Zioncheck, T.F. et al., Expression and characterization of heptocyte growth factor receptor-IgG fusion proteins: effects of mutations in the potential proteolytic cleavage site on processing and ligand binding, *J. Biol. Chem.* 267, 26166–26171, 1992; Bristol, J.A., Freedman, S.J., Furie, B.C., and Furie, B., Profactor IX: the propeptide inhibits binding to membrane surfaces and activation by factor IXa., *Biochemistry* 33, 14136–14143, 1994; Groskreutz, D.J., Sliwkowski, M.X., and Gorman, C.M., Genetically engineering proinsulin constitutively processed and secreted as mature, active insulin, *J. Biol. Chem.* 269, 6241–6245, 1994; Lind, P., Larsson, K., Spira, J. et al., Novel forms of B-domain deleted recombinant factor VIII molecules. Construction and biochemical characterization, *Eur. J. Biochem.* 232, 19–27, 1995; Ayoubi, T.A., Meulemans, S.M., Roebroek, A.J., and Van de Ven, W.J., Production of recombinant proteins in Chinese hamster ovary cells overexpressing the subtilisin-like proprotein converting enzyme furin, *Mol. Biol. Rep.* 23, 87–95, 1996; Chiron, M.F., Fryling, C.M., and Fitzgerald, D., Furin-mediated cleavage of *Pseudomonas* exo-toxin-derived chimeric toxins, *J. Biol. Chem.* 272, 31707–31711, 1997). Furins are functionally related to secretases in being protein precursor processing enzymes (Anders, L., Mertins, P., Lammich, S. et al., Furin-, ADAM 10-, and γ-secretase-mediated cleavage of a receptor tyrosine phosphatase and regulation of β-cateinin's transcriptional activity, *Mol. Cell. Biol.* 26, 3917–3924, 2006). There has been some work on the possible role of furin in the processing of β-secretase (Bennett, B.D., Denis, P., Haniu, M. et al., A furin-like convertase mediates propeptide cleavage of BACE, the Alzheimer's β-secretase, *J. Biol. Chem.* 275, 37712–37717, 2000; Creemers, J.W.M., Dominguez, D.I., Plets, E. et al., Processing of β-secretase by furin and other members of the proprotein convertase family, *J. Biol. Chem.* 276, 4211–4217, 2001; Pinnix, I., Council, J.E., Roseberry, B. et al., Convertases other than furin cleave β-secretase to its mature form, *FASEB J.* 15, 1810–1812, 2001).

G Protein   A heteromeric protein that functions in signal transduction via modulation by G-protein-coupled receptors (GPCRs). See Spiegel, A.M. and Downs, R.W., Jr., Guanine nucleotides: key regulators of hormone receptor-adenylate cyclase, *Endocr. Rev.* 2, 275–305, 1981; Cooper, D.M. and Londos, C., GTP-stimulation and inhibition of adenylate cyclase, *Horiz. Biochem. Biophys.* 6, 309–333, 1982; Poste, G., New insights into receptor regulation, *J. Appl. Physiol.* 57, 1297–1305, 1984; Cuatrecasas, P., Hormone receptors, membrane phospholipids, and protein kinases, *Harvey Lect.* 80, 89–128, 1984–1985; Neer, E.J., Guanine nucleotide-binding proteins involved in transmembrane signaling, *Symp. Fundam. Cancer Res.* 39, 123–136, 1986; Spiegel, A.M., Signal transduction by guanine nucleotide binding proteins, *Mol. Cell. Endocrinol.* 49, 1–16, 1987; Bockaert, J., Homburger, V., and Rouot, B., GTP binding proteins: a key role in cellular communication, *Biochimie* 69, 329–338, 1987; Houslay, M.D. and Milligan, G., *G-Proteins as Mediators of Cellular Signalling Processes,*

Wiley, Chichester, UK, 1990; Naccache, P.H., *G Proteins and Calcium Signaling,* CRC Press, Boca Raton, FL, 1990; Johnson, R.A. and Corbin, J.D., *Adenyl Cyclase, G Proteins, and Guanylyl Cyclase,* Academic Press, San Diego, CA, 1991; Ravi, I., *Heterotrimeric G Proteins*, Academic Press, San Diego, CA, 1994; Watson, S.P., and Arkinstall, S., *The G-Protein Linked Receptor Factbooks,* Academic Press, London, 1994; Siderovski, D.P., *G Proteins and Calcium Signaling,* Elsevier, Amsterdam, 2004; Zhang, Z., Melia, T.J., He, F. et al., How a G protein binds a membrane, *J. Biol. Chem.* 279, 33937–33945, 2004; Gavi, S., Shumay, E., Wang, H.Y., and Malbon, C.C., G-protein-coupled receptors and tyrosine kinases: crossroads in cell signaling and regulation, *Trends Endocrinol. Metab.* 17, 48–54, 2006; Sato, M., Blumer, J.B., Simon, V., and Lanier, S.M., Accessory proteins for G proteins: partners in signaling, *Annu. Rev. Pharmcol. Toxicol.* 46, 151–187, 2006.

Gamma(γ)-Secretase   A membrane-associated regulatory protease responsible for the cleavage of amyloid precursor protein and Notch protein. Gamma(γ)-secretase is composed of four subunits: presenilin, nicastrin, Aph-1, and Pen-2. Presenilin is responsible for the catalytic gamma(γ)-secretase activity and nicastrin and Aph-2 have a function in substrate recognition and complex stabilization while Pen-2 assists in catalytic function. Gamma(γ)-secretase is a therapeutic treatment for Alzheimer's disease. See Mundy, D.L., Identification of the multicatalytic enzyme as a possible γ-secretase for the amyloid precursor protein, *Biochem. Biophys. Res. Commun.* 204, 333–341, 1994; Wolfe, M.S. and Haass, C., The role of presenilins in gamma-secretase activity, *J. Biol. Chem.* 276, 5413–5416, 2001; Sisodia, S.S. and St. George-Hyslop, P.H., γ-secretase, Notch, Aβ and Alzheimer's disease: where do the presenilins fit in? *Nat. Rev. Neurosci.* 3, 281–290, 2002; Kimberly, W.T. and Wolfe, M.S., Identity and function of gamma-secretases, *J. Neurosci. Res.* 74, 353–360, 2003; Iwatsubo, T., The gamma secretase complex: machinery for intramembrane proteolysis, *Curr. Opin. Neurobiol.* 14, 379–383, 2004; Raemakers, T., Esselens, C., and Annaert, W., Presenilin 1: more than just gamma-secretase, *Biochem. Soc. Trans.* 33, 559–562, 2005; De Strooper, B., Nicastrin: gatekeeper of the gamma-secretase complex, *Cell* 122, 318–320, 2005; Churcher, I. and Beher, D., Gamma-secretase as a therapeutic target for the treatment of Alzheimer's disease, *Curr. Pharm. Des.* 11, 3363–3382, 2005; Barten, D.M., Meredith, J.E., Jr., Zaczek, R. et al., Gamma-secretase inhibitors for Alzheimer's disease: balancing efficacy and toxicity, *Drugs R & D* 7, 87–97, 2006; Wolfe, M.S., The γ-secretase complex: membrane-embedded proteolytic ensemble, *Biochemistry* 45, 7931–7939, 2006.

Gelsolin; Gelsolinlike Domains   Gelsolin is a signature protein for a family of proteins that interact with actin and influence the structure of the cytoskeleton. Gelsolin is a calcium-dependent, actin-binding protein that modulates actin filament length. See Yin, H.L., Hartwig, J.H., Maruyama, K., and Stossel, T.P., Ca^{2+} control of actin filament length. Effects of macrophage gelsolin on actin polymerization, *J. Biol. Chem.* 256, 9693–9697, 1981; Matasudaira, P., Jakes, R., and Walker, J.E., A gelsolin-like Ca^{2+}-dependent actin-binding domain in villin, *Nature* 315, 248–250, 1985; Dixon, R.A.F., Kobilka, B.K., and Strader, D.J., Cloning of the gene and cDNA for mammalian β-adrenergic receptor and homology with rhodopsin, *Nature* 321, 75–79, 1986; Libert, F., Parmentier, M., Lefort, A. et al., Selective amplification and cloning of four new members of the G protein-coupled receptor family, *Science* 244, 569–572, 1989; Yu, F.X., Zhou, D.M., and Yin, H.L., Chimeric and truncated gCap39 elucidate the requirements for actin filament severing and end capping by the gelsolin family of proteins, *J. Biol. Chem.* 266, 19269–19275, 1991; Wen, D., Corina, K., Chow, E.P. et al., The plasma and cytoplasmic forms of human gelsolin differ in disulfide structure, *Biochemistry* 35, 9700–9709, 1996; Isaacson, R.L., Weeds, A.G., and Fersht, A.R., Equilibria and kinetics of folding of gelsolin domain 2 and mutants involved in familial amyloidosis-Finnish type, *Proc. Natl. Acad. Sci. USA* 96, 11247–11252, 1996; Liu, Y.T. and Yin, H.L., Identification of the binding partners for flightless I, a novel protein bridging the leucine-rich repeat and the gelsolin superfamilies, *J. Biol. Chem.* 273, 7920–7927, 1998; Benyamini, H., Gunasekaran, K., Wolfson, H., and Nussinov, R., Conservation and amyloid formation: a study of the gelsolin-like family, *Proteins* 51, 266–282, 2003; Uruno, T., Remmert, K., and Hammer, J.A., III, CARMIL is a potent capping protein antagonist: identification of a conserved CARMIL domain that inhibits the activity of capping protein and uncaps capped actin filaments, *J. Biol. Chem.* 281, 10635–10650, 2006.

Gene Expression Domain   A genomic region that contains a gene and all of the *cis*-acting elements that are required to obtain the homeostatic level and timing of gene expression *in vivo*. Gene expression domains are generally defined by their ability to function independently of the site of integration into a transgene.

General Transcription Factors   A group of *trans*-acting factors that have a central role in the initiation of transcription by RNA polymerase II (pol II). The components are likely similar to the earlier described basal transcription factors. See Greenblatt, J., RNA polymerase-associated transcription factors, *Trends Biochem. Sci.* 16, 408–411, 1991; Corden, J.L., RNA polymerase II transcription cycles, *Curr. Opin. Genet. Dev.* 3, 213–218, 1993; Travers, A., Transcription: building an initiation machine, *Curr. Biol.* 6, 401–403, 1996; Reese, J.C., Basal transcription factors, *Curr. Opin. Genet. Dev.* 13, 114–118, 2003; Asturias, F.J., RNA polymerase II structure, and organization of the preinitiation complex, *Curr. Opin. Struct. Biol.* 14, 121–129, 2004; Boeger, H., Bushnell, D.A., Davis, R. et al., Structural basis of eukaryotic gene transcription, *FEBS Lett.* 579, 899–903, 2005; Szutarisz, H., Dillon, N., and Tora, L., The role of enhancers as centres for general transcription factor recruitment, *Trends Biochem. Sci.* 30, 593–599, 2005; Gross, P. and Oelgeschlager, T., Core promoter-selective RNA polymerase II transcription, *Biochem. Soc. Symp.* 73, 225–236, 2006.

Generic Drug   A generic drug is the same as a brand name drug in dosage, safety, strength, administration, quality, and intended use. The suitability of a generic drug is based on "therapeutic equivilance." By law, a generic product must contain the identical amount of the same active ingredient(s) as the brand name product. See Verbeeck, R.K., Kanfer, I., and Walker, R.B., Generic substitution: the use of medicinal products containings different salts and implications for safety and efficacy, *Eur. J. Pharm.Sci.* 28, 1–6, 2006; Devine, J.W., Cline, R.R., and Farley, J.F., Follow-on biologics: competition in the biopharmaceutical marketplace, *J. Am. Pharm. Assoc.* 46, 193–201, 2006.

Genome   The complete gene complement of any organism, contained in a set of chromosomes in eukaryotes, a single chromosome in bacteria, or a DNA or RNA molecule in viruses; the complete set of genes inside the cell or virus. Singer, M. and Berg, P., *Genes & Genomes: A Changing Perspective*, University Science, Mill Valley, CA, 1991; Murray, T.H. and Rothstein, R.A., *The Human Genome Project and the Future of Health Care,* Indiana University Press, Bloomington, 1996; Brown, T.A., *Genome*, Bios/Wiley-Liss, New York, 1999; Ridley, M., *Genome: The Autobiography of a Species of 23 Chapters*, HarperCollins, New York, 1999.

Genome-Based Proteomics   Gene-based analysis of the proteome; analytical strategies based on the knowledge of the genome. See Rosamond, J. and Allsop, A., Harnessing the power of the genome in the search for new antibiotics, *Science* 287, 1973–1976, 2000; Agaton, C., Uhlen, M., and Hober, S., Genome-based proteomics, *Electrophoresis* 25, 1280–1288, 2004; Wisz, M.S., Suarez, M.K., Holmes, M.R., and Giddings, M.C., GFSWeb: a web tool for genome-based identification of proteins from mass spectrometric samples, *J. Proteome Res.* 3, 1292–1295, 2004; Romero, P., Wagg, J., Green, M.L. et al., Computational prediction of human metabolic pathways from the complete human genome, *Genome Biol.* 6, R2, 2005; Ek. S., Adreasson, U., Hober, S. et al., From gene expression analysis to tissue microarrays: a rational approach to identify therapeutic and diagnostic targets in lymphoid malignancies, *Mol. Cell. Proteomics* 5, 1072–1081, 2006.

Genomic Databases   See Baxevaris, A.D., Using genomic databases for sequence-based biological discovery, *Molec. Med.* 9, 185–192, 2003.

Genomics   The study of the structure and function of the genome, including information about sequence, mapping, and expression, and how genes and their products work in the organism; the study of the genetic composition of organisms.

Genotype   The internally coded, inheritable information carried by all living organisms; the genetic constitution of an organism.

Glass/Glasses   A large inhomogenous class of materials with highly variable mechanical and optical properties that solidify from the molten state without crystallization. The cooling of the melt must occur without crystallization. Glasses are most frequently derived from silicates by fusing with boric oxide, aluminum oxide, or phosphorus pentoxide. Glasses are generally hard, brittle, and transparent or translucent, and are considered to be supercooled liquids rather than true solids. See Santoro, M., Gorelli, F.A., Bini, F. et al., Amorphous silica-like carbon dioxide, *Nature* 441, 857–860, 2006; Huang, W., Day, D.E., Kittiratanapiboon, K., and Rahaman, M.N., Kinetics and mechanisms of the conversion of silicate (45S5), borate, and borosilicate glasses to hydroxyapatite in dilute phosphate solutions, *J. Mater. Sci. Mater. Med.* 17, 583–596, 2006; Abraham, S., Mallia, V.A., Ratheesh, K.V. et al., Reversible thermal and photochemical switching of liquid crystalline phases and luminescence in diphenylbutadiene-based mesogenic dimers, *J. Am. Chem. Soc.* 128, 7692–7698, 2006; Lehner, A., Corbineau, F., and Bailly, C., Changes in lipid status and glass properties in cotyledons of developing sunflower seeds, *Plant Cell Physiol.*, 47, 818–828, 2006; Chang, R. and Yethiraj, A., Dynamics of chain molecules in disordered materials, *Phys. Rev. Lett.* 96, 107802, 2006; Katritzky, A.R., Singh, S., Kirichenko, K. et al., In search of ionic liquids incorporating azolate anions, *Chemistry* 12, 4630–4641, 2006.

Glass Transition/ Glass Transition Temperature   The glass transition generally refers to the change of a polymer from an amorphous material to a brittle material. The glass transition of a non-crystalline material is the critical temperature at which the material changes its behavior from being a glass or brittle material to being an amorphous rubberlike material. For lyophilization, it is a critical temperature during the drying cycle that is important to the final product cake. See MacKenzie, A.P., Non-equilibrium freezing behavior of aqueous systems, *Philos. Trans. R. Soc. Lond. B Biol. Sci.* 278, 167–189, 1977; Schenz, T.W., Israel, B., and Rosolen, M.A., Thermal analysis of water-containing systems, *Adv. Exp. Med. Biol.* 302, 199–214, 1991; Craig, D.Q., Royall, P.G., Kett, V.L., and Hopton, M.L., The relevance of the amorphous state to pharmaceutical dosage forms: glassy drugs and freeze dried systems, *Int. J. Pharm.* 179, 179–207, 1999; Oliver, A.E., Hincha, D.K., and Crowe, J.H., Looking beyond sugars: the role of amphiphilic solutes in preventing adventitious reactions in anhydrobiotes at low water contents, *Comp. Biochem. Physiol. A Mol. Integr. Physiol.* 131, 515–525, 2002; Nail, S.L., Jiang, S., Chongprasert, S., and Knopp, S.A., Fundamentals of freeze-drying, *Pharm. Biotechnol.* 14, 281–360, 2002; Franks, F., Scientific and technological aspects of aqueous glasses, *Biophys. Chem.* 105, 251–261, 2003; Vranic, E., Amorphous pharmaceutical solids, *Bosn. J. Basic Med. Sci.* 4, 35–39, 2004; Hilden, L.R. and Morris, K.R., Physics of amorphous solids, *J. Pharm. Sci.* 93, 3–12, 2004.

Global Proteomics
Analysis of all proteins in a cell or tissue of an organism. See Hancock, W.S., Wu, S.L., Stanley, R.R., and Gombocz, E.W., Publishing large proteome datasets: scientific policy meets emerging technologies, *Trends Biotechnol.* 20 (Suppl. 12), S39–S44, 2002; Godovac-Zimmermann, J. and Brown, L.R., Proteomics approaches to elucidation of signal transduction pathways, *Curr. Opin. Mol. Ther.* 5, 241–249, 2003; Kumar, G.K. and Klein, J.B., Analysis of expression and posttranslational modification of proteins during hypoxia, *J. Appl. Physiol.* 96, 1178–1186, 2004; Hoskisson, P.A. and Hobbs, G., Continuous culture — making a comeback? *Microbiology* 151, 3153–3159, 2005.

Globulin
A classic definition of proteins that are insoluble in water and soluble in dilute salt solutions and migrate more slowly than albumin in an electrophoretic system (Cooper, G.R., Electrophoretic and ultracentrifugal analysis of normal human serum, in *The Plasma Proteins*, Putnam, F.W., Ed., Academic Press, New York, 1960, pp. 51–103). The globulins were separated into several fractions including the the γ-globulins, which contain the various immunoglobulin fractions and were defined as the most slowly moving protein fraction on electrophoresis at pH 8.6 (Porter, H.R., γ-globulins and antibodies, in *The Plasma Proteins*, Putnam, F.W., Ed., Academic Press, New York, 1960, pp. 241–277). See Gehrke, C.W., Oh, Y.H., and Freeark, C.W., Chemical fractionation and starch gel-urea electrophoretic characterization of albumins, globulins, gliadins, and glutenins in soft wheat, *Anal. Biochem.* 7, 439–460, 1964; Nilsson, U.R. and Mueller-Eberhard, H.J., Isolation of beta IF-globulin form human serum and its characterization as the fifth component of complement, *J. Exp. Med.* 122, 277–298, 1965; Sun, S.M. and Hall, T.C., Solubility characteristics of globulins from *Phaseolus* seeds in regard to their isolation and characterization, *J. Agric. Food Chem.* 23, 184–189, 1975; Hauptman, S.P., Macromolecular insoluble cold globulin (MICG): a novel protein from mouse lymphocytes — I. Isolation and characterization, *Immunochemistry* 15, 415–422, 1978.

Glucose Oxidase
A flavoprotein (FAD) enzyme (EC 1.1.3.4; β-D-glucose:oxygen 1-oxidoreductase), which catalyzes the oxidation of β-D-glucose to glucolactone/

gluconic acid and hydrogen peroxide. The enzyme is highly specific for this form of glucose (Keilin, D. and Hartree, E.F., The use of glucose oxidase [Notatin] for the determination of glucose in biological material and for the study of glucose-producing systems by mannometric methods, *Biochem. J.* 42, 230–238, 1942; Sols, A. and de la Fuente, G., On the substrate specificity of glucose oxidase, *Biochim. Biophys. Acta* 24, 206–207, 1957; Wurster, B. and Hess, B., Anomeric specificity of enzymes for D-glucose metabolism, *FEBS Lett.* 40 (Suppl.), S112–S118, 1974) and is the basis of most of the assays for glucose in blood and bioreactors. The vast majority of assays measure the hydrogen peroxide released in the reaction (Kiang, S.W., Kuan, J.W., Kuan, S.S., and Guilbault, G.G., Measurement of glucose in plasma, with use of immobililized glucose oxidase and peroxidase, *Clin. Chem.* 22, 1378–1382, 1976; Chua, K.S. and Tan, I.K., Plasma glucose measurement with the Yellow Springs glucose analyzer, *Clin. Chem.* 24, 150–152, 1978; Artiss, J.D., Strandbergh, D.R., and Zak, B., On the use of a sensitive indicator reaction for the automated glucose oxidase-peroxidase coupled reaction, *Clin. Biochem.* 1, 334–337, 1983; Burtis, C.A., Ashwood, E.R., and Bruns, D.F., Eds., *Tietz Textbook of Clinical Chemistry and Molecular Diagnostics*, 4th ed., Elsevier-Saunders, St. Louis, MO, 2006). Glucose oxidase was discovered in the early 1900s and originally described as an antibacterial factor derived from molds such as *Pencillium notatum* and *Aspergillus niger* (Coulthard, C.E., Michaealis, R., Short, W.F. et al., Notatin: an antibacterial glucose aerodehydrogenase from *Penicillium notatum* and *Penicillium resitculosum* sp. nov, *Biochem. J.* 39, 24–36, 1945). Glucose oxidase has subsequently been identified as the antibacterial/antibiotic activity in honey (White, J.W., Jr., Subers, M.H., and Schepartz, A.I., The identification of inhibine, the antibacterial factor in honey, as hydrogen peroxide and its origin in a honey glucose-oxidase system, *Biochim. Biophys. Acta* 73, 57–70, 1963; Schepartz, A.T. and Subers, M.H., The glucose oxidase of honey. I. Purification and some general properties of the enzyme, *Biochim. Biophys. Acta* 85, 228–237, 1964; Bang, L.M., Bunting, C., and Molan, P., The effect of dilution on the rate of hydrogen peroxide production in honey and its implications for wound healing, *J. Alternative Complementary Med.* 9, 267–273, 2003; Badawy, O.F., Shafii, S.S., Tharwat, E.E., and Kamal, A.M., Antibacterial activity of bee honey and its therapeutic usefulness against *Escherichia coli* 0157:H7 and *Salmonella typhimurium* infection, *Rev. Sci. Tech.* 23, 1011–1022, 2004) and as a critical component of the honey bee invertebrate immune system (Xang, X. and Cox-Foster, D.L., Impact of an ectoparasite on the immunity and pathology of an invertebrate: evidence for host immunosuppression and viral amplification, *Proc. Natl. Acad. Sci. USA* 102, 7470–7475, 2005). Glucose oxidase is also involved in herbivore offense in plants (Musser, R.O., Cipollini, D.F., Hum-Musser, S.M. et al., Evidence that the caterpillar salivary enzyme glucose oxidase provides herbivore offense in solanaceous plants, *Archs. Insect Biochem. Physiol.* 58, 128–137, 2005).

Glucose-Regulated Protein, 78kD   Grp78; glucose-regulated protein, identical with BiP, a chaperonelike protein that was also described as the immunoglobulin heavy-chain-binding protein. See Munro, S. and Pelham, H.R., An Hsp70-like protein in the ER: identity with the 78 kd glucose-regulated protein and immunoglobulin heavy chain binding protein, *Cell* 46, 291–300, 1986; Hendershot, L.M., Ting, J., and

Lee, A.S., Identity of the immunoglobulin heavy-chain-binding protein with the 78,000 dalton glucose-regulated protein and the role of posttranslational modifications in its binding function, *Mol. Cell Biol.* 8, 4250–4256, 1988; Haas, I.G., BiP (Grp78), an essential hsp70 resident protein in the endoplasmic reticulum, *Experentia* 50, 1012–1020, 1994; Kleizen, B. and Braakman, I., Protein folding and quality control in the endoplasmic reticulum, *Curr. Opin. Cell Biol.* 16, 343–349, 2004; Okudo, H., Kato, H., Arakaki, Y., and Urade, R., Cooperation of ER-60 and BiP in the oxidative refolding of denatured proteins *in vitro*, *J. Biochem.* 138, 773–780, 2005; Sorgjerd, K., Ghafouri, B., Jonsson, B.H. et al., Retention of misfolded mutant transthyretin by the chaperone BiP/GRP78 mitigates amyloidogenesis, *J. Mol. Biol.* 356, 469–482, 2006; Panayi, G.S., and Corrigall, V.M., BiP regulates autoimmune inflammation and tissue damage, *Autoimmune Rev.* 5, 140–142, 2006; Li, J. and Lee, A.S., Stress induction of GRP78/BiP and its roles in cancer, *Curr. Mol. Med.* 6, 45–54, 2006; Tajima, H. and Koizumi, N., Induction of BiP by sugar independent of a *cis*-element for the unfolded protein response in *Arabidopsis thaliana*, *Biochem. Biophys. Res. Commun.* 346, 926–930, 2006.

Glucosyl-transferase A glycosyltransferase specific for the transfer of glucosides. See Doyle, R.J. and Ciardi, J.E., *Glucosyltransferases, Glucans, Sucrose, and Dental Caries*, IRL Press, Washington, DC, 1983; Bleicher, R.J. and Cabot, M.C., Glucosylceramide synthesis and apoptosis, *Biochim. Biophys. Acta* 1585, 172–178, 2002; Yang, J., Hoffmeister, D., Liu, L. et al., Natural product glycorandomization, *Bioorg. Med. Chem.* 12, 1577–1584, 2004; Lorenc-Kukula, K., Korobczak, A., Aksamit-Stachurska, A. et al., Glucosyltransferase: the gene arrangement and enzyme function, *Cell Mol. Biol. Lett.* 9, 935–946, 2004; Trombetta, E.S. and Parodi, A.J., Glycoprotein reglucosylation, *Methods* 35, 328–337, 2005.

GLUT A family of membrane transporters that mediate the uptake of hexoses in mammalian cells. See Gould, G.W. and Holman G.D., The glucose transporter family — structure, function, and tissue-specific expression, *Biochem. J.* 295, 329–341, 1993; Yang, J., Dowden, J., Tatibouet, A., Hatanaka, Y., and Holman, G.D., Development of high-affinity ligands and photoaffinity labels for the D-fructose transporter GLUT5, *Biochem. J.* 367, 533–539, 2002.

Glycome The total carbohydrates within an organism. See Feizi, T., Progress in deciphering the information content of the "glycome" — a crescendo in the closing years of the millennium, *Glycoconj. J.* 17, 553–565, 2001; Hirabayashi, J., Arata, Y., and Kasai, K., Glycome project: concept, strategy, and preliminary application to *Caenorhabditis elegans*, *Proteomics* 1, 295–303, 2001; Loel, A., Glycome: a medical paradigm, *Adv. Exp. Biol. Med.* 546, 445–451, 2004; Hsu, K.L., Pilobello, K.T., and Mahal, L.K., Analyzing the dynamic bacterial glycome with a lectin microarray approach, *Nat. Chem. Biol.* 2, 125–126, 2006; Freeze, H.H., Genetic defects in the human glycome, *Nat. Rev. Genet.* 7, 537–551, 2006.

Glycomics The study of the structure, function, and interactions of carbohydrates within the gycome. See Drickhamer, K. and Taylor, M.E., Glycan arrays for functional glycomics, *Genome Biol.* 3, 1034, 2002; Love, K.R. and Seeberger, P.H., Carbohydrate arrays as tools for glycomics, *Angew. Chem. Int. Ed. Engl.* 41, 3583–3586, 2002; Hirabayashi, J., Oligosaccharide microarrays for glycomics, *Trends Biotechnol.* 21, 141–143, 2003; Feizi, T., Fazio, F., Chai, W., and Wong, C.H., Carbohydrate microarrays — a new set of technologies at the frontiers of glycomics, *Curr. Opin. Struct. Biol.* 13, 637–645, 2003; Morelle, W. and Michalski, J.C., Glycomics and mass spectrometry, *Curr. Pharm. Des.* 11, 2615–2645, 2005; Raman, R., Raguram, S., Venkataraman, G. et al., Glycomics: an integrated systems approach to structure-function relationships of glycans, *Nat. Methods* 2, 817–824, 2005.

Glycosidase   An enzyme that hydrolyzes glycosidic bonds, most often in oligosaccharides and polysaccharides. See Allen, H.J. and Kisailus, E.C., *Glycoconjugates: Composition, Structure, and Function,* Dekker, New York, 1992; Lennarz, W.J. and Hart, G.W., Eds., *Guide to Techniques in Glycobiology,* Academic Press, San Diego, CA, 1994; Bucke, C., *Carbohydrate Biotechnology Protocols*, Humana Press, Totowa, NJ, 1999; Himmel, M.E. and Baker, J.O., *Glycosyl Hydrolases for Biomass Conversion*, American Chemical Society, Washington, DC, 2001.

Glycosyl-   An enzyme that synthesizes compounds with glycosidic bonds by catalyzing
transferase   the transfer of glycosyl groups. See Carib, E., Carbohydrate metabolism, *Annu. Rev. Biochem.* 32, 321–354, 1963; Heath, E.C., Complex polysaccharides, *Annu. Rev. Biochem.* 40, 29–56, 1971; Honjo, T. and Hayashi, O., Enzymatic ADP-ribosylation of proteins and regulation of cellular activity, *Curr. Top. Cell Regul.* 7, 87–127, 1973; Alavi, A. and Axford, J.S., Eds., *Glycoimmunology*, Plenum Press, New York, 1995; Fukuda, M. and Hindsgaul, O., Eds., *Molecular Glycobiology*, Oxford University Press, Oxford, UK, 1994; Endo, T., Aberrant glycosylation of alpha-dystroglycan and congenital muscular dystrophies, *Acta Myol.* 24, 64–69, 2005; Serafini-Cessi, F., Monti, A., and Cavallone, D., *N*-glycans carried by Tamm-Horsfall glycoprotein have a crucial role in the defense against urinary tract diseases, *Glycoconj. J.* 22, 383–394, 2005; Milewski, S., Gabriel, I., and Olchowy, J., Enzymes of UDP-GlcNAc in yeast, *Yeast* 23, 1–14, 2006; Millar, C.M. and Brown, S.A., Oligosaccharide structures of von Willebrand factor and their potential role in von Willebrand disease, *Blood Rev.* 20, 83–92, 2006; Koch-Nolte, F., Adriouch, S., Bannas, P. et al., ADP-ribosylation of membrane proteins: unveiling the secrets of a crucial regulatory mechanism in mammalian cells, *Ann. Med.* 38, 189–199, 2006.

Goblet Cell   A type of cell found in the epithelium with high occurrence in respiratory/digestive tracts that secrete mucus. See Rogers, D.F., Motor control of airway goblet cells and glands, *Respir. Physiol.* 125, 129–144, 2001; Jeffery, P. and Zhu, J., Mucin-producing elements and inflammatory cells, *Novartis Found. Symp.* 248, 51–68, 2002; Rogers, D.F., The airway goblet cell, *Int. J. Biochem. Cell. Biol.* 35, 1–6, 2003; Kim, S. and Nadel, J.A., Role of neutrophils in mucus hypersecretion in COPD and implications for therapy, *Treat. Respir. Med.* 3, 147–159, 2004; Bai, T.R. and Knight, D.A., Structural changes in the airways in asthma: observations and consequences, *Clin. Sci.* 108, 463–477, 2005; Rose, M.C., and Voynow, J.A., Respiratory tract mucin genes and mucin glycoproteins in health and disease, *Physiol. Rev.* 86, 245–278, 2006; Lievin-Le Moal, V. and Servin, A.L., The front line of enteric host defense against unwelcome intrusion of harmful microorganisms: mucins, antimicrobial peptides, and microbiota, *Clin. Microbiol. Rev.* 19, 315–337, 2006.

Golgi Apparatus   A subcellular organelle consisting of a series of membrane structures; the Golgi apparatus can be considered a single membrane structure containing a number of membrane-bound vesicles. The Golgi apparatus functions in the protein secretory pathway by transporting and packing proteins for distribution elsewhere in the cell. The Golgi has a *cis*-side facing the endoplasmic reticulum and a *trans*-side that interfaces with the plasma membrane and components of the endocytotic pathway. See Northcote, D.H., The Golgi apparatus, *Endeavor* 30, 26–33, 1971; Shnitka, T.K. and Seligman, A.M., Ultrastructural localization of enzymes, *Annu. Rev. Biochem.* 40, 375–396, 1971; Schachter, H., The subcellular sites of glycosylation, *Biochem. Soc. Trans.* 40, 47–71, 1974; Whaley, W.B., *The Golgi Apparatus*, Springer-Verlag, New York, 1975; Novikoff, A.B., The endoplasmic reticulum: a cytochemist's view, *Proc. Natl. Acad. Sci. USA* 73, 2781–2787, 1976; Pavelka, M., *Functional Morphology of the Golgi Apparatus*, Springer-Verlag, Berlin, 1987; Loh, Y.P., *Mechanisms of Intracellular Trafficking and Processing of Preproteins*, CRC Press, Boca Raton, FL, 1993; Rothblatt, J. and Novak, P., Eds., *Guidebook to the Secretory Pathway*, Oxford University Press, Oxford, UK, 1997; Berger, E.G. and Roth, J., Eds., *The Golgi Apparatus*, Birkhäuser Verlag, Basel, 1997; Robinson, D.G., *The Golgi Apparatus and the Plant Secretory Pathway,*, CRC Press, Boca Raton, FL, 2003; Hawes, C. and Satiat-Jeunemailtre, B., The plant Golgi apparatus — going with the flow, *Biochim. Biophys. Acta* 1744, 466–480, 2005; Meyer, H.H., Golgi reassembly afer mitosis: the AAA family meets the ubiquitin family, *Biochim. Biophys. Acta* 1744, 481–492, 2005; Toivola, D.M., Tao, G.Z., Hbtezion, A., Liao, J., and Omary, M.B., Cellular integrity plus: organelle-related and protein-targeting functions of intermediate filaments, *Trends Cell Biol.* 15, 608–617, 2005; Jolliffe, N.A., Craddock, C.P., and Frigerio, L., Pathways for protein transport to see storage granules, *Biochem. Soc. Trans.* 33, 1016–1018, 2005; Ungar, D., Oka, T., Kreiger, M., and Hughson, F.M., Retrograde transport on the COG railway, *Trends Cell Biol.* 16, 113–120, 2006; Quatela, S.E. and Phillips, M.R., Ras signaling on the Golgi, *Curr. Opin. Cell Biol.* 18, 162–167, 2006; D'Souza-Schorey, C. and Chavrier, P., ARF proteins: roles in membrane traffic and beyond, *Nat. Rev. Mol. Cell Biol.* 7, 347–358, 2006.

Golgins   A family of proteins found in the Golgi apparatus. The members of this protein family are characterized by the presence of a long region of coiled-coil segments that have a tendency to form long rodlike structures. See Fritzler, M.J., Hamel, J.C., Ocha, R.L., and Chan, E.K., Molecular characterization of two human autoantigens: unique cDNAs encoding 95- and 160-kD proteins of a putative family in the Golgi complex, *J. Exp. Med.* 178, 49–62, 1993; Kjer-Nielsen, L., Teasdale, R.D., van Vliet, C., and Gleeson, P.A., A novel Golgi-localization domain shared by a class of coiled-coil peripheral membrane proteins, *Curr. Biol.* 9, 385–388, 1999; Munro, S. and Nichols, B.J., The GRIP domain — a novel Golgi-targeting domain found in several coiled-coil proteins, *Curr. Biol.* 9, 377–380, 1999; Pfeffer, S.R., Constructing a Golgi complex, *J. Cell Biol.* 155, 873–883, 2001; Barr, F.A. and Short, B., Golgins in the structure and dynamics of the Golgi apparatus, *Curr. Opin. Cell Biol.* 15, 405–413, 2003; Darby, M.C., van Vliet, C., Brown, D. et al., Mammalian GRIP domain proteins differ in their membrane binding properties and are recruited to distinct domains of the TGN, *J. Cell Biol.* 177, 5865–5874, 2004; Fridmann-Sirkis, Y., Siniossoglou, S., and Pelham, H.R., TMF is a golgin that binds Rab6 and influences Golgi morphology, *BMC Cell Biol.* 5, 18, 2004; Malsam,

J., Satch, A., Pelletier, L., and Warren, G., Golgin tethers define subpopulations of COPI vesicles, *Science* 307, 1095–1098, 2005; Short, B., Haas, A., and Barr, F.A., Golgins and GTPases, giving identity and structure to the Golgi apparatus, *Biochim. Biophys. Acta* 1744, 383–395, 2005; Satoh, A., Beard, M., and Warren, G., Preparation and characterization of recombinant golgin tethers, *Methods Enzymol.* 404, 279–296, 2005.

G-Protein-Coupled Receptor (GPCR)   A membrane receptor that is functional, linked to the activation of a trimeric G protein complex characterized by the presence of seven transmembrane segments.

Gα Protein   The alpha-subunit of the heterotrimeric G protein, which separates into a Gα-protein-GTP complex when GTP replaces GDP. See Albert, P.R. and Robillard, L., G protein specificity: traffic direction required, *Cell Signalling* 14, 407–418, 2002; Kurose, H., $Gα_{12}$ and $Gα_{13}$ as key regulatory mediators in signal transduction, *Life Sci.* 74, 155–161, 2003; Kostenis, E., Waelbroeck, M., and Milligan, G., Techniques: promiscuous Gα proteins in basic research and drug discovery, *Trends Pharmacol. Sci.* 26, 595–602, 2005; Herrman, R., Heck, M., Henklein, P. et al., Signal transfer from GPCRs to G proteins: role of the Gα N-terminal region in rhodopsin-transducin coupling, *J. Biol. Chem.*, 281, 30234–30241, 2006.

Granzyme   Granzymes are exogenous serine proteases that are contained in cytoplasmic granules in cytotoxic T-cells and natural killer cells. Granzymes enter the target cell through pores created by perforin and induce apoptosis through a variety of mechanisms including caspace-dependent and caspace-independent pathways. See Jenne, D.E. and Tchopp, J., Granzymes, a family of serine proteases released from granules of cytolytic T lymphocytes upon T-cell receptor stimulation, *Immunol. Rev.* 103, 53–71, 1988; Smyth, M.J. and Trapani, J.A., Granzymes: exogenous proteinases that induce target cell apoptosis, *Immunol. Today* 16, 202–206, 1995; Lieberman, J. and Fan, Z., Nuclear war: the granzyme A-bomb, *Curr. Opin. Immunol.* 15, 553–559, 2003; Andrade, F., Casciola-Rosen, L.A., and Rosen, A., Granzyme B-induced cell death, *Acta Haematol.* 111, 28–41, 2004; Waterhouse, N.J., Clarke, C.J., Sedelies, K.A., Teng, M.W., and Trapani, J.A., Cytotoxic lymphocytes; instigators of dramatic target cell death, *Biochem. Pharmacol.* 68, 1033–1040, 2004; Ashton-Rickardt, P.G., The granule pathway of programmed cell death, *Crit. Rev. Immunol.* 25, 161–182, 2005; Bleackely, R.C., A molecular view of cytotoxic T lymphocyte induced killing, *Biochem. Cell Biol.* 83, 747–751, 2005.

Growth   Can be defined as weight or mass increase with age in a multiplicative way (from Medewar, P., Size, shape and age, in *Essays in Growth and Form Presented to D'Arcy Wentworth Thompson*, Clarendon Press, Oxford, UK, 1945, p. 708, as cited by Smith, R.W. and Ottema, C., Growth, oxygen consumption and protein and RNA synthesis rates in the yolk sac larvae of the African catfish (*Clarias gariepinos*), *Comp. Biochem. Physiol. Part A* 143, 315–325, 2006).

GTP-Binding Protein   Intracellular proteins that bind GTP and have a wide variety of functions including signal transduction and in turn protein synthesis and cell proliferation. These proteins are "active" when GTP is bound; on hydrolysis of the GTP to GDP, "activity" is lost. See Rouot, B., Brabet, P., Homberger, V. et al., Go, a major brain GTP-binding protein in search of a function: purification, immunological, and biochemical characterization, *Biochimie* 69, 339–349, 1987; Obar, P.A., Shpetner, H.S., and Vallee, R.B., Dynamin:

a microtubule-associated GTP-binding protein, *J. Cell Sci.* 14 (Suppl.), 143–145, 1991; Lillie, T.H. and Gomperts, B.D., A cell-physiological description of GE, a GTP-binding protein that mediates exocytosis, *Ciba Found. Symp.* 176, 164–179, 1993; Kjeldgaard, M., Nyborg, J., and Clark, B.G., The GTP-binding motif: variations on a theme, *FASEB J.* 10, 1347–1386, 1996; Im, M.J., Russell, M.A., and Feng, J.F., Transglutaminase II: a new class of GTP-binding protein with new biological functions, *Cell Signal.* 9, 477–482, 1997; Ridley, A.J., The GTP-binding protein Rho, *Int. J. Biochem. Cell Biol.* 29, 1225–1229, 1997; Sugden, P.H. and Clerk, A., Activation of the small GTP-binding protein Ras in the heart by hypertrophic agonists, *Trends Cardiovasc. Med.* 10, 1–8, 2000; Caron, E., Cellular functions of the Rap1 GTP-binding protein: a pattern emerges, *J. Cell Sci.* 116, 435–440, 2003; Gasper, R., Scrima, A., and Wittinghofer, A., Structural insights into HypB, a GTP-binding protein that regulates metal binding, *J. Biol. Chem.*, 281, 27492–27502, 2006.

Haber-Weiss Reaction

A cycle consisting of the reaction of hydroxyl radicals with hydrogen peroxide, generating the superoxide with the subsequent reaction of superoxide with peroxide generating hydroxyl anion and hydroxyl radical; it is possible that this second reaction is catalyzed by ferric ion. See Kehrer, J.P., The Haber-Weiss reaction and mechanisms of toxicity, *Toxicology* 149, 43–50, 2000; Koppenol, W.H., The Haber-Weiss cycle — 70 years later, *Redox Rep.* 6, 229–234, 2001.

Heat Capacity (Cρ)

The quantity of thermal energy needed to raise the temperature of an object by 1°C; Cρ = mass × specific heat; see *Specific Heat*. Heat capacity in proteins is measured with techniques such as differential scanning calorimetry and isothermal titration calorimetry. An understanding of heat capacity is important in understanding the glass transition in the lyophilization of proteins. See Cooper, A., Heat capacity effects in protein folding and ligand binding: a re-evaluation of the role of water in biomolecular thermodynamics, *Biophys. Chem.* 115, 89–97, 2005; Prabhu, N.V. and Sharp, K.A., Heat capacity I proteins, *Annu. Rev. Phys. Chem.* 56, 521–548, 2005; van Teeffelen, A.M., Melinders, M.B., and de Jongh, H.H., Identification of pitfalls in the analysis of heat capacity changes of beta-lactoglobulin A, *Int. J. Biol. Macromol.* 30, 28–34, 2005; Kozlov, A.G. and Lohman, T.M., Effects of monovalent anions on a temperature-dependent heat capacity change for *Escherichia coli* SSB tetramer binding to single-stranded DNA, *Biochemistry* 45, 5190–5205, 2006; Gribenko, A.V., Keiffer, T.R., and Makhatadze, G.I., Amino acid substitutions affecting protein dynamics in eglin C do not affect heat capacity change upon unfolding, *Proteins* 64, 295–300, 2006; Lemaster, D.M., Heat capacity–independent determination of differential free energy of stability between structurally homologous proteins, *Biophys. Chem.* 119, 94–100, 2006.

Heat-Shock Proteins

Heat-shock proteins (HSP) are a family of proteins with chaperone activity. Heat-shock proteins are involved in protein synthesis and folding, vesicular trafficking, and antigen presentation. Glucose-regulated protein 78 kDA (GRP78), which is also known as immunoglobulin heavy chain-binding protein (BiP) is one of the better-known members of this family and is constitutively expressed in the endoplasmic reticulum (ER) in a wide variety of cell types. Heat-shock proteins were first described as part of the response of the cell to heat shock and other stress situations such as hypoxia. See Tissières, A., Mitchell, H.K., and Tracy, U.M., Protein synthesis in salivary glands of *Drosophila melanogaster*: relation to chromosome puffs, *J. Mol. Biol.* 84, 389–398, 1974; Schedl, P., Artavanis-Tsakonas, S., Steward, R. et al., Two hybrid plasmids with *D. melanogaster* DNA sequences complementary to mRNA coding for the major heat shock protein, *Cell* 14, 921–929, 1978; Artavanis-Tsakonas, S., Schedl, P., Mirault, M.E. et al., Genes for the 70,000 dalton heat shock protein in two cloned *D. melanogaster* DNA segments, *Cell* 17, 9–18, 1979; McAlister, L. and Finklestein, D.B., Heat shock proteins and thermal resistance in yeast, *Biochem. Biophys. Res. Commun.* 93, 819–824, 1980; Wang, C., Gomer, R.H., and Lazarides, E., Heat shock proteins are methylated in avian and mammalian cells, *Proc. Natl. Acad. Sci. USA* 78, 3531–3535, 1981; Roccheri, M.C., Di Bernardo, M.G., and Giudice, G., Synthesis of heat-shock proteins in developing sea urchins, *Dev. Biol.* 83, 173–177, 1981; Lindquist, S., Regulation of protein synthesis during heat shock, *Nature* 283, 311–314, 1981; Loomis, W.F., Wheeler, S., and Schmidt, J.A., Phosphorylation of the major heat shock protein of *Dictyostelium discoideum*, *Mol. Cell Biol.* 2, 484–489, 1982; Neidhardt, F.C., VanBogelen, R.A., and Vaughn, V., The genetics and regulation of heat-shock proteins, *Annu. Rev. Genet.* 18, 295–329, 1984; Schlesinger, M.J., Heat shock proteins: the search for functions, *J. Cell Biol.* 103, 321–325, 1986; Lanks, K.W., Modulators of the eukaryotic heat shock response, *Exp. Cell Res.* 165, 1–10, 1986; Lindquist, S. and Craig, E.A., The heat-shock proteins, *Annu. Rev. Genet.* 22, 631–677, 1988; Tanguay, R.M., Transcriptional activation of heat-shock genes in eukaryotes, *Biochem. Cell Biol.* 66, 584–593, 1988; Pelham, H.R., Control of protein exit from the endoplasmic reticulum, *Annu. Rev. Cell Biol.* 5, 1–23, 1989; Bukau, B., Weissman, J., and Horwich, A., Molecular chaperones and protein quality control, *Cell* 125, 443–451, 2006; Panyai, G.S. and Corrigal, V.W., BiP regulates autoimmune inflammation and tissue damage, *Autoimmun. Rev.* 5, 140–142, 2006. More recently, there has been interest in heat-shock proteins as therapeutic targets in oncology. See Dai, C. and Whitesell, L., HSP90: a rising star on the horizon of anticancer targets, *Future Oncol.* 1, 529–540, 2005; Li, J. and Lee, A.S., Stress induction of GRP78/BiP and its role in cancer, *Curr. Mol. Med.* 6, 45–54, 2006; Kim, Y., Lillo, A.M., Steiniger, S.C.J. et al., Targeting heat shock proteins on cancer cells: selection, characterization, and cell-penetrating properties of a peptidic GRP78 ligand, *Biochemistry* 45, 9434–9444, 2006.

Hedgehog   A family of proteins important in tissue formation during embryonic development; generally expressed on cell exteriors and bind to receptors on adjacent cells. Sonic hedgehog is a glycoprotein important as a signal molecule during differentiation. See Echelard, Y., Epstein, D.J., St-Jacques, B. et al., Sonic hedgehog, a member of a family of putative signaling molecules, is implicated in the regulation of CNS polarity, *Cell* 75, 1417–1430, 1993; Johnson, R.L., Riddle, R.D., Laufer, E., and Tabin, C., Sonic hedgehog: a key mediator of anterior-posterior patterning of the limb and dorso-ventral patterning of axial embryonic structures, *Biochem. Soc. Trans.* 22, 569–574, 1994; Bumcrot, D.A. and McMahon, A.P., Somite differentiation. Sonic signals somites, *Curr. Biol.* 5, 612–614, 1995; Lum, L. and Beachy, P.A., The Hedgehog response network: sensors, switches, and routers, *Science* 304, 1755–1759, 2004; Ishibashi, M., Saitsu, H., Komada, M., and Shiota, K., Signaling cascade coordinating growth of dorsal and ventral tissues of the vertebrate brain, with special reference to the involvement of sonic hedgehog signaling, *Anat. Sci. Int.* 80, 30–36, 2005; Hooper, J.A. and Scott, M.P., Communicating with hedgehogs, *Nat. Rev. Mol. Cell Biol.* 6, 206–317, 2005; Kalderon, D., The mechanism of hedgehog signal transduction, *Biochem. Soc. Trans.* 33, 1509–1512, 2005; Davy-Grosjean, L. and Couve-Privat, S., Sonic hedgehog signaling in basel cell carcinomas, *Cancer Lett.* 225, 181–192, 2005.

Heterochromatin   "Condensed" or modified chromatin not conducive to gene transcription. See Hyde, B.B., Ultrastructure in chromatin, *Prog. Biophys. Mol. Biol.* 15, 129–148, 1965; Brown, S.W., Heterochromatin, *Science* 151, 417–425, 1966; Back, F., The variable condition of h euchromatin and heterochromatin, *Int. Rev. Cytol.* 45, 25–64, 1976; Lewis, J. and Bird, A., DNA methylation and chromatin structure, *FEBS Lett.* 285, 155–159, 1991; Wu, C.T., Transvection, nuclear structure, and chromatin proteins, *J. Cell. Biol.* 120, 587–590, 1993; Karpen, G.H., Position-effect variegation and the new biology of heterochromatin, *Curr. Opin. Genet. Dev.* 4, 281–291, 1994; Kornberg, R.D. and Lorch, Y., Interplay between chromatin structure and transcription, *Curr. Opin. Cell Biol.* 7, 371–375, 1995; Zhimulev, I.F., Polytene chromosomes, heterochromatin, and position effect variegation. *Adv. Genet.* 37, 1–566, 1998; Martin, C. and Zhang, Y., The diverse functions of histone, lysine methylation, *Nat. Rev. Mol. Cell Biol.* 6, 838–849, 2005; Wallace, J.A. and Orr-Weaver, T.L., Replication of heterochromatin: insights into mechanisms of epigenetic inheritance, *Chromosoma* 114, 389–402, 2005; Hiragami, K. and Festenstein, R., Heterochromatin protein 1: a pervasive controlling influence, *Cell. Mol. Life Sci.* 62, 2711–2726, 2005.

Heterolytic   An uneven division of a molecule such as $HCl \rightarrow H^+ + Cl^{-1}$, which usually
Cleavage,   generates ions. The hydrogenase reaction and the oxygen radical oxidation
Heterolysis   of fatty acids are examples of heterolytic cleavages. See Gardner, H.W., Oxygen radical chemistry of polyunsaturated fatty acids, *Free Radic. Biol. Med.* 7, 65–86, 1989; Fontecilla-Camps, J.C., Frey, M., Garcin, E. et al., Hydrogenase: a hydrogen-metabolizing enzyme. What do the crystal structures tell us about its mode of action? *Biochimie* 79, 661–666, 1997; Richard, J.P. and Amyes, T.L., Proton transfer at carbon, *Curr. Opin. Chem. Biol.* 5, 626–633, 2001; Solomon, E.I., Decker, A., and Lehnert, N., Non-heme iron enzymes: contrasts to heme catalysis, *Proc. Natl. Acad. Sci. USA* 100, 3589–3594, 2003; Zampella, G., Bruschi, M., Fantucci, P., and De Gioia, L., Investigation of H_2 activation by [M(NHPnPr3)('S3')] (M = Ni, Pd). Insight into key factors relevant to the design of hydrogenase functional models, *J. Amer. Chem. Soc.* 127, 13180–13189, 2005.

His-Tag   Generally a hexahistidine sequence that can be attached to the carboxyl-terminal or amino-terminal end of an expressed protein. This tag can be used for the affinity purification or separation of a protein by binding to an IMAC (immobilized metal affinity chromatography) column. The tag can also be used to provide an affinity site for interaction with another molecule in solution. See Hengen, P., Purification of His-Tag fusion proteins from *Escherichia coli*, *Trends Biochem. Sci.* 20, 285–286, 1995; Sigal, G.B., Bamdad, C., Barberis, A., Strominger, J., and Whitesides, G.M., A self-assembled monolayer for the binding and study histidine-tagged proteins by surface plasmon resonance, *Anal. Chem.* 68, 490–497, 1996; Müller, K.M., Arndt, K.M., Bauer, K., and Plückthun, A., Tandem immobilized metal-ion affinity chromatography/immunoaffinity purification of His-tagged proteins — evaluation of two anti-His-tag monoclonal antibodies, *Analyt. Biochem.* 259, 54–61, 1998; Altendorf, K., Stalz, W., Greie, J., and Deckers-Hebestreit, G., Structure and function of the F(o) complex of the ATP synthase from *Escherichia coli*, *J. Exptl. Biol.* 203, 19–28, 2000; Terpe, K., Overview of tag protein fusions: from molecular and biochemical fundamentals to commercial systems, *Appl. Microbiol. Biotechnol.* 60, 523–533, 2003; Jenny, R.J., Mann, K.G., and Lundblad, R.L., A critical review of the methods for cleavage of fusion proteins with thrombin and factor Xa, *Protein Expr. Purif.* 31, 1–11, 2003; Meredith, G.D., Wu, H.Y., and Albritton, N.L., Targeted protein functionalization using His tags, *Bioconjugate Chem.* 15, 969–982, 2004; Zhao, Y., Benita, Y., Lok, M. et al., Multi-antigen immunization using IgG binding domain ZZ as carrier, *Vaccine* 23, 5082–5090, 2005.

Hofmeister Series   Also known as the lyotropic; the order of certain ions to "salt out" or precipitate certain hydrophilic materials from aqueous solution. Polyvalent anions such as citrate and sulfate tend to precipitate while monovalent anions such as chloride and thiocyanate tend to solubilize. A similar series exists for cations. It is thought that this phenomenon is related to the ability of the various ions to bind water — hence the term "salting out." See Cacace, M.G., Landau, E.M., and Ramsden, J.J., The Hofmeister series: salt and solvent effects on interfacial phenomena, *Quart. J. Biophys.* 30, 241–277, 1997; Boström, M., Tavares, F.W., Finet, S. et al., Why forces between proteins follow different Hofmeister series for pH above and below PI, *Biophys. Chem.* 117, 217–224, 2005.

Holliday Junction   A transient structure formed between two adjoining DNA molecules during homologous recombination, which provides for the transfer of DNA sequences between the adjacent strands. See Symington, L.S. and Kolodner, R., Partial purification of an enzyme from *Saccharomyces cerevisiae* that cleaves Holliday junctions, *Proc. Natl. Acad. Sci. USA* 82, 7247–7251, 1985; Churchill, M.E., Tullius, T.D., Kallenbach, N.R., and Seeman, N.C., Holliday recombinanation intermediate is twofold symmetric, *Proc. Natl. Acad. Sci. USA* 85, 4653–4656, 1988; Dukett, D.R., Murchie, A.I., Diekmann, S. et al., The structure of the Holliday junction and its resolution, *Cell* 55, 79–89, 1988; Jeyaseelan, R. and Shanmugam, G., Human placental endonuclease cleaves Holliday junctions, *Biochem. Biophys. Res. Commun.* 156, 1054–1060, 1988; Sharples, G.J., Ingleston, S.M., and Lloyd, R.G., Holliday junction processing in bacteria: insights from the evolutionary conservation of RuvABC, RecG, and RusA, *J. Bacteriol.* 181, 5543–5550, 1999; Sharples, G.J., The X philes: structure-specific endonuclease that resolve Holliday junctions, *Mol. Microbiol.* 39, 823–834, 2001; Ho, P.S. and Eichman, B.F., The crystal structures of DNA Holliday junctions, *Curr. Opin. Struct. Biol.* 11, 302–308, 2001; Heyer, W.D., Ehmsen, K.T., and Solinger, J.A., Holliday junctions in the eukaryotic nucleus: resolution in sight? *Trends Biochem. Sci.* 28, 548–557, 2003; Heyer, W.D., Recombination: Holliday junction resolution and crossover formation, *Curr. Biol.* 14, R56–R58, 2004; Khuu, P.A., Voth, A.R., Hays, F.A., and Ho, P.S., The stacked-X DNA Holliday junction and protein recognition, *J. Mol. Recognit.* 19, 234–242, 2006.

Holoenzyme   The intact function enzyme unit that could consist of a protein, metal ions, coenzymes, and other protein components. This term was originally used to describe the combination of a coenzyme or other low-molecular weight cofactor such as a metal ion with a protein component designated as the apoenzyme to form the holoenzyme. More recently, the term holoenzyme has been used to describe DNA and RNA polymerases. See Hokin, L.E., Purification and molecular properties of te (sodium + potassium)-adenosinetriphosphatase and reconstitution of coupled sodium and potassium transport in phospholipid vesicles containing purified enzyme, *J. Exp. Zool.* 194, 197–205, 1975; Dalziel, K., McFerran, N.V., and Wonacott, A.J., Glyceraldehyde-3-phosphate dehydrogenase, *Philos. Trans. R. Soc. Lond. B Biol. Sci.* 293, 105–118, 1981; McHenry, C.S., DNA polymerase III holoenzyme. Components, structure, and mechanism of a true replicative complex, *J. Biol. Chem.* 266, 19127–19130, 1991; Ishihama, A., A multifunctional enzyme with RNA polymerase and RNase activities: molecular anatomy of influenza virus RNA polymerase, *Biochimie* 78, 1097–1102, 1996; Greenblatt, J., RNA polymerase II holoenzyme and transcriptional regulation, *Curr. Opin. Cell Biol.* 9, 310–319, 1997; Amieux, P.S. and McKnight, G.S., The essential role of RI alpha in the maintenance of regulated PKA activity, *Ann. N.Y. Acad. Sci.* 968, 75–95, 2002; Taggart, A.K. and Zakian, V.A., Telomerase: what are the est proteins doing? *Curr. Opin. Cell Biol.* 15, 275–280, 2003; Borukhov, S. and Nudler, E., RNA polymerase holoenyzme: structure, function, and biological significance, *Curr. Opin. Microbiol.* 6, 93–100, 2003; McHenry, C.S., Chromosomal replicases as asymmetric dimers: studies of subunit arrangement and functional consequences, *Mol. Microbiol.* 49, 1157–1165, 2003.

Holotype   The single specimen or illustration designated as the type for naming a species or subspecies when no type was specified. See Crickmore, N., Zeigler, D.R., Feitelson, J. et al., Revision of the nomenclature for the *Bacillus thuringiensis* pesticidal crystal proteins, *Microbiol. Mol. Biol. Rev.* 62, 807–813, 1998; Pecher, W.T., Robledo, J.A., and Vasta, G.R., Identification of a second rRNA gene unit in the *Parkinsus andrewsi* genome, *J. Eukaryot. Microbiol.* 51, 234–245, 2004.

Homeobox   A brief sequence of nucleotides whose base sequence is virtually identical in all the genes that contain said sequence. Originally described in *Drosphila*, it has now been found in many organisms including *Homo sapiens*. In the fruit fly, a homeobox appears to determine when particular groups of genes are expressed during development. Homeobox regions encode proteins containing homeodomain regions. See Gehring, W.J. and Hiromi, Y., Homeotic genes and the homeobox, *Annu. Rev. Genet.* 20, 147–173, 1986; Stern, C.D. and Keynes, R.J., Spatial patterns of homeobox gene expression in the developing mammalian CNS, *Trends Neurosci.* 11, 190–192, 1988; Kappen, C., Schughart, K., and Ruddle, F.H., Organization and expression of homeobox genes in mouse and man, *Ann. N.Y. Acad. Sci.* 567, 243–252, 1989; Wray, G.A., Transcriptional regulation and the evolution of development, *Int. J. Dev. Biol.* 47, 675–684, 2003; Del Bene, F. and Wittbrodt, J., Cell cycle control by homeobox genes in development and disease, *Semin. Cell Dev. Biol.* 16, 449–460, 2005; Samuel, S. and Naora, H., Homeobox gene expression in cancer: insights from developmental regulation and deregulation, *Eur. J. Cancer* 41, 2428–2437, 2005.

Homeodomain   A domain in a protein that is encoded for by a homeobox; these proteins are transcription factors. Homeodomains are approximately 60 amino acids in length and are composed of three α-helices and bind DNA. See Scott, M.P., Tamkun, J.W., and Hartzell, G.W., III, The structure and function of the homeodomain, *Biochim. Biophys. Acta* 989, 25–48, 1989; Affolter, M., Schier, A., and Gehring, W.J., Homeodomain proteins and the regulation of gene expression, *Curr. Opin. Cell Biol.* 2, 485–495, 1990; Izpisua-Belmonte, J.C. and Deboule, D., Homeobox genes and pattern formation in the vertebrate limb, *Dev. Biol.* 152, 26–36, 1992; Yates, A. and Chambers, I., The homeodomain protein Nanog and pluripotency in mouse embryonic stem cells, *Biochem. Soc. Trans.* 33, 1518–1521, 2005; Towle, H.C., Glucose as regulator of eukaryotic gene transcription, *Trends in Endocrinol. Metab.* 16, 489–494, 2005.

Homeotic
A shift in structural development as in a major shift in the developmental fate of an organ or body. See Dessain, S. and McGinnis, W., Regulating the expression and function of homeotic genes, *Curr. Opin. Genet. Dev.* 1, 275–282, 1991; Morata, G., Homeotic genes of *Drosophila*, *Curr. Opin. Genet. Dev.* 3, 606–614, 1993; Doboule, D. and Morata, G., Colinearity and functional hierarchy among genes of the homeotic complexes, *Trends Genet.* 10, 358–364, 1994; Mann, R.S., The specificity of homeotic gene function, *Bioessays* 17, 855–863, 1995; Duncan, I., How do single homeotic genes control multiple segment identities? *Bioessays* 18, 91–94, 1996; Graba, Y., Aragnol, D., and Pradel, J., *Drosophila Hox* complex downstream targets and the function of homeotic genes, *Bioessays* 19, 379–388, 1997; Reichert, H. and Simone, A., Conserved usage of gap and homeotic genes in patterning the CNS, *Curr. Opin. Neurobiol.* 9, 589–595, 1999; Irish, V.F., The evolution of floral homeotic gene function, *Bioessays* 25, 637–646, 2003; Zubko, M.K., Mitochondrial tuning fork in nuclear homeotic functions, *Trends Plant Sci.* 9, 61–64, 2004. See also *HOX Genes.*

Homolytic
Cleavage,
Homolysis
An even division of a molecule such as HCl → H· + Cl,· which generates free radicals. The decomposition of a precursor molecule can proceed via either a homolytic pathway, a heterolytic pathway, or both. See White, R.E., Sligar, S.G., and Coon, M.J., Evidence for a hemolytic mechanism of peroxide oxygen–oxygen bond cleavage during substrate hydroxylation by cytochrome P-450, *J. Biol. Chem.* 255, 11108–11011, 1980; Yang, G., Candy, T.E., Boaro, M. et al., Free radical yields from the homolysis of peroxynitrous acid, *Free Radic. Biol. Med.* 12, 327–330, 1992; Correia, M.A., Yao, K., Allentoff, A.J. et al., Interactions of peroxy quinols with cytochromes P450 2B1, 3A1, and 3A5: influence of the apoprotein on heterocyclic versus hemolytic O–O bond cleavage, *Arch. Biochem. Biophys.* 317, 471–478, 1995; Barr, D.P., Martin, M.V., Guengerich, F.P., and Mason, R.P., Reaction of cytochrome P450 with cumene hydroperoxide: ESR spin-trapping evidence for the hemolytic scission of the peroxide O–O bond by ferric cytochrome P450 1A2, *Chem. Res. Toxicol.* 9, 318–325, 1996; Marsh, E.N. and Ballou, D.P., Coupling of cobalt–carbon bond homolysis and hydrogen atom abstraction in adenosylcobalamin-dependent glutamate mutase, *Biochemistry* 37, 11864–11872, 1998; Licht, S.S., Booker, S., and Stubbe, J.,. Studies on the catalysis of carbon–cobalt bond homolysis by ribonucleoside triphosphate reductase: evidence for concerted carbon–cobalt bond homolysis and thiyl radical formation, *Biochemistry* 38, 1221–1233, 1999; Vlasie, M.D. and Banerjee, R., Tyrosine 89 accelerates Co-carbon bond homolysis in methylmalonyl-CoA mutase,

J. Am. Chem. Soc. 125, 5431–5435, 2003; Lymar, S.V., Khairutdinov, R.F., and Hurst, J.K., Hydroxyl radical formation by O–O bond homolysis in peroxynitrous acid, *Inorg. Chem.* 42, 5259–5266, 2003; Rees, M.D. and Davies, M.J., Heparan sulfate degradation via reductive homolysis of its *N*-chloro derivatives, *J. Am. Chem. Soc.* 128, 3085–3097, 2006.

Homotype A structure having the same general function as another, which may or may not be opposing. For example, the left arm is a homotype of the right arm. A selectin can be a homotype of another selectin. Homotypic is a descriptor referring to homotype. See Rouhandeh, H., Yau, T., and Lang, P.A., Homotypic and heterotypic interference among picornovirus ribonucleic acids, *Arch. Gesamte Virusforsch.* 27, 236–243, 1969; Bendini, C., Lanfranchi, A., Nobili, R., and Miyake, A., Ultrastructure of meiosis-inducing (heterotypic) and noninducing (homotypic) cell unions in conjugation of *Blepharisma, J. Cell Sci.* 32, 31–43, 1978; Daunter, B., Immune response: tissue specific T-lymphocytes, *Med. Hypotheses* 37, 76–84, 1992; Wagner, M.C., Molnar, E.E., Molitoris, B.A., and Goebl, M.G., Loss of the homotypic fusion and vacuole protein sorting or Golgi-associated retrograde protein vesicle tethering complexes results in gentamicin sensitivity in the yeast *Saccharmyces cerevesiae, Antimicrob. Agents Chemother.* 50, 587–595, 2006; Karaulanov, E.E., Bottcher, R.T., and Niehrs, C., A role for fibronectin-leucine-rich transmembrane cell-surface proteins in homotypic cell adhesion, *EMBO Rep.* 7, 283–290, 2006; Brandhorst, D., Zwilling, D., Rizzoli, S.O. et al., Homotypic fusion of early endosomes: SNAREs do not determine fusion specificity, *Proc. Natl. Acad. Sci. USA* 103, 2701–2706, 2006; Decker, B.L. and Wickner, W.T., Enolase activates homotypic vacuole fusion and protein transport to the vacuole in yeast, *J. Biol. Chem.,* 281, 14523–14528, 2006; Stroupe, C., Collins, K.M., Fratti, R.A., and Wickner, W., Purification of active HOPS complex reveals its affinities for phosphoinositides and the SNARE Vam7p, *EMBO J.* 25, 1579–1589, 2006; Brereton, H.C., Carvell, M.J., Asare-Anane, H. et al., Homotypic cell contact enhances insulin but not glucagon secretion, *Biochem. Biophys. Res. Commun.* 344, 995–1000, 2006.

Hoogsteen Bond The hydrogen bonds formed in the hybridization of DNA chains to form a triple helix. See Searle, M.S. and Wickham, G., Hoogsteen versus Watson-Crick A-T basepairing in DNA complexes of a new group of "quinomycin-like" antibiotics, *FEBS Lett.* 272, 171–174, 1990; Raghunathan, G., Miles, H.T., and Sasisekharan, V., Symmetry and structure of RNA and DNA triple helices, *Biopolymers* 36, 333–343, 1995; Soliva, R., Luque, F.J., and Orozco, M., Can G-C Hoogsteen-wooble pairs contribute to the stability of d(G, C-C) triplexes? *Nucleic Acids Res.* 27, 2248–2255, 1999; Li, J.S., Shikiya, R., Marky, L.A., and Gold, B., Triple helix forming TRIPside molecules that target mixed purine/pyrimidine DNA sequences, *Biochemistry* 43, 1440–1448, 2004.

Hormonology The study of hormones; it has been proposed as a substitute for endocrinology. See Ross, J.W., Hormonology in obstetrics, *J. Natl. Med. Assoc.* 46, 19–21, 1954; Swain, C.T., Hormonology, *N. Engl. J. Med.* 280, 388–389, 1969; Kulinskii, V.I. and Kolesnichenko, L.S., Current aspects of hormonology, *Biochemistry (Mosc.)* 62, 1171–1173, 1997; Holland, M.A., Occam's razor applied to hormonology (are cytokines produced by plants?), *Plant Physiol.* 115, 865–868, 1997; Hadden, D.R., One hundred years of hormonology: a view from No. 1 Wimpole Street, *J. R. Soc. Med.*

98, 325–326, 2005; Hsueh, A.J.W., Bouchard, P., and Ben-Shlomo, I., Hormonology: a genomic perspective on hormonal research, *J. Endocrinol.* 187, 333–338, 2005.

HOX Genes Encodes a family of transcription factors, *Hox* proteins. See Sekimoto, T., Yoshinobu, K., Yoshida, M. et al., Region-specific expression of murine *Hox* genes implies the *Hox* code-mediated patterning of the digestive tract, *Genes to Cells* 3, 51–64, 1998; Hoegg, S. and Meyer, A., *Hox* clusters as models for vertebrate genome evolution, *Trends Genet.* 21, 421–424, 2005; Morgan, S., *Hox* genes: a continuation of embryonic patterning? *Trends Genet.* 22, 67–69, 2006.

Hydrogels An easily deformed pseudo-solid mass formed from largely hydrophilic colloids dispersed in an aqueous medium (dispersion medium or continuous phase). There is considerable interest in the use of hydrogels for drug delivery. See Jhon, M.S. and Andrade, J.D., Water and hydrogels, *J. Biomed. Mat. Res.* 7, 509–522, 1973; Dusek, K., *Reponsive Gels; Volume Transitions*, Springer-Verlag, Berlin, 1993; Roorda, W., Do hydrogels contain different classes of water? *J. Biomater. Sci. Polym. Ed.* 5, 383–395, 1994; Dumitriu, S., *Polymeric Biomaterials*, Marcel Dekker, New York, 1994; Zrinyl, N., *Gels*, Springer, Darmstadt, Germany, 1996; McCormick, C.L., *Stimuli-Responsive Water Soluble and Amphiphilic Polymers*, American Chemical Society, Washington, DC, 2001; Dumitriu, S., *Polymeric Biomaterials*, Marcel Dekker, New York, 2002; Omidian, H., Rocca, J.G., and Park, K., Advances in superporous hydrogels, *J. Control. Release* 102, 3–12, 2005; Frokjaer, S. and Otzen, D.E., Protein drug stability: a formulation challenge, *Nat. Rev. Drug Discov.* 4, 298–306, 2005; Kashyap, N., Kumar, N., and Kumar, K.N., Hydrogels for pharmaceutical and biomedical applications, *Crit. Rev. Ther. Drug Carrier Syst.* 22, 107–149, 2005; Fairman, R. and Akerfeldt, K.S., Peptides as novel smart materials, *Curr. Opin. Struct. Biol.* 15, 453–463, 2005; Young, S., Wong, M., Tabata, Y., and Mikos, A.G., Gelatin as a delivery vehicle for the controlled release of bioactive molecules, *J. Control. Release* 109, 256–274, 2005.

Hydrophobic, Hydrophobic Effect, Hydrophobic Forces Tendency of a molecular structure to avoid water, which results in an association or clustering of hydrophobic groups. The term nonpolar is frequently used to describe such groups or molecules. Polar and nonpolar groups or functions can exist in the same molecule; for example, the ε-amino group of lysine is polar, but the methylene carbon chain between the ε-amino group and the α-carbon is nonpolar. See Kauzmann, W., Some forces in the interpretation of protein denaturation, *Adv. Prot. Chem.* 14, 1–63, 1959; Tanford, C., The hydrophobic effect and the organization of living matter, *Science* 200, 1012–1018, 1978; Kumar, S. and Nussinov, R., Close-range electrostatic interactions in proteins, *ChemBioChem* 3, 604–617, 2002; Kyte, J., The basis of the hydrophobic effect, *Biophys. Chem.* 100, 193–203, 2003; Lesk, A.M., Hydrophobicity-getting into hot water, *Biophys. Chem.* 105, 179–182, 2003; Seelig, J., Thermodynamics of lipid-peptide interactions, *Biochim. Biophys. Acta* 1666, 40–50, 2004; Hofinger, S. and Zerbetto, F., Simple models for hydrophobic hydration, *Chem. Soc. Rev.* 34, 1012–1020, 2005; Chander, D., Interfaces and the driving force of hydrophobic assembly, *Nature* 437, 640–647, 2005.

Hydrophobins Hydrophobins are secreted proteins functioning in fungal growth and development. Hydrophobins self-assemble at hydrophilic/hydrophobic interfaces forming amphipathic membranes. See Wessels, J., De Vries, O.,

Asgeirsdottir, S.A., and Schuren, F., Hydrophobin genes involved in formation of aerial hyphae and fruit bodies in *Schizophyllum*, *Plant Cell* 3, 793–799, 1991; Wessels, J.G., Hydrophobins: proteins that change the nature of the fungal surface, *Adv. Microb. Physiol.* 38, 1–45, 1997; Ebbole, D.J., Hydrophobins and fungal infection of plants and animals, *Trends Microbiol.* 5, 405–408, 1997; Wosten, H.A., Hydrophobins: multipurpose proteins, *Annu. Rev. Microbiol.* 55, 625–646, 2001; Linder, M.B., Szilvay, G.R., Nakari-Setala, T., and Penttila, M.E., Hydrophobins: the protein-amphiphiles of filamentous fungi, *FEMS Microbiol. Rev.* 29, 877–896, 2005.

Hypsochromic   A shift of light absorption or emission to a shorter wavelength ($\lambda < \lambda_o$); a "blue" shift. See Crescitelli, F. and Karvaly, B., The gecko visual pigment: the anion hypsochromic shift, *Vision Res.* 31, 945–950, 1991; Zalis, S., Sieger, M., Greulich, S. et al., Replacement of the 2,2′-bipyridine by 1,4-diazabutadiene acceptor ligands: why the bathochromic shift for [N empty set N)IrCl(C5Me5)] + complexes but the hypsochromic shift for (N empty set N)Ir(C5Me5)? *Inorg. Chem.* 42, 5185–5191, 2003; Meier, H., Gerold, J., Kolshrn, H., and Muhling, B., Extension of conjugation leading to bathochromic or hypsochromic effects in OPV series, *Chemistry* 23, 360–370, 2004; de Garcia Ventrini, C., Andreaus, J., Machado, V.G., and Machado, C., Solvent effects in the interaction of methyl-β-cyclodextrin with solvatochromic merocyanine dyes, *Org. Biomol. Chem.* 3, 1751–1756, 2005; Kidman, G. and Northrop, D.B., Effect of pressure on nucleotide binding to yeast alcohol dehydrogenase, *Protein Pept. Lett.* 12, 495–497, 2005; Li, Y., He, W., Dong, Y. et al., Human serum albumin interaction with formononetin studied using fluorescence anisotropy, FT-IR spectroscopy, and molecular modeling methods, *Bioorg. Med. Chem.* 14, 1431–1436, 2006; Schonefeld, K., Ludwig, R., and Feller, K.H., Fluorescence studies of host-guest interaction of a dansyl amide labeled calyx[6]arene, *J. Fluoresc.* 16, 449–454, 2006; Correa, N.M. and Levinger, N.E., What can you learn from a molecular probe? New insights on the behavior of C343 in homogeneous solutions and AOT reverse micelles, *J. Phys. Chem. B Condens. Matter Surf. Interfaces Biophys.* 110, 13050–13061, 2006.

Idiotypic   Refers to an idiotype where the idiotype is that portion of the variable region of an antibody that confers specificity. See Bigazzi, P.E., Regulation of autoimmunity and the idiotypic network, *Immunol. Ser.* 54, 39–64, 1991; Schoenfeld, Y., Idiotypic induction of autoimmunity: do we need an autoantigen? *Clin. Exptl. Rheumatol.* 12 (Suppl. 11), S37–S40, 1994; Schoenfeld, Y. and George, J., Induction of autoimmunity. A role for the idiotypic network, *Ann. N.Y. Acad. Sci.* 815, 342–349, 1997; Bianchi, A. and Massaia, M., Idiotypic vaccination in B-cell malignancies, *Mol. Med. Today* 3, 435–441, 1997; Lacroix-Desmazes, S., Bayry, J., Misra, N. et al., The concept of idiotypic vaccination against factor VIII inhibitors in haemophilia A, *Haemophilia* 8 (Suppl. 2), 55–59, 2002; Coutinho, A., Will the idiotypic network help to solve natural tolerance? *Trends Immunol.* 24, 53–54, 2003.

IMAC   A chromatographic fractionation procedure that uses a matrix consisting of
(Immobilized   a metal ion tightly bound to a matrix. Nickel is the most common metal
Metal Affinity   ion used but there is use of copper and other transition metals. See Porath,
Chromatography) J. and Olin, B., Immobilized metal ion affinity adsorption and immobilized
                     metal ion affinity chromatography of biomaterials. Serum protein affinities

for gel-immobilized iron and nickel ions, *Biochemistry* 23, 1621–1630, 1982; Porath, J., Immobilized metal ion affinity chromatography, *Protein Expr. Purif.* 3, 263–281, 1992; Skerra, A., Engineered protein scaffolds for molecular recognition, *J. Mol. Recognit.* 13, 167–187, 2000; Gaberc-Proekar, V. and Menart, V., Perspectives of immobilized-metal affinity chromatography, *J. Biochem. Biophys. Methods* 49, 335–360, 2001; Ueda, E.K., Gout, P.W., and Morganti, L., Current and prospective applications of metal ion-protein binding, *J. Chromatog. A* 988, 1–23, 2003.

Imino Sugars
A class of carbohydrate mimetics that contain nitrogen in the place of oxygen in the ring. These sugars inhibit glycosylation reactions by acting as transition state analogues. See Paulsen, H. and Brockhausen, I., From imino sugars to cancer glycoproteins, *Glycoconjugate J.* 18, 867–870, 2001; Dwek, R.A., Butters, T.D., Platt, F.M., and Zitzmann, N., Targeting glycosylation as a therapeutic approach, *Nat. Rev. Drug Disc.* 1, 65–75, 2002; El-Ashry, E.-S.H., and El Nemr, A., Synthesis of mono- and di-hydroxylated prolines and 2-hydroxymethylpyrrolidines from non-carbohydrate precursors, *Carbohyd. Res.* 338, 2265–2290, 2003; Butters, T.D., Dwek, R.A., and Platt, F.M., New therapeutics for the treatment of glycosphingolipid lysosomal storage diseases, *Adv. Exp. Med. Biol.* 535, 219–226, 2003; Butters, T.D., Dwek, R.A., and Platt, F.M., Imino sugar inhibitors for treating the lysosomal glycosphingolipidoses, *Glycobiology* 14, 43R–52R(epub), 2005.

Immunoblotting
A technique for the identification of immunoreactive substances such as proteins. Most frequently, detection by immunoblotting first involves a gel-based electrophoretic separation step followed by electrophoretic transfer to another matrix such as nitrocellulose or PVDF in a manner such that the original separation pattern is retained. The separated proteins are measured by reaction with a primary probe such as an antibody labeled with an enzyme or other signal; it is also possible to use a secondary probe that would react with the primary probe. A secondary probe could be an antibody with a signal such as an enzyme. This latter situation is similar to an indirect ELISA. See Bjerrum, O.J. and Heegaard, N.H.H., Eds., *CRC Handbook of Immunoblotting of Proteins*, CRC Press, Boca Raton, FL, 1988; Harlow, E. and Lane, D., Eds., *Antibodies: A Laboratory Manual*, Cold Spring Harbor, NY, 1988; Manchenko, G.P., *Handbook of Detection of Enzymes on Electrophoresis*, CRC Press, Boca Raton, FL, 1994; Stot, D.I., Immunoblotting, dot-blotting, and ELISPOT assay: methods and applications, in *Immunochemistry*, Van Oss, C.J. and van Regenmortel, M.H.V., Eds., Marcel Dekker, New York, 1994, pp. 925–948; Burns, R., *Immunochemical Protocols*, Humana Press, Totowa, NJ, 2005. Western blotting is a form of immunoblotting. See *ELISA, Elispot, Western Blotting*.

Immunoglobulin
A group of plasma proteins (Ig) that are synthesized by plasma cells, which are formed from B-cells. There are five general classes of immunoglobulins: IgA, IgE, IgD, IgG, and IgM. With the exception of some unique immunoglobulins such as camelids, immunoglobulins are based on a structure of dimers or heterodimers where the heterodimers are composed of a light chain and a heavy chain. IgM is a pentamer of this basic building block while IgA can be a monomer, dimer, or trimer of the basic building block. The basic building block is bivalent in that each heterodimer can

bind an antigen; IgA may be bivalent, tetravalent, or hexavalent while IgM is decavalent. See Pernis, B. and Vogel, H.J., *Cells of Immunoglobulin Synthesis*, Academic Press, New York, 1979; Calabi, F. and Neuberger, M.S., *Molecular Genetics of Immunoglobulin,* Elsevier, Amsterdam, 1987; Langone, J.J., *Antibodies, Antigens, and Molecular Mimicry,* Academic Press, San Diego, CA, 1989; Kuby, J., *Immunology,* W.H. Freeman, New York, 1992; Cruse, J.M. and Lewis, R.E., *Atlas of Immunology,* CRC Press, Boca Raton, FL, 1999.

Immunoglobulin Superfamily   A family of cell surface glycoproteins that contain an extracellular domain homologous to immunoglobulin (Ig), a transmembrane component, and a cytoplasmic extension, and which interact with other cell adhesion molecules such as integrins in homotypic interactions. See Anderson, P., Morimoto, C., Breitmeyer, J.B., and Schlossman, S.F., Regulatory interactions between members of the immunoglobulin superfamily, *Immunol. Today* 9, 199–203, 1988; Hunkapiller, T. and Hood, L., Diversity of the immunoglobulin gene superfamily, *Adv. Immunol.* 44, 1–63, 1989; Barclay, A.N., Membrane proteins with immunoglobulin-like domains — a master superfamily of interaction molecules, *Semin. Immunol.* 15, 215–223, 2003; Naka, Y., Bucciarelli, L.G., Wendt, T. et al., RAGE axis: animal models and novel insights into the vascular complications of diabetes, *Arterioscler. Thromb. Vasc. Biol.* 24, 1342–1349, 2004; Mittler, R.S., Foell, J., McCausland, M. et al., Anti-CD137 antibodies in the treatment of autoimmune disease and cancer, *Immunol. Res.* 29, 197–208, 2004; Du Pasquier, L., Zucchetti, I., and De Santis, R., Immunoglobulin superfamily receptors in protochordates: before RAG time, *Immunol. Rev.* 198, 233–248, 2004; Peggs, K.S. and Allison, J.P., Co-stimulatory pathways in lymphocyte regulation: the immunoglobulin superfamily, *Br. J. Haematol.* 130, 809–824, 2005.

Immunomics   Study of the molecular functions associated with all immune-related coding and noncoding mRNA transcripts. See Maecker, B., von Bergwelt-Baildon, M., Anderson, K.S., Vonderheide, R.H., and Schultze, J.L., Linking genomics to immunotherapy by reverse immunology — "immunomics" in the new millennium, *Curr. Mol. Med.* 1, 609–619, 2001; Schonbach, C., From immunogenetics to immunomics: functional prospecting of genes and transcripts, *Novartis Found. Symp.* 254, 177–188, 2003.

Immuno Proteasome   A type of proteasome that is involved in processing proteins for MHC class I antigen presentation. See Aki, M., Shimbara, N., Takashina, M. et al., Interferon-gamma induces different subunit organizations and functional diversity of proteasomes, *J. Biochem.* 115, 257–269, 1994; Dahlmann, B., Ruppert, T., Kuehn, L. et al., Different proteasome subtypes in a single tissue exhibit different enzymatic properties, *J. Mol. Biol.* 303, 643–653, 2000; Tenzer, S., Stoltze, L., Schonfisch, B. et al., Quantitative analysis of prion-protein degradation by constitutive and immuno-20S proteasomes indicates differences correlated with disease susceptibility, *J. Immunol.* 172, 1083–1091, 2004; Dahlmann, B., Proteasomes, *Essays Biochem.* 41, 31–48, 2005; Ishii, K., Hisaeda, H., Duan, X. et al., The involvement of immunoproteomics in induction of MHC class I-restricted immunity targeting Toxoplasma SAG1, *Microbes Infect.* 8, 1045–1053, 2006.

Immuno- proteomics   Definition is a work in progress varying from the screening of two-dimensional gels for reactive antibodies to the use of mass spectrometry to study targets of the immune system; in general, the use of proteomics to study the cellular and humoral immune systems. See Haas, G., Karaali, G., Ebermayer,

K. et al., Immunoproteomics of *Helicobacter pylori* infection and relation to gastric disease, *Proteomics* 2, 313–324, 2002; Purcell, A.W. and Gorman, J.J., Immunoproteomics: mass spectrometry–based methods to study the targets of the immune response, *Mol. Cell. Proteomics* 3, 193–208, 2004; Chen, Z., Peng, B., Wang, S., and Peng, X., Rapid screening of highly efficient vaccine candidates by immunoproteomics, *Proteomics* 4, 3203–3213, 2004; Hess, J.L., Blazer, L., Romer, T. et al., Immunoproteomics, *J. Chromatog. B Analyt. Technol Biomed. Life Sci.* 815, 65–75, 2005; Paul-Satyaseela, M., Karched, M., Bian, Z. et al., Immunoproteomics of *Actinobacillus actinomycetemcomitans* outer-membrane proteins reveal a highly immunoreactive peptidoglycan-associated lipoprotein, *J. Med. Microbiol.* 55, 931–942, 2006; Falisse-Poirrier, N., Ruelle, V., Elmoualij, B. et al., Advances in immunoproteomics for serological characterization of microbial antigens, *J. Microbiol. Methods*, 67, 593–596, 2006.

Immunostimula-tory Sequence (ISS); Immuno-stimulatory Sequence Oligodeoxy-nucleotide (ISS-ODN) A specific sequence nonmethylated DNA containing cytosine and guanine. It has immunostimulatory properties and can serve as an adjuvant. See Horner, A.A., Ronaghy, A., Cheng, P.M. et al., Immunostimulatory DNA is a potent mucosal adjuvant, *Cell. Immunol.* 190, 77–82, 1998; Miyazaki, D., Liu, G., Clark, L., and Ono, S.J., Prevention of acute allergic conjunctivitis and late-phase inflammation with immunostimulatory DNA sequences, *Invest. Ophthalmol.* 41, 3850–3855, 2000; Horner, A.A. and Raz, E., Immunostimulatory sequence oligodeoxynucleotide: a novel mucosal adjuvant, *Clin. Immunol.* 95, S19–S29, 2000; Marshall, J.D., Abtahi, S., Eiden, J.J. et al., Immunostimulatory sequence DNA linked to the Amb a 1 allergen promotes T(H) 1 cytokine expression while downregulating T(H)2 cytokine expression in PBMCs from human patients with ragweed allergy, *J. Allergy Clin. Immunol.* 108, 191–197, 2001; Horner, A.A. and Raz, E., Immunostimulatory sequence oligodeoxynucleotide-based vaccination and immunomodulation: two unique but complementary strategies for the treatment of allergic diseases, *J. Allergy Clin. Immunol.* 110, 706–712, 2002; Teleshova, N., Kenney, J., Williams, V. et al., CpG ISS-ODN activation of blood-derived B-cells from healthy and chronic immunodeficiency virus–infected macaques, *J. Leukoc. Biol.* 79, 257–267, 2006.

Industrial Plantation A large-scale, usually single-crop forestry or agricultural enterprise.

Infrared Spectroscopy The common range for infrared spectroscopy is 10–12,800 cm^{-1} (780–10^6 nm). Absorption spectra are described as a function of the wavenumber of the incident; the wavenumber ( ) is the reciprocal of the wavelength and has the advantage of being linear with energy. The infrared region can be divided into near-infrared, mid-infrared, and far-infrared regions.

Inhibin Inhibin is a dimeric glycoprotein secreted by the follicular cells of the ovary and the Sertoli cells of the testes; inhibin regulates secretion of follicle-stimulating hormones from the anterior pituitary. Inhibin has received recent attention as a biomarker for ovarian cancer. See Chari, S., Chemistry and physiology of inhibin — a review, *Endokrinologie* 70, 99–107, 1977; Grady, R.R., Charlesworth, M.C., and Schwartz, N.B., Characterization of the FSH-suppressing activity in follicular fluids, *Recent Prog. Horm. Res.* 38, 409–456, 1982; Schwartz, N.B., Role of ovarian inhibin (folliculostatin) in regulating FSH secretion in the female rat, *Adv. Exp. Med. Biol.* 147, 15–36, 1982; Burger, H.G. and Igarashi, M., Inhibin: definition and nomenclature, including related substances, *Mol. Endocrinol.* 2,

391–392, 1988; Robertson, D.M., Stephenson, T., Cahir, N. et al., Development of an inhibin subunit ELISA with broad specificity, *Mol. Cell. Endocrinol.* 180, 79–86, 2001; Robertson, D.M., Stephenson, T., Pruysers, E. et al., Inhibins/activins as diagnostic markers for ovarian cancers, *Mol. Cell. Endocrinol.* 191, 97–103, 2002; Khosravi, J., Krishna, R.G., Khaja, N., Bodani, U., and Diamandi, A., Enzyme-linked immunosorbent assay of total inhibin: direct determination based on inhibin subunit-specific monoclonal antibodies, *Clin. Biochem.* 37, 370–376, 2004; Cook, R.W., Thompson, T.B., Jardtzky, T.S., and Woodruff, T.K., Molecular biology of inhibin action, *Semin. Reprod. Med.* 22, 269–276, 2004.

Insulin Receptor   A heterotetramer consisting of two extracellular alpha subunits that bind insulin and two transmembrane beta subunits that have tyrosine kinase activity. See Chang, L., Chiang, S.-H., and Saltiel, A.R., Insulin signaling and the regulation of glucose transport, *Molec. Med.* 10, 65–71, 2004; Kanzaki, M., Insulin receptor signals regulating GLUT4 translocation and actin dynamics, *Endocr. J.* 53, 267–293, 2006; Martinez, S.C., Cras-Meneur, C., Bernal-Mizrachi, E., and Permutt, M.A., Glucose regulates Fox01 through insulin receptor signaling in the pancreatic islet β-cells, *Diabetes* 55, 1581–1591, 2006; Marine, S., Zamiara, E., Todd Smith, S. et al., A miniaturized cell-based fluorescence resonance energy transfer assay for insulin-receptor activation, *Anal. Biochem.* 355, 267–277, 2006; Hao, C., Whittaker, L., and Whittaker, J., Characterization of a second ligand binding site of the insulin receptor, *Biochem. Biophys. Res. Commun.* 347, 334–339, 2006; Sisci, D., Morelli, C., and Garofalo, C., Expression of nuclear insulin receptor substrate 1 (IRS-1) in breast cancer, *J. Clin. Pathol.*, in press, 2006.

Integrin-   Pathway involved in the membrane transport of ferric iron. See Conrad, M.E,
Mobilferrin   Umbreit, J.N., Peterson, R.D. et al., Function of integrin in duodenal mucosal
Pathway (IMP)   uptake of iron, *Blood* 81, 517–521, 1993; Wolf, G. and Wessling-Resnick, M., An integrin-mobilferrin iron transport pathway in intestine and hematopoietic cells, *Nutr. Rev.* 52, 387–389, 1994; Conrad, M.E. and Umbreit, J.N., Iron absorption and transport — an update, *Amer. J. Hematol.* 64, 287–298, 2000; Umbreit, J.N., Conrad, M.E., Hainsworth, L.N., and Simovich, M., The ferrireductase paraferritin contains divalent metal transporter as well as mobilferrin, *Am. J. Physiol. Gastrointest. Liver Physiol.* 282, G534–G539, 2002.

Integrins   Cell membrane glycoproteins that function as receptor for extracellular matrix components. Integrins are heterodimers containing an α-subunit and a β-subunit. The β-subunit contains RGD sequences that "recognize" ligands such as fibronectin, platelet glycoprotein II b/IIIa, and extracellular matrix components or structural analogues or homologues. See Akiyama, S.K., Yamada, K.M., and Hayashi, J., The structure of fibronectin and its role in cellular adhesion, *J. Supramol. Struct. Cell. Biochem.* 16, 345–348, 1981; Mosher, D.F., Physiology of fibronectin, *Annu. Rev. Med.* 35, 561–575, 1984; Hynes, R.O., Integrins: a family of cell surface receptors, *Cell* 48, 549–554, 1987; Malech, H.L. and Gallin, J.I., Current concepts: immunology. Neutrophils in human diseases, *N. Engl. J. Med.* 317, 687–694, 1987; Bennett, J.S., Structure and function of the platelet integrin II3, *J. Clin. Invest.* 115, 3363–3369, 2005; Serini, G., Valdembri, D., and Bussolino, F., Integrins and angiogenesis: a sticky business, *Exp. Cell Res.* 312, 651–658, 2005; Caswell, P.T. and Norman, J.C., Integrin trafficking and the control of cell migration, *Traffic* 7, 14–21, 2006; Legate, K.R., Montanez, E., Kudlacek, O., and Fassler, R., ILK, PINCH, and parvin: the tIPP of integrin signalling, *Nat. Rev. Mol. Cell Biol.* 7, 20–31, 2006.

Intein   Intervening protein sequences that are removed by posttranslational self-splicing; analogous to exon splicing. Intein regions are surrounded by an *N*-terminal extein and a *C*-terminal extein. Intein splicing has proved useful for the preparation of *N*-terminal cysteine residues, which can be coupled to a matrix. See Colston, M.J. and Davies, E.O., The ins and outs of protein splicing elements, *Mol. Microbiol.* 12, 359–363, 1994; Cooper, A.A. and Stevens, T.H., Protein splicing: self-splicing of genetically mobile elements at the protein level, *Trends Biochem. Sci.* 20, 351–356, 1995; Paulus, H., Protein splicing and related forms of protein autoprocessing, *Annu. Rev. Biochem.* 69, 447–496, 2000; Xu, M.Q. and Evans, T.C., Jr., Intein-mediated ligation and cyclization of expressed proteins, *Methods* 24, 257–277, 2001; Durek, T. and Becker, C.F., Protein semi-synthesis: new problems for functional and structural studies, *Biomol. Eng.* 22, 153–172, 2005; Tan, L.P. and Yao, S.Q., Intein-mediated, *in vitro* and *in vivo* protein modifications with small molecules, *Protein Pep. Lett.* 12, 769–775, 2005; Anderson, L.L., Marshall, G.R., and Baranski, T.J., Expressed protein ligation to study protein interactions: semi-synthesis of the G-protein alpha subunit, *Protein Pep. Lett.* 12, 783–787, 2005; Eckenroth, B., Harris, K., Turanov, A.A. et al., Semisynthesis and characterization of mammalian thioredoxin reductase, *Biochemistry* 45, 5158–5170, 2006; Hackenberger, C.P., Chen, M.M., and Imperiali, B., Expression of *N*-terminal cys-protein fragments using an intein refolding strategy, *Bioorg. Med. Chem.*, 14, 5043–5048, 2006; Sharma, S.S., Chong, S., and Harcum, S.W., Intein-mediated protein purification of fusion proteins expressed under high-cell density conditions in *E.coli*, *J. Biotechnol.*, 125, 48–56, 2006; Kwon, Y., Coleman, M.A., and Camarero, J.A., Selective immobilization of proteins onto solid supports through split-intein-mediated protein trans-splicing, *Angew. Chem. Int. Ed. Engl.* 45, 1726–1729, 2006.

Interactome   The protein–protein interactions within a proteome. See Ito, T., Chiba, T., and Yoshida, M., Exploring the protein interactome using comprehensive two-hybrid projects, *Trends Biotechnol.* 19 (Suppl. 10), S23–S27, 2001; Ito, T., Ota, K., Kubota, H. et al., Roles for the two-hybrid system in exploration of the yeast protein interactome, *Mol. Cell. Proteomics* 1, 561–566, 2002; Vidal, M., Interactome modeling, *FEBS Lett.* 579, 1834–1838, 2005; Cusick, M.E., Klitgord, N., Vidal, M., and Hill, D.E., Interactome: gateway into systems biology, *Hum. Mol. Genetics* 15 (14 Spec. No. 2), R171–R181, 2005; Ghavidel, A., Cagney, G., and Emili, A., A skeleton on the human protein interactome, *Cell* 122, 830–832, 2005.

Intercalation   The insertion of a molecule, usually planer, between adjacent base pairs of DNA.

Interleukin   A functionally defined group of small proteins that "communicate" between various immune cell types (inter + leukocytes = interleukin) (Aardem, L.A., Brunner, T.K., Creottini, J.C. et al., Revised nomenclature for antigen-nonspecific T-cell proliferation and helper factors, *J. Immunol.* 123, 2928–2929, 1979; Paul, W.E., Kishimoto, T., Melchers, F. et al., Nomenclature for secreted regulatory proteins of the immune system [interleukins], *Clin. Immunol. Immunopathol.* 64, 3–4, 1992; IUIS/WHO Standing Committee on Interleukin Designation, Nomenclature for secreted regulatory proteins of the immune system [interleukins]: update, *Bull. World*

Health Org. 75, 175, 1997). This term was developed to rationalize the nomenclature for these materials as the different terms/names were selected on the basis of activity in a particular assay system rather than an intrinsic physical or biological property; this situation is not unlike that which occurred in blood coagulation somewhat earlier. Thus, lymphocyte-activating factor (LAF; mitogenic protein, B-cell differentiation factor) is IL-1, while thymocyte-stimulating factor (TSF, T-cell growth factor, killer cell helper factor) is IL-2 (see Watson, J. and Mochizuki, D., Interleukin 2: a class of T-cell growth factors, *Immunol. Rev.* 51, 287–278, 1980; Mizel, S.B., Interleukin 1 and T-cell activation, *Immunol. Rev.* 63, 51–72, 1982; Wagner, H., Hardt, C., Heeg, K. et al., The *in vivo* effects of interleukin 2 (TCGF), *Immunobiology* 161, 139–156, 1982; Farrar, J.J., Benjamin, W.R., Hilfiker, M.L. et al., The biochemistry, biology, and role of interleukin 2 in the induction of cytotoxic T-cell and antibody-forming B-cell responses, *Immunol. Rev.* 63, 129–166, 1982; Gillis, S., Interleukin 2: biology and biochemistry, *J. Clin. Immunol.* 3, 1–13, 1983; Durum, S.K., Schmidt, J.A., and Oppenheim, J.J., Interleukin 1: an immunological perspective, *Annu. Rev. Immunol.* 3, 263–287, 1985). Work on the interleukins is usually considered within the greater area of cytokines. See Porter, J.R. and Jezová, D., *Circulating Regulatory Factors and Neuroendocrine Function,* Plenum Press, New York, 1990; Kimball, E.S., *Cytokines and Inflammation,* CRC Press, Boca Raton, FL, 1991; Kishimoto, T., *Interleukins: Molecular Biology and Immunology,* Karger, Basal, 1992; Thrompson, A.W., *The Cytokine Handbook,* Academic Press, London, 1994; Austen, K.F., *Therapeutic Immunology,* Blackwell Science, Malden, MA, 2001; Janeway, C.A., Travers, P., Walport, M., and Shlomchik, M., *Immunobiology 5: The Immune System in Health and Disease,* Garland Publishing/Taylor & Francis, New York, 2001; Cruse, J.M., Lewis, R.F., and Wang, H., *Immunology Guidebook,* Elsevier, Amsterdam, 2004.

Internal Standard   A compound or material that is not an analyte but is included in an unknown or standard to correct for issues in the processing or analysis of an analyte or analytes; an internal standard is not a calibration standard. See Julka, S. and Regnier, F., Quantification in proteomics through stable isotope coding: a review, *J. Proteome Res.* 3, 350–363, 2004; Bronstrup, M., Absolute quantification strategies in proteomics based on mass spectrometry, *Expert Rev. Proteomics* 1, 503–512, 2004; Coleman, D. and Vanatta, L., Statistics in analytical chemistry, part 19-internal standards, *American Laboratory,* December 2005.

Intrabodies   Intrabodies are intracellular antibodies or functional antibody fragments. Intrabodies can be expressed as intracellular antibodies using recombinant DNA technology and used for the study of intracellular pathways and protein–protein interactions using two-hybrid technology. See Lobato, M.N. and Rabbitts, T.H., Intracellular antibodies and challenges facing their use as therapeutic agents, *Trends Mol. Med.* 9, 390–396, 2003; Stocks, M.R., Intrabodies: production and promise, *Drug Discov. Today* 9, 960–966, 2004; Visintin, M., Meli, G.A., Cannistraci, I., and Cattaneo, A., Intracellular antibodies for proteomics, *J. Immunol. Methods* 290, 135–153, 2004; Miller, T.W. and Messer, A., Intrabody applications in neurological disorders: progress and future prospects, *Mol. Ther.* 12, 394–401, 2005; Stocks, M., Intrabodies as drug discovery tools and therapeutics, *Curr. Opin. Chem. Biol.* 9, 359–365, 2005. Antibodies or antibody fragments can also be introduced into the cell through the use of cell-penetrating peptides (Zhao, Y., Lou, D., Burkett, J., and Kohler, H., Chemical engineering of cell-penetrating antibodies, *J. Immunol. Methods* 254, 137–145, 2001; Gupta, B., Levchenko, T.S., and Torchilin, V.P., Intracellular delivery of large molecules and small particles by cell-penetrating proteins and peptides, *Adv. Drug Deliv. Rev.* 57, 637–651, 2005; De Coupade, C., Fittipaldi, A., Chagnas, V. et al., Novel human-derived cell-penetrating peptides for specific subcellular delivery of therapeutic biomolecules, *Biochem. J.* 390, 407–418, 2005; Gupta, B. and Torchilin, V.P., Transactivating transcriptional activator-mediated drug delivery, *Expert Opin. Drug Deliv.* 3, 177–190, 2006).

Intron A segment of DNA that is not transcribed into messenger RNA and is designated as non-coding DNA. This is different from DNA that is imprinted to preclude transcription. See Stone, E.M. and Schwartz, R.J., *Intervening Sequences in Evolution and Development,* Oxford University Press, New York, NY, USA, 1990; Gesteland, R.F. and Cech, T., *The RNA World: The Nature of Modern RNA Suggests a Prebiotic RNA World,* Cold Spring Harbor Laboratory Press, Cold Spring Harbor, NY, USA, 2006; Mahler, H.R. The exon:intron structure of some mitochondrial genes and its relation to mitochondrial evolution, *Int. Rev. Cytol.* 82, 1–98, 1983; Patthy, L., Intron-dependent evolution: preferred types of exons and introns, *FEBS Lett.* 214, 1–7, 1987; Hawkins, J.D., A survey on intron and exon lengths, *Nuc. Acids Res.* 16, 9893–9908, 1988; Long, M., de Souza, S.J., and Gilbert, W., Evolution of the intron–exon structure of eukaryotic genes, *Curr. Opin. Genet. Dev.* 5, 774–778, 1995; Roy, S.W., Intron-rich ancestors, *Trends Genet.* 22, 468–471, 2006.

Intron Density Average number of introns per gene over an entire genome. See Grover, D., Mukerji, M., Bhatnagar, P., Kannan, K., and Brahmachari, S.K., Alu repeat analysis in the complete human genome: trends and variations with respect to genomic composition, *Bioinformatics* 20, 813–827, 2004; Niu, D.K., Hou, W.R., and Li, S.W., mRNA-mediated intron losses: evidence from extraordinary large exons, *Mol. Biol. Evol.* 22, 1475–1481, 2005; Sironi, M., Menozzi, G., Comi, G.P. et al., Analysis of intronic conserved elements indicates that functional complexity might represent a major source of negative selection on non-coding sequences, *Hum. Mol. Genet.* 14, 2533–2546, 2005; Toyoda, T. and Shinozaki, K., Tiling array-driven elucidation of transcriptional structures based on maximum-likelihood and Markov models, *Plant J.* 43, 611–621, 2005; Keeling, P.J. and Slamovits, C.H., Causes and effects of nuclear genome reduction, *Curr. Opin. Genet. Dev.* 15, 601–608, 2005; Jeffares, D.C., Mourier, T., and Penny, D., The biology of intron gain and loss, *Trends Genet.* 22, 16–22, 2006; de Cambiare, J.C., Otis, C., Lemieux, C., and Turmel, M., The complete chloroplast genome sequence of the chlorophycean green alga *Scenedesmus obliquus* reveals a compact gene organization and a biased distribution of genes on the two DNA strands, *BMC Evol. Biol.* 6, 37, 2006.

Ion Channels Integral membrane proteins providing for the regulated transport of ions across a membrane via the formation of a porelike structure. See Schonherr, R., Clinical relevance of ion channels for diagnosis and therapy of cancer, *J. Membr. Biol.* 205, 175–184, 2005; Yu, F.H., Yarov-Yarovoy, V., Gutman, G.A., and Catterall, W.A., Overview of molecular relationships in the voltage-gated ion channel superfamily, *Pharmacol. Rev.* 57, 387–395, 2005;

Clapham, D.E., Julius, D., Montell, C., and Schultz, G., International Union of Pharmacology XLIX. Nomenclature and structure-function relationships of transient receptor potential channels, *Pharmacol. Rev.* 57, 427–450, 2005.

Ionization
Potential
Energy required to remove a given electron from its atomic orbital; value in electron volts (eV).

Ionophore
A chemical compound that binds ions and provides transport across a biological membrane. More recent work has led to the development of ion-specific electrodes and other sensors. One of the most common examples is A23187, which functions as a calcium ionophore (Haynes, D.H., Detection of ionophore-cation complexes on phospholipid membranes, *Biochim. Biophys. Acta* 255, 406–410, 1972; Scarpa, A. and Inesi, G., Ionophore-mediated equilibration of calcium ion gradients in fragmented-sarcoplasmic reticulum, *FEBS Lett.* 22, 273–276, 1972; Reed, P.W. and Lardy, H.A., A23187: a divalent cation ionophore, *J. Biol. Chem.* 247, 6970–6977, 1972; Chaney, M.O., Demarco, P.V., Jones, N.D., and Occolowitz, J.L., The structure of A23187, a divalent cation ionophore, *J. Am. Chem. Soc.* 96, 1932–1933, 1974; Ferreira, H.G. and Lew, V.L., Use of ionophore A23187 to measure cytoplasmic Ca buffering and activation of the Ca pump by internal Ca, *Nature* 259, 47–49, 1976; Estensen, R.D., Reusch, M.E., Epstein, M.L., and Hill, H.R., Role of Ca^{2+} and Mg^{2+} in some human neutrophil functions as indicated by ionophore A23187, *Infect. Immun.* 13, 146–151, 1976; Painter, G.R. and Pressman, B.C., Dynamic aspects of ionophore-mediated membrane transport, *Top. Curr. Chem.* 101, 83–110, 1982). See Shampsipur, M., Avenes, A., Javanbakht, M. et al., A 9,10-anthraquinone derivative having two propenyl arms as a neutral ionophore for highly selective and sensitive membrane sensors for Copper(II) ion, *Anal. Sci.* 18, 875–879, 2002; Benco, J.S., Nienaber, H.A., and McGimpsey, W.G., Synthesis of an ammonium ionophore and its application in a planar ion-selective electrode, *Anal. Chem.* 75, 152–156, 2003; Kim, Y.K., Lee, Y.H., Lee, H.Y. et al., Molecular recognition of anions through hydrogen bonding stabilization of anion–ionophore adducts: a novel trifluoroacetatophenone-based binding motif, *Org. Lett.* 5, 4003–4006, 2003; Grote, Z., Lehaire, M.L., Scopelliti, R., and Severin, K., Selective complexation of Li^+ in water at neutral pH using a self-assembled ionophore, *J. Am. Chem. Soc.* 125, 13638–13639, 2003; Dhungana, S., White, P.S., and Crumbliss, A.L., Crystal and molecular structures of ionophore-siderophore host-guest supramolecular assemblies relevant to molecular recognition, *J. Am. Chem. Soc.* 125, 14760–14767, 2003; Mahajan, R.K., Kaur, I., Kaur, R. et al., Lipophilic lanthanide tris(beta-diketonate) complexes as an ionophore for Cl⁻ anion-selective electrodes, *Anal. Chem.* 76, 7354–7359, 2004; Fisher, A.E., Lau, G., and Naughton, D.P., Lipophilic ionophore complexes as superoxide dismutase mimetics, *Biochem. Biophys. Res. Commun.* 329, 930–933, 2005; Zhang, Y.L., Dunlop, J, Phung, T. et al., Supported bilayer lipid membranes modified with a phosphate ionophore, *Biosens. Bioelectron.* 21, 2311–2314, 2006; Shirai, O., Yoshida, Y., and Kihara, S., Voltammetric study on ion transport across a bilayer lipid membrane in the presence of a hydrophobic ion or ionophore, *Anal. Bioanal. Chem.*, 386, 494–504, 2006; Rose, L and Jenkins, A.T., The effect of the ionophore valinomycin on biomimetic solid supported lipid DPPTE/EPC membranes, *Bioelectrochemistry*, in press, 2006.

IQ Motif   A linear sequence of amino acids that binds calmodulin and calmodulinlike proteins where IQ are the first conserved amino acids. See Greeves, M.A. and Holmes, K.C., Structural basis of muscle contraction, *Ann. Rev. Biochem.* 68, 687–728, 1999; Bähler, M. and Rhoads, A., Calmodulin signaling via the IQ motif, *FEBS Lett.* 513, 107–113, 2002.

Isobaric   Having the same molecular mass but different chemical properties and structure; such compounds are called isobars (the term isobar also has a meaning in atmospheric science). Also, a process or reaction can be considered isobaric if performed under constant pressure within either space or time. See Uline, M.J. and Corti, D.S., Molecular dynamics in the isothermal-isobaric ensemble: the requirement of a "shell" molecule. II. Simulation results, *J. Chem. Phys.* 123, 164102, 2005; Rosgen, J. and Hinz, H.J., Pressure-modulated differential scanning calorimetry: theoretical background. *Anal. Chem.* 78, 991–996, 2006; Wu, W.W., Wang, G., Baek, S.J., and Shen, R.F., Comparative study of the three proteomic quantitative methods, DIGE, cICAT, and iTRAQ, using 2D gel or LC-MALDI TOF/TOF, *J. Proteome Res.* 5, 651–658, 2006; Sachon, E., Mohammed, S., Bache, N., and Jensen, O.N., Phosphopeptide quantitation using amine-reactive isobaric tagging reagents and tandem mass spectrometry: application to protein isolated by gel electrophoresis, *Rapid Commun. Mass Spectrom.* 20, 1127–1134, 2006; Langrock, T., Czihal, P., and Hoffman, R., Amino acid analysis by hydrophilic interaction chromatography coupled on-line to electrospray ionization mass spectrometry, *Amino Acids*, 30, 291–297, 2006.

Isocratic   A term used in chromatography to describe a stepwise elution process as opposed to a gradient elution. The term isocratic also refers to a governing system with equality. See Wang, N.W., Ion exchange in purification, *Bioprocess Technol.* 9, 359–400, 1990; Frey, D.D., Feedback regulation in preparative elution chromatography, *Biotechnol. Prog.* 7, 213–224, 1991; Coffman, J.L., Roper, D.K., and Lightfoot, E.N., High-resolution chromatography of proteins in short columns and adsorptive membranes, *Bioseparation* 4, 183–200, 1994; Hajos, P. and Nagy, L., Retention behaviours and separation of carboxylic acids by ion-exchange chromatography, *J. Chromatog. B Biomed. Sci. Appl.* 717, 27–38, 1998; Marsh, A., Clark, B.J., and Altria, K.D., A review of the background, operating parameters, and applications of microemulsion liquid chromatography, *J. Sep. Sci.* 28, 2023–2032, 2005.

Isoelectric Focusing (IEF)   An electrophoretic method for separating amphoteric molecules in pH gradients. Isoelectric focusing is an integral part of the two-dimensional analysis of proteins/peptides in proteomics using immobilized pH gradients (IPG). See Righetti, P.G. and Drysdale, J.W., Isoelectric focusing in polyacrylamide gels, *Biochim. Biophys. Acta* 236, 17–28, 1971; Haglund, H., Isoelectric focusing in pH gradients — a technique for fractionation and characterization of ampholytes, *Methods Biochem. Anal.* 19, 1–104, 1971; Righetti, P.G. and Drysdale, J.W., Small-scale fractionation of proteins and nucleic acids by isoelectric focusing in polyacrylamide gels, *Ann. N.Y. Acad. Sci.* 209, 163–186, 1973; Righetti, P.G., Molarity and ionic strength of focused carrier ampholytes in isoelectric focusing, *J. Chromatog.* 190, 275–282, 1980; Righetti, P.G., Tudor, G., and Gianazza, E., Effect of 2-mercaptoethanol on pH gradients in isoelectric focusing, *J. Biochem. Biophys. Methods* 6, 219–227, 1982; Righetti, P.G., Isoelectric focusing as the crows flies, *J. Biochem. Biophys. Methods* 16, 99–108, 1988;

Strege, M.A., and Lagu, A.L., Capillary electrophoresis of biotechnology-derived proteins, *Electrophoresis* 18, 2343–2352, 1997; Korlach, J., Hagedorn, R., and Fuhr, G., pH-Regulated electroretention chromatography: towards a new method for the separation of proteins according to their isoelectric points, *Electrophoresis* 19, 1135–1139, 1998; Molloy, M.P., Two-dimensional electrophoresis of membrane proteins using immobilized pH gradients, *Anal. Biochem* 280, 1–10, 2000; Kilar, F., Recent applications of capillary isoelectric focusing, *Electrophoresis* 23, 3908–3916, 2003; Righetti, P.G., Determination of the isoelectric point of proteins by capillary isoelectric focusing, *J. Chromatog. A* 1037, 491–499, 2004; Stastna, M., Travnicek, M., and Slais, K., New azo dyes as colored point markers for isoelectric focusing in the acidic pH region, *Electrophoresis* 26, 53–59, 2005; Kelly, R.T. and Woolley, A.T., Electric field gradient focusing, *J. Sep. Sci.* 28, 1985–1993, 2005; Righetti, P.G., The alpher, bethe, gamow of isoelectric focusing, the alpha-Centaury of electrokinetic methodologies. Part I, *Electrophoresis* 27, 923–938, 2006.

Isoelectric Point (I_p) The pH at which an amphoteric molecule such as a protein has a net charge of zero. It is, however, possible for a protein at the isoelectric point to have localized areas or patches of positivity or negativity. See Ingram, V.M., Isoelectric point of chymotrypsinogen by a Donnan equilibrium method, *Nature* 170, 250–251, 1952; Harden, V.P. and Harris, J.O., The isoelectric point of bacterial cells, *J. Bacteriol.* 65, 198–202, 1953; Sophianopoulos, A.J. and Sasse, E.A., Isoelectric point of proteins by differential conductimetry, *J. Biol. Chem.* 240, PC1864–PC1866, 1965; Bishop, W.H. and Richards, F.M., Isoelectric point of a protein in the crosslinked crystallized state, *J. Mol. Biol.* 33, 415–421, 1968; McDonagh, P.F. and Williams, S.K., The preparation and use of fluorescent-protein conjugates for microvascular research, *Microvasc. Res.* 27, 14–27, 1984; Palant, C.E., Bonitati, J., Bartholomew, W.R. et al., Nodular glomerulosclerosis associated with multiple myeloma. Role of light chain isoelectric point, *Am. J. Med.* 80, 98–102, 1986; Karpinska, B., Karlsson, M., Schinkel, H. et al., A novel superoxide dismutase with a high isoelectric point in higher plants. Expression, regulation, and protein localization, *Plant Physiol.* 126, 1668–1677, 2001; Lim, T.K., Imai, S., and Matsunaga, T., Miniaturized amperometric flow immunoassay using a glass fiber membrane modified with anion, *Biotechnol. Bioeng.* 77, 758–763, 2002; Cargile, B.J. and Stephenson, J.L., Jr., An alternative to tandem mass spectrometry: isoelectric point and accurate mass for the identification of peptides, *Anal. Chem.* 76, 267–275, 2004; Shi, Q., Zhou, Y., and Sun, Y., Influence of pH and ionic strength on the steric mass-action model parameters around the isoelectric point of protein, *Biotechnol. Prog.* 21, 516–523, 2005; Sillero, A. and Maldonado, A., Isoelectric point determination of proteins and other macromoecules: oscillating method, *Comput. Biol. Med.* 36, 157–166, 2006. Proteins are usually least soluble at the isoelectric point and this has been suggested as a useful tool in crystallization (see Kantaardjieff, K.A. and Rupp, B., Protein isoelectric point as a predictor for increased crystallization screening efficiency, *Bioinformatics* 20, 2162–2168, 2004; Canaves, J.M., Page, R., Wilson, I.A., and Stevens, R.C., Protein biophysical properties that correlate with crystallization success in *Thermotoga maritime*: maximum clustering strategy for structural genomics, *J. Mol. Biol.* 344, 977–991, 2004).

Isopeptide Bond An amide bond between a carboxyl group of one amino acid and an amino group of another amino acid where either the carboxyl or amino group or both is or are not α in position; for example, the peptide bond formed between glutamine and lysine in transamidation reaction or the peptide bond formed with the β-carboxyl group of aspartic acid and the proximate amino group under acid conditions in peptides and proteins; also the bond between ubiquitin and ubiquitinlike modifiers and substrate proteins. Recently, isopeptide bonds have been described from the reaction of homocysteine lactone with ε-amino group I proteins. See Di Donato, A., Ciardiello, M.A., de Nigris, M. et al., Selective demidation of ribonuclease A. Isolation and characterization of the resulting isoaspartyl and aspartyl derivatives, *J. Biol. Chem.* 268, 4745–4751, 1993; Chen, J.S. and Mehta, K., Tissue transglutaminase: an enzyme with a split personality, *Int. J. Biochem. Cell Biol.* 31, 817–836, 1999; Pickart, C.M., Mechanisms underlying ubiquitination, *Annu. Rev. Biochem.* 70, 502–533, 2001; Perna, A.F., Capasso, R., Lombardi, C. et al., Hyperhomocysteinemia and macromolecule modifications in uremic patients, *Clin. Chem. Lab. Med.* 43, 1032–1038, 2005.

Isosteres   Chemical compounds with the same number of valence electrons but different numbers and types of atoms. See Showell, G.A. and Mills, J.S., Chemistry challenges in lead optimization: silicon isosteres in drug discovery, *Drug Discov. Today* 8, 551–556, 2003; Venkatesan, N. and Kim, B.H., Synthesis and enzyme inhibitory activities of novel peptide isosteres, *Curr. Med. Chem.* 9, 2243–2270, 2002; Roy, R. and Baek, M.G., Glycodendrimers: novel glycotope isosteres unmasking sugar coding. Case study with T-antigen markers from breast cancer MUC1 glycoprotein, *J. Biotechnol.* 90, 291–309, 2002; Rye, C.S. and Baell, J.B., Phosphate isosteres in medicinal chemistry, *Curr. Med. Chem.* 12, 3127–3141, 2005.

Isotherm   For chromatography, an arithmetic function that describes the partitioning of chromatographic solute (adsorbate) between solvent and the matrix/adsorbent. The Langmuir isotherm is an empirical isotherm based on a postulated kinetic mechanism describing an equilibrium process for the process of adsorption based on several assumptions; assumptions include absolute uniformity of the adsorbent service, all adsorption occurs by the same mechanism, and adsorbate molecules adsorb in a uniform monolayer on the adsorbent. See Jacobson, J., Frenz, J., and Horvath, C., Measurement of adsorption isotherms by liquid chromatography, *J. Chromatog.* 316, 53–68, 1984; Chang, C. and Lenhoff, A.M., Comparison of protein adsorption isotherms and uptake rates in preparative cation-exchange materials, *J. Chromatog. A*, 827, 281–293, 1998; Di Giovanni, O., Mazzotti, M., Morbidell, M. et al., Supercritical fluid simulated moving bed chromatography II. Langmuir isotherm, *J. Chromatog. A* 919, 1–12, 2001; Grajek, H., Comparison of the differential isosteric adsorption enthalpies and entropies calculated from chromatographic data, *J. Chromatog.* 986, 89–99, 2003; Xia, F., Nagrath, D., and Cramer, S.M., Modeling of adsorption in hydrophobic interaction chromatography systems using a preferential interaction quadratic isotherm, *J. Chromatog. A*, 989, 47–54, 2003; Piatkowski, W., Antos, D., Gritti, F., and Guiochon, G., Study of the competitive isotherm model and the mass transfer kinetics for a BET binary system, *J. Chromatog. A*, 1003, 73–89, 2003; Lapizco-Encinas, B.H. and Pinto, N.G., Determination of adsorption isotherms of proteins by H-root method: comparison between open micro-channels and convential packed column, *J. Chromatog. A* 1070, 201–205, 2005; Cecchi, T., Use of lipophilic ion adsorption isotherms to determine the surface area and the monolayer capacity of a chromatographic packing, as well as the thermodynamic equilibrium constant for its adsorption, *J. Chromatog. A* 1072, 201–206, 2005; Cano, T., Offringa, N.D., and Willson, R.C., Competitive ion-exchange adsorption of proteins: competitive isotherms with controlled competitor concentration, *J. Chromatog. A* 1079, 116–126, 2005; Zhang, W., Shan, Y., and Seidel-Morgenstern, A., Breakthrough curves and elution profiles of single solutes in case of adsorption isotherms with two inflection points, *J. Chromatog. A* 1107, 215–225, 2006. This concept is also represented by the distribution coefficient (see *British Pharmacopoeia*, 2004). The distribution coefficient, or partitioning coefficient, is also used in countercurrent distribution.

Isothermal Titration Calorimetry (ITC) A physical method that directly measures the heat of interaction of two or more substances. Changes in temperature are measured as one substance is added to another and molar heat (kcal/mol) is determined as function of the amount of material added. This information is used to calculate changes in enthalpy (ΔH). This can be applied to large molecule (ligand-receptor) and small molecule interactions. See Rudolph, M.G., Luz, J.G., and Wilson, I.A., Structural and thermodynamic correlates of T-cell signaling, *Ann. Rev. Biophys. Biomol. Struct.* 31, 121–149, 2002; Velazquez-Campoy, A., Leavitt, S.A., and Freire, E., Characterization of protein–protein interactions by isothermal titration calorimetry, *Methods Mol. Biol.* 261, 35–54, 2004; Ciulli, A. and Abell, C., Biophysical tools to monitor enzyme-ligand interactions of enzymes involved in vitamin biosynthesis, *Biochem. Soc. Trans.* 33, 767–771, 2005; Holdgate, G.A. and Ward, W.H.J., Measurements of binding thermodynamics in drug discovery, *Drug Discov. Today* 10, 1543–1550, 2005.

Isotropy A physical measurement such as the melting point is identical when measured in different principal directions; antonym, anisotropy.

Isotype *Iso* (Gr. equal); isotype usually refers to the immunoglobulin subclasses as defined by the chemical and antigenic characteristics of their constant regions. In biological terms, an isotype is a biological specimen that is a duplicate of a holotype (*holo*, Gr. complete).

Isotype Switching The process where antibody class expression changes as in the rearrangement of genes in B-cells resulting from the exposure of the B-cells to its antigens. Naive B-cells express IgA (secretory immunoglobulin) while stimulated or exposed B-cells may express other immunoglobulin isotypes including IgG and IgE. Isotype switching is also referred to as antibody class switching. This is a process separate from that of somatic hypermutation, which involves the variable regions of the immunoglobulins and is responsible for antibody functional diversity. See Rothman, P., Li, S.C., and Alt, F.W., The molecular events in heavy chain class-switching, *Semin. Immunol.* 1, 65–77, 1989; Whitmore, A.C., Haughton, G., and Arnold, L.W., Isotype switching in CD5 B-cells, *Ann. N.Y. Acad. Sci.* 651, 143–151, 1992; Vercelli, D. and Geha, R.S., Regulation of isotype switching, *Curr. Opin. Immunol.* 4, 794–797, 1992; Rothman, P., Interleukin 4 targeting of immunoglobulin heavy chain class-switch recombination, *Res. Immunol.* 144, 579–583, 1993; Snapper, C.. and Mond, J.J., Towards a comprehensive view of immunoglobulin class switching, *Immunol. Today* 14, 15–17, 1993; Diamant, E. and Melamed, D., Class switch recombination in B lymphopoiesis: a potential pathway for B-cell autoimmunity, *Autoimmun. Rev.* 3, 464–469, 2004; Frasca, D., Riley, R.L., and Blomberg,

B.B., *Crit. Rev. Immunol.* 24, 297–320, 2004; Fiset, P.O., Cameron, L., and Hamid, Q., Local isotype switching to IgE in airway mucosa, *J. Allergy Clin. Immunol.* 116, 233–236, 2005; Min, I.M. and Selsing, E., Antibody class switch recombination: roles for switch seqences and mismatch repair proteins, *Adv. Immunol.* 87, 297–328, 2005; Aplan, P.D., Causes of oncogenic chromosomal translocation, *Trends Genet.* 22, 46–55, 2006.

JAK (Janus-Associated Kinase)
A family of tyrosine kinases involved in signal transduction through cytokine receptors. There are four JAK family members: JAK1, JAK2, JAK3, and TYK2. JAK1 and JAK2 are involved in type II interferon (interferon-gamma) signaling, whereas JAK1 and TYK2 are involved in type I interferon signaling. The term Janus is derived from the Roman god for doors and pathways, who is frequently depicted with two faces looking in opposite directions. See Karnitz, L.M. and Abraham, R.T., Cytokine receptor signaling mechanisms, *Curr. Opin. Immunol.* 7, 320–326, 1995; Ihle, J.N., The Janus protein tyrosine kinase family and its role in cytokine signaling, *Adv. Immunol.* 60, 1–35, 1995; Yamaoka, K., Saharinen., P., Pesu, M. et al., The Janus kinases (Jaks), *Genome Biol.* 5, 253 (epub), 2004; Gilmour, K.C. and Reich, N.C., Signal transduction and activation of gene transcription by interferon, *Gene Expr.* 5, 1–18, 1995; Gao, B., Cytokines, STATs, and liver disease, *Cell. Mol. Immunol.* 2, 92–100, 2005.

JASPAR
A database for transcription factor binding. See Sandelin, A., Alkema, W., Engstrom P., et al., JASPAR: an open-access database for eukaryotic transcription factor binding profiles, *Nuc. Acid Res.* 32, D91–D94, 2004.

Karyology
Study of the nucleus of the cell, specifically the chromosomes. Used in the characterization of master cell banks and working cell banks for recombinant DNA products. See Chiarelli, A.B. and Koen, A.L., *Comparative Karyology of Primates*, Moulton, The Hague, 1979; Macgregor, H.C., *An Introduction to Animal Cytogenetics*, Chapman & Hall, London, 1993; Petricciani, J.C. and Horaud, F.N., Karyology and tumorigenicity testing requirements: past, present, and future, *Dev. Biol. Stand.* 93, **5**, 5–13, 1998.

Katal
An international standard (SI; Systems International d'Unites) unit for enzyme activity. A katal (kat) is defined as 1 mol/s. A unit for enzyme activity is defined by the International Union of Biochemistry and Molecular Biology (IUBMB) as 1 μmol/min; then one unit of enzyme activity is equal to 16.67×10^{-9} kat or 16.67 nkat. This term is used in clinical chemistry more than in basic biomedical investigation. See Dybkær, R., Problems of quantities and units in enzymology, *Enzyme* 20, 46–64, 1975; Lehmann, H.P., Metrication of clinical laboratory data in SI units, *Am. J. Clin. Pathol.* 65, 2–18, 1976; Lehman, H.P., SI units, *CRC Crit. Rev. Clin. Lab. Sci.* 10, 147–170, 1979; Bowers, G.N., Jr. and McComb, R.B., A unifying reference system for clinical enzymology: aspartate aminotransferase and the International Clinical Enzyme Scale, *Clin. Chem.* 39, 1128–1136, 1984; Powsner, E.R., SI quantities and units for American Medicine, *JAMA* 252, 1737–1741, 1984; van Assendelft, O.W., The international system of units (SI) in historical perspective, *Am. J. Public Health* 77, 1400–1403, 1987; Dybkær, R. and Storring, P.L., Application of IUPAC-IFCC recommendations on quantities and units to WHO biological reference materials for diagnostic use. International Union of Pure and Applied Chemistry (IUPAC) and International Federation of Clinical Chemistry (IFCC), *Eur. J. Clin. Chem. Clin. Biochem.* 33, 623–625, 1995; Dybkær, R., The tortuous road to the adoption of katal for the expression

	of catalytic activity by the general conference on weights and measures, *Clin. Chem.* 48, 586–590, 2002.
Keratin	A fibrous protein found in skin, hair, and surface hard tissue such as fingernails. Keratins are characterized by a relatively high content of sulphur-containing amino acids. See Crewther, W.G., Fraser, R.D., Lennox, F.G., and Lindley, H., The chemistry of keratins, *Adv. Protein Chem.* 20, 191–346, 1965; Roe, D.A., Sulphur metabolism in relation to cutaneous disease, *Br. J. Dermatol.* 81 (Suppl. 2), 49–60, 1969; Bradbury, J.H., Keratin and its formation, *Curr. Probl. Dermatol.* 6, 34–86, 1976; Fuchs, E. and Green, H., Multiple keratins of cultured human epidermal cells are translated from different mRNA molecules, *Cell* 17, 573–582, 1979; Sun, T.T., Eichner, R., Nelson, W.C. et al., Keratin classes: molecular markers for different types of epithelial differentiation, *J. Invest. Dermatol.* 81 (Suppl. 1), 109s–115s, 1983; Steinert, P.M., Jones, J.C., and Goldman, R.D., Intermediate filaments, *J. Cell Biol.* 99, 22s–27s, 1984; Virtanen, I., Miettinen, M., Lehto, V.P. et al., Diagnostic application of monoclonal antibodies to intermediate filaments, *Ann. N.Y. Acad. Sci.* 455, 635–648, 1985; Dale, B.A., Resing, K.A., and Lonsdale-Eccles, J.D., Filaggrein: a keratin filament-associated protein, *Ann. N.Y. Acad. Sci.* 455, 330–342, 1985; Fuchs, E., Keratin genes, epidermal differentiation, and animal models for the study of human skin diseases, *Biochem. Soc. Trans.* 19, 1112–1115, 1991; Oshima, R.G., Intermediate filament molecular biology, *Curr. Opin. Cell Biol.* 4, 110–116, 1992; Coulombe, P.A., The cellular and molecular biology of keratins: beginning a new era, *Curr. Opin. Cell Biol.* 5, 17–29, 1993; Liao, J., Ku, N.O., and Omary, M.B., Keratins and the keratinocyte activation cycle, *J. Invest. Dermatol.* 116, 633–640, 2001; Kierszenbaum, A.L., Keratins: unraveling the coordinated construction of scaffolds in spermatogenesic cells, *Mol. Reprod. Dev.* 61, 1–2, 2002; Lane, E.B. and McLean, W.H., Keratins and skin disorders, *J. Pathol.* 204, 355–366, 2004; Zatloukal, K., Stumpter, C., Fuchsbichler, A. et al., The keratin cytoskeleton in liver disease, *J. Pathol.* 204, 367–376, 2004; Gupta, R. and Ramnani, P., Microbial keratinases and their prospective applications: an overview, *Appl. Microbiol. Biotechnol.* 70, 21–33, 2006.
Kinome	The protein kinases in a proteome of an organism. See Manning, G., Whyle, D.B., Martinez, R., Hunter, T., and Sudarsanam, S., The protein kinase complement of the human genome, *Science* 298, 596–601, 2000; ter Haar, E., Walters, W.P., Pazhanisamy, S. et al., Kinase chemogenomics: targeting the human kinome for target validation and drug discovery, *Mini Rev. Med. Chem.* 4, 235–253, 2004.
Kinomics	Analysis of all kinases in the proteome of a given organism. See Vieth, M., Sutherland, J.J., Robertson, D.H., and Campbell, R.M., Kinomics: characterizing the therapeutically validated kinase space, *Drug Discov. Today* 10, 839–846, 2005; Johnson, S.A. and Hunter, T., Kinomics: methods for deciphering the kinome, *Nat. Methods* 2, 17–25, 2005.
Knockdown	This term was originally used to describe the incapacitation of insects such as mosquitoes by insecticides (Asher, K.R., Preferential knockdown action of cetyl bromoacetate for certain laboratory-reared resistant stains of houseflies, *Bull. World Health Organ.* 18, 675–677, 1958; Cohan, F.M. and Hoffmann A.A., Genetic divergence under uniform selection. II. Different responses to selection for knockdown resistance to ethanol among *Drosophila melanogaster* populations and their replicate lines, *Genetics* 114, 145–164, 1986; Bloomquist, J.R. and Miller, TA., Sodium channel

neurotoxins as probes of the knockdown resistance mechanism, *Neurotoxicity* 7, 217–223, 1986) but has seen increased use for the phenomena of the inhibition of transcription by the process of RNA interference/RNA silencing or by the use of antisense oligonucleotides. See Nasevicius, A. and Ekker, S.C., Effective targeted gene 'knockdown' in zebrafish, *Nat. Genet.* 26, 216–220, 2000; Araki, I. and Brand, M., Morpholino-induced knockdown of fgf8 efficiently phenocopies the acerebellar (ace) phenotype, *Genesis* 30, 157–159, 2001; Dick, J.M., Van Molle, W., Libert, C., and Lefebvre, R.A., Antisense knockdown of inducible nitric oxide synthase inhibits the relaxant effect of VIP in isolated smooth muscle cells of the mouse gastric fundus, *Br. J. Pharmacol.* 134, 425–433, 2001; Scherer, L.J. and Rossi, J.J., Approaches for the sequence-specific knockdown of mRNA, *Nat. Biotechnol.* 21, 1457–1465, 2003; Achenbach, T.V., Brunner, B., Heermeier, K., Oligonucleotide-based knockdown technologies: antisense versus RNA interference, *Chem Bio Chem* 4, 928–935, 2003; Voorhaeve, P.M. and Agami, R., Knockdown stands up, *Trends Biotechnol.* 21, 2–4, 2003; Tiscornia, G., Singer, O., Ikawa, M., and Verma, I.M., A general method for gene knockdown in mice using lentiviral vectors expressing small interfering RNA, *Proc. Nat. Acad. Sci. USA* 100, 1844–1888, 2004; Manfredsson, F.P., Lewis, A.S., and Mandel, R.J., RNA knockdown as a potential therapeutic strategy in Parkinson's disease, *Gene Ther.* 13, 517–524, 2006.

Krüppel-Like Factor A family of zinc finger transcription factors; the name is derived from the *Drosophila* Krüppel embryonic pattern regulator. See Sugawara, M., Scholl, T., Ponath, P.D., and Strominger, J.L., A factor that regulates the class II major histocompatibility complex gene DPA is a member of a subfamily of zinc finger proteins that includes a *Drosophila* developmental control protein, *Mol. Cell. Biol.* 14, 8438–8450, 1994; Kaczynski, J., Cook, T., and Urrutia, R., SpI- and Krüppel-like transcription factors, *Genome Biology* 4, article 206, 2003.

Labile Zinc Zinc is an essential mineral for most organisms. Zn is either labile or fixed. Fixed Zn is that Zn tightly bound to metalloproteins while labile zinc is bound loosely to proteins or low molecular thiols such as glutathione. Total cellular Zn is measured by atomic absorption analysis while labile Zn can be measured, for example, with fluorophoric reagents. See Pattison, S.E. and Cousins, R.J., Zinc uptake and metabolism by hepatocytes, *Fed. Proc.* 45, 2805–2809, 1986; Truong-Tran, A.Q., Ho, L.H., Chai, F., and Zalewski, P.D., Cellular zine fluxes and the regulation of apoptosis/gene-directed cell death, *J. Nutr.* 130 (Suppl. 5S), 1459S–1466S, 2000; Paski, S.C. and Xu, Z., Growth factor–stimulated cell proliferation is accompanied by an elevated labile intracellular pool of zinc in 3T3 cells, *Can. J. Physiol. Pharmacol.* 80, 790–795, 2002; Eide, D.J., Multiple regulatory mechanisms maintain zinc homeostasis in *Saccharomyces cerevisiae*, *J. Nutr.* 133 (5 Suppl. 1), 1532S–1535S, 2003; Sauer, G.R., Smith, D.M., Cahalane, M., Wu, L.N., and Wuthier, R.E., Intracellular zinc fluxes associated with apoptosis in growth plate chondrocytes, *J. Cell. Biochem.* 88, 954–969, 2003; Roschitzki, B. and Vasak, M., Redox labile site in a Zn4 cluster of Cu4, Zn4-metallothionein-3, *Biochemistry* 42, 9822–9828, 2003; Ho, L.H., Ruffin, R.E., Murgia, C. et al., Labile zinc and zinc transporter ZnT4 in mast cell granules: role in regulation of caspase activation and NF-B translocation, *J. Immunol.* 172, 7750–7760, 2004;

Atsriku, C., Scott, G.K., Benz, C.C., and Baldwin, M.A., Reactivity of zinc finger cysteines: chemical modification within labile zinc fingers in estrogen receptors, *J. Am. Soc. Mass Spectrom.* 16, 2017–2026, 2005; Lee, J.Y., Hwang, J.J., Park, M.H., and Koh, J.Y., Cytosolic labile zinc: a marker for apoptosis in the developing rat brain, *Eur. J. Neurosci.* 23, 435–442, 2006; Zalewski, P., Truong-Tran, A., Lincoln, S. et al., Use of a zinc fluorophore to measure labile pools of zinc in body fluids and cell-conditioned media, *BioTechniques* 40, 509–520, 2006; Haase, H., Hebel, S., Engelhardt, G., and Rink, L., Flow cytometric measurement of labile zinc in peripheral blood mononuclear cells, *Analyt. Biochem.* 352, 222–230, 2006.

Lactoferrin   Lactoferrin is an iron-binding protein of very high affinity originally described in milk and other secreted biological fluids such as saliva (see Weinberg, E.D., The therapeutic potential of lactoferrin, *Expert Opin. Investig. Drugs* 12, 841–851, 2003; Van Nieuw Amerongen, A., Bolscher, J.G., and Veerman, E.C., Salivary proteins: protective and diagnostic value in cariology? *Caries Res.* 38, 247–253, 2004). Lactoferrin is also found in specific granules of neutrophils. Lactoferrin is considered to play an important role in the nonspecific defense process by sequestering iron required for bacterial growth. See Goldman, A.S. and Smith, C.W., Host resistance factors in human milk, *J. Pediatr.* 82, 1082–1090, 1973; Bullen, J.J., Rogers, H.J., and Griffiths, E., Role of iron in bacterial infection, *Curr. Top. Microbiol. Immunol.* 80, 1–35, 1978; Reiter, B., The biological significance of lactoferrin, *Int. J. Tissue React.* 5, 87–96, 1983; Birgens, H.S., The biological significance of lactoferrin in haematology, *Scand. J. Haematol.* 33, 225–230, 1984; De Sousa, M., Breedvelt, F., Dynesius-Trentham, R., and Lum, J., Iron, iron-binding proteins, and immune system cells, *Ann. N.Y. Acad. Sci.* 526, 310–322, 1988; Levay, P.F. and Viljoen, M., Lactoferrin: a general review, *Haematologica* 80, 252–267, 1995; Legrand, D., Elass, E., Pierce, A., and Mazurier, J., Lactoferrin and host defense: an overview of its immuno-modulating and anti-inflammatory properties, *Biometals* 17, 225–229, 2004; Yalcin, A.S., Emerging therapeutic potential of whey proteins and peptides, *Curr. Pharm. Des.* 12, 1637–1643, 2006.

Latarcins   A newly defined group of antimicrobial and cytolytic peptides from spider venom. See Kozlov, S.A., Vassilevski, A.A., Feofanov, A.V. et al., Latarcins, antimicrobial and cytolytic peptides from the venom of the spider *Lachesana tarabaevi* (Zodariidae) that exemplify biomolecular diversity, *J. Biol. Chem.* 281, 20983–20992, 2006.

Lectin   A protein that selectively binds carbohydrates. Lectin affinity columns can be used for the purification of carbohydrate chains and glycoproteins. Lectins are also used in histochemistry and cytochemistry. See Cohen, E., *Recognition Proteins, Receptors, and Probes: Invertebrates: Proceedings of a Symposium Entitled Recognition and Receptor Display, Lectin Cell Surface Receptors and Probes*, A.R. Liss, New York, 1984; Gabius, H.J. and Gabius, S., *Lectins and Glycobiology*, Springer-Verlag, Berlin, 1993; Fukuda, M. and Kobata, A., *Glycobiology: A Practical Approach*, IRL Press at Oxford University Press, Oxford, UK, 1993; Doyle, R.J. and Shifkin, M., *Lectin-Microorganism Interactions*, Marcel Dekker, New York, 1994; Brooks, S.A., Leathern, A.J.C., and Schumacher, L., *Lectin Histochemistry: A Concise Practical Handbook,* BIOS Scientific, Oxford, UK, 1997.

Linkage Group   A group of genes inherited as a unit so that they are described as linked, such that disparate phenotypic expressions are also described as linked. See Lamm, L.U. and Petersen, G.B., The HLA genetic linkage group, *Transplant Proc.* 11, 1692–1696, 1979; Campbell, R.D., Dunham, I., and Sargent, C.R., Molecular mapping of the HLA-linked complement genes and the RCA linkage group, *Exp. Clin. Immunogenet.* 5, 81–98, 1988; Haig, D., A brief history of human autosomes, *Philos. Trans. R. Soc. Lond. B Biol. Sci.* 354, 1447–1470, 1999.

Lipofection   Originally described as cellular membrane translocation of DNA for gene therapy via the use of cationic lipids as micelles. More generally, the membrane translocation of RNA or DNA encapsulated in a lipid micelle and is used now for RNAi studies. The liposome and its cargo are referred to as a lipoplex. See Zuhorn, I.S., Kalicharan, R., and Hoekstra, D., Lipoplex-mediated transfection of mammalian cells occurs through the cholesterol-dependent clathrin-mediated pathway of endocytosis, *J. Biol. Chem.* 277, 18021–18028, 2002; Hart, S.L., Lipid carriers for gene therapy, *Curr. Drug Deliv.* 2, 423–438, 2005.

Lipophilic   Affinity for hydrophobic materials such as lipids; compounds that will dissolve in organic/nonpolar solvents such as benzene or cyclohexane but not in water; also hydrophobicity. This quality is frequently measured by distribution or partitioning in an octanol-water system and can be assigned a value such as log P. Lipophilicity can also be measured by retention on an HPLC column with a suitable matrix or on thin-layer chromatography. See Markuszewski, M.J., Wiczling, P., and Kaliszan, R., High-throughput evaluation of lipophilicity and acidity by new gradient HPLC methods, *Comb. Chem. High Throughput Screen.* 7, 281–289, 2004; Klopman, G. and Zhu, H., Recent methodologies for the estimation of *n*-octanol/water partition coefficients and their use in the prediction of membrane transport properties of drugs, *Mini Rev. Med. Chem.* 5, 127–133, 2005; Mannhold, R., The impact of lipophilicity in drug research: a case report on beta-blockers, *Mini Rev. Med. Chem.* 5, 197–205, 2005; Gocan, S., Cimpan, G., and Comer, J., Lipophilicity measurements by liquid chromatography, *Adv. Chromatog.* 44, 79–176, 2006.

Liposomes   A relatively large (nano to micro) micelle composed of polar lipids. There is considerable interest in liposomes as models for biological membranes and for drug delivery. See Bangham, A.D., Lipid bilayers and biomembranes, *Annu. Rev. Biochem.* 41, 753–776, 1972; Gulik-Krzywicki, T., Structural studies: the association between biological membrane components, *Biochim. Biophys. Acta* 415, 1–28, 1975; Pressman, B.C., Biological applications of ionophores, *Annu. Rev. Biochem.* 45, 501–530, 1976; Schreier, S., Polnaszek, C.F., and Smith, I.C., Spin labels in membranes. Problems in practice, *Biochim. Biophys. Acta* 515, 395–436, 1978; Hart, S.L., Lipid carriers for gene therapy, *Curr. Drug Deliv.* 2, 423–428, 2005; Zamboni, W.C., Liposomal, nanoparticle, and conjugated formulations of anticancer agents, *Clin. Chem. Res.* 11, 8230–8234, 2005; Taylor, T.M., Davidson, P.M., Bruce, B.D., and Weiss, J., Liposomal nanocapsules in food science and agriculture, *Crit. Rev. Food Sci. Nutr.* 45, 587–605, 2005; Kshirsager, N.A., Pandya, S.K., Kirodian, G.B., and Sanath, S., Liposomal drug delivery system from laboratory to clinic, *J. Postgrad. Med.* 51 (Suppl. 1), S5–S15, 2005; Paleos, C.M. and Tsiourvas, D., Interaction between complementary liposomes: a process leading to multicompartment systems formation, *J. Mol. Recognit.* 19, 60–67, 2006.

Localized Surface Plasmon Resonance

The use of noble metal (Ag, Au) nanoparticles as sensors for macromolecular interactions. This phenomenon is related to surface plasmon resonance and is based on the spectral properties of nanoparticles as compared to the bulk metal. See Haes, A.J. and Van Duyne, R.P., A unified view of propagating and localized surface plasmon resonance biosensors, *Anal. BioAnal. Chem.* 370, 920–930, 2004; Haes, A.J., Stuart, D.A., Nie, S., and Van Duyne, R.P., Using solution-phase nanoparticles, surface-confined nanoparticle arrays, and single nanoparticles as biological sensing platforms, *J. Fluoresc.* 14, 355–367, 2004; Dahlin, A., Zach, M., Rindzevicius, T. et al., Localized surface plasmon resonance sensing of lipid-membrane–mediated biorecognition events, *J. Am. Chem. Soc.* 127, 5043–5048, 2005; Endo, T., Kerman, K., Nagatani, N., Takamura, Y., and Tamiya, E., Label-free detection of peptide nucleic acid-DNA hybridization using localized surface plasmon resonance-based optical biosensor, *Anal. Chem.* 77, 6976–6984, 2005; Wang, Y., Chen, H., Dong, S., and Wang, E., Surface-enhanced Raman scattering of *p*-aminothiophenol self-assembled monolayers in sandwich structure fabricated on glass, *J. Chem. Phys.* 124, 74709, 2006.

Locus Control Region

A *cis*-acting region on a gene locus, which controls the expression of a transgene *in vivo*; a region in a gene sequence that regulates the independent expression of a gene in transgenic mice; a regulatory element critical for globin gene expression. See Festenstein, R. and Kioussis, D., Locus control regions and epigenetic chromatin modifiers, *Curr. Opin. Genet. Dev.* 10, 199–203, 2000; Levings, P.P. and Bungert, J., The human β-globin locus region. A center of attraction, *Eur. J. Biochem.* 269, 1589–1599, 2002; Dekker, J., A closer look at long-range chromosomal interactions, *Trends in Biochem. Sci.* 28, 277–280, 2003; Harrow, F. and Ortiz, B.D., The TCR locus control region specifies thymic, but not peripheral, patterns of TCR gene expression, *J. Immunol.* 175, 6659–6667, 2005; Dean, A., On a chromosome far, far away: LCRs and gene expression, *Trends Genet.* 22, 38–45, 2006.

Longin Domain

An *N*-terminal domain in nonsyntaxin soluble *N*-ethylmaleimide-sensitive factor (NSF) attachment protein receptors (SNARE), which have been suggested to regulate membrane trafficking processes. Such proteins are described as longins (longin domains are also found in YKT-like proteins). The term "longin" was suggested to differentiate such domains from "brevins," which are shorter *N*-terminal regions found in vesicle-associated membrane proteins. See Filippini, F., Rossi, V., Galli, T. et al., Longins: a new evolutionary conserved VAMP family sharing a novel SNARE domain, *Trends Biochem. Sci.* 26, 407–409, 2001; Dietrich, L.E.P., Boeddinghaus, C., LaGrassa, T.J., and Ungermann, C., Control of eukaryotic membrane fusion by N-terminal domain of SNARE proteins, *Biochim. Biophys. Acta* 1641, 111–119, 2003; Martinez-Arca, S., Rudge, R., Vacca, M. et al., A dual mechanism controlling the localization and function of exocytic v-SNAREs, *Proc. Natl. Acad. Sci. USA* 100, 9011–9016, 2003; Rossi, V., Picco, R., Vacca, M. et al., VAMP subfamilies identified by specific R-SNARE motifs, *Biology of the Cell* 96, 251–256, 2004; Uemura, T., Sato, M.H., and Takeyasu, K., The longin domain regulates subcellular targeting of VAMP7 in *Arabidopsis thaliana*, *FEBS Lett.* 579, 2842–2846, 2005; Schlenker, O., Hendricks, A., Sinning, I., and Wild, K.,

The structure of the mammalian signal recognition particle (SRP) receptor as prototype for the interaction of small GTPases with longin domains, *J. Biol. Chem.* 281, 8898–8906, 2006.

Luminescence   The emission of electromagnetic radiation from an excited molecule when an electron returns to ground state from an excited state (other than from thermal energy changes). Fluorescence, phosphorescence, and chemiluminescence are forms of luminescence. The decay time for fluorescence is shorter than phosphorescence and both phenomena result from irradiation of compounds with electromagnetic irradiation while chemiluminescence is the result of a chemical reaction, usually oxidation. Phosphorescence also shows a larger shift in the wavelength of emitted light than fluorescence. Electrochemiluminescence differs from all of the above in that chemiluminescence is generated at the surface of an electrode. See Abrams, B.L. and Hollaway, P.H., Role of the surface in luminescent processes, *Chem. Rev.* 104, 5783–5801, 2004; Richter, M.M., Electrochemiluminescence, *Chem. Rev.* 104, 3003–3036, 2004; Hemmila, I. and Laitala, V., Progress in lanthanides as luminescent probes, *J. Fluoresc.* 15, 529–542, 2005; Tsuji, F.I., Role of molecular oxygen in the bioluminescence of the firefly squid, *Watasenia scintillans*, *Biochem. Biophys. Res. Commun.* 338, 250–253, 2005; Aslan, K., Lakowicz, J.R., and Geddes, C.D., Plasmon light scattering in biology and medicine: new sensing approaches, visions, and perspectives, *Curr. Opin. Chem. Biol.* 9, 538–544, 2005; Medlycott, E.A. and Hanan, G.S., Designing tridentate ligands for ruthenium(II) complexes with prolonged room temperature luminescence lifetimes, *Chem. Soc. Rev.* 34, 133–142, 2005.

Lutheran
Glycoprotein   The Lutheran glycoprotein is one of the two components of the Lutheran blood group system, a family of red blood cell antigens. The Lutheran glycoprotein and the basal cell adhesion molecule (together recognized as CD239) are members of the immunoglobulin superfamily. See Telen, M.J., Lutheran antigens, CD44-related antigens, and Lutheran regulatory genes, *Transfus. Clin. Biol.* 2, 291–301, 1995; Daniels, G. and Crew, V., The molecular basis of the Lutheran blood group antigens, *Vox. Sang.* 83 (Suppl. 1), 189–192, 2002; Kikkawa, Y. and Miner, J.H., Review: Lutheran/B-CAM: a laminin receptor on red blood cells and in various tissues, *Connect. Tissue Res.* 46, 193–199, 2005; Eyler, C.E. and Telen, M.J., The Lutheran glycoprotein; a multifunction adhesion receptor, *Transfusion* 46, 668–667, 2006.

Lysosomes   An intracellular organelle in a eukaryotic cell responsible for *controlled* intracellular digestion. Lysosomal protein degradation is also involved in MHC class II antigen preparation. Lysosomes contain a variety of hydrolytic enzymes having an acid pH optimum. See Gatti, E. and Pierre, P., Understanding the cell biology of antigen presentation: the dendritic cell contribution, *Curr. Opin. Cell Biol.* 15, 468–473, 2003; Honey, K. and Rudensky, A.Y., Lysomal cysteine proteases regulate antigen presentation, *Nat. Rev. Immunol.* 3, 472–482, 2003; Wolf, D.H., From lysosome to proteosome: the power of yeast in the dissection of proteinase function in cellular regulation and waste disposal, *Mol. Life Sci.* 61, 1601–1614, 2004; Luzio, J.P., Pryor, P.R., Gray, S.R. et al., Membrane traffic to and from lysosomes, *Biochem. Soc. Symp.* 72, 77–86, 2005; Bagshaw, R.D., Mahuran, D.J., and Callahan, J.W., Lysosomal membrane proteomics and biogenesis of lysosomes, *Mol. Neurobiol.* 32, 27–41, 2005; Hsing, L.C. and

Rudensky, A.Y., The lysosomal cysteine proteases in MHC class II antigen presentation, *Immunol. Rev.* 207, 229–241, 2005; Hideshima, T., Bradner, J.E., Chauhan, D., and Anderson, K.C., Intracellular protein degradation and its therapeutic implications, *Clin. Cancer Res.* 11, 8530–8533, 2005; Mellman, T., Antigen processing and presentation by dendritic cells: cell biological mechanisms, *Adv. Exp. Med. Biol.* 560, 63–67, 2005; Lip, P., Gregg, J.L., Wang, N. et al., Compartmentalization of class II antigen presentation: contribution of cytoplasmic and endosomal processing, *Immunol. Rev.* 207, 206–217, 2005.

Macrolide A large ring structure with many functional groups. An example is provided by the kabiramides. See Petchprayoon, C., Swanbonriux, K., Tanaka, J. et al., Fluorescent kabiramides: new probes to quantify actin *in vitro* and *in vivo*, *Bioconugate Chem.* 16, 1382–1389, 2005.

Macrophage An immune system cell that is derived from a monocyte after passage through the endothelium and is a phagocytic cell and an antigen-presenting cell. See Pearsall, N.N. and Weiser, R.S., Eds., *The Macrophage*, Lea & Febiger, Philadelphia, 1970; Carr, I., *The Macrophage: A Review of Ultrastructure and Function*, Academic Press, London, 1973; Nelson, D.S. and Alexander, P., *Immunobiology of the Macrophage*, Academic Press, New York, 1976; Horst, M., *The Human Macrophage System: Activity and Functional Morphology*, Karger, Basel, 1988; Russell, S.W. and Gordon, S., Eds., *Macrophage Biology and Activation*, Springer-Verlag, Berlin, 1992; Zwilling, B.S. and Eisenstein, T.K., Eds., *Macrophage-Pathogen Interactions*, Marcel Dekker, New York, 1992; Bernard, B. and Lewis, C.E., Eds., *The Macrophage*, Oxford University Press, Oxford, UK, 2002. Foam cells, which are derived from macrophages, play an important role in the pathogenesis of atherosclerosis (Schwartz, C.J., Valente, A.J., Sprague, E.A. et al., The pathogenesis of atherosclerosis: an overview, *Clin. Cardiol.* 14 [2 Suppl. 1], I1–I16, 1991; Osterud, B. and Bjorklid, E., Role of monocytes in atherogenesis, *Physiol. Rev.* 83, 1069–1112, 2003; Linton, M.F. and Fazio, S., Macrophages, inflammation, and atherosclerosis, *Int. J. Obes. Relat. Metab. Disord.* 27 [Suppl. 3], S35–S40, 2003; Shashkin, P., Dragulev, B., and Ley, K., Macrophage differentiation to foam cells, *Curr. Pharm. Des.* 11, 3061–3072, 2005; Bobryshev, Y.V., Monocyte recruitment and foam cell formation in atherosclerosis, *Micron* 37, 208–222, 2006).

Macropinocytosis Actin-mediated endocytotic process possessing unique ultrastructural features such as the formation of the macropinosome. See Pratten, M.K. and Lloyd, J.B., Effects of temperature, metabolic inhibitors, and some other factors on fluid-phase and adsorptive pinocytosis by rat peritoneal macrophages, *Biochem. J.* 180, 567–571, 1979; Sallusto, F., Cella, M., Danieli, C., and Lanzavecchia, A., Dendritic cells use macro-pinocytosis and the mannose receptor to concentrate macromolecules in the major histocompatibilty complex class II compartment: downregulation by cytokines and bacterial products, *J. Exp. Med.* 182, 389–400, 1995; Kaplan, I.M., Wadia, J.S., and Dowdy, S.F., Cationic TAT peptide transduction domain enters cells by macropinocytosis, *J. Control. Release* 192, 247–253, 2005; Chia, C.P., Gomathinayagam, S., Schmaltz, R.J., and Smoyer, L.K., Glycoprotein gp130 of dictyostelium discodeum influences macropinocytosis and adhesion, *Mol. Biol. Cell* 16, 2681–2693, 2005; Melikov, K. and Chernomordik, L.V., Arginine-rich cell-penetrating peptides: from endosomal uptake to nuclear delivery, *Cell. Mol. Life Sci.* 62, 2739–2749, 2005; Kirkham, M.

and Parton, R.G., Clathrin-independent endocytosis: new insights into cave-olae and non-caveolar lipid raft carriers, *Biochim. Biophys. Acta* 1746, 349–363, 2005; Zaro, J.L., Rajapaksa, T.E., Okamoto, C.T., and Shen, W.C., Membrane transduction of oligoarginine in HeLa cells is not mediated by macropinocytosis, *Mol. Pharm.* 3, 181–186, 2006; Mettlen, M., Platek, A., Van Der Smissen, P. et al., Src triggers circular ruffling and macropinocytosis at the apical surface of polarized MDCK cells, *Traffic* 7, 589–603, 2006; von Delwig, A., Hilkens, C.M., Altmann, D.M. et al., Inhibition of macro-pinocytosis blocks antigen presentation of type II collagen *in vitro* and *in vivo* in HLA-DR1 transgenic mice, *Arthritis Res. Ther.* 8, R93, 2006.

Madin-Darby Canine Kidney (MDCK)   Used in reference to a mammalian cell line (MDCK cells) used to represent polarized epithelial cells. See Hidalgo, I.J., Assessing the absorption of new pharmaceuticals, *Curr. Top. Med. Chem.* 1, 385–401, 2001; Cohen, D. and Musch, A., Apical surface formation in MDCK cells: regulation by the serine/threonine kinase EMK1, *Methods* 30, 69–276, 2003; Sidorenko, Y. and Reichl, U., Structured model of influenza virus replication in MDCK cells, *Biotechnol. Bioeng.* 88, 1–14, 2004; Urquhart, P., Pang, S., and Hooper, N.M., *N*-glycans as apical targeting signals in polarized epithelial cells, *Biochem. Soc. Symp.* 72, 39–45, 2005.

Maillard Reaction   A reaction (named after Louis-Camille Maillard) between a protein amino group, usually the epsilon-amino group of lysine and a reducing sugar/aldose. There is an initial condensation reaction to form a Schiff base, which undergoes rearrangement to form an Amadori product. This process can initiate a chain reaction, resulting in protein crosslinking and the formation of complex chemicals referred to as advanced glycation endproducts (AGE). Reaction with chemicals such as glucose and meth-ylglyoxal results in Maillard reactions. The browning of foods and the tanning of animal skin are examples of the Maillard reaction. See Hodge, J.E., Chemistry of the browning reactions in model systems, *J. Agric. Food Chem.* 1, 928–943, 1953; Waller, G.R. and Feather, M.S., *The Maillard Reaction in Foods and Nutrition,* American Chemical Society, Washington, DC, 1983; Njoroge, F.G. and Monnier, V.M., The chemistry of the Maillard reaction under physiological conditions: a review, *Prog. Clin. Biol. Res.* 304, 85–107, 1989; Kaanane, A. and Labuza, T.P., The Maillard reaction in foods, *Prog. Clin. Biol. Res.* 304, 301–327, 1989; Labuza, T.P., *Maillard Reactions in Chemistry, Food, and Nutrition,* Royal Society of Chemistry, Cambridge, UK, 1994; Ikan, R., *The Maillard Reaction: Consequences for the Chemical and Life Sciences*, Wiley, Chichester, UK, 1996; Lederer, M.O., Gerum, F., and Severin, T., Cross-linking of proteins by Maillard processes-model reactions of D-glucose or methylglyoxal with butylamine and guanidine derivatives, *Bioorg. Med. Chem.* 6, 993–1002, 1998; Oya, T., Hattori, N., Mizuno, Y. et al., Methylglyoxal modification of protein. Chemical and immunochemical characterization of methylglyoxal-arginine adducts, *J. Biol. Chem.* 274, 18492–18502, 1999; Fayle, S.E. and Gerrard, J.A., *The Maillard Reaction*, Royal Society of Chemistry, Cambridge, UK, 2002; Marko, D., Habermeyer, M., Keméy, M. et al., Maillard reaction products modulating the growth of human tumor cells *in vitro*, *Chem. Res. Toxicol.* 16, 48–55, 2003; Takeguchi, M., Yamagishi, S., Iwaki, M. et al., Advanced glycation end product (age) inhbitors and their therapeutic implications in diseases, *Int. J. Clin. Pharmacol. Res.* 24, 95–1010, 2004; Jing, H. and Nakamura, S., Production and use of Maillard products as oxidative stress modulators, *J. Med. Food* 8, 291–298, 2005; Nursten, H., *The Maillard Reaction. Chemistry, Biochemistry, and Implications*, Royal Society of Chemistry, Cambridge, UK, 2005; Sell, D.R., Biemel, K.M., and Reihl, O., Glucosepane is a major protein cross-link of the senescent human extracellular matrix. Relationship with diabetes, *J. Biol. Chem.* 280, 12310–12315, 2005.

Major Groove and Minor Groove — DNA Structure

The channels or grooves formed when two complementary strands of DNA hybridize with each other to form a double helix. The major groove is ~22 Å wide and the minor groove ~11 Å wide. See Jovin, T.M., McIntosh, L.P., Anrdt-Jovin, D.J. et al., Left-handing DNA: from synthetic polymers to chromosomes, *J. Biomol. Struct. Dyn.* 1, 21–57, 1983; Uberbacher, E.C., and Bunick. G.J., Structure of the nucleosome core particle at 8 Å resolution, *J. Biomol. Struct. Dyn.* 7, 1–18, 1989; Reddy, B.S., Sondhi, S.M., and Lown, J.W., Synthetic DNA minor groove-binding drugs, *Pharmacol. Ther. Sci.*, 1–111, 1999; Haq, I., Thermodynamics of drug-DNA interactions, *Arch. Biochem. Biophys.* 403, 1–15, 2002; Sundaralingam, M. and Pan, B., Hydrogen and hydration of DNA and RNA oligonucleotides, *Biophys. Chem.* 95, 273–282, 2002; Susbielle, G., Blatters, R., Brevet, V., Monod, C., and Kas, E., Target practice: aiming at satellite repeats with DNA minor groove binders, *Curr. Med. Chem. Anticancer Agents* 5, 409–420, 2005; Dragan, A.I., Li, Z., Makeyeva, E.I. et al., Forces driving the binding of homeodomains to DNA, *Biochemistry* 45, 141–151, 2006; Horton, J.R., Zhang, X., Maunus, R. et al., DNA nicking by HikP1I endonuclease: bending, based flipping, and minor groove expansion, *Nucleic Acids Res.* 34, 939–948, 2006; Lamoureux, J.S. and Glover, J.N., Principles of protein-DNA recognition revealed in the structural analysis of Ndt80-MSE DNA complexes, *Structure* 14, 555–565, 2006.

Major Histocompatibility Complex (MHC)

The major histocompatibility complex (MHC) or locus is a cluster of genes on chromosome 6 (chromosome 17 in the mouse) that encodes a family of membrane glycoproteins referred to as MHC proteins or molecules. There are some nonmembrane proteins such as HLA-DM and HLA-DO, which are encoded by the MHC and function in the processing of peptides for delivery to the MHC membrane glycoproteins. MHC membrane glycoproteins are found in antibody-presenting cells (APCs; professional antigen-presenting cells, dendritic cells, macrophages, and B-cells) and "present" antigens to effector CD-4 and CD-8 T-cells. MHC membrane glycoproteins are divided into two groups: MHC class I and MHC class II. MHC class I proteins present peptides generated in the cytoplasm to CD-8 T-cells stimulating the formation of cytotoxic T-cells. Peptides for MHC class I receptors are processed by proteosomes and can be derived from virus-infected cells. MHC class II receptors present peptides to CD-4 T-cells, which activate B-cells to form plasma cells, which synthesize and secrete antibody. Peptides for MHC class II receptors are derived from the action of lysosomal proteases on endosomes. See Janeway, C.A., Jr., Travers, P., Walport, M., and Capra, J.D., *Immunobiology: The Immune System in Health and Disease,* 4th ed., Garland, New York, 1999; Lyczak, J.B., The major histocompatibility complex, in *Immunology, Infection, and Immunity*, Pier, G.B., Lyczak, J.B., and Wetzler, L.M., Eds., ASM Press, Washington, DC, 2004, pp. 261–282; Drozina, G., Kohoutek, J., Jabrane-Ferrat, N., and Peterlin, B.M., Expression of MHC II genes, *Curr. Top. Microbiol. Immunol.* 290, 147–170, 2005; Wucherpfennig, K.W., The structural interactions between T-cell receptors and MHC-peptide complexes place physical limits on self-nonself discrimination, *Curr. Top. Microbiol. Immunol.* 296, 19–37, 2005; Krawczyk, M. and Reith, W., Regulation of MHC class II expression, a unique regulatory system identified by the study of a primary immunodeficiency disease, *Tissue Antigens* 67, 183–197, 2006; Serrano, N.C., Millan, P., and Paez, M.C., Non-HLC associations with autoimmune diseases, *Autoimmune Rev.* 5, 209–214, 2006; Hoglund, P., Induced peripheral regulatory T-cells: the family grows larger, *Eur. J. Immunol.* 36, 264–266, 2006.

MAPPER Search engine for the identification of putative transcription factor–binding sites (TFBSs). See Marinescu, V.D., Kohane, I.S., and Riva, A., MAPPER: a search engine for the computational identification of putative transcription factor binding sites in multiple genomes, *BMC Informatics* 6, 79, 2005.

Mass Spectrometer A device that assigns mass-to-charge ratios to ions based on their momentum, cyclotron frequency, time-of-flight, or other parameters. See Kiser, R.W., *Introduction to Mass Spectrometry and Its Applications*, Prentice-Hall, Englewood Cliffs, NJ, 1965; Blauth, E.W., *Dynamic Mass Spectrometers*, Elsevier, Amsterdam, 1966; Majer, J.R., *The Mass Spectrometer*, Wykeham, London, 1977; Herbert, C.G. and Johnstone, R.A.W., *Mass Spectrometer Basics*, CRC Press, Boca Raton, FL, 2003.

Mast Cell A type of leukocyte found in tissues with large basophilic secretory granules containing histamine and other physiological materials such as heparin during inflammatory responses. See Puxeddu, H., Ribetti, D., Crivellato, E., and Levi-Schaffer, F., Mast cells and eosinophils: a novel link between inflammation and angiogenesis in allergic diseases, *J. Allergy Clin. Immunol.* 116, 531–536, 2005; Galli, S.J., Kalesnikoff, J., Grimbaldeston, M.A. et al., Mast cells as "tunable" effector and immunoregulatory cells: recent advances, *Annu. Rev. Immunol.* 23, 749–786, 2005; Saito, H., Role of mast cell proteases in tissue remodeling, *Chem. Immunol. Allergy* 87, 80–84, 2005; Vliagoftis, H. and Befus, A.D., Mast cells at mucosal frontiers, *Curr. Mol. Med.* 5, 573–589, 2005; Krishnaswamy, G., Ajitawi, O., and Chi, D.S., The human mast cell: an overview, *Methods Mol. Biol.* 315, 13–34, 2006.

Metabolipidomics The study of the metabolism of lipids in the proteome. See Bleijerveld, O.B., Howeling, M., Thomas, M.J., and Cui, Z., Metabolipidomics: profiling metabolism of glycerophospholipid species by stable isotopic precursors and tandem mass spectrometry, *Anal. Biochem.* 352, 1–14, 2006.

Metabolite/Metabolic Profiling Identification/quantification of a select group of metabolites, generally part of a specific pathway such as glycolysis or fatty acid synthesis. See Dunn, W.B. and Ellis, D.I., Metabolomics: current analytical platforms and methodologies, *Trends Anal. Chem.* 24, 285–294, 2005.

Metabolome The total metabolites produced by the products of the genome. See Tweeddale, H., Notley-McRobb, L., and Perenci, T., Effect of slow growth on metabolism of *Escherichia coli*, as revealed by global metabolite pool ("metabolome") analysis, *J. Bacteriol.* 180, 5109–5116, 1998; Tweeddale, H., Notley-McRobb, L., and Ferenci, T., Assessing the effect of reactive oxygen species on *Escherichia coli* using a metabolome approach, *Redox. Rep.* 4, 237–241, 1999; ter Kuile, B.H. and Westerhoff, H.V., Transcriptome meets metabolome: hierarchical and metabolic regulation of the glycolytic pathway, *FEBS Lett.* 500, 169–171, 2001; Mendes, P., Emerging bioinformatics for the metabolome, *Brief Bioinform.* 3, 134–145, 2002; Mazurek, S., Grimm, H., Boschek, C.B., Vaupel, P., and Eingebrodt, E., Pyruvate kinase type?

Phytochemistry 62, 837–849, 2003; Nobeli, I., Ponstingl, H., Krissinel, E.B., and Thornton, J.M., A structure-based anatomy of the *E.coli* metabolome, *J. Mol. Biol.* 334, 697–719, 2003; Parsons, L. and Orban, J., Structural genomics and the metabolome: combining computational and NMR methods to identify target ligands, *Curr. Opin. Drug Discov. Devel.* 7, 62–68, 2004; Soloviev, M. and Finch, P., Peptidomics: bridging the gap between proteome and metabolome, *Proteomics* 6, 744–747, 2006.

Metabolomics   The study of the metabolome; ideally, the nonbiased identification/quantification of all metabolites in a biological system such as a cell or organism. See Reo, N.V., NMR-based metabolomics, *Drug Chem. Toxicol.* 25, 375–382, 2002; German, J.B., Roberts, M.A., Fay, L., and Watkins, S.M., Metabolomics and individual metabolic assessment: the next great challenge for nutrition, *J. Nutr.* 132, 2486–2487, 2002; Watkins, S.M. and German, S.B., Metabolomics and biochemical profiling in drug discovery and development, *Curr. Opin. Mol. Ther.* 4, 224–228, 2002; Phelps, T.J., Palumbo, A.V., and Beliaev, A.S., Metabolomics and microarrays for improved understanding of phenotypic characteristics controlled by both genomics and environmental constraints, *Curr. Opin. Biotechnol.* 13, 20–24, 2002; Grivet, J.P., Delort, A.M., and Portais, J.C., NMR and microbiology: from physiology to metabolomics, *Biochemie* 85, 823–840, 2003; Brown, S.C., Kruppa, G., and Dasseux, J.L., Metabolomics applications of FT-ICR mass spectrometry, *Mass Spectrom. Rev.* 24, 223–231, 2005; Bhalla, R., Narasimhan, K., and Swarup, S., Metabolomics and its role in understanding cellular responses in plants, *Plant Cell Rep.* 24, 562–571, 2005; Fridman, E. and Pichersky, E., Metabolomics, genomics, proteomics, and the identification of enzymes and their substrates and products, *Curr. Opin. Plant Biol.* 8, 242–248, 2005; Rochfort, S., Metabolomics reviewed: a new "omics" platform technology for systems biology and implications for natural product research, *J. Nat. Prod.* 68, 1813–1820, 2005; Griffin, J.L., The Cinderella story of metabolic profiling: does metabolomics get to go to the functional genomics ball? *Philos. Trans. R. Soc. Lond. B Biol. Sci.* 361, 147–161, 2006.

Metagenomics   The genomic analysis of a mixed microbial population. See Schloss, P.D. and Handelsman, J., Biotechnological prospects from metagenomics, *Curr. Opin. Biotechnol.* 14, 303–310, 2003; Riesenfled, C.S., Schloss, P.D., and Handelsman, J., Metagenomics: genomic analysis of microbial communities, *Annu. Rev. Genet.* 38, 525–552, 2004; Steele, H.L. and Streit, W.R., Metagenomics: advances in ecology and biotechnology, *FEMS Microbiol. Lett.* 247, 105–111, 2005.

Metaproteomics   The proteomic analysis of mixed microbial communities. See Wilmes, P. and Bond, P.L., The application of two-dimensional polyacrylamide gel electrophoresis and downstream analyses to a mixed community of prokaryotic microorganisms, *Environ. Microbiol.* 6, 911–920, 2004; Kan, J., Hanson, T.E., Ginter, J.M., Wang, K., and Chen, F., Metaproteomic analysis of Chesapeake Bay microbial communities, *Saline Systems* 1, 7, 2005; Wilmes, P. and Bond, P.L, Metaproteomics: studying functional gene expression in microbial ecosystems, *Trends Microbiol.* 14, 92–97, 2006; Valenzuela, L., Chi, A., Beard, S. et al., Genomics, metagenomics, and proteomics in biomining microorganisms, *Biotechnol. Adv.* 24, 195–209, 2006; Ward, N., New directions and interactions in metagenomics research, *FEMS Microbiol. Ecol.* 55, 331–338, 2006.

Metazoan   Animals with differentiated cells and tissues and usually a discrete digestive tract with specialized cells.

Micelles   Small (nanoscale) particles composed of individual molecules, considered an aggregate. In biochemistry and molecular biology, micelle usually refers to a structure of polar lipids in aqueous solution, which can form an amphipathic layer with the polar groups directed toward solvent and nonpolar groups clustered toward the interior. However, a micelle can be composed of proteins or other organic materials. Casein polymers have been extensively studied because of their presence in milk. The term "critical micellar concentration" (CMC) defines the concentration when, for example, a lipid would move from solution phase to micelle. Micelles are used for drug delivery. Micelles also refer to small particles of materials used in humus soil formulations. See also *Liposomes, Nanotechnology.* See Hartley, G.S., *Aqueous Solutions of Paraffin-Chain Salts: A Study in Micelle Formation*, Hermann and Cie, Paris, 1936; Mukerjee, P., *Critical Micelle Concentrations of Aqueous Surfactant Systems*, U.S. National Bureau of Standards, U.S. Government Printing Office, Washington, DC, 1971; Bloomfield, V.A. and Mead, R.J., Jr., Structure and stability of casein micelles, *J. Dairy Sci.* 58, 592–601, 1975; Kreuter, J., Nanoparticles and nanocapsules — new dosage forms in the nanometer size range, *Pharm. Acta Helv.* 53, 33–39, 1978; Furth, A.J., Removing unbound detergent from hydrophobic proteins, *Anal. Biochem.* 109, 207–215, 1980; Rosen, M.J., *Surfactants and Interfacial Phenomena* , Wiley-Interscience, Hoboken, NJ, 2004; Bagchi, B., Water dynamics in the hydration layer around proteins and micelles, *Chem. Rev.* 105, 3197–3219, 2005; Chandler, D., Interfaces and the driving force of hydrophobic assembly, *Nature* 437, 640–647, 2005; Gentle, I. and Barnes, G., *Interfacial Science: An Introduction*, Oxford University Press, Oxford, UK, 2005; Aliabadi, H.M. and Lavasanifar, A., Polymeric micelles for drug delivery, *Expert Opin. Drug Deliv.* 3, 139–162, 2006.

Microarray   Generally referring to an array of probes displayed on a matrix similar to a microscope slide. Microarrays are used to analyze complex mixtures for specific analytes. The sample is usually labeled with a signal such as a fluorescent dye. Both sample preparation and analysis are complex. See *Tissue Microarray.* See Schena, M., Heller, R.A., Theriault, T.P. et al., Microarrays: Biotechnology's discovery platform for functional genomics, *Trends Biotechnol.* 16, 301–306, 1998; Gerhold, D., Rushmore, T., and Caskey, C.T., DNA chips: promising toys have become powerful tools, *Trends Biochem. Sci.* 24, 168–173, 1999; Ness, S.A., Basic microarray analysis: strategies for successful experiments, *Methods Mol. Biol.* 316, 13–33, 2006; Wang, S. and Cheng, Q., Microarray analysis in drug discovery and clinical applications, *Methods Mol. Biol.* 316, 49–65, 2006; Kozarova, A., Petrinac, S., Ali, A., and Hudson, J.W., Array of informatics: applications in modern research, *J. Proteome Res.* 5, 10551–10559, 2006; Sievertzon, M., Nilsson, P., and Lundeberg, J., Improving reliability and performance of DNA microarrays, *Expert Rev. Mol. Diagn.* 6, 481–492, 2006; Sobek, J., Bartscherer, K., Jacob, A., Hoheisel, J.D., and Angenendt, P., Microarray technology as a universal tool for high-throughput analysis of biological systems, *Comb. Chem. High Throughput Screen.* 9, 365–380, 2006; Pedroso, S. and Guillen, I.A., Microarray and nanotechnology applications of functional nanoparticles, *Comb. Chem. High Throughput Screen.* 9, 389–397,

2006. Microarray analysis can also be applied to the study of multiple tissue specimens with tissue microarrays (Rimm, D.L., Camp, R.L, Charette, L.A. et al., Tissue microarray: a new technology for amplification of tissue resources, *Cancer J.* 7, 24–31, 2001; Bubendorf, L., Nocito, A., Moch, K., and Sauter, G., Tissue microarray [TMA] technology: minaturized pathology archives for high-throughput *in situ* studies, *J. Pathol.* 195, 72–79, 2001; Fedor, H.L. and De Marzo, A.M., Practical methods for tissue microarray construction, *Methods Mol. Biol.* 103, 89–101, 2005).

Microdialysis   A process for sampling low-molecular weight metabolites in the extracellular space of tissues. This technique is used for the study of tissue metabolism and pharmacokinetic studies. Microdialysis is accomplished through the use of a probe constructed as a concentric tube which is implanted into a tissue and a perfusion fluid (a physiological solution such as Hank's Balanced Salt Solution) enters through an inner tube flowing toward the distal end and, entering the space between the inner tube and the outer dialysis membrane, flows back toward the proximal end of the probe. Dialysis takes place during the passage of fluid toward the proximal end and the exiting fluid is sampled for the analyte in question. It is viewed as a noninvasive method of evaluated tissue metabolism. See Lonnroth, P. and Smith, U., Microdialysis — a novel technique for clinical investigations, *J. Intern. Med.* 227, 295–300, 1990; Ungerstedt, U., Microdialysis — principles and applications for studies in animals and man, *J. Intern. Med.* 230, 365–373, 1991; Parsons, L.H. and Justice, J.B., Jr., Quantitative approaches to *in vivo* brain microdialysis, *Crit. Rev. Neurobiol.* 9, 189–220, 1994; Schuck, V.J., Rinas, I., and Derendorf, H., *In vitro* microdialysis sampling of docetaxel, *J. Pharm. Biomed. Anal.* 36, 607–613, 2004; Hocht, C., Opezzo, J.A., and Taira, C.A., Microdialysis in drug discovery, *Curr. Drug. Discov. Technol.* 1, 269–285, 2004; Rooyackeres, O., Thorell, A., Nygren, J., and Ljungqvist, O., Microdialysis method for measuring human metabolism, *Curr. Opin. Clin. Nutr. Metab. Care* 7, 515–552, 2004; Cano-Cebrian, M.J., Zornoza, T., Polache, A., and Granero, L., Quantitative *in vivo* microdialysis in pharmacokinetic studies: some reminders, *Curr. Drug Metab.* 6, 83–90, 2005; Abrahamsson, P. and Winso, O., An assessment of calibration and performance of the microdialysis system, *J. Pharm. Biomed. Anal.* 39, 730–734, 2005; Ao, X. and Stenken, J.A., Microdialysis sample of cytokines, *Methods* 38, 331–341, 2006. Successful interpretation of microdialysis experiments will require a thorough understanding of the factor in fluxing membrane of the specific analyte or analytes.

Microsatellite   Tandem repeats of one to six nucleotides; sometimes referred to as simple sequence repeats or simple sequence lengths. These sites are used for genetic mapping and show variability in oncological disorders. Changes are measured by simple sequence-length polymorphisms. See Tautz, D. and Schlotterer, C., Simple sequences, *Curr. Opin. Genet. Dev.* 4, 832–837, 1994; Albrecht, A. and Mundlos, S., The other trinucleotide repeat: polyalanine expansion disorders, *Curr. Opin. Genet. Dev.* 15, 285–293, 2005; Di Prospero, N.A. and Fischbeck, K.H., Therapeutics development for triplet repeat expansion diseases, *Nat. Rev. Genet.* 6, 756–765, 2005; Zheng, H.T., Peng, Z.H., Li, S., and He, L., Loss of heterozygosity analyzed by single nucleotide polymorphism array in cancer, *World J. Gastroenterol.* 11, 6740–6744, 2005.

Minicollagen   Minicollagen is a miniprotein derivative of collagen. See Mazzorana, M., Snellman, A., Kivirikkov, K.I., van der Rest, M., and Pihlajaniemi, T., Involvement of prolyl-4-hydroxylase in the assembly of trimeric minicollagen XII. Study in a baculovirus expression system, *J. Biol. Chem.* 271, 29003–29008, 1996; Milbradt, A.G., Boulegue, C., Moroder, L., and Renner, C., The two cysteine-rich head domains of minicollagen from *Hydra* nematocysts differ in their cystine framework and overall fold despite an identical cysteine sequence pattern, *J. Mol. Biol.* 354, 591–600, 2005.

Miniprotein   The smallest portion of a protein, which can fold into a unique three-dimensional structure thus differing from a peptide. See Degrado, W.F. and Sosnick, T.R., Protein minimization: downsizing through mutation, *Proc. Natl. Acad. Sci. USA* 93, 5680–5681, 1996; Vita, C., Vizzavona, J., Drakopoulou, E. et al., Novel miniproteins engineered by the transfer of active sites to small nature scaffolds, *Biopolymers* 47, 93–100, 1998; Neuweiler, H., Doose, S., and Sauer, M., A microscopic view of miniprotein folding: enhanced folding efficiency through formation of an intermediate, *Proc. Natl. Acad. Sci. USA* 102, 16650–16655, 2005.

Modulus of Elasticity   The stress required to produce unit strain, causing a change in length (Young's modulus) or a twist or shear (shear modulus) or a change volume (bulk modulus); units are dynes/cm^2. See Nash, G.B. and Gratzer, W.B., Structural determinants of the rigidity of the red cell membrane, *Biorheology* 30, 397–407, 1993; Urry, D.W. and Pattanaik, A., Elastic protein-based materials in tissue reconstruction, *Ann. N.Y. Acad. Sci.* 831, 32–46, 1997; Ambrosio, L., De Santis, R., and Nicolais, L., Composite hydrogels for implants, *Proc. Inst. Mech. Eng.* 212, 93–99, 1998; Roberts, R.J. and Rowe, R.C., Relationships between the modulus or elasticity and tensile strength for pharmaceutical drugs and excipients, *J. Pharm. Pharmcol.* 51, 975–977, 1999; Hegner, M. and Grange, W., Mechanics and imaging of single DNA molecules, *J. Muscle Res. Cell. Motil.* 23, 367–375, 2002; Carr, M.E., Jr., Development of platelet contractile force as a research and clinical measure of platelet function, *Cell Biochem. Biophys.* 38, 55–78, 2003; Balshakova, A.V., Kiselyova, O.I., and Yaminsky, I.V, Microbial surfaces investigated using atomic force microscopy, *Biotechnol. Prog.* 20, 1615–1622, 2004; Zhang, G., Evaluating the viscoelastic properties of biological tissues in a new way, *J. Musculoskelet. Neuronal Interact.* 5, 85–90, 2005; Seal, B.L. and Panitch A., Physical matrices stabilized by enzymatically sensitive covalent crosslinks, *Acta Biomater.* 2, 241–251, 2006.

Molecular Beacons   DNA probes that are chain-loop-and-stem structures. A fluorophore at the 5′ terminus is in proximity to a quencher at the 3′ terminus and is released on the binding of the probes to a specific DNA sequence. See Piatek, A.S., Tyagi, S., Pol, A.C. et al., Molecular beacon sequence analysis for detecting drug resistance in *Mycobacterium tuberculosis*, *Nature Biotechnol.* 16, 35–363, 1998; Tan, W., Fang, X., Li, J., and Liu, X., Molecular beacons: a novel DNA probe for nucleic acid and protein studies, *Chem. Eur. J.* 6, 1107–1111, 2000; Fang, X., Mi, Y., Li, J.J. et al., Molecular beacons: fluorogenic probes for living cell study, *Cell Biochem. Biophys.* 37, 71–81, 2002; Stöhr, K., Häfner, B., Nolte, O. et al., Species-specific identification of myobacterial 16S rRNA PCR amplicons using smart probes, *Anal. Chem.* 77, 7195–7203, 2005.

Molecular Clock   Places timescales on evolutionary events. See Zuckerkandl, E., On the molecular evolutionary clock, *J. Mol. Evol.* 26, 34–46, 1987; Easteal, S., A mammalian molecular clock, *Bioessays* 14, 415–419, 1992; Seoighe, C., Turning the clock back on ancient genome duplication, *Curr. Opin. Genet. Dev.* 13, 636–643, 2003; Freitas, C., Rodrigues, S., Saude, L., and Palmeirim, I., Running after the clock, *Int. J. Dev. Biol.* 49, 317–324, 2005; Renner, S.S., Relaxed molecular clocks for dating historical plant dispersal events, *Trends Plant Sci.* 10, 550–558, 2005; Ho, S.Y.W. and Larson, G., Molecular clocks: when times are a-changing, *Trends Genet.* 22, 79–83, 2006.

Monocot (Monocotyledon)   Flowering plants that have embryos with only one cotyledon. See Iyer, L.M., Kumpatla, S.P., Chadrasekharan, M.B., and Hall, T.C., Transgene silencing in monocots, *Plant Mol. Biol.* 43, 323–346, 2000; Agrawal, G.K., Iwahashi, H., and Rakwal, R., Rice MAPKs, *Biochem. Biophys. Res. Commun.* 302, 171–180, 2003.

Monocyte   Circulating white blood cell; an immune system cell that is a precursor to macrophage. See Nelson, D.S. and Alexander, P., *Immunobiology of the Macrophage*, Academic Press, New York, 1976; Schmalzl, F. and Huhn, D., *Disorders of the Monocyte Macrophage System: Pathophysiological and Clinical Aspects*, Springer-Verlag, Berlin, 1981; Russell, S.W. and Siamon, G., *Macrophage Biology and Activation*, Springer-Verlag, Berlin, 1992; Zwilling, B.S. and Eisenstein, T.K., *Macrophage-Pathogen Interactions*, Marcel Dekker, New York, 1994; Mire-Sluis, A.R. and Thorpe, R., *Cytokines*, Academic Press, San Diego, CA, 1998; Bellosta, S. and Bernini, F., Modulation of macrophage function and metabolism, *Handb. Exp. Pharmacol.* 170, 665–695, 2005; Condeelis, J. and Pollard, J.W., Macrophages: obligate partners for tumor cell migration, invasion, and metastasis, *Cell* 124, 263–266, 2006; Noda, M., Current topics in pharmacological research on bone metabolism: regulation of bone mass by the function of endogenous modulators of bone morphogenetic protein in adult stage, *J. Pharmcol. Sci.* 100, 211–214, 2006; Lews, C.E. and Pollard, J.W., Distinct role of macrophages in different tumor microenvironments, *Cancer Res.* 66, 605–612, 2006; Bobryshev, Y.V., Monocyte recruitment and foam cell formation in atherosclerosis, *Micron* 37, 208–222, 2006; Cathelin, S., Rébe, C., Haddaoui, L. et al., Identification of proteins cleaved downstream of caspases activation in monocytes undergoing macrophage differentiation, *J. Biol. Chem.* 281, 17779–17788, 2006.

Morpho-proteomics   Morphoproteomics combines the technical approaches of histopathology, molecular biology, and proteomics for the study of cell biology and systems biology. See Brown, R.E., Morphoproteomic portrait of the mTOR pathway in mesenchymal chondrosarcoma, *Ann. Clin. Lab. Sci.* 34, 397–399, 2004; Brown, R.E., Morphoproteomics: exposing protein circuitries in tumors to identify potential therapeutic targets in cancer patients, *Expert Rev. Proteomics* 2, 337–348, 2005.

Mucus/Mucins   A viscous biological fluid containing mucins. Mucins are very large, highly glycosylated proteins with considerable asymmetry. The carbohydrate moiety is sulfated, resulting in a high negative charge that contributes to the extended conformation. It is the physical properties of mucin that are responsible for the viscosity of mucus. See Gerken, T.A., Biophysical approaches to salivary mucin structure, conformation, and dynamics, *Crit. Rev. Oral Biol. Med.* 4, 261–270, 1993; Dodd, S., Place, G.A., Hall, R.L., and Harding, S.E., Hydrodynamic properties of mucins secreted by primary cultures of guinea-pig tracheal epithelial cells: determination of diffusion coefficients by analytical ultracentrifugation and kinetic analysis of mucus gel hydration and dissolution, *Eur. Biophys. J.* 28, 38–47, 1999; Brockhausen, I., Sulphotransferases acting on mucin-type oligosaccharides, *Biochem. Soc. Trans.* 31, 318–325, 2003; Lafitte, G., Thuresson, K., and Soderman, O., Mixtures of mucin and oppositely charged surfactant aggregates with varying charge density. Phase behavior, association, and dynamics, *Langmuir* 21, 7097–7104, 2005; Rose, M.C. and Voynow, J.A., Respiratory tract mucin genes and mucin glycoproteins in health and disease, *Physiol. Rev.* 86, 245–278, 2006.

Mullerian-Inhibiting Substance (Mullerian-Inhibiting Hormone)   A peptide growth factor that is a member of the TGF family, which was originally described as a large glycoprotein secreted by neonatal/fetal testes responsible for regression of the Mullerian duct during the process of sexual differentiation. See Budzik, G.P., Powell, S.M., Kamagata, S., and Donahoe, P.K., Mullerian-inhibiting substance fractionation of dye affinity chromatography, *Cell* 34, 307–314, 1983; Visser, J.A. and Themmen, A.P.N., Anti-Müllerian hormone and folliculogenesis, *Mol. Cell. Endocrinol.* 234, 81–86, 2005.

Mutation Rate   The instantaneous rate at which nucleotide changes occur in the genome. Lethal or near-lethal mutations are often ignored in the calculation of maturation rate. See Ellegren, H., Smith, N.G., and Webster, M.T., Mutation rate variation in the mammalian genome, *Curr. Opin. Genet. Dev.* 13, 562–568, 2003; Sniegowski, P., Evolution: bacterial mutation in stationary phase, *Curr. Biol.* 14, R245–R246, 2004; Wang, C.L. and Wabl, M., Precise dosage of an endogenous mutagen in the immune system, *Cell Cycle* 3, 983–985, 2004; Pakendorf, B. and Stoneking, M., Mitochondrial DNA and human evolution, *Annu. Rev. Genomics Hum. Genet.* 6, 165–183, 2005.

Myeloid Progenitor   Precursor of granulocyte, macrophages, and mast cells; involved in chronic myeloid leukemia (CML). Development regulated by colony-simulating factors (CSFs). See Burgess, A.W. and Metcalf, D., The nature and action of granulocyte-macrophage colony stimulating factors, *Blood* 56, 947–958, 1980; Islam, A., Haemopoietic stem cells: a new concept, *Leuk. Res.* 9, 1415–1432, 1985; Cannistra, S.A. and Griffin, J.D., Regulation of the production and function of granulocytes and monocytes, *Semin. Hematol.* 25, 173–188, 1988; Grimwade, D. and Enver, T., Acute promyelocytic leukemia: where does it stem from? *Leukemia* 19, 375–384, 2004; Coulombel, L., Identification of hematopoietic stem/progenitor cells: strength and drawbacks of functional assays, *Oncogene* 23, 7210–7223, 2004; Rosmarain, A.G., Yang, Z., and Resendes, K.K., Transcriptional regulation in myelopoiesis: hematopoietic fate choice, myeloid differentiation, and leukemogenesis, *Exp. Hematol.* 33, 131–143, 2005; Marley, S.B., and Gordon, M.Y., Chronic myeloid leukemia: stem cell derived but progenitor cell driven, *Clin. Sci.* 109, 13–25, 2005.

Nanofiltration   Filtration of small (nano) particles from solvent using a filter with extremely small pores (0.001–0.010 micron); finer than ultrafiltration, not as fine as reverse osmosis. Used for the removal of viruses from plasma protein products. See Yaroshchuk, A.E., Dielectric exclusion of ions from membranes, *Adv. Colloid Interface Sci.* 85, 193–230, 2000; Rossano, R., D'Elia, A., and Riccio, P., One-step separation from lactose: recovery and purification of major cheese-whey proteins by hydroxyapatite — a flexible procedure suitable for small- and medium-scale preparations, *Protein Expr. Purif.* 21, 165–169, 2001; Burnouf, T. and Radosevich, M., Nanofiltration of plasma-derived biopharmaceutical products, *Haemophilia* 9, 24–37, 2003; Bhanushali, D. and Bhattacharyya, D., Advances in solvent-resistant nanofiltration membranes: experimental observations and application, *Ann. N.Y. Acad. Sci.* 984, 159–177, 2003; Weber, R., Chmiel, H., and Mavrov, V., Characteristics and application of ceramic nanofiltration membranes, *Ann. N.Y. Acad. Sci.* 984, 178–193, 2003; Tieke, B., Toutianoush, A., and Jin, W., Selective transport of ions and molecules across layer-by-layer assembled membranes of polyelectrolytes, *p*-sulfonato-calixin[n]arenes, and Prussian Blue–type complex salts, *Adv. Colloid Interface Sci.* 116, 121–131, 2005; Berot, S., Compoint, J.P., Larre, C., Malabat, C., and Gueguen, J., Large-scale purification of rapeseed proteins (*Brassica napus L.*), *J. Chromatog. B Analyt. Technol. Biomed. Life Sci.* 818, 35–42, 2005; Zhao, K. and Li, Y., Dielectric characterization of a nanofiltration membrane in electrolyte solutions: its double-layer structure and ion permeation, *J. Phys. Chem. B Condens. Matter Mater. Surf. Interfaces Biophys.* 110, 2755–2763, 2006; Bulut, M., Gevers, L.E., Paul, J.S., Vankelecom, I.F., and Jacobs, P.A., Directed development of high-performance membranes via high-throughput and combinatorial strategies, *J. Comb. Chem.* 8, 168–173, 2006.

Nanog   Nanog is a homeodomain transcription factor that is found in undifferentiated embryonic stem cells and is considered important for the maintenance of pluripotency. See Mitsui, K., Tokuzawa, Y., Itoh, H. et al., The homeoprotein Nanog is required for maintenance of pluripotency in mouse epiblast and ES cells, *Cell* 113, 631–642, 2003; Chambers, I., Colby, D., Robertson, M. et al., Functional expression of Nanog, a pluripotency sustaining factor in embryonic stem cells, *Cell* 113, 643–655, 2003; Oh, J.H., Do, H.J., Yang, H.M. et al., Identification of a putative transactivation domain in human Nanog, *Exp. Mol. Med.* 37, 250–254, 2005; Xu, Y., A new role for p53 in maintaining genetic stability in embryonic stem cells, *Cell Cycle* 4, 363–364, 2005; Ralston, A. and Rossant, J., Genetic regulation of stem cells origins in the mouse embryo, *Clin. Genet.* 68, 106–112, 2005; Yates, A. and Chambers, I., The homeodomain protein Nanog and pluripotency in mouse embryonic stem cells, *Biochem. Soc. Trans.* 33, 1518–1521, 2005; Silva, J., Chambers, I., Pollard, S., and Smith, A., Nanog promotes transfer of pluripotency afer cell fusion, *Nature* 441, 997–1001, 2006.

Nanotechnology   The study of particles, devices, and substances having physical dimensions of 1–100 nanometers (0.001 to 0.100 micrometers [microns]). This includes the study of liposomes for drug delivery, nanofiltration, micelles, dendrimers, and quantum dots. See Ciofalo, M., Collins, M.W., and Hennessy, T.R., Modeling nanoscale fluid dynamics and transport in physiological flows, *Med. Eng. Phys.* 18, 437–451, 1996; Seeman, N.C., DNA nanotechnology: novel DNA constructions, *Annu. Rev. Biophys. Biomol. Struct.* 27, 225–248, 1998; Melo, E.P., Aires-Barros, M.R., and Cabral, J.M., Reverse micelles and protein biotechnology, *Biotechnol. Annu. Rev.* 7, 87–129, 2001; Wilson, M., *Nanotechnology: Basic Science and Emerging Technologies*, Chapman & Hall, Boca Raton, FL, 2002; Poole, C.P. and Owens, F.J., *Introduction to Nanotechnology*, John Wiley & Sons, Hoboken, NJ, 2003; Di Ventra, M. and Evoy, S., *Introduction to Nanoscale Science and Technology*, Kluwer Academic, Boston, MA, 2004; Bhushan, B., *Springer Handbook of Nanotechnology*, Springer, Berlin, 2004; Williams,

D.J. and Sebastine, I.M., Tissue engineering and regenerative medicine: manufacturing challenges, *IEE Proc. Nanobiotechnol.* 152, 207–210, 2005; Bhattacharya, D. and Gupta, N.K., Nanotechnology and potential of microorganisms, *Crit. Rev. Biotechnol.* 25, 199–204, 2005; Eijkel, J.C. and van den Berg, A., The promise of nanotechnology for separation devices — from a top-down approach to nature-inspired separation devices, *Electrophoresis* 27, 677–685, 2006; Lange, C.F. and Finlay, W.H., Liquid atomizing: nebulizing and other methods of producing aerosols, *J. Aerosol. Med.* 18, 28–35, 2006; Langford, R.M., Focused ion beam nano-fabrication: a comparison with conventional processing techniques, *J. Nanosci. Nanotechnol.* 6, 661–668, 2006; Wang, J. and Ren, J., Lumines-cent quantum dots: a very attractive and promising tool in biomedicine, *Curr. Med. Chem.* 13, 897–909, 2006.

NAR Gene, *Nar* Operon, *Nar* Promoter Promoter region of the *nar* operon, which encodes nitrate reductase in *Escherichia coli*. The promoter is generally only maximally induced under anaerobic conditions. It has been shown that the *nar* promoter in some strains of *Escherichia coli* can be induced under conditions of very low oxygen tension in the presence of nitrate. This observation has been used to develop some useful processes for recombinant protein expression in *Escherichia coli*. See Li, S.F. and DeMoss, J.A., Promoter region of the *nar* operon of *Escherichia coli*: nucleotide sequence and transcription initiation signals, *J. Bacteriol.* 169, 4614–4620, 1987; Han, S.J., Chang, H.N., and Lee, J., Characterization of an oxygen-dependent inducible promoter, the *nar* promoter of *Escherichia coli*, to utilize in metabolic engineering, *Biotechnol. Bioeng.* 72, 573–577, 2001; Lee, K.H., Cho, M.H., Chung, T. et al., Characterization of an oxygen-dependent inducible promoter: the *Escherichia coli nar* promoter, in gram-negative host strains, *Biotechnol. Bioeng.* 82, 271–277, 2003.

Nascent Peptide Exit Tunnel A "tunnel"/pore starting at the peptidyl transferase center on the ribosome and ending on the solvent side on the large ribosomal subunit. See Gabashvili, I.S., Gregory, S.T., Valle, M. et al., The polypeptide tunnel system in the ribosome and its gating in erythromycin resistance mutants of L4 and L22, *Mol. Cell* 8, 181–188, 2001; Tenson, T. and Ehrenberg, M., Regulatory nascent peptides in the ribosomal tunnel, *Cell* 108, 591–594, 2002; Jenni, S. and Ban, N., The chemistry of protein synthesis and voyage through the ribosomal tunnel, *Curr. Opin. Struct. Biol.* 13, 212–219, 2003; Vimberg, V., Ziong, L., Bailey, M., Tenson, T., and Mankin, A., Peptide-mediated macrolide resistance reveals possible specific interactions in the nascent peptide exit tunnel, *Mol. Microbiol.* 54, 376–385, 2004; Baram, D. and Yonath, A., From peptide-bond formation to cotranslational folding: dynamic, regulatory, and evolutionary aspects, *FEBS Lett.* 579, 948–994, 2005; Egea, P.F., Stroud, R.M., and Walter, P., Targeting proteins to mem-branes: structure of the signal recognition particle, *Curr. Opin. Struct. Biol.* 15, 213–220, 2005; Markin, A.S., Nascent peptide in the "birth canal" of the ribosome, *Trends Biochem. Sci.* 31, 11–16, 2006.

Nephelometry Detection of electromagnetic wave scattering in a direction different from the direct path of the transmitted light; for example, electromagnetic energy scattered at a 90° angle from the incident radiation. Nephelometry is widely used for the measurement of immune complexes. See Deverilli, I. and Reeves, W.G., Light scattering and absorption — developments in immunology, *J. Immunol. Methods* 38, 191–204, 1980; Blackstock, R.,

In vitro methods for detection of circulating immune complexes and other solution protein–protein interactions, *Ann. Clin. Lab. Sci.* 11, 262–268, 1981; Steinberg, K.K., Cooper, G.R., Graiser, S.R., and Rosseneu, M., Some considerations of methodology and standardization of apolipoprotein A-I immunoassays, *Clin. Chem.* 29, 415–426, 1983; Price, C.P., Spencer, K., and Whicher, J., Light-scattering immunoassay of specific proteins: a review, *Ann. Clin. Biochem.* 20, 1–14, 1983; Brinkman, J.W., Bakker, S.J., Gansevoort, R.T. et al., Which method for quantifying urinary albumin excretion gives what outcome? A comparison of immunonephelometry with HPLC, *Kidney Int.* 92 (Suppl.), S69–S75, 2004; Yeh, A.T. and Hirshburg, J., Molecular interactions of exogenous chemical agents with collagen — implications for tissue optical clearing, *J. Biomed. Opt.* 11, 014003, 2006.

NF-κB
Nuclear factor kappa B — a transcription factor that is thought to have a major role in the growth of malignant cells. See Zingarelli, B., Nuclear factor kappaB, *Crit. Care Med.* 33 (Suppl. 12), S414–S416, 2005; Jimi, E. and Ghosh, S., Role of the nuclear factor-kappaB in the immune system and bone, *Immunol. Rev.* 208, 80–87, 2005; Bubici, C., Papa, S., Pham, C.G., Zazzeroni, F., and Froanzoso, G., The NF-kappaB-mediated control of the ROS and JNK signaling, *Histol. Histopathol.* 21, 69–80, 2006; Kim, H.J., Hawke, N., and Baldwin, A.S., NF-kappaB and IKK as therapeutic targets in cancer, *Cell Death Differ.* 13, 738–747, 2006.

Nonidet P-40™
Nonidet is a popular nonionic (polyoxythelene glycol derivative) detergent that has been used for membrane protein solubilization. See Dunkley, P.R., Holmes, R., and Rodnight, R., Phosphorylation of synaptic-membrane proteins from ox cerebral cortex *in vitro*. Preparation of fractions enriched in phosphorylated proteins by using extraction with detergents urea, and gel filtration, *Biochem. J.* 163, 369–378, 1977; Sharma, C.B., Lehle, L., and Tanner, W., *N*-glycosylation of yeast proteins. Characterization of the solubilized oligosaccharyl transferase, *Eur. J. Biochem.* 116, 101–108, 1981; Perez-Machin, R., Henriquez-Hernandez, L., Perez-Luzardo, O. et al., Solubilization and photoaffinity labeling identification of glucorticoid binding peptides in endoplasmic reticulum from rat liver, *J. Steroid Biochem. Mol. Biol.* 84, 245–253, 2003; Shiozaki, A., Tsuji, T., and Kohno, R., Proteome analysis of brain proteins in Alzheimer's disease: subproteomics following sequentially extracted protein preparation, *J. Alzheimers Dis.* 6, 257–268, 2004; Zintl, A., Pennington, S.R., and Mulcahy, G., Comparison of different methods for the solubilization of *Neospora caninum* (Phylum Apicomplexa) antigen, *Vet. Paristol.* 135, 205–213, 2006; Kalabis, J., Rosenberg, I., and Podolskyi, D.K., Vangil protein acts as a downstream effector of intestinal trefoil factor (ITF/TFF3) signaling and regulates wound healing of intestinal epithelium, *J. Biol. Chem.* 281, 6434–6441, 2006. Used in renaturing allergens after blotting (Muro, M.D., Fernandez, C., and Moneo, I., Renaturation of blotting allergens increases the sensitivity of specific IgE detection, *J. Investig. Allergol. Clin. Immunol.* 6, 166–171, 1996) and has been described as being used in an aqueous two-phase separation system (Sanchez-Ferrer, A., Bru, R., and Garcia-Carmona, F., Phase separation of biomolecules in polyoxyethylene glycol nonionic detergents, *Crit. Rev. Biochem. Mol. Biol.* 29, 275–313, 1994).

Northern Blot
A technique similar to the southern blot. RNA separated by electrophoresis is transferred to a PDVF membrane. Specific RNA sequences are detected

with a labeled cDNA probe. See Hayes, P.C., Wolf, C.R., and Hayes, J.D., Blotting techniques for the study of DNA, RNA, and proteins, *BMJ* 299, 965–968, 1989; Dallman, M.J., Montgomery, R.A., Larsen, C.P., Wanders, A., and Wells, A.F., Cytokine gene expression: analysis using northern blotting, polymerase chain reaction and *in situ* hybridization, *Immunol. Rev.* 119, 163–179, 1991; Mengod, G., Goudsmit, E., Probst, A., and Palacios, J.M., *In situ* hybridization histochemistry in the human hypothalamus, *Prog. Brain Res.* 93, 45–55, 1992; Pajor, A.M., Hirayama, B.A., and Wright, E.M., Molecular biology approaches to comparative study of Na(+)–glucose cotransport, *Am. J. Physiol.* 263, R489–R495, 1992; Kroczek, R.A., Southern and northern analysis, *J. Chromatog.* 618, 133–145, 1993; Farrell, R.E., *RNA Methodologies: A Laboratory Guide for Isolation and Characterization,* Academic Press, San Diego, CA, 1993; Raval, P., Qualitative and quantitative determination of mRNA, *J. Pharmacol. Toxicol. Methods* 32, 125–127, 1994; Durrant, I., Enhanced chemiluminescent detection of horseradish peroxidase labeled probes, *Methods Mol. Biol.* 31, 147–161, 1994; Darling, D.C. and Brickell, P.M., *Nucleic Acid Blotting:The Basics*, Oxford University Press, Oxford, UK, 1994; Aravin, A. and Tuschi, T., Identification and characterization of small RNAs involved in RNA silencing, *FEBS Lett.* 579, 5830–5840, 2005.

Northwestern Blot　A protein blotting technique related to the various other blotting techniques such as the western blot and the northern blot. In the northwestern blot, the protein mixtures are separated by gel electrophoresis and transferred by electrophoresis to a PVDF or nitrocellulose membrane. Specific proteins are identified through the binding of radiolabeled or fluorophore-labeled RNA oligomers; (double-stranded RNA) dsRNA is used to identify dsRNA binding proteins. See Schiff, L.A., Nibert, M.L., Co, M.S., Brown, E.G., and Fields, B.N., Distinct binding sites for zinc and double-stranded RNA in the reovirus outer capsid protein sigma 3, *Mol. Cell. Biol.* 8, 273–283, 1988; Chen, X., Sadlock, J., and Schon, E.A., RNA-binding patterns in total human tissue proteins: analysis by northwestern blotting, *Biochem. Biophys. Res. Commun.* 191, 18–25, 1993; Kumar, A., Kim, H.R., Sobol, R.W. et al., Mapping of nucleic acid binding in proteolytic domains of HIV-1 reverse transcriptase, *Biochemistry* 32, 7466–7474, 1993; Lin, G.Y., Paterson, R.G., and Lamb, R.A., The RNA binding region of the paramyxovirus SV5 V and P proteins, *Virology* 238, 460–469, 1997; Zhao, S.L., Liang, C.Y., Zhang, W.J., Tang, X.C., and Peng, H.Y., Characterization of the RNA-binding domain in the *Decrolimus punctatus* cytoplasmic polyhedrosis virus nonstructural protein p44, *Virus Res.* 114, 80–88, 2005; Sekiya, S., Noda, K., Nishikawa, F. et al., Characterization and application of a novel RNA aptamers against the mouse prion protein, *J. Biochem.* 139, 383–390, 2006.

Notch　A receptor class that regulates cell differentiation and development. Notch was first identified in *Drosophila*, where it is thought to be involved in long-term memory and neuronal plasticity. See Artavanis-Tsakonas, S., The molecular biology of the Notch locus and the fine tuning of differentiation in *Drosophila*, *Trends Genet.* 4, 95–100, 1988; Jones, P.A., Epithelial stem cells, *Bioessays* 19, 683–690, 1997; Lai, E.C., Notch signaling: control of cell communication and cell fate, *Development* 131, 965–973, 2004; Wilkin, M.B. and Baron, M., Endocytic regulation of North activation and down-regulation, *Membrane Molec. Biol.* 22, 279–289, 2005.

Nuclear   A technique that detects nuclear-spin orientation in an applied magnetic field;
Magnetic   detection of a nuclear magnetic moment, usually measured as the chemical
Resonance   shift from a standard such as tetramethyl silane for hydrogen and trichloro-
fluoromethane for fluorine. Coupling constants (spin-spin coupling, J)
are also measured in two-dimensional analyses. See Roberts, J.D., *Nuclear
Magnetic Resonance: Applications to Organic Chemistry*, McGraw-Hill,
New York, 1959; Pople, J.A., *High-Resolution Nuclear Magnetic Reso-
nance*, McGraw-Hill, New York, 1959; Dyer, J.R., *Applications of Absorp-
tion Spectroscopy of Organic Compounds,* Prentice-Hall, Englewood
Cliffs, NJ, 1965; Knowles, P.R., March, D., and Rattle, H.W.E., *Magnetic
Resonance of Biomolecules: An Introduction to the Theory and Practice
of NMR and ESR in Biological Systems*, Wiley, New York, 1976; Leyden,
D.E. and Cox, R.H., *Analytical Applications of NMR*, Wiley, New York,
1977; Jardetzky, O. and Roberts, G.C.K., *NMR in Molecular Biology*,
Academic Press, New York, 1981; Wüthrich, K., *NMR of Proteins and
Nucleic Acids*, Wiley, New York, 1986; Paudler, W.W., *Nuclear Magnetic
Resonance: General Concepts and Applications*, Wiley, New York, 1987;
Schrami, J. and Bellama, J.M., *Two-Dimensional NMR Spectroscopy*,
Wiley, New York, 1988; Sanders, J.K.M. and Hunter, B.K., *Modern NMR
Spectroscopy: A Guide for Chemists*, Oxford University Press, Oxford,
UK, 1993; Hore, P.J. and Jones, J.A., *NMR, The Tookit,* Oxford University
Press, Oxford, UK, 2000; James, T.L. and Schmitz, U., *Nuclear Magnetic
Resonance of Biological Molecules,* Academic Press, San Diego, CA,
2001; Lambert, J.B and Mazzola, E.P., *Nuclear Magnetic Resonance Spec-
troscopy: An Introduction to Principles, Applications, and Experimental
Methods*, Pearson/Prentice-Hall, Upper Saddle River, NJ, 2004; Mitchell,
T.N. and Costisella, B., *NMR — From Spectra to Structures; An Experi-
mental Approach*, Springer, Berlin, 2004; Friebolin, H., *Basic One- and
Two-Dimensional NMR Spectroscopy*, Wiley-VCH, Weinheim, Germany,
2005.

Nuclear Pore A large transporter that spans the nuclear envelope (nuclear membrane). This
Complex   structure forms a channel between the inner and outer nuclear membranes,
providing for the transport of materials to and from the nucleus and
cytoplasm; all transport mechanisms in and out of the nucleus, active and
passive, occur through a tubular element in this pore structure. The
Karyopherin β family of proteins is involved in these transport processes.
See Faberge, A.C., The nuclear pore complex: its free existence and an
hypothesis as to its origin, *Cell Tissue Res.* 151, 403–415, 1974; Maul,
G.G., Nuclear pore complexes. Elimination and reconstruction during
mitosis, *J. Cell Biol.* 74, 492–500, 1977; Wozniak, R. and Clarke, P.R.,
Nuclear pores: sowing the seeds of assembly on the chromatin landscape,
Curr. Biol. 13, R970–R972, 2003; Rabut, G., Lenart, P., and Ellenberg,
J., Dynamics of nuclear pore complex organization through the cell cycle,
Curr. Opin. Cell Biol. 16, 314–321, 2004; Sazer, S., Nuclear envelope:
nuclear pore complexity, *Curr. Biol* 15, R23–R26, 2005; Peters, R.,
Translocation through the nuclear pore complex: selectivity and speed
by reduction-of-dimensionality, *Traffic* 6, 421–427, 2005; Devos, D.,
Dokudovskaya, S., Williams, R. et al., Simple fold composition and
modular architecture of the nuclear pore complex, *Proc. Natl. Acad. Sci.
USA* 103, 2172–2177, 2006; van der Aa, M.A.E.M., Mastrobattista, E.,

Oosting, R.S. et al., The nuclear pore complex: the gateway to successful nonviral gene delivery, *Pharmaceut. Res.* 23, 447–459, 2006.

Nucleic Acid Testing The use of PCR technology to test for the presence of nucleic acid sequences in biological materials. This approach is receiving attention in theranostics and the screening of blood for viral pathogens. See Tabor, E. and Epstein, J.S., NAT screening of blood and plasma donations: evolution of technology and regulatory policy, *Transfusion* 42, 1230–1237, 2002; Valentine-Thon, E., Quality control in nucleic acid testing — where do we stand? *J. Clin. Virol.* 25, S13–S21, 2002; Dimech, W., Bowden, D.S., Brestovac, B. et al., Validation of assembled nucleic acid–based tests in diagnostic microbiology laboratories, *Pathology* 36, 45–50, 2004.

Nucleosome An octomer of histone proteins associated with an approximate 140 bp DNA; the octomer is composed of two each of H2A, H2B, H3, and H4. See Kornberg, R.D. and Lorch, Y., Irresistible force meets immovable object: transcription and the nucleosome, *Cell* 67, 833–836, 1991; Turner, B.M., Decoding the nucleosome, *Cell* 75, 5–8, 1993; Sivolob, A. and Prunell, A., Nucleosome conformational flexibility and implications for chromatin dynamics, *Philos. Transact. A Math. Phys. Eng. Sci.* 362, 1519–1547, 2004; Lieb, J.D. and Clarke, N.D., Control of transcription through intragenic patterns of nucleosome composition, *Cell* 123, 1187–1190, 2005; Decker, P., Nucleosome autoantibodies, *Clin. Chim. Acta* 366, 48–60, 2006; Stockdale, C., Bruno, M., Ferreira, H. et al., Nucleosome dynamics, *Biochem. Soc. Symp.* 73, 109–119, 2006; Reinberg, D. and Sims, R.J., III, de FACTo nucleosome dynamics, *J. Biol. Chem.* 281, 23297–23301, 2006; Segal, E., Fodufe-Mittendorf, Y., Chen, L. et al., A genomic code for nucleosome positioning, *Nature* 442, 772–778, 2006; Bash, R., Wang, H., Anderson, C. et al., AFM imaging of protein movements: histone H2A-H2B release during nucleosome remodeling, *FEBS Lett.* 580, 4757–5761, 2006; Pisano, S., Pascucci, E., Cacchione, S. et al., AFM imaging and theoretical modeling studies of sequence-dependent nucleosome positioning, *Biophys. Chem.*, 124, 81–89, 2006.

Nutrigenomics Genomics of nutrition. The science of nutrigenomics seeks to provide a molecular understanding for how common dietary chemicals (i.e., nutrition) affect health by altering the expression or structure of an individual's genetic makeup. See van Ommen, B. and Stierum, R., Nutrigenomics: exploiting system biology in the nutrition and health arena, *Curr. Opin. Biotechnol.* 13, 517–721, 2002; Muller, M. and Kersten, S., Nutrigenomics: goals and strategies, *Nat. Rev. Genet.* 4, 315–322, 2003; Bauer, M., Hamm, A., and Pankratz, M.J., Linking nutrition to genomics, *Biol. Chem.* 385, 593–596, 2004; Davis, C.D. and Milner, J., Frontiers in nutrigenomics, proteomics, metabolomics, and cancer prevention, *Mutat. Res.* 551, 51–64, 2004; van Ommen, B., Nutrigenomics: exploiting systems biology in the nutrition and health arenas, *Nutrition* 20, 4–8, 2004; Mutch, D.M., Wahli, W., and Williamson, G., Nutrigenomics and nutrigenetics: the emerging faces of nutrition, *FASEB J.* 19, 1602–1616, 2005; Corthesy-Theulaz, I., den Dunnen, J.T., Ferre, P. et al., Nutrigenomics: the impact of biomics technology on nutrition research, *Ann. Nutr. Metab.* 49, 355–365, 2005; Trujillo, E., Davis, C., and Milner, J., Nutrigenomics, proteomics, metabolomics, and the practice of dietetics, *J. Am. Diet. Assos.* 106, 403–413, 2006; Afman, L. and Muller, M., Nutrigenomics: from molecular nutrition to prevention of disease, *J. Am. Diet. Assoc.* 106, 569–576, 2006; http://nutrigenomics.ucdavis.edu.

Ogston Effect   A model for the electrophoretic migration of a polymer within a fiber network, which treats the fiber network or soluble polymer network as a molecular sieve and the migrating solute as an undeformable particle. The reptation or biased-reptation model treats the migrating solute as a flexible material that can "snake" through the network. There has been interest in the application of this model to the electrophoresis of large DNA molecules. See Ogston, A.G., The spaces in a uniform random suspension of fibers, *Trans. Faraday Soc.* 54, 1754–1757, 1958; Grossman, P.D. and Soane, D.S., Experimental and theoretical studies of DNA separations by capillary electrophoresis in entangled polymer solutions, *Biopolymers* 31, 1221–1228, 1991; Kotaka, T., Adachi, S., and Shikata, T., Biased sinusoidal field gel electrophoresis for the separation of large DNA, *Electrophoresis* 14, 313–321, 1993; Guttma, A., Lengyel, T., Szoke, M., and Sasvari-Szekely, M., Ultra-thin-layer agarose gel electrophoresis II. Separation of DNA fragments on composite agarose-linear polymer matrices, *J. Chromatog. A* 871, 289–298, 2000; Labrie, J., Merdcier, J.F., and Slater, G.W., An exactly solvable Ogston model of gel electrophoresis. V. Attractive gel-analyte interactions and their effects on the Ferguson plot, *Electrophoresis* 21, 823–833, 2000; Slater, G.W., A theoretical study of an empirical function for the mobility of DNA fragments in sieving matrices, *Electrophoresis* 23, 1410–1416, 2002; Mercier, J.-F. and Slater, G.W., Universal interpolating function for the dispersion coefficient of DNA fragments in sieving matrices, *Electrophoresis* 27, 1453–1461, 2006.

Okazaki Fragment   Smaller fragments of DNA that are synthesized and then incorporated into larger DNA molecules, showing that replication can be a discontinuous process. See Okazaki, R., Okazaki, T., Sakabe, K., Sugimoto, K., and Sugino, A., Mechanism of DNA chain growth. I. Possible discontinuity and unusual secondary structure of newly synthesized chains, *Proc. Natl. Acad. Sci. USA* 59, 598–605, 1968; Hyodo, M. and Suzuki, K., Chain elongation of DNA and joining of DNA intermediates in intact and permeabilized mouse cells, *J. Biochem.* 88, 17–25, 1980; Alberts, B.M., Prokaryotic DNA replication mechanisms, *Philos. Trans. R. Soc. Lond. B Biol. Sci.* 317, 395–420, 1987; Nethanel, T., Reisfeld, S., Dinter-Gottlieb, G., and Kaufmann, G., An Okazaki piece of simian virus 40 may be synthesized by ligation of shorter precursor chains, *J. Virol.* 62, 2867–2873, 1988; Egli, M., Usman, N., Zhang, S.G., and Rich, A., Crystal structure of an Okazaki fragment at 2–Å resolution, *Proc. Natl. Acad. Sci. USA* 89, 534–538, 1992; Kim, J.H., Kang, Y.H., Kang, H.J. et al., *In vivo* and *in vitro* studies of Mgsl suggest a link between genome instability and Okazaki fragment processing, *Nucleic Acids Res.* 33, 6137–6150, 2005; Sporbert, A., Domaing, P., Leonhardt, H., and Cardoso, M.C., PCNA acts as a stationary loading platform for transiently interacting Okazaki fragment maturation proteins, *Nucleic Acids Res.* 33, 3521–3526, 2005.

OMP85   A protein found in gram-negative bacteria, which integrates proteins into bacterial outer membranes. See Gentle, I.E., Burri, L., and Littigow, T., Molecular architecture and function of the Omp85 family of proteins, *Molecular Microbiol.* 58, 1216–1225, 2005.

Oncogene   A gene that transforms normal cells into cancerous tumor cells, especially a viral gene that transforms a host cell into a tumor cell; a gene that encodes a protein product, which will stimulate uncontrolled cellular proliferation. Oncogenes are derivatives of normal cellular genes. See also *Proto-Oncogenes*. See Wiman, K.G. and Hayward, W.S., Rearrangement and activation of the *c-myc* gene in avian and human B-cell lymphomas, *Tumour Biol.* 5, 211–219, 1984; Balmain, A., Transforming *ras* oncogenes and multistage carcinogenesis, *Br. J. Cancer* 51, 1–7, 1985; Newbold, R.F., Malignant transformation of mammalian cells in culture: delineation of stages and role of cellular oncogenes activation, *IARC Sci. Publ.* 67, 31–53, 1985; Ratner, L., Josephs, S.F., and Wong-Staal, F., Oncogenes: their role in neoplastic transformation, *Annu. Rev. Microbiol.* 39, 419–449, 1985; Giehl, K., Oncogenic *Ras* in tumour progression and metastasis, *Biol. Chem.* 386, 193–205, 2005; Sanchez, P. Clement, V., Ruis, I., and Altaba, A., Therapeutic targeting of the Hedgehog-GLI pathway in prostate cancer, *Cancer Res.* 65, 2990–2992, 2005; Bellacosa, A., Kumar, C.C., Di Cristafano, A., and Testa, J.R., Activation of AKT kinases in cancer: implications for therapeutic targeting, *Adv. Cancer Res.* 94, 29–86, 2005; Kranenburg, O., The *KRAS* oncogenes: past, present, and future, *Biochim. Biophys. Acta* 1756, 81–82, 2005.

Oncogenomics   The use of molecular medicine tools such as DNA microarray and proteomics to study the oncology process, cancer genomics; study of oncogenes. See Sakamoto, K.M., Oncogenomics: dissecting cancer through genome research, *IDrugs* 4, 392–393, 2001; Rosell, R., Monzo, M., O'Brate, A., and Taron, M., Translational oncogenomics: toward rational therapeutic decision-making, *Curr. Opin. Oncol.* 14, 171–179, 2002; Strausberg, R.L., Simpson, A.J., Old, L.J., and Riggins, G.J., Oncogenomics and the development of new cancer therapies, *Nature* 429, 469–474, 2004; Jain, K.K., Role of oncoproteomics in the personalized management of cancer, *Expert Rev. Proteomics* 1, 49–55, 2004; Lam, S.H. and Gong, Z., Modeling liver cancer using zebrafish: a comparative oncogenomics approach, *Cell Cycle* 5, 573–577, 2006.

Onconase   Onconase is a ribonuclease isolated from amphibia. Onconase is homologous to pancreatic ribonuclease and is used in clinical trials as a biopharmaceutical. See Ardelt, W., Mikulski, S.M., and Shogen, K., Amino acid sequence of an anti-tumor protein from *Rana Pipiens* oocytes and early embryos. Homology to pancreatic ribonuclease, *J. Biol. Chem.* 266, 245–251, 1991; Wu, Y., Mikulski, S.M., Ardelt, W. et al., A cytotoxic ribonuclease. Study of the mechanism of onconase cytotoxicity, *J. Biol. Chem.* 268, 10686–10693, 1993; Leland, P.A., Schultz, L.W., Kim, B.W., and Raines, R.T., Ribonuclease A variants with potent cytotoxic activity, *Proc. Natl. Acad. Sci. USA* 95, 10407–10412, 1998; Notomista, E., Catanzano, F., Graziano, G. et al., Onconase: an unusually stable protein, *Biochemistry* 39, 8711–8718, 2000; Bosch, M., Benito, A., Ribo, M. et al., A nuclear localization sequence endows human pancreatic ribonuclease with cytotoxic activity, *Biochemistry* 43, 2167–2177, 2004; Kim, B.-M., Kim, H., Raines, R.T. et al., Glycosylation of onconase increases its conformational stability and toxicity for cancer cells, *Biochem. Biophys. Res. Commun.* 315, 976–983, 2004; Tafech, A., Bassett, T., Sparanese, D., and Lee, C.H., Destroying RNA as a therapeutic approach, *Curr. Med. Chem.* 13, 863–881, 2006; Suhasini, A.N. and Sirdeshmukh, R., Transfer RNA cleavages by onconase reveal unusual cleavage sites, *J. Biol. Chem.* 281, 12201–12209, 2006.

Opsonization   The process by which an antigen, usually a bacterial cell, is coated with an antibody (an opsonin) and then destroyed by the subsequent process of phagocytosis. The process of opsonization uses the Fab' portion of the antibody-recognized antigen and the Fc domain for complement activation and interaction in phagocytic cells such as neutrophils. See Peterson, P.K., Kim, Y., Schemling, D. et al., Complement-mediated phagocytosis of *Pseudomonas aeruginosa, J. Lab. Clin. Med.* 92, 883–894, 1978; Cunnion, K.M., Hair, P.S., and Buescher, E.S., Cleavage of complement C3b to iC3b on the surface of *Staphyloccus aureus* is mediated by serum complement factor I, *Infect. Immun.* 72, 2858–2863, 2004; Mueller-Ortiz, S.L., Drouin, S.M., and Wetsel, R.A., The alternative activation pathway and complement component C3 are critical for a protective immune response against *Pseudomonas aeruginosa* in a murine model of pneumonia, *Infect. Immun.* 72, 2899–2906, 2004; Coban, E., Ozdogan, M., Tuncer, M., Bozcuk, H., and Ersoy, F., The treatment of low-dose intraperitoneal immunoglobulin administration in the treatment of peritoneal dialysis-related peritonitis, *J. Nephrol.* 17, 427–430, 2004; Blasi, E., Mucci, A., Neglia, R. et al., Biological importance of the two Toll-like receptors, TLR2 and TLR4, in macrophage response to infection with *Candida albicans, FEMS Immunol. Med. Microbiol.* 46, 69–79, 2005; Tosi, M.F., Innate immune responses to infection, *J. Allergy Clin. Immunol.* 116, 241–249, 2005; Rus, H., Cudrici, C., and Niculescu, F., The role of the complement system innate immunity, *Immunol. Res.* 33, 103–112, 2005; Foster, T.J., Immune invasion by staphylococci, *Nat. Rev. Microbiol.* 3, 948–958, 2005; Arbo, A., Pavia-Ruz, N., and Santos, J.I., Opsonic requirements for the respiratory burst of neutrophils against *Giardia lamblia* trophozoites, *Archs. Med. Res.* 37, 465–473, 2006.

Optical Activity   The ability of chemical compounds to change the plane of polarization of polarized light; compounds may be dextrorotatory (*d*) or levorotatory (*l*). These descriptions have, in part, been replaced with R and S to indicate right and left, respectively. The optical activity of a chemical compound is a chemical property and an index of stereochemical purity. Optical rotatory dispersion and circular dichroism are measurements of optical activity.

Optical Rotatory   The measurement of the differential change in the velocity of light- and
Dispersion   right-circularly polarized light. This technique has been used to study the conformation of molecules. See McKenzie, H.A. and Frier, R.D., The behavior of R-ovalbumin and its individual components A1, A2, and A3 in urea solution: kinetics and equilibria, *J. Prot. Chem.* 22, 207–214, 2003; Chen, E., Kumita, J.R., Woolley, G.A., and Kliger, D.S., The kinetics of helix unfolding of an azobenzene cross-linked peptide probed by nanosecond time-resolved optical rotatory dispersion, *J. Amer. Chem. Soc.* 125, 12443–12449, 2003; Chen, E., Goldbeck, R.A., and Kliger, D.S., The earliest events in protein folding: a structural requirement for ultrafast folding in cytochrome C, *J. Amer. Chem. Soc.* 126, 11175–11181, 2004; Giorgio, E., Viglione, R.G., Zanasi, R., and Rosini, C., Ab initio calculation of optical rotatory dispersion (ORD) curves: a simple and reliable approach to the assignment of the molecular absolute configuration, *J. Amer. Chem. Soc.* 126, 12968–12976, 2004.

Optical Switches   In telecommunication, an optical switch is a switch that enables signals in optical fibers or integrated optical circuits (IOCs) to be selectively switched from one circuit to another. In biology, there are several definitions — one is whether an optical switch is a chemical probe that undergoes a spectral transition in response to light, where such a probe competes for a specific binding partner or ligand when in one but not the other spectral state (Sakata, T., Yan, Y., and Marriott, G., Family of site-selective molecular optical switches, *J. Org. Chem.* 70, 2009–2013, 2005). Another definition is of a spectral probe that is sensitive to a specific intra-cellular biological event (Graves, E.E., Weissleder, R., and Ntziachristos, V., Fluorescence molecular imaging of small animal tumors, *Curr. Mol. Med.* 4, 419–430, 2004). The term "optical switch" is also used in conjunction with optical scissors (Feringa, B.L., In control of motion: from molecular switches to molecular motors, *Acc. Chem. Res.* 34, 504–513, 2001; Capitano, M, Vanzi, F., Broggio, C. et al., Exploring molecular motors and molecular switches at the single-molecule level, *Microsc. Res. Tech.* 65, 194–204, 2004).

ORFeome   The total number of protein-coding open reading frames in an organism. See Rual, J.F., Hill, D.E., and Vidal, M., ORFeome projects: gateway between genomics and omics, *Curr. Opin. Chem. Biol.* 8, 20–25, 2004; Brasch, M.A., Hartlety, J.L., and Vidal, M., ORFeome cloning and systems biology: standardized mass production of the parts from a parts list, *Genome Res.* 14, 2001–2009, 2004; Uetz, P., Rajagopala, S.V., Dong, Y.A., and Haas, J., From ORFeomes to protein interaction maps in viruses, *Genome Res.* 14, 2029–2033, 2004; Johnson, N.M., Behm, C.A., and Trowell, S.C., Heritable and inducible gene knockdown in *C. elegans* using Wormgate and the ORFeome, *Gene* 359, 26–34, 2005; Schroeder, B.K, House, B.L., Mortimer, M.W. et al., Development of a functional genomics platform for *Sinorhizobium meliloti* construction of an ORFeome, *Appl. Environ. Microbiol.* 71, 5858–5864, 2005.

Organelle   Analysis of subcellular organelles such as mitochondria, the nucleus, and
Proteomics   the endocytotic apparatus by proteomic techniques. See Jan van Wijk, K., Proteomics or the chloroplast: experimentation and prediction, *Trends Plant Sci.* 5, 420–425, 2000; Taylor, S.W., Fahy, E., and Ghosh, S.S., Global organellar proteomics, *Trends Biotechnol.* 21, 82–88, 2003; Huber, L.A., Pfaller, K., and Vistor, I., Organelle proteomics: implications for subcellular fractionation in proteomics, *Circ. Res.* 92, 962–968, 2003; Dreger, M., Subcellular proteomics, *Mass Spectrom. Rev.* 22, 27–56, 2003; Brunet, S., Thibault, P.. Gagnon, E. et al., *Trends Cell Biol.* 12, 629–638, 2003; Jarvis, P., Organelle proteomics: chloroplasts in the spotlight, *Curr. Biol.* 14, R317–R319, 2004; Warnock, D.E., Fahy, E., and Taylor, S.W., Identification of protein associations in organelles, using mass spectrometry-based pro-teomics, *Mass Spectrom. Rev.* 23, 259–280, 2004; van Wijk, K.J., Plastid proteomics, *Plant Physiol. Biochem.* 42, 963–977, 2004; Yates, J.R., III, Gilchrist, A., Howell, K.E., and Bergeron, J.J., Proteomics of organelles and large cellular structures, *Nat. Rev. Mol. Cell Biol.* 6, 702–714, 2005.

Orthogonal   Two lines intersecting at right angles (in mathematics). The term is derived from the Greek *orthos* meaning straight, upright, vertical. Orthogonal has been used to describe a variety of activities in biochemistry and molecular biology including protein purification and analysis. More recently, orthogonal has been used to describe tRNA/tRNA synthase pairs that will react with each other and not with other pairs in *Escherichia coli*, permitting the incorporation of unnatural amino acids into proteins. See Liu, D.R., Magliery, T.J., Pastrnak, M., and Schultz, P.G., Engineering a tRNA and aminoacyl-tRNA synthetase for the site-specific incorporation of unnatural amino acids into protein *in vivo*, *Proc. Natl. Acad. Sci. USA* 94, 10092–10097, 1997; Guilhaus, M., Selby, D., and Miynski, V., Orthogonal acceleration time-of-flight mass spectrometry, *Mass Spectrom. Rev.* 19, 65–107, 2000; Chin, J.V., Martin, A.B., King, D.S., Wang, L., and Schultz, P.G., Addition of a photocrosslinking amino acid to the genetic code of *Escherichia coli*, *Proc. Natl. Acad. Sci. USA* 99, 11020–11024, 2002; Köhrer, C., Sullivan, E.L., and RajBhandary, U.L., Complete set of orthogonal 21st aminoacyl-tRNA synthetase-amber, ochre, and opal suppressor tRNA pairs: concomitant suppression of three different termination codons in an mRNA in mammalian cells, *Nucl. Acids Res.* 21, 6200–6211, 2004; Evans, C.R. and Jorgenson, J.W., Multidimension LC-LC and LC-CE for high-resolution separations of biologicals, *Anal. BioAnal. Chem.* 378, 1952–1961, 2004; Righetti, P.G., Bioanalysis: its past, present, and future, *Electrophoresis* 25, 2111–2127, 2004; Speers, A.E. and Cravatt, B.F., A tandem orthogonal proteolysis strategy for high-content chemical proteomics, *J. Amer. Chem. Soc.* 127, 10018–10019, 2005.

Orthologues   Genes in different organisms that have similar functions. See Lovejoy, D.A., Peptide hormone evolution: functional heterogeneity within GnRN and CRF families, *Biochem. Cell Biol.* 74, 1–7, 1996; Cole, C.N., mRNA export: the long and winding road, *Nat. Cell Biol.* 2, E55–E58, 2000; Trowsdale, J., Barten, R., Haude, A. et al., The genomic context of natural killer receptor extended gene families, *Immunol. Rev.* 181, 20–38, 2001; Lieschke, G.J., Zebrafish — an emerging genetic model for the study of cytokines and hematopoiesis in the era of functional gneomics, *Int. J. Hematol.* 73, 25–31, 2001; Lamotagne, B., Larose, S., Boulanger, J., and Elela, S.A., The RNase III family: a conserved structure and expanding functions in eukaryotic dsRNA metabolism, *Curr. Issues Mol. Biol.* 3, 71–78, 2001; Stothard, P. and Pilgrim, D., Sex-determination gene and pathway evolution in nematodes, *Bioessays* 25, 221–231, 2003; Chen, T.Y., Structure and function of clc channels, *Annu. Rev. Physiol.* 67, 809–839, 2005; Nair, V. and Zavolan, M., Virus-encoded microRNAs: novel regulators of gene expression, *Trends Microbiol.* 14, 169–175, 2006.

Osmosensor   A molecular system sensing osmotic stress and regulating pathways involved in preservation of osmotic equilibrium. See Wurgler-Murphey, S.M. and Saito, H., Two-component signal transducers and MAPK cascades, *Trends Biochem. Sci.* 22, 172–176, 1997; Urao, T., Yamaguchi-Shinozaki, K., and Shinozaki, K., Plant histidine kinases: an emerging picture of two-component single transduction in hormone and environmental responses, *Sci. STKE 2001* 109, RE18, 2001; Reiser, V., Raitt, D.C., and Saito, H., Yeast osmosensor Sln1 and plant cytokine receptor Cre1 respond to changes in turgor pressure, *J. Cell Biol.* 161, 1035–1040, 2003; Poolman, B., Spitzer, J.J., and Wood, J.M., Bacterial osmosensing: roles of membrane structure and electrostatics in lipid–protein and protein–protein interactions, *Biochim. Biophys. Acta* 1666, 88–104, 2004; Liedtke, W., TRPV4 as osmosensor: a transgenic approach, *Pflügers Arch.* 451, 176–180, 2005; Schiller, D., Ott, V., Kramer, R., and Morbach, S., Influence of membrane composition on osmosensitivity by the betaine carrier BetP from *Corynebacterium glutamicum*, *J. Biol. Chem.* 281, 7737–7746, 2006.

Osteoclast   Bone-degrading cells, derived from macrophages; opposite function to osteoblasts. See Rifkin, B.R. and Gay, C.V., *Biology and Physiology of* *the Osteoclast*, CRC Press, Boca Raton, FL, 1992; Abou-Samra, A.-B. and Mundy, G.R., *Physiology and Pharmacology of Bone*, Springer-Verlag, Berlin, 1993; Massaro, E.J. and Rogers, J.M., *The Skeleton: Biochemical, Genetic, and Molecular Interactions in Development and Homeostasis*, Humana Press, Totowa, NJ, 2004; Bronner, F. and Carson, M.C., *Bone Resorption*, Springer, London, 2005; Li, Z., Kong, K., and Qi, W., Osteoclast and its roles in calcium metabolism and bone development and remodeling, *Biochem. Biophys. Res. Commun.* 343, 345–350, 2006; Fukumoto, S., Iwamoto, T., Sakai, E. et al., Current topics in pharmacological research of bone metabolism: osteoclasts differentiation regulated by glycospingolipids, *J. Pharmacol. Sci.* 100, 195–200, 2006; Wada, T., Nakashima, T., Hiroshi, N., and Penninger, J.M., RANKL-RNAK signaling in osteoclastogenesis and bone disease, *Trends Mol. Med.* 12, 17–25, 2006.

Osteoprotegerin (OPG)   A protein that suppresses the production of osteoclasts; functions in combination with RANKL-regulated bone resorption. See Hofbauer, L.C. and Heufelder, A.E., Osteoprotegerin: a novel local player in bone metabolism, *Eur. J. Endocrinol.* 137, 345–346, 1997; Kong, Y.Y., Boyle, W.J., and Penninger, J.M., Osteoprotegerin ligand: a regulator of immune responses and bone physiology, *Immunol. Today* 21, 495–502, 2000; Hofbauer, L.C., Khosla, S., Dunstan, C.R. et al., The roles of osteoprotegerin and osteoprotegerin ligand in the paracrine regulation of bone resorption, *J. Bone Miner. Res.* 15, 2–12, 2000; Theoleyre, S., Wittrant, Y., Tat, S.K. et al., The molecular triad OPG/RANK/RANKL: involvement in the orchestration of pathophysiological bone remodeling, *Cytokine Growth Factor Rev.* 15, 457–475, 2004; Bezerra, M.C., Carvalho, J.F., Prokopowitsch, A.S., and Pereira, R.M., RANK, RANKL, and osteoprotegerin in arthritic bone loss, *Braz. J. Med. Biol. Res.* 38, 161–170, 2005; Feng, X., RANKing intracellular signaling in osteoclasts, *IUBMB Life* 57, 389–395, 2005; Kostenuik, P.J., Osteoprotegerin and RANKL regulate bone resorption, density, geometry, and strength, *Curr. Opin. Pharmacol.* 5, 618–625, 2005; Neumann, E., Gay, S., and Muller-Ladner, U., The RANK/RANKL/osteoprotegerin system in rheumatoid arthritis: new insights from animal models, *Arthritis Rheum.* 52, 2960–2967, 2005; Wada, T., Nakashima, T., Hiroshi, N., and Penninger, J.M., RANKL-RANK signaling in osteoclastogenesis and bone disease, *Trends Mol. Med.* 12, 17–25, 2006.

Oxyanion Hole   A feature of the active sites of hydrolytic enzymes such as lipases or chymotrypsin where the acyl carbonyl of the acyl-enzyme intermediate is stabilized by hydrogen bonding to peptide amide nitrogens on the enzyme. See Menard, R. and Storer, A.C., Oxyanion hole interactions in serine cysteine proteases, *Biol. Chem. Hoppe Seyler* 373, 393–400, 1992; Whiting, A.K. and Peticolas, W.L., Details of the acyl-enzyme intermediate and the oxyanion hole in serine protease catalysis, *Biochemistry* 33, 552–561, 1994; Johal, S.S., White, A.J., and Wharton, C.W., Effect of specificity on ligand conformation in acyl-chymotrypsins, *Biochem. J.* 297, 281–287, 1994; Cui, J., Marankan, F., Fu, W. et al., An oxyanion-hole selective serine protease inhibitor in complex with trypsin, *Bioorg. Med. Chem.* 10, 41–46, 2002; Lee, L.C., Lee, Y.L., Leu, R.J., and Shaw, J.F., Functional role of catalytic triad and oxyanion hole-forming residues on enzyme activity of *Escherichia coli* thioesterase I/protease I/phospholipase L1, *Biochem. J.* 397, 69–76, 2006.

Palindrome   A sequence that reads the same forwards and backwards; usually refers to a nucleic acid sequence where opposing strands read the same; that is the $3'\rightarrow5'$ sequence in one strand is the same as the $5'\rightarrow3'$ sequence in the opposing strand. Palindromic sequences are frequently present at the sites of restriction class II enzyme cleavages. See Leach, D.R., Long DNA palindromes, cruciform structures, genetic instability, and secondary structure repair, *Bioessays* 16, 893–900, 1994; Beato, M., Chavez, S., and Truss, M., Transcriptional regulation by steroid hormones, *Steroids* 61, 240–251, 1996; Cho-Chung, Y.S., CRE-palindrome oligonucleotide as a transcription factor decoy and an inhibitor of tumor growth, *Antisense Nucleic Acid Drug Dev.* 8, 167–170, 1998.

Paracrine   Usually in reference to a hormone or other biological effector such as a peptide growth factor or cytokine, which has an effect on the cell or tissue immediately surrounding the cell or tissue responsible for the synthesis of the given compound. See Franchimont, P., *Paracrine Control*, Saunders, Philadelphia, 1986; Piva, F., *Cell to Cell Communication in Endocrinology*, Raven Press, New York, 1988; Krey, L.C. and Gulyas, B.J., Eds., *Autocrine and Paracrine Mechanisms in Reproductive Endocrinology*, Plenum Press, New York, 1989; Habenicht, A., *Growth Factors, Differentiation Factors and Cytokines*, Springer-Verlag, Berlin, 1990; Hardie, D.G., *Biochemical Messengers: Hormones, Neurotransmitters, and Growth Factors*, Chapman & Hall, London, 1991; Vallesi, A., Giuli, G., Bradshaw, R.A., and Luporini, P., Autocrine mitogenic activity of pheromones produced by the protozoan ciliate *Euplotes raikovi*, *Nature* 376, 522–524, 1995.

Paralogues   Genes within the same genome that have evolved by gene duplication. See Ohno, S., The one-to-four rule and paralogues of sex-determining genes, *Cell. Mol. Life Sci.* 55, 824–830, 1999; Forterre, P., Displacement of cellular proteins by functional analogues from plasmids or viruses could explain puzzling phylogenies of many DNA informational proteins, *Mol. Microbiol.* 33, 457–465, 1999; Gilbert, J.M., The evolution of engrailed genes after duplication and speciation events, *Dev. Gene. Evol.* 212, 307–318, 2002; Ferrier, D.E., Hox genes: Did the vertebrate ancestor have a Hox14? *Curr. Biol.* 14, R210–R211, 2004; Tsuru, T., Kawai, M., Mizutani-Ui, Y., Uchiyama, I., and Kobayashi, I., Evolution of paralogous genes: reconstruction of genome rearrangements through comparison of multiple genomes with *Staphylococcus aureus*, *Mol. Biol. Evol.*, 23, 1269–1285, 2006.

PDZ Domain   A protein domain involved in protein–protein interactions with a preference for C-terminal regions. PDZ proteins are recognized as components of biological scaffolds. The acronym is derived from the homology of a motif in P̲SD-95, the *Drosophila* D̲iscs-Large septate junction protein, and the epithelial junction protein Z̲O-1. See Ranganathan, R. and Ross, E.M., PDZ domain proteins; scaffolds for signaling complexes, *Curr. Biol.* 7, R770–R773, 1997; Harris, B.Z. and Lim, W.A., Mechanism and role of PDZ domains in signaling complex assembly, *J. Cell Sci.* 114, 3219–3231, 2001; Dev, K.K., Making protein interactions druggable: targeting PDZ domains, *Nat. Rev. Drug Disc.* 3, 1047–1056, 2004; Schlieker, C., Mogk, A., and Bukau, B., A PDZ switch for a cellular stress response, *Cell* 117, 417–419, 2004; Brone, B. and Eggermont, J., PDZ proteins retain and regulate membrane transporters in polarized epithelial cell membranes, *Am. J. Physiol. Cell Physiol.* 3, 1047–1056, 2005.

Pedigree Rate   Estimate of the mutation rate determined by calculating the number of nucleotide changes observed over a known number of reproductive events. See Howell, N., Smejkal, C.B., Mackey, D.A. et al., The pedigree rate of sequence divergence in the human mitochondrial genome: there is a difference between phylogenetic and pedigree rates, *Am. J. Hum. Genet.* 72, 659–670, 2003.

Pepducins   Cell-penetrating lipidated (palmitic acid) membrane-tethered peptides, which act as G-protein-coupled receptors. See Covic, L., Gresser, A.L., Talavera, J. et al., Activation and inhibition of G-protein-coupled receptors by cell-penetrating membrane-tethered peptides, *Proc. Natl. Acad. Sci. USA* 99, 643–648, 2002; Lomas-Neira, J. and Ayala, A., Pepducins: an effective means to inhibit GPCR signaling by neutrophils, *Trends Immunol.* 26, 619–621, 2005. See also *Cell-Penetrating Peptides*.

Peptergents   Peptide detergents; small, self-assembling peptides with detergent properties. These peptide detergents appear quite useful for the study of membrane proteins. See Yeh, J.I., Du, S., Tortajada, A., Paulo, J., and Zhang, S., Peptergents: peptide detergents that improve stability and functionality of a membrane protein, Glycerol-3-phosphate dehydrogenase, *Biochemistry* 44, 16912–16919, 2005; Kiley, P., Zhao, X., Vaughn, M. et al., Self-assembling peptide detergents stabilize isolated photosystem I on a dry surface for an extended time, *PLoS Biol.* 3, e230, 2005.

Peptidome   The peptide complement of a genome. Peptidomics is the study of the peptidome. See Schrader, M., and Schulz-Knappe, P., Peptidomics technologies for human body fluids, *Trends Biotechnol.* 19 (Suppl. 10), S55–S60, 2001; Jurgens, M. and Schrader, M., Peptidomic approaches in proteomic research, *Curr. Opin. Mol. Ther.* 4, 236–241, 2002; Schulze-Knappe, M., Schrader, M., and Zucht, H.D., The peptidomics concept, *Comb. Chem. High Throughput. Screen.* 8, 697–704, 2005; Zheng, X., Baker, H., and Hancock, W.S., Analysis of the low molecular weight serum peptidome using ultrafiltration and a hybrid ion trap-Fourier transform mass spectrometry, *J. Chromatog. A*, 1120, 173–184, 2006; Hortin, G.L., The MALDI TOF mass spectrometric view of the plasma proteome and peptidome, *Clin. Chem.*, 52, 1223–1237, 2006.

Perforin   Perforin is a protein located in the granules of CD8 T-cells (cytotoxic T-cells) and natural killer cells. Upon degranulation of these cells, perforin inserts itself into the target cell's plasma membrane, forming a pore resulting in lysis of the target cell. See Catalfamo, M., and Henkart, P.A., Perforin and the granule exocytosis cytotoxicity pathway, *Curr. Opin. Immunol.* 15, 522–527, 2003; Smith, M.J., Cretney, E., Kelly, J.M. et al., Activation of NK cell cytotoxicity, *Mol. Immunol.* 42, 501–510, 2005; Ashton-Rickardt, P.G., The granule pathway of programmed cell death, *Crit. Rev. Immunol.* 25, 161–182, 2005; Yoon, J.W. and Jun, H.S., Autoimmune destruction of pancreatic beta cells, *Am. J. Ther.* 12, 580–591, 2005; Andersen, M.H., Schrama, D., Thor Straten, P., and Becker, J.C., Cytotoxic T-cells, *J. Invest. Dermatol.* 126, 32–41, 2006.

Peroxiredoxins   Group of antioxidant thioredoxin-dependent enzymes with a catalytic function in the detoxification of cellular-toxic peroxides. See Claiborne, A., Ross, R.P., and Parsonage, D., Flavin-linked peroxide reductases: protein-sulfenic acids and the oxidative stress response, *Trends Biochem. Sci.* 17, 183–186, 1992; Dietz, K-J., Horhing, F., König, J., and Baien, M., The function of chloroplast 2-cysteine peroxiredoxin I peroxide detoxification and its regulation, *J. Expt. Bot.* 53, 1321–1329, 2002; Immenschuh, S.

and Baumgart-Vogt, E., Peroxiredoxins, oxidative stress, and cell proliferation, *Antioxid. Redox. Signal.* 7, 768–777, 2005; Rouhier, N. and Jacquot, J.P., The plant multigenic family of thiol peroxidases, *Free Rad. Biol. Med.* 38, 1413–1421, 2005; Rhee, S.G., Chae, H.Z., and Kim, K., Peroxiredoxins: a historical overview and speculative preview of novel mechanisms and emerging concepts in cell signaling, *Free Radic. Biol. Med.* 38, 1543–1552, 2005.

Peroxynitrite   An oxidizing/nitrating agent derived from the reaction of nitric oxide and superoxide, which reacts with proteins, lipids, and nucleic acids. The reactions are complex and in addition to oxidation reactions such as carbonyl formation and disulfide formation, there are reactions such as nitrosylation of cysteine and the nitration of tyrosine. See Beckman, J.S. and Crow, J.P., Pathological implications of nitric oxide, superoxide, and peroxynitrite formation, *Biochem. Soc. Trans.* 21, 330–334, 1993; Pryor, W.A. and Squadrito, G.L., The chemistry of peroxynitrite: a product from the reaction of nitric oxide with superoxide, *Am. J. Physiol.* 268, L699–L722, 1995; Uppu, R.M., Squadrito, G.L., Cueto, R., and Pryor, W.A., Synthesis of peroxynitrite by azide-ozone reaction, *Methods Enzymol.* 269, 311–321, 1996; Beckman, J.S. and Koppenol, W.H., Nitric oxide, superoxide, and peroxynitrite: the good, the bad, and ugly, *Am. J. Physiol.* 271, C1424–C1437, 1996; Girotti, A.W., Lipid hydroperoxide generation, turnover, and effector action in biological systems, *J. Lipid Res.* 39, 1529–1542, 1998; Radi, R., Denicola, A., and Freeman, B.A., Peroxynitrite reactions with carbon dioxide-bicarbonate, *Methods Enzymol.* 301, 353–357, 1999; Groves, J.T., Peroxynitrite: reactive, invasive and enigmatic, *Curr. Opin. Chem. Biol.* 3, 226–235, 1999; Halliwell, B., Zhao, K., and Whiteman, M., Nitric oxide and peroxynitrite. The ugly, the uglier, and the not so good: a personal view of the recent controversies, *Free Radic. Res.* 31, 651–669, 1999; Estevez, A.G. and Jordan, J., Nitric oxide and superoxide, a deadly cocktail, *Ann. N.Y. Acad. Sci.* 962, 207–211, 2002; Ohmori, H., and Kanayama, N., Immunogenicity of an inflammation-associated product, tyrosine nitrated self-proteins, *Autoimmun. Rev.* 4, 224–229, 2005; Hurd, T.R., Filipovska, A., Costa, N.J. et al., Disulphide formation on mitochondrial protein thiols, *Biochem. Soc. Trans.* 44, 1390–1393, 2005; Sawa, T. and Ohshima, H., Nitrative DNA damage in inflammation and its possible role in carcinogenesis, *Nitric Oxide* 14, 91–100, 2006; Niles, J.C., Wishnok, J.S., and Tannenbaum, S.R., Peroxynitrite-induced oxidation and nitration products of guanine and 8-oxoguanine: structures and mechanisms of product formation, *Nitric Oxide* 14, 109–121, 2006; Uppu, R.M., Synthesis of peroxynitrite using isoamyl nitrite and hydrogen peroxide in a homogeneous solvent system, *Anal. Biochem.* 354, 165–168, 2006. The reaction of tyrosine with peroxynitrite is sensitive to solvent environment with nitration favored in a hydrophobic environment as opposed to oxidation (Zhang, H., Joseph, J., Feix, J. et al., Nitration and oxidation of a hydrophobic tyrosine probe by peroxynitrite in membranes: comparison with nitration and oxidation of tyrosine by peroxynitrite in aqueous solution, *Biochemistry* 40, 7675–7686, 2001).

Pescadillo   A nuclear protein originally demonstrated in zebrafish. Pescadillo is thought to be important in a variety of nuclear activities including DNA replication, ribosome formation, and cell cycle control. See Allende, M.L., Amsterdam, A., Becker, T. et al., Insertional mutagenesis in zebrafish identifies two novel genes, pescadillo and dead eye, essential for embryonic development, *Genes Dev.* 10, 3141–3155, 1996; Haque, J., Boger, S., Li, J., and Duncan, D.A., The murine Pes1 gene encodes a nuclear protein containing a BRCT domain, *Genomics* 70, 201–210, 2000; Kinoshita, Y., Jarell, A.D., Flaman, J.M. et al., Pescadillo, a novel cell cycle regulatory protein abnormally expressed in malignant cells, *J. Biol. Chem.* 276, 6656–6665, 2001; Maiorana, A., Tu. X., Cheng, G., and Baserga, R., Role of pescadillo in the transformation and immortalization of mammalian cells, *Oncogene* 23, 7116–7124, 2004; Killian, A., Le Meur, N., Sesboue, R. et al., Inactivation of the RRB1-Pescadillo pathway involved in ribosome biogenesis induces chromosomal instability, *Oncogene* 23, 8597–8602, 2004; Sikorski, E.M., Uo, T., Morrison, R.S., and Agarwal, A., Pescadillo interacts with the cadmium response element of the human heme oxygenase-1 promoter in renal epithelial cells, *J. Biol. Chem.* 281, 24423–24430, 2006.

Pfam
Protein family; used to describe a protein family database. See Persson, B., Bioinformatics in protein analysis, *EXS* 88, 215–231, 2000; Bateman, A., Birney, E., Cerruti, L. et al., The Pfam protein families database, *Nucl. Acids Res.* 30, 276–280, 2002; Lubec, G., Afjehi-Sadat, L., Yang, J.W., and John, J.P., Searching for hypothetical proteins: theory and practice based upon original data and literature, *Prog. Neurobiol.* 77, 90–127, 2005; Anderston, J.N., Del Vecchio, R.L., Kannan, N. et al., Computational analysis of protein tyrosine phosphatases: practical guide to bioinformatics and data resources, *Methods* 35, 90–114, 2005.

Pharmaceutical Equivalence
Drug products can be considered to be pharmaceutical equivalents if such products (1) contain the same active ingredients, (2) are of the same dosage form and route of administration, and (3) are identical in strength and concentration. The term therapeutic equivalence is also used to describe pharmaceutical equivalence. Pharmaceutically equivalent drug products may differ in attributes such as shape, color, excipients, and release mechanisms. Pharmaceutical equivalence is of importance in the development of generic drugs. See http://www.fda.gov.

Pharmaco-genomics
The use of genomics to study the development and utilization of drugs. Pharmacogenomics will be essential to the development of predictive medicine/personalized medicine. See Robson, B. and Mushlin, R., Genomic messaging system and DNA mark-up language for information-based personalized medicine with clinical and proteome research applications, *J. Proteome Res.* 3, 930–948, 2004; Wilke, R.A., Reif, D.M., and Moore, J.H., Combinatorial pharmacogenetics, *Nat. Rev. Drug Discov.* 4, 911–918, 2005; Ginsburg, G.S., Konstance, R.P., Allsbrook, J.S., and Schulman, K.A., Implications of pharmacogenomics for drug development and clinical practice, *Arch. Intern. Med.* 165, 2331–2336, 2005; Siest, G., Marteau, J.B., Maumus, S. et al., Pharmacogenomics and cardiovascular drugs: need for integrated biological system with phenotypes and proteomic markers, *Eur. J. Pharmacol.* 527, 1–22, 2005.

Pharmacophore
The collection of structural features of a compound, which provides for pharmacological properties. Uses information derived from QSAR studies. See Guner, O., Clent, O., and Kurogi, Y., Pharmacophore modeling and three dimensional database searching for drug design using catalyst: recent advances, *Curr. Med. Chem.* 11, 2991–3005, 2004; Glennon, R.A., Pharmacophore identification for sigma-1 (sigma1) receptor binding: application of the "deconstruction-reconstruction-elaboration" approach, *Mini*

	Rev. Med. Chem. 5, 927–940, 2005; Guner, O.F, The impact of pharmacophore modeling in drug design, *IDrugs* 9, 567–572, 2005.
Pharmaco-proteomics	The use of proteomics to predict individual reaction to a drug or drugs; also the use of proteomics for drug discovery and development; related to personalized medicine, theranostics, pharmacogenomics. See Witzmann, F.A. and Grant, R.A., Pharmacoproteomics in drug development, *Pharmacogenomics J.*, 3, 69–76, 2003; Di Paolo, A., Danesi, R., and Del Tacca, M., Pharmacogenetics of neoplastic diseases: new trends, *Pharmacol. Res.* 49, 331–342, 2004; Jain, K.K., Role of pharmacoproteomics in the development of personalized medicine, *Pharmacogenomics J.* 5, 331–336, 2004.
Phase Diagram(s)	A graph showing the relationship between phases (i.e., solid/gas/liquid) over a range of physical conditions (usually temperature and pressure). A phase is generally defined as a homogenous part of a heterogeneous system that is clearly separated from other phases by a physical boundary; the separation between ice and water is an example of a boundary between phases. See Ohgushi, M. and Wada, A., Liquid-like state of side chains at the intermediate stage of protein denaturation, *Adv. Biophys.* 18, 75–90, 1984; Dorfler, H.D., Mixing behavior of binary insoluble phospholipid monolayers. Analysis of the mixing properties of binary lecithin and cephalin systems by application of several surface and spreading techniques, *Adv. Colloid Interface Sci.* 31, 1–110, 1990; Diamond, A.D. and Hsu, J.T., Aqueous two-phase systems for biomolecule separation, *Adv. Biochem. Eng. Biotechnol.* 47, 89–135, 1992; Mason, J.T., Investigation of phase transitions in bilayer monolayers, *Methods Enzymol.* 295, 468–494, 1998; Crowe, J.H., Tablin, F., Tsvetkova, N. et al., Are lipid phase transitions responsible for chilling damage in human platelets? *Cryobiology* 38, 180–191, 1999; Smeller, L, Pressure-temperature phase diagrams of biomolecules, *Biochim. Biophys. Acta* 1595, 11–29, 2002; Dill, K.A., Truskett, T.M., Vlachy, V., and Hribar-Lee, B., Modeling water, the hydrophobic effect, and ion salvation, *Annu. Rev. Biochem. Biomol. Struct.* 34, 173–199, 2005; Scharnagl, C., Reif, M., and Friedrich, J., Stability of proteins: temperature, pressure and the role of the solvent, *Biochim. Biophys. Acta* 1749, 187–213, 2005.
Phenotype	The physical manifestation of the genes of an organism; the collection of structure and function expressed by the genotype of an organism; the visible properties of an organism that are produced by the interaction of a genotype and the environment. See Padykula, H.A., *Control Mechanisms in the Expression of Cellular Phenotypes*, Academic Press, New York, 1970; Levine, A.J., *The Transformed Phenotype*, Cold Spring Harbor Laboratory Press, Cold Spring Harbor, NY, 1984; Dewitt, T.J. and Scheiner, S.M., *Phenotypic Plasticity Function and Conceptual Approaches*, Oxford University Press, Oxford, UK, 2004; Pigliucci, M. and Preston, K., *Phenotypic Integration: Studying the Ecology and Evolution of Complex Phenotypes*, Oxford University Press, Oxford, UK, 2004.
Phospholipase C	A family of intracellular enzymes central to many signal transduction pathways via effects on Ca^{2+} and protein kinase C. Phospholipase C catalyzes the hydrolysis of phosphoinositol 4,5-bisphosphate to yield 1,4,5-inositol triphosphate and diacylglycerol. See Irvine, R.F., The enzymology of stimulated inositol lipid turnover, *Cell Calcium* 3, 295–309, 1982; Farese, R.V., Phospholipids as intermediates in hormone action, *Mol. Cell Endocrinol.* 35, 1–14, 1984; Majerus, P.W., The production

of phosphoinositide-derived messenger molecules, *Harvey Lect.* 82, 145–155, 1986–87; Litosch, I. and Fain, J.N., Regulation of phosphoinositide breakdown by guanine nucleotides, *Life Sci.* 39, 187–194, 1986; Putney, J.W., Jr., Formation and actions of calcium-mobilizing messenger, inositol 1,4,5-trisphosphate, *Am. J. Physiol.* 252, G149–G157, 1987; Lemmon, M.A., Pleckstrin homology domains: two halves make a hole? *Cell* 120, 574–576, 2005; Malbon, C.C., G proteins in development, *Nat. Rev. Mol. Cell Biol.* 6, 689–701, 2005; Gilfillan, A.M. and Tkaczyk, C., Integrated signaling pathways for mast-cell activation, *Nat. Rev. Immunol.* 6, 218–230, 2006; Drin, G., Dougnet, D., and Scarlata, S., The pleckstrin homology domain of phospholipase C transmits enzymatic activation through modulation of the membrane-domain orientation, *Biochem.* 45, 5712–5724, 2006.

Cis-**Phosphorylation (or *Cis*-Autophosphorylation)** An autophosphorylation event where the kinase catalyzes the phosphorylation of itself as opposed to another molecule of the same kinase. See Frattali, A.L., Treadway, J.L., and Pessin, J.E., Transmembrane signaling by the human insulin receptor kinase. Relationship between intramolecular beta subunit *trans*- and *cis*-autophorylation and substrate kinase activation, *J. Biol. Chem.* 267, 19521–19528, 1992; Cann, A.D. and Kohanski, R.A., *Cis*-autophosphorylation of juxtamembrane tyrosines in the insulin receptor kinase domain, *Biochemistry* 36, 7681–7689, 1997; Cann, A.D., Bishop, S.M., Ablooglu, A.J., and Kohanski, R.A., Partial activation of the insulin receptor kinase domain by juxtamembrane autophosphorylation, *Biochemistry* 37, 11289–11300, 1998; Leu, T.H. and Maa, M.C., Tyr-863 phosphorylation enhances focal adhesion kinase autophosphorylation at Tyr-397, *Oncogene* 21, 6992–7000, 2002; Iyer, G.H., Moore, M.J., and Taylor, S.S., Consequences of lysine 72 mutation on the phosphorylation and activation state of cAMP-dependent kinase, *J. Biol. Chem.* 280, 8800–8807, 2005; Yang, K., Kim, J.H., Kim, H.J. et al., Tyrosine 740 phosphorylation of discoidin domain receptor 2 by Src stimulates intramolecular autophosphorylation and Shc signaling complex formation, *J. Biol. Chem.* 280, 39058–39066, 2005.

Trans-**Phosphorylation (or *Trans*-Autophosphorylation)** An autophosphorylation event where the kinase catalyzes the phosphorylation of another molecule of the same kinase as opposed to *cis*-phosphorylation where the kinase phosphorylates itself. A more generic definition is the transfer of a phosphoryl function from one site to another site (transfer of a phosphate function). See Wei, L., Hubbard, S.R., Hendrickson, W.A., and Ellis, L., Expression, characterization, and crystallization of the catalytic core of the human insulin receptor protein-tyrosine kinase domain, *J. Biol. Chem.* 270, 8122–8130, 1995; McKeehan, W.L., Wang, F., and Kan, M., The heparin sulfate-fibroblast growth factor family: diversity of structure and function, *Prog. Nucleic Acid Res. Mol. Biol.* 59, 135–176, 1998; Klint, P. and Claesson-Welsh, L., Signal transduction by fibroblast growth factor receptors, *Front. Biosci.* 4, D165–D177, 1999; DiMaio, D., Lai, C.C., and Mattoon, D., The platelet-derived growth factor beta receptor as a target of the bovine papillomavirus E5 protein, *Cytokine Growth Factor Rev.* 11, 283–293, 2000; DiMaio, D. and Matoon, D., Mechanisms of cell transformation by papillomavirus E5 proteins, *Oncogene* 20, 7866–7873, 2001; Schwarz, J.K., Lovly, C.M., and Piwnica-Worms, H., Regulation of the Chk2 protein kinase by oligomerization-mediated *cis*- and *trans*-phosphorylation, *Mol. Cancer Res.* 1,

598–609, 2003; Wu, S. and Kaufman, R.J., *trans*-autophosphorylation by the isolated kinase domain is not sufficient for dimerization of activation of the dsRNA-activated protein kinase PKR, *Biochemistry* 43, 11027–11034, 2004; Shi, G.W., Chen, J., Concepcion, F. et al., Light causes phosphorylation of nonactivated visual pigments in intact mouse rod photoreceptor cells, *J. Biol. Chem.* 280, 41184–41191, 2005; Gao, X. and Harris, T.K., Steady-state kinetic mechanism of PDK1, *J. Biol. Chem.*, 281, 21670–21681, 2006.

Phylogenetic Rate Estimate of the substitution rate calculated by comparing the molecular sequence data obtained from different species. See Heyer, E., Zietkiewicz, E., Rochowski, A. et al., Phylogenetic and familial estimates of mitochondrial substitution rates: study of control region mutations in deep-rooting pedigrees, *Am. J. Human Genet.* 69, 1113–1126, 2001; Ritchie, P.A., Miller, C.D., Gibb, G.C., Baroni, C., and Lambert, D.M., Ancient DNA enables timing of the pleistocene origin and holocene expansion of two adelie penguin lineages in Antarctica, *Mol. Biol. Evol.* 21, 240–248, 2004.

Phytoremediation Use of plants for remediation of toxic chemicals. This can be separated into phytoremediation/rhizofiltration, which describes the removal of toxic materials from water, and phytoextraction, which describes the removal of toxic materials from soil by plants. See Arthur, E.L., Rice, P.J., Rice, P.J. et al., Phytoremediation — an overview, *Crit. Rev. Plant Sci.* 24, 109–122, 2005.

Pinocytosis Transcytosis across an endothelial cell through endosomic vesicles and/or a tubovesicular pathway. See Chapman-Andresen, C., *Studies on Pinocytosin Amoebae*, Danish Science Press, Copenhagen, 1962; LaBella, F.S., *Pinocytosis*, MSS Information Group, New York, 1973; Josefsson, J.-O., *Pinocytosis*, University of Lund, Sweden, 1973; Stossel, T.P., Contractile proteins in cell structure and function, *Annu. Rev. Med.* 29, 427–457, 1978; Lloyd, J.B., Insights into mechanisms of interacellular protein turnover from studies on pinocytosis, *Ciba Found. Symp.* 75, 151–165, 1979; Barondes, S.H., Lectins: their multiple endogenous cellular function, *Ann. Rev. Biochem.* 50, 207–231, 1981; Besterman, J.M., and Low, R.B., Endocytosis: a review of mechanisms and plasma membrane dynamics, *Biochem. J.* 210, 1–13, 1983; Mansilla, A.O., Arguero, R.S., Rico, F.G., and Alba, C.C., Cellular receptors, acceptors, and clinical implications, *Arch. Med. Res.* 24, 127–137, 1993; Meier, O. and Greber, U.F., Adenovirus endocytosis, *J. Gene. Med.* 5, 451–462, 2003; Batahori, G., Cervenak, L., and Karadi, I., Caveolae — an alternative endocytotic pathway for targeted drug delivery, *Crit. Rev. Ther. Drug Carrier Syst.* 21, 67–95, 2004.

Piranha Solution A mixture of concentrated sulfuric acid and hydrogen peroxide (as an example, a 7:3[v/v] ratio of 98% H_2SO_4 [concentrated sulfuric acid] and 30% [w/v] H_2O_2) that is used to clean glass and other surfaces. See Seeboth, A. and Hettrich, W., Spatial orientation of highly ordered self-assembled silane monolayers or glass surfaces, *J. Adhesion Sci. Technol.* 11, 495–505, 1997; Gray, D.E., Case-Green, S.C., Fell, T.S., Dobson, P.J., and Southern, E.M., Ellipsometric and interferometric characterization of DNA probes immobilized on a combinatorial array, *Langmuir* 13, 2833–2842, 1997; Steiner, G., Möller, H., Savchuk, O. et al., Characterization of ultra-thin polymer films by polarization modulation FITR spectroscopy, *J. Mol. Struct.* 563–564, 273–277, 2001; Guo, W. and Ruckenstein, E.,

Crosslinked glass fiber affinity membrane chromatography and its application to fibronectin separation, *J. Chromatog. B Technol. Biomed. Life Sci.* 795, 61–72, 2003; Ziegler, K.J., Gu, Z., Peng, H. et al., Controlled oxidative cutting of single-walled carbon nanotubes, *J. Amer. Chem. Soc.* 127, 1541–1547, 2005; Wang, M., Liechti, K.M., Wang, Q., and White, J.M., Self-assembled monolayers: fabrication with nanoscale uniformity, *Langmuir* 21, 1848–1857, 2005; Szuneritz, S. and Boukherroub, R., Preparation and characterization of thin films of SiO(x) on gold substrates for surface plasmon resonance studies, *Langmuir* 22, 1660–1663, 2006; Petrovykh, D.Y., Kimura-Suda, H., Opdahl, A. et al., Alkanethiols on platinum: multicomponent self-assembled monolayers, *Langmuir* 14, 2578–2587, 2006.

Plant Cell   Resulting from the capture of a cyanobacterium by a eukaryotic, mitochondria-possessing cell; the endosymbiont (cyanobacter) lost its identity and became a chloroplast. See Martin, W., Rujan, T., Richly, E. et al., Evolutionary analysis of Aribidopsis, cyanobacterial, and chloroplast genomes reveals plastid phylogeny and thousands of cyanobacterial genes in the nucleus, *Proc. Natl. Acad. Sci. USA* 99, 12246–12251, 2002; Grevich, J.J. and Daniell, H., Chloroplast genetic engineering: recent advances and future perspectives, *Crit. Rev. Plant Sci.* 23, 84–107, 2005.

Plasma/Serum   The identification and characterization of the proteins in the blood plasma/serum.
Proteome   See Lathrop, J.T., Anderson, N.L., Anderson, N.G., and Hammond, D.J., Therapeutic potential of the plasma proteome, *Curr. Opin. Mol. Ther.* 5, 250–257, 2003; Veenstra, T.D., Conrads, T.P., Hood, B.L. et al., Biomarkers: mining the biofluid proteome, *Mol. Cell. Proteomics* 4, 409–418, 2005.

Plastid   Usually refers to a chloroplast before the development of chlorophyll but can refer to any small intracellular pigmented vacuole. Plastids have been described in some bacteria. See Granick, S. and Gibor, A., The DNA of chloroplasts, mitochondria, and centrioles, *Prog. Nucleic Acid Res. Mol. Biol.* 6, 143–186, 1967; Gibor, A. and Ganick, S., Plastids and mitochondria: inheritable systems, *Science* 145, 890–897, 1964; Lopez-Juez, E. and Pyke, K.A., Plastids unleashed: their development and their integration in plant development, *Int. J. Dev. Biol.* 49, 557–577, 2005; Mackenzie, S.A., Plant organellar protein targeting: a traffic plan still under construction, *Trends Cell. Biol.* 15, 548–554, 2005; Miyagishima, S.Y., Origin and evolution of the chloroplast division machinery, *J. Plant Res.* 118, 295–306, 2005; Beck, C.F., Signaling pathways for the chloroplast to the nucleus, *Planta* 222, 743–756, 2005; Toyoshima, Y., Onda, Y., Shiina, T., and Nakahira, Y., Plastid transcription in higher plants, *Crit. Rev. Plant Sci.* 24, 59–81, 2005; Pilon, M., Abdel-Ghany, S.E., van Hoewyk, D., Ye, H., and Pilon-Smits, E.A., Biogenesis of iron-sulfur cluster proteins in plastids, *Genet. Eng.* 27, 101–117, 2006.

Plate Number   In chromatography, a plate is a separation instance or moment that a solute encounters during passage through a chromatographic column. The higher the number of plates, the more possibility for high resolution, but such resolution depends on the individual behavior of solutes (see *Resolution*). Plates may be theoretical or effective plates. The efficiency of a column is measured in the number of plates referred to as plate number (N). One equation for plate number (N), $N = 5.54(t_r/W_{1/2})^2$, where t_r is band retention time and $W_{1/2}$ is peak width at peak half-height. See Anspach, B., Gierlich, H.U., and Unger, K.K., Comparative study of Zorbax Bio series GF 250 and GF 450 and Tsk-Gel 3000 SW and SWXL columns in size-exclusion chromatography of proteins, *J. Chromatog.* 443, 45–54, 1988; Boyes, B.E. and Kirkland, J.J., Rapid, high-resolution HPLC separation of peptides using small particles at elevated temperatures, *Pept. Res.* 6, 249–258, 1993; Palsson, E., Axelsson, A., and Larsson, P.O., Theories of chromatographic efficiency applied to expanded base, *J. Chromatog. A* 912, 235–248, 2001; Mahesan, B. and Lai, W., Optimization of selected chromatographic responses using a designed experiment at the fine-tuning stage in reversed-phase high-performance liquid chromatographic method development, *Drug Dev. Ind. Pharm.* 27, 585–590, 2001; Ishizuka, N., Kobayashi, H., Minakuchi, H. et al., Monolithic silica columns for high-efficiency separations by high-performance liquid chromatography, *J. Chromatog. A* 960, 85–96, 2002; Jandera, P., Halama, M., and Novotna, K., Stationary-phase effects in gradient high-performance liquid chromatography, *J. Chromatog. A* 1030, 33–41, 2004; Lim, L.W., Hirose, K., Tatsumi, S. et al., Sample enrichment by using monolithic precolumns in microcolumn liquid chromatography, *J. Chromatog. A* 1033, 205–212, 2004; Okanda, F.M. and Rassi, Z., Capillary electrochromatography with monolithic stationary phases. 4. Preparation of neutral stearyl-acrylate monoliths and their evaluation in capillary electrochromatography of neutral and charged small species as well as peptides and proteins, *Electrophoresis* 26, 1988–1995, 2005; Berezkin, V.G. and Lapin, A.B., Ultra-short open capillary columns in gas-liquid chromatography, *J. Chromatog. A* 1075, 197–203, 2005; Donohoe, E., Denaturing high-performance liquid chromatography using the WAVE DNA fragment analysis system, *Methods Mol. Med.* 108, 173–187, 2005; Lohrmann, M., Schulte, M., and Strube, J., Generic method for systematic phase selection and method development of biochromatographic processes. Part I. Selection of a suitable cation-exchanger for the purification of a pharmaceutical protein, *J. Chromatog. A* 1092, 89–100, 2005; Chester, T.L. and Teremmi, S.O., A virtual-modeling and multivariate-optimization examination of HPLC parameter interactions and opportunities for saving analysis time, *J. Chromatog. A* 1096, 16–27, 2005.

Pleckstrin Homology Domain A protein domain consisting of approximately 100 amino acids, which binds phosphoinositide and other activators such as heterotrimeric G proteins and participates in the process of signal transduction. The name was derived from the platelet protein pleckstrin (platelet and leukocyte C kinase substrate) identified as a substrate for protein kinase C. See Tyers, M., Rachubinski, R.A., Stewart, M.I. et al., Molecular cloning and expression of the major protein kinase C substrate of platelets, *Nature* 333, 470–473, 1988; Mayer, B.J., Ren, R., and Clark, K.L., A putative modular domain present in diverse signaling proteins, *Cell* 73, 629–630, 1993; Musacchio, A., Gibson, T., Rice, P., Thompson, J., and Saraste, M., The PH domain: a common piece in the structural patchwork of signaling proteins, *Trends Biochem. Sci.* 18, 343–348, 1993; Ingley, E. and Hemmings, B.A., Pleckstrin homology (PH) domains in signal transduction, *J. Cell. Biochem.* 56, 436–443, 1994; Lemmon, M.A., Ferguson, K.M., and Abrams, C.S., Pleckstrin homology domains and the cytoskeleton, *FEBS Lett.* 513, 71–76, 2002; Philip, F., Guo, Y., and Scarlata, S., Multiple roles of pleckstrin homology domains in the phospholipase Cbeta function, *FEBS Lett.* 531, 29–32, 2002; Lemmon, M.A., Phosphoinositide recognition domains, *Traffic* 4, 201–213, 2003; Cozier, G.E., Carlton, J., Bouyoucef, D., and

Cullen, P.J., Membrane targeting by pleckstrin homology domains, *Curr. Top. Microbiol. Immunol.* 282, 49–88, 2004; Balla, T., Inositol-lipid binding motifs: signal integrators through protein–lipid and protein–protein interaction, *J. Cell Sci.* 119, 2093–2104, 2005; Perry, R.J. and Ridgway, N.D., Molecular mechanisms and regulation of ceramide transport, *Biochim. Biophys. Acta* 1734, 220–234, 2005.

Pleiotropic   Having more than one phenotypic expression of a gene; an effector molecule associated with more than a single event depending on the stimulation or, in the case of regulatory proteins and peptides, more than one target receptor. See Takeda, Y., Pleiotropic actions of aldosterone and the effects of eplerenone, a selective mineralocorticoid receptor antagonist, *Hypertens. Res.* 27, 781–789, 2004; Wilkie, A.O., Bad bones, absent smell, selfish testes: the pleiotropic consequences of human FGF receptor mutations, *Cytokine Growth Factor Rev.* 16, 187–203, 2005; Staels, B. and Fruchart, J.C., Therapeutic roles of peroxisome proliferator-activated receptor agonists, *Diabetes* 54, 2460–2470, 2005; Russo, V.C., Gluckman, P.D., Feldman, E.L., and Werther, G.A., The insulin-like growth factor system and its pleiotropic functions in brain, *Endocrine Rev.* 26, 916–943, 2005; Carrillo-Vico, A., Guerrero, J.M., Lardone, P.J., and Reiter, R.J., A review of the multiple actions of melatonin on the immune system, *Endocrine* 27, 189–200, 2005.

Podosome   A specialized cell-matrix contact point, which is structurally distinct from focal adhesion complexes. See Linder, S. and Aepfelbacher, M., Podosomes: adhesion hot-spots of invasive cells, *Trends Cell Biol.* 13, 376–385, 2003; McNiver, M.A., Baldassarre, M., and Buccione, R., The role of dynamin in the assembly and function of podosomes and invadopodia, *Front. Biosci.* 9, 1944–1953, 2004; Linder, S., and Kopp, P., Podosomes at a glance, *J. Cell Sci.* 118, 2079–2082, 2005.

Poisson Distribution   A probability density function that is an approximation to the biomodal distribution and is characterized by its mean being equal to its variance. See Mezei, L.M., *Practical Spreadsheet Statistics and Curve Fitting for Scientists and Engineers*, Prentice-Hall, Englewood Cliffs, NJ, 1990; Dowdy, S.M. and Wearden, S., *Statistics for Research*, Wiley, New York, 1991; Balakrishnan, N. and Nevzorov, V.B., *A Primer on Statistical Distributions*, Wiley, Hoboken, NJ, 2003.

Polyadenylation   The attachment of 200 adenyl residues to the 3′ end of messenger RNA, protecting the mRNA from degradation by nucleases and aiding in transfer of mRNA from nucleus to cytoplasm. The polyadenylation follows a specific cleavage at the termination of transcription. See Wilt, F.H., Polyadenylation of material RNA of sea urchin eggs after fertilization, *Proc. Natl. Acad. Sci. USA* 70, 2345–2349, 1973; Cooper, D.L. and Marzluff, W.F., Polyadenylation of RNA in a cell-free system from mouse myeloma cells, *J. Biol. Chem.* 253, 8375–8380, 1978; Bernstein, P. and Ross, J., Poly(A), poly(A) binding protein, and the regulation of mRNA stability, *Trends in Biochem. Sci.* 14, 373–377, 1989; Manley, J.L., Polyadenylation of mRNA precursors, *Biochim. Biophys. Acta* 950, 1–12, 1988; Buratowski, S., Connections between mRNA 3′ end processing and transcription termination, *Curr. Opin. Cell Biol.* 17, 257–261, 2005.

Polymerase Chain Reaction (PCR)   A method for synthesizing and amplifying a specific DNA sequence based on the use of specific oligonucleotide primers and unique DNA polymerases such as the thermostable DNA polymerase from *Thermus aquaticus*

(Taq polymerase). PCR amplicons from the amplified sequence are analyzed by size or sequence. See Kleppe, K., Ohtsuka, E., Kleppe, R. et al., Studies on polynucleotides XCVI. Repair replication of short synthetic DNAs as catalyzed by DNA polymerases, *J. Mol. Biol.* 56, 341–346, 1971; Saiki, R.K., Scharf, S., Faloona, F. et al., Enzymatic amplification of beta-globin genomic sequences and restriction site analysis for diagnosis of sickle cell anemia, *Science* 230, 1350–1354, 1985; Saiki, R.K., Gelfand, D.H., Stoffel, S. et al., Primer-directed enzymatic amplification of DNA with a thermostable DNA polymerase, *Science* 239, 487–491, 1988; Vosberg, H.P., The polymerase chain reaction: an improved method for the analysis of nucleic acids, *Hum. Genet.* 83, 1–15, 1989; Mullis, K.B., The unusual origin of the polymerase chain reaction, *Sci. Amer.* 262, 56–61, 1990; Innes, M.A., Ed., *PCR Protocols: A Guide to Methods and Applications*, Academic Press, San Diego, CA, 1990; White, B.A., Ed., *PCR Protocols: Current Methods and Applications*, Humana Press, Totowa, NJ, 1993; Dieffenbacah, C.W. and Dveksler, G.S., *PCR Primer: A Laboratory Manual*, Cold Spring Harbor Laboratory Press, Cold Spring Harbor, NY, 1995; Taylor, G.R. and Robinson, P., The polymerase chain reaction: from functional genomics to high-school practical classes, *Curr. Opin. Biotechnol.* 9, 35–42, 1998; Sninsky, J.J and Innes, M.A., Eds., *PCR Applications: Protocols for Functional Genomics*, Academic Press, San Diego, CA, 1999. See *Gene Expression Domain*; *Real-Time RT-PCR*; *Reverse Transcriptase–Polymerase Chain Reaction (RT-PCR)*.

Polyvinyl-pyrrolidone

Polyvinylpyrrolidone (PVP) is a polymer similar to poly(ethylene)glycol (PEG) in that it is readily soluble in water and is used for the stabilization of proteins. Unlike PEG, PVP is useful in the lyophilization of proteins. PVP is synthesized by the free-radical polymerization of *N*-vinylpyrrolidinone (1-vinyl-2-pyrrolidinone). The final size of the polymer is controlled by choice of experimental conditions. It has a wide application in biotechnology. See Antonsen, K.P., Gombotz, W.R., and Hoffman, A.S., Attempts to stabilize a monoclonal antibody with water soluble synthetic polymers of varying hydrophobicity, *J. Biomater. Sci. Polym. Ed.* 6, 55–65, 1994; Gombotz, W.R., Pankey, S.C., Phan, D. et al., The stabilization of a human IgM monoclonal antibody with poly(vinylpyrrolidone), *Pharm. Res.* 11, 624–632, 1994; Gibson, T.D., Protein stabilization using additives based on multiple electrostatic interactions, *Dev. Biol. Stand.* 87, 207–217, 1996; Anchordoquy, T.J. and Carpenter, J.F., Polymers protect lactate dehydrogenase during freeze-drying by inhibiting dissociation in the frozen state, *Arch. Biochem. Biophys.* 332, 231–238, 1996; Yoshioka, S., Aso, Y., and Kojima, S., The effect of excipients on the molecular mobility of lyophilized formulations, as measured by glass transition temperature and NMR relaxation-based critical mobility temperature, *Pharm. Res.* 16, 135–140, 1999; Sharp, J.M. and Doran, P.M., Strategies for enhancing monoclonal antibody accumulation in plant cell and organ cultures, *Biotechnol. Prog.* 17, 979–992, 2001. PVP has some direct therapeutic use (Kaneda, Y., Tsutsumi, Y., Yoshioka, Y. et al., The use of PVP as a polymeric carrier to improve the plasma half-life of drugs, *Biomaterials* 25, 3259–3266, 2004) and as a carrier for iodine as a disinfectant (Art, G., Combination povidone-iodine and alcohol formulations more effective, more convenient versus formulations containing either iodine or alcohol alone: a review of the literature, *J. Infus. Nurs.* 28, 314–320, 2005). An HPLC method for the analysis of PVP in pharmaceutical products has been developed (Jones, S.A., Martin, G.P., and Brown, M.B., Determination of polyvinylpyrrolidone using high-performance liquid chromatography, *J. Pharm. Biomed. Anal.* 35, 621–624, 2004).

Posttranslational Modification   A covalent modification of a protein following translation of the RNA to form the polypeptide chain. Such modification may or may not be enzyme catalyzed (γ-carboxylation vs. nitration) and may or may not be reversible (phosphorylation vs. γ-carboxylation).

Pre-initiation Complex   A complex of general transcription factors (GTFs) that are formed at each core promoter prior to transcriptional activation and required for the action of RNA polymerase II. Recent work suggests that core promoter elements may not be an absolute requirement. See Svejstrup, J.Q., The RNA polymerase II transcription cycle: cycling though chromatin, *Biochim. Biophys. Acta* 1677, 64–73, 2004; Govind, C.K., Yoon, S., Qiu, H., Govind, S., and Hinnebusch, A.G., Simultaneous recruitment of coactivators by Gcn4p stimulates multiple steps of transcription *in vivo*, *Mol. Cell. Biol.* 25, 5626–5638, 2005; George, A.A., Sharma, M., Singh, B.N., Sahoo, N.C., and Rao, K.V., Transcription regulation from a TATA and INR-less promoter: spatial segregation of promoter function, *EMBO J.*, 25, 811–821, 2006; Maag, D., Algire, M.A., and Lorsch, J.R., Communication between eukaryotic translation initiation factors 5 and 1A within the ribosomal pre-initiation complex plays a role in start site selection, *J. Mol. Biol.* 356, 724–737, 2006.

Primase   Primase is an enzyme that catalyzes polymerization of ribonucleoside 5′-triphosphates to form RNA primers in a sequence that is directed by a DNA template. See Foiani, M., Lucchini, G., and Plevani, P., The DNA polymerase alpha-primase complex couples DNA replication, cell-cycle progression, and DNA-damage response, *Trends Biochem. Sci.* 22, 424–427, 1997; Arezi, B. and Kuchta, R.D., Eurkaryotic DNA primase, *Trends Biochem. Sci.* 25, 572–576, 2000; Frick, D.N. and Richardson, C.C., DNA primases, *Annu. Rev. Biochem.* 70, 39–80, 2001; Benkovic, S.J., Valentine, A.M., and Salinas, F., Replisome-mediated DNA replication, *Annu. Rev. Biochem.* 70, 181–208, 2001; MacNeil, S.A., DNA replication: partners in the Okazaki two-step, *Curr. Biol.* 11, F842–F844, 2001; Kleymann, G., Helicase primase: targeting the Achilles heel of herpes simplex viruses, *Antivir. Chem. Chemother.* 15, 135–140, 2004; Lao-Sirieix, S.H., Pellegrini, L., and Bell, S.D., The promiscuous primase, *Trends Genet.* 21, 568–572, 2005; Lao-Sirieix, S.H., Nookala, R.K., Roversi, P. et al., Structure of the heterodimeric core primase, *Nat. Struct. Mol. Biol.* 12, 1137–1144, 2005; Shutt, T.E. and Gray, M.W., Twinkle, the mitochondrial replicative DNA helicase, is widespread in the eukaryotic radiation and may also be the mitochondrial DNA primase in most eukaryotes, *J. Mol. Evol.* 62, 588–599, 2006; Rodina, A. and Godson, G.N., Role of conserved amino acids in the catalytic activity of *Escherichia coli* primase, *J. Bacteriol.* 188, 3614–3621, 2006.

Promoter Elements   A region of a segment of DNA (usually *cis*), which regulate the transcription (mRNA synthesis) from information encoded on that segment of DNA. These are elements regulating the nuclear transcription process and binding transcription factors and other regulatory factors. See Kingston, R.E., Baldwin, A.S., and Sharp, P.A., Transcription control by oncogenes, *Cell* 41, 3–5, 1985; Wasylk, B., Transcription elements and factors of RNA polymerase B promoters of higher eukaryotes, *CRC Crit. Rev. Biochem.* 23, 77–120, 1988; Khokha, M.K. and Loots, G.G., Strategies for characterizing *cis*-regulatory elements in *Xenopus*, *Brief Funct. Genomic Proteomic.* 4, 58–68, 2005; Sipos, L. and Gyurkovics, H., Long-distance interactions between enhancers and promoters, *FEBS J.* 272, 3253–3259, 2005; Fukuchi, M., Tabuchi, A., and Tsuda, M., Transcriptional regulation of neuronal genes and its effect on neural functions: cumulative mRNA expression of PACAP and BDNP genes controlled by calcium and cAMP signals in neurons, *J. Pharmacol. Sci* 98, 212–218, 2005; Anderson, S.K., Transcriptional regulation of NK cell receptors, *Curr. Top. Microbiol. Immunol.* 298, 59–75, 2006.

Protamines A family of basic proteins associated with the chromatin in the nucleus of the cell. Protamines are characterized by a high content of arginine and replace histones in the process of spermiogenesis. See Lewis, S.D. and Ausió, J., Protamine-like proteins: evidence for a novel chromatin structure, *Biochem. Cell Biol.* 80, 353–361, 2002; Meistrich, M.L., Mohapatra, B., Shirley, C.R., and Zhao, M., Roles of transition nuclear proteins in spermiogenesis, *Chromasoma* 111, 483–488, 2003; Aoki, V.W. and Carrell, D.T., Human protamines and the developing spermatid: their structure, function, expression, and relationship with male infertility, *Asian J. Androl.* 5, 315–324, 2003; Erin-López, J.M., Frehlich L.J., and Ausió, J., Protamines, in the footsteps of linker histone evolution, *J. Biol. Chem.* 281, 1–4, 2006.

Protease A protease/proteolytic enzyme catalyzes the hydrolysis of a peptide bond in a protein. A simple classification of proteases divides these enzymes into two functional categories and four chemical categories. The functional categories are regulatory and digestive. Examples of regulatory proteolysis is proprotein processing by furin and blood coagulation while digestive enzymes include enzymes like pepsin, trypsin, and chymotrypsin found in mammalian digestive systems. Chemical categories describe the functional groups at enzyme active sites and include serine proteases such as trypsin or chymotrypin, cysteine proteases such as papain and the caspaces, aspartic acid proteases such as pepsin, and metalloproteinases such as ADAM proteases and matrix metalloproteinase (MMP). See Magnusson, S., Ed., *Regulatory Proteolytic Enzymes and Their Inhibitors,* Pergammon Press, Oxford, UK, 1978; Barrett, A.J. and McDonald, J.K., Eds., *Mammalian Proteases: A Glossary and Bibliography,* Academic Press, New York, 1980; Polgár, L., *Mechanism of Protease Action,* CRC Press, Boca Raton, FL, 1989; Dunn, B.M., Ed., *Proteases of Infectious Agents,* Academic Press, San Diego, CA, 1999; Zwickl, P. and Baumeister, W., Eds., *The Proteosome-Ubiquitin Protein Degradation Pathway,* Springer, Berlin, 2002; Saklatvala, J. and Nagase, H., Eds., *Proteases and the Regulation of Biological Processes,* Portland Press, London, 2003.

Protease-Activated Receptor (PAR) Protease-activated receptors (PARs) are a family of G-protein-coupled receptors in which the (tethered) ligand is intrinsic to the receptor protein and is exposed by proteolysis in the *N*-terminal external region. These receptors may also be activated by peptides where the sequence is identical to or related to the ligand sequence in the receptor. The protease-activated receptor was first described in platelets by Coughlin and colleagues (Vu, T.K., Hung, D.T., Wheaton, V.I., and Coughlin, S.R., Molecular cloning of a functional thrombin receptor reveals a novel proteolytic mechanism of receptor activation, *Cell* 64, 1057–1068, 1991; Coughlin, S.R., Vu, T.K., Hung, D.T., and Wheaton, V.I., Expression cloning and characterization of a functional thrombin receptor reveals a novel proteolytic mechanism of receptor activation, *Semin. Thromb. Hemost.* 18, 161–166, 1992) and by another group in France (Rasmussen, U.B., Vouret-Cravieri, V., Jallet, S. et al., cDNA cloning and expression of a hamster alpha-thrombin receptor coupled to Ca^{2+} mobilization, *FEBS Lett.* 288, 123–128, 1991) using fibroblasts. Since the original work, four PARs have been described on a wide variety of cell types. See Chen, J., Bernstein, H.S., Chen, M. et al., Tethered ligand library for discovery of peptide agonists, *J. Biol. Chem.* 270, 23398–23401, 1995; Santulli, R.J., Derian, C.K., Darrow, A.L. et al., Evidence for the presence of a proteinase-activated receptor distinct from the thrombin receptor in human keratinocytes, *Proc. Natl. Acad. Sci. USA* 92, 9151–9155, 1995; Ishihara, H., Connolly, A.J., Zeng, D. et al., Protease-activated receptor 3 is a second thrombin receptor in humans, *Nature* 386, 502–506, 1997; Brass, L.F. and Molino, M., Protease-activated G protein coupled receptors on human platelets and endothelial cells, *Thromb. Haemostas.* 78, 234–241, 1997; Brass, L.F., Thrombin receptor antagonists: a work in progress, *Coron. Artery Dis.* 8, 49–58, 1997; Niclou, S.P., Suidan, H.S., Pavlik, A., Vejsada, R., and Monard, D., Changes in the expression of protease-activated receptor 1 and protease nexin-1 mRNA during rat nervous system development and after nerve lesion, *Eur. J. Neurosci.* 10, 1590–1607, 1998; Hou, L., Howells, G.L., Kapas, S., and Macey, M.G., The protease-activated receptors and their cellular expression and function in blood-related cells, *Br. J. Haematol.* 101, 1–9, 1998; Coughlin, S.R., Protease-activated receptors and platelet function, *Thromb. Haemost.* 82, 353–356, 1999; Cooks, T.M. and Moffatt, J.D., Protease-activated receptors: sentries for inflammation, *Trends Pharmacol. Sci.* 21, 103–108, 2000; Macfarlane, S.R., Seatter, M.J., Kanke, T., Hunter, G.D., and Plevin, R., Proteinase-activated receptors, *Pharmacol. Rev.* 53, 245–282, 2001; Bucci, M., Roviezzo, F., and Cirino, G., Protease-activated receptor-2 (PAR2) in cardiovascular system, *Vascul. Pharmacol.* 43, 247–253, 2005; Wang, P. and Defea, K.A., Protease-activated receptor-2 simultaneously directs beta-arrestin-1-dependent inhibition and Gαq-dependent activation of phosphatidylinositol 3-kinase, *Biochemistry* 45, 9374–9385, 2006; Oikonomopoulou, K., Hansen, K.K., Saifeddine, M. et al., Proteinase-activated receptors (PARs): targets for kallikrein signalling, *J. Biol. Chem.*, 281, 32095–32112, 2006; Wang, L., Luo, J., Fu, Y., and He, S., Induction of interleukin-8 secretion and activation of ERK1/2, p38 MAPK signaling pathways by thrombin in dermal fibroblasts, *Int. J. Biochem. Cell Biol.* 38, 1571–1583, 2006; Page, K., Hughes, V.S., Bennett, G.W., and Wong, H.R., German cockroach proteases regulate matrix metalloproteinase-9 in human bronchial epithelial cells, *Allergy* 61, 988–995, 2006.

Protease Inhibitor Cocktail   A mixture of protease inhibitors, which is used to preserve protein integrity during the processing of samples for subsequent analysis. The term "cocktail" refers to a mixture of components. A protease inhibitor cocktail is composed of a broad spectrum of protease inhibitors and intends to inhibit the diverse proteolytic enzymes found in tissue extracts and biological fluids. See Pringle, J.R., Methods for avoiding proteolytic artifacts in studies with enzymes and other proteins from yeasts, *Methods Cell Biol.* 12, 149–184, 1975; Drubin, D.G., Miller, K.G., and Botstein, D., Yeast actin-binding proteins: evidence for a role in morphogenesis, *J. Cell Biol.* 107, 2551–2561, 1988; Nanoff, C., Jacobson, C.A., and Stiles, G.L., The A2 adenosine receptor: guanine nucleotide modulation of agonist binding is enhanced by proteolysis, *Mol. Pharmacol.* 39, 130–135, 1991; Palmer, T.M., Jacobson, K.A., and Stiles, G.L., Immunological identification of A2 adenosine receptors by two antipeptide antibody preparations, *Mol. Pharmacol.* 42, 391–397, 1992; Pyle, L.E., Barton, P., Fujiwara, Y., Mitchell, A., and Fidge, N., Secretion of biologically active human proapolipoprotein A-1 in a baculovirus-insect cell system: protection from degradation by protease inhibitors, *J. Lipid Res.* 36, 2355–2361, 1995; Weidner, M.-F., Grenier, D., and Mayrand, D., Proteolytic artifacts in SDS-PAGE analysis of selected periodontal pathogens, *Oral Microbiol. Immunol.* 11, 103–108, 1996; Hassel, M., Klenk, G., and Frohme, M., Prevention of unwanted proteolysis during extraction of proteins from protease-rich tissue, *Anal. Biochem.* 242, 274–275, 1996; Salvesen, G. and Nagase, H., Inhibition of proteolytic enzymes, in *Proteolytic Enzymes: Practical Approaches,* 2nd ed., Benyon, R. and Bond, J.S., Eds., Oxford University Press, Oxford, UK, 2001, pp. 105–130; North, M.J. and Benyon, R.J., Prevention of unwanted proteolysis, in *Proteolytic Enzymes: Practical Approaches,* 2nd ed., Benyon, R. and Bond, J.S., Eds., Oxford University Press, Oxford, UK, 2001, pp. 211–232; Castellanos-Serra, L. and Paz-Lago, D., Inhibition of unwanted proteolysis during sample preparation: evaluation of its efficiency in challenge experiments, *Electrophoresis* 23, 1745–1753, 2002; Kikuchi, S., Hirohashi, T., and Nakai, M.,. Characterization of the preprotein translocon at the outer envelope membrane of chloroplasts by blue native PAGE, *Plant Cell Physiol.* 47, 363–371, 2006. The term "protease cocktail" also refers to the combination of therapeutic protease inhibitors used in AIDS therapy. See Tamamura, H. and Fujii, N., Two orthogonal approaches to overcome multi-drug resistant HIV-1s: development of protease inhibitors and entry inhibitors based on CXCR4 antagonists, *Curr. Drug Targets Infect. Disord.* 3, 103–110, 2004; Wicovsky, A., Siegmund, D., and Wajant, H., Interferons induce proteolytic degradation of TRAILR4, *Biochem. Biophys. Res. Commun.* 337, 184–190, 2005. The term "cocktail" is also used to describe the combination of chemicals and solvent used for liquid scintillation counting of radioisotopes. See Kobayashi, Y. and Maudsely, D.V., Practical aspects of liquid scintillation counting, *Methods Biochem. Anal.* 17, 55–133, 1969; Wood, K.J., McElroy, R.G., Surette, R.A., and Brown, R.M., Tritium sampling and measurement, *Health Phys.* 65, 610–627, 1993; Jaubert, F., Tartes, I., and Cassette, P., Quality control of liquid scintillation counting, *Appl. Radiat. Isot.*, 64, 1163–1170, 2006.

Proteasome   A multisubunit complex that functions in the degradation of intracellular proteins in eukaryotic cells. It is composed of catalytic subunits with different specificity and regulatory subunits. In eukaryotic cells, proteins are "marked" for proteasomal degradation by ubiquitinylation. There is a specialized proteasome that functions in MHC I antigen presentation. See Arrigo, A.P., Tanaka, K., Goldberg, A.L., and Welch, W.J., Identity of the 19S "prosome" particle with the large multifunctional protease complex of mammalian cells (the proteasome), *Nature* 331, 192–194, 1988; Falkenberg, P.E., and Kloetael, P.M., Identification and characterization of three different subpopulations of the *Drosophila* multicatalytic proteinase (proteasome), *J. Biol. Chem.* 264, 6660–6666, 1989; Dahlmann, B., Kopp, F., Kuehn, L. et al., The multicatalytic proteinase (prosome, proteasome): comparison of the eukaryotic and archaebacterial enzyme, *Biomed. Biochim. Acta* 50, 465–469, 1991; Demartino, G.W., Orth, K., McCullough, M.L. et al., The primary structure of four subunits of the human, high-molecular weight proteinase, macropain (proteasome), are distinct but homologous, *Biochim. Biophys. Acta* 1079, 29–38, 1991; Wlodawer, A., Proteasome: a complex protease with a new fold and a distinct mechanism, *Structure* 3, 417–420, 1995; Baumeister, W., Cejka, Z., Kania, M., and Seemuller, E., The proteasome: a macromolecular assembly designed to confine proteolysis to a nanocompartment, *Biol. Chem.* 378, 121–130, 1997; Grune, T., Merker, K., Sandig, G., and Davies, K.J., Selective degradation of oxidatively modified protein substrates by the proteasome, *Biochem. Biophys. Res. Commun.* 305, 709–718, 2003; Hartmann-Petersen, R. and Gordon, C., Proteins interacting with 26S proteasome, *Cell. Mol. Life Sci.* 61, 1589–1595, 2004; Smalle, J. and Veirstra, R.D., The ubiquitin 26S proteasome proteolytic pathway, *Ann. Rev. Plant Biol.* 55, 555–590, 2004; Dalton, W.S., The proteasome, *Semin. Oncol.* 31 (6 Suppl. 16), 3–9, 2004; Qureshi, N., Vogel, S.N., Van Way, C., III et al., The proteasome: a central regulator of inflammation and macrophage function, *Immunol. Res.* 31, 243–260, 2005; Glickman, M.H. and Raveh, D., Proteasome plasticity, *FEBS Lett.* 579, 3214–3223, 2005; Ye, Y., The role of the ubiquitin-proteasome system in ER quality control, *Essays Biochem.* 41, 99–112, 2005; Gao, G. and Luo, H., The ubiquitin-proteasome pathway in viral infections, *Can. J. Physiol. Pharmacol.* 84, 5–14, 2006. There is interest in the proteasome as a drug target in oncology. See Montagut, C., Rovira, A., and Albanell, J., The proteasome: a novel target for anticancer therapy, *Clin. Transl. Oncol.* 8, 313–317, 2006. There is a variation in the proteasomes involved in the presentation of MHC I antigens; see *Immuno Proteasome.*

Protein Classification   There are a number of approaches to protein classification. One simple approach is based on environmental conditions and divides proteins into three different groups (see Finkelstein, A.V. and Ptitsyn, O.B., *Protein Physics: A Course of Lectures*, Academic Press, London, 2002). Fibrous proteins are usually in nonaqueous environments and usually form high, regular hydrogen-bonded structures such as those seen in cartilage; membrane proteins are also found in nonaqueous environments; water-soluble proteins are found in the cytoplasm and extracellular fluids. Water-soluble proteins can be divided into albumins and globulins on the basis of solubility properties. Water-soluble proteins can form three-dimensional structures maintained by a variety of forces including hydrogen bonds and van der Waals forces. Water-soluble proteins can have effects on enzyme activity separate from their intrinsic activity (see Derham, B.K. and Harding, J.J., The effect of the presence of globular proteins and elongated polymers on enzyme activity, *Biochim. Biophys. Acta* 1764, 1000–1006, 2006).

Protein Disulfide Isomerase/ Ero1p   Protein disulfide isomerase and Ero1p are enzymes that help form disulfide bonds within the endoplasmic reticulum. These factors are critical for the normal formation of disulfide bonds during protein folding. See Lodi, T., Neglia, B., and Donnini, C., Secretion of human serum albumin by *Kluyveromyces lactis* overexpressing *KlPDIUl* and *KlERO1*, *Appl. Environ.*

Microbiol. 71, 4359–4364, 2005; Kulp, M.S., Frickel, E.-M., Ellgaard, L., and Weissman, J.S., Domain architecture of protein-disulfide isomerase facilitates its dual role as an oxidase and an isomerase in Ero1p-mediated disulfide formation, *J. Biol. Chem.* 281, 876–884, 2006; Gross, E., Sevier, C.S., Heldman, N. et al., Generating disulfides enzymatically: reaction products and electron acceptors of the endoplasmic reticulum thiol oxidase Ero1p, *J. Biol. Chem.* 281, 299–304, 2006.

Protein Profiling The use of algorithms to determine the relationship of multiple proteins as determined by proteomic analysis such as protein microarray technology, shotgun proteomics, or SELDI-TOF-MS. See Tomlinson, I.M. and Holt, L.J., Protein profiling comes of age, *Genome Biol.* 2, 1004, 2001; Kingamore, S.F. and Patel, D.D., Multiplexed protein profiling on antibody-based microarrays by rolling circle amplification, *Curr. Opin. Biotechnol.* 14, 74–81, 2003; Jessani, N. and Cravatt, B.F., The development and application of methods for activity-based protein profiling, *Curr. Opin. Chem. Biol.* 8, 54–59, 2004; Berger, A.B., Vitorino, P.M., and Bogyo, M., Activity-based protein profiling: applications to biomarker discovery, *in vitro* imaging, and drug discovery, *Am. J. Pharmacogenomics* 4, 371–381, 2004; Steel, L.F., Haab, B.B., and Hanash, S.M., Methods of comparative proteomic profiling for disease diagnostics, *J. Chromatog. B Analyt. Technol. Biomed. Life Sci.* 815, 275–284, 2005; Kislinger, T. and Emili, A., Multidimensional protein identification technology: current status and future prospects, *Expert Rev. Proteomics* 2, 27–39, 2005; Katz, J.E., Mallick, P., and Agus, D.B., A perspective on protein profiling of blood, *BJU Int.* 96, 477–482, 2005; Bons, J.A., Wodzig, W.K., and van Dieijen-Visser, M.P., Protein profiling as a diagnostic tool in clinical chemistry: a review, *Clin. Chem. Lab. Med.* 43, 1281–1290, 2005.

Protein Tyrosine Phosphatases A family of hydrolytic enzymes that catalyze the dephosphorylation of protein-bound *O*-tyrosine phosphate. Dephosphorylation of tyrosine residues can modulate biological activity and may be a specific or nonspecific process. See Fischer, E.H., Tonks, N.K., Charbonneau, H. et al., Protein tyrosine phosphatases: a novel family of enzymes involved in transmembrane signalling, *Adv. Second Messenger Phosphoprotein Res.* 24, 272–279, 1990; Calya, X., Goris, J., Hermann, J. et al., Phosphotyrosyl phosphatase activity of the polycation-stimulated protein phosphatases and involvement of dephosphorylation in cell cycle regulation, *Adv. Enzyme Reg.* 39, 265–285, 1990; Saito, H. and Streuli, M., Molecular characterization of protein tyrosine phosphatases, *Cell Growth Differ.* 2, 59–65, 1991; Tonks, N.K., Yang, Q., and Guida, P., Jr., Structure, regulation, and function of protein tyrosine phosphatases, *Cold Spring Harbor Symp. Quant. Biol.* 56, 265–273, 1991; Lawrence, D.S., Signaling protein inhibitors via the combinatorial modification of peptide scaffolds, *Biochim. Biophys. Acta* 1754, 50–57, 2005; Boutros, R., Dozier, C., and Ducommun, B., The when and wheres of CDC25 phosphatases, *Curr. Opin. Cell Biol.* 18, 185–191, 2006; Ostman, A., Hellberg, C., and Bohmer, F.D., Protein-tyrosine phosphatases and cancer, *Nat. Rev. Cancer* 6, 307–320, 2006; Burridge, K., Sastry, S.K., and Salfee, J.L., Regulation of cell adhesion by protein-tyrosine phosphatases. I. Cell-matrix adhesion, *J. Biol. Chem.* 281, 15593–15596, 2006; Sallee, J.L., Wittchen, E.S., and Burridge, K., Regulation of cell adhesion by protein-tyrosine phosphatases. II. Cell–cell adhesion, *J. Biol. Chem.* 281, 16189–16192, 2006.

Proteome   The total expressed protein content of a genome. See Wasinger, V.C., Cordwell, S.J., Cerpa-Poljak, A. et al., Progress with gene-product mapping of the mollicutes: *Mycoplasma genitalium*, *Electrophoresis* 16, 1090–1094, 1995; Kahn, P., From genome to proteome: looking at a cell's proteins, *Science* 270, 369–370, 1995; Wilkens, M.R., Sanchez, J.C., Gooley, A.A. et al., Progress with proteome projects: why all proteins expressed by a genome should be identified and how to do it, *Biotechnol. Genet. Eng. Rev.* 13, 19–50, 1996; Figeys, D., Gygi, S.P., Zhang, Y. et al., Electrophoresis combined with novel mass spectrometry techniques: powerful tools for the analysis of proteins and proteomics, *Electrophoresis* 19, 1811–1818, 1998; Blackstock, W.P. and Weir, M.P., Proteomics: quantitative and physical mapping of cellular proteins, *Trends Biotechnol.* 17, 121–127, 1999; Bradshaw, R.A., Proteomics — boom or bust? *Mol. Cell. Proteomics* 1, 177–178, 2002; Bradshaw, R.A. and Burlingame, A.L., From proteins to proteomics, *IUBMB Life* 57, 267–272, 2005; Domon, B. and Aebersold, R., Mass spectrometry and protein analysis, *Science* 312, 212–217, 2006.

Proteometabolism Metabolism of the proteome.

Proteomics   The study of the proteome; not technology limited; the qualitative and quantitative study of the proteome under various conditions including protein expression, modification, localization, function, and protein–protein interactions, as a means of understanding biological processes.

Proto-Oncogenes   Normal cellular genes whose activation or modification to an oncogene is linked to malignant transformation; progenitors of oncogenes; proto-oncogenes can become oncogenes either by transduction into a virus or by a "disturbance" such as chromosomal translocation, amplification, or point mutation at the location in a chromosome. *c-Myc* is one of the most studied of the proto-oncogenes. See Bishop, J.M., Oncogenes and proto-oncogenes, *J. Cell. Physiol. Suppl.* 4, 1–5, 1986; Cory, S., Activation of cellular oncogenes in hemapoietic cells by chromosome translocation, *Adv. Cancer Res.* 47, 189–243, 1986; Bishop, J.M., and Hannfusa, W., Proto-oncogenes in normal and neoplastic cells, in *Scientific American Molecular Oncology*, Bishop, J.M. and Weinberg, R.A., Eds., Scientific American, New York, 1996, pp. 61–83; Shachaf, C.M. and Felsher, D.W., Rehabilitation of cancer through oncogene inactivation, *Trends Mol. Med.* 11, 316–321, 2005; Barry, E.L., Baron, J.A., Grau, M.V., Wallace, K., and Haile, R.W., K-ras mutations in incident sporadic colorectal adenomas, *Cancer* 106, 1036–1040, 2006.

ProtParam   A program that allows the calculation of a number of physical and chemical properties for a protein from the known amino acid sequence. See http://www.expasy.ch/tools/protparam.html.

Proximal Promoter Element   A region located 30–200 bp upstream from the transcription start site; region usually contains multiple transcription factor-binding sites. See van de Klundert, F.A., Jansen, H.J., and Bloemendal, H., A proximal promoter element in the hamster desmin upstream regulatory region is responsible for activation by myogenic determination factors, *J. Biol. Chem.* 269, 220–225, 1994; Petrovic, N., Black, T.A., Fabian, J.R. et al., Role of proximal promoter elements in regulation of rennin gene transcription, *J. Biol. Chem.* 271, 22499–22505, 1996; Mori, A., Kaminuma, O., Ogawa, K., Okudaira, H. and Akiyama, K., Transcriptional regulation of IL-5 gene by nontransformed human T-cells through the proximal promoter element,

Intern. Med. 39, 618–625, 2000; Ghosh-Choudhury, N., Choudhury, G.G., Harris, M.A. et al., Autoregulation of mouse BMP-2 gene transcription is directed by the proximal promoter element, *Biochem. Biophys. Res. Commun.* 286, 101–108, 2001; Rentsendorj, O., Nagy, A., Sinko, I. et al., Highly conserved proximal promoter element harbouring paired Sox9-binding sites contributes to the tissue- and developmental stage-specific activity of the matrilin-1 gene, *Biochem. J.* 389, 705–716, 2005.

Pseudogenes (Retropseudogenes) Copies of cellular RNA that have been reverse transcribed and inserted into the genome. See Vanin, E.F., Processed pseudogenes: characteristics and evolution, *Annu. Rev. Genet.* 19, 253–272, 1985; Weiner, A.M., Deininger, P.L., and Efstratiadis, A., Nonvirial retroposons, genes, pseudogenes, and transposable elements generated by the reverse flow of genetic information, *Annu. Rev. Biochem.* 55, 631–661, 1986; Pascual, V. and Capra, J.D., Human immunoglobulin heavy-chain variable region genes: organization, polymorphism, and expression, *Adv. Immunol.* 49, 1–74, 1991; King, C.C., Modular transposition and the dynamical structure of eurkaryote regulatory evolution, *Genetica* 86, 127–142, 1992; D'Errico, I., Gadaleta, G., and Saccone, C., Pseudogenes in metazoan: origin and features, *Brief Funct. Genomic Proteomic.* 3, 157–167, 2004; Rodin, S.N., Parkhomchuk, D.V., Rodin, A.S., Holmquist, G.P., and Riggs, A.D., Repositioning-dependent fate of duplicate genes, *DNA Cell Biol.* 24, 529–542, 2005; Pavlicek, A., Gentles, A.J., Paes, J. et al., Retroposition of processed pseudogenes: the impact of RNA stability and translational control, *Trends Genet.* 22, 69–73, 2006.

Psychogenomics The process of applying the tools of genomics, transcriptomics, and proteomics to understand the molecular basis of behavioral abnormalities.

Psychrophilic Functioning more efficiently at cold temperatures. See Feller, G. and Gerday, C., Psychrophilic enzymes: hot topics in cold adaptation, *Nat. Rev. Microbiol.* 1, 200–208, 2003; Bolter, M., Ecophysiology of psychrophilic and psychrotolerant microorganisms, *Cell. Mol. Biol.* 50, 563–573, 2004; Zecchinon, L., Oriol, A., Netzel, U. et al., Stability domains, substrate-induced comformational changes, and hinge-bending motions in a psychrophilic phosphoglycerate kinase. A microcalorimetric study, *J. Biol. Chem.* 280, 41307–41314, 2005.

"Pull-Down" The process of the capture of a protein, a protein complex, or other biological by binding to an immobilized capture reagent such as an antibody. See Cavailles, V., Dauvois, S., Danielian, P.S., and Parker, M.G., Interaction of proteins with transcriptionally active estrogen receptors, *Proc. Natl. Acad. Sci. USA* 91, 10009–10013, 1994; Magnaghi-Jaulin, L., Masutani, H., Robin, P., Lipinski, M., and Harel-Bellan, A., SRE elements are binding sites for the fusion protein EWS-FLI-1, *Nucleic Acids Res.* 24, 1052–1058, 1996; Dombrosky-Ferlan, P.M. and Corey, S.J., Yeast two-hybrid *in vivo* association of the Src kinase Lyn with the proto-oncogene product Cbl but not with the p85 subunit of PI 3-kinase, *Oncogene* 14, 2019–2024, 1997; Graves, P.R. and Haystead, T.A., A functional proteomics approach to signal transduction, *Recent Prog. Horm. Res.* 58, 1–24, 2003.

Pullulanase Enzyme degrading pullulan, a branched starch; pullulanase catalyzes the hydrolysis of the α-1,6-glucosidic linkage in α-glucans. Pullulanase preferentially hydrolyzes pullulan while isoamylase has a preference for glycogen and amylopectin. See Wallenfels, K., Bender, H., and Rached, J.R., Pullulanase from *Aerobacter aerogenes*; production in a cell-bound state. Purification and properties of the enzymes, *Biochem. Biophys. Res.*

Commun. 22, 254–261, 1966; Hardie, D.G. and Manners, D.J., A visco-metric assay for pullulanase-type, debranching enzymes, *Carbohdr. Res.* 36, 207–210, 1974; Harada, T., Special bacterial polysaccharides and polysaccharidases, *Biochem. Soc. Symp.* 48, 97–116, 1983; Vihinen, M. and Mantsala, P., Microbial amylolytic enzymes, *Crit. Rev. Biochem. Mol. Biol.* 24, 329–418, 1989; Doman-Pytka, M. and Bardowski, J., Pullalan degrading enzymes of bacterial origin, *Crit. Rev. Microbiol.* 30, 107–121, 2004; Lammerts van Bueren, A., Finn, R., Ausio, J., and Boraston, A.B., Alpha-glucan recognition by a new family of carbohydrate-binding mod-ules found primarily in bacterial pathogens, *Biochemistry* 43, 15633–15642, 2004; Mikami, B., Iwamoto, H., Malle, D. et al., Crystal structure of pullulanase: evidence for parallel binding of oligosaccharides in the active site, *J. Mol. Biol.* 359, 690–707, 2006; Hytonen, J., Haataja, S., and Finne, J., Use of flow cytometry for the adhesion analysis of *Streptococcus pyogenes* mutant strains to epithelial cells: investigation of the possible role of surface pullulanase and cysteine protease, and the transcriptional regulation Rgg, *BMC Microbiol.* 6, 18, 2006.

Pulsed-Field Gel Electrophoresis   A gel electrophoretic technique for the analysis of very large DNA molecules. It usually uses an agarose gel matrix with alternating current in that the direction of the electric field is changed (or pulsed) periodically for sep-aration. See Cantor, C.R., Smith, C.L., and Mathew, M.K., Pulsed-field gel electrophoresis of very large DNA molecules, *Ann. Rev. Biophys. Biophys. Chem.* 17, 287–304, 1988; Lat, E., Birren, B.W., Clark, S.M., Simon, M.I., and Hood, L., Pulsed-field gel electrophoresis, *Biotechniques* 7, 34–42, 1989; Olson, M.V., Separation of large DNA molecules by pulsed-field gel electrophoresis. A review of the basic phenomenology, *J. Chromatog.* 470, 377–383, 1989; Aires de Sousa, M. and de Lencastre, H., Bridges from hospitals to the laboratory: genetic portraits of methicillin-resistant *Staphyllococcus aureus* clones, *FEMS Immunol. Med. Microbiol.* 40, 101–111, 2004; Dukhin, A.S. and Dukhin, S.S., Aperiodic capillary electrophoresis method using an alternating current electric field for sep-aration of macromolecules, *Electrophoresis* 26, 2149–2153, 2005.

Pulse Radiolysis   A technique related to flash photolysis; pulse radiolysis uses very short (nanosecond) intense pulses of ionizing radiation to generate transient high concentrations of reactive species. See Salmon, G.A. and Sykes, A.G., Pulse radiolysis, *Methods Enzymol.* 227, 522–534, 1993; Maleknia, S.D., Kieselar, J.G., and Downard, K.M., Hydroxyl radical probe of the surface of lysozyme by synchrotron radiolysis and mass spectrometry, *Rapid Commun. Mass Spectrom.* 16, 53–61, 2002; Nakuna, B.N., Sun, G., and Anderson, V.E., Hydroxyl radical oxidation of cytochrome c by aerobic radiolysis, *Free Radic. Biol. Med.* 37, 1203–1213, 2004; Bataille, C., Baldacchino, G., Cosson, R.P. et al., Effect of pressure on pulse radiolysis reduction of proteins, *Biochim. Biophys. Acta* 1724, 432–439, 2005.

Quadrupole Mass Spectrometry   Mass spectrometry where only electric fields are used to separate ions on the basis of mass as they pass along the central axis of four parallel rods having an applied DC charge and alternative voltage applied (Herbert, C.G. and Johnstone, R.A.W., *Mass Spectrometry Basics*, CRC Press, Boca Raton, FL, Chapter 25, 2003). These instruments are generally referred to as quadrupole/time-of-flight mass spectrometers. See Horning, E.C., Carroll, D.I., Dzidic, I. et al., Development and use of analytical systems based on mass spectrometry, *Clin. Chem.* 23, 13–21, 1977; Yost, R.A. and

Boyd, R.K., Tandem mass spectrometry: quadrupole and hybrid instruments, *Methods Enzymol.* 193, 154–200, 1990; Jonscher, K.R. and Yates, J.R., III, The quadrupole ion trap mass spectrometry — a small solution to a big challenge, *Anal. Biochem.* 244, 1–15, 1997; Chernushevich, I.V., Loboda, A.V., and Thomson, B.A., An introduction to quadrupole-time-of-flight mass spectrometry, *J. Mass Spectrom.* 36, 849–865, 2001; Ens, W. and Standing, K.G., Hybrid quadrupole/time-of-flight mass spectrometers for analysis of biomolecules, *Methods Enzymol.* 402, 49–78, 2005; Payne, A.H. and Glish, G.L., Tandem mass spectrometry in quadrupole ion trap and ion cyclotron resonance mass spectrometers, *Methods Enzymol.* 402, 109–148, 2005.

Quantum Dots   Fluorescent semiconducting (usually CdSe surrounded by a passivation shell) nanocrystals used in the imaging of cells and subcellular particles. It is considered to have considerable advantage over other fluorescent imaging approaches. See Penner, R.M., Hybrid electrochemical/chemical synthesis of Q dots, *Acc. Chem. Res.* 33, 78–86, 2000; Lidke, D.S. and Arndt-Jovin, D.J., Imaging takes a quantum leap, *Physiology* 19, 322–325, 2004; Arya, H., Kaul, Z., Wadhwa, R. et al., Quantum dots in bio-imaging: revolution by the small, *Biochem. Biophys. Res. Commun.* 378, 1173–1177, 2005; Bentzen, E.L., Tomlinson, I.D., Mason, J. et al., Surface modification to reduce nonspecific binding of quantum dots in live cell assays, *Bioconjugate Chem.* 16, 1488–1494, 2005.

Quantum Yield   Efficiency of fluorescence; percentage of incident energy emitted after absorption. The higher the quantum yield, the greater the intensity of the fluorescence, luminescence, or phosphorescence. See Papp, S. and Vanderkooi, J.M., Tryptophan phosphorescence at room temperature as a tool to study protein structure and dynamics, *Photochem. Photobiol.* 49, 775–784, 1989; Plasek, J. and Sigler, K., Slow fluorescent indicators of membrane potential: a survey of different approaches to probe response analysis, *J. Photochem. Photobiol.* 33, 101–124, 1996; Vladimirov, Y.A., Free radicals in primary photobiological processes, *Membr. Cell Biol.* 12, 645–663, 1998; Maeda, M., New label enzymes for bioluminescent enzyme immunoassay, *J. Pharm. Biomed. Anal.* 30, 1725–1734, 2003; Imahori, H., Porphyrin-fullerene linked systems as artificial photosynthetic mimics, *Org. Biomol. Chem.* 2, 1425–1433, 2004; Katerinopoulos, H.E., The coumarin moiety as chromophore of fluorescent ion indicators in biological systems, *Curr. Pharm. Des.* 10, 3835–3852, 2004.

Quelling   A term used to describe the forceful suppression of a political uprising; to reduce to submission. In biology, quelling is suggested to uniquely describe posttranslational gene silencing in *Neurospora* and, by extension, to other fungi. Quelling has some characteristics similar to RNA interference (RNAi) and cosuppression (posttranslational gene silencing) in plants. Quelling involves the silencing of gene expression by segments of DNA in express of the normal number. See Morel, J.B. and Vaucheret, H., Post-transcriptional gene silencing mutants, *Plant Mol. Biol.* 43, 275–284, 2000; Fagard, M., Boutet, S., Morel, J.B. et al., AGO1, QDE-2, and RDE-1 are related proteins required for post-transcriptional gene silencing in plants, quelling in fungi, and RNA interference in animals, *Proc. Natl. Acad. Sci. USA* 97, 11650–11654, 2000; Shiu, P.K., Raju, N.B., Zickler, D., and Metzenberg, R.L., Meiotic silencing by unpaired DNA, *Cell* 107, 905–916, 2001; Pickford, A.S., Catalanotto, C., Cogoni, C., and

Macino, G., Quelling in *Neurosporo crassa*, *Adv. Genet.* 46, 277–303, 2002; Goldoni, M., Azzalin, G., Macino, G., and Cogoni, C., Efficient gene silencing by expression of double-stranded RNA in *Neurospora crassa*, *Fungal Genet. Biol.* 4, 1016–1024, 2004; Nakayashi, H., RNA silencing in fungi: mechanisms and applications, *FEBS Lett.* 579, 5950–5957, 2005.

Raman Spectroscopy (Raman Scattering)
A form of spectroscopy that uses inelastic light scattering, which provides information on molecular vibrations. It is similar to infrared spectroscopy but can be used for aqueous solutions. See Warshel, A., Interpretation of resonance Raman spectra of biological molecules, *Annu. Rev. Biophys. Bioeng.* 6, 273–300, 1977; Mathlouthi, M. and Koenig, J.L., Vibrational spectra of carbohydrates, *Adv. Carbohydr. Chem. Biochem.* 44, 7–89, 1986; Ghomi, M., Letellier, R., Liquier, J., and Taillandier, E., Interpretation of DNA vibrational spectra by normal coordinate analysis, *Int. J. Biochem.* 22, 691–699, 1990; Kitagawa, T., Investigation of higher order structures of proteins by ultraviolet resonance Raman spectroscopy, *Prog. Biophys. Mol. Biol.* 58, 1–18, 1992; Loehr, T.M. and Sanders-Loehr, J., Techniques for obtaining resonance Raman spectra of metalloproteins, *Methods Enzymol.* 226, 431–470, 1993; Barron, L.D., Hecht, L., Blanch, E.W., and Bell, A.F., Solution structure and dynamics of biomolecules from Raman optical activity, *Prog. Biophys. Mol. Biol.* 73, 1–49, 2000; Blanch, E.W., Hecht, L., and Barron, L.D., Vibrational Raman optical activity of proteins, nucleic acids, and viruses, *Methods* 29, 196–209, 2003; Spiro, T.G. and Wasbotten, I.H., CD as a vibrational probe of heme protein active sites, *J. Inorgan. Biochem.* 99, 34–44, 2005; Scheidt, W.R., Durbin, S.M., and Sage, J.T., Nuclear resonance vibrational spectroscopy — NRVS, *J. Inorgan. Biochem.* 99, 60–71, 2005; Aroca, R.F., Alvarez-Puebla, R.A., Pieczonka, N., Sanchez-Cortez, S., and Garcia-Ramos, J.V., Surface-enhanced Raman scattering on colloidal nanostructures, *Adv. Colloid Interface Sci.* 116, 45–61, 2005; Hammond, B.R. and Wooten, B.R., Resonance Raman spectroscopy measurement of carotenoids in the skin and retina, *J. Biomed. Opt.* 10, 054002, 2005; Owen, C.A., Selvakumaran, J., Notingher, I. et al., *In vitro* toxicology evaluation of pharmaceuticals using Raman micro-spectroscopy, *J. Cell. Biochem.*, 99, 178–186, 2006; Vandenabeele, P. and Moens, L., Introducing students to Raman spectroscopy, *Anal. BioAnal. Chem.* 385, 209–211, 2006.

Randomization
An unbiased process by which individual sample units (e.g., wells in microplate, experimental subjects) are assigned to experimental classes. An example is the assignment of subjects to two or more treatment groups. See Lachin, J.M., Statistical properties of randomization in clinical trials, *Control. Clin. Trials* 9, 289–311, 1988; Greenland, S., Randomization, statistics, and casual interference, *Epidemiology* 1, 421–429, 1990; Kernan, W.N., Viscoli, C.M., Makuch, R.W. et al., Stratified randomization for clinical trials, *J. Clin. Epidemiol.* 52, 19–26, 1999; Abel, U. and Koch, A., The role of randomization in clinical trials: myths and beliefs, *J. Clin. Epidemiol.* 52, 487–497, 1999. Mendelian randomization refers to the randomization of genes that are transferred from a parent to offspring at the time of gamete formation (Nitsch, D., Molokhia, M., Smeeth, L. et al., Limits to causal inference based on Mendelian randomization: a comparison with randomized controlled trials, *Am. J. Epidemiol.* 163, 397–403, 2006; Zoccali, C., Testa, A., Spoto, B. et al., Mendelian randomization: a new approach to studying epidemiology in ESRD, *Am. J. Kidney Dis.* 47, 332–341, 2006).

Real-Time PCR; Real-Time RT-PCR

Real-time PCR permits the assay of the rate of amplicon formation during replication in the PCR reaction. Conventional PCR amplicons are measured either by size analysis or by sequence analysis and while there is a relation of amplicon number to target number in the early phases, such a quantitative relationship is lost at high levels of amplification. The use of a fluorescent compound such as SYBR Green I to bind to double-stranded DNA increases fluorescence. Another approach uses FRET with a donor/acceptor pair. The use of fluorescence to measure the synthesis of amplicons permits the measurement of amplification in real time with the use of appropriate instrumentations. Real-time RT-PCR is an approach to quantitative use of the reverse transcriptase-polymerase chain reaction (RT-PCR) to measure messenger RNA and viral pathogen RNA. This is an adaptation of techniques that were based on the use of fluorescent tags to measure PCR amplicons in real time and has proved useful for the study of gene expression where real-time RT-PCR is used to "validate" other approaches to gene expression analysis such as the use of DNA microarrays. See Lie, Y.S. and Petropoulos, C.J., Advances in quantitative PCR technology: 5′ nuclease assays, *Curr. Opin. Biotechnol.* 9, 43–48, 1998; Edwards, K.J. and Saunders, K.A., Real-time PCR used to measure stress-induced changes in the expression of the genes of the aliginate pathway of *Pseudomonas aeruginoses, J. Appl. Microbiol.* 91, 29–37, 2001; Brechtbuehl, K., Whalley, S.H., Dusheiko, G.M., and Saunders, N.A., A rapid real-time quantitative polymerase chain reaction for hepatitis B virus, *J. Virol. Methods* 93, 105–113, 2001; Giulietti, A., Overbergh, L., Valckx, D. et al., An overview of real-time quantitative PCR: applications to quantify cytokine gene expression, *Methods* 25, 386–401, 2001; Klein, D., Quantification using real-time PCR technology: applications and limitations, *Trends Mol. Med.* 8, 257–260, 2002; Mackay, I.M., Arden, K.E., and Nitsche, A., Real-time PCR in virology, *Nucleic Acids Res.* 30, 1292–1305, 2002; Edwards, K., Logan, J., and Saunders, N., Eds., *Real Time PCR: An Essential Guide*, Horizon Biosciences, Wymandham, Norfolk, UK, 2004; Bustin, S.A., Benes, V., Nolan, T., and Pfaffi, M.W., Quantitative real-time RT-PCR — a perspective, *J. Mol. Endocrinol.* 34, 597–601, 2005; Bustin, S.A. and Mueller, R., Real-time reverse transcription PRC (qRT-PCR) and its potential use in clinical diagnosis, *Clin. Sci.* 109, 365–379, 2005; Delenda, C. and Gaillard, C., Real-time quantitative PCR for the design of lentiviral vector analytical assays, *Gene Ther.* 12 (Suppl. 1), S36–S50, 2005; Kubista, M., Andrade, J.M., Bengtsson, M. et al., The real-time polymerase chain reaction, *Mol. Aspects Med.* 27, 95–125, 2006; Kuypers, J., Wright, N., Ferrenberg, J. et al., Comparison of real-time PCR assays with fluorescent-antibody assays for diagnosis of respiratory virus infections in children, *J. Clin. Microbiol.* 44, 2382–2388, 2006; Diederen, B.M., de Jong, C.M., Kluytmans, J.A. et al., Detection and quantification of Legionella pneumonia DNA in serum: case reports and review of the literature, *J. Med. Microbiol.* 55, 639–642, 2006; Peano, C., Severgnini, M., Cifola I. et al., Transcriptome amplification methods in gene expression profiling, *Expert Rev. Mol. Diag.* 6, 465–480, 2006; Leong, W.F. and Chow, W.T.K., Transcriptomic and proteomic analyses of rhabdomyosarcoma cells reveal differential cellular gene expression in response to enterovirus 71 infection, *Cell. Microbiol.* 8, 565–580, 2006.

Receptor Activator of NF-κB (RANK)	A member of the neuroregulin/tumor necrosis factor superfamily. See Roundy, K., Smith, R., Weis, J.J., and Weis, J.H., Overexpression of RANKL (receptor-activator of NF-κB ligand) implicates IFN-beta-mediated elimination of B-cell precursors in the osteopetrotic bone of microphthalmic mice, *J. Bone Miner. Res.* 18, 278–288, 2003; Huang, W., Drissi, M.H., O'Keefe, R.J., and Schwarz, E.M., A rapid multiparameter approach to study factors that regulate osteoclastogenesis: demonstration of the combinatorial dominant effects of TNF-alpha and TGF-ss in RANKL-mediated osteoclastogenesis, *Calcif. Tissue Int.* 73, 584–593, 2003; Neumann, E., Gay, S., and Muller-Ladner, U., The RANK/RANKL/osteoprotegerin system in rheumatoid arthritis: new insights from animal models, *Arthritis Rheum.* 2, 3257–3268, 2005; Hamdy, N.A., Osteoprotegerin as a potential therapy for osteoporosis, *Curr. Osteoporos. Rep.* 3, 121–125, 2005; Wada, T., Nakashima, T., Hiroshi, N., and Penniger, J.M., RANKL-RANK signaling in osteoclastogenesis and bone disease, *Trends Mol. Med.* 12, 17–25, 2006.
Receptor Activity Modifying Proteins (RAMP)	Receptor activity modifying proteins (RAMPs) were identified as part of an effort to clone calcitonin gene-related peptides. There are three members of the family and they have been demonstrated to modulate the activity of G-protein-coupled receptors. See McLatchie, L.M., Fraser, M.J., Main, M.J. et al., RAMPs regulate the transport and ligand specificity of the calcitonin-receptor-like receptors, *Nature* 393, 333–339, 1998; Foord, S.M. and Marshall, F.H., RAMPs: accessory proteins for seven transmembrane domain receptors, *Trends Pharmacol. Sci.* 20, 184–187, 1999; Sexton, P.M., Abiston, A., Morfis, M. et al., Receptor activity modifying proteins, *Cell Signal.* 13, 73–82, 2001; Fischer, J.A., Muff, R., and Born, W., Functional relevance of G-protein-coupled-receptor-associated proteins, exemplified by receptor-activity-modifying proteins (RAMPs), *Biochem. Soc. Trans.* 30, 455–460, 2002; Morfis, M., Christopolous, A., and Sexton, P.M., RAMPs: 5 years on. Where to now? *Trends Pharmacol. Sci.* 34, 596–601, 2003; Hay, D.L., Conner, A.C., Howitt, S.G. et al., The pharmacology of CGRP-responsive receptors in cultured and transfected cells, *Peptides* 25, 2019–2026, 2004; Udawela, M., Hay, D.L., and Sexton, P.M., The receptor activity modifying protein family of G-protein-coupled receptor accessory proteins, *Sem. Cell Dev. Biol.* 15, 299–308, 2004; Young, A., Receptor pharmacology, *Adv. Pharmacol.* 52, 47–65, 2005.
Receptorome	That portion of the proteome that functions via ligand recognition. This category is subject to subdivision by receptor type as the GPCR receptorome. See Setola, V., Hufeisne, S.J., Grande-Allen, K.J. et al., 3,4-methylene-dioxymethamphetamine (MDMA, "Ecstasy") induces fenfluoramine-like proliferative actions on human cardiac valvular interstitial cells *in vitro*, *Mol. Pharmacol.* 62, 1223–1229, 2003; Armbruster, B.N. and Roth, B.L., Mining the receptorome, *J. Biol. Chem.* 280, 5129–5132, 2005; Roth, B.L., Receptor systems: will mining the receptorome yield novel targets for pharmacotherapy? *Pharmacol. Ther.* 108, 59–64, 2005.
Receptors for AGE (RAGE)	Cell-surface receptors for advanced glycation endproducts (AGE). These receptors are members of the immunoglobulin superfamily and are involved in the processes of inflammation and are suggested to be involved in the pathogenesis of diseases such as diabetes and neurogenerative diseases such as Alzheimer's disease. It is also noted that RAGE are also receptors for S100/calgranulin. See Bucciarelli, L.G., Wendt, T., Rong, L. et al., RAGE is a multiligand receptor of the immunoglobulin superfamily: implications

for homeostasis and chronic disease, *Cell. Mol. Life Sci.* 59, 1117–1128, 2002; Yan, S.F., Ramasamy, R., Naka, Y., and Schmidt, A.M., Glycation, inflammation, and RAGE. A scaffold for the macrovascular complications of diabetes and beyond, *Circ. Res.* 93, 1159–1169, 2003; Ramasamy, R., Vannucci, S.J., Yan, S.S. et al., Advanced glycation endproducts and RAGE: a common thread in aging, diabetes, neurodegeneration, and inflammation, *Glycobiology* 15, 16R–28R, 2005; Jensen, L.J., Ostergaard, J., and Flyvbjerg, A., AGE-RAGE and AGE crosslink interaction: important players in the pathogenesis of diabetic kidney disease, *Horm. Metab. Res.* 37 (Suppl. 1), 26–34, 2005; Bierhaus, A., Humpert, P.M., Stern, D.M., Arnold, B., and Nawroth, P.P., Advanced glycation endproduct receptor-mediated cellular dysfunction, *Ann. N.Y. Acad. Sci.* 1043, 676–680, 2005; Ding, Q. and Keller, J.N., Evaluation of rage isoforms, ligands, and signaling in the brain, *Biochim. Biophys. Acta* 1746, 18–27, 2005.

Receptor Tyrosine Kinase   A relatively simple transmembrane protein with an extracellular ligand-binding domain and an intracellular protein kinase domain. Receptor tyrosine kinases are coupled with receptors such as epidermal growth factor receptor, insulin receptor, etc. See Gourley, D.R., Isolation and characterization of membrane drug receptors, *Prog. Drug Res.* 20, 323–346, 1976; Adamson, E.D. and Rees, A.R., Epidermal growth factor receptors, *Mol. Cell. Biochem.* 34, 129–152, 1981; Carpenter, G., The biochemistry and physiology of the receptor-tyrosine kinase for epidermal growth factor, *Mol. Cell. Endocrinol.* 31, 1–19, 1983; Alaoui-Jamali, M.A., Paterson, J., Al Moustafa, A.E., and Yen, L., The role of Erb-2 tyrosine kinase receptor in cellular intrinsic chemoresistance: mechanisms and implications, *Biochem. Cell Biol.* 75, 315–325, 1997; Smit, L. and Borst, J., The Cb1 family of signal transduction molecules, *Crit. Rev. Oncog.* 8, 359–379, 1997; Elchebly, M., Cheng, A., and Tremblay, M.L., Modulation of insulin signaling by protein tyrosine phosphatases, *J. Mol. Med.* 78, 473–482, 2000; Carraway, K.L, Ramsauer, V.P., Haq, B., and Carrothers Carraway, C.A., Cell signaling through membrane mucins, *Bioessays* 25, 66–71, 2003; Murai, K.K. and Pasquale, E.B., Eph receptors, ephrins, and synaptic function, *Neuroscientist* 10, 304–314, 2004; Monteiro, H.P., Rocha Oliveira, C.J., Curcio, M.F., Morales, M.S., and Arai, R.J., Tyrosine phosphorylation in nitric oxide-mediated signaling events, *Methods Enzymol.* 396, 350–358, 2005; Heroult, M., Schaffner, F., and Augustin, H.G., Eph receptor and ephrin ligand-mediated interactions during angiogenesis and tumor progression, *Exp. Cell Res.* 312, 642–650, 2006; Perona, R., Cell signaling: growth factors and tyrosine kinase receptors, *Clin. Transl. Oncol.* 8, 77–82, 2006; Li, E. and Hristova, K., Role of receptor tyrosine kinase transmembrane domains in cell signaling and human pathologies, *Biochemistry* 45, 6242–6251, 2006.

Receptosome   An intracellular organelle resulting from receptor-mediated endocytosis of a ligand. See Willingham, M.C. and Pastan, I., The receptosome: an intermediate organelle of receptor-mediated endocytosis in cultured fibroblasts, *Cell* 21, 67–77, 1980; Pastan, I.L. and Willingham, M.C., Journal of the center of the cell: role of the receptosome, *Science* 214, 504–509, 1981; Chitambar, C.R. and Zivkovic-Gilgenbach, Z., Role of the acidic receptosome in the uptake and retention of 67Ga by human leukemic HL60 cells, *Cancer Res.* 50, 1484–1487, 1990. While this specific term has not seen extensive use, there is interest in receptor-mediated endocytosis coupled with "specific" vesicular transport (Sano, H., Higashi, T.,

Matsumoto, K. et al., Insulin enhances macrophage scavenger receptor-mediated endocytotic uptake of advanced glycation endproducts, *J. Biol. Chem.* 273, 8630–8637, 1998). There is some interest in receptor-mediated endocytosis for drug delivery (Selbo, P.K., Hogset, A., Prasmickaite, L., and Berg, K., Photochemical internalization: a novel drug delivery system, *Tumour Biol.* 23, 102–112, 2002).

Refractive Index (Index of Refraction)   Ratio of wavelength or phase velocity of an electromagnetic wave in a vacuum to that in a substance. Changes in the refractive index of solutions have been used to measure solute concentration in techniques such as analytical ultracentrifugation and chromatography. Techniques based on refractive index have been used to study cells. More recently, refractive index has provided the basis for measurement of macromolecules on surfaces. See Hawkes, J.B. and Astheimer, R.W., Thermal coefficient of the refractive index of water, *Science* 110, 717, 1949; Barer, R. and Tkaczyk, S., Refractive index of concentrated protein solutions, *Nature* 173, 821–822, 1954; Barer, R. and Dick, D.A., Interferometry and refractometry of cells in tissue culture, *Exp. Cell Res.* 13 (Suppl. 4), 103–135, 1957; Fishman, H.A., Greenwald, D.R., and Zare, Z.N., Biosensors in chemical separations, *Annu. Rev. Biophys. Biomol. Struct.* 27, 165–198, 1998; Van Regenmortel, M.H., Altschuh, D., Chatellier, J. et al., Measurement of antigen–antibody interactions with biosensors, *J. Mol. Recognit.* 11, 163–167, 1998; Eremeeva, T., Size-exclusion chromatography of enzymatically treated cellulose and related polysaccharides, *J. Biochem. Biophys. Methods* 56, 253–264, 2003; Mogridge, J., Using light-scattering to determine the stoichiometry of protein complexes, *Methods Mol. Biol.* 261, 113–118, 2004; Hut, T.S., Biophysical methods for monitoring cell-substrate interactions in drug discovery, *Assay Drug Dev. Technol.* 1, 479–488, 2003; Haes, A.J. and Van Duyne, R.P., A unified view of propagating and localized surface plasmon resonance, *Anal. Bioanal. Chem.* 379, 920–930, 2004; Stuart, D.A., Haes, A.J., Yonzon, C.R., Hicks, E.M., and Van Duyne, R.P., Biological applications of localized surface plasmonic phenomenae, *IEE Proc. Nanobiotechnol.* 152, 13–32, 2005; Yuk, J.S., Hong, D.G., Jung, J.W. et al., Sensitivity enhancement of spectral surface plasmon resonance biosensors for the analysis of protein arrays, *Eur. Biophys. J.*, 35, 469–476, 2006; Ogusu, K., Suzuki, K., and Nishio, H., Simple and accurate measurement of the absorption coefficient of an absorbing plate by use of the Brewster angle, *Opt. Lett.* 31, 909–911, 2006; Cardenas-Valencia, A.M, Dlutowski, J., Fries, D., and Langdebrake, L., Spectrometric determination of the refractive index of optical wave guiding materials used in lab-on-a-chip applications, *Appl. Spectrosc.* 60, 322–329, 2006; Coelho Neto, J., Agero, U., Gazzinelli, R.T., and Mesquita, O.N., Measuring optical and mechanical properties of a living cell with defocusing microscopy, *Biophys. J.*, 91, 1108–1115, 2006; Friebel, M. and Meinke, M., Model function to calculate the refractive index of native hemoglobin in the wavelength range of 250–1100 nm, *Appl. Opt.* 45, 2838–2842, 2006.

Regulators of G-Protein Signaling (RGS)   Regulators of G-protein signaling are a family of proteins that bind to the activated α-subunit of the heterotrimer G-protein complex, where GDP has been replaced by GTP, and block signal transmission. There is increasing evidence for a broader role for RGS proteins in cell function. See Dohlman, H.G. and Thorner, J., RGS proteins and signaling by heterotrimeric

G proteins, *J. Biol. Chem.* 272, 3871–3874, 1997; Berman, D.M. and Gilman, A.G., Mammalial RGS proteins: barbarians at the gate, *J. Biol. Chem.* 273, 1269–1272, 1998; Hepler, J.R., Emerging roles for RGS proteins in cell signaling, *Trends Pharmacol. Sci.* 20, 376–382, 1999; Hepler, J.R., Emerging roles for RGS proteins in cell signaling, *Trends Pharmacol. Sci.* 20, 376–382, 1999; Burchett, S.A., Regulators of G-protein signaling: a bestiary of modular protein binding domains, *J. Neurochem.* 75, 1335–1351, 2000; Hollinger, S. and Hepler, J.R., Cellular regulation of RGS proteins: modulators and intergrators of G-protein signaling, *Pharmacol. Rev.* 54, 527–559, 2002; Kehrl, J.H., G-protein-coupled receptor signaling, RGS proteins, and lymphocyte function, *Crit. Rev. Immunol.* 24, 409–423, 2004; Wilkie, T.M. and Kinch, L., New roles for G and RGS proteins: communication continues despite pulling sisters apart, *Curr. Biol.* 15, R843–R854, 2005.

Regulatory Transcription Factors
A *trans*-acting component, usually a protein or protein complex, which interacts with a *cis*-regulatory region on a gene distant from the transcription initiation site to enhance or suppress the rate of transcription. A regulatory transcription factor is not considered a part of the basal transcription apparatus. See Fujita, T., Kimura, Y., Miyamoto, M. et al., Induction of endogenous IFN-alpha and IFN-beta genes by a regulatory transcription factor, IRF-1, *Nature* 337, 270–272, 1989; Wingender, E., *Gene Regulation in Eukaryotes*, VCH, Weinheim, Germany, 1993; Gopakrishnan, R.V., Dolle, P., Mattei, M.G. et al., Genomic structure and developmental expression of the mouse cell cycle regulatory transcription factor DP1, *Oncogene* 13, 2671–2680, 1996; Bachmaier, K., Neu, N., Pummerer, C. et al., iNOS expression and nitrotyrosine formation in the myocardium in response to inflammation is controlled by the interferon regulatory transcription factor 1, *Circulation* 96, 585–591, 1997; Larochelle, O., Stewart, G., Moffatt, P. et al., Characterization of the mouse metal-regulatory-element-binding proteins, metal element protein-1 and metal regulatory transcription factor-1, *Biochem. J.* 353, 591–601, 2001; Courey, A.J., Regulatory transcription factors and regulatory regions, in *Transcription Factors,* Locker, J., Ed., Academic Press, San Diego, CA, 2001, pp. 17–34; Willmore, W.G., Control of transcription in eukaryotic cells, in *Functional Metabolism: Regulation and Adaptation*, Storey, K.B., Ed., Wiley-Liss, Hoboken, NJ, 2004, pp. 153–187.

Resolution (Chromatographic)
The chromatographic separation between components in a mixture of components. A number of equations can be developed for the expression of resolution, such as $R = 2(t_2 - t_1)/W_1 + W_2$, where t_1 is the elution time of component 1 having peak width of W_1, and t_2 is the elution time of component 2 having peak width of W_2. See Hearn, M.T., General strategies in the separation of proteins by high-performance liquid chromatographic methods, *J. Chromatog.* 418, 3–26, 1987; Feibush, B. and Santasania, C.T., Hydrophilic shielding of hydrophobic, cation- and anion-exchange phases for separation of small analytes: direct injection of biological fluids onto high-performance liquid chromatographic columns, *J. Chromatog.* 544, 41–49, 1991; Hagan, R.L., High-performance liquid chromatography for small-scale studies of drug stability, *Am. J. Hosp. Pharm.* 51, 2162–2175, 1994; Coffman, J.L., Roper, D.K., and Lightfoot, E.N., High-resolution chromatography of proteins in short columns and adsorptive membranes, *Bioseparation* 4, 183–200, 1994; Myher, J.J., and Kuksis, A.,

General strategies in chromatographic analysis of lipids, *J. Chromatog. B* 671, 3–33, 1995; Chen, H. and Horvath, C., High-speed high-performance liquid chromatography of peptides and proteins, *J. Chromatog. A* 705, 3–20, 1995; Bojarski, J. and Aboul-Enein, H.Y., Recent applications of chromatographic resolution of enantiomers in pharmaceutical analysis, *Biomed. Chromatog.* 10, 297–302, 1996; Dolan, J.W., Temperature selectivity in reversed-phase high-performance liquid chromatography, *J. Chromatog. A* 965, 195–205, 2002; Jupille, T.H., Dolan, J.W., Snyder, L.R., and Molnar, I., Two-dimensional optimization using different pairs of variables for the reversed-phase high-performance liquid chromatographic separation of a mixture of acidic compounds, *J. Chromatog. A* 948, 35–41, 2002; Pellett, J., Lukulay, P., Mao, Y. et al., "Orthogonal" separations for reversed-phase liquid chromatography, *J. Chromatog. A* 1101, 122–135, 2006; Nageswara Rao, R., Narasa Raju, A., and Nagaraju, D., Development and validation of a liquid chromatographic method for determination of enantiomeric purity of citalopram in bulk drugs and pharmaceuticals, *J. Pharm. Biomed. Anal.* 41, 280–285, 2006.

Resurrection Plants   Usually found in arid regions, these are plants that adopt a compact shape during water deprivation and change shape upon rehydration. See Kranner, I., Beckett, R.P., Wornik, S., Zorn, M., and Preifhofer, H.W., Revival of a resurrection plant correlates with its antioxidant status, *Plant J.* 31, 13–24, 2002; Schluepmann, H., Pellny, T., van Dijken, A., Smeeken, S., and Paul, M., Trehalose 6-phosphate is indispensable for carbohydrate utilization and growth in *Arabidopsis thaliana*, *Proc. Natl. Acad. Sci. USA* 100, 6849–6854, 2003; Jones, L. and McQueen-Mason, S., A role for expansins in dehydration and rehydration of the resurrection plant, *Craterostigma plantagineum*, *FEBS Lett.* 559, 61–65, 2004; Helseth, L.E. and Fischer, T.M., Physical mechanisms of rehydration in *Polypodium polypodioides*, a resurrection plant, *Phys. Rev. E. Stat. Nonlin. Soft Matter. Phys.* 71 (6 Pt. 1), 061903, epub, 2005.

Retention Time   For chromatography, the retention time (t_r) is the time from injection of solute to the apex (zenith) of the peak of the respective solute. For planer chromatography such as thin-layer chromatography or paper chromatography, the retardation factor (R_f) is the ratio of the distance traveled by the solvent (solvent front) and the distance traveled by the solute. See Palmblad, M., Ramstrom, M., Bailey, C.G. et al., Protein identification by liquid chromatography-mass spectrometry using retention time prediction, *J. Chromatog. B Analyt. Technol. Biomed. Life Sci.* 803, 131–135, 2004; Joutovsky, A., Hadzi-Nesic, J., and Nardi, M.A., HPLC retention time as a diagnostic tool for hemoglobin variants and hemoglobinopathies: a study of 60,000 samples in a clinical diagnostic laboratory, *Clin. Chem.* 50, 1736–1747, 2004; Pierce, K.M., Wood, L.F., Wright, B.W., and Synovec, R.E., A comprehensive two-dimensional retention time alignment algorithm to enhance chemometric analysis of comprehensive two-dimensional separation data, *Anal. Chem.* 77, 7735–7743, 2005. The term "retention time" is also used to describe the period of time that a material resides in the digestive tract (Bernard, L. and Doreau, M., Use of rare earth elements as external markers for mean retention time measurements in ruminants, *Reprod. Nutr. Dev.* 40, 89–101, 2000; Pearson, R.A., Archibald, R.F., and Muirhead, R.H., A comparison of the effect of forage type and level of feeding on the digestibility and gastrointestinal mean retention time of dry forages given to cattle, sheep, ponies, and donkeys, *Br. J. Nutr.* 95, 88–98, 2006); retention in filtration systems (Lee, Y.W., Chung, J., Jeong, Y.D. et al., Backwash-based methodology for the estimation of solids retention time in biological aerated filter, *Environ. Technol.* 27, 777–787, 2006), and retention time of solid waste in a bioreactor (Maase, A., Sperandio, M., and Cabassud, C., Comparison of sludge characteristics and performance of a submerged membrane bioreactor and an activated sludge process at high solids retention time, *Water Res.* 40, 2405–2415, 2006).

Retention Volume   For chromatography, the retention volume is a function of the flow rate of the mobile phase and the retention time (Frigon, R.P., Leypoldt, J.K., Uyeji, S., and Henderson, L.W., Disparity between Stokes radii of dextrans and proteins as determined by retention volume in gel permeation chromatography, *Anal. Chem.* 55, 1349–1354, 1983; Dyr, J.E. and Suttnar, J., On the increased retention volume of human hemoglobin in high-performance gel filtration, *J. Chromatog.* 408, 303–307, 1987; Griotti, F. and Guiochon, G., Influence of the pressure on the properties of chromatographic columns. III. Retention time of thiourea, hold-up volume, and compressibility of the C_{18}-bonded layer, *J. Chromatog. A* 1075, 117–126, 2005). The term is also used to refer to urine retention (Dutkiewicz, S., Witeska, A., and Stepien, K., Relation between prostate-specific antigen, prostate volume, retention volume, and age in benign prostatic hypertrophy [BPH], *Int. Urol. Nephrol.* 27, 762–768, 1995; Demaria, F., Amar, N., Blau, D. et al., Prospective 3-D ultrasonographic evaluation of immediate postpartum urine retention volume in 100 women who delivered vaginally, *Int. Urogynecol. J.Pelvic Floor Dysfunct.* 15, 281–285, 2004).

Retromer   A multiprotein complex thought to function in endosome-Golgi retrieval. See Pfeffer, S.R., Membrane transport: retromer to the rescue, *Curr. Biol.* 11, R109–R111, 2001; Seaman, M.N.J., Recycle your receptors with retromer, *Trends Cell Biol.* 15, 68–75, 2005; Griffin, C.T., Trejo, J., and Magnuson, T., Genetic evidence for a mammalian retromer complex containing sorting nexins 1 and 2, *Proc. Natl. Acad. Sci. USA* 102, 15173–15177, 2005; Gullapalli, A., Wolfe, B.L., Griffin, C.T., Magnuson, T., and Trejo, J., An essential role of SNX1 in lysosomal sorting of protease-activated receptor-1: evidence for retromer-, Hrs-, and Tsg101-independent functions of sorting nexins, *Mol. Biol. Cell*, 17, 1228–1238, 2006.

Retro-Translocation (Retrotranslocation)   A process by which misfolded proteins or other incorrect translation products are transported from the lumen of the endoplasmic reticulum to the cytoplasm for subsequent degradation by the proteosome. See Johnson, A.E. and Haigh, N.G., The ER translocon and retrotranslocation: is the shift into reverse manual or automatic? *Cell* 102, 709–712, 2000; Svedine, S., Wang, T., Halaban, R., and Herbert, D.N., Carbohydrates act as sorting determinants in ER-associated degradation of tyrosinase, *J. Cell Sci.* 117, 2937–2949, 2004; Schulze, A., Sandera, S., Buerger, E. et al., The ubiquitin- domain protein HERP forms a complex with components of the endoplasmic reticulum–associated degradation pathway, *J. Mol. Biol.* 354, 1021–1027, 2005.

Reverse Immunology   Prediction of antigen structure based on peptide reactivity with cytotoxic T-cell MHC proteins; most frequently used in the study of tumor antigens. See Boon, T. and van der Bruggen, P., Human tumor antigens recognized by T lymphocytes, *J. Exptl. Med.* 183, 725–729, 1996; Maecker, B., von Bergwelt-Baildon, M.S., Anderson, K.S. et al., Linking genomics to immunotherapy by reverse immunology — "immunomics" in the new millennium, *Curr. Mol. Med.* 1, 609–619, 2001; Anderson, K.S. and LaBaer, J., The sentinel within: exploiting the immune system for cancer biomarkers, *J. Proteome Res.* 4, 1123–1133, 2005. See *SEREX*.

Reverse Micelle   A reverse micelle or inverted micelle is a stable assembly of a surfactant around an aqueous core where the lipophilic part of the surfactant is directed toward the exterior, which is a nonpolar solvent and the charged portion is directed toward the aqueous core. Reverse micelles have been used for the stabilization of proteins in organic solvents, for protein purification, and for drug delivery. See Bernert, J.T., Jr. and Sprecher, H., Solubilization and partial purification of an enzyme involved in rat liver microsomal fatty acid chain elongation: beta-hydroxyacyl-CoA dehydrase, *J. Biol. Chem.* 254, 11584–11590, 1979; Grandi, C., Smith, R.E., and Luisi, P.L., Micellar solubilization of biopolymers in organic solvents. Activity and conformation of lysozyme in isooctane reverse micelles, *J. Biol. Chem.* 256, 837–843, 1981; Leser, M.E., Wei, G., Luisi, P., and Maestro, M., Application of reverse micelles for the extraction of proteins, *Biochem. Biophys. Res. Commun.* 135, 629–635, 1986; Luisi, P.L., and Magid, L.J., Solubilization of enzymes and nucleic acids in hydrocarbon micellar solutions, *CRC Crit. Rev. Biochem.* 20, 409–474, 1986; Huruguen, J.P. and Pileni, M.P., Drastic change of reverse micellar structure by protein or enzyme addition, *Eur. Biophys. J.* 19, 103–107, 1991; Bru, R., Sanchez-Ferrer, A., and Garcia-Caroma, F., Kinetic models in reverse micelles, *Biochem. J.* 310, 721–739, 1995; Nicot, C. and Waks, M., Proteins as invited guests of reverse micelles: conformational effects, significance, applications, *Biotechnol. Genet. Eng. Rev.* 13, 267–314, 1996; Tuena de Gomez-Puyou, M. and Gomez-Puyou, A., Enzymes in low-water systems, *CRC Rev. Biochem. Mol. Biol.* 33, 53–89, 1998; Orlich, B. and Schomacker, R., Enzyme catalysis in reverse micelles, *Adv. Biochem. Eng. Biotechnol.* 75, 185–208, 2002; Krishna, S.H., Srinivas, N.D., Ragnavarao, K.S., and Karanth, N.G., Reverse micellar extraction for downstream processing of proteins/enzymes, *Adv. Biochem. Eng. Biotechnol.* 75, 119–183, 2002; Marhuenda-Egea, F.C. and Bonete, M.J., Extreme halophilic enzymes in organic solvents, *Curr. Opin. Biotechnol.* 13, 385–389, 2002; Muller-Goymann, C.C., Physicochemical characterization of colloidal drug delivery systems such as reverse micelles, vesicles, liquid crystals, and nanoparticles for topical administration, *Eur. J. Pharm. Biopharm.* 58, 343–356, 2004.

Reverse Proteomics   Proteomic analysis where genomic sequence information is used to predict the resulting proteome providing the basis for experiment design. See Lamesch, P., Milstein, S., Hao, T. et al., *C. elegans* ORFeome version 3.1: increasing the coverage of ORFeome resources with improved gene production, *Genome Res.* 14, 2064–2069, 2004; Rual, J.F., Hirozane-Kishikawa, T., Hao, T. et al., Human ORFeom version 1.1: a platform for reverse proteomics, *Genome Res.* 14, 2128–2135, 2004; Gillette, W.K., Esposito, D., Frank, P.H. et al., Pooled ORF expression technology (POET), *Mol. Cell. Proteom.* 4, 1647–1652, 2005.

Reverse Transcriptase   An enzyme that catalyzes the formation of DNA from an RNA template. This is an enzyme critical for the replication of RNA viruses such as HIV and is a major drug target for AIDS and other RNA viral diseases. See O'Conner, T.E., Reverse transcriptase-progress, problems, and prospects, *Bibl. Haematol.* 39, 1165–1181, 1973; Wu, A.M. and Gallo, R.C., Reverse transcriptase, *CRC Crit. Rev. Biochem.* 3, 289–347, 1975; Verma, I.M., The reverse transcriptase, *Biochim. Biophys. Acta* 473, 1–38, 1977; Chandra, P., Immunological characterization of reverse transcriptase from human tumor tissues, *Surv. Immunol. Res.* 2, 170–177, 1983; Lim, D. and Maas, W.K., Reverse transcriptase in bacteria, *Mol. Microbiol.* 3, 1141–1144, 1989; Barber, A.M., Hizi, A., Maizel, J.V., Jr., and Hughes, S.H., HIV-1 reverse transcriptase: structure predictions for the polymerase domain, *AIDS Res. Hum. Retroviruses* 6, 1061–1072, 1990; Durantel, D., Brunelle, M.N., Gros, E. et al., Resistance of human hepatitis B virus to reverse transcriptase inhibitors: from genotypic to phenotypic testing, *J. Clin. Vitrol.* 34 (Suppl. 1), S34–S43, 2005; Menendez-Arias, L., Matamoros, T., and Cases-Gonzalez, C.E., Insertions and deletions in HIV-1 reverse transcriptase: consequences for drug resistance and viral fitness, *Curr. Pharm. Des.* 12, 1811–1825, 2006; Srivastava, S., Sluis-Cremer, N., and Tachedjian, G., Dimerization of human immunodeficiency virus type 1 reverse transcriptase as an antiviral target, *Curr. Pharm. Des.* 12, 1879–1894, 2006.

Reverse Transcriptase– Polymerase Chain Reaction (RT-PCR)   A variation of the PCR technique in which cDNA is made from RNA via reverse transcription. The cDNA is then amplified using standard PCR protocols. See Mocharla, H., Mocharla, R., and Hodes, M.E., Coupled reverse transcription–polymerase chain reaction (RT-PCR) as a sensitive and rapid method for isozyme genotyping, *Gene* 93, 271–275, 1990; Weis, J.H., Tan, S.S., Martin, B.K., and Willwer, C.T., Detection of rare mRNAs via quantitative RT-PCR, *Trends Genet.* 8, 263–264, 1992; Akoury, D.A., Seo, J.J., James, C.D., and Zaki, S.R., RT-PCR detection of mRNA recovered from archival glass slide smears, *Mol. Pathol.* 6, 195–200, 1993; Silver, J., Maudru, T., Fujita, K., and Repaske, R., An RT-PCR assay for the enzyme activity of reverse transcriptase capable of detecting single virions, *Nucleic Acids Res.* 21, 3593–3594, 1993; Taniguchi, A., Kohsaka, H., and Carson, D.A., Competitive RT-PCR ELISA: a rapid, sensitive, and nonradioactive method to quantitate cytokine mRNA, *J. Immunol. Methods* 169, 101–109, 1994; Prediger, E.A., Quantitating mRNAs with relative and competitive RT-PCR, *Methods Mol. Biol.* 160, 49–63, 2001; Lion, T., Current recommendations for positive controls in RT-PCR assays, *Leukemia* 15, 1033–1037, 2001; Joyce, C., Quantitative RT-PCR. A review of current methodologies, *Methods Mol. Biol.* 193, 83–92, 2002; Ransick, A., Detection of mRNA by *in situ* hybridization and RT-PCR, *Methods Mol. Biol.* 74, 601–620, 2004; Tallini, G. and Brandao, G., Assessment of RET/PTC oncogene activation in thyroid nodules utilizing laser microdissection followed by nested RT-PCR, *Methods Mol. Biol.* 293, 103–111, 2005; Ooi, C.P., Rohani, A., Zamree, I., and Lee, H.L., Temperature-related storage evaluation of an RT-PCR test kit for the detection of dengue infection in mosquitoes, *Trop. Biomed.* 22, 73–76, 2005.

Rho Factor   A ring-shaped homohexameric bacterial protein encoded by the *rho* gene, which regulates RNA polymerase. See Lathe, R., RNA polymerase of *Escherichia coli*, *Curr. Top. Microbiol. Immunol.* 83, 37–91, 1978; Adhya, S. and Gottesman, M., Control of transcription termination, *Annu. Rev. Biochem.* 47, 967–996, 1978; Aktories, K., Schmidt, G., and Just, I., Rho GTPases as targets of bacterial protein toxins, *Biol. Chem.* 381, 421–426, 2000; Anston, A.A., Single-stranded-RNA binding proteins, *Curr. Opin. Struct. Biol.* 10, 87–94, 2000; Richardson, J.P., Rho-dependent termination and ATPases in transcript termination, *Biochim. Biophys. Acta* 1577, 251–260, 2002; Banerjee, S., Chalissery, J., Bandey, I., and Sen, R.,

Rho-dependent transcription termination: more questions than answers, *J. Microbiol.* 44, 11–22, 2006.

Rhomboid   A family of transmembrane proteins with proteolytic activity and considered to be a regulatory of EGF signaling. Rhomboid was originally described in *Drosophila* as protease-cleaving Spitz, a membrane-bound EGF. See Noll, R., Sturtevant, M.A., Gollapudi, R.R., and Bier, E., New functions of the *Drosophila* rhomboid gene during embryonic and adult development are revealed by a novel genetic method, enhancer piracy, *Development* 120, 2329–2338, 1994; Lage, P., Yan, Y.N., and Jarman, A.P., Requirement for EGF receptor signaling in neural recruitment during formation of *Drosophila* chordotonal sense organ clusters, *Curr. Biol.* 7, 166–175, 1997; Sturtevant, M.A., Roark, M., and Bier, E., The *Drosophila* rhomboid gene mediates the localized formation of wing veins and interacts genetically with components of the EGF-R signaling pathway, *Genes Dev.* 7, 961–973, 1993; Klambt, C., EGF receptor signaling: the importance of presentation, *Curr. Biol.* 10, R388–R391, 2000; Guichard, A., Roark, M., Ronshaugen, M., and Bier, E., Brother of rhomboid, a rhomboid-related gene expressed during early *Drosophila* oogenesis, promotes EGF-R/MAPK signaling, *Dev. Biol.* 226, 255–266, 2000; Urban, S., Lee, J.R., and Freeman, M., *Drosophila* rhomboid-1 defines a family of putative intramembrane serine proteases, *Cell* 107, 173–182, 2001; Urban, S., Lee, J.R., and Freeman, M., A family of rhomboid intramembrane proteases activates all *Drosophila* membrane-tethered EGF ligands, *EMBO J.* 21, 4277–4286, 2002; Jaszai, J. and Brand, M., Cloning and expression of Ventrhoid, a novel vertebrate homologue of the *Drosophila* EGF pathway gene rhomboid, *Mech. Dev.* 113, 73–77, 2002; Zhou, X.W., Blackman, M.J., Howell, S.A., and Carruthers, V.B., Proteomic analysis of cleavage events reveals a dynamic two-step mechanism for proteolysis of a key parasite adhesive complex, *Mol. Cell. Proteomics* 3, 565–576, 2004; Sik, A., Passer, B.J., Koonin, E.V., and Pellegrini, L., Self-regulated cleavage of the mitochondrial intramembrane-cleaving protease PARL yields Pbeta, a nuclear-targeted peptide, *J. Biol. Chem.* 279, 15323–15329, 2004; Kanaoko, M.M., Urban, S., Freeman, M., and Okada, K., An *Arabidopsis* rhomboid homolog is an intermediate protease in plants, *FEBS Lett.* 579, 5723–5728, 2005; Howell, S.A., Hackett, F., Johgco, A.M. et al., Distinct mechanisms govern proteolytic shedding of a key invasion protein in apicomplexan pathogens, *Mol. Microbiol.* 57, 1342–1356, 2005; Uban, S. and Wolfe, M.S., Reconstitution of intramembrane proteoysis *in vitro* reveals that pure rhomboid is sufficient for catalysis and specificity, *Proc. Natl. Acad. Sci. USA* 102, 1883–1888, 2005; Nakagawa, T., Guichard, A., Castro, C.P. et al., Characterization of a human rhomboid homolog, p100hRho/RHBDF1, which interacts with TGF-alpha family ligands, *Dev. Dyn.* 233, 1315–1331, 2005. "Rhomboid" also describes a geometric shape such as a parallelogram or rhombus and, as such, has been used to describe intracellular crystal formation (Machhi, J., Kouzova, M., Komorowski, D.J. et al., Crystals of alveolar soft part sarcoma in a fine needle aspiration biopsy cytology smear. A case report, *Acta Cytol.* 46, 904–908, 2002; Duan, X., Bruneval, P., Hammadeh, R. et al., Metastatic juxtaglomerular cell tumor in a 52-year-old man, *Am. J. Surg. Pathol.* 28, 1098–1102, 2004; Stewart, C.J. and Spagnolo, D.V., Crystalline plasma cell inclusions in Helicobacter-associated gastritis, *J. Clin. Pathol.*, 59, 851–854, 2006).

Rhomboid is also an anatomical term (Dong, H.W. and Swanson, L.W., Organization of axonal projections from the anterolateral area of the bed nuclei of the stria terminalis, *J. Comp. Neurol.* 468, 277–298, 2004).

Riboswitch A discrete RNA sequence in the leader sequences (UTR regions) of certain mRNAs, which encode enzymes involved in metabolism. Earlier described as the *RFN* element (Gefland, M.A., Mironov, A.A., Jomantas, J. et al., A conserved RNA structure element involved in the regulation of bacterial riboflavin synthesis genes, *Trends Genet.* 15, 439–442, 1999). Riboswitches are conceptually similar to aptamers where there is specific binding of a ligand. Binding of a ligand to a specific riboswitch influences the expression of the cognate gene at both the transcriptional and translational level. See Winkler, W., Nahvi, A., and Breaker, R.R., Thiamine derivatives bind messenger RNAs directly to regulate bacterial gene expression, *Nature* 419, 952–956, 2002; Winkler, W.C., Cohen-Chalamish, S., and Breaker, R.R., An mRNA structure that controls gene expression by binding FMN, *Proc. Natl. Acad. Sci. USA* 99, 15908–15913, 2002; Sudarsan, N., Wickiser, J.K., Nakamura, S. et al., An mRNA structure in bacteria that controls gene expression by binding lysine, *Gene Dev.* 17, 2688–2697, 2003; Nudler, E. and Mironov, A.S., The riboswitch control of bacterial metabolism, *Trends Biochem. Sci.* 29, 11–17, 2004; Batey, R.T., Gilbert, S.D., and Montange, R.K., Structure of a natural guanine-responsive riboswitch complexed with the metabolite hypoxanthine, *Nature* 432, 411–415, 2004; Winkler, W.C., Riboswitches and the role of noncoding RNAs in bacterial metabolic control, *Curr. Opin. Chem. Biol.* 9, 594–602, 2005; Montange, R.K. and Batey, R.T., Structure of the *S*-adenosylmethionine riboswitch regulatory mRNA element, *Nature* 441, 1172–1175, 2006.

Ring-Finger Proteins/ Ring-Finger Domains Ring-finger, which is related to zinc finger, describes a family of proteins defined as a zinc-binding ring-finger motif. This motif was first described in RING1 protein but occurs in a wide variety of proteins including proteins involved in ubiquitinylation and c-Cbl oncoprotein. See Lovering, R., Hanson, I.M., Borden, K.L. et al.., Identification and preliminary characterization of a protein motif related to the zinc finger, *Proc. Natl. Acad. Sci. USA* 90, 2112–2116, 1993; Fremont, P.S., The RING finger. A novel protein sequence motif related to the zinc finger, *Ann. N.Y. Acad. Sci.* 684, 174–192, 1993; Hu, H.M., O'Rourke, K., Boguski, M.S., and Dixit, V.M., A novel RING finger protein interacts with the cytoplasmic domain of CD40, *J. Biol. Chem.* 269, 30069–30072, 1994; Borden, K.L. and Freemont, P.S., The RING finger domain: a recent example of a sequence-structure family, *Curr. Opin. Struct. Biol.* 6, 395–401, 1996; Smit, L. and Borst, J., The Cbl family of signal transduction molecules, *Crit. Rev. Oncol.* 8, 359–379, 1997; Jackson, P.K., Eldridge, A.G., Freed, E. et al., The lore of the RINGs: substrate recognition and catalysis by ubiquitin ligases, *Trends Cell Biol.* 10, 429–439, 2000; Gregorio, C.C., Perry, C.N., and McElhinny, A.S., Functional properties of the titin/connectin-associated proteins, the muscle-specific RING finger proteins (MURFs), in striated muscle, *J. Muscle Res. Cell Motil.* 14, 1–12, 2006.

RNA-Induced Silencing Complex (RISC) The functional complex formed from the interaction of interfering RNA (RNA interference, RNAi) from small interfering RNAs (siRNA) or microRNAs (miRNAs) with mRNA-protein (Argonaut protein) to form a complex that results in posttranscriptional gene silencing because of mRNA cleavage from the siRNA/miRNA. See Sontheimer, E.J., Assembly and function of RNA silencing complexes, *Nat. Rev. Mol. Cell Biol.* 6, 127–138, 2005; Tang, G., siRNA and miRNA: an insight into RISCs, *Trends Biochem. Sci.* 30, 106–114, 2005; Filipowicz, W., RNAi: the nuts and bolts of the RISC machine, *Cell* 122, 17–20, 2005; Hutvagner, G., Small RNA asymmetry in RNAi: function in RISC assemble and gene regulation, *FEBS Lett.* 579, 5850–5857, 2005; Hammond, S.M., Dicing and slicing: the core machinery of the RNA interference pathway, *FEBS Lett.* 579, 5822–5829, 2005; Gilmore, I.R., Fox, S.P., Hollins, A.J., and Akhtar, S., Delivery strategies for siRNA-mediated gene silencing, *Curr. Drug Deliv.* 3, 147–155, 2006.

RNA Interference (RNAi)   The inhibition of gene transcription mediated through the production of small interfering RNA fragments (siRNA) and binding of these fragments and protein to messenger RNA. RNA interference is also referred to as RNA silencing. See Fire, A., Xu, S., Montgomery, M.K. et al., Potent and specific genetic interference by double-stranded RNA in *Caenorhabditis elegans*, *Nature* 391, 806–811, 1998; Bosher, J.M., Dufourcq, P., Sookhareea, S., and Labousse, M., RNA interference can target pre-mRNA: consequences for gene expression in a *Caenorhabditis elegans* operon, *Genetics* 153, 1245–1256, 1999; Tabara, H., Sarkissian, M., Kelly, W.G. et al., The rde-1 gene, RNA interference, and transposon silencing in *C. elegans, Cell* 99, 123–132, 1999; Plasterk, R.H. and Ketting, R.F., The silence of the genes, *Curr. Opin. Genet. Dev.* 10, 562–567, 2000; Barstead, R., Genome-wide RNAi, *Curr. Opin. Chem. Biol.* 5, 63–66, 2001; Tuschi, T., RNA interference and small interfering RNAs, *Chembiochem* 2, 239–245, 2001; Baulcombe, D., RNA silencing, *Trends Biochem. Sci.* 30, 290–293, 2005; Filipowicz, W., Jaskiewicz, L., Kolb, F.A., and Pillai, R.S., Posttranscriptional gene silencing by siRNAs and miRNAs, *Curr. Opin. Struct. Biol.* 15, 331–341, 2005; Shearwin, K.E., Callen, B.P., and Egan, J.B., Transcriptional interference — a crash course, *Trends Genet.* 21, 339–345, 2005; Sarov, M., and Stewart, A.F., The best control for the specificity of RNAi, *Trends Biotechnol.* 23, 446–448, 2005; Collins, R.E. and Cheng, X., Structural domains in RNAi, *FEBS Lett.* 579, 541–549, 2005; Zamore, P.D. and Haley, B., Ribo-genome: the big world of small RNAs, *Science* 309, 1519–1524, 2005; Yeung, M.L., Bennasser, Y., Le, S.Y., and Jeang, K.T., siRNA, miRNA, and HIV: promises and challenges, *Cell. Res.* 15, 935–946, 2005; Carmichael, G.C., Ed., *RNA Silencing: Methods and Protocols*, Humana Press, Totowa, NJ, 2005; Galun, E., *RNA Silencing*, World Scientific Publishing, Singapore, 2005; Fanning, G.C. and Symonds, G., Gene-expressed RNA as a therapeutic: issues to consider, using ribozymes and small hairpin RNA as specific examples, *Handb. Exp. Pharmacol.* 173, 289–303, 2006.

RNA Isolation ("Tri-Reagents")   Study of gene expression can involve the isolation of mRNA from cells and tissues for analysis by microarray technology, RT-PCR, and northern blot analysis. A variety of approaches are involved including the treatment of water with diethylpyrocarbonate (ethoxyformic anhydride), the use of RNAse inhibitors, and various extraction technologies having the prefix "tri" such as Tri Reagent® and TRIzol®. These reagents use a solution of guanidine isothiocyanate and phenol (see Chomczynski, P. and Sacchi, N., Single-step method of RNA isolation by acid guanidinium thiocyanate-phenol-chloroform extraction, *Anal. Biochem.* 162, 156–159, 1987) for extraction of cells and tissue followed by a phenol extraction, which yields an aqueous phase with RNA. A variety of other technologies are available and there are several excellent comparative studies. See Verhofstede, C., Fransen, K., Marissens, D. et al., Isolation of HIV-1 RNA from plasma: evaluation of eight different extraction methods, *J. Virol. Methods* 60, 155–159, 1996; Chadderton, T., Wilson, C., Bewick, M., and Gluck, S., Evaluation of three rapid RNA extraction reagents: relevance for use in RT-PCRs and measurement of low-level gene expression in clinical samples, *Cell. Mol. Biol.* 43, 1227–1234, 1997; Weber, K., Bolander, M.E., and Sarkar, G., PIG-B: a homemade monophasic cocktail for the extraction of RNA, *Mol. Biotechnol.* 73–77, 1998; Mannhalter, C., Koizar, D., and Mitterbauer, G., Evaluation of RNA isolation methods and reference genes for RT-PCR analyses of rare target RNA, *Clin. Chem. Lab. Med.* 38, 171–177, 2000; Deng, M.Y., Wang, H., Ward, G.B., Beckham, T.R., and McKenna, T.S., Comparison of six RNA extraction methods for the detection of classical swine fever virus by real-time and conventional reverse transcription-PCR, *J. Vet. Diagn. Invest.* 17, 574–578, 2005; Culley, D.E., Kovacik, W.P., Jr., Brockman, F.J., and Zhang, W., Optimization of RNA isolation from the archaebacterium *Methanosacrcina barkeri* and validation for oligonucleotide microarray analysis, *J. Microbiol. Methods*, 67, 36–43, 2006; Prezeau, N., Silvy, M., Gabert, J., and Picard, C., Assessment of a new RNA stabilizing reagent (Tempus GLood RNA) for minimal residual disease in onco-hematology usng the EAC-protocol, *Leuk. Res.* 30, 569–574, 2006.

RNA Polymerase The enzymes responsible for the biosynthesis of DNA-directed RNA synthesis. RNA polymerase is a nucleotide transferase that synthesizes RNA from ribonucleotides. In bacteria there is only one RNA polymerase (see Lathe, R., RNA polymerase of *Escherichia coli*, *Curr. Top. Microbiol. Immunol.* 83, 37–91, 1978); arachaea also has a single RNA polymerase (Geiduschek, E.P. and Ouhammouch, M., Archaeal transcription and its regulators, *Mol. Microl.* 56, 1397–1407, 2005). Eukaryotic cells have three RNA polymerases: RNA polymerase I (polI) catalyzes the synthesis of ribosomal RNA species in the form of a precursor pre-rRNA (45S), which is processed into other species such as 28S and 18S RNAs; RNA polymerase III (polIII) synthesizes tRNA (transfer RNAs) and other smaller RNA species. RNA polymerase II (polI) (Hahn, S., Structure and mechanism of the RNA polymerase II transcription machinery, *Nature Struct. Mol. Biol.* 11, 394–403, 2004) is responsible for the synthesis of the various pre-mRNAs (messenger RNAs), which mature into the mRNA species responsible for the direction of protein biosynthesis. Viral RNA polymerases appear to be different from other RNA polymerases and appear to be derived from DNA polymerases. See Losick, R. and Chamberlin, M., Eds., *RNA Polymerase*, Cold Spring Harbor Laboratory Press, Cold Spring Harbor, NY, 1976; Kuo, L.C. and Olsen, D.B., Eds., *Viral Polymerases and Related Proteins*, Academic Press, San Diego, CA, 1996; Adhya, S.L., *RNA Polymerase and Associated Factors*, Academic Press, San Diego, CA, 1996; Goodbourn, S., Ed., *Eukaryotic Gene Transcription*, IRL Press at Oxford University Press, Oxford, UK, 1996; Richter, J.D., *mRNA Formation and Function*, Academic Press, San Diego, CA, 1997; Borukhov, S. and Nudler, E., RNA polymerase holoenzyme: structure, function, and biological implications, *Curr. Opin. Microbiol.* 6,

93–100, 2003; Murakami, K.S. and Darst, S.A., Bacterial RNA polymerases: the whole story, *Curr. Opin. Struct. Biol.* 13, 31–39, 2003; Studitsky, V.M., Walter, W., Kireev, M. et al., Chromatin remodeling by RNA polymerases, *Trends Biochem. Sci.* 29, 127–136, 2004; Bartlett, M.S., Determinants of transcription initiation by archaeal RNA polymerase, *Curr. Opin. Microbiol.* 8, 677–684, 2005; Boeger, H., Bushnell, D.A., Davis, R. et al., Structural basis of eukaryotic gene transcription, *FEBS Lett.* 579, 899–903, 2005; Gralla, J.D., *Escherichia coli* ribosomal RNA transcription: regulatory roles for ppGpp, NTPs, architectural proteins, and a polymerase-binding protein, *Molec. Microbiol.* 55, 973–977, 2005; Banerjee, S., Chalissery, J., Bandey, I., and Sen, R., Rho-dependent transcription: more questions than answers, *J. Microbiol.* 44, 11–22, 2006.

RNA Splicing   The removal of introns from the sequence of an mRNA following transcription to form an uninterrupted coding sequence. During this process, introns or intervening regions are removed and the remaining regions, exons, join together in a splicing process to form the mature RNA transcript. See Sharp, P.A., The discovery of split genes and RNA splicing, *Trends Biochem. Sci.* 30, 279–281, 2005; Matlin, A., Clark, F., and Smith, C.W., Understanding alternative splicing: toward a cellular code, *Nat. Rev. Mol. Cell Biol.* 6, 386–398, 2005; Stetefeld, J. and Ruegg, M.A., Structural and functional diversity generated by alternative mRNA splicing, *Trends Biochem. Sci.* 30, 510–521, 2005.

(RNAse III)   A family of ribonucleases that is involved in RNA silencing or RNA
Ribonuclease III   interference. See Conrad, C. and Rauhut, R., Ribonuclease III: new sense from nuisance, *Int. J. Biochem.* 34, 116–129, 2002.

Rolling   The initial interaction between a leukocyte and the endothelium; also known as margination.

RTX Toxins   RTX family of bacterial toxins, which are a group of cytolysins and cytotoxins. Hemolysin (HlyA) is often quoted as the model for RTX toxins. See Coote, J.G., Structural and functional relationship among the RTX toxin determinants of Gram-negative bacteria, *FEMS Microbiol. Rev.* 8, 137–161, 1992.

S100 Proteins   A multifunctional family of intracellular proteins distinguished by solution in saturated ammonium sulfate (100% saturation) and their interactions with calcium ions. The seminal member of the family, S100 protein, was described as a protein unique to the nervous system. See Moore, B.W., A soluble protein characteristic of the nervous system, *Biochem. Biophys. Res. Commun.* 19, 739–744, 1965; Donato, R., Perspectives in S-100 protein biology, *Cell Calcium* 12, 713–726, 1991; Passey, R.J., Xu, K., Hume, D.A., and Geczy, C.L., S100A8: emerging functions and regulations, *J. Leukocyte Biol.* 66, 549–556, 1999; Heizmann, C.W., The multifunctional S100 protein family, *Methods Mol. Biol.* 172, 69–80, 2002; Emberley, E.D., Murphy, L.C., and Watson, P.H., S100 proteins and their influence on pro-survival pathways in cancer, *Biochem. Cell Biol.* 82, 508–515, 2004.

Saposins   Sphingolipid activator proteins; small heat-stable proteins that appear to be cofactors in the hydrolysis of sphingolipids. There are four saposins: A, B, C, and D, which are generated from a common precursor, prosaposin. See Vaccaro, A.M., Salivioli, R., Tatti, M., and Ciaffoni, F., Saposins and their interactions with lipids, *Neurochemical Res.* 24, 307–314, 1999.

Scaffold   In combinatorial chemistry or parallel synthetic strategy, it is the common
                    platform that serves as the core for synthesis of individual scaffold family
                    members; also the matrix for tissue development such as bone. See
                    Hollister, S.J., Porous scaffold design for tissue engineering, *Nat. Mater.*
                    4, 518–524, 2005; Hosse, R.J., Rothe, A., and Power, B.E., A new
                    generation of protein display scaffolds for molecular recognition, *Protein
                    Science* 15, 14–27, 2006; Hammond, J.S., Beckingham, I.J., and
                    Shakesheff, K.M., Scaffolds for liver tissue engineering, *Expert Rev.
                    Med. Devices* 3, 21–27, 2006; Li, C., Vepari, C., Jin, H.J., Kim, H.J.,
                    and Kaplan, D.L., Electrospun silk-BMP-2 scaffolds for bone tissue
                    engineering, *Biomaterials*, 27, 3115–3124, 2006; van Lieshout, M.I.,
                    Vaz, C.M., Rutten, M.C., Peters, G.W., and Baaijens, F.P., Electrospining
                    versus knitting: two scaffolds for tissue engineering of the aortic valve,
                    J. Biomater. Sci. Polym. Ed. 17, 77–89, 2006.

Scattering   As in light scattering. Scattering may be elastic where energy is conserved
                    and the scattered electromagnetic waves are of the same frequency as in
                    incident electromagnetic radiation; when the frequency of the scattered
                    radiation is different from the incident radiation, the scattering is inelastic.
                    Reflection and refraction are types of light scattering. Turbidimetry and
                    nephelometry are applications of light scattering. Raman spectroscopy is
                    an example of inelastic light scattering.

Sec-Dependent   Secretory protein translocation. See Stephenson, K., Sec-dependent protein
                    translocation across biological membranes. Evolutionary conservation of an
                    essential protein transport pathway, *Mol. Membrane Biol.* 22, 17–28, 2005.

Secretase   This term describes those proteolytic activities involved in the processing
                    of amyloid precursor protein (APP) to yield the soluble circulating amy-
                    loid protein. See Stephens, B.J. and Austen, B.M., Characterization of
                    beta-secretase, *Biochem. Soc. Trans.* 26, 500–504, 1998; Wolfe, M.S.,
                    Secretase targets for Alzheimer's disease: identification and therapeutic
                    potential, *J. Med. Chem.* 44, 2039–2060, 2001; Vassar, R., The beta-
                    secretase, BACE: a prime target for Alzheimer's disease, *J. Mol. Neuro-
                    sci.* 17, 157–170, 2001; Hooper, N.M. and Turner, A.J., The search for
                    alpha-secretase and its potential as a therapeutic approach to Alzheimer's
                    disease, *Curr. Med. Chem.* 9, 1107–1119, 2002; Pollack, S.J. and Lewis,
                    H., Secretase inhibitors for Alzheimer's disease: challenges of a promis-
                    cuous protease, *Curr. Opin. Invest. Drugs* 6, 35–47, 2005. There is a
                    more general definition of secretase as a "sheddase" responsible for the
                    proteolysis of type I and type II membrane proteins (Hooper, N.M.,
                    Karran, E.H., and Turner, A.J., Membrane protein secretases, *Biochem.
                    J.* 321, 265–279, 1997; Wolfe, M.S., and Kopan, R., Intramembrane
                    proteolysis, *Science* 305, 1119–1123, 2004). The term "secretase" has
                    been used to describe the activity responsible for the release of TNF from
                    membranes (Mezyk, R., Browska, M., and Bereta, J., Structure and
                    functions of tumor necrosis factor-alpha converting enzyme, *Acta Bio-
                    chim. Pol.* 50, 625–645, 2003). There is a relationship between ADAM
                    proteases and secretases (Fahrenholz, F., Gilbert, S., Kojro, E. et al.,
                    Alpha-secretase activity of the disintegrin metalloprotease ADAM 10.
                    Influence of domain structure, *Ann. N.Y. Acad. Sci.* 920, 215–222, 2000;
                    Higashiyama, S. and Nanba, D., ADAM-mediated ectodomain shedding
                    of HB-EGF in receptor cross-talk, *Biochim. Biophys. Acta* 1751,
                    111–117, 2005). See also *Gamma(γ)-Secretase.*

SELDI   Surface-enhanced laser/desorption ionization mass spectrometry; Protein-Chip®. See Tang, N., Tornatore, P., and Weinberger, S.R., Current developments in SELDI affinity technology, *Mass Spectrom. Rev.* 23a, 34–44, 2004.

Selectivity and Selectivity Factor   The discrimination shown by a compound in reacting with two or more positions on the same compound or several compounds. It is quantitatively expressed by ratios of rate constants of the competing reactions or by the decadic logarithms of such ratios. It also refers to the differential affinity of compounds to a chromatographic matrix. Chromatographic selectivity is a determining factor in resolution; the use of "selectivity" is discouraged in favor of the use of the term "separation factors."

SELEX   Systematic evolution of nucleic acid ligands (aptamers) by exponential enrichment. This represents an approach to the development of nucleic acid ligands for affinity chromatography and therapeutic aptamers by selection from combinatorial oligonucleotide libraries directed against a putative target. See Klug, S.J. and Famulok, M., All you wanted to know about SELEX, *Mol. Biol. Rep.* 20, 97–107, 1994; Joyce, G.J., *In vitro* evolution of nucleic acids, *Curr. Opin. Struct. Biol.* 4, 331–336, 1994; Gold, L., Brown, D., He, Y. et al., From oligonucleotide shapes to genomic SELEX: novel biological regulatory loops, *Proc. Natl. Acad. Sci. USA* 94, 59–64, 1997; Jayasena, S.D., Aptamers: an emerging class of molecules that rival antibodies in diagnostics, *Clin. Chem.* 45, 1628–1650, 1999; Clark, S.L. and Remcho, V.T., Aptamers as analytical reagents, *Electrophoresis* 23, 1335–1340, 2002; Tuerk, C. and Gold, L., Systematic evolution of ligands by expotential enrichment: RNA ligands to bacteriophage T4 DNA polymerase, *Science* 249, 505–510, 1990; Liu, J. and Stormo, G.D., Combining SELEX with quantitative assays to rapidly obtain accurate models of protein–DNA interactions, *Nuc. Acid Res.* 33, e141, 2005; Guthrie, J.W., Hamula, C.L., Zhang, H. et al., Assays for cytokines using aptamers, *Methods* 39, 324–330, 2006; Ulrich, H., RNA aptamers: from basic science towards therapy, *Handb. Exp. Pharmacol.* 173, 305–326, 2006.

Separase   A regulatory protease that initiates the metaphase–anaphase transition by cleavage of the Sec1 subunit of cohesion, a chromosomal protein complex. This is a process regulated by shugoshin ("guardian spirit"). See Yanagida, M., Cell cycle mechanisms of sister chromatid separation; roles of Cut1/separin and Cut2/securin, *Genes Cells* 5, 1–8, 2000; Amon, A., Together until separin do us part, *Nat. Cell Biol.* 3, E12–E14, 2001; Uhlmann, F., Secured cutting: controlling separase at the metaphase to anaphase transition, *EMBO Rep.* 2, 487–492, 2001; Hearing, C.H. and Nasmyth, K., Building and breaking bridges between sister chromatids, *Bioessays* 25, 1178–1191, 2003; Uhlmann, F., The mechanism of sister chromatid cohesion, *Exp. Cell Res.* 296, 80–85, 2004; Watanabe, Y. and Kitajima, T.S., Shugoshin protects cohesion complexes at centromeres, *Philos. Tran. R. Soc. Lond. B Biol. Sci.* 360, 515–521, 2005; Watanabe, Y., Shugoshin: guardian spirit at the centromere, *Curr. Opin. Cell Biol.* 17, 590–595, 2005.

Separation Factor   Designated by the term α and refers to the relative affinity of two components for a chromatographic matrix and related to the resolution. By definition the separation factor is larger than one and could be described by the following expression: $\alpha = t_2/t_1$, where t_2 is the elution time for the apex of the more slowly moving solute, and t_1 is the elution time for the apex of the more rapidly moving solute. See Chen, Y., Kele, M., Quinones, I.,

Sellergren, B., and Guiochon, G., Influence of the pH on the behavior of an imprinted polymeric stationary phase — supporting evidence for a binding site model, *J. Chromatog. A* 927, 1–17, 2001; Avramescu, M.E., Borneman, Z., and Wessling, M., Mixed-matrix membrane adsorbers for protein separation, *J. Chromatog. A* 1006, 171–183, 2003; Ziomek, G., Kaspereit, M., Jezowski, J., Seidel-Morgenstern, A., and Antos, D., Effect of mobile phase composition on the SMB processes efficiency. Stochastic optimization of isocratic and gradient operation, *J. Chromatog. A* 1070, 111–124, 2005; Lesellier, E. and Tchapla, A., A simple subcritical chromatographic test for an extended ODS high-performance liquid chromatography column classification, *J. Chromatog. A* 1100, 45–59, 2005; Lapointe, J.F., Gauthier, S.F., Pouliot, Y., and Bouchard, C., Selective separation of cationic peptides from a tryptic hydrolyzate of beta-lactoglobulin by electrofiltration, *Biotechnol. Bioeng.* 94, 223–233, 2006.

SERCA
Sarcoplasmic reticulum Ca^{2+} ATPase, responsible for calcium ion transport. See Martonosi, A.N. and Pikula, S., The structure of the Ca^{2+}-ATPase of sarcoplasmic reticulum, *Acta Biochim. Pol.* 50, 337–365, 2003; Strehler, E.E. and Treiman, M., Calcium pumps of plasma membrane and cell interior, *Curr. Mol. Med.* 4, 323–335, 2004.

SEREX
Serological identification of antigens by recombinant expression cloning. See Sahin, U., Tureci, O., Schmitt, H. et al., Human neoplasms elicit multiple specific immune responses in the autologous host, *Proc. Natl. Acad. Sci. USA* 92, 11810–11813, 1995; Chen, Y.-T., Scanlan, M.J., Sahin, U. et al., A testicular antigen aberrantly expressed in human cancers detected by autologous antibody screening, *Proc. Natl. Acad. Sci. USA*, 94, 1914–1918, 1997; Fernandez, M.F., Tang, N., Alansari, H. et al., Improved approach to identify cancer-associated autoantigens, *Autoimmun. Rev.* 4, 230–235, 2005; www.licr.org/SEREX.html; www2.licr.org/CancerImmunomeDB/.

Serial Lectin Affinity Chromatography
The use of a series of two or more lectin affinity chromatography columns of known specificity for the fractionation of oligosaccharides, glycoproteins, or glycopeptides into structurally distinct groups. See Cummings, R.D. and Kornfeld, S., Fractionation of asparagine-linked oligosaccharides by serial lectin-agarose affinity chromatography. A rapid, sensitive, and specific technique, *J. Biol. Chem.* 257, 11235–11240, 1982; Qiu, R. and Regnier, F.E., Comparative glycoproteomics of N-linked complex-type glycoforms containing sialic acid in human serum, *Anal. Chem.* 77, 7725–7231, 2005.

Serpin
A term now in its own right. It was developed as an acronym for serine protease inhibitor (Carroll, R.W. and Travis, J., α-1-antitrypsin and the serpins: variation and countervariation, *Trends Biochem. Sci.* 10, 20–24, 1985). It is considered to be a structurally homologous superfamily (Hunt, L.T. and Dayhoff, M.O., A surprising new protein superfamily containing ovalbumin, antithrombin-III, and α1-proteinase inhibitor, *Biochem. Biophys. Res. Commun.* 95, 864–871, 1980) of proteins having masses in the range of 40 kDa to 100 kDa. See Gettins, P., Patson, P.A., and Schapira, M., The role of conformational change in serpin structure and function, *Bioessays* 15, 461–467, 1993; Schulze, A.J., Huber, R., Bode, W., and Engh, R.A., Structural aspects of serpin inhibition, *FEBS Lett.* 344, 117–124, 1994; Potempa, J., Korzus, E., and Travis, J., The serpin superfamily of proteinase inhibitors: structure, function, and regulation, *J.*

Biol. Chem. 269, 15957–15960, 1994; Lawrence, D.A., The role of reactive-center loop mobility in the serpin inhibitory mechanism, *Adv. Exp. Med. Biol.* 425, 99–108, 1997; Whisstock, J., Skinner, R., and Lesk, A.M., An atlas of serpin conformations, *Trends Biochem. Sci.* 23, 63–67, 1998; Gettins, P.G., Serpin structure, mechanism, and function, *Chem. Rev.* 102, 4751–4804, 2002; Huntington, J.A., Shape-shifting serpins — advantages of a mobile mechanism, *Trends Biochem. Sci.* 31, 427–435, 2006.

Shotgun Proteomics
Identification of peptides (usually by mass spectrometry) obtained by the enzymatic or chemical digestion of the entire proteome. A naturally occurring protein mixture such as cell extract, blood plasma, or other biological fluid is reduced, alkylated, and subjected to tryptic hydrolysis. The tryptic hydrolysis is fractionated by liquid chromatography and analyzed by mass spectrophotometry. See Wolters, D.A., Washburn, M.P., and Yates, J.R., III, An automated multidimensional protein identification technology for shotgun proteomics, *Anal. Chem.* 73, 5683–5690, 2001; Liu, H., Sadygov, R.G., and Yates, J.R., III, A model for random sampling and estimation of relative protein abundance in shotgun proteomics, *Anal. Chem.* 76, 4193–4201, 2004.

Shugoshin
A protein family having a role in the centromeric protection of cohesion; protects the centromeric cohesion at meiosis I by inhibiting the action of separase on cohesion. See Salic, A., Waters, J.C., and Mitchison, T.J., Vertebrate shugoshin links sister centromere cohesion and kinetochore microtubule stability in mitosis, *Cell* 118, 567–578, 2004; Kitajima, T.S., Kawashima, S.A., and Watanabe, Y., The conserved kinetochore protein shugoshin protects centromeric cohesion during meiosis, *Nature* 427, 510–517, 2005; Goulding, S.E. and Earnshaw, W.C., Shugoshin: a centromeric guardian senses tension, *Bioessays* 27, 588–591, 2005; Watanabe, Y., Shugoshin: guardian spirit at the centromere, *Curr. Opin. Cell Biol.* 17, 590–595, 2005; Stemmann, O., Boos, D., and Gorr, I.H., Rephrasing anaphase: separase FEARs shugoshin, *Chromosoma* 113, 409–417, 2005; Mcgee, P., Molecular biology: chromosome guardian on duty, *Nature* 441, 35–37, 2006.

Sigma Factor
A factor that binds to RNA polymerase and provides specificity for the transcriptional process. It also provides for DNA strand separation during the transcriptional process. Sigma factors could be considered subunits of the RNA polymerase enzyme. See Kazmierczak, M.J., Wiedmann, M., and Boor, K.J., Alternative sigma factors and their roles in bacterial virulence, *Microbiol. Mol. Biol. Rev.* 69, 527–543, 2005; Mooney, R.A., Darst, S.A., and Landick, R., Sigma and RNA polymerase: an on-again, off-again relationship? *Mol. Cell* 20, 335–345, 2005; Kill, K., Binnewies, T.T., Sicheritz-Ponten, T. et al., Genome update: sigma factors in 240 bacterial genomes, *Microbiology* 151, 3147–3150, 2005; Wigneshweraraj, S.R., Burrows, P.C., Bordes, P. et al., The second paradigm for activation of transcription, *Prog. Nucl. Acid Res. Mol. Biol.* 79, 339–369, 2005.

Signalosome
An endosome with an active signaling component, which is transported to a juxtanuclear position. See Perret, E., Lakkaraju, A., Deborde, S., Schreiner, R., and Rodriguez-Boulan, E., Evolving endosomes: how many varieties and why?, *Curr. Opin. Cell Biol.* 17, 423–434, 2005.

Signal Recognition Particle
A targeting chaperone involved in the transmembrane transport of proteins; involves the recognition of the signal peptide. See Pool, M.R, Signal recognition particles in chloroplasts, bacteria, yeast, and mammals, *Mol. Membrane Biol.* 22, 3–15, 2004.

Signature Domain

An amino acid sequence that is closely conserved within a group of proteins and is considered unique to that group of proteins, which is also called a protein family. The sequences may or may not have homologous function (see Khuri, S., Bakker, F.T., and Dunwell, J.M., Phylogeny, function, and evolution of the cupins, a structurally conserved, functionally diverse superfamily of proteins, *Mol. Biol. Evolution* 18, 593–605, 2001). In this sense, the use of the term "signature" is related to historical use to describe a physical property or feature of a plant or other natural object as an indication of pharmacological impact because of relation of such a feature to the body part (see *Oxford English Dictionary*, Oxford University Press, Oxford, UK, 1989; *Webster's Third International Dictionary, unabridged*, Merriam-Webster, Springfield, MA, 1996). One of the most studied examples is the C1q domain (see Bérubé, N.G., Swanson, X.H., Bertram, M.J. et al., Cloning and characterization of CRF, a novel C1q-related factor, expressed in areas of the brain involved in motor function, *Mol. Brain Res.* 63, 233–240, 1999; Kishore, U., Gaboriaud, C., Waters, P. et al., C1q and tumor necrosis factor superfamily: modularity and versatility, *Trends Immunol.* 25, 551–561, 2004). For general considerations, see Tousidou, E., Nanopoulos, A., and Manolopoulos, Y., Improved methods for signature-tree construction, *Comput. J.* 43, 301–314, 2000; Ye, Y. and Godzik, A., Comparative analysis of protein domain organization, *Genome Res.* 14, 343–353, 2004.

Single-Chain Fv Fragment (scFv)

A synthetic (usually recombinant) peptide/protein composed of the V_L and V_H domains of an antibody linked by a peptide. It is relatively small (30 kDa) and as a single peptide chain is easily expressed in bacterial systems. It is possible to express the scFv inside the cell (intracellular expression) as intrabodies for analytical and therapeutic purposes. It is also possible to increase the avidity of these engineered fragments by dimerization to form diabodies and higher order polymers. Also on occasion referred to as minibodies. See Plückthun, A. and Pack, P., New protein engineering approaches to multivalent and bispecific antibody fragments, *Immunotechnology* 3, 93–105, 1997; Kerschbaumer, R.J., Hirschl, S., Kaufmann, A. et al., Single-chain Fv fusion proteins suitable as coating and detecting reagents in a double antibody sandwich enzyme-linked immunosorbent assay, *Anal. Biochem.* 249, 219–227, 1997; Hudson, P.J. and Kortt, A.A., High avidity scFv multimers; diabodies and triabodies, *J. Immunol. Methods* 231, 177–189, 1999; Chadd, H.E. and Chamow, S.M., Therapeutic antibody expression technology, *Curr. Opin. Biotechnol.* 12, 188–194, 2001; Krebs, B., Rauchenberger, R., Reiffert, S. et al., High-throughput generation and engineering of recombinant human antibodies, *J. Immunol. Methods* 254, 67–84, 2001; de Graaf, M., van der Meulen-Mulleman, I.H., Pinedo, H.M., and Haisma, H.J., Expression of scFvs and scFv fusion proteins in eukaryotic cells, *Methods Mol. Biol.* 178, 379–387, 2002; Lennard, S., Standard protocols for the construction of scFv libraries, *Methods Mol. Biol.* 178, 59–71, 2002; Fong, R.B., Ding, Z., Hoffman, A.S., and Stayton, P.S., Affinity separation using an Fv antibody fragment — "smart" polymer conjugate, *Biotechnol. Bioengin.* 79, 271–276, 2002; Sinacola, J.R. and Robinson, A.S., Rapid refolding and polishing of single-chain antibodies from *Escherichia coli* inclusion bodies, *Protein Express. Purif.* 26, 301–308, 2002; Lunde, E., Lauvrak, V., Rasmussen, I.B. et al.,

Troybodies and pepbodies, *Biochem. Soc. Trans.* 30, 500–506, 2002; Kim, S.-E., Expression and purification of recombinant immunotoxin — a fusion protein stabilizes a single-chain Fv (scFv) in denaturing condition, *Prot. Express. Purif.* 27, 85–89, 2003; Leath, C.A., III, Douglas, J.T., Curiel, D.T., and Alvarez, R.D., Single-chain antibodies: a therapeutic modality for cancer gene therapy, *Int. J. Oncol.* 24, 765–771, 2004; Visintin, M., Meli, G.A., Cannistraci, I., and Cattnaeo, A., Intracellular antibodies for proteomics, *J. Immunol. Methods* 290, 135–153, 2004; Lobato, M.N. and Rabbitts, T.H., Intracellular antibodies as specific reagents for function ablation: future therapeutic molecules, *Curr. Mol. Med.* 4, 519–528, 2004; Holliger, P. and Hudson, P.J., Engineered antibody fragments and the rise of single domains, *Nat. Biotechnol.* 23, 1126–1136, 2005; Röthlisberger, D. Honengger, A., and Plückthun, A., Domain interactions in the Fab fragment: a comparative evaluation of the single-chain Fv and Fab format engineered with variable domains of different stability, *J. Mol. Biol.* 347, 773–789, 2005.

Small Interfering RNA (siRNA)	A short-length double-stranded RNA (21–27 nucleotides in length) derived from intracellular double-stranded RNA by the action of specific endonucleases such as RNAse III (see *Dicer, Drosha*). The siRNA stimulates the cellular machinery to cut up messenger RNA, thus inhibiting the process of transcription; this process is called knockdown. See Bass, B.L., Double-stranded RNA as a template for gene silencing, *Cell* 101, 235–238, 2000; Myers, J.W. and Ferrell, J.E., Jr., Silencing gene expression with Dicer-generated siRNA pools, in Carmichael, G.G., Ed., *RNA Silencing. Methods and Protocols*, Humana Press, Totowa, NJ, 2005, pp. 93–196; Aravin, A. and Tuschi, T., Identification and characterization of small RNAs involved in RNA silencing, *FEBS Lett.* 579, 5830–5840, 2005; Kim, V.N., Small RNAs: classification, biogenesis, and function, *Mol. Cell.* 19, 1–15, 2005.
Small Nuclear Ribonucleoprotein (RNA Plus Protein) Particle	Component of the spliceosome, the intron-removing apparatus in eukaryotic nuclei. See Graveley, B.R., Sorting out the complexity of SR protein functions, *RNA* 6, 1197–1211, 2000; Will, C.L. and Luhrmann, R., Spliceosomal UsnRNP biogenesis, structure, and function, *Curr. Opin. Cell Biol.* 13, 290–301, 2001; Turner, I.A., Norman, C.R., Churcher, M.J., and Newman, A.J., Roles of the U5 snRNP in spliceosome dyanamics and catalysis, *Biochem. Soc. Trans.* 32, 928–931, 2004.
Small Temporal RNA	Messenger RNAs that are expressed only at a specific stage in development and encode proteins involved in specific developmental timing events. See Pasquinelli, A.E., Reinhart, B.J., Slack, R. et al., Conservation of the sequence and temporal expression of *let-7* heterochronic regulatory RNA, *Nature* 408, 86–89, 2000; Moss, E.G., RNA interference: it's a small RNA world, *Curr. Biol.* 11, R772–R775, 2001.
Smart Probes	Usually a nucleic acid probe that emits a signal only when bound to a specific target. An example is a molecular beacon. See Stöhr, K., Häfner, B., Nolte., O. et al., Species-specific identification of mycobacterial 16S rRNA PCR amplicons using smart probes, *Anal. Chem.* 77, 7195–7203, 2005. There are other examples of smart probes including proteins (Wunder, A., Tung, C.-H., Müller-Ladner, U., Weissleder, R., and Mahmood, U., *In vivo* imaging of protease activity in arthritis, a novel approach for monitoring treatment response, *Arthritis & Rheumatism* 50, 2459–2465, 2004) and chiral compounds (Tsukube, H. and Shinoda, S., Lanthanide complexes as smart

CD probes for chirality sensing of biological substrates, *Enantiomer* 5, 13–22, 2000). "Smart" contrast reagents have also been developed for magnetic resonance studies (Lowe, M.P., Activated MR contrast reagents, *Curr. Pharm. Biotechnol.* 5, 519–528, 2004).

SNARE Proteins SNAREs (soluble NSF attachment protein receptors) participate in eukaryotic membrane fusion. It is suggested that vesicle SNARE proteins fuse with target SNARE proteins during processes such as exocytosis. Most SNARE proteins have a C-terminal transmembrane domain, a substantial cytosolic domain, and a variable N-terminal domain (brevin domain, longin domain, YKT-domain), which regulate membrane fusion reactions. SNARE proteins can be classified as Q-SNAREs or R-SNAREs depending on amino acid sequence homology. There are other classification systems as well. See Ferro-Novick, S. and Jahn, R., Vesicle fusion from yeast to man, *Nature* 370, 191–193, 1994; Rothman, J.E. and Warren, G., Implications of the SNARE hypothesis for intracellular membrane topology and dynamics, *Curr. Biol.* 4, 220–233, 1994; Morgan, A., Exocytosis, *Essays Biochem.* 30, 77–95, 1995; Burgoyne, R.D., Morgan, A., Barnard, A.J. et al., SNAPs and SNAREs in exocytosis in chromaffin cells, *Biochem. Soc. Trans.* 24, 653–657, 1996; Wilson, M.C., Mehta, P.P., and Hess, E.J., SNAP-25, ensnared in neurotransmission and regulation of behavior, *Biochem. Soc. Trans.* 24, 670–676, 1996; Hya, J.C. and Scheller, R.H., SNAREs and NSF in targeted membrane fusion, *Curr. Opin. Cell Biol.* 9, 505–512, 1997; Pelham, H.R., SNAREs and the secretory pathway — lessons from yeast, *Exp. Cell Res.* 247, 1–8, 1997; Whiteheart, S.W., Schraw, T., and Matleeva, E.A., N-ethylmaleimide sensitive factor (NSF) structure and function, *Int. Rev. Cytol.* 207, 71–112, 2001; Hay, J.C., SNARE complex structure and function, *Exp. Cell Res.* 271, 10–21, 2001; Dietrich, L.E.P., Boedinghaus, C., LaGrassa, J.T., and Ungermann, C., Control of eukaryotic membrane fusion by N-terminal domains of SNARE proteins, *Biochim. Biophys. Acta* 1641, 111–119, 2003; Hong, W., SNAREs and traffic, *Biochim. Biophys. Acta* 1744, 493–517, 2005; Montecucco, C., Schiavo, G., and Pantano, S., SNARE complexes and neuroexocytosis: how many, how close? *Trends Biochem. Sci.* 30, 367–372, 2005.

Soft Ionization Ionization techniques such as fast atom bombardment (FAD), electrospray ionization (ESI), or matrix-assisted laser desorption/ionization (MALDI) that initiate the desorption and ionization of nonvolatile thermally labile compounds such as proteins or peptides. See Fenn, J.B., Mann, M., Meng, C.K. et al., Electrospray ionization for mass spectrometry of large biomolecules, *Science* 246, 64–71, 1989; Reinhold, V.N., Reinhold, B.B., and Costello, C.B., Carbohydrate molecular weight profiling, sequence, linkage, and branching data: ES-MS and CID, *Anal. Chem.* 67, 1772–1784, 1995; Griffiths, W.J., Jonsson, A.P., Liu, S., Rai, D.K., and Wang, Y., Electrospray and tandem mass spectrometry in biochemistry, *Biochem. J.* 355, 545–561, 2001; Schalley, C.A., Molecular recognition and supramolecular chemistry in the gas phase, *Mass Spectrom. Rev.* 20, 253–309, 2001; Kislinger, T., Humeny, A., and Pischetsrider, M., Analysis of protein glycation products by matrix-assisted laser desorption ionization time-of-flight mass spectrometry, *Curr. Med. Chem.* 11, 2185–2193, 2004; Laskin, J. and Futrell, J.H., Activation of large ions in FT-ICR mass spectrometry, *Mass Spectrom. Rev.* 24, 135–167, 2005; Bolbach, G., Matrix-assisted laser desorption/ionization analysis of non-covalent complexes: funda-mentals and applications, *Curr. Pharm. Des.* 11, 2535–2557, 2005; Bald-win, M.A., Mass spectrometers for the analysis of biomolecules, *Methods Enzymol.* 402, 3–48, 2005.

Somatic   The increased mutation in the variable region of immunoglobulin genes,
Hypermutation   which allows for diversity of immune recognition. See Steele, E.J., Rothen-fluh, H.S., and Both, G.W., Defining the nucleic acid substrate for somatic hypermutation, *Immunol. Cell Biol.* 70, 129–144, 1992; Jacob, J., Miller, C., and Kelsoe, G., *In situ* studies of the antigen-driven somatic hyper-mutation of immunoglobulin genes, *Immunol. Cell Biol.* 70, 145–152, 1992; George, J. and Clafin, L., Selection of B-cell clones and memory B-cells, *Semin. Immunol.* 4, 11–17, 1992; Neuberger, M.S. and Milstein, C.S., Somatic hypermutation, *Curr. Opin. Immunol.* 7, 24–254, 1995; Hengstschlager, M., Maizels, N., and Leung, H., Targeting and regulation of immunoglobulin gene somatic hypermutation and isotype switch recombination, *Prog. Nucleic Acid Res. Mol. Biol.* 50, 67–99, 1995; Steele, E.J., Rothenflug, H.S., and Blanden, R.V., Mechanism of antigen-driven somatic hypermutation of rearranged immunoglobulin V(D)J genes in the mouse, *Immunol. Cell Biol.* 75, 82–95, 1997; Rajewsky, K., Clonal selection and learning in the antibody system, *Nature* 381, 751–758, 1996; Storb, U., Peters, A., Klotz, E., et al., *Cis*-acting sequences that affect somatic hypermutation of Ig genes, *Immunol. Rev.* 162, 153–160, 1998; Neuberger, M.S., Ehrenstein, M.R., Klix, N. et al., Monitoring and interpreting the intrinsic features of somatic hypermuta-tion, *Immunol. Rev.* 162, 107–116, 1998; Kuppers, R., Goossens, T., and Klein, U., The role of somatic hypermutation in the generation of dele-tions and duplications in human Ig V region genes and chromosomal translocations, *Curr. Top. Microbiol. Immunol.* 246, 193–198, 1999; Harris, R.S., Kong, Q., and Maizels, N., Somatic hypermutation and the three Rs: repair, replication, and recombination, *Mutat. Res.* 436, 157–178, 1999; Jacobs, H. and Bross, L., Towards an understanding of somatic hypermutation, *Curr. Opin. Immunol.* 13, 208–218, 2001; Seki, M., Gearhart, P.J., and Wood, R.D., DNA polymerases and somatic hyper-mutation of immunoglobulin genes, *EMBO Rep.* 6, 1143–1148, 2005; Neuberger, M.S., Di Noia, J.M., Beale, R.C. et al., Somatic hypermutation at A–T pairs: polymerase error versus dUTP incorporation, *Nat. Rev. Immunol.* 5, 171–178, 2005.

Southern Blotting The use of a complement oligonucleotide/polynucleotide to identify dena-tured DNA transferred by absorption from an agarose gel to another matrix, such as a nitrocellulose membrane. See Southern, E.M., Detection of specific sequences among DNA fragments separated by gel electro-phoresis, *J. Mol. Biol.* 98, 503–517, 1975; Darbre, P.D., *Introduction to Practical Molecular Biology*, Wiley, Chichester, UK, 1988; Southern, E.M., Detection of specific sequences among DNA fragments separated by gel electrophoresis, *Biotechnology* 24, 122–139, 1992; Issac, P.G., *Protocols for Nucleic Acid Analysis by Nonradioactive Probes*, Humana Press, Totowa, NJ, 1994; Darling, D.C. and Brickell, P.M., *Nucleic Acid Blotting: The Basics,* Oxford University Press, Oxford, UK, 1994; Kelly, K.F., Southern blotting, *Proc. Nutr. Soc.* 55, 591–597, 1996; Keichle, F.L., DNA technology in the clinical laboratory, *Arch. Pathol. Lab. Med.* 123, 1151–1153, 1999; Southern, E.M., Blotting at 25, *Trends in Biochem. Sci.*

25, 585–588, 2000; Porchet, N. and Aubert, J.P., Southern blot analysis of large DNA fragments, *Methods Mol. Biol.* 125, 313–321, 2000; Voswinkel, J. and Gause, A., From immunoglobulin gene fingerprinting to motif-specific hybridization: advances in the analysis of B lymphoid clonality in rheumatic diseases, *Arthritis Res.* 4, 1–4, 2002; Rose, M.G., Degar, B.A., and Berliner, N., Molecular diagnostics of malignant disorders, *Clin. Adv. Hematol. Oncol.* 2, 650–660, 2004; Wong, L.J. and Boles, R.G., Mitochondrial DNA analysis in clinical laboratory diagnostics, *Clin. Chim. Acta* 354, 1–20, 2005.

Southwestern Blotting
An analytical procedure used to identify the specific binding of a deoxyribonucleic acid sequence to a protein that uses a technical approach similar to southern blot and western blot. A protein mixture is separated by electrophoresis and the resulting electrophoretograms are transferred to a PVDF membrane electrophoresis. The proteins are renatured on the membrane and a ^{32}P-labeled oligodeoxyribonucleotide probe of defined sequence is used to identify the specific binding protein(s). Other labels such as cyanine dyes or fluorescein can be used for the oligonucleotide probe. See Zhu, Q., Andrisani, O.M., Pot, D.A., and Dixon, J.E., Purification and characterization of a 43-kDa transcription factor required for rat somatostatin gene expression, *J. Biol. Chem.* 264, 6550–6556, 1989; Ogura, M., Takatori, T., and Tsuro, T., Purification and characterization of NF-R1 that regulates the expression of the human multidrug resistance (MDR1) gene, *Nucleic Acids Res.* 20, 5811–5817, 1992; Kwast-Welfeld, J., de Belle, I., Walker, P.R., Whitfield, J.F., and Sikorska, M., Identification of a new cAMP response element-binding factor by southwestern blotting, *J. Biol. Chem.* 268, 19551–19585, 1993; Liu, Z. and Jacob, S.T., Characterization of a protein that interacts with the rat ribosomal gene promoter and modulates RNA polymerase I transcription, *J. Biol. Chem.* 269, 16618–16625, 1994; Handen, J.S. and Rosenberg, H.F., An improved method for southwestern blotting, *Front. Biosci.* 2, c9–c11, 1997; Coffman, J.A. and Yuh, C.H., Identification of sequence-specific DNA binding proteins, *Methods Cell Biol.* 74, 653–675, 2004; Fedorov, A.V., Lukyanov, D.V., and Podgornaya, O.T., Identification of the proteins specifically binding to the rat LINE1 promoter, *Biochem. Biophys. Res. Commun.* 340, 553–559, 2006. There is a southwestern approach used for histochemistry (Hishikawa, Y., Damavandi, E., Izumi, S., and Koji, T., Molecular histochemical analysis of estrogen receptor alpha and beta expression in the mouse ovary: *in situ* hybridization and southwestern histochemistry, *Med. Electron Microsc.* 36, 67–73, 2003) and for ELISA (Fukuda, I., Nishiumi, S., Yabushita, Y. et al., A new southwestern chemistry-based ELISA for detection of aryl hydrocarbon receptor transformation: application to the screening of its receptor agonists and antagonists, *J. Immunol. Methods* 287, 187–201, 2004).

Specific Heat
The amount of heat required to raise the temperature of one gram of a substance by 1°C; specific heat of water is one calorie (4.184 joule). *Heat of fusion* is the amount of thermal energy to melt one mole of a substance at the melting point; also referred to as latent heat of fusion, kcal/mole, or kJ/mole. *Heat of vaporization* is the amount of energy required to convert one mole of a substance to vapor at the boiling point; also referred to as the latent heat of vaporization, kcal/mole, or kJ/mole.

Specificity
In assay validation, the ability of an assay to recognize a single analyte in a sample, which might contain closely related species; for example, in

DNA microarray assays, specificity would be the ability of a probe to bind to a unique target sequence and produce a signal proportional to the amount of that specific target sequence only. Also referred to as selectivity. In statistics, specificity is the proportion of negative tests to the total number of negative tests.

Spectroscopy   The interaction of electromagnetic radiation with materials including scattering, absorption, and emission. It does not include chemical effects such as bond formation or free radical formation. It does include some aspects of photochemistry, which is a specialized form of energy transduction. See Campbell, I.D. and Dwek, R.A., *Biological Spectroscopy*, Benjamin Cummings, Menlo Park, CA, 1984; Stuart, B., *Infrared Spectroscopy*, John Wiley & Sons, Chichester, UK, 2004.

Spectrum   A pattern of emissions from a particle following the application of energy. The emissions may be in the form of electromagnetic waves such as observed in spectroscopy or in the form of a mass such as that observed in mass spectrometry.

Sp1-Like Transcription Factors   A family of zinc-finger transcription factors in mammalian cells, which bind to GC-rich promoter elements. Originally described for SV40 virus. See Dynan, W.S. and Tjian, R., Isolation of transcription factors that discriminate between different promoters recognized by RNA polymerase II, *Cell* 32, 669–680, 1983; Lomberk, G. and Urrutia, R., The family feud: turning off Sp1 by Sp1-like KLF proteins, *Biochem. J.* 392, 1–11, 2005.

Spliced-Leader *Trans*-Splicing   Spliced-leader *trans*-splicing is a process mediated by a spliceosome where a short RNA sequence derived from the 5′ end of a non-mRNA to an acceptor site (3′ splice acceptor site) on a pre-RNA molecule. As a result, a diverse group of mRNA molecules in an organism acquires a common 5′ sequence. This process appears to occur in the same nuclear location as *cis*-splicing. See Murphey, W.J., Watkins, K.P., and Agabian, N., Identification of a novel Y branch structure as an intermediate in trypanosome mRNA processing: evidence for *trans* splicing, *Cell* 47, 517–525, 1986; Bruzik, J.P., Van Doren, K., Hirsh, D., and Steitz, J.A., *Trans* splicing involves a novel form of small nuclear ribonucleoprotein particles, *Nature* 335, 559–562, 1988; Layden, R.E. and Eisen, H., Alternate *trans* splicing in *Trypanosoma equiperdum*: implication for splice site selection, *Mol. Cell. Biol.* 8, 1352–1360, 1988; Hastings, K.E.M., SL *trans*-splicing: easy come or easy go? *Trends Genet.* 21, 240–247, 2005.

Spliceosome   A complex of RNA and protein components that assists the process of RNA splicing in ribosomes. Prokaryote RNA mRNA is less complex than eukaryotic mRNA and are not subject to RNA splicing. Eukaryotic RNA species that participate in spliceosome function include U1, U2, U4, U5, and U6. These RNA species are rich in uridine, which recognizes species sequences at the 5′ and 3′ sites on the pre-mRNA. The regions between these specific sites are excised so that the two remaining exons are joined together in the splicing process. See Robash, M. and Seraphin, B., Who's on first? The U1 snRNP-5′ splice site interaction and splicing, *Trends Biochem. Sci.* 16, 187–190, 1991; Garcia-Blanco, M.A., Messenger RNA reprogramming by spliceosome-mediated RNA *trans*-splicing, *J. Clin. Invest.* 112, 474–480, 2003; Kramer, A., Frefoglia, F., Huang, C.J. et al., Structure-function analysis of the U2 snRNP-associated splicing factor SF3a, *Biochem. Soc. Trans.* 33, 439–442, 2005.

Splicing Silencers Weakly interacting *cis* and *trans* factors that repress constitutive and alternative splicing during mRNA processing. There are exonic splicing silencers (ESS) and intronic splicing silencers (ISS). This is distinct from transcriptional silencing and is also known as transcriptional repression. See Staffa, A. and Cochrane, A., Identification of positive and negative splicing regulatory elements within the terminal tat-rev exon of human immunodeficiency virus type 1, *Mol. Cell. Biol.* 15, 4597–4605, 1995; Amendt, B.A., Si, Z.H., and Stoltzfus, C.M., Presence of exon splicing silencers with human immunodeficiency virus type 1 tat-exon 2 and tat-rev exon 3: evidence for inhibition mediated by cellular factors, *Mol. Cell. Biol.* 15, 4606–4615, 1995; Chew, S.L., Baginsky, L., and Eperon, I.C., An exonic splicing silencer in the testes-specific DNA ligase III beta exon, *Nucleic Acids Res.* 28, 402–410, 2000; Puzzoli, U. and Sironi, M., Silencers regulate both constitutive and alternative splicing events in mammals, *Cell. Mol. Life Sci.* 62, 1579–1604, 2005; Paca-Uccaralertkun, S., Damgaard, C.K., Auewarakul, P. et al., The effect of a single nucleotide substitution in the splicing silencer in the tat/rev intron on HIV type 1 envelope expression, *AIDS Res. Human Retroviruses* 22, 76–82, 2006.

SR Family of Proteins A family of phylogenetically conserved proteins that are essential cofactors in the splicing that occurs during the maturation of messenger RNA. SR proteins are essential for both constitutive and alternative splicing events. SR proteins are characterized by the presence of an N-terminal RNA recognition motif or motifs and a C-terminal region characterized by repeated arginine/serine residues. See Birney, E., Kumar, S., and Krainer, A.R., Analysis of the RNA-recognition motif and RS and RBB domains: conservation in metazoan pre-mRNA splicing factors, *Nuc. Acids Res.* 25, 503–5816, 1993; Ramchatesingh, J., Zahler, A.M., Neugebauer, K.M., Roth, M.B., and Cooper, T.A., A subset of SR proteins activates splicing of the cardiac troponin T alternative exon by direct interactions with an exonic enhancer, *Mol. Cell. Biol.* 15, 4898–4907, 1995; McNally, L.M. and McNally, M.T., SR protein splicing factors interact with the Rous sarcoma virus negative regulator of splicing elements, *J. Virol.* 70, 1163–1172, 1996; Zahler, A.M., Purification of SR protein splicing factors, *Methods. Mol. Biol.* 118, 419–432, 1999; Katsarou, M.E., Papakyriakou, A., Katsaros, N., and Scorilas, A., Expression of the C-terminal domain of novel human SR-A1 protein: interaction with the CTD domain of RNA polymerase II, *Biochem. Biophys. Res. Commun.* 334, 61–68, 2005; Sanford, J.R., Ellis, J., and Cáceres, J.F., Multiple roles of arginine/serine-rich splicing factors in RNA processing, *Biochem. Soc. Trans.* 33, 443–446, 2005; Rasheva, V.I., Knight, D., Borko, P., Marsh, K., and Frolov, M.V., Specific role of the SR protein splicing factors B52 in cell cycle control in *Drosophila*, *Mol. Cell. Biol.* 26, 3468–3477, 2006.

Staining A process by which contrast is introduced into a sample such as a tissue section or electrophoretograms. In positive staining, the item of interest is "staining" (absorbs the stain); in negative staining, the item of interest is unreactive and the background absorbs the stain. See Horne, R.W., Some recent applications of negative-staining methods to the study of biological structure in the electron microscope, *J. R. Micros. Soc.* 83, 169–177, 104; Kasten, F.H., Cytochemical studies with acridine orange and the influence of dye contaminants in the staining of nucleic acids, *Int. Rev. Cytol.* 21,

141–202, 1967; Biswas, B.B., Basu, P.S., and Pal, M.K., Gram staining and its molecular mechanism, *Int. Rev. Cytol.* 29, 1–27, 1970; Rabilloud, T., Mechanisms of protein silver staining in polyacrylamide gels: a 10-year synthesis, *Electrophoresis* 11, 785–794, 1990; Kiselev, N.A., Sherman, M.B. and Tsuprun, V.L., Negative staining of proteins, *Electron Microsc. Rev.* 3, 43–72, 1990; Gabriel, O. and Gersten, D.M., Staining for enzymatic activity after gel electrophoresis, *Anal. Biochem.* 203, 1–21, 1992; Lyon, R.O., Dye purity and dye standardization for biological staining, *Biotech. Histochem.* 77, 57–80, 2002; Hardy, E. and Castellanos-Serra, L.R., "Reverse-staining" of biomolecules in electrophoresis gesl: analytical and micropreparative applications, *Anal. Biochem.* 328, 1–13, 2004.

Standard Conditions (Standard State)	1 atm, 25°C (298°K).
Standard Electrode Potential	The value ($E°$) for the standard electromotive force of a cell in which hydrogen under standard conditions is oxidized to hydronium ions (solvated protons) at the left-hand electrode. This value is used as a standard to measure electrode potentials.
Standard Free Energy	A thermodynamic function designated G (after Walter Gibbs, frequently referred to as the Gibbs free energy). The change in G (ΔG) for a given reaction provides the information on the amount of energy derived from the reaction and is a product of the changes in enthalpy and entropy: $\Delta G = \Delta H - T\Delta S$. The standard free energy designated $\Delta G°$ indicates the values are those obtained for standard conditions. ΔG is negative for a thermodynamically favorable reaction. See *Enthalpy* and *Entropy*.
Stark Effect	The effect of an electrical field on the absorption/emission of spectra of a probe such as fluorescein or a coumarin derivative. It is derived from the interaction of the induced dipole(s) in the probe interacting with the charged group. See Sitkoff, D., Lockhart, D.J., Sharp, K.A., and Honig, B., Calculation of electrostatic effects at the amino terminal of an helix, *Biophys. J.* 67, 2251–2260, 1994; Pierce, D.W. and Boxer, S.A., Stark effect spectroscopy of tryptophan, *Biophys. J.* 68, 1583–1591, 1995; Klymchenko, A.S., Avilov, S.V., and Demchenko, A.P., Resolution of Cys and Lys labeling of α-crystallin with site-sensitive fluorescent 3-hydroxyflavone dye, *Anal. Biochem.* 329, 43–57, 2004.
Statistical Power	The probability of detecting a true effect of a particular size; equal to 1-false negative rate. Power increases as the random error of a procedure decreases.
Steroid Hormone Receptor (SHR)	Receptors for steroid hormones located in the nucleus. These are ligand-activated transcription factors. See Lavery, D.N. and McEwan, I.J., Structure and functions of steroid receptor AF1 transactivation domains: induction of active conformations, *Biochem. J.* 391, 449–464, 2005.
Stochastic	Involving or containing random errors.
Stochastic Process	A process consisting of a series of random variables (x_t), where t assumes values in a certain range of T. See Everitt, B.S., Ed., *The Cambridge Dictionary of Statistics*, Cambridge University Press, Cambridge, UK, 1998.
Structural Biology	Study of the secondary, tertiary, and higher structures of proteins in the proteome including, but limited to, the use of crystallography, nuclear magnetic resonance, and electron microscopy. See Smith, C.U.M., *Molecular*

Biology: A Structural Approach, MIT Press, Cambridge, MA, 1968; Devons, S., *Biology and the Physical Sciences*, Columbia University Press, New York, 1969; Rhodes, D. and Schwabe, J.W., Structural biology. Complex behavior, *Nature* 352, 478–479, 1991; Riddihough, G., Structural biology. Picture an enzyme at work, *Nature* 362, 793, 1993; Diamond, R., *Molecular Structures in Biology*, Oxford University Press, Oxford, UK, 1993; Kendrew, J.C. and Lawrence, E., Eds., *The Encyclopedia of Molecular Biology*. Blackwell Science, Cambridge, MA, 1994; Waksman, G. and Caparon, M., *Structural Biology of Bacterial Pathogenesis*, ASM Press, Washington, DC, 2005; Weiner, S., Sagi, I., and Addadi, L., Structural biology. Choosing the crystallization path less traveled, *Science* 309, 1027–1028, 2005; Sundstrom, S. and Martin, N., *Structural Genomics and High Throughput Structural Biology*, Taylor & Francis, Boca Raton, FL, 2006; Chiu, W., Baker, M.L., and Almo, S.C., Structural biology of cellular machines, *Trends Cell Biol.* 16, 144–150, 2006; Aravind, L., Iyer, L.M., and Koonin, E.V., Comparative genomics and structural biology of the molecular innovations of eukaryotes, *Curr. Opin. Struct. Biol.* 16, 409–419, 2006.

Structural Genomics Focuses on the physical aspects of the genome through the construction and comparison of gene maps and sequences, as well as gene discovery, localization, and characterization; determination of the three-dimensional structures of gene products using x-ray crystallography and NMR; known in a previous life as crystallography. See Kim, S.H., Shining a light on structural genomics, *Nat. Struct. Biol.* 5 (Suppl.), 643–645, 1998; Skolnick, J., Fetrow, J.S., and Kolinski, A., Structural genomics and its importance for gene function analysis, *Nat. Biotechnol.* 18, 283–287, 2000; Burley, S.K. and Bonnano, J.B., Structural genomics of proteins from conserved biochemical pathways and processes, *Curr. Opin. Struct. Biol.* 12, 383–391, 2002; Bourne, P.E. and Weissig, H., *Structural Bioinformatics*, Wiley-Liss, Hoboken, NJ, 2003; Staunton, D., Owen, J., and Cambell, I.D., NMR and structural genomics, *Acc. Chem. Res.* 36, 207–214, 2003; Burley, S.K. and Bonanno, J.B., Structural genomics, *Methods Biochem. Anal.* 44, 591–612, 2003; Goldsmith-Fishman, S. and Honig, B., Structural genomics: computational methods for structure analysis, *Protein Sci.* 12, 1813–1821, 2003; Bernardi, G., *Structural and Evolutionary Genomics: Natural Selection in Genome Evolution,* Elsevier, Amsterdam, 2004; Schmid, M.B., Seeing is believing: the impact of structural genomics on antimicrobial drug discovery, *Nat. Rev. Microbiol.* 2, 739–746, 2004; Lundstrom, K., Structural genomics of GPCRs, *Trends Biotechnol.* 12, 103–108, 2005; Sundstrom, M. and Norin, M., *Structural Genomics and High Throughput Structural Biology*, Taylor & Francis, Boca Raton, FL, 2006; Chandonia, J.M. and Brenner, S.A., The impact of structural genomics: expectations and outcomes, *Science* 311, 347–351, 2006.

Structural Proteomics Study of the primary, secondary, and tertiary structure of the proteins in a proteome; functional predictions from primary structure. See Norin, M. and Sundstrom, M., Structural proteomics: developments in structure-to-function predictions, *Trends Biotechnol.* 20, 79–84, 2002; Mylvagenam, S.E., Prahbakaran, M., Tudor, S.S. et al., Structural proteomics: methods in deriving protein structural information and issues in data management, *Biotechniques* (March Suppl.) 42–46, 2002; Sali, A., Glaseser, R., Earnest,

T., and Baumeister, W., From words to literature in structural proteomics, *Nature* 422, 216–225, 2003; Lefkovits, I., Functional and structural proteomics: a critical appraisal, *J. Chromatog. B Analyt. Technol. Biomed. Life Sci.* 787, 1–10, 2003; Smith, R.D., and Veenstra, T.D., *Proteome Characterization and Proteomics*, Academic Press, San Diego, CA, 2003; Kamp, R.M. and Calvete, J.J., *Methods in Proteome and Protein Analysis*, Springer, Berlin, 2004; Jung, J.W. and Lee, W., Structure-based functional discovery of proteins: structural proteomics, *J. Biochem. Mol. Biol.* 37, 28–34, 2004; Yakunin, A.F., Yee, A.A., Savchenko, A. et al., Structural proteomics: a tool of genome annotation, *Curr. Opin. Chem. Biol.* 9, 42–48, 2004; Chan, K. and Fernandez, D., Patent prosecution in structural proteomics, *Assay Drug Dev. Technol.* 2, 313–319, 2004; Liu, H.L. and Hsu, J.P., Recent developments in structural proteomics for protein structure determination, *Proteomics* 5, 2056–2068, 2005; Vinarov, D.A. and Markley, J.L., High-throughput automated platform for nuclear magnetic resonance-based structural proteomics, *Expert Rev. Proteomics* 2, 49–55, 2005.

SUMOylation The modification of proteins with the small ubiquitinlike modifier (SUMO). SUMO are ubiquitinlike proteins such as Rub1, Apg9, and Apg12 and are separate from ubiquitin domain proteins such as RAD23 and DSK2. Unlike modification with ubiquitin, SUMOylation does not signal protein degradation but rather appears to enhance stability and/or specific transport. See Müller, S., Hoege, C., Pyrowolakis, G., and Jenisch, S., SUMO, ubiquitin's mysterious cousin, *Nat. Rev. Molec. Cell Biol.* 2, 202–210, 2001; Watts, F.Z., SUMO modification of proteins other than transcription factors, *Semin. Cell Dev. Biol.* 15, 211–220, 2004; Gill, G., SUMO and ubiquitin in the nucleus: different functions, similar mechanisms? *Genes Dev.* 18, 2046–2059, 2004; Navotchova, M., Budhiraja, R., Coupland, G., Eisenhaber, F., and Bachmair, A., SUMO conjugation in plants, *Planta* 220, 1–8, 2004; Bossis, G. and Melchior, F., Regulation of SUMOylation by reversible oxidation and SUMO-conjugating enzymes, *Mol. Cell* 21, 349–357, 2006.

Surface Plasmon Resonance A technique that uses affinity binding to measure analytes in solution. Conceptually, surface plasmon resonance is related to other binding assays such as ELISA assays. In surface plasmon resonance, binding is measured by the increase in mass on a target probe, which is bound to a surface. Frequently, gold is the surface. Incident light is refracted from the surface and measured as reflectance (surface plasmon resonance). See also *Localized Surface Plasmon Resonance*. See Englebienne, P., Van Hoonacker, A.S., and Verhas, M., Surface plasmon resonance: principles, methods, and applications in biomedical sciences, *Spectroscopy* 17, 255–273, 2003; Smith, E.A. and Corn, R.M., Surface plasmon resonance imaging as a tool to monitor biomolecular interactions in an array-based format, *Appl. Spectros.* 57, 320A–332A, 2003; Lee, J.H., Yan, Y., Marriott, G., and Corn, R.M., Quantitative functional analysis of protein complexes on surfaces, *J. Physiol.* 563, 61–71, 2005; Piehler, J., New methodologies for measuring protein interactions *in vivo* and *in vitro*, *Curr. Opin. Struct. Biol.* 15, 4–14, 2005; Buijs, J. and Franklin, G.C., SPR-MS in functional proteomics, *Brief Funct. Genomic Proteomics* 4, 39–47, 2005; Pattnaik, P., Surface plasmon resonance: applications in understanding receptor–ligand interaction, *Appl. Biochem. Biotechnol.* 126, 76–92, 2005; Homola, J.,

Vaisocherova, H., Dostalek, J., and Piliarik, M., Multi-analyte surface plasmon resonance biosensing, *Methods* 37, 26–36, 2005.

Surface Tension A phenomenon that occurs when two fluids are in contact with each other due to molecular attraction between molecules of the two liquids at the surface of separation.

Surfactant The term "surfactant" dates to the 1950s when it was developed as a shortened version of "surface-active agent." Surfactants are amphipathic/amphiphilic molecules that tend to migrate to surfaces or interfaces in solutions (at equilibrium, the concentration of a surfactant is higher at the interface than the concentration in bulk solution). The term "detergent" is sometimes used interchangeably with "surfactant"; the purist might consider that "detergency" reflects on cleansing, which is one of the several properties of surfactants. Surfactants can also be described as dispersing agents, emulsifiers, foaming agents, stabilizers, solubilizers, or wetting agents depending on their performance activity and final effects. Surfactants can be divided into four broad chemical categories: anionic compounds such as soaps, which are sodium salts of long-chain alkyl carboxylic acids (alkanoic acids); cationic compounds such as alkyl amine derivatives such as Triton™ RW; amphoteric derivatives; and nonionic surfactants such as alkylphenol ethoxylates (Igepal™) and anhydrosorbitol esters (Tween derivatives). See Schick, M.J., Ed., *Nonionic Surfactants*, Marcel Dekker, New York, 1966; Attwood, D. and Florence, A.T., *Surfactant Systems: Their Chemistry, Pharmacy, and Biology*, Chapman and Hall, London, 1983; *Kirk-Othmer Encyclopedia of Chemical Technology*, 3rd ed., Vol. 22, Wiley-Interscience, New York, 1983; Cross, J., *Anionic Surfactants Analytical Chemistry*, Marcel Dekker, New York, 1998; van Oss, N.M., *Nonionic Surfactants Organic Chemistry*, Marcel Dekker, New York, 1998; Holmberg, K., *Novel Surfactants Preparation, Applications, and Biodegradabilty*, Marcel Dekker, New York, 1998; Kwak, J.C.T., *Polymer-Surfactant Systems*, Marcel Dekker, New York, 1998; Hus, J.-P., *Interfacial Forces and Fields Theory and Applications*, Marcel Dekker, New York, 1999; Pefferkorn, E., *Interfacial Phenomena in Chromatography*, Marcel Dekker, New York, 1999; Myers, D., *Surfaces, Interfaces, and Colloids: Principles and Applications*, Wiley-VCH, New York, 1999; Broze, G., *Handbook of Detergents*, Marcel Dekker, New York, 1999; Marangani, A.G. and Narine, S.S., Eds., *Physical Properties of Lipids*, Marcel-Dekker, New York, 2002. Surfactants are used extensively in the solubilization of membranes and phospholipid (Lichtenberg, D., Robson, R.J., and Dennis, E.A., Solubilization of phospholipids by detergents. Structural and kinetic aspects, *Biochim. Biophys. Acta* 737, 285–304, 1983; Dennis, E.A., Micellization and solubilization of phospholipid by surfactants, *Adv. Colloid Interface Sci.* 26, 155–175, 1986; Silvius, J.R., Solublization and functional reconstitution of biomembrane components, *Annu. Rev. Biophys. Biomol. Struct.* 21, 323–348, 1992; Henry, G.D. and Sykes, B.D., Methods to study membrane protein structure in solution, *Methods Enyzmol.* 239, 515–535, 1994; Bowie, J.H., Stabilizing membrane proteins, *Curr. Opin. Struct. Biol.* 11, 397–402, 2001; Seddon, A.M., Curow, P., and Booth, B.J., Membrane proteins, lipids, and detergents: not just a soap opera, *Biochim. Biophys. Acta* 1666, 105–117, 2004). Nonionic surfactants have an effect (drag reduction) on fluid flow at low concentrations (Jacobs, E.W., Anderson, G.W., Smith, C.A. et al., Drag reduction using high molecular weight fractions of poly-ethylene oxide, in Sellin, R.H.J. and Moses, R.T., Eds., *Drag Reduction in Fluid Flows: Techniques for Friction Control*, Ellis-Horwood, Chichester, UK, 1989; Drappier, J., Divoux, T., Amarouchene, Y. et al., Turbulent drag reduction by surfactants, *Europhysics Lett.* 74, 362–368, 2006).

Surrogate Marker   A biomarker that can be used in place of a clinical indication for diagnosis or a clinical endpoint for prognosis; also referred to as surrogate endpoint. See Hilsenbeck, S.G. and Clark, G.M., Surrogate endpoints in chemoprevention of breast cancer: guidelines for evaluation of new biomarkers, *J. Cell. Biochem.* (Suppl. 17G), 205–211, 1993; Morrish, P.K., How valid is dopamine transporter imaging as a surrogate marker in research trials on Parkinson's disease? *Mov. Disord.* 18 (Suppl. 7), S63–S70, 2003; Kluft, C., Principles of use of surrogate markers and endpoints, *Maturitas* 47, 293–298, 2004; Bowdish, M.E., Arcasoy, S.M., Wilt, J.S. et al., Surrogate markers and risk factors for chronic lung allograft dysfunction, *Am. J. Transplant.* 4, 1171–1178, 2004; Lieberman, R., Evidence-based medical perspectives: the evolving role of PSA for early detection, monitoring of treatment response, and as a surrogate endpoint of efficacy for intervention in men with different clinical risk states for the prevention and progression of prostate cancer, *Am. J. Ther.* 11, 501–506, 2004; Li, Z., Chines, A.A., and Meredith, M.P., Statistical validation of surrogate endpoints: is bone density a valid surrogate for fracture? *J. Musculoskelet. Neuronal Interact.* 4, 64–74, 2004; Kantarci, K., and Jack, C.R., Jr., Quantitative magnetic resonance techniques as surrogate markers of Alzheimer's disease, *NeuroRx* 1, 196–205, 2004; Wier, C.J. and Walley, R.J., Statistical evaluation of biomarkers as surrogate endpoints: a literature review, *Stat. Med.* 25, 183–203, 2006.

Synovial Proteome   The total protein content of synovial fluid; results obtained from the proteomic analysis of synovial fluid. See Dasuri, K., Antonovici, M., Chen, K. et al., The synovial proteome: analysis of fibroblast-like synoviocytes, *Arthritis Res. Ther.* 6, R161–R168, 2004; Romeo, M.J., Espina, V., Lowenthal, M. et al., CSF proteome: a protein repository for potential biomarker idenfication, *Expert. Rev. Proteomics* 2, 57–70, 2005; Hueber, W., Kidd, B.A., Tomooka, B.H. et al., Antigen microarray profiling of autoantibodies in rheumatoid arthritis, *Arthritis Rheum.* 52, 2645–2655, 2005.

Systems Biology   The integration of data at the genomic, transcriptomic, proteomic, and metabolomic levels including functional and structural data to understand biological function, which can be described by a mathematical function. See Ideker, T., Galitski, T., and Hood, L., A new approach to decoding life: systems biology, *Annu. Rev. Genomics Hum. Genet.* 2, 343–372, 2001; Levesque, M.P. and Benfey, P.N., Systems biology, *Curr. Biol.* 14, R179–R189, 2004; Weston, A.D. and Hood, L.J., Systems biology, proteomics, and the future of health care: toward predictive, preventative, and personalized medicine, *J. Proteome Res.* 3, 179–196, 2004; Benner, S.A. and Ricardo, A., Planetary systems biology, *Mol. Cell* 17, 471–472, 2005; Friboulet, A. and Thomas, D., Systems biology — an interdisciplinary approach, *Biosens. Bioelectron.* 20, 2404–2407, 2005; Aderem, A., Systems biology: its practices and challenges, *Cell* 121, 511–513, 2005; Alberghina, L. and Westerhoff, H.V., Eds., *Systems Biology: Definitions and Perspectives*, Springer, Berlin, 2005; Theilgaard-Monch, K., Porse, B.T., and Borregaard, N., Systems biology of neutrophil differentiation and immune response, *Curr. Opin. Immunol.* 18, 54–60, 2006; Mustacchi, R., Hohmann, S., and Nielsen, J., Yeast systems biology to unravel the network of life, *Yeast* 23, 227–238, 2006; Baker, M.D., Wolanin, P.M., and Stock, J.B., Systems biology of bacterial chemotaxis, *Curr. Opin. Microbiol.* 9, 187–192, 2006; Philippi, S. and Kohler, J., Addressing the problems with life-science databases for traditional uses and systems biology, *Nat. Rev. Genet.* 7, 482–488, 2006; Palsson, B., *Systems Biology: Properties of Reconstructuced Networks*, Cambridge University Press, Cambridge, UK, 2006.

Targeted Proteomics
Analysis of a defined portion of a proteome such as a glycoproteome, phosphoproteome, or ribosomal proteins. See Knepper, M.A. and Masilamani, S., Targeted proteomics in the kidney using ensembles of antibodies, *Acta Physiol. Scand.* 173, 11–21, 2001; Warcheid, B. and Fenselau, C., A targeted proteomics approach to the rapid identification of bacterial cell mixtures by matrix-assisted laser desorption/ionization mass spectrometry, *Proteomics* 4, 2877–2892, 2004; Ecelbarger, C.A., Targeted proteomics using immunoblotting technique for studying dysregulation of ion transporters in renal disorders, *Expert Rev. Proteomics* 1, 219–227, 2004; Immler, D., Greven, S., and Reinemer, P., Targeted proteomics in biomarker validation: detection and quantification of proteins using a multidimensional peptide separation strategy, *Proteomics* 6, 2947–2958, 2006.

Target of Rapomycin (TOR, mTOR)
A highly conserved serine/threonine kinase that regulates cell growth and which is a target for immunosuppression in kidney transplantation with therapeutics such as sirolimus and everolimus. See Hall, M.N., The TOR signaling pathway and growth control in yeast, *Biochem. Soc. Trans.* 24, 234–239, 1996; Dennis, P.B., Fumagalli, S., and Thomas, G., Target of rapamycin (TOR): balancing the opposing forces of protein synthesis and degradation, *Curr. Opin. Genet. Dev.* 9, 49–54, 1999; Bestard, O., Cruzado, J.M., and Grinyo, J.M., Calcineurin-inhibitor-sparing immunosuppressive protocols, *Transplant Proc.* 37, 3729–3732, 2005; Mabasa, V.H. and Ensom, M.H., The role of therapeutic monitoring of everolimus in solid organ transplantation, *Ther. Drug Monit.* 27, 666–676, 2005; Avruch, J., Lin, Y., Long, X., Murthy, S., and Ortiz-Vega, S., Recent advances in the regulation of the TOR pathway by insulin and nutrients, *Curr. Opin. Nutr. Metab. Care,* 8, 67–72, 2005; Abraham, R.T., TOR signaling: an odyssey from cellular stress to the cell growth machinery, *Curr. Biol.* 15, R139–R141, 2005; Cesareni, G., Ceol, A., Gavrila, C. et al., Comparative interactomics, *FEBS Lett.* 579, 1828–1833, 2005; Martin, D.E. and Hall, M.N., The expanding TOR signaling network, *Curr. Opin. Cell Biol.* 17, 158–166, 2005; Neuberger, J. and Jothimani, D., Long-term immunosuppression for prevention of nonviral disease recurrence, *Transplant. Proc.* 37, 1671–1674, 2005; dos Sarbassov, D., Ali, S.M., and Sabatini, D.M., Growing roles for the mTOR pathway, *Curr. Opin. Cell Biol.* 17, 596–603, 2005; Wullschleger, S., Loewith, R., and Hall, M.N., TOR signaling in growth and metabolism, *Cell* 124, 471–484, 2006.

Telomerase
A ribonucleoprotein complex that catalyzes the synthesis of DNA at the ends of chromosomes and confers replicative immortality to cells. It is considered to be important in the growth of cancer cells and is a therapeutic target. See Blackburn, E.H., Greider, C.W., Henderson, E. et al., Recognition and elongation of telomeres by telomerase, *Genome* 31, 553–560, 1989; Lamond, A.I., Tetrahymena telomerase contains an internal RNA

template, *Trends Biochem. Sci.* 14, 202–204, 1989; Greider, C.W., Telomeres, telomerase, and senescence, *Bioessays* 12, 363–369, 1990; Shippen-Lentz, D. and Blackburn, E.H., Functional evidence for an RNA template in telomerase, *Science* 247, 546–552, 1990; Romero, D.P. and Blackburn, E.H., A conserved secondary structure for telomerase RNA, *Cell* 67, 343–353, 1991; Greider, C.W., Telomerase and telomere-length regulation: lessons from small eukaryotes to mammals, *Cold Spring Harb. Symp. Quant. Biol.* 58, 719–723, 1993; Harley, C.B., Kim, N.W., Prowse, K.R. et al., Telomerase, cell immortality, and cancer, *Cold Spring Harb. Symp. Quant. Biol.* 59, 307–315, 1994; Rhyu, M.S., Telomeres, telomerase, and immortality, *J. Natl. Cancer Inst.* 87, 884–894, 1995; Buchkovich, K.J., Telomeres, telomerase, and the cell cycle, *Prog. Cell Cycle Res.* 2, 187–195, 1996; Shay, J.W. and Wright, W.E., Use of telomerase to create bioengineered tissues, *Ann. N.Y. Acad. Sci.* 1057, 479–491, 2005; Wirth, T., Kuhnel, F., and Kubicka, S., Telomerase-dependent gene therapy, *Curr. Mol. Med.* 5, 243–251, 2005; Hahn, W.C., Telomere and telomerase dynamics in human cells, *Curr. Mol. Med.* 5, 227–231, 2005; Blackburn, E.H., Telomeres and telomerase: their mechanisms of action and the effects of altering their functions, *FEBS Lett.* 579, 859–862, 2005; Shin, J.S., Hong, A., Solomon, M.J., and Lee, C.S., The role of telomeres and telomerase in the pathology of human cancer and aging, *Pathology* 38, 103–113, 2006; Flores, I., Benetti, R., and Blasco, M.A., Telomerase regulation and stem cell behavior, *Curr. Opin. Cell Biol.* 18, 254–260, 2006.

Tetramer   Using a protein molecule containing four either identical or nonidentical subunits. In the case of nonidentical subunits, the tetramer is frequently a dimer of dimers such as hemoglobin (Verzili, D., Rosato, N., Ascoli, F., and Chiancone, E., Aromatic amino acids and subunit assembly in the hemoglobins from *Scapharca inaequivalvis*; a fluorescence and Cd study of the apoproteins, *Biochim. Biophys. Acta* 954, 108–113, 1988; Ackers, G.K. and Holt, J.M., Asmmetric cooperativity in a symetric tetramer: human hemoglobin, *J. Biol. Chem.* 281, 11441–11443, 2006) and MHC II receptors (Schafer, P.H., Pierce, S.K., and Jardetzky, T.S., The structure of MHC class II: a role for dimer of dimers, *Semin. Immunol.* 7, 389–398, 1995; Yadati, S., Nydam, T., Demian, D. et al., Salt bridge residues between I-Ak dimer of dimers alpha-chains modulate antigen presentation, *Immunol. Lett.* 67, 47–55, 1999; Lindstedt, R., Monk, N., Lombardi, G., and Lechler, R., Amino acid substitutions in the putative MHC class II "dimer of dimers" interface inhibit CD4+ T-cell activation, *J. Immunol.* 166, 800–808, 2001) although there are examples of tetrameric proteins such as lactose dehydrogenase where there can be varying subunit composition (Pesce, A., McKay, R.H., Stolzenbach, F. et al., The comparative enzymology of lactic dehydrogenases. I. Properties of the crystalline beef and chicken enzymes, *J. Biol. Chem.* 239, 1753–1761, 1964; Maekawa, M., Lactate dehydrogenase isoenzymes, *J. Chromatog.* 429, 373–398, 1988). See Kosaka, M., Iishi, Y., Okagawa, K. et al., Tetramer Bence Jones protein in the immunoproliferative diseases. Angioimmunoblastic lymphadenopathy, primary amyloidosis, and multiple myeloma, *Am. J. Clin. Pathol.* 91, 639–646, 1989; Meissner, G., Ligand binding and cooperative interactions among the subunits of the tetrameric Ca^{2+} release channel complex of sarcoplasmic reticulum, *Adv. Exp. Med. Biol.* 311, 277–287, 1992; Furey, W., Arjunan, P., Chen, L. et al., Structure–function relationships and flexible tetramer assembly in pyruvate decarboxylase revealed by analysis of crystal structures, *Biochim. Biophys. Acta* 1385, 253–270, 1998; Grotzinger, J., Kerneback, T., Kallen, K.J., and Rose-John, S., IL-6 type cytokine receptor complexes: hexamer, tetramer, or both? *Biol. Chem.* 380, 803–813, 1999; Kang, H.M., Choi, K.S., Kassam, G. et al., Role of annexin II tetramer in plasminogen activation, *Trends Cardiovasc. Med.* 9, 92–102, 1999; Hamiche, A. and Richard-Foy, H., Characterization of specific nucleosomal states by use of selective substitution reagents in model octamer and tetramer structures, *Methods* 19, 457–464, 1999; Xu, X.N. and Screaton, G.R., MHC/peptide tetramer studies of T-cell function, *J. Immunol. Methods* 268, 21–28, 2002; Zimmerman, A.L., Two B or not two B? Questioning the rotational symmetry of tetrameric ion channels, *Neuron* 36, 997–999, 2002; Kita, H., He, X.S., and Gershwin, M.E., Application of tetramer technology in studies on autoimmune diseases, *Autoimmun. Rev.* 2, 43–49, 2003.

Tetrose   A four-carbon monosaccharide including *meso*-erythritol (erythritol), D-erythrose, L-erythrulose, and D-threose. See Horecker, B.L., Smyrniotis, P.Z., Hiatt, H.H, and Marks, P.A., Tetrose phosphate and the formation of aldoheptulose diphosphate, *J. Biol. Chem.* 212, 827–836, 1955; Batt, R.D., Dickens, F., and Williamson, D.H., Tetrose metabolism. 1. The preparation and degradation of specifically labelled [^{14}C] tetroses and [^{14}C] tetritols, *Biochem. J.* 77, 272–281, 1960; Taylor, G.A. and Ballou, C.E., D-glycero-tetrulose 1,4-diphosphate (D-erythulose 1,4-diphosphate), *Biochemistry* 2, 553–555, 1963; Giudici, T.A. and Fluharty, A.L., A specific method for the determination of the aldotetroses, *Anal. Biochem.* 13, 448–457, 1965; Lai, C.Y., Martinez-de Dretz, G., Bacila, M. et al., Labeling of the active site of aldolase with glyceraldehyde 3-phosphate and erythrose 4-phosphate, *Biochem. Biophys. Res. Commun.* 27, 665–672, 1968; Benov, L. and Fridoich, I., Superoxide dependence of the toxicity of short chain sugars, *J. Biol. Chem.* 273, 25741–25744, 1998; Lehman, T.D. and Ortwerth, B.J., Inhibition of advanced glycation end product-associated protein crosslinking, *Biochim. Biophys. Acta* 1535, 110–119, 2001; Chaput, J.C. and Szostak, J.W., TNA synthesis by DNA polymerases, *J. Am. Chem. Soc.* 125, 9274–9275, 2003; Kempeneers, V., Vastmans, K., Rozenski, J., and Herdewijn, P., Recognition of threosyl nucleotides by DNA and RNA polymerases, *Nucleic Acids Res.* 31, 6221–6226, 2003; Ichida, J.K., Zou, K., Horhota, A. et al., An *in vitro* selection system for TNA, *J. Am. Chem. Soc.* 127, 2802–2803, 2005; Horhota, A., Zou, K., Ichida, J.K. et al., Kinetic analysis of an efficient DNA-dependent TNA polymerase, *J. Am. Chem. Soc.* 127, 7427–7434, 2005; Wamelink, M.M, Struys, E.A., Huck, J.H. et al., Quantification of sugar phosphate intermediates of the pentose phosphate pathway by LC-MS/MS: application to two new inherited defects of metabolism, *J. Chromatog. B Analyt. Technol. Biomed. Life Sci.* 823, 18–25, 2005; Weber, A.L. and Pizzarello, S., The peptide-catalyzed stereospecific synthesis of tetroses: a possible model for prebiotic molecular evolution, *Proc. Natl. Acad. Sci. USA*, 103, 12713–12717, 2006.

Theragnostic   As in theragnostic imaging, described as the combined use of molecular and functional imaging to prescribe the distribution of radiations in three-dimensional space as a function of time in the process of radiation therapy. See Bentzen, S.M., Theragnostic imaging for radiation oncology: dose-painting by numbers, *Lancer Oncol.* 6, 112–117, 2005.

Theranostics   The use of diagnostic laboratory tests to guide therapeutic outcomes. Current use has emphasized "real-time" PCR assays for the identification of pathogens. See Picard, F.J. and Bergeron, M.G., Rapid molecular theranostics in infectious disease, *Drug Discov. Today* 7, 1092–1101, 2002.

Therapeutic Equivalence (TE)   Drug products including biologics can be considered to be therapeutically equivalent if such products can be substituted for brand products/prescribed products/originator products with the full expectation that such substituted products will produce the same clinical effect and safety as the brand products/prescribed products/originator products. See Patnaik, R., Hauck, W.W. et al., An individual bioequivalence criterion: regulatory considerations, *Stat. Med.* 19, 2821–2842, 2000; Meyer, M.C., United States Food and Drug Administration requirements for approval of generic drug products, *J. Clin. Psychiatry* 62 (Suppl. 5), 4–9, 2001; Temple, R., Policy developments in regulatory approval, *Stat. Med.* 21, 2939–3048, 2002; Gould, A.L, Substantial evidence of effect, *J. Biopharm. Stat.* 12, 53–77, 2002; Chen, M.L., Panhard, X., and Mentre, F., Evaluation by simulation of tests based on nonlinear mixed-effects models in pharmacokinetic interaction and bioequivalence cross-over clinical trials, *Stat. Med.* 24, 1509–1524, 2005; Bolton, S., Bioequivalence studies for levothyroxine, *AAPS J.* 7, E47–E53, 2005.

Thermal Conductivity   Rate of heat transfer by conduction through unit thickness, across unit area for unit difference of temperature; measured as calories/second/cm^2/cm (thickness) and a temperature difference of 1°C; unit is cal/cm sec °K or W/cm °K. See Harting, R. and Pfeiffenberger, U., Thermal conductivity of bovine and pig retina: an experimental study, *Grafes Arch. Clin. Exp. Ophthalmol.* 219, 290–291, 1982; Miller, J.H., Wilson, W.E., Swenberg, C.E. et al., Stochastic model of free radical yields in oriented DNA exposed to densely ionizing radiation at 77K, *Int. J. Radiat. Biol. Relat. Stud. Phys. Chem. Med.* 53, 901–907, 1988; Arkin, H., Holmes, K.R., and Chen, M.M., A technique for measuring the thermal conductivity and evaluating the "apparent conductivity" concept in biomaterials, *J. Biomech. Eng.* 111, 276–282, 1989; Cheng, J., Shoffner, M.A., Mitchelson, K.R. et al., Analysis of ligase chain reaction products amplified in a silicon-glass chip using capillary electrophoresis, *J. Chromatog. A* 732, 151–158, 1996; Bhattacharya, A. and Mahajan, R.L., Temperature dependence of thermal conductivity of biological tissues, *Physiol. Meas.* 24, 769–783, 2003; Rodriguez, I., Lesaicherre, M., Tie, Y. et al., Practical integration of polymerase chain reaction amplification and electrophoretic analysis in microfluidic devices for genetic analysis, *Electrophoresis* 24, 172–178, 2003.

Thermophilic   Term used to describe an organism that grows at elevated temperatures. The most famous thermophilic organism is the bacteria (*Thermus aquaticus*) responsible for TAQ polymerase, which is used in the PCR. See Friedman, S.M., Protein-synthesizing machinery of thermophilic bacteria, *Bacteriol. Rev.* 32, 27–38, 1968; Singleton, R., Jr. and Amelunxen, R.E., Protein from thermophilic microorganisms, *Baceriol. Rev.* 37, 320–342, 1973; Lasa, I. and Berenguer, J., Thermophilic enzymes and their biotechnology potential, *Microbiologica* 9, 77–89, 1993; Kelly, R.M., Peeples, T.L., Halio, S.B. et al., Extremely thermophilic microorganisms. Metabolic strategies, genetic characteristics, and biotechnological potential, *Ann. N.Y. Acad. Sci.* 745, 409–425, 1994; Russell, R.J. and Taylor, G.L., Engineering thermostability: lessons from the thermophilic proteins, *Curr. Opin. Biotechnol.* 6, 370–374, 1995; Kumar, S. and Nussinov, R., How do thermophilic proteins deal with heat? *Cell. Mol. Life Sci.* 58, 1216–1233, 2001; Sambongi, Y., Uchiyama, S., Kobayashi, Y. et al., Cytochrome c from a thermophilic bacterium has provided insights into the mechanisms of protein maturation, folding, and stability, *Eur. J. Biochem.* 269, 3355–3361, 2002; Radianingtyas, H. and Wright, P.C., Alcohol dehydrogenases from thermophilic and hyperthermophilic archae and bacteria, *FEMS Microbiol. Rev.* 27, 593–616, 2003; McMullen, G., Christie, J.M., Rahman, T.J. et al., Habitat, applications, and genomics of the aerobic, thermophilic genus *Geobacillus*, *Biochem. Soc. Trans.* 32, 214–217, 2004; Averhoff, B., DNA transport and natural transformation in mesophilic and thermophilic bacteria, *J. Bioenerg. Biomembr.* 36, 25–33, 2004; Pereira, M.M., Bandeiras, T.M., Fernandes, A.S. et al., Respiratory chains from aerobic thermophilic prokaryotes, *J. Bioenerg. Biomembr.* 36, 93–105, 2004; Nishida, C.R. and Ortiz de Montellano, P.R., Thermophilic cytochrome P450 enzymes, *Biochem. Biophys. Res. Commun.* 338, 437–445, 2005; Egorova, K. and Antranikian, G., Industrial relevance of thermophilic archaea, *Curr. Opin. Microbiol.* 8, 649–655, 2005.

Thiolase Superfamily
A family of condensing enzymes with diverse functions such as the formation of carbon–carbon bonds in the synthesis of fatty acids and polyketides. The expression of peroxisome thiolases is regulated in the PPAR receptors. See Swartzman, E.E., Viswanathan, M.N., and Thorner, J., The PAL1 gene product is a peroxisomal ATP-binding cassette transporter in the yeast *Saccharomyces cerevisiae*, *J. Cell Biol.* 132, 549–563, 1996; Aoyama, T., Peters, J.M., Iritani, N. et al., Altered constitutive expression of fatty acid-metabolizing enzymes in mice lacking the peroxisome proliferator-activated receptor alpha (PPARalpha), *J. Biol. Chem.* 273, 5678–5684, 1998; Latruffe, N., Chekauoui, M.M., Nicholas-Frances, V. et al., Regulation of the peroxisomal beta-oxidation dependent by peroxisome proliferator-activated receptor alpha and kinases, *Biochem. Pharmacol.* 60, 1027–1032, 2000; Rabus, R., Kube, M., Beck, A., Widdel, F., and Reinhardt, R., Genes involved in the anaerobic degradation of ethyl benzene in a denitrifying bacterium, strain EbN1, *Arch. Microbiol.* 178, 506–516, 2002; Kursula, P., Sikkila, H., Fukao, T., Kondo, N., and Wierenga, R.K., High-resolution crystal structures of human cytosolic thiolase (CT): a comparison of the active sites of human CT, bacterial thiolase, and bacterial KAS I., *J. Mol. Biol.* 347, 189–201, 2005; Haapalinen, A.M., Merilainen, G., and Wierenga, R.K., The thiolase superfamily: condensing enzymes with diverse reaction specifications, *Trends Biochem. Sci.* 31, 64–71, 2006.

Thioredoxin
A small protein that functions as a hydrogen donor/reducing agent in biological systems; considered to be a major regulator of redox reactions in the cell. See Holmgren, A., Thioredoxin, *Annu. Rev. Biochem.* 54, 237–271, 1985; Brot, N. and Weissbach, H., Biochemisty of methionine sulfoxide residues in proteins, *Biofactors* 3, 91–96, 1991; Martin, J.L., Thioredoxin — a fold for all reasons, *Structure* 3, 245–250, 1995; Aslund, F. and Beckwith, J., The thioredoxin superfamily: redundancy, specificity, and gray-area genomics, *J. Bacteriol.* 181, 1375–1379, 1999; Arrigo, A.P., Gene expression and the thiol redox state, *Free Radic. Biol. Med.* 27, 936–944, 1999; Holmgren, A., Antioxidant function of thioredoxin and gluaredoxin systems, *Antioxid. Redox Signal.* 2, 811–820,

2000; Arner, E.S.J. and Holmgren, A., Physiological functions of thioredoxins and thioredoxin reductase, *Eur. J. Biochem.* 267, 6102–6109, 2000; Pearson, W.R., Phylogenes of glutathione transferase families, *Methods Enzymol.* 401, 186–204, 2005; Burke-Gaffney, A., Callister, M.E., and Nakamura, H., Thioredoxin: friend or foe in human disease? *Trends Pharmacol. Sci.* 26, 398–404, 2005; Stefankova, P., Kollarova, M., and Barak, I., Thioredoxin — structural and functional complexity, *Gen. Physiol. Biophys.* 24, 3–11, 2005; Koc, A., Mathews, C.K., Wheeler, L.J., Gross, M.K., and Merrill, G.F., Thioredoxin is required for deoxyribonucleotide pool maintenance during S phase, *J. Biol. Chem.* 281, 15058–15063, 2006.

Thymosins   A group of peptide hormones (alpha, beta, and gamma) originally described in the thymus, which have a broad range of functions including the modulation of actin polymerization and control of lymphoid tissue function in the development of T-cells and the maturation of B-cells to form plasma cells. See Goldstein, A.L. and White, A., Role of thymosin and other thymic factors in the development, maturation, and functions of lymphoid tissue, *Curr. Top. Exp. Endocrinol.* 1, 121–149, 1971; Bach, J.F. and Carnaud, C., Thymic factors, *Prog. Allergy* 21, 342–408, 1976; Low, T.L., Thurman, G.B., Chincarini, C. et al., Current status of thymosin research: evidence for the existence of a family of thymic factors that control T-cell maturation, *Ann. N.Y. Acad. Sci.* 332, 33–48, 1979; Ampe, C. and Vandekerchove, J., Actin–actin binding protein interfaces, *Semin. Cell Biol.* 5, 175–182, 1994; Chen, C., Li, M., Yang, H. et al., Roles of thymosins in cancers and other organ systems, *World J. Surg.* 29, 264–270, 2005; Goldstein, A.L., Hannappel, E., and Kleinman, H.K., Thymosin beta4: actin-sequestering protein moonlights to repair injured tissues, *Trends Mol. Med.* 11, 421–429, 2005.

Time-of-Flight   The term "time-of-flight" designates techniques and apparatus that depend on the time taken by particles to traverse a set distance, for example the separation of ions according to their mass in mass spectrometry. Time-of-flight mass spectrometry (TOF-MS) measures flight time of ions; lighter ions travel a greater distance than heavier ions; mass proportional to time squared; converted to m/z by calibration with standards (Cotter, R.J., Time-of-flight mass spectrometry: an increasing role in the life sciences, *Biomed. Environ. Mass Spectrom.* 18, 513–532, 1989; Guilhaus, M., Selby, D., and Mlynski, V., Orthogonal acceleration time-of-flight mass spectrometry, *Mass Spectrom. Rev.* 19, 65–107, 2000; Belu, A.M., Graham, D.J., and Castner, D.G., Time-of-flight secondary ion mass spectrometry: techniques and applications for the characterization of biomaterial surfaces, *Biomaterials* 24, 3635–3653, 2003; Seibert, V., Wiesner, A., Buschmann, T., and Meuer, J., Surface-enhanced laser desorption ionization time-of-flight mass spectrometry [SELDI TOF-MS] and ProteinChip technology in proteomics research, *Pathol. Res. Pract.* 200, 83–94, 2004; Vestal, M.L., and Campbell, J.M., Tandem time-of-flight mass spectrometry, *Methods Enzymol.* 402, 79–108, 2005). However, the technique is used in other applications such as measurement of blood flow (Kochhar, R., Khandelwal, N., Singh, P., and Suri, S., Arterial contamination: a useful indirect sign of cerebral sino-venous thrombosis, *Acta Neurol. Scand.* 114, 139–142, 2006; Han, S., Granwehr, J., Garcia, S. et al., Auxillary probe design adaptable to existing probes for remote detection NMR, MRI, and time-of-flight tracing, *J. Magn. Reson.*, in press, 2006).

Tissue   A microarray consisting of cores (0.6 mm in diameter, for example) of
Microarray   tissue embedded in a paraffin block. Samples are taken from existing
                    paraffin-block sections This technology allows multiple samples to be
                    processed at the same time under the same conditions. See Moch, H.,
                    Schrami, P., Bubendorf, L. et al., High-throughput tissue microarray anal-
                    ysis to evaluate genes uncovered by cDNA microarray screening in renal
                    cell carcinoma, *Am. J. Pathol.* 154, 981–986, 1999; Kallioniemi, O.P.,
                    Wagner, U., Kononen, J., and Sauter, G., Tissue microarray technology
                    for high-throughput molecular profiling of cancer, *Hum. Mol. Genet.* 10,
                    657–662, 2001; Rao, J., Seligson, D., and Hemstreet, G.P., Protein expres-
                    sion analysis using quantitative fluorescence image analysis on tissue
                    microarray slides, *BioTechniques* 32, 928–930, 2002; Rubin, M.A., Dunn,
                    R., Strawderman, M., and Pienta, K.J., Tissue microarray sampling strat-
                    egy for prostate cancer biomarker analysis, *Am. J. Surg. Pathol.* 26,
                    212–219, 2002; Hedvat, C.V., Hedge, A., Chaganti, R.S. et al., Application
                    of tissue microarray technology to the study of non-Hodgkins and
                    Hodgkins lymphoma, *Hum. Pathol.* 33, 968–974, 2002; Kim, W.H., Rubin,
                    M.A., and Dunn, R.L., High-density tissue microarray, *Am. J. Surg. Pathol.*
                    26, 1236–1238, 2002; Parker, R.L., Huntsman, D.G., Lesack, D.W. et al.,
                    Assessment of interlaboratory variation in the immunohistochemical
                    determination of estrogen receptor status using a breast cancer tissue
                    microarray, *Am. J. Clin. Pathol.* 117, 723–728, 2002; Giltnane, J.M. and
                    Rimm, D.L., Technology insight: identification of biomarkers with tissue
                    microarray technology, *Nat. Clin. Pract. Oncol.* 1, 104–111, 2004;
                    Zimpfer, A., Schonberg, S., Lugli, A. et al., Construction and validation
                    of a bone marrow tissue microarray, *J. Clin. Pathol.*, 60, 57–61, 2007.
Titin   A very large protein (mass of approximately 3 million daltons), the third
                    most abundant protein in vertebrate striated muscle; it has a role in pro-
                    viding sarcomeric alignment and recoil. Titin is a single-chain protein that
                    extends from the M line to the Z line, forming a thick filament. See
                    Maruyama, K., Connectin, an elastic filamentous protein of striated mus-
                    cle, *Int. Rev. Cytol.* 104, 81–114, 1986; Trinick, J., Elastic filaments and
                    giant proteins in muscle, *Curr. Opin. Cell Biol.* 3, 112–119, 1991; Fulton,
                    A.B., and Isaacs, W.B., Titin, a huge, elastic sarcomeric protein with a
                    probable role in morphogenesis, *Bioessays* 13, 157–161, 1991; Trinick,
                    J., Understanding the functions of titin and nebulin, *FEBS Lett.* 307,
                    44–48, 1992; Kellermeyer, M.S. and Grama, L., Stretching and visualizing
                    titin molecules: combining structure, dynamics, and mechanics, *J. Muscle
                    Res. Cell Motil.* 23, 499–511, 2002; Tskhovrebova, L. and Trinick, J.,
                    Titin: properties and family relationships, *Nat. Rev. Mol. Cell Biol.* 4,
                    679–689, 2003; Granzier, H.L. and Labeit, S., The giant protein titin: a
                    major player in myocardial mechanics, signaling, and disease, *Circ. Res.*
                    94, 284–295, 2004; Tskhovrebova, L. and Trinick, J., Properties of titin
                    immunoglobulin and fibronectin-3 domains, *J. Biol. Chem.* 279,
                    46351–46354, 2005; Samori, B., Zuccheri, G., and Baschieri, R., Protein
                    unfolding and refolding under force: methodologies for nanomechanics,
                    Chemphyschem. 6, 29–34, 2005; Lange, S., Ehler, E., and Gautel, M.,
                    From A to Z and back? Multicompartment proteins in the sarcomere,
                    Trends Cell Biol. 16, 11–18, 2006; Ferrari, M.B., Podugu, S., and Eskew,
                    J.D., Assembling the myofibril: coordinating contractile cable construction
                    with calcium, *Cell Biochem. Biophys.* 45, 317–337, 2006.

Toll-Like Receptor The Toll-like receptor is derived from the relationship of these proteins to the Toll receptor in *Drosophila* (Kuno, K. and Matsushima, K., The IL-1 receptor signaling pathway, *J. Leukoc. Biol.* 56, 542–547, 1994; Meister, M., Lemaitre, B., and Hoffman, J.A., Antimicrobial peptide defense in *Drosophila*, *Bioessays* 19, 1019–1026, 1997; Dushay, M.S. and Eldon, E.D., *Drosophila* immune responses as models for human immunity, *Am. J. Hum. Genet.* 62, 10–14, 1998; O'Neill, L.A. and Greene, C., Signal transduction pathways activated by the IL-1 receptor family: ancient signaling machinery in mammals, insects, and plants, *J. Leukoc. Biol.* 63, 650–657, 1998). Toll-like receptors in mammals are immune cell receptors that recognize infectious agents and activate the adaptive immune system. See Aderem, A. and Ulevitch, R.J., Toll-like receptors in the induction of the innate immune response, *Nature* 406, 782–787, 2000; Aderem, A., Role of the Toll-like receptors in inflammatory response in macrophages, *Crit. Care Med.* 29 (Suppl. 7), S16–S18, 2001; Beutler, B., Sepsis begins at the interface of pathogen and host, *Biochem. Soc. Trans.* 29, 853–859, 2001; Beutler, B. and Rietschel, E.T., Innate immune sensing and its roots: the story of endotoxin, *Nat. Rev. Immunol.* 3, 169–176, 2003; Philpott, D.J. and Girardin, S.E., The role of Toll-like receptors and Nod proteins in bacterial infection, *Mol. Immunol.* 41, 1099–1108, 2004; Jenner, R.G. and Young, R.A., Insights into host responses against pathogens from transcriptional profiling, *Nat. Rev. Microbiol.* 3, 281–294, 2005; Pasare, C. and Medzhitov, R., Toll-like receptors: linking innate and adaptive immunity, *Adv. Exp. Med. Biol.* 560, 11–18, 2005; O'Neill, L.A., How Toll-like receptors signal: what we know and what we don't know, *Curr. Opin. Immunol.* 18, 3–9, 2006; Kreig, A.M., Therapeutic potential of Toll-like receptor 9 activation, *Nat. Rev. Drug Discov.* 5, 471–484, 2006; Turvey, S.E. and Hawn, T.R., Towards subtlety: understanding the role of Toll-like receptor signaling in susceptibility to human infections, *Clin. Immunol.* 120, 1–9, 2006.

TonB A membrane-spanning protein that functions in receptors. See Ferguson, A.D. and Deisenhofer, J., TonB-dependent receptors — structural perspectives, *Biochim. Biophys. Acta* 1565, 318–332, 2002; Koebnik, R., TonB-dependent trans-envelope signaling: the exception or the rule? *Trends Microbiol.* 13, 343–347, 2005.

Tonoplast A membrane surrounding an intracellular structure or vacuole. See Barbeir-Brygoo, H., Renaudin, J.P., and Guern, J., The vacuolar membrane of plant cells: a newcomer in the field of biological membranes, *Biochimie* 68, 417–425, 1986; Bertl, A. and Slayman, C.L., Complex modulation of cation channels in the tonoplast and plasma membrane of *Saccharomyces cerevisieae*: single-channel studies, *J. Exp. Biol.* 172, 271–287, 1992; Neuhaus, J.M. and Rogers, J.C., Sorting of proteins to vacuoles in plant cells, *Plant Mol. Biol.* 38, 127–144, 1998; Luttge, U., The tonoplast functioning as the master switch for circadian regulation of crassulacean acid metabolism, *Planta* 211, 761–769, 2000.

Top-Down Proteomics Mass spectrometric analysis of intact proteins as opposed to bottom-up proteomics, where mass spectrometry analyzes peptides derived from the proteolytic enzyme digests of proteins. Successful top-down proteomics generally requires highly sophisticated instrumental approaches such as Fourier transform ion cyclotron resonance (FITCR) mass spectrometry. See Ge, Y., Lawhorn, G., ElNagger, M. et al., Top-down characterization of larger proteins (45 kDa) by electron capture dissociation mass spectrometry, *J. Am. Chem. Soc.* 124, 672–678, 2002; Nemeth-Cawley, J.F., Tangarone, B.S., and Rouse, J.C., "Top-down" characterization is a complementary technique to peptide sequencing for identifying protein species in complex mixtures, *J. Proteome Res.* 2, 495–505, 2003; Hirano, H., Islam, N., and Kawaski, H., Technical aspects of functional proteomics in plants, *Phytochemistry* 65, 1487–1498, 2004; Vaidyanathan, S., Kell, D.B., and Goodacre, R., Selective detection of proteins in mixtures using electrospray ionization mass spectrometry: influence of instrumental settings and implications for proteomics, *Anal. Chem.* 76, 5024–5032, 2004; Zhang, S. and van Pelt, C.K., Chip-based nanoelectrospray mass spectrometry for protein characterization, *Expert. Rev. Proteomics* 1, 449–468, 2004; Copper, H.J., Hakansson, K., and Marshall, A.G., The role of electron capture dissociation in biomolecular analysis, *Mass Spectrom. Rev.* 24, 201–222, 2005; Godovac-Zimmerman, J., Kleiner, O., Brown, L.R., and Drukier, A.K., Perspectives in splicing up proteomics with splicing, *Proteomics* 5, 699–709, 2005; Kaiser, N.K., Anderson, G.A., and Bruce, J.E., Improved mass accuracy for tandem mass spectrometry, *J. Am. Soc. Mass Spectrom.* 16, 463–470, 2005; Williams, T.L., Monday, S.R., Edelson-Mammel, S., Buchanan, R., and Musser, S.M., A top-down proteomics approach for differentiating thermal resistant strains of *Enterobacter sakazakii*, *Proteomics* 5, 4161–4169, 2005; Demirev, P.A., Feldman, A.B., Kowalski, P., and Lin, J.S., Top-down proteomics for rapid identification of intact microorganisms, *Anal. Chem.* 77, 7455–7461, 2005; Du, Y., Parks, B.A., Sohn, S., Kwast, K.E., and Kelleher, N.L., Top-down approaches for measuring expression ratios of intact yeast proteins using Fourier transform mass spectrometry, *Anal. Chem.* 78, 686–694, 2006.

Topoisomerase   A family of enzymes which alters the topology of DNA by catalyzing relaxation and unknotting the double-stranded DNA complex. This is accomplished through the alteration of the supercoiling of the DNA helix. Topoisomerase I cleaves one strand of DNA while topoisomerase II cleaves both strands of the DNA helix and there are isoforms of human topoisomerase II (Kondapi, A.K., Satyanarayana, N., and Saikrishna, A.D., A study of the topoisomerase II activity in HIV-1 replication using the ferrocene derivatives as probes, *Arch. Biochem. Biophys.* 450, 123–132, 2006). This process is critical for the transcription of DNA to yield messenger RNA and for the replication process. There are two other classes of topoisomerases: topoisomerase III, which appears to be involved in recombination; topoisomerase IV appears to be involved in the separation of chromosomes. Topoisomerases are targets for cancer chemotherapy. See Glisson, B.S. and Ross, W.E., DNA topoisomerase II: a primer on the enzyme and its unique role as a multidrug target in cancer chemotherapy, *Pharmacol. Ther.* 32, 89–106, 1987; Osheroff, N., Biochemical basis for the interactions of type I and type II topoisomerases with DNA, *Pharmacol. Ther.* 41, 223–241, 1989; Gmeiner, W.H., Yu, S., Pon, R.T., Pourquier, P., and Pommier, Y., Structural basis for topoisomerase I inhibition by nucleoside analogs, *Nucleosides, Nucleotides, Nucleic Acids* 22, 653–658, 2003; Porter, A.C. and Farr, C.J., Topoisomerase II: untangling its contribution at the centromere, *Chromosome Res.* 12, 569–583, 2004; Pindur, U., Jansen, M., and Lemster, T., Advances in DNA-ligand with groove binding, intercalating, and/or alkylating activity: chemistry, DNA-binding, and biology, *Curr. Med. Chem.* 12, 2805–2847, 2005; Yangida, M., Basic mechanism of eukaryotic chromosome segregation, *Philos. Trans. R. Soc. London B Biol. Sci.* 360, 609–621, 2005; Martincic, D. and Hande, K.R., Topoisomerase II inhibitors, *Cancer Chemother. Biol. Response Modif.* 22, 101–121, 2005.

Topological Proteomics A technology that analyzes proteins and protein interactions on a single-cell level (MELK technology); study of the toponome; analysis of protein networks. See Owens, J., Topological proteomics: a new approach to drug discovery, *Drug Discov. Today* 6, 1081–1082, 2001; Shubert, W., Topological proteomics, toponomics, MELK technology, *Adv. Biochem. Eng. Biotechnol.* 82, 189–209, 2003; Han, J.D., Dupuy, D., Bertin, N., Cusick, M.E., and Vidal, M., Effect of sampling on topology predictions of protein–protein interaction networks, *Nat. Biotechnol.* 23, 839–844, 2005; Stelzl, U., Worm, U., Lalowski, M. et al., A human protein–protein interaction network: a resource for annotating the proteome, *Cell* 122, 957–968, 2005.

***Trans*-Activation** Enhancement of transcription (transcriptional activation) by a transcription factor binding to a *cis*-factor in DNA and influencing the activity of RNA polymerase. See Roizman, B., Kristie, T., McKnight, J.L. et al. The *trans*-activation of herpes simplex virus gene expression: comparison of two factors and their *cis* sites, *Biochimie* 70, 1031–1043, 1988; Nevins, J.R., Mechanisms of viral-mediated *trans*-activation of transcription, *Adv. Virus Res.* 37, 35–83, 1989; Green, N.M, Cellular and viral transcriptional activators, *Harvey Lect.* 88, 67–96, 1992–1993; de Folter, S. and Angenent, G.C., *Trans* meets *cis* in MADS science, *Trends Plant Sci.* 11, 223–231, 2006; Gomez-Roman, N., Felteon-Edkins, Z.A., Kenneth, N.S. et al., Activation by c-Myc or transcription by RNA polymerases I, II, and III, *Biochem. Soc. Symp.* 73, 141–154, 2006; Campbell, K.J. and Perkins, N.D., Regulation of NF-B function, *Biochem. Soc. Symp.* 73, 165–180, 2006; Belakvadi, M. and Fondell, J.D., Role of the mediator complex in nuclear hormone receptor signaling, *Rev. Physiol. Biochem. Pharmacol.* 156, 23–43, 2006.

Trans- Inactivation (Transinactivation) Gene or transgene silencing mediated by heterochromatin; gene inactivation by *trans*-inactivation is considered to be an epigenetic event. See Sabl, J.F. and Laird, C.D., Epigene conversion: a proposal with implications for gene mapping in humans, *Am. J. Hum. Genet.* 50, 1171–1177, 1992; Opsahl, M.L., Springbett, A., Lathe, R. et al., Mono-allelic expression of variegating transgene locus in the mouse, *Transgenic Res.* 12, 661–669, 2003; Sage, B.T., Jones, J.L., Holmes, A.L., Wu, M.D., and Csink, A.K., Sequence elements in *cis* influence heterochromatic silencing in *trans*, *Mol. Cell Biol.* 25, 377–388, 2005. It also is the suppression of the *trans*-phosphorylation of receptors and resulting signaling pathways. See Graness, A., Hanke, S., Boehmer, F.D., Presek, P., and Liebmann, C., Protein-tyrosine-phosphatase-mediated epidermal growth factor (EGF) receptor transinactivation and EGF receptor-independent stimulation of mitogen-activated protein kinase by bradykinin in A431 cells, *Biochem. J.* 347, 441–447, 2000; Elbaz, N., Bedecs, K., Masson, M. et al., Functional *trans*-inactivation of insulin receptor kinase by growth-inhibitory angiotensis II AT2 receptor, *Mol. Endocrinol.* 14, 795–804, 2000; Nouet, S., Amzallag, N., Li, J.M. et al., *Trans*-inactivation of receptor tyrosine kinases by novel angiotensin II AT2 receptor-interacting protein, ATIP, *J. Biol. Chem.* 279, 28989–28997, 2004.

Transcription   The process by which genetic information is transferred from DNA to RNA.
                             See Hames, B.D. and Glover, D.M., *Transcription and Splicing*, IRL Press,
                             Oxford, UK, 1988; Neidle, S., *DNA Structure and Recognition*, IRL Press,
                             Oxford, UK, 1994; Baumann, P., Qureshi, S.A., and Jackson, S.P., Tran-
                             scription: new insights from studies on Archaea, *Trends Genet.* 11,
                             279–283, 1995; Singer, M. and Berg, P., *Exploring Genetic Mechanisms*,
                             University Science Books, Sausalito, CA, 1997; Lewin, B., *Genes VII*,
                             Oxford University Press, Oxford, UK, 2000; Lodish, H.F., *Molecular Cell
                             Biology*, W.H. Freeman, New York, 2000; Brown, W.M., and Brown, P.M.,
                             Transcription, Taylor & Francis, London, 2002; Alton, G., Schwanborn,
                             K., Satoh, Y., and Westwick, J.K., Therapeutic modulation of inflammatory
                             gene transcription by kinase inhibitors, *Expert. Opin. Biol. Ther.* 2,
                             621–632, 2002; Lee, D.K., Seol, W., and Kim, J.S., Custom DNA-binding
                             proteins and artificial transcription factors, *Curr. Top. Med. Chem.* 3,
                             645–657, 2003; Mondal, N. and Parvin, J.D., Transcription from the per-
                             spective of the DNA: twists and bumps in the road, *Crit. Rev. Eukaryot.
                             Gene Expr.* 13, 1–8, 2003; Olson, M.O.J., *The Nucleolus*, Landes Bio-
                             science, Georgetown, TX, 2004; Sausville, E.A. and Holbeck, S.L., Tran-
                             scription profiling of gene expression in drug discovery and development:
                             the NCI experience, *Eur. J. Cancer* 40, 2544–2549, 2004; Uesugi, M.,
                             Synthetic molecules that modulate transcription and differentiation: hints
                             for future drug discovery, *Comb. Chem. High Throughput Screen.* 7,
                             653–659, 2004.

Transcription   These are *trans* factors, which are proteins or protein complexes that bind to
Factors   *cis* factors or regions that are intrinsic to the DNA sequence of the regulated
                             gene and control the process of transcription. Transcription can be general
                             transcription factors, which are required for the basal transcription appa-
                             ratus, or regulatory transcription factors, which may bind upstream or
                             downstream from the transcription initiation site and either enhance or
                             suppress the rate of transcription. See McKnight, S.L. and Yamamoto,
                             K.R., Eds., *Transcriptional Regulation*, Cold Spring Harbor Laboratory
                             Press, Cold Spring Harbor, NY, 1992; Goodbourn, S., *Eukaryotic Gene
                             Transcription,* IRL Press, Oxford, UK, 1996; Tymms, M.J., Ed., *Tran-
                             scription Factor Protocols*, Humana Press, Totowa, NJ, 2000; Locker, J.,
                             Ed., *Transcription Factors*, Bios, Oxford, UK, 2001; Michalik, L. and
                             Wahli, W., Involvement of PPAR nuclear receptors in tissue injury and
                             wound repair, *J. Clin. Invest.* 116, 598–606, 2006; Kikuchi, A., Kishida,
                             S., and Yamamoto, H., Regulation of Wnt signaling by protein–protein
                             interaction and posttranslational modification, *Exp. Mol. Med.* 28, 1–10,
                             2006; Sharrocks, A.D., PIAS proteins and transcriptional regulation —
                             more than just SUMO E3 ligases? *Genes Dev.* 20, 754–758, 2006; Camp-
                             bell, K.J. and Perkins, N.D., Regulation of NF-kappaB function, *Biochem.
                             Soc. Symp.* 73, 165–180, 2006; Russell, J. and Zomerdijk, J.C., The RNA
                             polymerase I transcription machinery, *Biochem. Soc. Symp.* 73, 203–216,
                             2006; Gross, P. and Oelgeschlarger, T., Core promoter-selective RNA
                             polymerase II transcription, *Biochem. Soc. Symp.* 73, 225–236, 2006. See
                             General Transcription Factors; *NF-B*; *Promoter Elements*; *Regulatory
                             Transcription Factors*; *RNA Polymerase.*

Transcriptional   Also known as transcription repression; results from the interaction of *cis-*
Silencing   and *trans-*components/sequences to inhibit the process of transcription of
                             mRNA. Distinct from splicing silencing. See Nasmyth, K. and Shore, D.,

Transcriptional regulation in the yeast life cycle, *Science* 237, 1162–1170, 1987; Pannell, D. and Ellis, J., Silencing of gene expression: implications for design of retrovirus vectors, *Rev. Med. Virol.* 11, 205–217, 2001; Wanzel, M., Herold, S., and Eilers, M., Transcriptional repression by Myc, *Trends Cell Biol.* 13, 146–250, 2003; Ellis, J. and Yao, S., Retrovirus silencing and vector design: relevance to normal and cancer stem cells, *Curr. Gene. Ther.* 5, 367–373, 2005; Baniahmad, A., Nuclear hormone receptor co-repressors, *J. Steroid Biochem. Mol. Biol.* 93, 89–97, 2005; Spellman, R. and Smith, C.W.J., Novel modes of splicing repression by PTB, *Trends Biochem. Sci.* 31, 73–76, 2006.

Transcriptomics The total RNA transcripts produced by a genome; the complete RNA messages coded from the DNA within a cell. See Betts, J.C., Transcriptomics and proteomics: tools for the identification of novel drug targets and vaccine candidates for tuberculosis, *IUBMB Life* 53, 239–242, 2002; Hegde, P.S., White, I.R., and Delbouck, C., Interplay of transcriptomics and proteomics, *Curr. Opin. Biotechnol.* 14, 647–651, 2003; Jansen, B.J. and Schalkwijk, J., Transcriptomics and proteomics of human skin, *Brief Funct. Genomic Proteomic* 1, 326–341, 2003; Hu, Y.F., Kaplow, J., and He, Y., From traditional biomarkers to transcriptome analysis in drug development, *Curr. Mol. Med.* 5, 29–38, 2005; Viguerie, N., Poitou, C., Cancello, R. et al., Transcriptomics applied to obesity and caloric restriction, *Biochemie* 87, 117–123, 2005; Seda, O., Tremblay, J., Sedova, L., and Hamet, P., Integrating genomics and transcriptomics with geo-ethnicity and the environment for the resolution of complex cardiovascular disease, *Curr. Opin. Mol. Ther.* 7, 583–587, 2005.

Transcytosis Movement through a cell (usually an endothelial cell and vascular wall transport) as opposed to junctional transport (paracellular pathway). Involves a combination of endocytotic and exocytotic pathways. See Patel, H.M., Transcytosis of drug carriers carrying peptides across epithelial barriers, *Biochem. Soc. Trans.* 17, 940–942, 1989; Mostov, K., The polymeric immunoglobulin receptor, *Semin. Cell Biol.* 2, 411–418, 1991; Michel, C.C., Transport of macromolecules through microvascular walls, *Cardiovasc. Res.* 32, 644–653, 1996; Caplan, M.J. and Rodriguez-Boulan, E., Epithelial cell polarity: challenges and methodologies, in *Handbook of Physiology. Section 14, Cell Physiology*, Hoffman, J.F. and Jamieson, J.D., Eds., Oxford University Press (for the American Physiological Society), New York, 1997; Florence, A.T. and Hussain, N., Transcytosis of nanoparticles and dendrimer delivery systems: evolving vistas, *Adv. Drug Deliv. Rev.* 50 (Suppl. 1), S69–S89, 2001; Vogel, S.M. and Malik, A.B., Albumin transcytosis in mesothelium: further evidence of a transcellular pathway in polarized cells, *Am. J. Physiol. Lung Cell Mol. Physiol.* 282, L1–L2, 2002; Ghetie, V. and Ward, E.S., Transcytosis and catabolism of antibody, *Immunol. Res.* 25, 97–113, 2002; Kreuter, J., Influence of the surface properties on nanoparticle-mediated transport of drugs to the brain, *J. Nanosci. Nanotechnol.* 4, 484–488, 2004; Rot, A., Contribution of Duffy antigen to chemokine function, *Cytokine Growth Factor Rev.* 16, 687–694, 2005.

Transformation Cell changes manifested by escape from control mechanisms, generally resulting in increased growth potential, alterations in the cell surface, and karyotypic abnormalities. Cell transformation generally occurs as a result of the acquisition of genetic information as by a virus entering the cell.

See Dulbecco, R., Transformation of cells *in vitro* by DNA-containing viruses, *JAMA* 190, 721–726, 1964; Enders, J.F., Cell transformation by viruses as illustrated by the response of human and hamster renal cells to Simian virus 40, *Harvey Lect.* 59, 113–153, 1965; Black, P.M., The oncogenic DNA viruses: a review of *in vitro* transformation studies, *Annu. Rev. Microbiol.* 22, 391–426, 1968; Hanafusa, H., Replication of oncogenic viruses in virus-induced tumor cells — their persistence and interaction with other viruses, *Adv. Cancer Res.* 12, 137–165, 1969; Berk, A.J., Recent lessons in gene expression, cell cycle control, and cell biology from adenovirus, *Oncogene* 24, 7673–7685, 2005; Gius, D., Bradbury, C.M., Sun, L. et al., The epigenome as a molecular marker target, *Cancer* 104, 1789–1793, 2005; Adhikary, S. and Eilers, M., Transcriptional regulation and transformation by Myc proteins, *Nat. Rev. Mol. Cell Biol.* 6, 635–645, 2005.

Transgene A piece or segment of DNA, usually coding DNA, which is introduced into a cell or organism to modify the genome. Derivative animals are referred to as transgenic. See Babinet, C., Morello, D., and Renard, J.P., Transgenic mice, *Genome* 31, 938–949, 1989; Dichek, D.A., Retroviral vector-mediated gene transfer into endothelial cells, *Mol. Biol. Med.* 8, 257–266, 1991; Grosveld, F.G. and Kollias, G.V., *Transgenic Animals*, Academic Press, San Diego, CA, 1992; Janne, J., Hyttinen, J.M., Peura, T. et al., Transgenic animals as bioproducers of therapeutic proteins, *Ann. Med.* 24, 273–280, 1992; Hiatt, A., *Transgenic Plants: Fundamentals and Applications,* Marcel Dekker, New York, 1993; Gluethmann, H. and Ohashi, P.S., *Transgenesis and Targeted Mutagenesis in Immunology*, Academic Press, San Diego, CA, 1994; Wright, D.C. and Wagner, T.E., Transgenic mice: a decade of progress in technology and research, *Mutat. Res.* 307, 429–440, 1994; Mittelstein Scheid, O., Transgene inactivation in *Aribidopsis thaliana*, *Curr. Top. Microbiol. Immunol.* 197, 29–42, 1995; Barry, M.A. and Johnston, S.A., Biological features of genetic immunization, *Vaccine* 15, 788–791, 1997; Patil, S.D., Rhodes, D.G., and Burgess, D.J., DNA-based therapeutics and DNA delivery systems: a comprehensive review, *AAPS J.* 7, E61–E77, 2005; Amsterdam, A. and Becker, T.S., Transgenes as screening tools to probe and manipulate the zebrafish genome, *Dev. Dyn.* 234, 255–268, 2005; Harrow, F. and Ortiz, B.D., The TCR locus control region specifies thymic, but not peripheral, patterns of TCR gene expression, *J. Immunol.* 175, 6659–6667, 2005; Peña, L., *Transgenic Plants: Methods and Protocols*, Humana Press, Totowa, NJ, 2005.

TRANSIL Porous silica beads that can be coated with a single phospholipid bilayer and used to study protein–lipid interactions. See Schmitz, A.A., Schleiff, E., Rohrig, C. et al., Interactions of myristoylated alanine-rich kinase substrates (MARCKS)-related protein with a novel solid-supported lipid membrane system (TRANSIL), *Analyt. Biochem.* 268, 343–353, 1999; Loidl-Stahlhofen, A., Hartmann, T., Schottner, M. et al., Multilamellar liposomes and solid-supported lipid membranes (TRANSIL): screening of lipid-water partitioning toward a high-throughput scale, *Pharm. Res.* 18, 1782–1788, 2001; Schuhmacher, J., Kohlsdorfer, C., Buhner, K. et al., High-throughput determination of the free fraction of drugs strongly bound to plasma proteins, *J. Pharm. Sci.* 93, 816–830, 2004.

Translation The process by which information is transferred from RNA to protein structure. See Ochoa, S., Translation of the genetic message, *Bull. Soc. Chim.*

Biol. 27, 721–737, 1967; Lewin, B., Units of transcription and translation: the relationship between heterogeneous nuclear RNA and messenger RNA, *Cell* 4, 11–20, 1975; Buetow, D.E. and Wood, W.M., The mitochondrial translation system, *Subcell. Biochem.* 5, 1–85, 1978; Phelps, C.S. and Arnstein, H.R.V., *Messenger RNA and Ribosomes in Protein Synthesis*, Biochemical Society, London, 1982; Arnstein, H.R.V. and Cox, R.A., *Protein Biosynthesis*, IRL Press, Oxford, UK, 1992; Belasco, J.G., and Brawerman, G., *Control of Messenger RNA Stability*, Academic Press, San Diego, CA, 1993; Ilan, J., *Translational Regulation of Gene Expression 2*, Plenum Press, New York, 1993; Kaufman, R.J., Control of gene expression at the level of translation initiation, *Curr. Opin. Biotechnol.* 5, 550–557, 1994; Tymms, M.J., In Vitro *Transcription and Translation Protocols*, Humana Press, Totowa, NJ, 1995; Weissman, S.M., *cDNA Preparation and Characterization*, Academic Press, San Diego, CA, 1999; Yarus, M., On translation by RNAs alone, *Cold Spring Harb. Symp. Quant. Biol.* 66, 207–215, 2001; Lapointe, J. and Brakier-Gingras, L., *Translation Mechanisms*, Landes Bioscience, Georgetown, TX, 2003; Frank, J., Towards an understanding of the structural basis of translation, *Genome Biol.* 4, 237, 2003; Schoenberg, D.R., *mRNA Processing and Metabolism: Methods and Protocols*, Humana Press, Totowa, NJ, 2004; Huang, Y.S. and Richter, J.D., Regulation of local mRNA translation, *Curr. Opin. Cell Biol.* 16, 308–313, 2004; Kapp, L.D. and Lorsch, J.R., The molecular mechanics of eukaryotic translation, *Annu. Rev. Biochem.* 73, 657–704, 2004; Piper, M. and Holt, C., RNA translation in axons, *Annu. Rev. Cell Dev. Biol.* 20, 505–523, 2004; Noller, H.F., The driving force for molecular evolution of translation, *RNA* 10, 1833–1837, 2004; Katz, L. and Ashley, G.W., Translation and protein synthesis: macrolides, *Chem. Rev.* 105, 499–528, 2005; Jackson, R.J., Alternative mechanisms of initiating translation of mammalian mRNAs, *Biochem. Soc. Trans.* 33, 1231–1241, 2005; Deana, A. and Belasco, J.G., Lost in translation: the influence of ribosomes on bacterial mRNA decay, *Genes Dev.* 19, 2526–2533, 2005; Pique, M., Lopez, J.M., and Mendez, R., Cytoplasmic mRNA polyadenylation and translation assays, *Methods Mol. Biol.* 322, 183–198, 2006; Schuman, E.M., Dynes, J.L., and Steward, O., Synaptic regulation of translation of dendritic mRNAs, *J. Neurosci.* 26, 7143–7146, 2006.

Translocation   The movement of a ribosome along mRNA during protein synthesis: this process involves the participation of elongation factor (EF-G) and is accompanied by GTP hydrolysis. Translocation also refers to the process of protein transport across membranes, which may be assisted by a chaperone. Protein secretion from the cell also is described as translocation; type II secretion (the general secretory pathway) involves a multiprotein complex referred to as the translocon. See Egae, P.F., Stroud, P.W., and Walter, P., Targeting proteins to membranes: structure of the signal recognition particle, *Curr. Opin. Struct. Biol.* 15, 213–220, 2005; Collinson, I., The structure of the bacterial translocation complex, SecYEG, *Biochem. Soc. Trans.* 33, 1225–1230, 2005; Chavan, M. and Lennarz, W., The molecular basis of coupling of translocation and *N*-glycosylation, *Trends Biochem. Sci.* 31, 17–20, 2006. Translocation also refers to the movement of water and solutes in a plant, in particular from the roots to the shoots. See Kutchan, T.M., A role for intra- and intercellular translocation in natural products, *Curr. Opin. Plant Biol.* 8, 292–300, 2005; Yang, X., Feng,

Y., He, Z., and Stoffells, P.J., Molecular mechanisms of heavy metal hyperaccumulation and phytoremediation, *J. Trace Elem. Med. Biol.* 18, 339–353, 2005; Mackenzie, S.A., Plant organellar protein targeting: a traffic plan still under construction, *Trends Cell Biol.* 15, 548–554, 2005; Thompson, M.V., Phloem: the long and the short of it, *Trends Plant Sci.* 11, 26–32, 2006; Takahashi, H., Yoshimoto, N., and Saito, K., Anionic nutrient transport in plants: the molecular basis of the sulfate transporter gene family, *Genet. Eng.* 27, 67–80, 2006.

Translocon   A multiprotein complex (composed of several ER proteins), which mediates protein transport (cotranslational protein translocation) across membranes; interacts with single recognition particle (SRP). See Johnson, A.E. and van Waes, M.A., The translocon: a dynamic gateway at the ER membrane, *Annu. Rev. Cell Dev. Biol.* 15, 799–842, 1999; May, T. and Soll, J., Chloroplast precursor protein translocon, *FEBS Lett.* 452, 52–56, 1999; Johnson, A.E. and Haigh, N.G., The ER translocon and retrotranslocation: is the shift into reverse manual or automatic? *Cell* 102, 709–712, 2000; White, S.H., Translocons, thermodynamics, and the folding of membrane proteins, *FEBS Lett.* 555, 116–221, 2003; Coombes, B.K. and Finlay, B.B., Insertion of the bacterial type III translocon: not your average needle stick, *Trends Microbiol.* 13, 92–95, 2006; Chavan, M. and Lennarz, W., The molecular basis of coupling of translocation and *N*-glycosylation, *Trends Biochem. Sci.* 31, 17–20, 2006.

***Trans*-Splicing**   A process that occurs with both nucleic acids and proteins. With nucleic acids, *trans*-splicing (transsplicing) occurs as part of pre-mRNA processing, increasing messenger diversity. The *trans*-splicing of pre-mRNA is not related to the removal of introns via *cis*-splicing. *Trans*-splicing transfers RNA segments from one RNA molecule to another while *cis*-splicing removes introns from the same RNA molecule. See Bonen, L., *Trans*-splicing of pre-mRNA in plants, animals, and protists, *FASEB J.* 7, 40–46, 1993; Nilsen, T.W., *Trans*-splicing: an update, *Mol. Biochem. Parasitol.* 73, 1–6, 1995; Adams, M.D., Rudner, D.Z., and Rio, D.C., Biochemistry and regulation of pre-mRNA splicing, *Curr. Opin. Cell Biol.* 8, 331–339, 1996; Frantz, C., Ebel, C., Paulus, F., and Imbault, P., Characterization of *trans*-splicing in Euglenoids, *Curr. Genet.* 37, 349–355, 2000; Garcia-Blanco, M.A., Messenger RNA reprogramming by spliceosome-mediated RNA *trans*-splicing, *J. Clin. Invest.* 112, 474–480, 2003; Kornblihtt, A.R., de la Maya, M., Fededa, J.P. et al., Multiple links between transcription and splicing, *RNA* 10, 1489–1498, 2004; Mitchell, L.G. and McGarrity, G.J., Gene therapy progress and prospects: reprogramming gene expression by *trans*-splicing, *Gene Ther.* 12, 1477–1485, 2005; Yang, Y. and Walsh, C.E., Spliceosome-mediated RNA *trans*-splicing, *Mol. Ther.* 12, 1006–1012, 2005; Cheng, G., Cohen, L., Ndegwa, D., and Davis, R.E., The flatworm spliced leader 3′-terminal AUG as a translation initiator methionine, *J. Biol. Chem.* 281, 733–743, 2006. SL (spliced leader) *trans*-splicing and alternative *trans*-splicing are special cases of *trans*-splicing for nucleic acids. *Trans*-splicing also occurs with proteins but is most often a technique to use intein chemistry for ligation. See Shi, J. and Muir, T.W., Development of a tandem protein *trans*-splicing system based on native and engineered split inteins, *J. Am. Chem. Soc.* 127, 6198–6206, 2005; Khan, M.S., Khalid, A.M., and Malik, K.A., Intein-mediated protein *trans*-splicing and transgene containment in plastids, *Trends Biotechnol.* 23,

217–220, 2005; Kwon, Y., Coleman, M.A., and Camarero, J.A., Selective immobilization of proteins onto solid supports through split-intein-mediated protein *trans*-splicing, *Angew. Chem. Int. Ed. Engl.* 45, 1726–1729, 2006; Iwai, H., Zuger, S., Jin, J., and Tam, P.H., Highly efficient protein *trans*-splicing by a naturally split DnaE intein from *Nostoc punctiforms, FEBS Lett.* 580, 1853–1858, 2006; Muralidharan, V. and Muir, T.W., Protein ligation: an enabling technology for the biophysical analysis of proteins, *Nat. Methods* 3, 429–438, 2006. See *Alternative Splicing*; *Intein*; *Spliced-Leader* Trans-*Splicing*.

Transportan   A 27 amino acid chimeric peptide with cell-penetrating properties. See *Cell-Penetrating Peptides*. See Pooga, M., Hällbrink, M., Zorko, M., and Langel, Ü., Cell penetration by transportan, *FASEB J.* 12, 67–77, 1998; Padiri, K., Säälik, P., Hansen, M. et al., Cell transduction pathways of transportans, *Biooconjugate Chem.* 16, 1399–1410, 2005.

Transvection   To carry over or to carry across. In mathematics, a linear function. In biology, where gene expression is influenced by *trans*-interactions between alleles depending on somatic pairing between homologous chromosome regions, it can result in partial complementation between mutant alleles. See Judd, B.H., Transvection: allelic cross talk, *Cell* 53, 841–843, 1988; Rassoulzadegan, M., Magliano, M., and Cuzin, F., Transvection effects involving DNA methylation during meiosis in the mouse, *EMBO J.* 21, 440–450, 2002; Duncan, I.W., Transvection effects in *Drosophila* 36, 521–556, 2002; Coulthard, A.B., Nolan, N., Bell, J.B., and Hilliker, A.J., Transvection at the vestigial locus of *Drosophila melanogaster, Genetics,* 170, 1711–1721, 2005.

Triabody   An noncovalent trimer formed with scFv fragments engineered with no linker between the V_H and V_L domains. The normal linker engineered between the V_H and V_L domains is 15 residues (usually glycine and serine to promote maximum flexibility) which yields as monomer; if the linker is reduced to 10 residues, a dimer (diabody) is formed while with no linker there is a trimer or higher order polymer. See Le Gall, E., Kipriyanov, S.M., Moldenhauer, G., and Little, M., Di-, tri- and tetrameric single chain Fv antibody fragments against human CD19: effect of valency on cell binding, *FEBS Lett.* 453, 164–168, 1999; Atwell, J.L., Breheney, K.A., Lawrence, L.J. et al., scFv multimers of the anti-neuraminidase antibody NC10: length of the linker between V_H and V_L domains dictates precisely the transition between diabodies and triabodies, *Protein Eng.* 12, 597–604, 1999; Todorovska, A., Roovers, R.C., Dolezal, O. et al., Design and application of diabodies, triabodies, and tetrabodies for cancer targeting, *J. Immunol. Methods* 248, 47–66, 2001. See also *Diabody*; *Single-Chain Fv Fragment*; *Tribody*. It has been observed that if the order of the variable regions are switched in scFv construct (V_L–V_H instead of V_H–V_L), the engineered scFv with a zero-length linker formed a dimer (diabody) instead of the expected trimer (see Arndt, M.A.E., Krauss, J., and Rybak, S.M., Antigen binding and stability properties of non-covalently linked anti-CD22 single-chain Fv dimers, *FEBS Lett.* 578, 257–261, 2004).

Tribody   A trivalent antibody construct with two scFv fragments attached to the C-terminal ends of a Fab fragment. See Schoonjans, R., Willems, A., Schoonooghe, S. et al., Fab chains as an efficient heterodimerization scaffold for the production of recombinant bispecific and trispecific antibody derivatives, *J. Immunol.* 165, 7050–7057, 2000; Willems, A., Leonen, J., Schoonooghe, S. et al., Optimizing expression and purification from cell culture of trispecific recombinant antibody derivatives, *J. Chromatog. B Analyt. Technol. Biomed. Life Sci.* 786, 161–176, 2003. See *Bibody*; *Triabody*.

"Tri-Reagents"   See *RNA Isolation*.

Tris-Lipidation   Linking a hydrophobic component to a peptide or protein to enhance membrane binding. The hydroxyl groups of Tris are esterified with fatty acids and subsequently coupled to a peptide or protein via the amino group. See Whittaker, R.G., Hayes, P.J., and Bender, V., A gentle method of linking Tris to amino acids and peptides, *Pept. Res.* 6, 125–128, 1993; Ali, M., Amon, M., Bender, V., and Manolis, N., Hydrophobic transmembrane-peptide lipid conjugation enhances membrane binding and functional activity in T-cells, *Bioconjugate Chem.* 16, 1556–1563, 2005.

Troybody   Antibody with specificity for APC, which has an antigenic sequence inserted into a constant domain region. See Lunde, E., Lauvrak, V., Rasmussen, I.B. et al., Troybodies and pepbodies, *Biochem. Soc. Trans.* 30, 500–506, 2002; Lunde, E., Western, K.H., Rasmussen, I.B., Sandlie, I., and Bogen, B., Efficient delivery of T-cell epitopes to APC by use of MHC class II-specific troybodies, *J. Immunol.* 168, 2154–2162, 2002; Tunheim, G., Schjetne, K.W., Fredrikson, A.B., Sandlie, I., and Bogen, G., Human CD14 is an efficient target for recombinant immunoglobulin vaccine constructs that deliver T-cell epitopes, *J. Leuk. Biol.* 77, 303–310, 2005.

Tubulin   A protein that polymerizes to form microtubules. Tubulin is a target for anticancer therapy. See Feit, H., Slusarek, L., and Shelanski, M.L., Heterogeneity of tubulin subunits, *Proc. Natl. Acad. Sci. USA* 68, 2028–2031, 1971; Fine, R.E., Heterogeneity of tubulin, *Nat. New Biol.* 233, 283–284, 1971; Rappaport, L., Leterrier, J.F., and Nunez, J., Non phosphorylation *in vitro* of 6 S tubulin from brain and thyroid tissue, *FEBS Lett.* 26, 239–352, 1972; Berry, R.W. and Shelanski, M.L., Interactions of tubulin with vinblastine and guanosine triphosphate, *J. Mol. Biol.* 71, 71–80, 1972; Hemminki, K., Relative turnover of tubulin subunits in rat brain, *Biochim. Biophys. Acta* 310, 285–288, 1973; Timasheff, S.N., Frigon, R.P., and Lee, J.C., A solution physical-chemical examination of the self-association of tubulin, *Fed. Proc.* 35, 1886–1891, 1976; Mohri, H., The function of tubulin in motile systems, *Biochim. Biophys. Acta* 456, 85–127, 1976; Caplow, M. and Zeeberg, B., Stoichiometry for guanine nucleotide binding to tubulin under polymerizing and nonpolymerizing conditions, *Arch. Biochem. Biophys.* 203, 404–411, 1980; Zeeberg, B., Cheek, J., and Caplow, M., Exchange of tubulin dimer into rings in microtubule assembly/disassembly, *Biochemistry* 19, 5078–5086, 1980; Cleveland, D.W., Treadmilling of tubulin and actin, *Cell* 28, 689–691, 1982; Sternlicht, H., Yaffe, M.B., and Farr, G.W., A model of the nucleotide-binding site in tubulin, *FEBS Lett.* 214, 226–235, 1987; Oakley, B.R., γ-tubulin, *Curr. Top. Dev. Biol.* 49, 27–54, 2000; Dutcher, S.K., Motile organelles: the importance of specific tubulin isoforms, *Curr. Biol.* 11, R419–R422, 2001; McKean, P.G., Vaughan, S., and Gull, K., The extended tubulin superfamily, *J. Cell Sci.* 114, 2723–2733, 2001; Cowan, N.J. and Lewis, S.A., Type II chaperonins, prefoldin, and the tubulin-specific chaperones, *Adv. Protein Chem.* 59, 73–104, 2001; Addinall, S.G. and Holland, B., The tubulin ancestor, FtsZ, draughtsman, designer, and driving force for bacterial cytokinesis, *J. Mol. Biol.* 318, 219–236, 2002; Szymanski, D., Tubulin folding cofactors: half a dozen for a dimer, *Curr. Biol.* 12, R767–R769, 2002; Dutcher,

S.K., Long-lost relatives reappear: identification of new members of the tubulin superfamily, *Curr. Opin. Microbiol.* 6, 634–640, 2003; Caplow, M. and Fee, L., Concerning the chemical nature of tubulin subunits that cap and stabilize microtubules, *Biochemistry* 42, 2122–2126, 2003; Nogales, E., Wang, H.W., and Niederstrasser, H., Tubulin rings: which way do they curve? *Curr. Opin. Struct. Biol.* 13, 256–261, 2003; Pellegrini, F. and Budman, D.R., Tubulin function, action of antitubulin drugs, and new drug development, *Cancer Invest.* 23, 264–273, 2005; Nogeles, E. and Wang, H.W., Structural mechanisms underlying nucleotide-dependent self-assembly of tubulin and its relatives, *Curr. Opin. Struct. Biol.* 16, 221–229, 2006.

Tumor Suppressor Gene
A gene responsible for the encoding of products that suppress the malignant phenotype. These genes were first identified in hybrid cells resulting from cell fusion. Loss of tumor suppressor genes results in cell cycle deregulation. p53 is one of the better-known tumor suppressor genes. See Wynford-Thomas, D., Oncogenes and anti-oncogenes; the molecular basis of tumor behavior, *J. Pathol.* 165, 187–201, 1991; Carbone, D.P., Oncogenes and tumor suppressor genes, *Hosp. Pract.* 28, 145–148, 1993; Skapek, S.X. and Chui, C.H., Cytogenetics and the biologic basis of sarcomas, *Curr. Opin. Oncol.* 12, 315–322, 2000; Lee, M.P., Genome-wide analysis of epigenetics in cancer, *Ann. N.Y. Acad. Sci.* 983, 101–109, 2003; Bocchetta, M. and Carbone, M., Epidemiology and molecular pathology at crossroads to establish causation: molecular mechanisms of malignant transformation, *Oncogene* 23, 6484–6491, 2004; Seth, A. and Watson, D.K., ETS transcription factors and their emerging roles in human cancer, *Eur. J. Cancer* 41, 2462–2478, 2005.

Turbidimetry
Turbidimetry is a measure of the light scattered from the direct path of the electromagnetic radiation; practically, it is the transmitted light. It represents electromagnetic radiation that is not absorbed as in spectroscopy but rather scattered. It is sometimes necessary to correct spectral measurements for turbidimetry, or more commonly, light scattering. The extent to which electromagnetic radiation is scattered and measured either by turbidimetry or nephelometry depends on the size of the particle and the wavelength of the incident radiation. Turbidimetry is used in clinical chemistry (Blirup-Jensen, S., Protein standardization III: method optimization basic principles for quantitative determination of human serum proteins on automated instruments based on turbidimetry or nephalometry, *Clin. Chem. Lab. Med.* 39, 1098–1109, 2001); platelet aggregation (Cruz, W.O., Platelet determination by turbidimetry, *Blood* 9, 920–926, 1954; Jarvis, G.E., Platelet aggregation: turbidimetric measurements, *Methods Mol. Biol.* 272, 65–76, 2004); and for the assay of some enzymes (Rapport, M.M., Meyer, K., and Linker, A., Correlation of reductimetric and turbidimetric methods for hyaluronidase, *J. Biol. Chem.* 186, 615–623, 1950; Houck, J.C., The turdimetric determination of deoxyribonuclease activity, *Arch. Biochem. Biophys.* 82, 135–144, 1959; Morsky, P., Turbidimetric determination of lysozyme with *Micrococcus lysodeikticus* cells: reexamination of reaction conditions, *Anal. Biochem.* 128, 77–85, 1983; Jenzano, J.W. and Lundblad, R.L., Effects of amines and polyamines on turbidimetric and lysoplate assays for lysozyme, *J. Clin. Microbiol.* 26, 34–37, 1988; Walker, M.B., Retzinger, A.C., and Retzinger, G.S., A turbidimetric method for measuring the activity of trypsin and its inhibition, *Anal.*

Biochem. 351, 114–121, 2006). See Zattoni, A., Loli Piccolomini, E., Torsi, G., and Rschiglian, P., Turbidimetric detection method in flow-assisted separation of dispersed samples, *Anal. Chem.* 75, 6469–6477, 2003; Mori, Y., Kitao, M., Tomita, N., and Natomi, T., Real-time turdimetry of LAMP reaction for quantifying template DNA, *J. Biochem. Biophys. Methods* 31, 145–157, 2004; Hianik, T., Rybar, P., Andreev, S.Y. et al., Detection of DNA hybridization on a liposome surface using ultrasound velocimetry and turbidimetry methods, *Bioorg. Med. Chem. Lett.* 14, 3897–3900, 2004; Gonzalez, V.D., Gugliotta, L.M., Vega, J.R., and Meira, G.R., Contamination by larger particles of two almost-uniform lattices: analysis by combined dynamic light scattering and turbidimetry, *J. Colloid Interface Sci.* 285, 581–589, 2005; Stano, P., Bufali, S., Damozou, A.S., and Luisi, P.L., Effect of tryptophan oligopeptides on the size distribution of POPC liposomes: a dynamic light scattering and turbidimetric study, *J. Liposome Res.* 15, 29–47, 2005; Mao, J., Kondu, S., Ji, H.F., and McShane, M.J., Study of the near-neutral pH-sensitivity of chitosan/gelatin hydrogels by turbidimetry and microcantilever deflection, *Biotechnol. Bioeng.*, 95, 333–341, 2006.

Tyrosine Kinase A large group of enzymes involved in intracellular signal transduction, which catalyzes the phosphorylation of tyrosine residues in target proteins. See Hardle, D.G., *Protein Phosphorylation: A Practical Approach*, Oxford University Press, Oxford, UK, 1993; Woodgett, J.R., *Protein Kinases,* IRL Press, Oxford, UK, 1994; Hardle, D.G. and Hanks, S., Eds., *The Protein Kinase Facts Book*, Academic Press, San Diego, CA, 1995; Krauss, G., Ed., *Protein Kinase Protocols,* Wiley-VCH, Weinheim, Germany, 2003.

Tyrphostins Inhibitors of protein tyrosine kinases. See Levitzki, A., Tyrphostins — potential antiproliferative agents and novel molecular tools, *Biochem. Pharmacol.* 40, 913–918, 1990; Lamb, D.J. and Shubhaba, S., Tyrphostins inhibit Sertoli cell-secreted growth factor stimulation of A431 cell growth, *Recent Prog. Homr. Res.* 48, 511–516, 1993; Wolbring, G., Hollenberg, M.D., and Schnetkamp, P.P., Inhibition of GTP-utilizing enzymes by tyrphostins, *J. Biol. Chem.* 269, 22470–22472, 1994; Holen, I., Stromhaug, P.E., Gordon, P.B. et al., Inhibition of autophagy and multiple steps in asialoglycoprotein endocytosis by inhibitors of tyrosine protein kinases (tyrphostins), *J. Biol. Chem.* 270, 12823–12831, 1995; Jaleel, M., Shenoy, A.R., and Visweswariah, S.S., Tyrphostins are inhibitors of guanylyl and adenylyl cyclases, *Biochemistry* 43, 8247–8255, 2004; Levitzki, A. and Mishani, E., Tyrphostins and other tyrosine kinase inhibitors, *Annu. Rev. Biochem.* 75, 93–109, 2006.

Ubiquitin Ubiquitin is a small intracellular protein that serves as a marker for protein degradation by the proteosome. This is a process of controlled proteolysis, which is an integral part of normal cell function. Some functions of the ubiquitin-proteosome system include the degradation of misfolded proteins and the production of peptides during MHC class I antigen presentation (Michalek, M.T., Grant, E.P., Gramm, C. et al., A role for the ubiquitin-dependent proteolytic pathway in MHC class I-restricted antigen presentation, *Nature* 363, 552–554, 1993). Ubiquitin is linked to a protein via an isopeptide bond in a process referred to as ubiquitinylation, which is catalyzed by a ubiquitin ligase (Pavletich, N.P., Structural biology of ubiquitin-protein ligases, *Harvey Lect.* 98, 65–102, 2002–2003; Robinson,

P.A. and Ardley, H.C., Ubiquitin-protein ligases, *J. Cell Sci.* 5191–5194, 2004). Ubiquitin is initially "activated" by the ubiquity ligase to form a high-energy thioester bond between the enzyme and the C-terminal glycine residue of ubiquitin; the ubiquitin is then transferred to a lysine residue on the target protein forming the isopeptide peptide. While the discovery of ubiquitin was based on its ability to target proteins to degradation, it is clear that there are other functions (Welchman, R.L., Gordon, C., and Mayer, R.J., Ubiquitin and ubiquitin-like proteins as multifunctional signals, *Nat. Rev. Mol. Cell Biol.* 6, 599–609, 2005; Hicke, L., Schubert, H.L., and Hill, C.P., Ubiquitin-binding domain, *Nature Rev. Mol. Cell Biol.* 6, 610–621, 2005; Chen, Z.J., Ubiquitin signaling in the NF-B pathway, *Nat. Cell Biol.* 7, 758–765, 2005). There is a ubiquitin family of proteins (Catic, A. and Ploegh, H.L., Ubiquitin — conserved protein or selfish gene? *Trends Biochem. Sci.* 30, 600–604, 2005) consisting of type I ubiquitinlike proteins and type II ubiquitinlike proteins (Pickart, C.M. and Eddins, M.J., Ubiquitin: structures, functions, mechanisms, *Biochim. Biophys. Acta* 1695, 55–72, 2004; Walters, K.J., Goh, A.M., Wang, Q. et al., Ubiquitin family proteins and their relationship to the proteosome: a structural perspective, *Biochim. Biophys. Acta* 1695, 73–87, 2004). While there are few type I family members, they are well known with Nedd8 and SUMO (Kroetz, M.B., SUMO: a ubiquitin-like protein modifier, *Yale J. Biol. Med.* 78, 197–201, 2005). See Rechsteiner, M., Ubiquitin-mediated pathways for intracellular proteolysis, *Annu. Rev. Cell Biol.* 3, 1–30, 1987; Ciechanover, A., Gonen, H., Elias, S., and Mayer, A., Degradation of proteins by the ubiquitin-mediated proteolytic pathway, *New Biol.* 2, 227–234, 1990; Smalle, J. and Vierstra, R.D., The ubiquitin 26S proteosome proteolytic pathway, *Annu. Rev. Plant Biol.* 55, 555–590, 2004; Denison, C., Kirkpatrick, D.S., and Gygi, S.P., Proteomic insights into ubiquitin-like proteins, *Curr. Opin. Chem. Biol.* 9, 69–75, 2005; Miller, J. and Gordon, C., The regulation of proteosome degradation by multi-ubiquitin chain-binding proteins, *FEBS Lett.* 579, 3224–3230, 2005; Ye, Y., The role of the ubiquitin-proteosome system in ER quality control, *Essays Biochem.* 41, 99–112, 2005; Salomens, F.A., Verhoef, L.G., and Dantuma, N.P., Fluorescent reporters of the ubiquitin-proteosome system, *Essays Biochem.* 41, 113–128, 2005; Nakayama, K.I. and Nakayama, K., Ubiquitin ligases: cell-cycle control and cancer, *Nat. Rev. Cancer* 6, 369–381, 2006.

UHF Dielectrometry Physical technique to study the state of protein-bound water. See Hackl, E.V., Gatash, S.V., and Nikalov, O.T., Using UHF-dielectrometry to study protein structural transitions, *J. Biochem. Biophys. Methods* 64, 127–148, 2005.

Ultraconserved Elements A class of conserved elements in genomes between orthologous domains that share 100% identity over at least 200 bp in mammalian genomes. See Berjano, G., Pheasant, M., Makunin, I. et al., Ultraconserved elements in the human genome, *Science* 304, 1321–1325, 2004.

Validity External validity refers to the extent to which a specific finding from an investigation or analytical process can be generalized beyond the context of the specific investigation or analytical process. For regulatory purposes such as the manufacture of drugs and therapeutic biologicals, validity can be considered to demonstrate the ability to repeat the process and/or assay. The validation process is the process by which an organization can demonstrate that the process is reproducible and, therefore, valid.

Variegation   The state of discrete, diversified coloration. In biology, this can refer to the
                       discrete coloration patterns in leaves or to the occurrence within a tissue
                       of sectors or clones of different phenotypes. In genetics, it is taken to
                       mean a chromosome position effect when particular loci are contiguous
                       with heterochromatin. See Baker, W.K., Position-effect variegation, *Adv.
                       Genet.* 14, 133–169, 1968; Henikoff, S., Position-effect variegation after
                       60 years, *Trends Genet.* 6, 422–426, 1990; Cook, K.R. and Karpen, G.H.,
                       A rosy future for heterochromatin, *Proc. Natl. Acad. Sci. USA* 91,
                       5219–5221, 1994; Martin, D.I. and Whitelaw, E., The vagaries of varie-
                       gating transgenes, *Bioessays* 18, 919–923, 1996; Klein, C.G. and Costa,
                       M., DNA methylation, heterochromatin, and epigenetic carcinogenesis,
                       Mutat. Res. 386, 163–180, 1997; Zhimulev, I.F., Polytene chromosomes,
                       heterochromatin, and position effect variegation, *Adv. Genet.* 37, 1–566,
                       1998; Hennig, W., Heterochromatin, *Chromosoma* 108, 1–9, 1999; Shotta,
                       G., Ebert, A., Dorn, R., and Reuter, G., Position-effect variegation and the
                       genetic dissection of chromatin regulation in *Drosophila, Semin. Cell Dev.
                       Biol.* 14, 67–75, 2003.

V(D)J   The process by which discontinuous regions of DNA become joined in
Recombination   lymphocytes, resulting in rearrangement of the DNA germline; the process
                       by which diversity is built into immunoglobulins. See Alt, F.W., Oltz,
                       E.M., Young, F. et al., V(D)J recombination, *Immunol. Today* 13, 306–314,
                       1992; Jung, D. and Alt, F.W., Unraveling V(D)J recombination; insights
                       into gene regulation, *Cell* 116, 299–311, 2004; Schatz, D.G., V(D)J recom-
                       bination, *Immunol. Rev.* 200, 5–11, 2004; Jones, J.M. and Gellert, M., The
                       taming of a transposon: V(D)J recombination and the immune system,
                       Immunol. Rev. 200, 233–248, 2004; Dudley, D.D., Chaudhuri, J., Bassing,
                       C.H., and Alt, F.W., Mechanism and control of V(D)J recombination
                       versus class switch recombination: similarities and differences, *Adv. Immu-
                       nol.* 86, 43–112, 2005; Aplon, P.D., Causes of oncogenic chromosomal
                       translocation, *Trends Genet.* 22, 46–55, 2006.

VGF   A neuronal protein involved in cell differentiation. See Salton, S.R.J., Fischber,
                       D.J., and Don, K.-W., Structure of the gene encoding VGF, a nervous
                       system-specific mRNA that is rapidly and selectively induced by nerve
                       growth factor in PC12 cells, *Mol. Cell Biol.* 11, 2335–2349, 1991.

VICKZ Proteins   A family of RNA-binding proteins recognizing specific *cis*-acting elements
                       acting on a variety of transcriptional processes involved in cell polarity
                       and migration. See Yisraeli, J.K., VICKZ proteins: a multi-talented family
                       of regulatory RNA-binding proteins, *Biol. Chem.* 97, 87–96, 2005.

Virulence Factors Factors elaborated by an organism, usually as bacteria, that are responsible
                       for the pathogenicity of the organism. An example would be a bacterial
                       exotoxin. See Evans, D.J., Jr. and Evans, D.G., Classification of pathogenic
                       Escherichia olie according to serotype and the production of virulence
                       factors, with special reference to colonization-factor antigens, *Rev. Infect.
                       Dis.* 5 (Suppl. 4), S692–S701, 1983; Lubran, M.M., Bacterial toxins, *Ann.
                       Clin. Lab. Sci.* 18, 58–71, 1988; Moxon, E.R. and Kroll, J.S., Type b
                       capsular polysaccharide as a virulence factor of *Haemophilis influenzae,
                       Vaccine* 6, 113–115, 1988; Pragman, A.A. and Schievert, P.M., Virulence
                       regulation in *Staphylococcus aureus*: the need for *in vivo* analysis of
                       virulence factor regulation, *FEMS Immuno. Med. Microbiol.* 42, 147–154,
                       2004; Walker, M.J., McArthur, J.D., McKay, F., and Ranson, M., Is plas-
                       minogen deployed as a *Streptococcus pyogenes* virulence factor? *Trends*

Microbiol. 13, 308–313, 2005; Lu, H., Yamaoka, Y., and Graham, D.Y., *Heliobacter pylori* virulence factors: fact and fantasies, *Curr. Opin. Gastroenterol.* 21, 653–659, 2005; Zaas, D.W., Duncan, M., Rae Wright, J., and Abraham, S.N., The role of lipid rafts in the pathogenesis of bacterial infections, *Biochim. Biophys. Acta* 1746, 305–313, 2005; Kazmierczak, M.J., Wiedmann, M., and Boor, K.J., Alternative sigma factors and their roles in bacterial virulence, *Microbiol. Mol. Biol. Rev.* 69, 527–543, 2005; Yates, S.P., Jørgensen, R., Andersen, G.R. et al., Stealth and mimicry by deadly bacterial toxins, *Trends Biochem. Sci.* 31, 123–133, 2006.

Viscosity
The property of a fluid indicating resistance to change in form or resistance to flow. There is considerable interest in the viscosity of blood as it is related to cardiovascular disease (Somer, T. and Meiselman, H.J., Disorders of blood viscosity, *Ann. Med.* 25, 31–39, 1993). See Kupke, D.W. and Crouch, T.H., Magnetic suspension: density-volume, viscosity, and osmotic pressure, *Methods Enzymol.* 48, 29–68, 1978; Ahmad, F. and McPhie, P., The intrinsic viscosity of glycoproteins, *Int. J. Biochem.* 11, 91–96, 1980; Harding, S.E., The intrinsic viscosity of biological macromolecules. Progress in measurement, interpretation, and application to structure in dilute solution, *Prog. Biophys. Mol. Biol.* 68, 207–262, 1997; Laghaei, R., Nasrabad, A.E., and Eu, B.C., Generic van der Waals equation of state, modified free volume theory of diffusion, and viscosity of simple liquids, *J. Phys. Chem. B Condens. Matter Mater. Surf. Interfaces Biophys.* 109, 5873–5883, 2005; Brookes, R., Davies, A., Ketwaroo, G., and Madden, P.A., Diffusion coefficients in ionic liquids: relationship to the viscosity, *J. Phys. Chem. B Condens. Matter Mater. Surf. Interfaces Biophys.* 109, 6485–6490, 2005; Stillinger, F.H. and Debenedetti, P.G., Alternative view of self-diffusion and shear viscosity, *J. Phys. Chem. B Condens. Matter Mater. Surf. Interfaces Biophys.* 109, 6605–6609, 2005; Ulker, P., Alexy, T., Meiselman, H.J., and Baskurt, O.K., Estimation of infused dextran plasma concentration via measurement of plasma viscosity, *Biorheology* 43, 161–166, 2006; Chopra, S., Lynch, R., Kim, S.H. et al., Effects of temperature and viscosity on R67 dihydrofolate reductase catalysis, *Biochemistry* 45, 6596–6605, 2006; Donoso, M. and Ghaly, E.S., Use of near-infrared for quantitative measurement of viscosity and concentration of active ingredients in pharmaceutical gel, *Pharm. Dev. Technol.* 11, 389–397, 2006; Haidekker, M.A., Akers, W.J., Fischer, D., and Theodorakis, E.A., Optical fiber-based fluorescent viscosity sensor, *Opt. Lett.* 31, 2529–2531, 2006.

Vitamers
Different chemical structural forms of a vitamin that have the same biological activity. See Bender, D.A., *Nutritional Biochemistry of the Vitamins*, 2nd ed., Cambridge University Press, Cambridge, UK, 2003; Voziyan, P.A. and Hudson, B.G., Pyridoxamine. The many virtues of a Maillard reaction inhibitor, *Ann. N.Y. Acad. Sci.* 1043, 807–816, 2005.

Walker A Motif
A motif described in SKN-1, a transcription factor in *Caenorhabditis elegans*. See Walker, A.K., See, R., Batchelder, C. et al., A conserved transcription motif suggesting functional parallels between *Caenorhabitis elegans* SKN-1 and Cap'n'Collar-related basic leucine zipper proteins, *J. Biol. Chem.* 275, 22166–22171, 2000.

Western Blotting
A method for identifying proteins after electrophoretic separation involving a specific probe, usually an antibody. It was derived from the earlier development of southern blotting and northern blotting. See also *Northwestern Blotting*; *Southwestern Blotting*. See Radka, S.F., Monoclonal

antibodies to human major histocompatibility complex class II antigens, *Crit. Rev. Immunol.* 8, 23–48, 1987; Heerman, K.H., Gultekin, H., and Gerlich, W.H., Protein blotting: techniques and application in virus hepatitis research, *Ric. Clin. Lab.* 18, 193–221, 1988; Hayes, P.C., Wolf, C.R., and Hayes, J.D., Blotting techniques for the study of DNA, RNA, and proteins, *BMJ* 299, 965–968, 1989; Baldo, B.A. and Tovey, E.R., *Protein Blotting: Methodology, Research, and Diagnostic Applications,* Karger, Basel, 1989; Harper, D.R., Kit, M.L., and Kangrok, H.O., Protein blotting: ten years on, *J. Virol. Methods* 30, 25–39, 1990; Dunn, M.J., Detection of proteins on blots using the avidin-biotin system, *Methods Mol. Biol.* 32, 227–232, 1994; Dunbar, B.S., *Protein Blotting: A Practical Approach,* IRL Press, Oxford, UK, 1994; Westermeier, R. and Marouga, R., Protein detection methods in proteomics research, *Biosci. Rep.* 25, 19–32, 2005.

WormBase   A public database for the genomics biology of *Caenorhabditis elegans* (a soil-dwelling nematode used extensively in biological research). See Chen, N., Harris, T.W., Antoschechkin, I. et al., WormBase: a comprehensive data resource for *Caenorhabditis* biology and genomics, *Nucleic Acids Res.* 33, D383–D389, 2005; O'Connell, K., There's no place like WormBase: an indispensable resource for *Caenorhabditis elegans* researchers, *Biol. Cell* 97, 867–872, 2005; Schwarz, E.M., Antoschechkin, I., Bastiani, C. et al., WormBase: better software, richer content, *Nucleic Acids Res.* 34, D475–D478, 2006.

Wormgate   A cloning system for the expression RNAai (hairpin RNA constructs) from the *C. elegans* ORFeome library. See Lamesch, P., Milstein, S., Hao, T. et al., *C. elegans* ORFeome version 3.1: increasing the coverage of ORFeome resource with improved gene prediction, *Genome Res.* 14, 2064–2069, 2004 (WormBase); Johnson, N.M., Behm, C.A., and Trowell, S.C., Heritable and inducible gene knockdown in *C. elegans* using Wormgate and the ORFeome, *Gene* 359, 26–34, 2005.

Xenobiotic   A chemical found in the body of an organism that is not the biosynthetic product of said organism and is therefore from an exogenous source. Benzene is an example of a xenobiotic compound. Organisms have utilized unique metabolic pathways for the metabolism/detoxification of xenobiotic compounds. See Paulson, G.D., Lamoureux, G.L., and Feil, V.J., Advances in methods and techniques for the identification of xenobiotic conjugates, *J. Toxicol. Clin. Toxicol.* 19, 571–608, 1982; Garattini, S., Notes on xenobiotic metabolism, *Ann. N.Y. Acad. Sci.* 407, 1–25, 1983; Glatt, H., Gemperlein, I., Turchi, G. et al., Search for cell culture systems with diverse xenobiotic-metabolizing activities and their use in toxicological studies, *Mol. Toxicol.* 1, 313–334, 1987–1988; Copley, S.D., Microbial dehalogenases: enzymes recruited to convert xenobiotic substrates, *Curr. Opin. Chem. Biol.* 2, 613–617, 1998; Gil, F. and Pla, A., Biomarkers as biological indicators of xenobiotic exposure, *J. Appl. Toxic.* 21, 245–255, 2001; Snyder, R., Xenobiotic metabolism and the mechanism(s) of benzene toxicity, *Drug Metab. Rev.* 36, 531–547, 2004; Pritchard, J.B. and Miller, D.S., *Toxicol. Appl. Pharmacol.* 204, 256–262, 2005; Dai, G. and Wan, Y.J., Animal models of xenobiotic receptors, *Curr. Drug Metab.* 6, 341–355, 2005; Cribb, A.E., Peyrou, M., Muruganandan, S., and Schneider, L., The endoplasmic reticulum in xenobiotic toxicity, *Drug Metab. Rev.* 37, 405–442, 2005; Janssen, D.B., Dinkla, I.J., Peolarends, G.J., and Terpstra, P., Bacterial degradation of xenobiotic compounds:

evolution and distribution of novel enzyme activities, *Environ. Microbiol.* 7, 1868–1882, 2005; Gong, H., Sinz, M.W., Feng, Y. et al., Animal models of xenobiotic receptors in drug metabolism and diseases, *Methods Enzymol.* 400, 598–618, 2005; Matsunaga, T., Shitani, S., and Hara, A., Multiplicity of mammalian reductases for xenobiotic carbonyl compounds, *Drug Metab. Pharmacokinet.* 21, 1–18, 2006.

Xerogel
Not a gel but rather a term used in reference to a dried, possibly open, gel; a gel in which the dispersing agent has been removed as opposed to a lyogel where the dispersing agent is still present as, for example, with a hydrogel that contains a substantial amount of water. There has been interest in xerogels as drug delivery vehicles. See Kortesuo, P., Ahola, M., Karlsson, S. et al., Sol-gel-processed sintered silica xerogel as a carrier in controlled drug delivery, *J. Biomed. Mat. Res.* 44, 162–167, 1999; Kortesuo, P., Ahola, M., Karlsson, S. et al., Silica xerogel as an implantable carrier for controlled drug delivery — evaluation of drug distribution and tissue effects after implantation, *Biomaterials* 21, 193–198, 2000; Shamansky, L.M., Luong, K.M., Han, D., and Chronister, E.L., Photoinduced kinetics of bacteriorhodopsin in a dried xerogel glass, *Biosens. Bioelectron.* 17, 227–231, 2002; Weng, K.C., Stalgren, J.J., Duval, D.J. et al., Fluid biomembranes supported on nanoporous aerogel/xerogel substrates, *Langmuir* 20, 7232–7239, 2004; Clifford, J.S. and Legge, R.L., Use of water to evaluate hydrophobicity of organically modified xerogel enzyme supports, *Biotechnol. Bioeng.* 92, 231–237, 2005; Oh, B.K., Robbins, M.E., Nablo, B.J., and Schoenfisch, M.H., Miniaturized glucose biosensor modified with a nitric oxide-releasing xerogel microarray, *Biosens. Bioelectron.* 21, 749–757, 2005; Copello, G.J., Teves, S., Degrossi, J. et al., Antimicrobial activity on glass materials subject to disinfectant xerogel coating, *J. Ind. Microbiol.* 33, 343–348, 2006; Xue, J.M., Tan, C.H., and Lukito, D., Biodegradable polymer-silica xerogel composite microspheres for controlled release of gentamicin, *J. Biomed. Mater. Res. B Appl. Biomater.* 78, 417–422, 2006.

Yeast Artificial Chromosomes
Yeast artificial chromosomes (YACs) are yeast DNA sequences that contain large segments of foreign recombinant DNA introduced by transformation. Yeast artificial chromosomes permit the cloning of large DNA fragments such as genes with flanking regulatory regions. See Schlessinger, D., Yeast artificial chromosomes: tools for mapping and analysis of complex genomes, *Trends Genet.* 6, 255–258, 1990; Huxley, C. and Gnirke, A., Transfer of yeast artificial chromosomes from yeast to mammalian cells, *Bioessays* 13, 545–550, 1991; Anand, R., Yeast artificial chromosomes (YACs) and the analysis of complex genomes, *Trends Biotechnol.* 10, 35–40, 1992; Huxley, C., Transfer of YACs to mammalian cells and transgenic mice, *Genet. Eng.* 16, 65–91, 1994; Schalkwyk, L.C., Francis, F., and Lehrach, H., Techniques in mammalian genome mapping, *Curr. Opin. Biotechnol.* 6, 37–43, 1995; Kouprina, N. and Larionov, V., Exploiting the yeast *Saccharomyces cerevisiae* for the study of the organization and evolution of complex genomes, *FEMS Microbiol. Rev.* 27, 629–649, 2003; Sasaki, T., Matsumoto, T., Antonio, B.A., and Nagamura, Y., From mapping to sequencing, post-sequencing, and beyond, *Plant Cell Physiol.* 46, 3–13, 2005.

Zebrafish
Zebrafish (*Danio rerio*) is a freshwater fish used for research in developmental biology. See http://zfin.org; http://www.neuro.uoregon.edu/k12/FAQs.html;

http://www.ncbi.nlm.nih.gov/genome/guide/zebrafish/. See also Kimmel, C.B., Genetics and early development of zebrafish, *Trends Genet.* 5, 283–288, 1989; Fulwiler, C. and Gilbert, W., Zebrafish embryology and neural development, *Curr. Opin. Cell Biol.* 3, 989–991, 1991; Driever, W., Stemple, D., Schier, A., and Solnica-Krezel, L., Zebrafish: genetic tools for studying vertebrate development, *Trends Genet.* 10, 152–159, 1994; Stemple, D.L. and Driever, W., Zebrafish: tools for investigating cellular differentiation, *Curr. Opin. Cell Biol.* 8, 858–864, 1996; Ingham, P.W. and Kim, H.R., Hedgehog signaling and the specification of muscle cell identity in the zebrafish embryo, *Exp. Cell Res.* 306, 336–342, 2005; Teh, C., Parinov, S., and Korzh, V., New ways to admire zebrafish: progress in functional genomics research methodology, *Biotechniques* 38, 897–906, 2005; Amsterdam, A. and Becker, T.S., Transgenes as screening tools to probe and manipulate the zebrafish genome, *Dev. Dyn.* 234, 255–268, 2005; Hsia, N. and Zon, L.I., Transcriptional regulation of hematopoietic stem cell development in zebrafish, *Exp. Hematol.* 33, 1007–1014, 2005; de Jong, J.L. and Zon, L.I., Use of the zebrafish to study primitive and definitive hematopoiesis, *Annu. Rev. Genet.* 39, 481–501, 2005; Alestrom, P., Holter, J.L., and Nourizadeh-Lillabadi, R., Zebrafish in functional genomics and aquatic biomedicine, *Trends Biotechnol.* 24, 15–21, 2006.

Zeolites
An aluminum silicate cagelike compound with a negative charge, which "captures" cations in the cavity. Zeolites are used as molecular sieves for drying solvents and gases (Mumpton, F.A., La roca magica: uses of natural zeolites in agriculture and industry, *Proc. Natl. Acad. Sci. USA* 96, 3463–3470, 1999; Kaiser, L.G., Meersmann, T., Logan, J.W., and Pines, A., Visualization of gas flow and diffusion in porous media, *Proc. Natl. Acad. Sci. USA* 97, 2414–2418, 2000; Kuznicki, S.M., Bell, V.A., Nair, S. et al., A titanosilicate molecular sieve with adjustable pores for size-selective adsorption of molecules, *Nature* 412, 720–724, 2001; Yan, A.X., Li, X.W., and Ye, Y.H., Recent progress on immobilization of enzymes on molecular sieves for reactions in organic solvents, *Appl. Biochem. Biotechnol.* 101, 113–129, 2002). There has been some interest in the specific adsorption of biopolymers such as proteins on zeolites (Matsui, M., Kiyozumi, Y., and Yamamoto, T., Selective adsorption of biopolymers on zeolites, *Chemistry* 7, 1555–1560, 2001; Chiku, H., Matsui, M., Murakami, S. et al., Zeolites as new chromatographic carriers for proteins — easy recovery of proteins adsorbed on zeolites by polyethylene glycol, *Anal. Biochem.* 318, 80–85, 2003; Sakaguchi, K., Matsui, M., and Mizukami, F., Applications of zeolite inorganic composites in biotechnology: current status and perspectives, *Appl. Microbiol. Biotechnol.* 67, 306–311, 2005). There are suggestions for the use of zeolites in health (Pavelic, K., Hadzija, M., and Bedrica, L., Natural zeolite clinoptilolite: new adjuvant in anticancer therapy, *J. Mol. Med.* 78, 708–720, 2001; Zarkovic, N., Zarkovic, K., Kralj, M. et al., Anticancer and antioxidative effects of micronized zeolite clinoptilolite, *Anticancer Res.* 23, 159–1595, 2003).

Zinc Finger Motifs
Motifs in DNA- and RNA-binding proteins whose amino acids are folded into a single structural unit around a zinc atom. In the classic zinc finger, one zinc atom is bound to two cysteines and two histidines. In between the cysteines and histidines are 12 residues that form a DNA-binding fingertip. By variations in the composition of the sequences in the fingertip

and the number and spacing of tandem repeats of the motif, zinc fingers can form a large number of different sequence-specific binding sites. Specificity of binding to the nucleic acid is achieved by recognition of an 18 bp sequence. See Schleif, R., DNA binding by proteins, *Science* 241, 1182–1187, 1988; Struhl, K., Helix-turn-helix, zinc-finger, and leucine-zipper motifs for eukaryotic transcriptional regulatory proteins, *Trends Biochem. Sci.* 14, 137–140, 1989; Gommans, W.M., Haisma, H.J., and Rots, M.G., Engineering zinc finger protein transcription factors: the therapeutic relevance of switching endogenous gene expression on or off at command, *J. Mol. Biol.* 354, 507–519, 2005; Durai, S., Mani, M., Kandavelou, K. et al., Zinc finger nucleases: custom-designed molecular scissors for genome engineering of plant and mammalian cells, *Nucl. Acid Res.* 26, 5978–5990, 2005; Chen, Y. and Varani, G., Protein families and RNA recognition, *FEBS J.* 272, 2088–2097, 2005.

Zinc Finger Nuclease
Zinc finger nucleases are engineered nucleases containing the zinc finger domain(s) fused to the nuclease domain from Fok1 restriction endonuclease. This nuclease domain is nonspecific, such that the sequence specificity cleavage of the zinc finger nucleases is provided from the zinc finger domain(s). See Urnov, F.D., Miller, J.C., Lee, Y.L. et al., Highly efficient endogenous human gene correction using zinc-finger nucleases, *Nature* 435, 646–651, 2005; Mani, M., Smith, J., Kandavelou, K., Berg, J.M., and Chandrasegaran, S., Binding of two zinc finger nuclease monomers to two specific sites is required for effective double-strand DNA cleavage, *Biochem. Biophys. Res. Commun.* 334, 1191–1197, 2005; Mani, M., Kandavelou, K., Dy, F.J., Durai, S., and Chandrasegaran, S., Design, engineering, and characterization of zinc finger nucleases, *Biochem. Biophys. Res. Commun.* 335, 447–457, 2005; Porteus, M.H., Mammalian gene targeting with designed zinc finger nucleases, *Mol. Ther.* 13, 438–446, 2006; Dhanasekaran, M., Negi, S., and Sugiura, Y., Designer zinc finger proteins: tools for creating artificial DNA-binding proteins, *Acc. Chem. Res.* 39, 45–52, 2006.

Zymography
A method for detecting enzyme activity on a matrix, usually a polyacrylamide gel or agarose gel after electrophoretic separation. See Frederiks, W.M. and Mook, O.R., Metabolic mapping of proteinase activity with emphasis on *in situ* zymography of gelatinases: review and protocols, *J. Histochem. Cytochem.* 52, 711–722, 2004; Lombard, C., Saulnier, J., and Wallach, J., Assays of matrix metalloproteinases (MMPs) activities: a review, *Biochemie* 87, 265–272, 2005.

Zymosan
An insoluble polysaccharide derived from the cell walls of fungi. More specifically, zymosan refers to a specific preparation from yeast that is used in models of inflammatory disease and multi-organ dysfunction. There is evidence for specific interaction with Toll receptors on macrophages. See Fitzpatrick, F.W. and DiCarlo, F.J., Zymosan, *Ann. N.Y. Acad. Sci.* 118, 233–262, 1964; Czop, J.K., Phagocytosis of particular activators of the alternative complement pathway: effects of fibronectin, *Adv. Immunol.* 38, 361–398, 1986; Stewart, J. and Weir, D.M., Carbohydrates as recognition molecules in macrophage activities, *J. Clin. Lab. Immunol.* 28, 103–108, 1989; Takeuchi, O. and Akira, S., Toll-like receptors: their physiological role and signal transduction system, *Int. Immunopharmacol.* 1, 625–635, 2001; Levitz, S.M., Interactions of Toll-like receptors with

fungi, *Microbes Infect.* 6, 1351–1355, 2004; Volman, T.J., Hendriks, T., and Goris, R.J., Zymosan-induced generalized inflammation: experimental studies into mechanisms leading to multiple organ dysfunction syndrome, *Shock* 23, 291–297, 2005; Ikeda, Y., Adachi, Y., Ishibashi, K., Miura, N., and Ohno, N., Activation of Toll-like receptor-mediated NF-kappa beta by zymosan-derived water-soluble fraction: possible contribution of endotoxinlike substances, *Immunopharmacol. Immunotoxicol.* 27, 285–298, 2005.

3 Chemicals Commonly Used in Biochemistry and Molecular Biology and Their Properties

Common Name	Chemical Name	M.W.	Properties and Comment
ACES	2-[2-amino-2-oxyethyl)-amino]ethanesulfonic Acid	182.20	One of the several "Good" buffers.

ACES, 2-[(2-amino-2-oxyethyl)amino]ethanesulfonic acid

Tunnicliff, G. and Smith, J.A., Competitive inhibition of gamma-aminobutyric acid receptor binding by N-hydroxyethylpiperazine-N′-2-ethanesulfonic acid and related buffers, *J. Neurochem.* 36, 1122–1126, 1981; Chappel, D.J., N-[(carbamoylmethyl) amino] ethanesulfonic acid improves phenotyping of α-1-antitrypsin by isoelectric focusing on agarose gel, *Clin. Chem.* 31, 1384–1386, 1985; Liu, Q., Li, X., and Sommer, S.S., pk-matched running buffers for gel electrophoresis, *Anal. Biochem.* 270, 112–122, 1999; Taha, M., Buffers for the physiological pH range: acidic dissociation constants of zwitterionic compounds in various hydroorganic media, *Ann. Chim.* 95, 105–109, 2005.

	Chemical Name	M.W.	Properties and Comment
Acetaldehyde	Acetaldehyde, Ethanal	44.05	Manufacturing intermediate; modification of amino groups; toxic chemical; first product in detoxification of ethanol.

Acetaldehyde   *gem*-diol form (approximately 60%)

Burton, R.M. and Stadtman, E.R., The oxidation of acetaldehyde to acetyl coenzyme A, *J. Biol. Chem.* 202, 873–890, 1953; Gruber, M. and Wesselius, J.C., Nature of the inhibition of yeast carboxylase by acetaldehyde, *Biochim. Biophys. Acta* 57, 171–173, 1962; Holzer, H., da Fonseca-Wollheim, F., Kohlhaw, G., and Woenckhaus, C.W., Active forms of acetaldehyde, pyruvate, and glycolic aldehyde, *Ann. N.Y. Acad. Sci.* 98, 453–465, 1962; Brooks, P.J. and Theruvathu, J.A., DNA adducts from acetaldehyde: implications for alcohol-related carcinogenesis, *Alcohol* 35, 187–193, 2005; Tyulina, O.V., Prokopieva, V.D., Boldyrev, A.A., and Johnson, P., Erthyrocyte and plasma protein modification in alcoholism: a possible role of acetaldehyde, *Biochim. Biophys. Acta* 1762, 558–563, 2006; Pluskota-Karwatka, D., Pawlowicz, A.J., and Kronberg, L., Formation of malonaldehyde-acetaldehyde conjugate adducts in calf thymus DNA, *Chem. Res. Toxicol.* 19, 921–926, 2006.

	Chemical Name	M.W.	Properties and Comment
Acetic Acid	Acetic Acid, Glacial	60.05	Solvent (particular use in the extraction of collagen from tissue), buffer component (used in urea-acetic acid electrophoresis). Use in endoscopy as mucous-resolving agent.

Acetic acid

Banfield, A.G., Age changes in the acetic acid-soluble collagen in human skin, *Arch. Pathol.* 68, 680–684, 1959; Steven, F.S. and Tristram, G.R., The denaturation of acetic acid-soluble calf-skin collagen. Changes in optical rotation, viscosity, and susceptibility towards enzymes during serial denaturation in solutions of urea, *Biochem. J.* 85, 207–210, 1962; Neumark, T.

and Marot, I., The formation of acetic-acid soluble collagen under polarization and electron microscrope, *Acta Histochem.* 23, 71–79, 1966; Valfleteren, J.R., Sequential two-dimensional and acetic acid/urea/Triton X-100 gel electrophoresis of proteins, *Anal. Biochem.* 177, 388–391, 1989; Smith, B.J., Acetic acid-urea polyacrylamide gel electrophoresis of proteins, *Methods Mol.Biol.* 32, 39–47, 1994; Banfield, W.G., MacKay, C.M., and Brindley, D.C., Quantitative changes in acetic acid-extractable collagen of hamster skin related to anatomical site and age, *Gerontologia* 12, 231–236, 1996; Lian, J.B., Morris, S., Faris, B. et al., The effects of acetic acid and pepsin on the crosslinkages and ultrastructure of corneal collagen, *Biochim. Biophys. Acta.* 328, 193–204, 1973; Canto, M.I., Chromoendoscopy and magnifying endoscopy for Barrett's esophagus, *Clin.Gastroenterol.Hepatol.* 3 (7 Suppl. 1), S12–S15, 2005; Sionkowska, A., Flash photolysis and pulse radiolysis studies on collagen Type I in acetic acid solution, *J. Photochem. Photobiol. B* 84, 38–45, 2006.

Acetic Anhydride

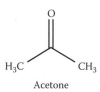

Acetic anhydride

Acetic Anhydride   102.07   Protein modification (trace labeling of amino groups); modification of amino groups and hydroxyl groups.

Jencks, W.P., Barley, F., Barnett, R., and Gilchrest, M., The free energy of hydrolysis of acetic anhydride, *J. Am. Chem. Soc.* 88, 4464–4467, 1966; Cromwell, L.D. and Stark, G.D., Determination of the carboxyl termini of proteins with ammonium thiocyanate and acetic anhydride, with direct identification of the thiohydantoins, *Biochemistry* 8, 4735–4740, 1969; Montelaro, R.C. and Rueckert, R.R., Radiolabeling of proteins and viruses *in vitro* by acetylation with radioactive acetic anhydride, *J. Biol. Chem.* 250, 1413–1421, 1975; Valente, A.J. and Walton, K.W., The binding of acetic anhydride- and citraconic anhydride-modified human low-density lipoprotein to mouse peritoneal macrophages. The evidence for separate binding sites, *Biochim. Biophys. Acta* 792, 16–24, 1984; Fojo, A.T., Reuben, P.M., Whitney, P.L., and Awad, W.M., Jr., Effect of glycerol on protein acetylation by acetic anhydride, *Arch. Biochem. Biophys.* 240, 43–50, 1985; Buechler, J.A., Vedvick, T.A., and Taylor, S.S., Differential labeling of the catalytic subunit of cAMP-dependent protein kinase with acetic anhydride: substrate-induced conformational changes, *Biochemistry* 28, 3018–3024, 1989; Baker, G.B., Coutts, R.T., and Holt, A., Derivatization with acetic anhydride: applications to the analysis of biogenic amines and psychiatric drugs by gas chromatography and mass spectrometry, *J. Pharmacol. Toxicol. Methods* 31, 141–148, 1994; Ohta, H., Ruan, F., Hakomori, S., and Igarashi, Y., Quantification of free Sphingosine in cultured cells by acetylation with radioactive acetic anhydride, *Anal. Biochem.* 222, 489–494, 1994; Yadav, S.P., Brew, K., and Puett, D., Holoprotein formation of human chorionic gonadotropin: differential trace labeling with acetic anhydride, *Mol. Endocrinol.* 8, 1547–1558, 1994; Miyazaki, K. and Tsugita, A., C-terminal sequencing method for peptides and proteins by the reaction with a vapor of perfluoric acid in acetic anhydride, *Proteomics* 4, 11–19, 2004.

Acetone

Acetone

Dimethyl Ketone;   58.08   Solvent, protein purification
2-propanone   (acetone powders); rare
                                 reaction with amino
                                 groups.

La Du, B., Jr. and Greenberg, D.M., The tyrosine oxidation system of liver. I. Extracts of rat liver acetone powder, *J. Biol. Chem.* 190, 245–255, 1951; Korn, E.D. and Payza, A.N., The degradation of heparin by bacterial enzymes. II. Acetone powder extracts, *J. Biol. Chem.* 223, 859–864, 1956; Ohtsuki, K., Taguchi, K., Sato, K., and Kawabata, M., Purification of ginger proteases by DEAE-Sepharose and isoelectric focusing, *Biochim. Biophys. Acta* 1243, 181–184, 1995; Selden, L.A., Kinosian, H.J., Estes, J.E., and Gershman, L.C., Crosslinked dimers with nucleating activity in actin prepared from muscle acetone powder, *Biochemistry* 39, 64–74, 2000; Abadir, W.F., Nakhla, V., and Chong, F., Removal of superglue from the external ear using acetone: case report and literature review, *J. Laryngol. Otol.* 109, 1219–1221, 1995; Jones, A.W., Elimination half-life of acetone in humans: case reports and review of the literature, *J. Anal. Toxicol.* 24, 8–10, 2000; Huang, L.P. and Guo, P., Use of acetone to attain highly active and soluble DNA packaging protein Gp16 of Phi29 for ATPase assay, *Virology* 312, 449–457, 2003; Paska, C., Bogi, K., Szilak, L. et al., Effect of formalin, acetone, and RNAlater fixatives on tissue preservation and different size amplicons by real-time PCR from paraffin-embedded tissues, *Diagn. Mol. Pathol.* 13, 234–240, 2004; Kuksis, A., Ravandi, A., and Schneider, M., Covalent binding of acetone to aminophospholipids *in vitro* and *in vivo*, *Ann. N.Y. Acad. Sci.* 1043, 417–439, 2005; Perera, A., Sokolic, F., Almasy, L. et al., On the evaluation of the

Kirkwood–Buff integrals of aqueous acetone mixtures, *J. Chem. Physics* 123, 23503, 2005; Zhou, J., Tao, G., Liu, Q. et al., Equilibrium yields of mono- and di-lauroyl mannoses through lipase-catalyzed condensation in acetone in the presence of molecular sieves, *Biotechnol. Lett.* 28, 395–400, 2006.

Acetonitrile   Ethenenitrile,   41.05   Chromatography solvent,
Methyl Cyanide   general solvent.

Acetonitrile

Hodgikinson, S.C. and Lowry, P.J., Hydrophobic-interaction chromatography and anion-exchange chromatography in the presence of acetonitrile. A two-step purification method for human prolactin, *Biochem. J.* 199, 619–627, 1981; Wolf-Coporda, A., Plavsic, F., and Vrhovac, B., Determination of biological equivalence of two atenolol preparations, *Int. J. Clin. Pharmacol. Ther. Toxicol.* 25, 567–571, 1987; Fischer, U., Zeitschel, U., and Jakubke, H.D., Chymotrypsin-catalyzed peptide synthesis in an acetonitrile-water-system: studies on the efficiency of nucleophiles, *Biomed. Biochim. Acta* 50, S131–S135, 1991; Haas, R. and Rosenberry, T.L., Protein denaturation by addition and removal of acetonitrile: application to tryptic digestion of acetylcholinesterase, *Anal. Biochem.* 224, 425–427, 1995; Joansson, A., Mosbach, K., and Mansson, M.O., Horse liver alcohol dehydrogenase can accept NADP$^+$ as coenzyme in high concentrations of acetonitrile, *Eur. J. Biochem.* 227, 551–555, 1995; Barbosa, J., Sanz-Nebot, V., and Toro, I., Solvatochromic parameter values and pH in acetonitrile-water mixtures. Optimization of mobile phase for the separation of peptides by high-performance liquid chromatography, *J. Chromatog. A* 725, 249–260, 1996; Barbosa, J., Hernandez-Cassou, S., Sanz-Nebot, V., and Toro, I., Variation of acidity constants of peptides in acetonitrile-water mixtures with solvent composition: effect of preferential salvation, *J. Pept. Res.* 50, 14–24, 1997; Badock, V., Steinhusen, U., Bommert, K., and Otto, A., Prefractionation of protein samples for proteome analysis using reversed-phase high-performance liquid chromatography, *Electrophoresis* 22, 2856–2864, 2001; Yoshida, T., Peptide separation by hydrophilic-interaction chromatography: a review, *J. Biochem. Biophys. Methods* 60, 265–280, 2004: Kamau, P. and Jordan, R.B., Complex formation constants for the aqueous copper(I)-acetonitrile system by a simple general method, *Inorg. Chem.* 40, 3879–3883, 2001; Nagy, P.I. and Erhardt, P.W., Monte Carlo simulations of the solution structure of simple alcohols in water-acetonitrile mixtures, *J. Phys. Chem. B Condens. Matter Mater. Surf. Interfaces Biophys.* 109, 5855–5872, 2005; Kutt, A., Leito, I., Kaljurand, I. et al., A comprehensive self-consistent spectrophotometric acidity scale of neutral Bronstad acids in acetonitrile, *J. Org. Chem.* 71, 2829–2938, 2006.

Acetyl Chloride   Ethanoyl Chloride   78.50   Acetylating agent.

Acetyl chloride

Hallaq, Y., Becker, T.C., Manno, C.S., and Laposata, M., Use of acetyl chloride/methanol for assumed selective methylation of plasma nonesterified fatty acids results in significant methylation of esterified fatty acids, *Lipids* 28, 355–360, 1993; Shenoy, N.R., Shively, J.E., and Bailey, J.M., Studies in C-terminal sequencing: new reagents for the synthesis of peptidylthiohydantoins, *J. Protein Chem.* 12, 195–205, 1993; Bosscher, G., Meetsma, A., and van De Grampel, J.C., Novel organo-substituted cyclophosphazenes via reaction of a monohydro cyclophosphazene and acetyl chloride, *Inorg. Chem.* 35, 6646–6650, 1996; Mo, B., Li, J., and Liang, S., A method for preparation of amino acid thiohydantoins from free amino acids activated by acetyl chloride for development of protein C-terminal sequencing, *Anal. Biochem.* 249, 207–211, 1997; Studer, J., Purdie, N., and Krouse, J.A., Friedel–Crafts acylation as a quality control assay for steroids, *Appl. Spectros.* 57, 791–796, 2003.

Acetylcysteine   *N*-acetyl-L-   163.2   Mild reducing agent for
cysteine   clinical chemistry (creatine
kinase); therapeutic use for
aminoacetophen
intoxication; some other
claimed indications.

N-acetylcysteine

Szasz, G., Gruber, W., and Bernt, E., Creatine kinase in serum. I. Determination of optimum reaction conditions, *Clin. Chem.* 22, 650–656, 1976; Holdiness, M.R., Clinical pharmacokinetics of *N*-acetylcysteine, *Clin. Pharmacokinet.* 20, 123–134, 1991; Kelley, G.S., Clinical applications of *N*-acetylcysteine, *Altern. Med. Rev.* 3, 114–127, 1998; Schumann, G., Bonora, R., Ceriotti, F. et al., IFCC primary reference procedures for the measurement of catalytic activity concentrations of enzymes at 37°C. Part 2. Reference procedure for the measurement of catalytic concentration of creatine kinase, *Clin. Chem. Lab. Med.* 40, 635–642, 2002; Zafarullah, M., Li, W.Q., Sylvester, J., and Ahmad, M., Molecular mechanisms of *N*-acetylcysteine actions, *Cell. Mol. Life Sci.* 60, 6–20, 2003; Marzullo, L., An update of *N*-acetylcysteine treatment for acute aminoacetophen toxicity in children, *Curr. Opin. Pediatr.* 17, 239–245, 2005; Aitio, M.L., *N*-acetylcysteine — passé-partout or much ado about nothing? *Br. J. Clin. Pharmacol.* 61, 5–15, 2006.

N-Acetylimidazole

1-acetyl-1*H*-imidazole   110.12   Reagent for modification of tyrosyl residues in proteins.

N-acetylimidazole

Lundblad, R.L., *Chemical Reagents for Protein Modification*, CRC Press, Boca Raton, FL, 2004; Gorbunoff, M.J., Exposure of tyrosine residues in proteins. 3. The reaction of cyanuric fluoride and *N*-acetylimidazole with ovalbumin, chymotrypsinogen, and trypsinogen, *Biochemistry* 44, 719–725, 1969; Houston, L.L. and Walsh, K.A., The transient inactivation of trypsin by mild acetylation with *N*-acetylimidazole, *Biochemistry* 9, 156–166, 1970; Shifrin, S. and Solis, B.G., Reaction of *N*-acetylimidazole with L-asparaginase, *Mol. Pharmacol.* 8, 561–564, 1972; Ota, Y., Nakamura, H., and Samejima, T., The change of stability and activity of thermolysin by acetylation with *N*-acetylimidazole, *J. Biochem.* 72, 521–527, 1972; Kasai, H., Takahashi, K., and Ando, T., Chemical modification of tyrosine residues in ribonuclease T1 with *N*-acetylimidazole and *p*-diazobenzenesulfonic acid, *J. Biochem.* 81, 1751–1758, 1977; Zhao, X., Gorewit, R.C., and Currie, W.B., Effects of *N*-acetylimidazole on oxytocin binding in bovine mammary tissue, *J. Recept. Res.* 10, 287–298, 1990; Wells, I. and Marnett, L.J., Acetylation of prostaglandin endoperoxide synthase by *N*-acetylimidazole: comparison to acetylation by aspirin, *Biochemistry* 31, 9520–9525, 1992; Cymes, G.D., Iglesias, M.M., and Wolfenstein-Todel, C., Chemical modification of ovine prolactin with *N*-acetylimidazole, *Int. J. Pept. Protein Res.* 42, 33–38, 1993; Zhang, F., Gao, J., Weng, J. et al., Structural and functional differences of three groups of tyrosine residues by acetylation of *N*-acetylimidazole in manganese-stabilizing protein, *Biochemistry* 44, 719–725, 2005.

Acetylsalicylic Acid

2-(acetoxy)benzoic   180.16   Analgesic, anti-
Acid; Aspirin   inflammatory; mild
                                             acetylating agent.

Acetylsalicylic acid (aspirin)
2–(acetoxy) benzoic acid
2–acetoxylbenzoic acid

Hawkins, D., Pinckard, R.N., and Farr, R.S., Acetylation of human serum albumin by acetylsalicylic acid, *Science* 160, 780–781, 1968; Kalatzis, E., Reactions of aminoacetophen in pharmaceutical dosage forms: its proposed acetylation by acetylsalicylic acid, *J. Pharm. Sci.* 59, 193–196, 1970; Pinckard, R.N., Hawkins, D., and Farr, R.S., The inhibitory effect of salicylate on the actylation of human albumin by acetylsalicylic acid, *Arthritis Rheum.* 13, 361–368, 1970; Van Der Ouderaa, F.J., Buytenhek, M., Nugteren, D.H., and Van Dorp, D.A., Acetylation of prostaglandin endoperoxide synthetase with acetylsalicylic acid, *Eur. J. Biochem.* 109, 1–8, 1980; Rainsford, K.D., Schweitzer, A., and Brune, K., Distribution of the acetyl compared with the salicyl moiety of acetylsalicylic acid. Acetylation of macromolecules in organs wherein side effects are manifest, *Biochem. Pharmacol.* 32, 1301–1308, 1983; Liu, L.R. and Parrott, E.L., Solid-state reaction between sulfadiazine and acetylsalicyclic acid, *J. Pharm. Sci.* 80, 564–566, 1991; Minchin, R.F., Ilett, K.F., Teitel, C.H. et al., Direct *O*-acetylation of *N*-hydroxy arylamines by acetylsalicylic acid to form carcinogen-DNA adducts, *Carcinogenesis* 13, 663–667, 1992.

Acrylamide   2-propenamide   71.08   Monomer unit of
polyacrylamide in gels,
hydrogels, hard polymers;
environmental carcinogen;
fluorescence quencher.

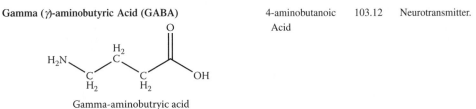

Eftink, M.R. and Ghiron, C.A., Fluorescence quenching studies with proteins, *Anal. Biochem.* 114, 199–227, 1981; Dearfield, K.L., Abernathy, C.O., Ottley, M.S. et al., Acrylamide: its metabolism, developmental and reproductive effects, *Mutat. Res.* 195, 45–77, 1988; Williams, L.R., Staining nucleic acids and proteins in electrophoresis gels, *Biotech. Histochem.* 76, 127–132, 2001; Hamden, M., Bordini, E., Galvani, M., and Righetti, P.G., Protein alkylation by acrylamide, its *N*-substituted derivatives and crosslinkers and its relevance to proteomics: a matrix-assisted laser desorption/ionization-time of flight-mass spectrometry study, *Electrophoresis* 22, 1633–1644, 2001; Cioni, P. and Strambini, G.B., Tryptophan phosphorescence and pressure effects on protein structure, *Biochim. Biophys. Acta* 1595, 116–130, 2002; Taeymans, D., Wood, J., Ashby, P. et al., A review of acrylamide: an industry perspective on research, analysis, formation, and control, *Crit. Rev. Food Sci. Nutr.* 44, 323–347, 2004; Rice, J.M., The carcinogenicity of acrylamide, *Mutat. Res.* 580, 3–20, 2005; Besaratinia, A. and Pfeifer, G.P., DNA adduction and mutagenic properties of acrylamide, *Mutat. Res.* 580, 31–40, 2005; Hoenicke, K. and Gaterman, R., Studies on the stability of acrylamide in food during storage, *J. AOAC Int.* 88, 268–273, 2005; Castle, L. and Ericksson, S., Analytical methods used to measure acrylamide concentrations in foods, *J. AOAC Int.* 88, 274–284, 2005; Stadler, R.H., Acrylamide formation in different foods and potential strategies for reduction, *Adv. Exp. Med. Biol.* 561, 157–169, 2005; Lopachin, R.M. and Decaprio, A.P., Protein adduct formation as a molecular mechanism in neurotoxicity, *Toxicol. Sci.* 86, 214–225, 2005.

Gamma (γ)-aminobutyric Acid (GABA)   4-aminobutanoic   103.12   Neurotransmitter.
Acid

Gamma-aminobutryic acid

Mandel, P. and DeFeudis, F.V., Eds., *GABA—Biochemistry and CNS Functions*, Plenum Press, New York, 1979; Costa, E. and Di Chiara, G., *GABA and Benzodiazepine Receptors,* Raven Press, New York, 1981; Racagni, G. and Donoso, A.O., *GABA and Endocrine Function*, Raven Press, New York, 1986; Squires, R.F., *GABA and Benzodiazepine Receptors*, CRC Press, Boca Raton, FL, 1988; Martin, D.L. and Olsen, R.W., *GABA in the Nervous System: The View at Fifty Years,* Lippincott, Williams & Wilkins, Philadelphia, PA, 2000.

Amiloride

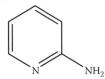

Amiloride

3,5-diamino-*N*-   229.63   Sodium ion channel blocker.
(amino-
iminomethyl)-6-
chloropyrazine-
carboxamide

Benos, D.J., A molecular probe of sodium transport in tissues and cells, *Am. J. Physiol.* 242, C131–C145, 1982; Garty, H., Molecular properties of epithelial, amiloride-blockable Na⁺ channels, *FASEB J.* 8, 522–528, 1994; Barbry, P. and Lazdunski, M., Structure and regulation of the amiloride-sensitive epithelial sodium channel, *Ion Channels* 4, 115–167, 1996; Kleyman, T.R., Sheng, S., Kosari, F., and Kieber-Emmons, T., Mechanism of action of amiloride: a molecular perspective, *Semin. Nephrol.* 19, 524–532, 1999; Alvarez de la Rosa, D., Canessa, C.M., Fyfe, G.K., and Zhang, P., Structure and regulation of amiloride-sensitive sodium channels, *Annu. Rev. Physiol.* 62, 573–594, 2000; Haddad, J.J., Amiloride and the regulation of NF-κβ: an unsung crosstalk and missing link between fluid dynamics and oxidative stress-related inflammation — controversy or pseudo-controversy, *Biochem. Biophys. Res. Commun.* 327, 373–381, 2005.

2-Aminopyridine

2-aminopyridine

α-aminopyridine   94.12   Precursor for synthesis of
pharmaceuticals and
reagents; used to derivatize
carbohydrates for analysis;
blocker of K⁺ channels.

Hase, S., Hara, S., and Matsushima, Y., Tagging of sugars with a fluorescent compound, 2-aminopyridine, *J. Biochem.* 85, 217–220, 1979; Hase, S., Ibuki, T., and Ikenaka, T., Reexamination of the pyridylamination used for fluorescence labeling of oligosaccharides and its application to glycoproteins, *J. Biochem.* 95, 197–203, 1984; Chen, C. and Zheng, X., Development of the new antimalarial drug pyronaridine: a review, *Biomed. Environ. Sci.* 5, 149–160, 1992; Hase, S., Analysis of sugar chains by pyridylamination, *Methods Mol. Biol.* 14, 69–80, 1993; Oefner, P.J. and Chiesa, C., Capillary electrophoresis of carbohydrates, *Glycobiology* 4, 397–412, 1994; Dyukova, V.I., Shilova, N.V., Galanina, O.E. et al., Design of carbohydrate multiarrays, *Biochim. Biophys. Acta* 1760, 603–609, 2006; Takegawa, Y., Deguchi, K., Keira, T. et al., Separation of isomeric 2-aminopyridine derivatized *N*-glycans and *N*-glycopeptides of human serum immunoglobulin G by using a zwitterionic type of hydrophilic-interaction chromatography, *J. Chromatog. A* 1113, 177–181, 2006; Suzuki, S., Fujimori, T., and Yodoshi, M., Recovery of free oligosaccharides from derivatives labeled by reductive amination, *Anal. Biochem.* 354, 94–103, 2006; Caballero, N.A., Melendez, F.J., Munoz-Caro, C., and Nino, A., Theoretical prediction of relative and absolute pK(a) values of aminopyridine, *Biophys. Chem.*, 124, 155–160, 2006.

Ammonium Bicarbonate

Ammonium bicarbonate, NH₄CO₃

Carbon dioxide(CO₂) + Water (H₂O)

Acid Ammonium   79.06   Volatile buffer salt.
Carbonate

Gibbons, G.R., Page, J.D., and Chaney, S.G., Treatment of DNA with ammonium bicarbonate or thiourea can lead to underestimation of platinum-DNA monoadducts, *Cancer Chemother. Pharmacol.* 29, 112–116, 1991; Sorenson, S.B., Sorenson, T.L., and Breddam, K., Fragmentation of protein by *S. aureus* strain V8 protease. Ammonium bicarbonate strongly inhibits the enzyme but does not improve the selectivity for glutamic acid, *FEBS Lett.* 294, 195–197, 1991; Fichtinger-Schepman, A.M., van Dijk-Knijnenburg, H.C., Dijt, F.J. et al., Effects of thiourea and ammonium bicarbonate on the formation and stability of bifunctional cisplatinin-DNA adducts: consequences for the accurate quantification of adducts in (cellular) DNA, *J. Inorg. Biochem.* 58, 177–191, 1995; Overcashier, D.E., Brooks, D.A., Costantino, H.R., and Hus, C.C., Preparation of excipient-free recombinant human tissue-type plasminogen activator by lyophilization from ammonium bicarbonate solution: an investigation of the two-stage sublimation process, *J. Pharm. Sci.* 86, 455–459, 1997.

ANS

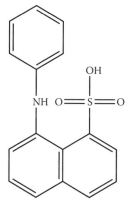

1-anilino-8-naphthalenesulfonate, ANS

| | 1-anilino-8-naphthalenesulfonate | 299.4 | Fluorescent probe for protein conformation; considered a hydrophobic probe; study of molten globules. |

Ferguson, R.N., Edelhoch, H., Saroff, H.A. et al., Negative cooperativity in the binding of thyroxine to human serum prealbumin. Preparation of tritium-labeled 8-anilino-1-naphthalenesulfonic acid, *Biochemistry* 14, 282–289, 1975; Ogasahara, K., Koike, K., Hamada, M., and Hiraoka, T., Interaction of hydrophobic probes with the apoenzyme of pig heart lipoamide dehydrogenase, *J. Biochem.* 79, 967–975, 1976; De Campos Vidal, B., The use of the fluorescence probe 8-anilinonaphthalene sulfate (ANS) for collagen and elastin histochemistry, *J. Histochem. Cytochem.* 26, 196–201, 1978; Royer, C.A., Fluorescence spectroscopy, *Methods Mol. Biol.* 40, 65–89, 1995; Celej, M.S., Dassie, S.A., Freire, E. et al., Ligand-induced thermostability in proteins: thermodynamic analysis of ANS-albumin interaction, *Biochim. Biophys. Acta* 1750, 122–133, 2005; Banerjee, T. and Kishore, N., Binding of 8-anilinonaphthalene sulfonate to dimeric and tetrameric concanavalin A: energetics and its implications on saccharide binding studied by isothermal titration calorimetry and spectroscopy, *J. Phys. Chem. B Condens. Matter Mater. Surf. Interfaces Biophys.* 110, 7022–7028, 2006; Sahu, K., Mondal, S.K., Ghosh, S. et al., Temperature dependence of salvation dynamics and anisotropy decay in a protein: ANS in bovine serum albumin, *J. Chem. Phys.* 124, 124909, 2006; Wang, G., Gao, Y., and Geng, M.L., Analysis of heterogeneous fluorescence decays in proteins. Using fluorescence lifetime of 8-anilino-1-naphthalenesulfonate to probe apomyoglobin unfolding at equilibrium, *Biochim. Biophys. Acta* 1760, 1125–1137, 2006; Greene, L.H., Wijesinha-Bettoni, R., and Redfield, C., Characterization of the molten globule of human serum retinol-binding protein using NMR spectroscopy, *Biochemistry* 45, 9475–9484, 2006.

Arachidonic Acid

COOH

COOH

Arachidonic acid

| | 5,8,11,14(all *cis*)-eicosotetraenoic Acid | 304.5 | Essential fatty acid; precursor of prostaglandins, thromboxanes, and leukotrienes. |

Moncada, S. and Vane, J.R., Interaction between anti-inflammatory drugs and inflammatory mediators. A reference to products of arachidonic acid metabolism, *Agents Actions Suppl.* 3, 141–149, 1977; Moncada, S. and Higgs, E.A., Metabolism of arachidonic acid, *Ann. N.Y. Acad. Sci.* 522, 454–463, 1988; Piomelli, D., Arachidonic acid in cell signaling, *Curr. Opin. Cell Biol.* 5, 274–280, 1993; Janssen-Timmen, U., Tomic, I., Specht, E. et al., The arachidonic acid cascade, eicosanoids, and signal transduction, *Ann. N.Y. Acad. Sci.* 733, 325–334, 1994; Wang, X. and Stocco, D.M., Cyclic AMP and arachidonic acid: a tale of two pathways, *Mol. Cell. Endocrinol.* 158, 7–12, 1999; Brash, A.R., Arachidonic acid as a bioactive molecule, *J. Clin. Invest.*

107, 1339–1345, 2001; Luo, M., Flamand, N., and Brock, T.G., Metabolism of arachidonic acid to eicosanoids within the nucleus, *Biochim. Biophys. Acta* 1761, 618–625, 2006; Balboa, M.A. and Balsinde, J., Oxidative stress and arachidonic acid mobilization, *Biochim. Biophys. Acta* 1761, 385–391, 2006.

Ascorbic Acid

Vitamin C; 3-oxo-L-gulofuranolactone   176.13   Nutrition, antioxidant (reducing agent); possible antimicrobial function.

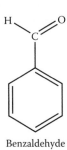

Ascorbic acid ⇌ Dehydroascorbic acid

Barnes, M.J. and Kodicek, E., Biological hydroxylations and ascorbic acid with special regard to collagen metabolism, *Vitam. Horm.* 30, 1–43, 1972; Leibovitz, B. and Siegel, B.V., Ascorbic acid and the immune response, *Adv. Exp. Med. Biol.* 135, 1–25, 1981; Englard, S. and Seifter, S., The biochemical functions of ascorbic acid, *Annu. Rev. Nutr.* 6, 365–406, 1986; Levine, M. and Hartzell, W., Ascorbic acid: the concept of optimum requirements, *Ann. N.Y. Acad. Sci.* 498, 424–444, 1987; Padh, H., Cellular functions of ascorbic acid, *Biochem. Cell Biol.* 68, 1166–1173, 1990; Meister, A., On the antioxidant effects of ascorbic acid and glutathione, *Biochem. Pharmacol.* 44, 1905–1915, 1992; Wolf, G., Uptake of ascorbic acid by human neutrophils, *Nutr. Rev.* 51, 337–338, 1993; Kimoto, E., Terada, S., and Yamaguchi, T., Analysis of ascorbic acid, dehydroascorbic acid, and transformation products by ion-pairing high-performance liquid chromatography with multiwavelength ultraviolet and electrochemical detection, *Methods Enzymol.* 279, 3–12, 1997; May, J.M., How does ascorbic acid prevent endothelial dysfunction? *Free Rad. Biol. Med.* 28, 1421–1429, 2000; Smirnoff, N. and Wheeler, G.L., Ascorbic acid in plants: biosynthesis and function, *Crit. Rev. Biochem. Mol. Biol.* 35, 291–314, 2000; Arrigoni, O. and De Tullio, M.C., Ascorbic acid: much more than just an antioxidant, *Biochim. Biophys. Acta* 1569, 1–9, 2002; Akyon, Y., Effect of antioxidant on the immune response of *Helicobacter pyrlori*, *Clin. Microbiol. Infect.* 8, 438–441, 2002; Takanaga, H., MacKenzie, B., and Hediger, M.A., Sodium-dependent ascorbic acid transporter family SLC23, *Pflügers Arch.* 447, 677–682, 2004.

Benzaldehyde

Benzoic Aldehyde; Essential Oil of Almond   106.12   Intermediate in manufacture of pharmaceuticals, flavors; reacts with amino groups, semicarbidizide.

Benzaldehyde

Chalmers, R.M., Keen, J.N., and Fewson, C.A., Comparison of benzyl alcohol dehydrogenases and benzaldehyde dehydrogenases from the benzyl alcohol and mandelate pathways in *Acinetobacter calcoaceticus* and the TOL-plasmid-encoded toluene pathway in *Pseudomonas putida*. N-terminal amino acid sequences, amino acid composition, and immunological cross-reactions, *Biochem. J.* 273, 99–107, 1991; Pettersen, E.O., Larsen, R.O., Borretzen, B. et al., Increased effect of benzaldehyde by exchanging the hydrogen in the formyl group with deuterium, *Anticancer Res.* 11, 369–373, 1991; Nierop Groot, M.N. and de Bont, J.A.M., Conversion of phenylalanine to benzaldehyde initiated by an aminotransferase in *Lactobacillus plantarum*, *Appl. Environ. Microbiol.* 64, 3009–3013, 1998; Podyminogin, M.A., Lukhtanov, E.A., and Reed, M.W., Attachment of benzaldehyde-modified oligodeoxynucleotide probes to semicarbazide-coated glass, *Nucleic Acids Res.* 29, 5090–5098, 2001; Kurchan, A.N. and Kutateladze, A.G., Amino acid-based dithiazines: synthesis and photofragmentation of their benzaldehyde adducts, *Org. Lett.* 4, 4129–4131, 2002; Kneen, M.M., Pogozheva, I.D., Kenyon, G.L., and McLeish, M.J., Exploring the active site of benzaldehyde lyase by modeling and mutagenesis, *Biochim. Biophys. Acta* 1753, 263–271, 2005; Mosbacher, T.G., Mueller, M., and Schultz, G.E., Structure and mechanism of the ThDP-dependent benzaldehyde lyase from *Pseudomonas fluorescens*, *FEBS J.* 272, 6067–6076, 2005; Sudareva, N.N. and Chubarova, E.V., Time-dependent conversion of benzyl alcohol to benzaldehyde and benzoic acid in aqueous solution, *J. Pharm. Biomed. Anal.* 41, 1380–1385, 2006.

Benzamidine HCl   156.61   Inhibitor of trypticlike
                                                                                 serine proteases.

Benzamidine

Ensinck, J.W., Shepard, C., Dudl, R.J., and Williams, R.H., Use of benzamidine as a proteolytic inhibitor in the radio-immunoassay of glucagon in plasma, *J. Clin. Endocrinol. Metab.* 35, 463–467, 1972; Bode, W. and Schwager, P., The refined crystal structure of bovine beta-trypsin at 1.8 Å resolution. II. Crystallographic refinement, calcium-binding site, benzamidine-binding site and active site at pH 7.0., *J. Mol. Biol.* 98, 693–717, 1975; Nastruzzi, C., Feriotto, G., Barbieri, R. et al., Differential effects of benzamidine derivatives on the expression of *c-myc* and HLA-DR alpha genes in a human B-lymphoid tumor cell line, *Cancer Lett.* 38, 297–305, 1988; Clement, B., Schmitt, S., and Zimmerman, M., Enzymatic reduction of benzamidoxime to benzamidine, *Arch. Pharm.* 321, 955–956, 1988; Clement, B., Immel, M., Schmitt, S., and Steinman, U., Biotransformation of benzamidine and benzamidoxime *in vivo*, *Arch. Pharm.* 326, 807–812, 1993; Renatus, M., Bode, W., Huber, R. et al., Structural and functional analysis of benzamidine-based inhibitors in complex with trypsin: implications for the inhibition of factor Xa, tPA, and urokinase, *J. Med. Chem.* 41, 5445–5456, 1998; Henriques, R.S., Fonseca, N., and Ramos, M.J., On the modeling of snake venom serine proteinase interactions with benzamidine-based thrombin inhibitors, *Protein Sci.* 13, 2355–2369, 2004; Gustavsson, J., Farenmark, J., and Johansson, B.L., Quantitative determination of the ligand content in benzamidine Sepharose® 4 Fast Flow media with ion-pair chromatography, *J. Chromatog. A* 1070, 103–109, 2005.

Benzene   Benzene   78.11   Solvent; a zenobiotic.

Benzene

Lovley, D.R., Anaerobic benzene degradation, *Biodegradation* 11, 107–116, 2000; Snyder, R., Xenobiotic metabolism and the mechanism(s) of benzene toxicity, *Drug Metab. Rev.* 36, 531–547, 2004; Rana, S.V. and Verma, Y., Biochemical toxicity of benzene, *J. Environ. Biol.* 26, 157–168, 2005; Lin, Y.S., McKelvey, W., Waidyanatha, S., and Rappaport, S.M., Variability of albumin adducts of 1,4-benzoquinone, a toxic metabolite of benzene, in human volunteers, *Biomarkers* 11, 14–27, 2006; Baron, M. and Kowalewski, V.J., The liquid water-benzene system, *J. Phys. Chem. A Mol. Spectrosc. Kinet. Environ. Gen. Theory* 100, 7122–7129, 2006; Chambers, D.M., McElprang, D.O., Waterhouse, M.G., and Blount, B.C., An improved approach for accurate quantiation of benzene, toluene, ethylbenzene, zylene, and styrene in blood, *Anal. Chem.* 78, 5375–5383, 2006.

Benzidine   *p*-benzidine; (1,1′- 184.24   Precursor for azo dyes;
                                           biphenyl)-4,4′-   mutagenic agent; forensic
                                           diamine   analysis for bloodstains
H₂N—   —NH₂   based on reactivity with
                                                                          hemoglobin.

p-benzidine

Ahlquist, D.A. and Schwartz, S., Use of leuco-dyes in the quantitative colorimetric microdetermination of hemoglobin and other heme compounds, *Clin. Chem.* 21, 362–369, 1975; Josephy, P.D., Benzidine: mechanisms of oxidative activation and mutagensis, *Fed. Proc.* 45, 2465–2470, 1986; Choudhary, G., Human health perspectives on environmental exposure to benzidine: a review, *Chemosphere* 32, 267–291, 1996; Madeira, P., Nunes, M.R., Borges, C. et al., Benzidine photodegradation: a mass spectrometry and UV spectroscopy combined study, *Rapid Commun. Mass Spectrom.* 19, 2015–2020, 2005; Saitoh, T., Yoshida, S., and Ichikawa, J., Naphthalene-1,8-diylbis(diphenylmethylium) as an organic two-electron oxidant: benzidine synthesis via oxidative self-coupling of *N,N*-dialkylanilines, *J. Org. Chem.* 71, 6414–6419, 2006.

BIG CHAP/Deoxy BIG CHAP

BigChap
N,N-bis-(3-gluconamidopropyl)cholamide

N,N-bis(3-d-gluconamido-propyl) cholamide/*N,N*-bis(3-d-gluconamido-propyl) deoxycholamide

878.1/
862.1

Nonionic detergents; protein solubilization, adenovirus gene transfer enhancement.

Bonelli, F.S. and Jonas, A., Reaction of lecithin: cholesterol acyltransferase with a water-soluble substrate: effects of surfactants, *Biochim. Biophys. Acta* 1166, 92–98, 1993; Aigner, A., Jager, M., Pasternack, R. et al., Purification and characterization of cysteine-*S*-conjugate *N*-acetyltransferase from pig kidney, *Biochem. J.* 317, 213–218, 1996; Mechref, Y. and Eirassi, Z., Micellar electrokinetic capillary chromatography with *in-situ* charged micelles. 4. Evaluation of novel chiral micelles consisting of steroidal glycoside surfactant borate complexes, *J. Chromatog. A* 724, 285–296, 1996; Abe, S., Kunii, S., Fujita, T., and Hiraiwa, K., Detection of human seminal gamma-glutamyl transpeptidase in stains using sandwich ELISA, *Forensic Sci. Int.* 91, 19–28, 1998; Akutsu, Y., Nakajima-Kambe, T., Nomura, N., and Nakahara, T., Purification and properties of a polyester polyurethane-degrading enzyme form *Comamonas acidovorans* TB-35, *Appl. Environ. Microbiol.* 64, 62–67, 1998: Connor, R.J., Engler, H., Machemer, T. et al., Identification of polyamides that enhance adenovirus-mediated gene expression in the urothelium, *Gene Therapy* 8, 41–48, 2001; Vajdos, F.F., Ultsch, M., Schaffer, M.L. et al., Crystal structure of human insulin-like growth factor-1: detergent binding inhibits binding protein interactions, *Biochemistry* 40, 11022–11029, 2001; Kuball, J., Wen, S.F., Leissner, J. et al., Successful adenovirus-mediated wild-type p53 gene transfer in patients with bladder cancer by intravesical vector instillation, *J. Clin. Oncol.* 20, 957–965, 2002; Susasara, K.M., Xia, F., Gronke, R.S., and Cramer, S.M., Application of hydrophobic interaction displacement chromatography for an industrial protein purification, *Biotechnol. Bioeng.* 82, 330–339, 2003; Ishibashi, A. and Nakashima, N., Individual dissolution of single-walled carbon nanotubes in aqueous solutions of steroid or sugar compounds and their Raman and near-IR spectral properties, *Chemistry*, 12, 7595–7602, 2006.

Biotin

Biotin

Coenzyme R

244.31

Coenzyme function in carboxylation reactions; growth factor; tight binding to avidin used for affinity interactions.

Knappe, J., Mechanism of biotin action, *Annu. Rev. Biochem.* 39, 757–776, 1970; Dunn, M.J., Detection of proteins on blots using the avidin-biotin system, *Methods Mol. Biol.* 32, 227–232, 1994; Wisdom, G.B., Enzyme and biotin labeling of antibody, *Methods Mol. Biol.* 32, 433–440, 1994; Wilbur, D.S., Pathare, P.M, Hamlin, D.K. et al., Development of new biotin/streptavidin reagents for pretargeting, *Biomol. Eng.* 16, 113–118, 1999; Jitrapakdee, S. and Wallace, J.C., The biotin enzyme family: conserved structural motifs and domain rearrangements, *Curr. Protein Pept. Sci.* 4, 217–229, 2003; Nikolau, B.J., Ohlrogge, J.B., and Wurtels, E.S., Plant biotin-containing carboxylases, *Arch. Biochem. Biophys.* 414, 211–222, 2003; Fernandez-Mejia, C., Pharmacological effects of biotin, *J. Nutri. Biochem.* 16, 424–427, 2005; Wilchek, M., Bayer, E.A., and Livnah, O., Essentials of biorecognition: the (strept)avidin-biotin system as a model for protein–protein and protein–ligand interactions, *Immunol. Lett.* 103, 27–32, 2006; Furuyama, T. and Henikoff, S., Biotin-tag affinity purification of a centromeric nucleosome assembly complex, *Cell Cycle* 5, 1269–1274, 2006; Streaker, E.D. and Beckett, D., Nonenzymatic biotinylation of a biotin carboxyl carrier protein: unusual reactivity of the physiological target lysine, *Protein Sci.* 15, 1928–1935, 2006; Raichur, A.M., Voros, J., Textor, M., and Fery, A., Adhesion of polyelectrolyte microcapsules through biotin-streptavidin specific interaction, *Biomacromolecules* 7, 2331–2336, 2006. For biotin switch assay, see Martinez-Ruiz, A. and Lamas, S., Detection and identification of *S*-nitrosylated proteins in endothelial cells, *Methods Enzymol.* 396, 131–139, 2005; Huang, B. and Chen, C., An ascorbate-dependent artifact that interferes with the interpretation of the biotin switch assay, *Free Radic. Biol. Med.* 41, 562–567, 2006; Gladwin, M.T., Wang, X., and Hogg, N., Methodological vexation about thiol oxidation versus *S*-nitrosation — a commentary on "An ascorbate-dependent artifact that interferes with the interpretation of the biotin-switch assay," *Free Radic. Biol. Med.* 41, 557–561, 2006.

Biuret

Biuret

Urea

Imidodicarbonic Diamide   103.08   Prepared by heating urea, reaction with cupric ions in base yields red-purple (the biuret reaction); nonprotein nitrogen (NPN) nutritional source.

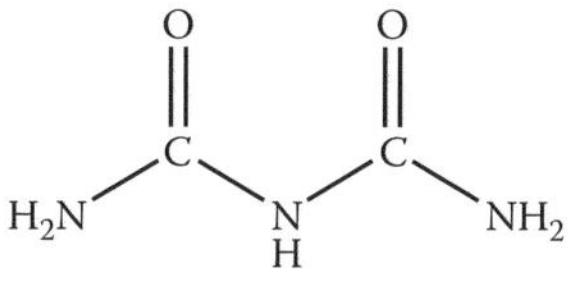

Jensen, H.L. and Schroder, M., Urea and biuret as nitrogen sources for *Rhizobium* spp., *J. Appl. Bacteriol.* 28, 473–478, 1965; Ronca, G., Competitive inhibition of adenosine deaminase by urea, guanidine, biuret, and guanylurea, *Biochim. Biophys. Acta* 132, 214–216, 1967; Oltjen, R.R., Slyter, L.L., Kozak, A.S., and Williams, E.E., Jr., Evaluation of urea, biuret, urea phosphate, and uric acid as NPN sources for cattle, *J. Nutr.* 94, 193–202, 1968; Tsai, H.Y. and Weber, S.G., Electrochemical detection of oligopeptides through the precolumn formation of biuret complexes, *J. Chromatog.* 542, 345–350, 1991; Gawron, A.J. and Lunte, S.M., Optimization of the conditions for biuret complex formation for the determination of peptides by capillary electrophoresis with ultraviolet detection, *Clin. Chem.* 51, 1411–1419, 2000; Roth, J., O'Leary, D.J., Wade, C.G. et al., Conformational analysis of alkylated biuret and triuret: evidence for helicity and helical inversion in oligoisocyates, *Org. Lett.* 2, 3063–3066, 2000; Hortin, G.L., and Mellinger, B., Cross-reactivity of amino acids and other compounds in the biuret reaction: interference with urinary peptide measurements, *Clin. Chem.* 51, 1411–1419, 2005.

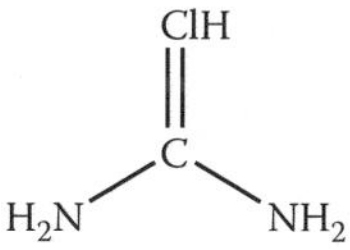

Blue Tetrazolium

Tetrazole form

Reduction

Formazan form

Tetrazolium Blue   727.65   Stain for cytotoxicity based on change to formazan on reduction. See nitro blue tetrazolium, which has similar chemistry and higher use.

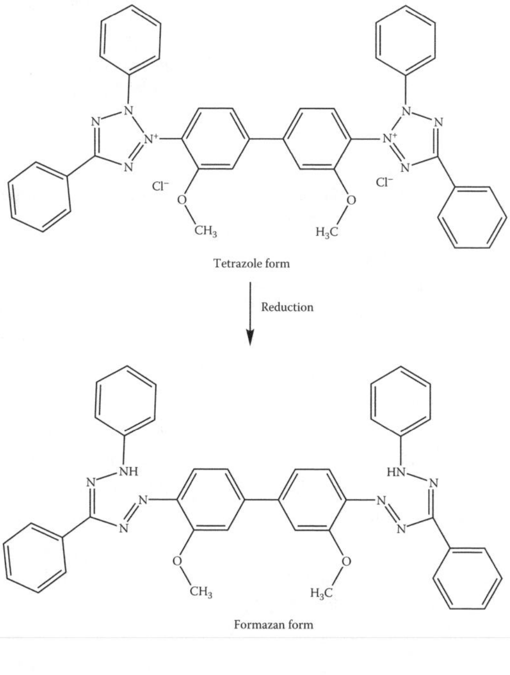

Tetrazolium Blue;
Blue Tetrazolium

Litteria, M. and Recknagel, R.O., A simplified blue tetrazolium reaction, *J. Lab. Clin. Med.* 48, 463–468, 1955; Sinsheimer, J.E. and Salim, E.F., Reactivity of blue tetrazolium with nonketol compounds, *Anal. Chem.* 37, 566–569, 1965; Graham, R.E., Biehl, E.R., Kenner, C.T. et al., Reduction of blue tetrazolium by corticosteroids, *J. Pharm. Sci.* 64, 226–230, 1975; Baba, N., Burtubise, P., and Myser, T., Immunofluorescence and immunoperoxidase observations of anti-lactic dehydrogenase-1 antibody, *J. Histochem. Cytochem.* 24, 572–577, 1976; Biehl, E., Wooten, R., Kenner, C.T., and Graham, R.E., Kinetic and mechanistic studies of blue tetrazolium reaction with phenylhydrazines, *J. Pharm. Sci.* 67, 927–930, 1978; Van Noorden, C.J., Tas, J., and Vogels, I.M., Cytophotometry of glucose-6-phosphate dehydrogenase activity in individual cells, *Histochem. J.* 15, 583–599, 1983; Maravelias, C., Dona, A., Athanaselis, S., and Koutselinis, A., The importance of performing *in vitro* cytotoxicity testing before immodulation evaluation, *Vet. Hum. Toxicol.* 42, 292–296, 2000; Reddy, R.M., Tsai, W.S., Ziauddin, M.F. et al., Cisplatin enhances apoptosis induced by a tumor-selective adenovirus-expressing tumor necrosis factor-related apoptosis-inducing ligand, *J. Thorac. Cardiovasc. Surg.* 128, 883–891, 2004.

Boric Acid *o*-boric Acid 61.83 Buffer salt, manufacturing; complexes with carbohydrates and other polyhydroxyl compounds; therapeutic use as a topic antibacterial/antifungal agent.

$$B(OH)_3 + 2H_2O \rightleftharpoons B(OH)_4^- + H_3O^+$$

Boric acid

Sciarra, J.J. and Monte Bovi, A.J., Study of the boric acid–glycerin complex. II. Formation of the complex at elevated temperature, *J. Pharm. Sci.* 51, 238–242, 1962; Walborg, E.F., Jr. and Lantz, R.S., Separation and quantitation of saccharides by ion-exchange chromatography utilizing boric acid–glycerol buffers, *Anal. Biochem.* 22, 123–133, 1968; Lerch, B. and Stegemann, H., Gel electrophoresis of proteins in borate buffer. Influence of some compounds complexing with boric acid, *Anal. Biochem.* 29, 76–83, 1969; Walborg, E.F., Jr., Ray, D.B., and Ohrberg, L.E., Ion-exchange chromatography of saccharides: an improved system utilizing boric acid/2,3-butanediol buffers, *Anal. Biochem.* 29, 433–440, 1969; Chen, F.T. and Sternberg, J.C., Characterization of proteins by capillary electrophoresis in fused-silica columns: review on serum protein anlaysis and application to immunoassays, *Electrophoresis* 15, 13–21, 1994; Allen, R.C. and Doktycz, M.J., Discontinuous electrophoresis revisited: a review of the process, *Appl. Theor. Electrophor.* 6, 1–9, 1996; Manoravi, P., Joseph, M., Sivakumar, N., and Balasubramanian, H., Determination of isotopic ratio of boron in boric acid using laser mass spectrometry, *Anal. Sci.* 21, 1453–1455, 2005; De Muynck, C., Beauprez, J., Soetaert, W., and Vandamme, E.J., Boric acid as a mobile phase additive for high-performance liquid chromatography separation of ribose, arabinose, and ribulose, *J. Chromatog. A* 1101, 115–121, 2006; Herrmannova, M., Kirvankova, L., Bartos, M., and Vytras, K., Direct simultaneous determination of eight sweeteners in foods by capillary isotachophoresis, *J. Sep. Sci.* 29, 1132–1137, 2006; Alencar de Queiroz, A.A., Abraham, G.A., Pires Camillo, M.A. et al., Physicochemical and antimicrobial properties of boron-complexed polyglycerol-chitosan dendrimers, *J. Biomater. Sci. Polym. Ed.* 17, 689–707, 2006; Ringdahl, E.N., Recurrent vulvovaginal candidiasis, *Mol. Med.* 103, 165–168, 2006.

BPNS-Skatole

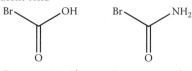

BPNS-Skatole

(2-[2′-nitrophenyl-sulfenyl]-3-methyl-3′-bromoindolenine	363.23	Tryptophan modification, peptide-bond cleavage; derived from skatole, which is also known as boar taint.

Boulanger, P., Lemay, P., Blair, G.E., and Russell, W.C., Characteriztion of adenovirus protein IX, *J. Gen. Virol.* 44, 783–800, 1979; Russell, J., Kathendler, J., Kowalski, K. et al., The single tryptophan residue of human placental lactogen. Effects of modification and cleavage on biological activity and protein conformation, *J. Biol. Chem.* 256, 304–307, 1981; Moskaitis, J.E. and Campagnoni, A.T., A comparison of the dodecyl sulfate-induced precipitation of the myelin basic protein with other water-soluble proteins, *Neurochem. Res.* 11, 299–315, 1986; Mahboub, S., Richard, C., Delacourte, A., and Han, K.K., Applications of chemical cleavage procedures to the peptide mapping of neurofilament triplet protein bands in sodium dodecyl sulfate-polyacrylamide gel electrophoresis, *Anal. Biochem.* 154, 171–182, 1986; Rahali, V. and Gueguen, J., Chemical cleavage of bovine beta-lactoglobulin by BPNS-skatole for preparative purposes: comparative study of hydrolytic procedure and peptide characterization, *J. Protein Chem.* 18, 1–12, 1999; Swamy, N., Addo, J., Vskokovic, M.R., and Ray, R., Probing the vitamin D sterol-binding pocket of human vitamin D-binding protein with bromoacetate affinity-labeling reagents containing the affinity probe at C-3, C-6, C-11, and C-19 positions of parent vitamin D sterols, *Arch. Biochem. Biophys.* 373, 471–478, 2000; Celestina, F. and Suryanarayana, T., Biochemical characterization and helix-stabilizing properties of HSNP-C' from the thermophilic archaeon *Sulfolobus acidocaldarius*, *Biochem. Biophys. Res. Commun.* 267, 614–618, 2000; Kibbey, M.M., Jameson, M.J., Eaton, E.M., and Rosenzweig, S.A., Insulinlike growth factor binding protein-2: contributions of the C-terminal domain to insulinlike growth factor-1 binding, *Mol. Pharmacol.* 69, 833–845, 2006.

Bromoacetic Acid

Bromoacetic acid   Bromoacetamide

Bromoacetic Acid	138.95	Alkylating agent; reacts with various nucleophiles.

Glick, D.M., Goren, H.J., and Barnard, E.A., Concurrent bromoacetate reaction at histidine and methionine residues in ribonuclease, *Biochem. J.* 102, 7C–10C, 1967; Goren, H.J. and Barnard, E.A., Relation of reactivity to structure in pancreatic ribonuclease. I. An analysis of the various reactions with bromoacetate in the pH range of 2–7, *Biochemistry* 9, 959–973, 1970; Goren, H.J. and Barnard, E.A., Relation of reactivity to structure in pancreatic ribonuclease. II. Positions of residues alkylated in certain conditions by bromoacetate, *Biochemistry* 9, 974–983, 1970; Lennette, E.P. and Plapp, B.V., Kinetics of carboxymethylation of histidine hydantoin, *Biochemistry* 18, 3933–3938, 1979; Adamczyk, M., Gebler, J.C., and Wu, J., A simple method to identify cysteine residues by isotopic labeling and ion trap mass spectrometry, *Rapid Commun. Mass Spectrom.* 13, 1813–1817, 1999; Schelte, P., Boeckler, C., Frisch, B., and Schuber, F., Differential reactivity of maleimide and bromoacetyl functions with thiols: application to the preparation of lysosomal diepitope constructs, *Bioconjug. Chem.* 11, 118–123, 2000; Filmon, R., Grizon, F., Basle, M.F., and Chappaard, D., Effects of negatively charged groups (carboxymethyl) on the calcification of poly(2-hydroxyethyl methacrylate), *Biomaterials* 23, 3053–3059, 2002; Barron, L. and Paull, B., Direct detection of trace haloacetates in drinking water using microbore ion chromatography. Improved detector sensitivity using a hydroxide gradient and a monolithic ion-exchange type suppressor, *J. Chromatog. A* 1047, 205–212, 2004; Zhang, L., Arnold, W.A., and Hozalski, R.M., Kinetics of haloacetic acid reactions with Fe(0), *Environ. Sci. Technol.* 38, 6881–6889, 2004; Lee, S. and Perez-Luna, V.H., Dextran-gold nanoparticle hybrid material for biomolecule immobilization and detection, *Anal. Chem.* 77, 7204–7211, 2005.

p-Bromophenacyl Bromide

p-bromophenacyl bromide

2-bromo-1-(4-bromophenyl)eth-anone; 4-bromophenacyl Bromide

277.04

Modification of various residues in proteins: reagent for identification of carboxylic acids; phospholipase A2 inhibitor.

Erlanger, B.F., Vratrsanos, S.M., Wasserman, N., and Cooper, A.G., A chemical investigation of the active center of pepsin, *Biochem. Biophys. Res. Commun.* 23, 243–245, 1966; Yang, C.C. and King, K., Chemical modification of the histidine residue in basic phospholipase A2 from the venom of *Naja nigricollis, Biochim. Biophys. Acta.* 614, 373–388, 1980; Darke, P.L., Jarvis, A.A., Deems, R.A., and Dennis, E.A., Further characterization and *N*-terminal sequence of cobra venom phospholipase A2, *Biochim. Biophys. Acta* 626, 154–161, 1980; Ackerman, S.K., Matter, L., and Douglas, S.D., Effects of acid proteinase inhibitors on human neutrophil chemotaxis and lysosomal enzyme release. II. Bromophenacyl bromide and 1,2-epoxy-3-(*p*-nitrophenoxy)propane, *Clin. Immunol. Immunopathol.* 26, 213–222, 1983; Carine, K. and Hudig, D., Assessment of a role for phospholipase A2 and arachidonic acid metabolism in human lymphocyte natural cytotoxicity, *Cell Immunol.* 87, 270–283, 1984; Duque, R.E., Fantone, J.C., Kramer, C. et al., Inhibition of neutrophil activation by *p*-bromophenacyl bromide and its effects on phospholipase A2, *Br. J. Pharmacol.* 88, 463–472, 1986; Zhukova, A., Gogvadze, G., and Gogvadze, V., *p*-bromophenacyl bromide prevents cumene hydroperoxide-induced mitochondrial permeability transition by inhibiting pyridine nucleotide oxidation, *Redox Rep.* 9, 117–121, 2004; Thommesen, L. and Laegreid, A., Distinct differences between TNF receptor 1- and TNR receptor 2-mediated activation of NF-κβ, *J. Biochem. Mol. Biol.* 38, 281–289, 2005; Yue, H.Y., Fujita, T., and Kumamoto, E., Phospholipase A2 activation by melittin enhances spontaneous glutamatergic excitatory transmission in rat substantia gelatinosa neurons, *Neuroscience* 135, 485–495, 2005; Costa-Junior, H.M., Hamaty, F.C., de Silva Farias, R. et al., Apoptosis-inducing factor of a cytotoxic T-cell line: involvement of a secretory phospholipase A(2), *Cell Tissue Res.* 324, 255–266, 2006; Marchi-Salvador, D.P., Fernandes, C.A., Amui, S.F. et al., Crystallization and preliminary X-ray diffraction analysis of a myotoxic Lys49-PLA2 from *Bothrops jararacussu* venom complexed with *p*-bromophenacyl bromide, *Acta Crystallograph. Sect. F Struct. Biol. Cryst. Commun.* 62, 600–603, 2006.

Bromophenol Blue

Bromphenol Blue

Bromophenol Blue

669.97

pH indicator; conformational probe for proteins; histochemical staining for basic proteins; some use as a vital stain.

Schilling, K. and Waldmann-Meyer, H., The interaction of bromophenol blue with serum albumin and gamma-globulin in acid medium, *Arch. Biochem. Biophys.* 64, 291–301, 1956; Cohen, A.H., Temperature jump studies of the binding of bromophenol blue to beta-lactoglobulin in the vicinity of the N–R transition, *J. Biol. Chem.* 245, 738–745, 1970; Harruff, R.C. and Jenkins, W.T., The binding of bromophenol blue to aspartate aminotransferase, *Arch. Biochem. Biophys.*

176, 206–213, 1976; Mitchell, J.P., Model system studies of staining procedures for lysine and arginine residues, *Histochemistry* 52, 151–157, 1977; Asao, T., Quantitative analysis of proteins by the use of SDS-polyacrylamide-gel electrophoresis, *Anal. Biochem.* 77, 321–331, 1977; Greenberg, C.S. and Craddock, P.R., Rapid single-step membrane protein assay, *Clin. Chem.* 28, 1725–1726, 1982; Bertsch, M. and Kassner, R.J., Selective staining of proteins with hydrophobic surface sites on a native electrophoretic gel, *J. Proteome Res.* 2, 469–475, 2003; Li, J., Chatterjee, K., Medek, A. et al., Acid-base characteristics of bromophenol blue-citrate buffer systems in the amorphous state, *J. Pharm. Sci.* 93, 697–712, 2004; Haritoglou, C., Yu, A., Freyer, W. et al., An evaluation of novel dyes for intraocular surgery, *Invest. Ophthalmol. Vis. Sci.* 46, 3315–3322, 2005; Haritoglou, C., Tadayoni, R., May, C.A. et al., Short-term *in vivo* evaluation of novel vital dyes for interocular surgery, *Retina* 26, 673–678, 2006; Schuettauf, F., Haritoglou, C., and May, C.A., Administration of novel dyes for intraocular surgery: an *in vivo* toxicity animal study, *Invest. Ophthalmol. Vis. Sci.* 47, 3573–3578, 2006; Zeroual, Y., Kim, B.S., Kim, C.S. et al., A comparative study on biosorption characteristics of certain fungi for bromophenol blue dye, *Appl. Biochem. Biotechnol.* 134, 51–60, 2006.

Cacodylic Acid	Dimethylarsinic Acid	138.10	Buffer salt in neutral pH range; largely replaced because of toxicity.

Cacodylic acid
Dimethylarsinic acid

McAlpine, J.C., Histochemical demonstration of the activation of rat acetylcholinesterase by sodium cacodylate and cacodylic acid using the thioacetic acid method, *J. R. Microsc. Soc.* 82, 95–106, 1963; Jacobson, K.B., Murphy, J.B., and Das Sarma, B., Reaction of cacodylic acid with organic thiols, *FEBS Lett.* 22, 80–82, 1972; Travers, F., Douzou, P., Pederson, T., and Gunsalus, I.C., Ternary solvents to investigate proteins at subzero temperature, *Biochimie* 57, 43–48, 1975; Young, C.W., Dessources, C., Hodas, S., and Bittar, E.S., Use of cationic disc electrophoresis near neutral pH in the evaluation of trace proteins in human plasma, *Cancer Res.* 35, 1991–1995, 1975; Chirpich, T.P., The effect of different buffers on terminal deoxynucleotidyl transferase activity, *Biochim. Biophys. Acta* 518, 535–538, 1978; Nunes, J.F., Aguas, A.P., and Soares, J.O., Growth of fungi in cacodylate buffer, *Stain Technol.* 55, 191–192, 1980; Caswell, A.H. and Bruschwig, J.P., Identification and extraction of proteins that compose the triad junction of skeletal muscle, *J. Cell Biol.* 99, 929–939, 1984; Parks, J.C. and Cohen, G.M., Glutaraldehyde fixatives for preserving the chick's inner ear, *Acta Otolaryngol.* 98, 72–80, 1984; Song, A.H. and Asher, S.A., Internal intensity standards for heme protein UV resonance Raman studies: excitation profiles of cacodylic acid and sodium selenate, *Biochemistry* 30, 1199–1205, 1991; Henney, P.J., Johnson, E.L., and Cothran, E.G., A new buffer system for acid PAGE typing of equine protease inhibitor, *Anim. Genet.* 25, 363–364, 1994; Jezewska, M.J., Rajendran, S., and Bujalowski, W., Interactions of the 8-kDa domain of rat DNA polymerase beta with DNA, *Biochemistry* 40, 3295–3307, 2001; Kenyon, E.M. and Hughes, M.F., A concise review of the toxicity and carcinogenicity of dimethylarsinic acid, *Toxicology* 160, 227–236, 2001; Cohen, S.M., Arnold, L.L., Eldan, M. et al., Methylated arsenicals: the implications of metabolism and carcinogenicity studies in rodents to human risk management, *Crit. Rev. Toxicol.* 99–133, 2006.

Calcium Chloride	$CaCl_2$; Various Hydrates	110.98	Anhydrous form as drying agent for organic solvents, variety of manufacturing uses; meat quality enhancement; therapeutic use in electrolyte replacement and bone cements; source of calcium ions for biological assays.

Barratt, J.O., Thrombin and calcium chloride in relation to coagulation, *Biochem. J.* 9, 511–543, 1915; Van der Meer, C., Effect of calcium chloride on choline esterase, *Nature* 171, 78–79, 1952; Bhat, R. and Ahluwalia, J.C., Effect of calcium chloride on the conformation of proteins. Thermodynamic studies of some model compounds, *Int. J. Pept. Protein Res.* 30, 145–152, 1987; Furihata, C., Sudo, K., and Matsushima, T., Calcium chloride inhibits stimulation of replicative DNA synthesis by sodium chloride in the pyloric mucosa of rat stomach, *Carcinogenesis* 10, 2135–2137, 1989; Ishikawa, K., Ueyama, Y., Mano, T. et al., Self-setting barrier membrane for guided tissue regeneration method: initial evaluation of alginate membrane made with sodium alginate and calcium chloride aqueous solutions, *J. Biomed. Mater. Res.* 47, 111–115, 1999; Vujevic, M., Vidakovic-Cifrek, Z., Tkalec, M. et al., Calcium chloride and calcium bromide aqueous solutions of technical and analytical grade in Lemna bioassay, *Chemosphere* 41, 1535–1542, 2000; Miyazaki, T., Ohtsuki, C., Kyomoto, M. et al., Bioactive PMMA bone cement prepared by modification with methacryloxypropyltrimethoxysilane and calcium chloride, *J. Biomed. Mater. Res. A* 67, 1417–1423, 2003; Harris, S.E., Huff-Lonegan, E., Lonergan, S.M. et al., Antioxidant status affects color stability and tenderness of calcium chloride-injected beef, *J. Anim. Sci.* 79, 666–677, 2001; Behrends, J.M., Goodson, K.J., Koohmaraie, M. et al., Beef customer satisfaction: factors affecting consumer evaluations of calcium chloride-injected top sirloin steaks when given instructions for preparation, *J. Anim. Sci.* 83, 2869–2875, 2005.

Cetyl Pyridinium Chloride

1-hexadecylpyridinium Chloride

350.01

Cationic detergent; precipitating agent and staining agent for glycosaminoglycans; antimicrobial agent.

Cetyl pyridinium chloride

Laurent, T.C. and Scott, J.E., Molecular weight fractionation of polyanions by cetylpyridinium chloride in salt solutions, *Nature* 202, 661–662, 1964; Kiss, A., Linss, W., and Geyer, G., CPC-PTA section staining of acid glycans, *Acta Histochem.* 64, 183–186, 1979; Khan, M.Y. and Newman, S.A., An assay for heparin by decrease in color yield (DECOY) of a protein-dye-binding reaction, *Anal. Biochem.* 187, 124–128, 1990; Chardin, H., Septier, D., and Goldberg, M., Visualization of glycosaminoglycans in rat incisor predentin and dentin with cetylpyridinium chloride-glutaraldehyde as fixative, *J. Histochem. Cytochem.* 38, 885–894, 1990; Chardin, H., Gokani, J.P., Septier, D. et al., Structural variations of different oral basement membranes revealed by cationic dyes and detergent added to aldehyde fixative solution, *Histochem. J.* 24, 375–382, 1992; Agren, U.M., Tammi, R., and Tammi, M., A dot-blot assay of metabolically radiolabeled hyaluronan, *Anal. Biochem.* 217, 311–315, 1994; Maccari, F. and Volpi, N., Glycosaminoglycan blotting on nitrocellulose membranes treated with cetylpyridinium chloride afer agarose-gel electrophoretic separation, *Electrophoresis* 23, 3270–3277, 2002; Maccari, F. and Volpi, N., Direct and specific recognition of glycosaminoglycans by antibodies after their separation by agarose gel electrophoresis and blotting on cetylpyridinium chloride-treated nitrocellulose membranes, *Electrophoresis* 24, 1347–1352, 2003.

CHAPS

3-[(3-cholamidopropyl)-dimethylammonio]-1-propanesulfonate

614.89

Detergent, solubilizing agent; extensive use for the solubilization of membrane proteins.

3[(3-cholamidopropyl)dimethylammonio]-1propanesulfonate (CHAPS)

nonoxynol; non-ionic detergent

Hjelmeland, L.M., A nondenaturing zwitterionic detergent for membrane biochemistry: design and synthesis, *Proc. Natl. Acad. Sci. USA* 77, 6368–6370, 1980; Giradot, J.M. and Johnson, B.C., A new detergent for the solubilization of the vitamin K–dependent carboxylation system from liver microsomes: comparison with triton X-100, *Anal. Biochem.* 121, 315–320, 1982; Liscia, D.S., Alhadi, T., and Vonderhaar, B.K., Solubilization of active prolactin receptors by a nondenaturing zwitterionic detergent, *J. Biol. Chem.* 257, 9401–9405, 1982; Womack, M.D., Kendall, D.A., and MacDonald, R.C., Detergent effects on enzyme activity and solubilization of lipid bilayer membranes, *Biochim. Biophys. Acta* 733, 210–215, 1983; Klaerke, D.A. and Jorgensen, P.L., Role of Ca^{2+}-activated K$^+$ channel in regulation of NaCl reabsorption in thick ascending limb of Henle's loop, *Comp. Biochem. Physiol. A* 90, 757–765, 1988; Kuriyama, K., Nakayasu, H., Mizutani, H. et al., Cerebral GABAB receptor: proposed mechanisms of action and purification procedures, *Neurochem. Res.* 18, 377–383, 1993; Koumanov, K.S., Wolf, C., and Quinn, P.J., Lipid composition of membrane domains, *Subcell. Biochem.* 37, 153–163, 2004.

Chloroform   Trichloromethane   177.38   Used for extraction of lipids,
                                                                                     usually in combination
                                                                                     with methanol.

Stevan, M.A. and Lyman, R.L., Investigations on extraction of rat plasma phospholipids, *Proc. Soc. Exp. Biol. Med.* 114, 16–20, 1963; Wells, M.A. and Dittmer, J.C., A microanalytical technique for the quantitative determination of twenty-four classes of brain lipids, *Biochemistry* 5, 3405–3418, 1966; Colacicco, G. and Rapaport, M.M., A simplified preparation of phosphatidyl inositol, *J. Lipid. Res.* 8, 513–515, 1967; Curtis, P.J., Solubility of mitochondrial membrane proteins in acidic organic solvents, *Biochim. Biophys. Acta* 183, 239–241, 1969; Privett, O.S., Dougherty, K.A., and Castell, J.D., Quantitative analysis of lipid classes, *Am. J. Clin. Nutr.* 24, 1265–1275, 1971; Claire, M., Jacotot, B., and Robert, L., Characterization of lipids associated with macromolecules of the intercellular matrix of human aorta, *Connect. Tissue Res.* 4, 61–71, 1976; St. John, L.C. and Bell, F.P., Extraction and fractionation of lipids from biological tissues, cells, organelles, and fluids, *Biotechniques* 7, 476–481, 1989; Dean, N.M. and Beaven, M.A., Methods for the analysis of inositol phosphates, *Anal. Biochem.* 183, 199–209, 1989; Singh, A.K. and Jiang, Y., Quantitative chromatographic analysis of inositol phospholipids and related compounds, *J. Chromatog. B Biomed. Appl.* 671, 255–280, 1995.

Cholesterol   386.66   The most common sterol in
                                                                                        man and other higher
                                                                                        animals. Cholesterol is
                                                                                        essential for the synthesis
                                                                                        of a variety of compounds
                                                                                        including estrogens and
                                                                                        vitamin D; also membrane
                                                                                        component.

Cholesterol

Doree, C., The occurrence and distribution of cholesterol and allied bodies in the animal kingdom, *Biochem. J.* 4, 72–106, 1909; Heilbron, I.M., Kamm, E.D., and Morton, R.A., The absorption spectrum of cholesterol and its biological significance with reference to vitamin D. Part I: Preliminary observations, *Biochem. J.* 21, 78–85, 1927; Cook, R.P., Ed., *Cholesterol: Chemistry, Biochemistry, and Pathology*, Academic Press, New York, 1958; Vahouny, G.V. and Treadwell, C.R., Enzymatic synthesis and hydrolysis of cholesterol esters, *Methods Biochem. Anal.* 16, 219–272, 1968; Heftmann, E., *Steroid Biochemistry*, Academic Press, New York, 1970; Nestel, P.J., Cholesterol turnover in man, *Adv. Lipid Res.* 8, 1–39, 1970; Dennick, R.G., The intracellular organization of cholesterol biosynthesis. A review, *Steroids Lipids Res.* 3, 236–256, 1972; J. Polonovski, Ed., *Cholesterol Metabolism and Lipolytic Enzymes*, Masson Publications, New York, 1977; Gibbons, G.F., Mitrooulos, K.A., and Myant, N.B., *Biochemistry of Cholesterol*, Elsevier, Amsterdam, 1982; Bittman, R., *Cholesterol: Its Functions and Metabolism in Biology and Medicine*, Plenum Press, New York, 1997; Oram, J.P. and Heinecke, J.W., ATP-binding cassette transporter A1: a cell cholesterol exporter that protects against cardiovascular disease, *Physiol. Rev.* 85, 1343–1372, 2005; Holtta-Vuori, M. and Ikonen, E., Endosomal cholesterol traffic: vesicular and non-vesicular mechanisms meet, *Biochem. Soc. Trans.* 34, 392–394, 2006; Cuchel, M. and Rader, D.J., Macrophage reverse cholesterol transport: key to the regression of atherosclerosis? *Circulation* 113, 2548–2555, 2006.

Cholic Acid   408.57   Component of bile;
                                                                                        detergent.

Cholic acid

Schreiber, A.J. and Simon, F.R., Overview of clinical aspects of bile salt physiology, *J. Pediatr. Gastroenterol. Nutr.* 2, 337–345, 1983; Chiang, J.Y., Regulation of bile acid synthesis, *Front. Biosci.* 3, dl176–dl193, 1998; Cybulsky, M.I., Lichtman, A.H., Hajra, L., and Iiyama, K., Leukocyte adhesion molecules in atherogenesis, *Clin. Chim. Acta* 286, 207–218, 1999.

Citraconic Anhydride

Methylmaleic Anhydride   112.1   Reversible modification of amino groups.

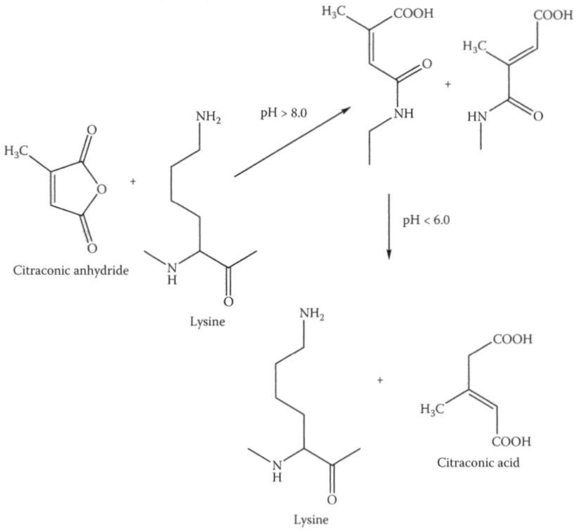

Dixon, H.B. and Perham, R.N., Reversible blocking of amino groups with citraconic anhydride, *Biochem. J.* 109, 312–314, 1968; Gibbons, I. and Perham, R.N., The reaction of aldolase with 2-methylmaleic anhydride, *Biochem. J.* 116, 843–849, 1970; Yankeelov, J.A., Jr. and Acree, D., Methylmaleic anhydride as a reversible blocking agent during specific arginine modification, *Biochem. Biophys. Res. Commun.* 42, 886–891, 1971; Takahashi, K., Specific modification of arginine residues in proteins with ninhydrin, *J. Biochem.* 80, 1173–1176, 1976; Brinegar, A.C. and Kinsella, J.E., Reversible modification of lysine in soybean proteins, using citraconic anhydride: characterization of physical and chemical changes in soy protein isolate, the 7S globulin, and lipoxygenase, *J. Agric. Food Chem.* 28, 818–824, 1980; Shetty, J.K. and Kensella, J.F., Ready separation of proteins from nucleoprotein complexes by reversible modification of lysine residues, *Biochem. J.* 191, 269–272, 1980; Yang, H. and Frey, P.A., Dimeric cluster with a single reactive amino group, *Biochemistry* 23, 3863–3868, 1984; Bindels, J.G., Misdom, L.W., and Hoenders, H.J., The reaction of citraconic anhydride with bovine alpha-crystallin lysine residues. Surface probing and dissociation-reassociation studies, *Biochim. Biophys. Acta* 828, 255–260, 1985; Al jamal, J.A., Characterization of different reactive lysines in bovine heart mitochondrial porin, *Biol. Chem.* 383, 1967–1970, 2002; Kadlik, V., Strohalm, M., and Kodicek, M., Citraconylation — a simple method for high protein sequence coverage in MALDI-TOF mass spectrometry, *Biochem. Biophys. Res. Commun.* 305, 1091–1093, 2003.

Congo Red

CI Direct Red 28;   696.68   pH indicator, histological stain for collagen, amyloid, elastin.

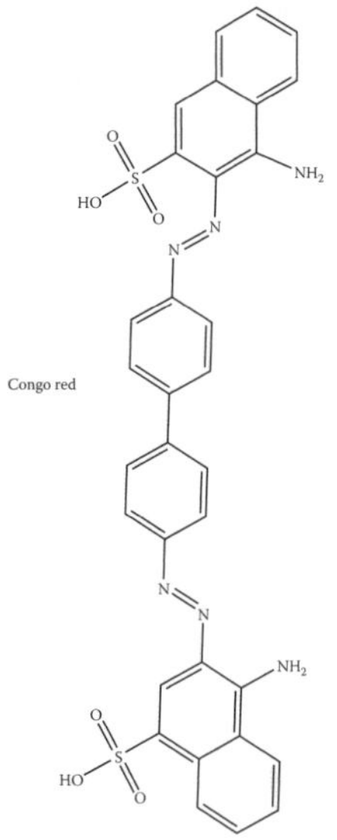

Sodium Diphenyldiazo-bis-naphthalamine-sulfonate

Mitchell, P., Crystallization of Congo red, *Nature* 165, 772–773, 1950; Helander, S., The distribution of Congo red in the tissues, *Acta. Med. Scand.* 138, 188–190, 1950; Hahn, N.J., The Congo red reaction in bacteria and its usefulness in the identification of rhizobia, *Can. J. Microbiol.* 12, 725–733, 1966; R.W. Horobin and J.A. Kiernan, Eds., *Conn's Biological Stains*, 10th ed., Bios Scientific Publishers, Oxford, UK, 2002; Inouye, H. and Kirschner, D.A., Alzheimer's beta-amyloid: insights into fibril formation and structure from Congo red binding, *Subcell. Biochem.* 38, 203–224, 2005; Inestrosa, N.C., Alvarez, A., Dinamarca, M.C. et al., Acetylcholinesterase-amyloid-beta-protein interaction: effect of Congo red and the role of the Wnt pathway, *Curr. Alzheimer Res.* 2, 301–306, 2005; Wu, X., Sun, S., Guo, C. et al., Resonance light scattering technique for the determination of proteins with Congo red and Triton X-100, *Luminescence* 21, 56–61, 2006; Halimi, M., Dayan-Amouyal, Y., Kariv-Inbal, Z. et al., Prion urine comprises a glycosaminoglycan-light chain IgG complex that can be stained by Congo red, *J. Virol. Methods* 133, 205–210, 2006; Bely, M. and Makovitzky, J., Sensitivity and specificity of Congo red staining according to Romhanyi. Comparison with Puchtler's or Bennhold's methods, *Acta Histochem.* 108, 175–180, 2006; McLaughlin, R.W., De Stigter, J.K., Sikkink, L.A. et al., The effects of sodium sulfate, glycosaminoglycans, and Congo red on the structure, stability, and amyloid formation of an immunoglobulin light-chain protein, *Protein Sci.* 15, 1710–1722, 2006; Cheung, S.T., Maheshwari, M.B., and Tan, C.Y., A comparative study of two Congo red stains for the detection of primary cutaneous amyloidosis, *J. Am. Acad. Dermatol.* 55, 363–364, 2006.

Coomassie Brilliant Blue G-250   CI Acid Blue 90   854   Most often used for the colorimetric determination of protein.

Coomassie brilliant blue R250

Bradford, M.M., A rapid and sensitive method for the quantitation of microgram quantities of protein utilizing the principle of protein-dye binding *Anal. Biochem.* 72, 248–254, 1976; Saleemuddin, M., Ahmad, H., and Husain, A., A simple, rapid, and sensitive procedure for the assay of endoproteases using Coomassie Brilliant Blue G-250, *Anal. Biochem.* 105, 202–206, 1980; van Wilgenburg, M.G., Werkman, E.M., van Gorkom, W.H., and Soons, J.B., Criticism of the use of Coomassie Brilliant Blue G-250 for the quantitative determination of proteins, *J.Clin. Chem.Clin. Biochem.* 19, 301–304, 1981; Mattoo, R.L., Ishaq, M., and Saleemuddin, M., Protein assay by Coomassie Brilliant Blue G-250-binding method is unsuitable for plant tissues rich in phenols and phenolases, *Anal. Biochem.* 163, 376–384, 1987; Lott, J.A., Stephan, V.A., and Pritchard, K.A., Jr., Evaluation of the Coomassie Brilliant Blue G-250 method for urinary proteins, *Clin. Chem.* 29, 1946–1950, 1983; Fanger, B.O., Adaptation of the Bradford protein assay to membrane-bound proteins by solubilizing in glucopyranoside detergents, *Anal. Biochem.* 162, 11–17, 1987; Marshall, T. and Williams, K.M., Recovery of proteins by Coomassie Brilliant Blue precipitation prior to electrophoresis, *Electrophoresis* 13, 887–888, 1992; Sapan, C.V., Lundblad, R.L., and Price, N.C., Colorimetric protein assay techniques, *Biotechnol. Appl. Biochem.* 29, 99–108, 1999.

Coomassie Brilliant Blue R-250   CI Acid Blue 83   826   Most often used for the detection of proteins on solid matrices such as polyacrylamide gels.

Coomassie brilliant blue R250

Vesterberg, O., Hansen, L., and Sjosten, A., Staining of proteins after isoelectric focusing in gels by a new procedure, *Biochim. Biophys. Acta* 491, 160–166, 1977; Micko, S. and Schlaepfer, W.W., Metachromasy of peripheral nerve collagen on polyacrylamide gels stained with Coomassie Brilliant Blue R-250, *Anal. Biochem.* 88, 566–572, 1978; Osset, M., Pinol, M., Fallon, M.J. et al., Interference of the carbohydrate moiety in Coomassie Brilliant Blue R-250 protein staining, *Electrophoresis* 10, 271–273, 1989; Pryor, J.L., Xu, W., and Hamilton, D.W., Immunodetection after complete destaining of Coomassie blue-stained proteins on immobilon-PVDF, *Anal. Biochem.* 202, 100–104, 1992; Metkar, S.S., Mahajan, S.K., and Sainis, J.K., Modified procedure for nonspecific protein staining on nitrocellulose paper using Coomassie Brilliant Blue R-250, *Anal. Biochem.* 227, 389–391, 1995; Kundu, S.K., Robey, W.G., Nabors, P. et al., Purification of commercial Coomassie Brilliant Blue R-250 and characterization of the chromogenic fractions, *Anal. Biochem.* 235, 134–140, 1996; Choi, J.K., Yoon, S.H., Hong, H.Y. et al., A modified Coomassie blue staining of proteins in polyacrylamide gels with Bismark brown R, *Anal. Biochem.* 236, 82–84, 1996; Moritz, R.L., Eddes, J.S., Reid, G.E., and Simpson, R.J., *S*-pyridylethylation of intact polyacrylamide gels and *in situ* digestion of electrophoretically separated proteins: a rapid mass spectrometric method for identifying cysteine-containing peptides, *Electrophoresis* 17, 907–917, 1996; Choi, J.K. and Yoo, G.S., Fast protein staining in sodium dodecyl sulfate polyacrylamide gel using counter ion-dyes, Coomassie Brilliant Blue R-250, and neutral red, *Arch. Pharm. Res.* 25, 704–708, 2002; Bonar, E., Dubin, A., Bierczynska-Krzysik, A. et al., Identification of major cellular proteins synthesized in response to interleukin-1 and interleukin-6 in human hepatoma HepG2 cells, *Cytokine* 33, 111–117, 2006.

Coomassie Brilliant Blue RL

CI Acid Blue 92;   695.6
Anazolene
Sodium

Cy 2

Fluorescent label used in proteomics and gene expression; use for internal standard.

Tonge, R., Shaw, J., Middleton, B. et al., Validation and development of fluorescence two-dimensional differential gel electrophoresis proteomics technology, *Proteomics* 1, 377–396, 2001; Chan, H.L., Gharbi, S., Gaffney, P.R. et al., Proteomic analysis of redox- and ErbB2-dependent changes in mammary luminal epithelial cells using cysteine- and lysine-labeling two-dimensional difference gel electrophoresis, *Proteomics* 5, 2908–2926, 2005; Misek, D.E., Kuick, R., Wang, H. et al., A wide range of protein isoforms in serum and plasma uncovered by a quantitative intact protein analysis system, *Proteomics* 5, 3343–3352, 2005; Doutette, P., Navet, R., Gerkens, P. et al., Steatosis-induced proteomic changes in liver mitochondria evidenced by two-dimensional differential in-gel electrophoresis, *J. Proteome Res.* 4, 2024–2031, 2005.

Cy 3

911.0   Fluorescent label used in proteomics and gene expression; in combination with Cy 5 is used for FRET-based assays.

Brismar, H. and Ulfake, B., Fluorescence lifetime measurements in confocal microscopy of neurons labeled with multiple fluorophores, *Nat. Biotechnol.* 15, 373–377, 1997; Strohmaier, A.R., Porwol, T., Acker, H., and Spiess, E., Tomography of cells by confocal laser scanning microscopy and computer-assisted three-dimensional image reconstruction: localization of cathepsin B in tumor cells penetrating collagen gels *in vitro, J. Histochem. Cytochem.* 45, 975–983, 1997; Alexandre, I., Hamels, S., Dufour, S. et al., Colorimetric silver detection of DNA microarrays, *Anal. Biochem.* 295, 1–8, 2001; Shaw, J., Rowlinson, R., Nickson, J. et al., Evaluation of saturation labeling two-dimensional difference gel electrophoresis fluorescent dyes, *Proteomics* 3, 1181–1195, 2003.

Cy 5

937.1   Fluorescent label used in proteomics and gene expression; also used in histochemistry.

Uchihara, T., Nakamura, A., Nagaoka, U. et al., Dual enhancement of double immunofluorescent signals by CARD: participation of ubiquitin during formation of neurofibrillary tangles, *Histochem. Cell Biol.* 114, 447–451, 2000; Duthie,

R.S., Kalve, I.M., Samols, S.B. et al., Novel cyanine dye-based dideoxynucleoside triphosphates for DNA sequencing, *Bioconjug. Chem.* 13, 699–706, 2002; Graves, E.E., Yessayan., D., Turner, G. et al., Validation of *in vivo* fluorochrome concentrations measured using fluorescence molecular tomography, *J. Biomed. Opt.* 10, 44019, 2005; Lapeyre, M., Leprince, J., Massonneau, M. et al., Aryldithioethyloxycarbonyl (Ardec): a new family of amine-protecting groups removable under mild reducing conditions and their applications to peptide synthesis, *Chemistry* 12, 3655–3671, 2006; Tang, X., Morris, S.L., Langone, J.J., and Bockstahler, L.E., Simple and effective method for generating single-stranded DNA targets and probes, *Biotechniques* 40, 759–763, 2006.

Cyanine Dye   (See glossary)   Cy 2, Cy 3, and Cy 5 are cyanine dye derivatives.

Cyanine

Carbocyanine trimethincyanine C₃

Pseudocyanine

Dicarbocyanine pentamethincyanine C₅

Selenocyanine

Thiacyaninine

α-Cyano-4-hydroxycinnamic Acid   4-HCCA; Cinnamate   189.2   Used as matrix substance for MALDI; transport inhibitor and enzyme inhibitor.

Alpha-cyano-4-hydroxycinnamic acid

Gobom, J., Schuerenberg, M., Mueller, M. et al., α-cyano-4-hydroxycinnamic acid affinity sample preparation. A protocol for MALDI-MS peptide analysis in proteomics, *Anal. Chem.* 73, 434–438, 2001; Zhu, X. and Papayannopoulos, I.A., Improvement in the detection of low concentration protein digests on a MALDI TOF/TOF workstation by reducing α-cyano-4-hydroxycinnamic acid adduct ions, *J. Biomol. Tech.* 14, 298–307, 2003; Neubert, H., Halket, J.M., Fernandez Ocana, M., and Patel, R.K., MALDI post-source decay and LIFT-TOF/TOF investigation of α-cyano-4-hydroxycinnamic acid cluster interferences, *J. Am. Soc. Mass Spectrom.* 15, 336–343, 2004; Kobayashi, T., Kawai, H., Suzuki, T. et al., Improved sensitivity for insulin in matrix-assisted laser desorption/ionization time-of-flight mass spectrometry by premixing α-cyano-4-hydroxycinnamic acid with transferrin, *Rapid Commun. Mass Spectrom.* 18, 1156–1160, 2004; Pshenichnyuk, S.A. and Asfandiarov, N.L., The role of free electrons in MALDI: electron capture by molecules of α-cyano-4-hydroxycinnamic acid, *Eur. J. Mass Spectrom.* 10, 477–486, 2004; Bogan, M.J., Bakhoum, S.F., and Agnes, G.R., Promotion of α-cyano-4-hydroxycinnamic acid and peptide cocrystallization within levitated droplets with net charge, *J. Am. Soc. Mass Spectrom.* 16, 254–262, 2005. As enzyme inhibitor: Clarke, P.D., Clift, D.L., Dooledeniya, M. et al., Effects of α-cyano-4-hydroxycinnamic acid on fatigue and recovery of isolated mouse muscle, *J. Muscle Res. Cell Motil.* 16, 611–617, 1995; Del Prete, E., Lutz, T.A., and Scharrer, E., Inhibition of glucose oxidation by α-cyano-4-hydroxycinnamic acid stimulates feeding in rats, *Physiol. Behav.* 80, 489–498, 2004; Briski, K.P. and Patil, G.D., Induction of Fox immunoreactivity labeling in rat forebrain metabolic loci by caudal fourth ventricular infusion of the monocarboxylate transporter inhibitor, α-cyano-4-hydroxycinnamic acid, *Neuroendocrinology* 82, 49–57, 2005.

Cyanogen C_2N_2;   53.03   Protein crosslinking at salt
                                                                 Ethanedinitrile   bridges.

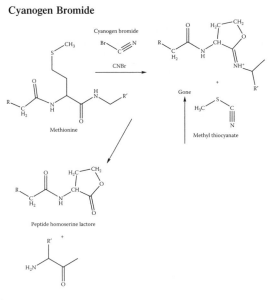

1-Cyano-4-dimethylaminopyridine

Cysteine   *S*-cyanocysteine

Ghenbot, G., Emge, T., and Day, R.A., Identification of the sites of modification of bovi carbonic anhydrase II (BCA II) by the salt bridge reagent cyanogen, C_2N_2, *Biochim. Biophys. Acta* 1161, 59–65, 1993; Karagozler, A.A., Ghenbot, G., and Day, R.A., Cyanogen as a selective probe for carbonic anhydrase hydrolase, *Biopolymers* 33, 687–692, 1993; Winters, M.S. and Day, R.A., Identification of amino acid residues participating in intermolecular salt bridges between self-associating proteins, *Anal. Biochem.* 309, 48–59, 2002; Winters, M.S. and Day, R.A., Detecting protein–protein interactions in the intact cell of *Bacillus subtilis*(ATCC 6633), *J. Bacteriol.* 185, 4268–4275, 2003.

Cyanogen Bromide   CNBr; Bromide   105.9   Protein modification;
                                                                 Cyanide   cleavage of peptide bonds;
                                                                                                                coupled nucleophiles to
                                                                                                                polyhydroxyl matrices;
                                                                                                                environmental toxicon
                                                                                                                derived from
                                                                                                                monobromamine and
                                                                                                                cyanide.

Cyanogen bromide

CNBr

Gone

Methionine   Methyl thiocyanate

Peptide homoserine lactore

Hofmann, T., The purification and properties of fragments of trypsinogen obtained by cyanogen bromide cleavage, *Biochemistry* 3, 356–364, 1964; Chu, R.C. and Yasunobu, K.T., The reaction of cyanogen bromide and *N*-bromosuccinimide with some cytochromes C, *Biochim. Biophys. Acta* 89, 148–149, 1964; Inglis, A.S. and Edman, P., Mechanism of cyanogen bromide reaction with methionine in peptides and proteins. I. Formation of imidate and methyl thiocyanate, *Anal. Biochem.* 37, 73–80, 1970; Kagedal, L. and Akerstrom, S., Binding of covalent proteins to polysaccharides by cyanogen bromide and organic cyanates. I. Preparation of soluble glycine-, insulin- and ampicillin-dextran, *Acta Chem. Scand.* 25, 1855–1899, 1971; Sipe, J.D. and Schaefer, F.V., Preparation of solid-phase immunosorbents by coupling human serum proteins to cyanogen bromide–activated agarose, *Appl. Microbiol.* 25, 880–884, 1973; March, S.C., Parikh, I., and Cuatrecasas, P., A simplified method for cyanogen bromide activation of agarose for affinity chromatography, *Anal. Biochem.* 60, 149–152, 1974; Boulware, D.W., Goldsworthy, P.D., Nardella, F.A., and Mannik, M., Cyanogen bromide cleaves Fc fragments of pooled human IgG at both methionine and tryptophan residues, *Mol. Immunol.* 22, 1317–1322, 1985; Jaggi, K.S. and Gangal, S.V., Monitoring of active groups of cyanogen bromide-activated paper discs used as allergosorbent, *Int. Arch. Allergy Appl. Immunol.* 89, 311–313, 1989; Villa, S., De Fazio, G., and Canosi, U., Cyanogen bromide cleavage at methionine residues of polypeptides containing disulfide bonds, *Anal. Biochem.* 177, 161–164, 1989; Luo, K.X., Hurley, T.R., and Sefton, B.M., Cyanogen bromide cleavage and proteolytic peptide mapping of proteins immobilized to membranes, *Methods Enzymol.* 201, 149–152, 1991; Jennissen, H.P., Cyanogen bromide and tresyl chloride chemistry revisited: the special reactivity of agarose as a chromatographic and biomaterial support for immobilizing novel chemical groups, *J. Mol. Recognit.* 8, 116–124, 1995; Kaiser, R. and Metzka, L., Enhancement of cyanogen bromide cleavage yields for methionyl-serine and methionyl-threonine peptide bonds, *Anal. Biochem.* 266, 1–8, 1999; Kraft, P., Mills, J., and Dratz, E., Mass spectrometric analysis of cyanogen bromide fragments of integral membrane proteins at the picomole level: application to rhodopsin, *Anal. Biochem.* 292, 76–86, 2001; Kuhn, K., Thompson, A., Prinz, T. et al., Isolation of *N*-terminal protein sequence tags from cyanogen bromide-cleaved proteins as a novel approach to investigate hydrophobic proteins, *J. Proteome Res.* 2, 598–609, 2003; Macmillan, D. and Arham, L., Cyanogen bromide cleavage generates fragments suitable for expressed protein and glycoprotein ligation, *J. Am. Chem. Soc.* 126, 9530–9531, 2004; Lei, H., Minear, R.A., and Marinas, B.J., Cyanogen bromide formation from the reactions of monobromamine and dibromamine with cyanide ions, *Environ. Sci. Technol.* 40, 2559–2564, 2006.

Cyanuric Chloride	2,4,6-trichloro-1,3,5-triazine	184.41	Coupling of carbohydrates to proteins; more recently for coupling of nucleic acid to microarray platforms.

Cyanuric chloride

Gray, B.M., ELISA methodology for polysaccharide antigens: protein coupling of polysaccharides for adsorption to plastic tubes, *J. Immunol. Methods* 28, 187–192, 1979: Horak, D., Rittich, B., Safar, J. et al., Properties of RNase A immobilized on magnetic poly(2-hydroxyethyl methacrylate) microspheres, *Biotechnol. Prog.* 17, 447–452, 2001; Lee, P.H., Sawan, S.P., Modrusan, Z. et al., An efficient binding chemistry for glass polynucleotide microarrays, *Bioconjug. Chem.* 13, 97–103, 2002; Steinberg, G., Stromsborg, K., Thomas, L. et al., Strategies for covalent attachment of DNA to beads, *Biopolymers* 73, 597–605, 2004; Abuknesha, R.A., Luk, C.Y., Griffith, H.H. et al., Efficient labeling of antibodies with horseradish peroxidase using cyanuric chloride, *J. Immunol. Methods* 306, 211–217, 2005.

1,2-Cyclohexylene-dinitrilotetraacetic Acid	Chelating agent suggested to have specificity for manganese ions; weaker for other metal ions such as ferric.

1, 2-Cyclohexylenedinitrilotetraacetic acid, CDTA

Tandon, S.K. and Singh, J., Removal of manganese by chelating agents from brain and liver of manganese, *Toxicology* 5, 237–241, 1975; Hazell, A.S., Normandin, L., Norenberg, M.D., Kennedy, G., and Yi, J.H., Alzheimer type II astrocyte changes following sub-acute exposure to manganese, *Neurosci. Lett.*, 396, 167–171, 2006; Hassler, C.S. and Twiss, M.R., Bioavailability of iron sensed by a phytoplanktonic Fe-bioreporter, *Environ. Sci. Tech.* 40, 2544–2551, 2006.

Dansyl Chloride	5-(dimethylamino)-1-naphthalene-sulfonyl Chloride	269.8	Fluorescent label for proteins; amino acid analysis.

Hill, R.D. and Laing, R.R., Specific reaction of dansyl chloride with one lysine residue in rennin, *Biochim. Biophys. Acta* 132, 188–190, 1967; Chen, R.F., Fluorescent protein-dye conjugates. I. Heterogeneity of sites on serum albumin labeled by dansyl chloride, *Arch. Biochem. Biophys.* 128, 163–175, 1968; Chen, R.F., Dansyl-labeled protein modified with dansyl chloride: activity effects and fluorescence properties, *Anal. Biochem.* 25, 412–416, 1968; Brown, C.S. and Cunningham, L.W., Reaction of reactive sulfhydryl groups of creatine kinase with dansyl chloride, *Biochemistry* 9, 3878–3885, 1970; Hsieh, W.T. and Matthews, K.S., Lactose repressor protein modified with dansyl chloride: activity effects and fluorescence properties,

Biochemistry 34, 3043–3049, 1985; Scouten, W.H., van den Tweel, W., Kranenburg, H., and Dekker, M., Colored sulfonyl chloride as an activated agent for hydroxylic matrices, *Methods Enzymol.* 135, 79–84, 1987; Martin, M.A., Lin, B., Del Castillo, B., The use of fluorescent probes in pharmaceutical analysis, *J. Pharm. Biomed. Anal.* 6, 573–583, 1988; Walker, J.M., The dansyl method for identifying *N*-terminal amino acids, *Methods Mol. Biol.* 32, 321–328, 1994; Walker, J.M., The dansyl-Edman method for peptide sequencing, *Methods Mol. Biol.* 32, 329–334, 1994; Pin, S. and Royer, C.A., High-pressure fluorescence methods for observing subunit dissociation in hemoglobin, *Methods Enzymol.* 323, 42–55, 1994; Rangarajan, B., Coons, L.S., and Scarnton, A.B., Characterization of hydrogels using luminescence spectroscopy, *Biomaterials* 17, 649–661, 1996; Kang, X., Xiao, J., Huang, X., and Gu, X., Optimization of dansyl derivatization and chromatographic conditions in the determination of neuroactive amino acids of biological samples, *Clin. Chim. Acta* 366, 352–356, 2006.

DCC

Dicyclohexylcarbodiimide

N,N′-dicyclohexyl-carbodiimide	206.33	Activates carboxyl groups to react with hydroxyl groups to form esters and with amines to form an amide bond; used to modify ion-transporting ATPases. Lack of water solubility has presented challenges.

Chau, A.S. and Terry, K., Analysis of pesticides by chemical derivatization. I. A new procedure for the formation of 2-chloroethyl esters of ten herbicidal acids, *J. Assoc. Off. Anal. Chem.* 58, 1294–1301, 1975; Patel, L. and Kaback, H.R., The role of the carbodiimide-reactive component of the adenosine-5′-triphosphatase complex in the proton permeability of *Escherichia coli* membrane vesicles, *Biochemistry* 15, 2741–2746, 1976; Esch, F.S., Bohlen, P., Otsuka, A.S. et al., Inactivation of the bovine mitochondrial F1-ATPase with dicyclohexyl[^{14}C]carbodiimide leads to the modification of a specific glutamic acid residue in the beta subunit, *J. Biol. Chem.* 256, 9084–9089, 1981; Hsu, C.M. and Rosen, B.P., Characterization of the catalytic subunit of an anion pump, *J. Biol. Chem.* 264, 17349–17354, 1989; Gurdag, S., Khandare, J., Stapels, S. et al., Activity of dendrimer-methotrexate conjugates on methotrexate-sensitive and -resistant cell lines, *Bioconjug. Chem.* 17, 275–283, 2006; Vgenopoulou, I., Gemperli, A.C., and Steuber, J., Specific modification of a Na[+] binding site in NADH: quinone oxidoreductase from *Klebsiella pneumoniae* with dicyclohexylcarbodiimide, *J. Bacteriol.* 188, 3264–3272, 2006; Ferguson, S.A., Keis, S., and Cook, G.M., Biochemical and molecular characterization of a Na[+]-translocating F1Fo-ATPase from the thermophilic bacterium *Clostridium paradoxum*, *J. Bacteriol.* 188, 5045–5054, 2006.

Deoxycholic Acid

Desoxycholic Acid	392.57	Detergent, nanoparticles.

Akare, S. and Martinez, J.D., Bile acid-induced hydrophobicity-dependent membrane alterations, *Biochim. Biophys. Acta* 1735, 59–67, 2005; Chae, S.Y., Son, S., Lee, M. et al., Deoxycholic acid-conjugated chitosan oligosaccharide nanoparticles for efficient gene carrier, *J. Control. Release* 109, 330–344, 2005; Dall'Agnol, M., Bernstein, C., Bernstein, H. et al., Identification of *S*-nitrosylated proteins after chronic exposure of colon epithelial cells to deoxycholate, *Proteomics* 6, 1654–1662, 2006; Dotis, J., Simitsopoulou, M., Dalakiouridou, M. et al. Effects of lipid formulations of amphotericin B on activity of human monocytes against *Aspergillus fumigatus*, *Antimicrob. Agents Chemother.* 128, 3490–3491, 2006; Darragh, J., Hunter, M., Pohler, E. et al., The calcium-binding domain of the stress protein SEP53 is required for survival in response to deoxycholic acid-mediated injury, *FEBS J.* 273, 1930–1947, 2006.

Deuterium Oxide

"Heavy Water"	20.03	Structural studies in proteins, enzyme kinetics; *in vivo* studies of metabolic flux.

Cohen, A.H., Wilkinson, R.R., and Fisher, H.F., Location of deuterium oxide solvent isotope effects in the glutamate dehydrogenase reaction, *J. Biol. Chem.* 250, 5343–5246, 1975; Rosenberry, T.L., Catalysis by acetylcholinesterase: evidence that the rate-limiting step for acylation with certain substrates precedes general acid-base catalysis, *Proc. Natl. Acad. Sci. USA* 72, 3834–3838, 1975; Viggiano, G., Ho, N.T., and Ho, C., Proton nuclear magnetic resonance and biochemical studies of oxygenation of human adult hemoglobin in deuterium oxide, *Biochemistry* 18, 5238–5247, 1979; Bonnete, F., Madern, D., and Zaccai, G., Stability against denaturation mechanisms in halophilic malate dehydrogenase "adapt" to solvent conditions, *J. Mol. Biol.* 244, 436–447, 1994; Thompson, J.F., Bush, K.J., and Nance, S.L., Pancreatic lipase activity in deuterium oxide, *Proc. Soc. Exp. Biol. Med.* 122, 502–505, 1996; Dufner, D. and Previs, S.F., Measuring *in vivo* metabolism using heavy water, *Curr. Opin. Clin. Nutr. Metab. Care* 6, 511–517, 2003; O'Donnell, A.H., Yao, X., and Byers, L.D., Solvent isotope effects on alpha-glucosidase, *Biochem. Biophys. Acta* 1703, 63–67, 2004; Hellerstein, M.K. and Murphy, E., Stable isotope-mass spectrometric measurements of molecular fluxes *in vivo*: emerging applications in drug development, *Curr. Opin. Mol. Ther.* 6, 249–264, 2004; Mazon, H., Marcillat, O., Forest, E., and Vial, C., Local dynamics measured by hydrogen/deuterium exchange and mass spectrometry of the creatine kinase digested by two proteases, *Biochimie* 87, 1101–1110, 2005; Carmieli, R., Papo, N., Zimmerman, H. et al., Utilizing ESEEM spectrscopy to locate the position of specific regions of membrane-active peptides within model membranes, *Biophys. J.* 90, 492–505, 2006.

DFP

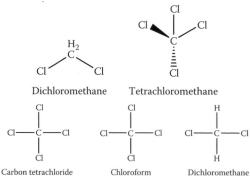

Phenylmethylsulfonyl fluoride (PMSF)
M.W. 174.2

3, 4-dichloroisocoumarin

4-(2-aminoethyl) benzenesulfonyl fluoride
Pefabloc SC′; M.W.

Diisopropylphosphorofluoridate

Diisopropylphos-phoro-fluoridate; Isofluorophate	184.15	Classic cholinesterase inhibitor; inhibitor of serine proteases, some nonspecific reaction tyrosine.

Baker, B.R., Factors in the design of active-site-directed irreversible inhibitors, *J. Pharm. Sci.* 53, 347–364, 1964; Dixon, G.H. and Schachter, H., The chemical modification of chymotrypsin, *Can. J. Biochem. Physiol.* 42, 695–714, 1964; Singer, S.J., Covalent labeling active site, *Adv. Protein Chem.* 22, 1–54, 1967; Kassell, B. and Kay, J., Zymogens of proteolytic enzymes, *Science* 180, 1022–1027, 1973; Fujino, T., Watanabe, K., Beppu, M. et al., Identification of oxidized protein hydrolase of human erythrocytes as acylpeptide hydrolase, *Biochim. Biophys. Acta* 1478, 102–112, 2000; Manco, G., Camardello, L., Febbraio, F. et al., Homology modeling and identification of serine 160 as nucleophile as the active site in a thermostable carboxylesterase from the archeon *Archaeoglobus fulgidus, Protein Eng.* 13, 197–200, 2000; Gopal, S., Rastogi, V., Ashman, W., and Mulbry, W., Mutagenesis of organophosphorous hydrolase to enhance hydrolysis of the nerve agent VX, *Biochem. Biophys. Res. Commun.* 279, 516–519, 2000; Yeung, D.T., Lenz, D.E., and Cerasoli, D.M., Analysis of active-site amino acid residues of human serum paraoxanse using competitive substrates, *FEBS J.* 272, 2225–2230, 2005; D'Souza, C.A., Wood, D.D., She, Y.M., and Moscarello, M.A., Autocatalytic cleavage of myelin basic protein: an alternative to molecular mimicry, *Biochemistry* 44, 12905–12913, 2005.

Dichloromethane

Dichloromethane   Tetrachloromethane

Carbon tetrachloride   Chloroform   Dichloromethane

Methylene Chloride	84.9	Lipid solvent; isolation of sterols, frequently used in combination with methanol.

Bouillon, R., Kerkhove, P.V., and De Moor, P., Measurement of 25-hydroxyvitamin D3 in serum, *Clin. Chem.* 22, 364–368, 1976; Redhwi, A.A., Anderson, D.C., and Smith, G.N., A simple method for the isolation of vitamin D metabolites from plasma extracts, *Steroids* 39, 149–154, 1982; Scholtz, R., Wackett, L.P., Egli, C. et al., Dichloromethane dehalogenase with improved catalytic activity isolated form a fast-growing dichloromethane-utilizing bacterium, *J. Bacteriol.* 170, 5698–5704, 1988; Russo, M.V., Goretti, G., and Liberti, A., Direct headspace gas chromatographic determination of dichloromethane in decaffeinated green and roasted coffee, *J. Chromatog.* 465, 429–433, 1989; Shimizu, M., Kamchi, S., Nishii, Y., and Yamada, S., Synthesis of a reagent for fluorescence-labeling of vitamin D and its use in assaying vitamin D metabolites, *Anal. Biochem.* 194, 77–81, 1991; Rodriguez-Palmero, M., de la Presa-Owens, S., Castellote-Bargallo, A.I. et al., Determination of sterol content in different food samples by capillary gas chromatography, *J. Chromatog. A* 672, 267–272, 1994; Raghuvanshi, R.S., Goyal, S., Singh, O., and Panda, A.K., Stabilization of dichloromethane-induced protein denaturation during microencapsulation, *Pharm. Dev. Technol.* 3, 269–276, 1998; El Jaber-Vazdekis, N., Gutierrez-Nicolas, F., Ravelo, A.G., and Zarate, R., Studies on tropane alkaloid extraction by volatile organic solvents: dichloromethane vs. chloroform, *Phytochem. Anal.* 17, 107–113, 2006.

Diethyldithiocarbamate

Diethyldithiocarbamate, sodium dithiocarb

Ditiocarb;	171.3	Chelating agent with
Dithiocarb; DTC	(Na)	particular affinity for Pb,
		Cu, Zn, Ni; colorimetric
		determination of Cu.

Matsuba, Y. and Takahashi, Y., Spectrophotometric determination of copper with *N,N,N′,N′*-tetraethylthiuram disulfide and an application of this method for studies on subcellular distribution of copper in rat brains, *Anal. Biochem.* 36, 182–191, 1970; Koutensky, J., Eybl, V., Koutenska, M. et al., Influence of sodium diethyldithiocarbamate on the toxicity and distribution of copper in mice, *Eur. J. Pharmacol.* 14, 389–392, 1971; Xu, H. and Mitchell, C.L., Chelation of zinc by diethyldithiocarbamate facilitates bursting induced by mixed antidromic plus orthodromic activation of mossy fibers in hippocampal slices, *Brain Res.* 624, 162–170, 1993; Liu, J., Shigenaga, M.K., Yan, L.J. et al., Antioxidant activity of diethyldithiocarbamate, *Free Radic. Res.* 24, 461–472, 1996; Zhang, Y., Wade, K.L., Prestera, T., and Talalav, P., Quantitative determination of isothiocyanates, dithiocarbamates, carbon disulfide, and related thiocarbonyl compounds by cyclocondensation with 1,2-benzenedithiol, *Anal. Biochem.* 239, 160–167, 1996; Shoener, D.F., Olsen, M.A., Cummings, P.G., and Basic, C., Electrospray ionization of neutral metal dithiocarbamate complexes using in-source oxidation, *J. Mass Spectrom.* 34, 1069–1078, 1999; Turner, B.J., Lopes, E.C., and Cheema, S.S., Inducible superoxide dismutase 1 aggregation in transgenic amyotrophic lateral sclerosis mouse fibroblasts, *J. Cell Biochem.* 91, 1074–1084, 2004; Xu, K.Y. and Kuppusamy, P., Dual effects of copper-zinc superoxide dismutase, *Biochem. Biophys. Res. Commun.* 336, 1190–1193, 2005; Jiang, X., Sun, S., Liang, A. et al., Luminescence properties of metal(II)-diethyldithiocarbamate chelate complex particles and its analytical application, *J. Fluoresc.* 15, 859–864, 2005; Wang, J.S. and Chiu, K.H., Mass balance of metal species in supercritical fluid extraction using sodium diethyldithiocarbamate and dibuylammonium dibutyldithiocarbamate, *Anal. Sci.* 22, 363–369, 2006.

Diethylpyrocarbonate (DEPC)

Ethoxyformic	162.1	Reagent for modification of
Anhydride		proteins and DNA; used as
		a sterilizing agent; RNAse
		inhibitor for RNA
		purification; preservative
		for wine and fruit fluids.

Wolf, B., Lesnaw, J.A., and Reichmann, M.E., A mechanism of the irreversible inactivation of bovine pancreatic ribonuclease by diethylpyrocarbonate. A general reaction of diethylpyrocarbonate with proteins, *Eur. J. Biochem.* 13, 519–525, 1970; Splittstoesser, D.F. and Wilkison, M., Some factors affecting the activity of diethylpyrocarbonate as a sterilant, *Appl. Microbiol.* 25, 853–857, 1973; Fedorcsak, I., Ehrenberg, L., and Solymosy, F., Diethylpyrocarbonate does not degrade RNA, *Biochem. Biophys. Res. Commun.* 65, 490–496, 1975; Berger, S.L., Diethylpyrocarbonate: an examination of its properties in buffered solutions with a new assay technique, *Anal. Biochem.* 67, 428–437, 1975; Lloyd, A.G. and Drake, J.J., Problems posed by essential food preservatives, *Br. Med. Bull.* 31, 214–219, 1975; Ehrenberg, L., Fedorcsak, I., and Solymosy, F., Diethylpyrocarbonate in nucleic acid research, *Prog. Nucleic Acid Res. Mol. Biol.* 16, 189–262, 1976; Saluz, H.P. and Jost, J.P., Approaches to characterize protein–DNA interactions *in vivo*, *Crit. Rev. Eurkaryot. Gene Expr.* 3, 1–29, 1993; Bailly, C. and Waring, M.J., Diethylpyrocarbonate and osmium tetroxide as probes for drug-induced changes in DNA conformation *in vitro*, *Methods Mol. Biol.* 90, 51–59, 1997; Mabic, S. and Kano, I., Impact of purified water quality on molecular biology experiments, *Clin. Chem. Lab. Med.* 41, 486–491, 2003; Colleluori, D.M., Reczkowski, R.S., Emig, F.A. et al., Probing the role of the hyper-reactive histidine residue of argininase, *Arch. Biochem. Biophys.* 444, 15–26, 2005; Wu, S.N. and Chang, H.D., Diethylpyrocarbonate, a histidine-modifying agent, directly stimulates activity of ATP-sensitive potassium channels in pituitary GH(3) cells, *Biochem. Pharmacol.* 71, 615–623, 2006.

Dimedone

Dimedone

5,5-dimethyl-1,3-cyclohexanedione	140.18	Originally described as reagent for assay of aldehydes; used as a specific modifier of sulfenic acid.

Bulmer, D., Dimedone as an aldehyde-blocking reagent to facilitate the histochemical determination of glycogen, *Stain Technol.* 34, 95–98, 1959; Sawicki, E. and Carnes, R.A., Spectrophotofluorimetric determination of aldehydes with dimedone and other reagents, *Mikrochem. Acta* 1, 95–98, 1968; Benitez, L.V. and Allison, W.S., The inactivation of the acyl phosphatase activity catalyzed by the sufenic acid form of glyceraldehyde 3-phosphate dehydrogenase by dimedone and olifins, *J. Biol. Chem.* 249, 6234–6243, 1974; Huszti, Z. and Tyihak, E., Formation of formaldehyde from S-adenosyl-L-[methyl-³H]methionine during enzymic transmethylation of histamine, *FEBS Lett.* 209, 362–366, 1986; Sardi, E. and Tyihak, E., Sample determination of formaldehyde in dimedone adduct form in biological samples by high-performance liquid chromatography, *Biomed. Chromatog.* 8, 313–314, 1994; Demaster, A.G., Quast, B.J., Redfern, B., and Nagasawa, H.T., Reaction of nitric oxide with the free sulfydryl group of human serum albumin yields a sulfenic acid and nitrous oxide, *Biochemistry* 34, 14494–14949, 1995; Rozylo, T.K., Siembida, R., and Tyihak, E., Measurement of formaldehyde as dimedone adduct and potential formaldehyde precursors in hard tissues of human teeth by overpressurized layer chromatography, *Biomed. Chromatog.* 13, 513–515, 1999; Percival, M.D., Ouellet, M., Campagnolo, C. et al., Inhibition of cathepsin K by nitric oxide donors: evidence for the formation of mixed disulfides and a sulfenic acid, *Biochemistry* 38, 13574–13583, 1999; Carballal, S., Radi, R., Kirk, M.C. et al., Sulfenic acid formation in human serum albumin by hydrogen peroxide and peroxynitrite, *Biochemistry* 42, 9906–9914, 2003; Poole, L.B., Zeng, B.-B., Knaggs, S.A., Yakuba, M., and King, S.B., Synthesis of chemical probes to map sulfenic acid modifications on proteins, *Bioconjugate Chem.* 16, 1624–1628, 2005; Kaiserov, K., Srivastava, S., Hoetker, J.D. et al., Redox activation of aldose reductase in the ischemic heart, *J. Biol. Chem.* 281, 15110–15120, 2006.

Dimethylformamide (DMF)

Dimethylformamide

N,N-dimethylformamide	73.09	Solvent.

Eliezer, N. and Silberberg, A., Structure of branched poly-alpha-amino acids in dimethylformamide. I. Light scattering, *Biopolymers* 5, 95–104, 1967; Bonner, O.D., Bednarek, J.M., and Arisman, R.K., Heat capacities of ureas and water in water and dimethylformamide, *J. Am. Chem. Soc.* 99, 2898–2902, 1977; Sasson, S. and Notides, A.C., The effects of dimethylformamide on the interaction of the estrogen receptor with estradiol, *J. Steroid Biochem.* 29, 491–495, 1988; Jeffers, R.J., Feng, R.Q., Fowlkes, J.B. et al., Dimethylformamide as an enhancer of cavitation-induced cell lysis *in vitro*, *J. Acoust. Soc. Am.* 97, 669–676, 1995; You, L. and Arnold, F.H., Directed evolution of subtilisin E in *Bacillus subtilis* to enhance total activity in aqueous dimethylformamide, *Protein Eng.* 9, 77–83, 1996; Szabo, P.T. and Kele, Z., Electrospray mass spectrometry of hydrophobic compounds using dimethyl sulfoxide and dimethylformamide, *Rapid Commun. Mass Spectrom.* 15, 2415–2419, 2001; Nishida, Y., Shingu, Y., Dohi, H., and Kobayashi, K., One-pot alpha-glycosylation method using Appel agents in *N,N*-dimethylformamide, *Org. Lett.* 5, 2377–2380, 2003; Shingu, Y., Miyachi, A., Miura, Y. et al., One-pot alpha-glycosylation pathway via the generation *in situ* of alpha-glycopyranosyl imidates I *N,N*-dimethylformamide, *Carbohydr. Res.* 340, 2236–2244, 2005; Porras, S.P. and Kenndler, E., Capillary electrophoresis in *N,N*-dimethylformamide, *Electrophoresis* 26, 3279–3291, 2005; Wei, Q., Zhang, H., Duan, C. et al., High sensitive fluorophotometric determination of nucleic acids with pyronine G sensitized by *N,N*-dimethylformamide, *Ann. Chim.* 96, 273–284, 2006.

Dimethyl Suberimidate (DMS)   Crosslinking agent.

Dimethylsuberimidate

Davies, G.E. and Stark, G.R., Use of dimethyl suberimidate, a crosslinking reagent, in studying the subunit structure of oligomeric proteins, *Proc. Natl. Acad. Sci. USA* 66, 651–656, 1970; Hassell, J. and Hand, A.R., Tissue fixation with diimidoesters as an alternative to aldehydes. I. Comparison of crosslinking and ultrastructure obtained with dimethylsuberimidate and glutaraldehyde, *J. Histochem. Cytochem.* 22, 223–229, 1974; Thomas, J.O., Chemical crosslinking of histones, *Methods Enzymol.* 170, 549–571, 1989; Roth, M.R., Avery, R.B., and Welti, R., Crosslinking of phosphatidylethanolamine neighbors with dimethylsuberimidate is sensitive to the lipid phase, *Biochim. Biophys. Acta* 986, 217–224, 1989; Redl, B., Walleczek, J., Soffler-Meilicke, M., and Stoffler, G., Immunoblotting analysis of protein–protein crosslinks within the 50S ribosomal subunit of *Escherichia coli*. A study using dimethylsuberimidate as crosslinking reagent, *Eur. J. Biochem.* 181, 351–256, 1989; Konig, S., Hubner, G., and Schellenberger, A., Crosslinking of pyruvate decarboxylase-characterization of the native and substrate-activated enzyme states, *Biomed. Biochim. Acta* 49, 465–471, 1990; Chen, J.C., von Lintig, F.C., Jones, S.B. et al., High-efficiency solid-phase capture using glass beads bonded to microcentrifuge tubes: immunoprecipitation of proteins from cell extracts and assessment of ras activation, *Anal. Biochem.* 302, 298–304, 2002; Dufes, C., Muller, J.M., Couet, W. et al., Anticancer drug delivery with transferrin-targeted polymeric chitosan vesicles, *Pharm. Res.* 21, 101–107, 2004; Levchenko, V. and Jackson, V., Histone release during transcription: NAP1 forms a complex with H2A and H2B and facilitates a topologically dependent release of H3 and H4 from the nucleosome, *Biochemistry* 43, 2358–2372, 2004; Jastrzebska, M., Barwinski, B., Mroz, I. et al., Atomic force microscopy investigation of chemically stabilized pericardium tissue, *Eur. Phys. J. E* 16, 381–388, 2005.

Dimethyl Sulfate

126.1   Methylating agent;
methylation of nucleic
acids; used for a process
called footprinting to
identify sites of
protein–nucleic acid
interaction.

Dimethylsulfate

Nielsen, P.E., *In vivo* footprinting: studies of protein–DNA interactions in gene regulation, *Bioessay* 11, 152–155, 1989; Saluz, H.P. and Jost, J.P., Approaches to characterize protein–DNA interactions *in vivo*, *Crit. Rev. Eurkaryot. Gene Expr.* 3, 1–29, 1993; Saluz, H.P. and Jost, J.P., *In vivo* DNA footprinting by linear amplification, *Methods Mol. Biol.* 31, 317–329, 1994; Paul, A.L. and Ferl, R.J., *In vivo* footprinting of protein–DNA interactions, *Methods Cell Biol.* 49, 391–400, 1995; Gregory, P.D., Barbaric, S., and Horz, W., Analyzing chromatin structure and transcription factor binding in yeast, *Methods* 15, 295–302, 1998; Simpson, R.T., *In vivo* to analyze chromatin structrure, *Curr. Opin. Genet. Dev.* 9, 225–229, 1999; Nawrocki, A.R., Goldring, C.E., Kostadinova, R.M. et al., *In vivo* footprinting of the human 11β-hydroxysteroid dehydrogenase type 2 promoter: evidence for cell-specific regulation by Sp1 and Sp3, *J. Biol. Chem.* 277, 14647–14656, 2002; McGarry, K.C., Ryan, V.T., Grimwade, J.E., and Leonard, A.C., Two discriminatory binding sites in the *Escherichia coli* replication origin are required for DNA stand opening by initiator DnaA-ATP, *Proc. Natl. Acad. Sci. USA* 101, 2811–2816, 2004; Kellersberger, K.A., Yu, E., Kruppa, G.H. et al., Two-down characterization of nucleic acids modified by structural probes using high-resolution tandem mass spectrometry and automated data interpretation, *Anal. Chem.* 76, 2438–2445, 2004; Matthews, D.H., Disney, M.D., Childs, J.L. et al., Incorporating chemical modification constraints into a dynamic programming algorithm for prediction of RNA secondary structure, *Proc. Natl. Acad. Sci. USA* 101, 7287–7292, 2004; Forstemann, K. and Lingner, J., Telomerase limits the extent of base pairing between template RNA and temomeric DNA, *EMBO Rep.* 6, 361–366, 2005; Kore, A.R. and Parmar, G., An industrial process for selective synthesis of 7-methyl guanosine 5′-diphosphate: versatile synthon for synthesis of mRNA cap analogues, *Nucleosides Nucleotides Nucleic Acids* 25, 337–340, 2006.

Dioxane

1,4-diethylene   88.1   Solvent.
Dioxide

1, 4-Dioxane

Sideri, C.N. and Osol, A., A note on the purification of dioxane for use in preparing nonaqueous titrants, *J. Am. Pharm. Am. Pharm. Assoc.* 42, 586, 1953; Martel, R.W. and Kraus, C.A., The association of ions in dioxane-water mixtures at 25 degrees, *Proc. Natl. Acad. Sci. USA* 41, 9–20, 1955; Mercier, P.L. and Kraus, C.A, The ion-pair equilibrium of electrolyte solutions in dioxane-water mixtures, *Proc. Natl. Acad. Sci. USA* 41, 1033–1041, 1995; Inagami, T., and Sturtevant, J.M., The trypsin-catalyzed hydrolysis of benzoyl-L-arginine ethyl ester. I. The kinetics in dioxane-water mixtures, *Biochim. Biophys. Acta* 38, 64–79, 1980; Zaeklj, A. and Gros, M., Electrophoresis of lipoprotein, prestained with Sudan Black B, dissolved in a mixture of dioxane and ethylene glycol, *Clin. Chim. Acta* 5, 947, 1960; Krasner, J. and McMenamy, R.H., The binding of indole compounds to bovine plasma albumin. Effects of potassium chloride, urea, dioxane, and glycine, *J. Biol. Chem.* 241, 4186–4196, 1966; Smith, R.R. and Canady, W.J., Solvation effects upon the thermodynamic substrate activity: correlation with the kinetics of enzyme-catalyzed reactions. II. More complex interactions of alpha-chymotrypsin with dioxane and acetone which are also competitive inhibitors, *Biophys. Chem.* 43, 189–195, 1992; Forti, F.L., Goissis, G., and Plepis, A.M., Modifications on collagen structures promoted by 1,4-dioxane improve thermal and biological properties of bovine pericardium as a biomaterial, *J. Biomater. Appl.* 20, 267–285, 2006.

Dithiothreitol

1,4-dithiothreitol; 154.3   Reducing agent.
DTT; Cleland's
Reagent; *threo*-
2,3-dihydroxy-
1,4-dithiolbutane

Dithiothreitol/Dithioerythritol

Cleland, W.W., Dithiothreitol, a new protective reagent for SH groups, *Biochemistry* 3, 480–482, 1964; Gorin, G., Fulford, R., and Deonier, R.C., Reaction of lysozyme with dithiothreitol and other mercaptans, *Experientia* 24, 26–27, 1968; Stanton, M. and Viswantha, T., Reduction of chymotryptin A by dithiothreitol, *Can. J. Biochem.* 49, 1233–1235, 1971; Warren, W.A., Activation of serum creatine kinase by dithiothreitol, *Clin. Chem.* 18, 473–475, 1972; Hase, S. and Walter, R., Symmetrical disulfide bonds as *S*-protecting groups and their cleavage by dithiothreitol: synthesis of oxytocin with high biological activity, *Int. J. Pept. Protein Res.* 5, 283–288, 1973; Fleisch, J.H., Krzan, M.C., and Titus, E., Alterations in pharmacologic receptor activity by dithiothreitol, *Am. J. Physiol.* 227, 1243–1248, 1974; Olsen, J. and Davis, L., The oxidation of dithiothreitol by peroxidases and oxygen, *Biochim. Biophys. Acta.* 445, 324–329, 1976; Chao, L.P., Spectrophotometric determination of choline acetyltransferase in the presence of dithiothreitol, *Anal. Biochem.* 85, 20–24, 1978; Fukada, H. and Takahashi, K., Calorimetric study of the oxidation of dithiothreitol, *J. Biochem.* 87, 1105–1110, 1980; Alliegro, M.C., Effects of dithiothreitol on protein activity unrelated to thiol-disulfide exchange: for consideration in the analysis of protein function with Cleland's reagent, *Anal. Biochem.* 282, 102–106, 2000; Rhee, S.S. and Burke, D.H., Tris(2-carboxyethyl)phosphine stabilization of RNA: comparison with dithiothreitol for use with nucleic acid and thiophosphoryl chemistry, *Anal. Biochem.* 325, 137–143, 2004; Pan, J.C., Cheng, Y., Hui, E.F., and Zhou, H.M., Implications of the role of reactive cysteine in arginine kinase: reactivation kinetics of 5,5'-dithiobis-(2-nitrobenzoic acid)-modified arginine kinase reactivated by dithiothreitol, *Biochem. Biophys. Res. Commun.* 317, 539–544, 2004; Thaxton, C.S., Hill, H.D., Georganopoulou, D.G. et al., A bio-barcode assay based upon dithiothreitol-induced oligonucleotide release, *Anal. Chem.* 77, 8174–8178, 2005.

DMSO

Dimethylsulfoxide   78.13   Solvent; suggested
therapeutic use; effect on
cellular function;
cyropreservative.

Dimethylsulfoxide

Huggins, C.E., Reversible agglomeration used to remove dimethylsulfoxide from large volumes of frozen blood, *Science* 139, 504–505, 1963; Yehle, A.V. and Doe, R.H., Stabilization of *Bacillus subtilis* phage with dimethylsulfoxide, *Can. J. Microbiol.* 11, 745–746, 1965; Fowler, A.V. and Zabin, I., Effects of dimethylsulfoxide on the lactose operon of *Escherichia coli*, *J. Bacteriol.* 92, 353–357, 1966; Williams, A.E. and Vinograd, J., The buoyant behavior of RNA and DNA in cesium sulfate solutions containing dimethylsulfoxide, *Biochim. Biophys. Acta* 228, 423–439, 1971; Levine, W.G., The effect of dimethylsulfoxide on the binding of 3-methylcholanthrene to rat liver fractions, *Res. Commum. Chem. Pathol. Pharmacol.* 4, 511–518, 1972; Fink, A.L, The trypsin-catalyzed hydrolysis of *N*-alpha-benzoyl-L-lysine *p*-nitrophenyl ester in dimethylsulfoxide at subzero temperatures, *J. Biol. Chem.* 249, 5072–5932, 1974; Hutton, J.R. and Wetmur, J.G., Activity of endonuclease S1 in denaturing solvents: dimethylsulfoxide, dimethylformamide, formamide, and formaldehyde, *Biochem. Biophys. Res. Commun.* 66, 942–948, 1975; Gal, A., De Groot, N., and Hochberg, A.A., The effect of dimethylsulfoxide on ribosomal fractions from rat liver, *FEBS Lett.* 94, 25–27, 1978; Barnett, R.E., The effects of dimethylsulfoxide and glycerol on Na+, K+-ATPase, and membrane structure, *Cryobiology* 15, 227–229, 1978; Borzini, P., Assali, G., Riva, M.R. et al., Platelet cryopreservation using dimethylsulfoxide/polyethylene glycol/sugar mixture as cryopreserving solution, *Vox Sang.* 64, 248–249, 1993; West, R.T., Garza, L.A., II, Winchester, W.R., and Walmsley, J.A., Conformation, hydrogen bonding, and aggregate formation of guanosine 5'-monophosphate and guanosine in dimethylsulfoxide, *Nucleic Acids Res.* 22, 5128–5134, 1994; Bhattacharjya, S. and Balarma, P., Effects of organic solvents on protein structures; observation of a structured helical core in hen egg-white lysozyme in aqueous dimethylsulfoxide, *Proteins* 29, 492–507, 1997; Simala-Grant, J.L. and Weiner, J.H., Modulation of the substrate specificity of *Escherichia coli* dimethylsulfoxide reductase, *Eur. J. Biochem.* 251, 510–515, 1998; Tsuzuki, W., Ue, A., and Kitamura, Y., Effect of dimethylsulfoxide on hydrolysis of lipase, *Biosci. Biotechnol. Biochem.* 65, 2078–2082, 2001; Pedersen, N.R., Halling, P.J., Pedersen, L.H. et al., Efficient transesterification of sucrose catalyzed by the metalloprotease thermolysin in dimethylsulfoxide, *FEBS Lett.* 519, 181–184, 2002; Fan, C., Lu, J., Zhang, W., and Li, G., Enhanced electron-transfer reactivity of cytochrome b5 by dimethylsulfoxide and *N,N'*-dimethylformamide, *Anal. Sci.* 18, 1031–1033, 2002; Tait, M.A. and Hik, D.S., Is dimethylsulfoxide a reliable solvent for extracting chlorophyll under field conditions? *Photosynth. Res.* 78, 87–91, 2003; Malinin, G.I. and

Malinin, T.I., Effects of dimethylsulfoxide on the ultrastructure of fixed cells, *Biotech. Histochem.* 79, 65–69, 2004; Clapisson, G., Salinas, C., Malacher, P. et al., Cryopreservation with hydroxyethylstarch (HES) + dimethylsulfoxide (DMSO) gives better results than DMSO alone, *Bull. Cancer* 91, E97–E102, 2004.

EDC

1-Cyclohexyl-2-(2-morpholinethyl)-carbodiimide

1, 3-Dicyclohexylcarbodiimide

1-ethyl-3-(3-dimethylaminopropyl)-carbodiimide

Glycine methyl ester

Carbodiimide

Protein carboxyl group

O-acylisourea

1-ethyl-(3-dimethylamino propyl)-carbodiimide; *N*-(3-dimethylamino-propyl)-*N'*-ethyl-carbodiimide

191.7 (HCl)

Water-soluble carbodiimide for the modification of carboxyl groups in proteins; zero-length crosslinking proteins; activation of carboxyl groups for amidation reactions, as for the coupling of amino-nucleotides to matrices for DNA microarrays.

Lin, T.Y. and Koshland, D.E., Jr., Carboxyl group modification and the activity of lysozyme, *J. Biol. Chem.* 244, 505–508, 1969; Carraway, K.L., Spoerl, P., and Koshland, D.E., Jr., Carboxyl group modification in chymotrypsin and chymotrypsinogen, *J. Mol. Biol.* 42, 133–137, 1969; Yamada, H., Imoto, T., Fujita, K. et al., Selective modification of aspartic acid-101 in lysozyme by carbodiimide reaction, *Biochemistry* 20, 4836–4842, 1981; Buisson, M. and Reboud, A.M., Carbodiimide-induced protein-RNA crosslinking in mammalian subunits, *FEBS Lett.* 148, 247–250, 1982; Millett, F., Darley-Usmar, V., and Capaldi, R.A., Cytochrome c is crosslinked to subunit II of cytochrome c oxidase by a water-soluble carbodiimide, *Biochemistry* 21, 3857–3862, 1982; Chen, S.C., Fluorometric determination of carbodiimides with trans-aconitic acid, *Anal. Biochem.* 132, 272–275, 1983; Davis, L.E., Roth, S.A., and Anderson, B., Antisera specificities to 1-ethyl-3-(3-dimethylaminopropyl) carbodiimide adducts of proteins, *Immunology* 53, 435–441, 1984; Ueda, T., Yamada, H., and Imoto, T., Highly controlled carbodiimide reaction for the modification of lysozyme. Modification of Leu129 or As119, *Protein Eng.* 1, 189–193, 1987; Ghosh, M.K., Kildsig, D.O., and Mitra, A.K., Preparation and characterization of methotrexate-immunoglobulin conjugates, *Drug. Des. Deliv.* 4, 13–25, 1989; Grabarek, Z. and Gergely, J., Zero-length crosslinking procedure with the use of active esters, *Anal. Biochem.* 185, 131–135, 1990; Gilles, M.A., Hudson, A.Q., and Borders, C.L., Jr., Stability of water-soluble carbodiimides in aqueous solutions, *Anal. Biochem.* 184, 244–248, 1990; Soinila, S., Mpitsos, G.J., and Soinila, J., Immunohistochemistry of enkephalins: model studies on hapten-carrier conjugates and fixation methods, *J. Histochem. Cytochem.* 40, 231–239, 1992; Soper, S.A., Hashimoto, M., Situma, C. et al., Fabrication of DNA microarrays onto polymer substrates using UV modification protocols with integration into microfluidic platforms for the sensing of low-abundant DNA point mutations, *Methods* 37, 103–113, 2005.

EDTA

EDTA, ethylenediaminetetraacetic acid

Ethylenediamine-tetraacetic Acid

292.24

Chelating agent; some metal ion-EDTA complexes (i.e., Fe^{2+}-EDTA) function as chemical nucleases.

Flaschka, H.A., *EDTA Titrations: An Introduction to Theory and Practice*, Pergammon Press, Oxford, UK, 1964; West, T.S., *Complexometry with EDTA and Related Reagents,* BDH Chemicals Ltd., Poole (Dorset), UK, 1969; Pribil, R., *Analytical Applications of EDTA and Related Compounds,* Pergammon Press, Oxford, UK, 1972; Papavassiliou, A.G., Chemical nucleases as probes for studying DNA–protein interactions, *Biochem. J.* 305, 345–357, 1995; Martell, A.E., and Hancock, R.D., *Metal Complexes in Aqueous Solutions*, Plenum Press, New York, 1996; Loizos, N. and Darst, S.A, Mapping protein–ligand interactions by footprinting, a radical idea, *Structure* 6, 691–695, 1998; Franklin, S.J., Lanthanide-mediated DNA hydrolysis, *Curr. Opin. Chem. Biol.* 5, 201–208, 2001; Heyduk, T., Baichoo, N., and Henduk, E., Hydroxyl radical footprinting of proteins using metal ion complexes, *Met. Ions Biol. Syst.* 38, 255–287, 2001; Orlikowsky, T.W., Neunhoeffer, F., Goelz, R. et al., Evaluation of IL-8-concentrations in plasma and lyszed EDTA-blood in healthy neonates and those with suspected early onset bacterial infection, *Pediatr. Res.* 56, 804–809, 2004; Matt, T., Martinez-Yamout, M.A., Dyson, H.J., and Wright, P.E., The CBP/p300 TAZ1 domain in its native state is not a binding partner of MDM2, *Biochem. J.* 381, 685–691, 2004; Nyborg, J.K. and Peersen, O.B., That zincing feeling: the effects of EDTA on the behavior of zinc-binding transcriptional regulators, *Biochem. J.* 381, e3–e4, 2004; Haberz, P., Rodriguez-Castanada, F., Junker, J. et al., Two new chiral EDTA-based metal chelates for weak alignment of proteins in solution, *Org. Lett.* 8, 1275–1278, 2006.

Ellman's Reagent

Ellman's reagent

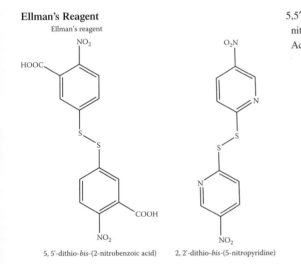

5, 5'-dithio-*bis*-(2-nitrobenzoic acid)   2, 2'-dithio-*bis*-(5-nitropyridine)

5,5'-dithio-*bis*-[2-   396.35   Reagent for determination
nitro-benzoic]   of sulfydryl groups/
Acid   disulfide bonds.

Ellman, G.L., Tissue sulfydryl groups, *Arch. Biochem. Biophys.* 82, 70–77, 1959; Boyne, A.F. and Ellman, G.L., A methodology for analysis of tissue sulfydryl components, *Anal. Biochem.* 46, 639–653, 1972; Brocklehurst, K., Kierstan, M., and Little, G., The reaction of papain with Ellman's reagent (5,5'-dithiobis-(2-nitrobenzoate), *Biochem. J.* 128, 811–816, 1972; Weitzman, P.D., A critical reexamination of the reaction of sulfite with DTNB, *Anal. Biochem.* 64, 628–630, 1975; Hull, H.H., Chang, R., and Kaplan, L.J., On the location of the sulfydryl group in bovine plasma albumin, *Biochim. Biophys. Acta* 400, 132–136, 1975; Banas, T., Banas, B., and Wolny, M., Kinetic studies of the reactivity of the sulfydryl groups of glyceraldehyde-3-phosphate dehydrogenase, *Eur. J. Biochem.* 68, 313–319, 1976; der Terrossian, E. and Kassab, R., Preparation and properties of *S*-cyano derivatives of creatine kinase, *Eur. J. Biochem.* 70, 623–628, 1976; Riddles, P.W., Blakeley, R.L., and Zerner, B., Ellman's reagent: 5,5'-dithiobis(2-nitrobenzoic acid) — a reexamination, *Anal. Biochem.* 94, 75–81, 1979; Luthra, N.P., Dunlap, R.B., and Odom, J.D., Characterization of a new sulfydryl group reagent: 6, 6'- diselenobis-(3-nitrobenzoic acid), a selenium analog of Ellman's reagent, *Anal. Biochem.* 117, 94–102, 1981; Di Simplicio, P., Tiezzi, A., Moscatelli, A. et al., The SH-SS exchange reaction between the Ellman's reagent and protein-containing SH groups as a method for determining conformational states: tubulin, *Ital. J. Biochem.* 38, 83–90, 1989; Woodward, J., Tate, J., Herrmann, P.C., and Evans, B.R., Comparison of Ellman's reagent with *N*-(1-pyrenyl)maleimide for the determination of free sulfydryl groups in reduced cellobiohydrolase I from *Trichoderma reesei, J. Biochem. Biophys. Methods* 26, 121–129, 1993; Berlich, M., Menge, S., Bruns, I. et al., Coumarins give misleading absorbance with Ellman's reagent suggestive of thiol conjugates, *Analyst* 127, 333–336, 2002; Riener, C.K., Kada, G., and Gruber, H.J., Quick measurement of protein sulfydryls of Ellman's reagents and with 4,4'-dithiopyridine, *Anal. Bio. Anal. Chem.* 373, 266–276, 2002; Zhu, J., Dhimitruka, I., and Pei, D., 5-(2-aminoethyl)dithio-2-nitrobenzoate as a more base-stable alternative to Ellman's reagent, *Org. Lett.* 6, 3809–3812, 2004; Owusu-Apenten, R., Colorimetric analysis of protein sulfydryl groups in milk: applications and processing effects, *Crit. Rev. Food Sci. Nutr.* 45, 1–23, 2005.

Ethanolamine

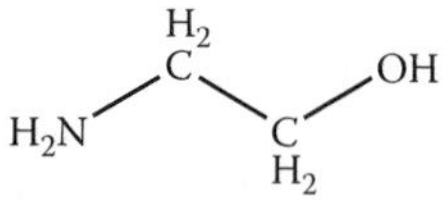

Ethanolamine

Glycinol   61.08   Buffer component;
component of a
phospholipid
(phosphatidyl
ethanolamine, PE).

Vance, D.E. and Ridgway, N.D., The methylation of phosophatidylethanolamine, *Prog. Lipid Res.* 27, 61–79, 1988; Louwagie, M., Rabilloud, T., and Garin, J., Use of ethanolamine for sample stacking in capillary electrophoresis, *Electrophoresis* 19, 2440–2444, 1998; de Nogales, V., Ruiz, R., Roses. M. et al., Background electrolytes in 50% methanol/water for the determination of acidity constants of basic drugs by capillary zone electrophoresis, *J. Chromatog. A* 1123, 113–120, 2006.

Ethidium Bromide

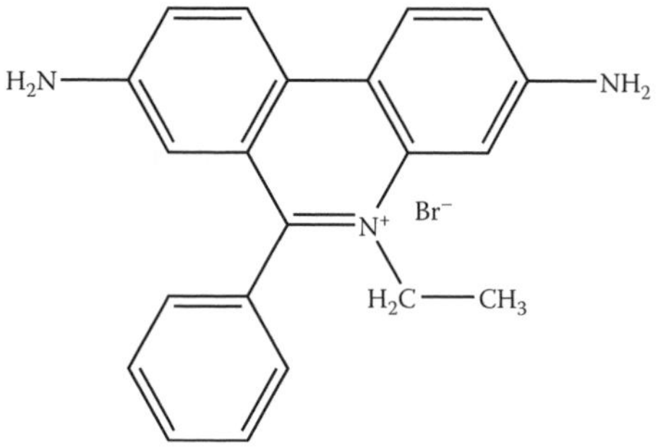

Ethidium bromide (Homidium Bromide)
3, 8-diamino-6-ethyl-5-phenylphenanthridium bromide

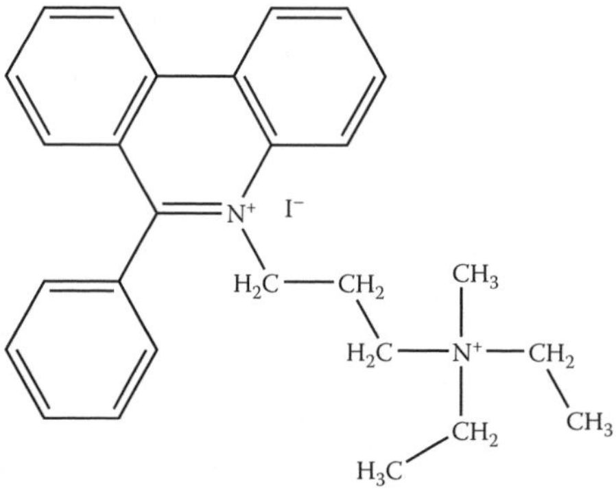

Propidium iodide

Sela, I., Fluorescence of nucleic acids with ethidium bromide: an indication of the configurative state of nucleic acids, *Biochim. Biophys. Acta* 190, 216–219, 1969; Le Pecq, J.B., Use of ethidium bromide for separation and determination of nucleic acids of various conformational forms and measurement of their associated enzymes, *Methods Biochem. Anal.* 20, 41–86, 1971; Borst, P., Ethidium DNA agarose gel electrophoresis: how it started, *IUBMB Life* 57, 745–747, 2005.

Ethyl Alcohol

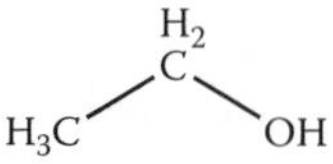

Ethanol

Ethanol   46.07   Solvent; used to adjust
solvent polarity; use in
plasma protein
fractionation.

Dufour, E., Bertrand-Harb, C., and Haertle, T., Reversible effects of medium dielectric constant on structural transformation of beta-lactoglobulin and its retinol binding, *Biopolymers* 33, 589–598, 1993; Escalera, J.B., Bustamante, P., and Martin, A., Predicting the solubility of drugs in solvent mixtures: multiple solubility maxima and the chameleonic effect, *J. Pharm. Pharmcol.* 46, 172–176, 1994; Gratzer, P.F., Pereira, C.A., and Lee, J.M., Solvent environment modulates effects of glutaraldehyde crosslinking on tissue-derived biomaterials, *J. Biomed. Mater. Res.* 31, 533–543, 1996; Sepulveda, M.R. and Mata, A.M., The interaction of ethanol with reconstituted synaptosomal plasma membrane Ca^{2+}, *Biochim. Biophys. Acta* 1665, 75–80, 2004; Ramos, A.S. and Techert, S., Influence of the water structure on the acetylcholinesterase efficiency, *Biophys. J.* 89, 1990–2003, 2005; Wehbi, Z., Perez, M.D., and Dalgalarrondo, M., Study of ethanol-induced conformation changes of holo and apo alpha-lactalbumin by spectroscopy anad limited proteolysis, *Mol. Nutr. Food Res.* 50, 34–43, 2006; Sasahara, K. and Nitta, K., Effect of ethanol on folding of hen egg-white lysozyme under acidic condition, *Proteins* 63, 127–135, 2006; Perham, M., Liao, J., and Wittung-Stafshede, P., Differential effects of alcohol on conformational switchovers in alpha-helical and beta-sheet protein models, *Biochemistry* 45, 7740–7749, 2006; Pena, M.A., Reillo, A., Escalera, B., and Bustamante, P., Solubility parameter of drugs for predicting the solubility profile type within a wide polarity range in solvent mixtures, *Int. J. Pharm.* 321, 155–161, 2006; Jenke, D., Odufu, A., and Poss, M., The effect of solvent polarity on the accumulation of leachables from pharmaceutical product containers, *Eur. J. Pharm. Sci.* 27, 133–142, 2006.

| **Ethylene Glycol** | 1,2-ethanediol | 62.07 | Solvent/cosolvent; increases viscosity (visogenic osmolyte); perturbant; cryopreservative. |

Ethylene glycol

Tanford, C., Buckley, C.E., III, De, P.K., and Lively, E.P., Effect of ethylene glycol on the conformation of gamma-globulin and beta-lactoglobulin, *J. Biol. Chem.* 237, 1168–1171, 1962; Kay, C.M. and Brahms, J., The influence of ethylene glycol on the enzymatic adenosine triphosphatase activity and molecular conformation of fibrous muscle proteins, *J. Biol. Chem.* 238, 2945–2949, 1963; Narayan, K.A., The interaction of ethylene glycol with rat-serum lipoproteins, *Biochim. Biophys. Acta* 137, 22–30, 1968; Bello, J., The state of the tyrosines of bovine pancreatic ribonuclease in ethylene glycol and glycerol, *Biochemistry* 8, 4535–4541, 1969; Lowe, C.R. and Mosbach, K., Biospecific affinity chromatography in aqueous-organic cosolvent mixtures. The effect of ethylene glycol on the binding of lactate dehydrogenase to an immobilized-AMP analogue, *Eur. J. Biochem.* 52, 99–105, 1975; Ghrunyk, B.A. and Matthews, C.R., Role of diffusion in the folding of the alpha subunit of tryptophan synthase from *Escherichia coli*, *Biochemistry* 29, 2149–2154, 1990; Silow, M. and Oliveberg, M., High concentrations of viscogens decrease the protein folding rate constant by prematurely collapsing the coil, *J. Mol. Biol.* 326, 263–271, 2003; Naseem, F. and Khan, R.H., Effect of ethylene glycol and polyethylene glycol on the acid-unfolded state of trypsinogen, *J. Protein Chem.* 22, 677–682, 2003; Hubalek, Z., Protectants used in the cyropreservation of microorganisms, *Cryobiology* 46, 205–229, 2003; Menezo, Y.J., Blastocyst freezing, *Eur. J. Obstet. Gynecol. Reprod. Biol.* 155 (Suppl. 1), S12–S15, 2004; Khodarahmi, R. and Yazdanparast, R., Refolding of chemically denatured alpha-amylase in dilution additive mode, *Biochim. Biophys. Acta.* 1674, 175–181, 2004; Zheng, M., Li, Z., and Huang, X., Ethylene glycol monolayer protected nanoparticles: synthesis, characterization, and interactions with biological molecules, *Langmuir* 20, 4226–4235, 2004; Bonincontro, A., Cinelli, S., Onori, G., and Stravato, A., Dielectric behavior of lysozyme and ferricytochrome-c in water/ethylene-glycol solutions, *Biophys. J.* 86, 1118–1123, 2004; Kozer, N. and Schreiber, G., Effect of crowding on protein–protein association rates: fundamental differences between low and high mass crowding agents, *J. Mol. Biol.* 336, 763–774, 2004; Levin, I., Meiri, G., Peretz, M. et al., The ternary complex of *Pseudomonas aeruginosa* dehydrogenase with NADH and ethylene glycol, *Protein Sci.* 13, 1547–1556, 2004; Stupishina, E.A., Khamidullin, R.N., Vylegzhanina, N.N. et al., Ethylene glycol and the thermostability of trypsin in a reverse micelle system, *Biochemistry* 71, 533–537, 2006; Nordstrom, L.J., Clark, C.A., Andersen, B. et al., Effect of ethylene glycol, urea, and *N*-methylated glycines on DNA thermal stability: the role of DNA base pair composition and hydration, *Biochemistry* 45, 9604–9614, 2006.

| **Ethyleneimine** | Aziridine | 43.07 | Modification of sulfydryl groups to produce amine functions; alkylating agent; reacts with carboxyl groups at acid pH; monomer unit for polyethylene amine, a versatile polymer. |

Ethyleneimine

Raftery, M.A. and Cole, R.D., On the aminoethylation of proteins, *J. Biol. Chem.* 241, 3457–3461, 1966; Fishbein, L., Detection and thin-layer chromatography of derivatives of ethyleneimine. I. *N*-carbamoyl and aziridines, *J. Chromatog.* 26, 522–526, 1967; Yamada, H., Imoto, T., and Noshita, S., Modification of catalytic groups in lysozyme with ethyleneimine, *Biochemistry* 21, 2187–2192, 1982; Okazaki, K., Yamada, H., and Imoto, T., A convenient *S*-2-aminoethylation of cysteinyl residues in reduced proteins, *Anal. Biochem.* 149, 516–520, 1985; Hemminki, K., Reactions of ethyleneimine with guanosine and deoxyguanosine, *Chem. Biol. Interact.* 48, 249–260, 1984; Whitney, P.L., Powell, J.T., and Sanford, G.L., Oxidation and chemical modification of lung beta-galactosidase-specific lectin, *Biochem. J.* 238, 683–689, 1986; Simpson, D.M., Elliston, J.F., and Katzenellenbogen, J.A., Desmethylnafoxidine aziridine: an electrophilic affinity label for the estrogen receptor with high efficiency and selectivity, *J. Steroid Biochem.* 28, 233–245, 1987; Musser, S.M., Pan, S.S., Egorin, M.J. et al., Alkylation of DNA with aziridine produced during the hydrolysis of *N,N′,N″*-triethylenethiophosphoramide, *Chem. Res. Toxicol.* 5, 95–99, 1992; Thorwirth, S., Muller, H.S., and Winnewisser, G., The millimeter- and submillimeter-wave spectrum and the dipole moment of ethyleneimine, *J. Mol. Spectroso.* 199, 116–123, 2000; Burrage, T., Kramer, E., and Brown, F., Inactivation of viruses by azirdines, *Dev. Biol.* (Basel) 102, 131–139, 2000; Brown, F., Inactivation of viruses by aziridines, *Vaccine* 20, 322–327, 2001; Sasaki, S., Active oligonucleotides incorporating alkylating agent as potential sequence- and base-selective modifier of gene expression, *Eur. J. Pharm. Sci.* 13, 43–51, 2001; Hou, X.L., Fan, R.H., and Dai, L.X., Tributylphosphine: a remarkable promoting reagent for the ring-opening reaction of aziridines, *J. Org. Chem.* 67, 5295–5300, 2002; Thevis, M., Loo, R.R.O., and Loo, J.A., In-gel derivatization of proteins for cysteine-specific cleavages and their analysis by mass spectrometry, *J. Proteome Res.* 2, 163–172, 2003; Sasaki, M., Dalili, S., and Yudin, A.K., *N*-arylation of aziridines, *J. Org. Chem.* 68, 2045–2047, 2003; Gao, G.Y., Harden, J.D., and Zhang, J.P., Cobalt-catalyzed efficient aziridination of alkenes, *Org. Lett.* 7, 3191–3193, 2005; Hopkins, C.E., Hernandez, G., Lee, J.P., and Tolan, D.R., Aminoethylation in model peptides reveals conditions for maximizing thiol specificity, *Arch. Biochem. Biophys.* 443, 1–10, 2005; Li, C. and Gershon, P.D., pK(a) of the mRNA cap-specific 2′-*O*-methyltransferase catalytic lysine by HSQC NMR detection of a two-carbon probe, *Biochemistry* 45, 907–917, 2006; Vicik R., Helten, H., Schirmeister, T., and Engels, B., Rational design of aziridine-containing cysteine protease inhibitors with improved potency: studies on inhibition mechanism, *ChemMedChem*, 1, 1021–1028, 2006.

Ethylene Oxide   Oxirane   44.05   Sterilizing agent; starting
                                                                                     material for ethylene
                                                                                     glycol and other products
                                                                                     such as nonionic
                                                                                     surfactants.

Ethylene oxide

Windmueller, H.G., Ackerman, C.J., and Engel, R.W., Reaction of ethylene oxide with histidine, methionine, and cysteine, *J. Biol. Chem.* 234, 895–899, 1959; Starbuck, W.C. and Busch, H., Hydroxyethylation of amino acids in plasma albumin with ethylene oxide, *Biochim. Biophys. Acta* 78, 594–605, 1963; Guengerich, F.P., Geiger, L.E., Hogy, L.L., and Wright,. P.L., *In vitro* metabolism of acrylonitrile to 2-cyanoethylene oxide, reaction with glutathione, and irreversible binding to proteins and nucleic acids, *Cancer Res.* 41, 4925–4933, 1981; Peter, H., Schwarz, M., Mathiasch, B. et al., A note on synthesis and reactivity towards DNA of glycidonitrile, the epoxide of acrylonitrile, *Carcinogenesis* 4, 235–237, 1983; Grammer, L.C. and Patterson, R., IgE against ethylene oxide-altered human serum albumin (ETO-HAS) as an etiologic agent in allergic reactions of hemodialysis patients, *Artif. Organs* 11, 97–99, 1987; Bolt, H.M., Peter, H., and Fost, U., Analysis of macromolecular ethylene oxide adducts, *Int. Arch. Occup. Environ. Health* 60, 141–144, 1988; Young, T.L., Habraken, Y., Ludlum, D.B., and Santella, R.M., Development of monoclonal antibodies recognizing 7-(2-hydroxyethyl) guanine and imidazole ring-opened 7-(2-hydroxyethyl) guanine, *Carcinogenesis* 11, 1685–1689, 1990; Walker, V.E., Fennell, T.R., Boucheron, J.A. et al., Macromolecular adducts of ethylene oxide: a literature review and a time-course study on the formation of 7-(2-hydroxyethyl)guanine following exposure of rats by inhalation, *Mutat. Res.* 233, 151–164, 1990; Framer, P.B., Bailey, E., Naylor, S. et al., Identification of endogenous electrophiles by means of mass spectrometric determination of protein and DNA adducts, *Environ. Health Perspect.* 99, 19–24, 1993; Tornqvist, M. and Kautianinen, A., Adducted proteins for identification of endogenous electrophiles, *Environ. Health Perspect.* 99, 39–44, 1993; Galaev, I. Yu. and Mattiasson, B., Thermoreactive water-soluble polymers, nonionic surfactants, and hydrogels as reagents in biotechnology, *Enzyme Microb. Technol.* 15, 354–366, 1993; Segerback, D., DNA alkylation by ethylene oxide and mono-substituted expoxides, *IARC Sci. Publ.* 125, 37–47, 1994; Phillips, D.H. and Farmer, P.B., Evidence for DNA and protein binding by styrene and styrene oxide, *Crit. Rev. Toxicol.* 24 (Suppl.), S35–S46, 1994; Marczynski, B., Marek, W., and Baur, X., Ethylene oxide as a major factor in DNA and RNA evolution, *Med. Hypotheses* 44, 97–100, 1995; Mosely, G.A. and Gillis, J.R., Factors affecting tailing in ethylene oxide sterilization part 1: when tailing is an artifact…and scientific deficiencies in ISO 11135 and EN 550, *PDA J. Pharm. Sci. Technol.* 58, 81–95, 2004.

N-Ethylmaleimide

H₂C—CH₃

1-ethyl-1*H*-pyrrole-2,5-dione

125.13

Modification of sulfydryl groups; basic building block for a number of reagents. Mechanism different from alkylating agent in that reaction involves a Michael addition.

N-Ethylmaleimide   Cysteine

Lundblad, R.L., *Chemical Reagent for Protein Modification*, 3rd ed., CRC Press, Boca Raton, FL, 2004; Bowes, T.J. and Gupta, R.S., Induction of mitochondrial fusion by cysteine-alkylators ethyacrynic acid and *N*-ethylmaleimide, *J. Cell Physiol.* 202, 796–804, 2005; Engberts, J.B., Fernandez, E., Garcia-Rio, L., and Leis, J.R., Water in oil microemulsions as reaction media for a Diels–Alder reaction between *N*-ethylmaleimide and cyclopentadiene, *J. Org. Chem.* 71, 4111–4117, 2006; Engberts, J.B., Fernandez, E., Garcia-Rio, L., and Leis, J.R, AOT-based microemulsions accelerate the 1,3-cycloaddition of benzonitrile oxide to *N*-ethylmaleimide, *J. Org. Chem.* 71, 6118–6123, 2006; de Jong, K. and Kuypers, F.A., Sulphydryl modifications alter scramblase activity in murine sickle cell disease, *Br. J. Haematol.* 133, 427–432, 2006; Martin, H.G., Henley, J.M., and Meyer, G., Novel putative targets of *N*-ethylmaleimide sensitive fusion proteins (NSF) and alpha/beta soluble NSF attachment proteins (SNAPs) include the Pak-binding nucleotide exchange factor betaPIX, *J. Cell. Biochem.*, 99, 1203–1215, 2006; Carrasco, M.R., Silva, O., Rawls, K.A. et al., Chemoselective alkylation of *N*-alkylaminooxy-containing peptides, *Org. Lett.* 8, 3529–3532, 2006; Pobbati, A.V., Stein, A., and Fasshauer, D., N- to C-terminal SNARE complex assembly promotes rapid membrane fusion, *Science* 313, 673–676, 2006; Mollinedo, F., Calafat, J., Janssen, H. et al., Combinatorial SNARE complexes modulate the secretion of cytoplasmic granules in human neutrophils, *J. Immunol.* 177, 2831–2841, 2006.

Fluorescein

HO

OH

332.31

Fluorescent dye that can be combined with a reactive function group such as fluorescein isothiocyanate (FITC); used for fluorescent angiography with emphasis on ophthalmology.

O

Fluorescein

NaOH/Zinc

O

COOH

Fluorescin

Chadwick, C.S., McEntegart, M.G., and Nairn, R.C., Fluorescent protein tracers; a simple alternative to fluorescein, *Lancet* 1(7017), 412–414, 1958; Holter, H. and Holtzer, H., Pinocytotic uptake of fluorescein-labeled proteins by various tissue cells, *Exp. Cell Res.* 18, 421–423, 1959; Schatz, H., *Interpretation of Fundus Fluorescein Angiography*, Mosby, St. Louis, MO, 1978; Voss, E.W., *Fluorescein Hapten: An Immunological Probe*, CRC Press, Boca Raton, FL, 1984; Katz, J.N., Gobetti, J.P., and Shipman, C., Jr., Fluorescein dye evaluation of glove integrity, *J. Am. Dent. Assoc.* 118, 327–331, 1989; Fan, J., Pope, L.E., Vitols, K.S., and Huennekens, F.M., Visualization of folate transport proteins by covalent labeling with fluorescein methotrexate, *Adv. Enzyme Regul.* 30, 3–12, 1990; Mauger, T.F. and Elson, C.L.,

Havener's Ocular Pharmacology, Mosby, St. Louis, MO, 1994; Isaac, P.G., *Protocols for Nucleic Acid Analysis by Nonradioactive Probes*, Humana Press, Totowa, NJ, 1994; Cavallerano, A.A., Ophthalmic fluorescein angiography, *Optom. Clin.* 5, 1–23, 1996; Mills, C.O., Milkiewicz, P., Saraswat, V., and Elias, E., Cholyllysyl fluorescein and related lysyl fluorescein conjugated bile acid analogues, *Yale J. Biol. Med.* 70, 447–457, 1997; Zhang, J., Malicka, J., Gryczynski, I., and Lakowicz, J.R., Surface-enhanced fluorescence of fluorescein-labeled oligonucleotides capping on silver nanoparticles, *J. Phys. Chem. B Condens. Matter Mater. Surf. Interfaces Biophys.* 109, 7643–7648, 2005; Goldsmith, C.R., Jaworski, J., Sheng, M., and Lippard, S.J., Selective labeling of extracellular proteins containing polyhistidine sequences by a fluorescein-nitrilotriacetic acid conjugate, *J. Am. Chem. Soc.* 128, 418–419, 2006; Sato, K. and Anzai, J., Fluorometric determination of sugars using fluorescein-labeled concanavalin A-glycogen conjugates, *Anal. Bio. Anal. Chem.* 384, 1297–1301, 2006; Maes, V., Hultsch, C., Kohl, S. et al., Fluorescein-labeled stable neurotensin derivatives, *J. Pept. Sci.* 12, 505–508, 2006.

Formaldehyde   Methanal   30.03   Tissue fixation; protein
   Formaldehyde   modification; zero-length
                                                                 crosslinking;
                                                                protein–nucleic acid
                                                                interactions.

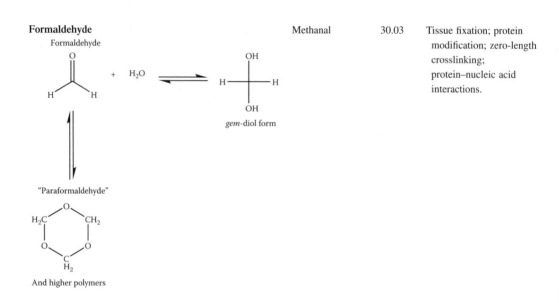

gem-diol form

"Paraformaldehyde"

And higher polymers

Feldman, M.Y., Reactions of nucleic acids and nucleoproteins with formaldehyde, *Prog. Nucleic Acid Res. Mol. Biol.* 13, 1–49, 1973; Russell, A.D. and Hopwood, D., The biological uses and importance of glutaraldehyde, *Prog. Med. Chem.* 13, 271–301, 1976; Means, G.E., Reductive alkylation of amino groups, *Methods Enzymol.* 47, 469–478, 1977; Winkelhake, J.L., Effects of chemical modification of antibodies on their clearance for the circulation. Addition of simple aliphatic compounds by reductive alkylation and carbodiimide-promoted amide formation, *J. Biol. Chem.* 252, 1865–1868, 1977; Yamazaki, Y. and Suzuki, H., A new method of chemical modification of N^6-amino group in adenine nucleotides with formaldehyde and a thiol and its application to preparing immobilized ADP and ATP, *Eur. J. Biochem.* 92, 197–207, 1978; Geoghegan, K.F., Cabacungan, J.C., Dixon, H.B., and Feeney, R.E., Alternative reducing agents for reductive methylation of amino groups in proteins, *Int. J. Pept. Protein Res.* 17, 345–352, 1981; Kunkel, G.R., Mehradian, M., and Martinson, H.G., Contact-site crosslinking agents, *Mol. Cell. Biochem.* 34, 3–13, 1981; Fox, C.H., Johnson, F.B., Whiting, J., and Roller, P.P., Formaldehyde fixation, *J. Histochem. Cytochem.* 33, 845–853, 1985; Conaway, C.C., Whysner, J., Verna, L.K., and Williams, G.M., Formaldehyde mechanistic data and risk assessment: endogenous protection from DNA adduct formation, *Pharmacol. Ther.* 71, 29–55, 1996; Masuda, N., Ohnishi, T., Kawamoto, S. et al., Analysis of chemical modifications of RNA from formalin-fixed samples and optimization of molecular biology applications for such samples, *Nucleic Acids Res.* 27, 4436–4443, 1999; Micard, V., Belamri, R., Morel, M., and Guilbert, S., Properties of chemically and physically treated wheat gluten films, *J. Agric. Food Chem.* 48, 2948–2953, 2000; Taylor, I.A. and Webb, M., Chemical modification of lysine by reductive methylation. A probe for residues involved in DNA binding, *Methods Mol. Biol.* 148, 301–314, 2001; Perzyna, A., Marty, C., Facopre, M. et al., Formaldehyde-induced DNA crosslink of indolizino[1,2-b]quinolines derived from the A-D rings of camptothecin, *J. Med. Chem.* 45, 5809–5812, 2002; Yurimoto, H., Hirai, R., Matsuno, N. et al., HxlR, a member of the DUF24 protein family, is a DNA-binding protein that acts as a positive regulator of the formaldehyde-inducible hx1AB operon in *Bacillus subtilis*, *Mol. Microbiol.* 57, 511–519, 2005.

Formic Acid   Methanoic Acid   46.03   Solvent; buffer component.

Formic acid   Formamide

Sarkar, P.B., Decomposition of formic acid by periodate, *Nature* 168, 122–123, 1951; Hass, P., Reactions of formic acid and its salts, *Nature* 167, 325, 1951; Smillie, L.B. and Neurath, H., Reversible inactivation of trypsin by anhydrous formic acid, *J. Biol. Chem* 234, 355–359, 1959; Hynninen, P.H. and Ellfolk, N., Use of the aqueous formic acid-chloroform-dimethylformamide solvent system for the purification of porphyrins and hemins, *Acta Chem. Scand.* 27, 1795–1806, 1973; Heukeshoven, J. and Dernick, R., Reversed-phase high-performance liquid chromatography of virus proteins and other large hydrophobic proteins in formic acid-containing solvents, *J. Chromatog.* 252, 241–254, 1982; Tarr, G.E. and Crabb, J.W., Reverse-phase high-performance liquid chromatography of hydrophobic proteins and fragments thereof, *Anal. Biochem.* 131, 99–107, 1983; Heukeshoven, J. and Dernick, R., Characterization of a solvent system for separation of water-insoluble poliovirus proteins by reversed-phase high-performance liquid chromatography, *J. Chromatog.* 326, 91–101, 1985; De Caballos, M.L., Taylor, M.D., and Jenner, P., Isocratic reverse-phase HPLC separation and RIA used in the analysis of neuropeptides in brain tissue, *Neuropeptides* 20, 201–209, 1991; Poll, D.J. and Harding, D.R., Formic acid as a milder alternative to trifluoroacetic acid and phosphoric acid in two-dimensional peptide mapping, *J. Chromatog.* 469, 231–239, 1989; Klunk W.E. and Pettegrew, J.W., Alzheimer's beta-amyloid protein is covalently modified when dissolved in formic acid, *J. Neurochem.* 54, 2050–2056, 1990; Erdjument-Bromage, H., Lui, M., Lacomis, L. et al., Examination of the micro-tip reversed phase liquid chromatographic extraction of peptide pools for mass spectrometric analysis, *J. Chromatog. A* 826, 167–181, 1998; Duewel, H.S. and Honek, J.F., CNBr/formic acid reactions of methionine- and trifluoromethionine-containing lambda lysozyme: probing chemical and positional reactivity and formylation side reactions of mass spectrometry, *J. Protein Chem.* 17, 337–350, 1998; Kaiser, R. and Metzka, L., Enhancement of cyanogen bromide cleavage yields for methionyl-serine and methionyl-threonine peptide bonds, *Anal. Biochem.* 266, 1–8, 1999; Rodriguez, J.C., Wong, L., and Jennings, P.A., The solvent in CNBr cleavage reactions determines the fragmentation efficiency of ketosteroid isomerase fusion proteins used in the production of recombinant peptides, *Protein Expr. Purif.* 28, 224–231, 2003; Zu, Y., Zhao, C., Li, C., and Zhang, L., A rapid and sensitive LC-MS/MS method for determination of coenzyme Q10 in tobacco (*Nicotiana tabacum* L.) leaves, *J. Sep. Sci.* 29, 1607–1612, 2006; Kalovidouris, M., Michalea, S., Robola, N. et al., Ultra-performance liquid chromatography/tandem mass spectrometry method for the determination of lercaidipine in human plasma, *Rapid Commun. Mass Spectrom.*, 20, 2939–2946, 2006; Wang, P.G., Wei, J.S., Kim, G. et al., Validation and application of a high-performance liquid chromatography-tandem mass spectrometric method for simultaneous quantification of lopinavir and ritonavir in human plasma using semi-automated 96-well liquid–liquid chromatography, *J. Chromatog. A*, 1130, 302–307, 2006.

Glutaraldehyde   Pentanedial   100.12   Protein modification; tissue
                                                                                               fixation; sterilization agent
                                                                                               approved by regulatory
                                                                                               agencies; use with albumin
                                                                                               as surgical sealant.

Hopwood, D., Theoretical and practical aspects of glutaraldehyde fixation, *Histochem. J.*, 4, 267–303, 1972; Hassell, J. and Hand, A.R., Tissue fixation with diimidoesters as an alternative to aldehydes. I. Comparison of crosslinking and ultrastructure obtained with dimethylsubserimidate and glutaraldehyde, *J. Histochem. Cytochem.* 22, 223–229, 1974; Russell, A.D. and Hopwood, D., The biological uses and importance of glutaraldehyde, *Prog. Med. Chem.* 13, 271–301, 1976; Woodroof, E.A., Use of glutaraldehyde and formaldehyde to process tissue heart valves, *J. Bioeng.* 2, 1–9, 1978; Heumann, H.G., Microwave-stimulated glutaraldehyde and osmium tetroxide fixation of plant tissue: ultrastructural preservation in seconds, *Histochemistry* 97, 341–347, 1992; Abbott, L., The use and effects of glutaraldehyde: a review, *Occup. Health* 47, 238–239, 1995; Jayakrishnan, A. and Jameela, S.R., Glutaraldehyde as a fixative in bioprosthesis and drug delivery matrices, *Biomaterials* 17, 471–484, 1996; Tagliaferro, P., Tandler, C.J., Ramos, A.J. et al., Immunofluorescence and glutaraldehyde fixation. A new procedure base on the Schiff-quenching method, *J. Neurosci. Methods* 77, 191–197, 1997; Cohen, R.J., Beales, M.P., and McNeal, J.E., Prostate secretory granules in normal and neoplastic prostate glands: a diagnostic aid to needle biopsy, *Hum. Pathol.* 31, 1515–1519, 2000; Chae, H.J., Kim, E.Y., and In, M., Improved immobilization yields by addition of protecting agents in glutaraldehyde-induced immobilization of protease, *J. Biosci. Bioeng.* 89, 377–379, 2000; Nimni, M.E., Glutaraldehyde fixation revisited, *J. Long Term Eff. Med. Implants* 11, 151–161, 2001; Fujiwara, K., Tanabe, T., Yabuchi, M. et al., A monoclonal antibody against the glutaraldehyde-conjugated polyamine, putrescine: application to immunocytochemistry, *Histochem. Cell Biol.* 115, 471–477, 2001; Chao, H.H. and Torchiana, D.F., Bioglue: albumin/glutaraldehyde sealant in cardiac surgergy, *J. Card. Surg.* 18, 500–503, 2003; Migneault, I., Dartiguenave, C., Bertrand, M.J., and Waldron, K.C., Glutaraldehyde: behavior in aqueous solution, reaction with proteins, and application to enzyme crosslinking, *Biotechniques* 37, 790–796, 2004; Jearanaikoon, S. and Abraham-Peskir, J.V., An x-ray microscopy perspective on the effect of glutaraldehyde fixation on cells, *J. Microsc.* 218, 185–192, 2005; Buehler, P.W., Boykins, R.A., Jia, Y. et al., Structural and functional characterization of glutaraldehyde-polymerized bovine hemoglobin and its isolated fractions, *Anal. Chem.* 77, 3466–3478, 2005; Kim, S.S., Lim, S.H., Cho, S.W. et al., Tissue engineering of heart valves by recellularization of glutaraldehyde-fixed porcine values using bone marrow-derived cells, *Exp. Mol. Med.* 38, 273–283, 2006.

Glutathione   γ-GluCysGly   307.32   Reducing agent;
                                                                                                        intermediate in phase II
                                                                                                        detoxification of
                                                                                                        xenobiotics.

Glutathione

Arias, I.M. and Jakoby, W.B., *Glutathione, Metabolism and Function,* Raven Press, New York, 1976; Meister, A., *Glutamate, Glutamine, Glutathione, and Related Compounds,* Academic Press, Orlando, FL, 1985; Sies, H. and Ketterer, B., *Glutathione Conjugation: Mechanisms and Biological Significance,* Academic Press, London, UK, 1988; Tsumoto, K., Shinoki, K., Kondo, H. et al., Highly efficient recovery of functional single-chain Fv fragments from inclusion bodies overexpressed in *Escherichia coli* by controlled introduction of oxidizing reagent — application to a human single-chain Fv fragment, *J. Immunol. Methods* 219, 119–129, 1998; Jiang, X., Ookubo, Y., Fujii, I. et al., Expression of Fab fragment of catalytic antibody 6D9 in an *Escherichia coli in vitro* coupled transcription/translation system, *FEBS Lett.* 514, 290–294, 2002; Sun, X.X., Vinci, C., Makmura, L. et al., Formation of disulfide bond in p53 correlates with inhibition of DNA binding and tetramerization, *Antioxid. Redox Signal.* 5, 655–665, 2003; Sies, H. and Packer, L., Eds., *Glutathione Transferases and Gamma-Glutamyl Transpeptidases,* Elsevier, Amsterdam, 2005; Smith, A.D. and Dawson, H., Glutathione is required for efficient production of infectious picornativur virions, *Virology,* 353, 258–267, 2006.

Glycine   Aminoacetic Acid   75.07   Buffer component; protein-
                                                                                                        precipitating agent,
                                                                                                        excipient for
                                                                                                        pharmaceutical
                                                                                                        formulation.

Glycine

Sarquis, J.L. and Adams, E.T., Jr., The temperature-dependent self-association of beta-lactoglobulin C in glycine buffers, *Arch. Biochem. Biophys.* 163, 442–452, 1974; Poduslo, J.F., Glycoprotein molecular-weight estimation using sodium dodecyl suflate-pore gradient electrophoresis: comparison of Tris-glycine and Tris-borate-EDTA buffer systems, *Anal. Biochem.* 114, 131–139, 1981; Patton, W.F., Chung-Welch, N., Lopez, M.F. et al., Tris-tricine and Tris-borate buffer systems provide better estimates of human mesothelial cell intermediate filament protein molecular weights than the standard Tris-glycine system, *Anal. Biochem.* 197, 25–33, 1991; Trasltas, G. and Ford, C.H., Cell membrane antigen-antibody complex dissociation by the widely used glycine-HC1 method: an unreliable procedure for studying antibody internalization, *Immunol. Invest.* 22, 1–12, 1993; Nail, S.L., Jiang, S., Chongprasert, S., and Knopp, S.A., Fundamentals of freeze-drying, *Pharm. Biotechnol.* 14, 281–360, 2002; Pyne, A., Chatterjee, K., and Suryanarayanan, R., Solute crystallization in mannitol-glycine systems — implications on protein stabilization in freeze-dried formulations, *J. Pharm. Sci.* 92, 2272–2283, 2003; Hasui, K., Takatsuka, T., Sakamoto, R. et al., Double immunostaining with glycine treatment, *J. Histochem. Cytochem.* 51, 1169–1176, 2003; Hachmann, J.P. and Amshey, J.W., Models of protein modification in Tris-glycine and neutral pH Bis-Tris gels during electrophoresis: effect of gel pH, *Anal. Biochem.* 342, 237–245, 2005.

Glyoxal   Ethanedial   58.04   Modification of proteins and nucleic acids; model for glycation reaction; fluorescent derivates formed with tryptophan.

Glyoxal

Nakaya, K., Takenaka, O., Horinishi, H., and Shibata, K., Reactions of glyoxal with nucleic acids. Nucleotides and their component bases, *Biochim. Biophys. Acta* 161, 23–31, 1968; Canella, M. and Sodini, G., The reaction of horse-liver alcohol dehydrogenase with glyoxal, *Eur. J. Biochem.* 59, 119–125, 1975; Kai, M., Kojima, E., Okhura, Y., and Iwaski, M., High-performance liquid chromatography of N-terminal tryptophan-containing peptides with precolumn fluorescence derivatization with glyoxal, *J. Chromatog. A.* 653, 235–250, 1993; Murata-Kamiya, N., Kamiya, H., Kayi, H., and Kasai, H., Glyoxal, a major product of DNA oxidation, induces mutations at G:C sites on a shuttle vector plasmid replicated in mammalian cells, *Nucleic Acids Res.* 25, 1897–1902, 1997; Leng, F., Graves, D., and Chaires, J.B., Chemical crosslinking of ethidium to DNA by glyoxal, *Biochim. Biophys. Acta* 1442, 71–81, 1998; Thrornalley, P.J., Langborg, A., and Minhas, H.S., Formation of glyoxal, methylglyoxal, and 3-deoxyglucosone in the glycation of proteins by glucose, *Biochem. J.* 344, 109–116, 1999; Sady, C., Jiang, C.L., Chellan, P. et al., Maillard reactions by alpha-oxoaldehydes: detection of glyoxal-modified proteins, *Biochim. Biophys. Acta* 1481, 255–264, 2000; Olsen, R., Molander P., Ovrebo, S. et al., Reaction of glyoxal with 2′-deoxyguanosine, 2′-deoxyadenosine, 2′-deoxycytidine, cytidine, thymidine, and calf thymus DNA: identification of the DNA adducts, *Chem. Res. Toxicol.* 18, 730–739, 2005; Manini, P., La Pietra, P., Panzella, L. et al., Glyoxal formation by Fenton-induced degradation of carbohydrates and related compounds, *Carbohydr. Res.* 341, 1828–1833, 2006.

Guanidine   Aminomethana-midine   59.07   Chaotropic agents; guanidine hydrochloride use for study of protein denaturation; GTIC is considered to be more effective than GuCl; GTIC used for nucleic acid extraction.

Guanidine Hydrochloride (GuCl)   95.53

Guanidine Thiocyanate (GTIC)   118.16

Guanidine   Guanidinium

Hill, R.L., Schwartz, H.C., and Smith, E.L., The effect of urea and guanidine hydrochloride on activity and optical rotation of crystalline papain, *J. Biol. Chem.* 234, 572–576, 1959; Appella, E. and Markert, C.L., Dissociation of lactate dehydrogenase into subunits with guanidine hydrochloride, *Biochem. Biophys. Res. Commun.* 6, 171–176, 1961; von Hippel, P.H. and Wong, K.-Y., On the conformational stability of globular proteins. The effects of various electrolytes and nonelectrolytes on the thermal transition ribonuclease transition, *J. Biol. Chem.* 240, 3909–3923, 1965; Katz, S., Partial molar volume and conformational changes produced by the denaturation of albumin by guanidine hydrochloride, *Biochim. Biophys. Acta* 154, 468–477, 1968; Shortle, D., Guanidine hydrochloride denaturation studies of mutant forms of staphylococcal nuclease, *J. Cell Biochem.* 30, 281–289, 1986; Lippke, J.A., Strzempko, M.N., Rai, F.F. et al., Isolation of intact high-molecular-weight DNA by using guanidine isothiocyanate, *Appl. Environ. Microbiol.* 53, 2588–2589, 1987; Alberti, S. and Fornaro, M., Higher transfection efficiency of genomic DNA purified with a guanidinium thiocyanate–based procedure, *Nucleic Acids Res.* 18, 351–353, 1990; Shirley, B.A., Urea and guanidine hydrochloride denaturation curves, *Methods Mol. Biol.* 40, 177–190, 1995; Cota, E. and Clarke, J., Folding of beta-sandwich proteins: three-state transition of a fibronectin type III module, *Protein Sci.* 9, 112–120, 2000; Kok, T., Wati, S., Bayly, B. et al., Comparison of six nucleic acid extraction methods for detection of viral DNA or RNA sequences in four different non-serum specimen types, *J. Clin. Virol.* 16, 59–63, 2000; Salamanca, S., Villegas, V., Vendrell, J. et al., The unfolding pathway of leech carboxypeptidase inhibitor, *J. Biol. Chem.* 277, 17538–17543, 2002; Bhuyan, A.K., Protein stabilization by urea and guanidine hydrochloride, *Biochemistry* 41, 13386–13394, 2002; Jankowska, E., Wiczk, W., and Grzonka, Z., Thermal and guanidine hydrochloride-induced denaturation of human cystatin C, *Eur. Biophys. J.* 33, 454–461, 2004; Fuertes, M.A., Perez, J.M., and Alonso, C., Small amounts of urea and guanidine hydrochloride can be detected by a far-UV spectrophotometric method in dialyzed protein solutions, *J. Biochem. Biophys. Methods* 59, 209–216, 2004; Berlinck, R.G., Natural guanidine derivatives, *Nat. Prod. Rep.* 22, 516–550, 2005; Rashid, F., Sharma, S., and Bano, B., Comparison of guanidine hydrochloride (GdnHCl) and urea denaturation on inactivation and unfolding of human placental cystatin (HPC), *Biophys. J.* 91, 686–693, 2006; Nolan, R.L. and Teller, J.K., Diethylamine extraction of proteins and peptides isolated with a mono-phasic solution of phenol and guanidine isothiocyanate, *J. Biochem. Biophys. Methods* 68, 127–131, 2006.

HEPES

4-(2-hydroxyethyl)-1-piperizineethane-sulfonic Acid

A "Good" buffer; reagent purity has been an issue; metal ion binding must be considered; there are buffer-specific effects that are poorly understood; component of tissue-fixing technique.

HEPES; 4-(2-hydroxyethyl)-1-piperazineethansulfonic acid oo

Good, N.E., Winget, G.D., Winter, W. et al., Hydrogen ion buffers for biological research, *Biochemistry* 5, 467–477, 1966; Turner, L.V. and Manchester, K.L., Interference of HEPES with the Lowry method, *Science* 170, 649, 1970; Chirpich, T.P., The effect of different buffers on terminal deoxynucleotidyl transferase activity, *Biochim. Biophys. Acta* 518, 535–538, 1978; Tadolini, B., Iron autoxidation in MOPS and HEPES buffers, *Free Radic. Res. Commun.* 4, 149–160, 1987; Simpson, J.A., Cheeseman, K.H., Smith, S.E., and Dean, R.T., Free-radical generation by copper ions and hydrogen peroxide. Stimulation by HEPES buffer, *Biochem. J.* 254, 519–523, 1988; Abas, L. and Guppy, M., Acetate: a contaminant in HEPES buffer, *Anal. Biochem.* 229, 131–140, 1995; Schmidt, K., Pfeiffer, S., and Mayer, B., Reaction of peroxynitrite with HEPES or MOPS results in the formation of nitric oxide donors, *Free Radic. Biol. Med.* 24, 859–862, 1998; Wiedorn, K.H., Olert, J., Stacy, R.A. et al., HOPE — a new fixing technique enables preservation and extraction of high molecular weight DNA and RNA of >20 kb from paraffin-embedded tissues. HEPES-glutamic acid buffer mediated organic solvent protection effect, *Pathol. Res. Pract.* 198, 735–740, 2002; Fulop, L., Szigeti, G., Magyar, J. et al., Differences in electrophysiological and contractile properties of mammalian cardiac tissues bathed in bicarbonate — and HEPES-buffered solutions, *Acta Physiol. Scand.* 178, 11–18, 2003; Mash, H.E., Chin, Y.P., Sigg, L. et al., Complexation of copper by zwitterionic amino-sulfonic (Good) buffers, *Anal. Chem.* 75, 671–677, 2003; Sokolowska, M. and Bal, W., Cu(II) complexation by "non-coordinating" *N*-2-hydroxyethylpiperazine-*N'*-ethanesulfonic acid (HEPES buffer), *J. Inorg. Biochem.* 99, 1653–1660, 2005; Zhao, G. and Chasteen, N.D., Oxidation of Good's buffers by hydrogen peroxide, *Anal. Biochem.* 349, 262–267, 2006; Hartman, R.F. and Rose, S.D., Kinetics and mechanism of the addition of nucleophiles to alpha,beta-unsaturated thiol esters, *J. Org. Chem.* 71, 6342–6350, 2006.

Hydrazine N_2H_4 32.05   Reducing agent;
                                                                                            modification of aldehydes
                                                                                            and carbohydrates;
                                                                                            hydrazinolysis used for
                                                                                            release of carbohydrates
                                                                                            from protein; derivatives
                                                                                            such as dinitrophenyl-
                                                                                            hydrazine used for analysis
                                                                                            of carbonyl groups in
                                                                                            oxidized proteins;
                                                                                            detection of acetyl and
                                                                                            formyl groups in proteins.

Schmer, G. and Kreil, G., Micro method for detection of formyl and acetyl groups in proteins, *Anal. Biochem.* 29, 186–192, 1969; Gershoni, J.M., Bayer, E.A., and Wilchek, M., Blot analyses of glycoconjugates: enzyme-hydrazine — a novel reagent for the detection of aldehydes, *Anal. Biochem.* 146, 59–63, 1985; O'Neill, R.A., Enzymatic release of oligosaccharides from glycoproteins for chromatographic and electrophoretic analysis, *J. Chromatog. A* 720, 201–215, 1996; Routier, F.H., Hounsell, E.F., and Rudd, P.M., Quantitation of the oligosaccharides of human serum IgG from patients with rheumatoid arthritis: a critical evaluation of different methods, *J. Immunol. Methods* 213, 113–130, 1998; Robinson, C.E., Keshavarzian, A., Pasco, D.S. et al., Determination of protein carbonyl groups by immunoblotting, *Anal. Biochem.* 266, 48–57, 1999; Merry, A.H., Neville, D.C., Royle, L. et al., Recovery of intact 2-aminobenzamide-labeled *O*-glycans released from glycoproteins by hydrazinolysis, *Anal. Biochem.* 304, 91–99, 2002; Vinograd, E., Lindner, B., and Seltmann, G., Lipopolysaccharides from *Serratia maracescens* possess one or two 4-amino-4-deoxy-L-arabinopyranose 1-phosphate residues in the lipid A and D-*glycero*-D-*talo*-Oct-ulopyranosonic acid in the inner core region, *Chemistry* 12, 6692–6700, 2006.

Hydrogen Peroxide H_2O_2 34.02   Oxidizing agent;
                                                                                            bacteriocidal agent.

Hydroxylamine H_3NO 33.03

8-Hydroxyquinoline   8-quinolinol   145.16   Metal chelator.

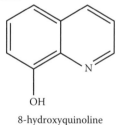

8-hydroxyquinoline

Imidazole   1,3-diazole   69.08   Buffer component.

Imidazole

Indole   2,3-benzyopyrrole   117.15

Indole

Indole-3-acetic Acid   Indoleacetic Acid;   175.19   Plant growth regulator.

Heteroauxin

Indoleacetic acid

Kawaguchi, M. and Syono, K., The excessive production of indole-3-acetic and its significance in studies of the biosynthesis of this regulator of plant growth and development, *Plant Cell Physiol.* 37, 1043–1048, 1996; Normanly, J. and Bartel, B., Redundancy as a way of life-IAA metabolism, *Curr. Opin. Plant Biol.* 2, 207–213, 1999; Leyser, O., Auxin signaling: the beginning, the middle, and the end, *Curr. Opin. Plant Biol.* 4, 382–386, 2001; Ljung, K., Hull, A.K., Kowalczyk, M. et al., Biosynthesis, conjugation, catabolism, and homeostasis of indole-3-acetic acid in *Arabidopsis thaliana, Plant Mol. Biol.* 49, 249–272, 2002; Kawano, T. Roles of the reactive oxygen species-generating peroxidase reactions in plant defense and growth induction, *Plant Cell Rep.* 21, 829–837, 2003; Aloni, R., Aloni, E., Langhans, M., and Ullrich, C.I., Role of cytokine and auxin in shaping root architecture: regulating vascular differentiation, laterial root initiation, root apical dominance, and root gravitropism*, Ann. Bot.* 97, 882–893, 2006.

Iodoacetamide   2-iodoacetamide   184.96   Alkylating agents that react with a variety of nucleophiles in proteins and nucleic acids. Reaction is more rapid than the bromo or chloro derivatives.

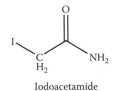

Iodoacetamide

Iodoacetic Acid   185.95

Iodoacetic acid

The amide is neutral and is not susceptible to either positive or negative influence from locally charged groups; iodoacetamide is frequently used to modify sulfydryl groups as part of reduction and carboxymethylation prior to structural analysis. Crestfield, A.M., Moore, S., and Stein, W.H., The preparation and enzymatic hydrolysis of reduced and *S*-carboxymethylated proteins, *J. Biol. Chem.* 238, 622–627, 1963; Watts, D.C., Rabin, B.R., and Crook, E.M., The reaction of iodoacetate and iodoacetamide with proteins as determined with a silver/silver iodide electrode, *Biochim. Biophys. Acta* 48, 380–388, 1961; Inagami, T., The alkylation of the active site of trypsin with iodoacetamide in the presence of alkylguanidines, *J. Biol. Chem.* 240, PC3453–PC3455, 1965; Fruchter, R.G. and Crestfield, A.M., The specific alkylation by iodoacetamide of histidine-12 in the active site of ribonuclease, *J. Biol. Chem.* 242, 5807–5812, 1967; Takahashi, K., The structure and function of ribonuclease T. X. Reactions of iodoacetate, iodoacetamide, and related alkylating reagents with ribonuclease T, *J. Biochem.* 68, 517–527, 1970; Whitney, P.L., Inhibition and modification of human carbonic anhydrase B with bromoacetate and iodoacetate, *Eur. J. Biochem.* 16, 126–135, 1970; Harada, M. and Irie, M., Alkylation of ribonuclease from *Aspirgillus saitoi* with iodoacetate and iodoacetamide, *J. Biochem.* 73, 705–716, 1973; Halasz, P. and Polgar, L., Effect of the immediate microenvironment on the reactivity of the essential SH group of papain, *Eur. J. Biochem.* 71, 571–575, 1976; Franzen, J.S., Ishman, P., and Feingold, D.S., Half-of-the-sites reactivity of bovine liver uridine diphosphoglucose dehydrogenase toward iodoacetate and iodoacetamide, *Biochemistry* 15, 5665–5671, 1976; David, M., Rasched, I.R., and Sund, H., Studies of glutamate dehydrogenase. Methionione-169: the preferentially carboxymethylated residue, *Eur. J. Biochem.* 74, 379–385, 1977; Ohgi, K., Watanabe, H., Emman, K. et al., Alkylation of a ribonuclease from *Streptomyces erthreus* with iodoacetate and iodoacetamide, *J. Biochem.* 90, 113–123, 1981; Dahl, K.H. and McKinley-McKee, J.S., Enzymatic catalysis in the affinity labeling of liver alcohol dehydrogenase with haloacids, *Eur. J. Biochem.*

118, 507–513, 1981; Syvertsen, C. and McKinley-McKee, J.S., Binding of ligands to the catalytic zinc ion in horse liver alcohol dehydrogenase, *Arch. Biochem. Biophys.* 228, 159–169, 1984; Communi, D. and Erneux, C., Identification of an active site cysteine residue in type Ins(1,4,5)P^{3}5-phosphatase by chemical modification and site-directed mutagenesis, *Biochem. J.* 320, 181–186, 1996; Sarkany, Z., Skern, T., and Polgar, L., Characterization of the active site thiol group of rhinovirus 21 proteinase, *FEBS Lett.* 481, 289–292, 2000; Lundblad, R.L., *Chemical Reagents for Protein Modification*, CRC Press, Boca Raton, FL, 2004.

2-Iminothiolane	Traut's Reagent	137.63	Introduction of sulfydryl group by modification of amino group; sulfydryl groups could then be oxidized to form cystine, which served as cleavable protein crosslink.
	(earlier as methyl-4-mercaptobutyrim-idate)		

Traut, R.R., Bollen, A., Sun, T.-T. et al., Methyl-4-mercaptobutyrimidate as a cleavable crosslinking reagent and its application to the *Escherichia coli* 30S ribosome, *Biochemistry* 12, 3266–3273, 1973; Schram, H.J. and Dulffer, T., The use of 2-iminothiolane as a protein crosslinking reagent, *Hoppe Seylers Z. Physiol.Chem.* 358, 137–139, 1977; Jue, R., Lambert, J.M., Pierce, L.R., and Traut, R.R., Addition of sulfyhryl groups *Escherichia coli* ribosomes by protein modification with 2-iminothiolane (methyl 4-mercaptobutyrimidate), *Biochemistry* 17, 5399–5406, 1978; Lambert, J.M., Jue, R., and Traut, R.R., Disulfide crosslinking of *Escherichia coli* ribosomal proteins with 2-iminothiolane (methyl 4-mercaptobutyrimidate): evidence that the crosslinked protein pairs are formed in the intact ribosomal subunit, *Biochemistry* 17, 5406–5416, 1978; Alagon, A.C. and King, T.P., Activation of polysaccharides with 2-iminothiolane and its use, *Biochemistry* 19, 4341–4345, 1980; Tolan, D.R. and Traut, R.R., Protein topography of the 40 S ribosomal subunit from rabbit reticulocytes shown by crosslinking with 2-iminothiolane, *J. Biol. Chem.* 256, 10129–10136, 1981; Boileau, G., Butler, P., Hershey, J.W., and Traut, R.R., Direct crosslinks between initiation factors 1, 2, and 3 and ribosomal proteins promoted by 2-iminothiolane, *Biochemistry* 22, 3162–3170, 1983; Kyriatsoulis, A., Maly, P., Greuer, B. et al., RNA-protein crosslinking in *Escherichia coli* ribosomal subunits: localization of sites on 16S RNA which are crosslinked to proteins S17 and S21 by treatment with 2-iminothiolane, *Nucleic Acids Res.* 14, 1171–1186, 1986; Uchiumi, T., Kikuchi, M., and Ogata, K., Crosslinking study on protein neighborhoods at the subunit interface of rat liver ribosomes with 2-iminothiolane, *J. Biol. Chem.* 261, 9663–9667, 1986; McCall, M.J., Diril, H., and Meares, C.F., Simplified method for conjugating macrocyclic bifunctional chelating agents to antibodies via 2-iminothiolane, *Bioconjug. Chem.* 1, 222–226, 1990; Tarentino, A.L., Phelan, A.W., and Plummer, T.H., Jr., 2-iminothiolane: a reagent for the introduction of sulphydryl groups into oligosaccharides derived from asaparagine-linked glycans, *Glycobiology* 3, 279–285, 1993; Singh, R., Kats, L., Blattler,

W.A., and Lambert, J.M., Formation of *N*-substituted 2-iminothiolanes when amino groups in proteins and peptides are modified by 2-iminothiolanes, *Anal. Biochem.* 236, 114–125, 1996; Hosono, M.N., Hosono, M., Mishra, A.K. et al., Rhenium-188-labeled anti-neural cell adhesion molecule antibodies with 2-iminothiolane modification for targeting small-cell lung cancer, *Ann. Nucl. Med.* 14, 173–179, 2000; Mokotoff, M., Mocarski, Y.M., Gentsch, B.L. et al., Caution in the use of 2-iminothiolane (Traut's reagent) as a crosslinking agent for peptides. The formation of *N*-peptidyl-2-iminothiolanes with bombesin (BN) antagonists (D-trp[6]-leu13-ψ[CH$_2$NH]-Phe[14]BN$_{6-14}$ and D-trp-gln-trp-NH$_2$, *J. Pept. Res.* 57, 383–389, 2001; Kuzuhara, A., Protein structural changes in keratin fibers induced by chemical modification using 2-iminothiolane hydrochloride: a Raman spectroscopic investigation, *Biopolymers* 79, 173–184, 2005.

Isatoic Anhydride

Isatoic anhydride

3,1-benzoxazine-2,4(1*H*)-dione   163.13   Fluorescent reagents for amines and sulfydryl groups; amine scavenger.

Gelb, M.H. and Abeles, R.H., Substituted isatoic anhydrides: selective inactivators of trypsinlike serine proteases, *J. Med. Chem.* 29, 585–589, 1986; Gravett, P.S., Viljoen, C.C., and Oosthuizen, M.M., Inactivation of arginine esterase E-1 of *Bitis gabonica* venom by irreversible inhibitors including a water-soluble carbodiimide, a chloromethyl ketone, and isatoic anhydride, *Int. J. Biochem.* 23, 1101–1110, 1991; Servillo, L., Balestrieri, C., Quagliuolo, L. et al., tRNA fluorescent labeling at 3′ end including an aminoacyl-tRNA-like behavior, *Eur. J. Biochem.* 213, 583–589, 1993; Churchich, J.E., Fluorescence properties of *o*-aminobenzoyl-labeled proteins, *Anal. Biochem.* 213, 229–233, 1993; Brown, A.D. and Powers, J.C., Rates of thrombin acylation and deacylation upon reaction with low molecular weight acylating agents, carbamylating agents, and carbonylating agents, *Bioorg. Med. Chem.* 3, 1091–1097, 1995; Matos, M.A., Miranda, M.S., Morais, V.M., and Liebman, J.F., Are isatin and isatoic anhydride antiaromatic and aromatic, respectively? A combined experimental and theoretic investigation, *Org. Biomol. Chem.* 1, 2566–2571, 2003; Matos, M.A., Miranda, M.S., Morais, V.M., and Liebman, J.F., The energetics of isomeric benzoxazine diones: isatoic anhydride revisited, *Org. Biomol. Chem.* 2, 1647–1650, 2004; Raturi, A., Vascratsis, P.O., Seslija, D. et al., A direct, continuous, sensitive assay for protein disulphide-isomerase based on fluorescence self-quenching, *Biochem. J.* 391, 351–357, 2005; Zhang, W., Lu, Y., and Nagashima, T., Plate-to-plate fluorous solid-phase extraction for solution-phase parallel synthesis, *J. Comb. Chem.* 7, 893–897, 2005.

Isoamyl Alcohol

Isoamyl alcohol

Isopentyl Alcohol; 3-methyl-1-butanol   88.15   Solvent.

Isopropanol

Isopropyl alcohol

2-propanol   60.10   Solvent; precipitation agent for purification of plasmid DNA; reagent in stability test for identification of abnormal hemoglobins.

Brosious, E.M., Morrison, B.Y., and Schmidt, R.M., Effects of hemoglobin F levels, KCN, and storage on the isopropanol precipitation test for unstable hemoglobins, *Am. J. Clin. Pathol.* 66, 878–882, 1976; Bensinger, T.A. and Beutler, E., Instability of the oxy form of sickle hemoglobin and of methemoglobin in isopropanol, *Am. J. Clin. Pathol.* 67, 180–183, 1977; Acree, W.E., Jr. and Bertrand, G.L., A cholesterol-isopropanol gel, *Nature* 269, 450, 1977; Naoum, P.C. Teixeira, U.A., de Abreu Machado, P.E., and Michelin, O.C., The denaturation of human oxyhemoglobin A, A2, and S by isopropanol/buffer method, *Rev. Bras. Pesqui. Med. Biol.* 11, 241–244, 1978; Ali, M.A., Quinlan, A., and Wong, S.C., Identification of hemoglobin E by the isopropanol solubility test, *Clin. Biochem.* 13, 146–148, 1980; Horer, O.L. and Enache, C., 2-propanol dependent RNA absorbances, *Virologie* 34, 257–272, 1983; De Vendittis, E., Masullo, M., and Bocchini, V., The elongation factor G carries a catalytic site for GTP hydrolysis, which is revealed by using 2-propanol in the absence of ribosomes, *J. Biol. Chem.* 261, 4445–4450, 1986; Wang, L., Hirayasu, K., Ishizawa, M., and Kobayashi, Y., Purification of genomic DNA from human whole blood by isopropanol-fractionation with concentrated NaI and SDS, *Nucleic Acids Res.* 22, 1774–1775, 1994; Dalhus, B. and Gorbitz, C.H., Glycyl-L-leucyl-L-tyrosine dehydrate 2-propanol solvate, *Acta Crystallogr. C* 52, 2087–2090, 1996; Freitas, S.S., Santos, J.A., and Prazeres, D.M., Optimization of isopropanol and ammonium sulfate precipitation steps in the purification of plasmid DNA, *Biotechnol. Prog.* 22, 1179–1186, 2006; Halano, B., Kubo, D., and Tagaya, H., Study on the reactivity of diarylmethane derivatives in supercritical alcohols media: reduction of diarylmethanols and diaryl ketones to diarylmethanes using supercritical 2-propanol, *Chem. Pharm. Bull.* 54, 1304–1307, 2006.

Isopropyl-β-D-thiogalactoside

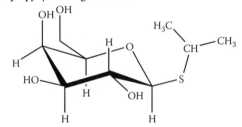

Isopropyl-β-D-thiogalactopyranoside; IPTG

IPTG, Isopropyl-β-D-thiogalactopyroanoside	238.3	"Gratuitous" inducer of the *lac* operon.

Cho, S., Scharpf, S., Franko, M., and Vermeulen, C.W., Effect of isopropyl-β-D-galactoside concentration on the level of *lac*-operon induction in steady state *Escherichia coli*, *Biochem. Biophys. Res. Commun.* 128, 1268–1273, 1985; Carlsson, U., Ferskgard, P.O., and Svensson, S.C., A simple and efficient synthesis of the induced IPTG made for inexpensive heterologous protein production using the *lac*-promoter, *Protein Eng.* 4, 1019–1020, 1991; Donovan, R.S., Robinson, C.W., and Glick, B.R., Review: optimizing inducer and culture conditions for expression of foreign proteins under control of the *lac* promoter, *J. Ind. Microbiol.* 16, 145–154, 1996; Hansen, L.H., Knudsen, S., and Sorensen, S.J., The effect of the lacy gene on the induction of IPTG-inducible promoters, studied in *Escherichia coli* and *Pseudomonas fluorescens, Curr. Microbiol.* 36, 341–347, 1998; Teich, A., Lin, H.Y., Andersson, L. et al., Amplification of ColE1 related plasmids in recombinant cultures of *Escherichia coli* after IPTG induction, *J. Biotechnol.* 64, 197–210, 1998; Ren, A. and Schaefer, T.S., Isopropyl-β-D-thiogalactoside (IPTG)-inducible tyrosine phosphorylation of proteins in *E. coli, Biotechniques* 31, 1254–1258, 2001; Ko, K.S., Kruse, J., and Pohl, N.L., Synthesis of isobutryl-C-galactoside (IBCG) as an isopropylthiogalactoside (IPTG) substitute for increased induction of protein expression, *Org. Lett.* 5, 1781–1783, 2003; Intasai, N., Arooncharus, P., Kasinrerk, W., and Tayapiwatana, C., Construction of high-density display of CD147 ectodomain on VCSM13 phage via gpVIII: effects of temperature, IPTG, and helper phage infection-period, *Protein Expr. Purif.* 32, 323–331, 2003; Faulkner, E., Barrett, M., Okor, S. et al., Use of fed-batch cultivation for achieving high cell densities for the pilot-scale production of a recombinant protein (phenylalanine dehydrogenase) in *Escherichia coli, Biotechnol. Prog.* 22, 889–897, 2006; Gardete, S., de Laencastre, H., and Tomasz, A., A link in transcription between the native pbpG and the acquired mecA gene in a strain of *Staphylococcus aureus, Microbiology* 152, 2549–2558, 2006; Hewitt, C.J., Onyeaka, H., Lewis, G. et al., A comparison of high cell density fed-batch fermentations involving both induced and noninduced recombinant *Escherichia coli* under well-mixed small-scale and simulated poorly mixed large-scale conditions, *Biotechnol. Bioeng.*, in press, 2006; Picaud, S., Olsson, M.E., and Brodelius, P.E., Improved conditions for production of recombinant plant sesquiterpene synthases in *Escherichia coli, Protein Expr. Purif.*, in press, 2006.

Maleic Anhydride

HO

2,5-furandione   98.06   Modification of amino groups in proteins. The dimethyl derivative (dimethylmaleic anhydride) is used for ribosome dissociation; monomer for polymer.

Maleic anhydride

+ NH₂   →(pH > 8.0)→   HN   →(pH < 3)→   NH₃⁺

Lysine   Lysine

Giese, R.W. and Vallee, B.L., Metallocenes. A novel class of reagents for protein modification. I. Maleic anhydride-iron tetracarbonyl, *J. Am. Chem. Soc.* 94, 6199–6200, 1972; Cantrell, M. and Craven, G.R., Chemical inactivation of *Escherichia coli* 30 S ribosomes with maleic anhydride: identification of the proteins involved in polyuridylic acid binding, *J. Mol. Biol.* 115, 389–402, 1977; Jordano, J., Montero, F., and Palacian, E., Relaxation of chromatin structure upon removal of histones H2A and H2B, *FEBS Lett.* 172, 70–74, 1984; Jordano, J., Montero, F., and Palacian, E., Rearrangement of nucleosomal components by modification of histone amino groups. Structural role of lysine residues, *Biochemistry* 23, 4280–4284, 1984; Palacian, E., Gonzalez, P.J., Pineiro, M., and Hernandez, F., Dicarboxylic acid anhydrides as dissociating agents of protein-containing structures, *Mol. Cell. Biochem.* 97, 101–111, 1990; Paetzel, M., Strynadka, N.C., Tschantz, W.R. et al., Use of site-directed chemical modification to study an essential lysine in *Escherichia coli* leader peptidase, *J. Biol. Chem.* 272, 9994–10003, 1997; Wink, M.R., Buffon, A., Bonan, C.D. et al., Effect of protein-modifying reagents on ecto-apyrase from rat brain, *Int. J. Biochem. Cell Biol.* 32, 105–113, 2000.

2-Mercaptoethanol

H₂
C
HS   C   OH
H₂

2-Mercaptoethanol

β-mercaptoethanol   78.13   Reducing agent; used frequently in the reduction and alkylation of proteins for structural analysis and for preservation of oxidation-sensitive enzymes.

Geren, C.R., Olomon, C.M., Jones, T.T., and Ebner, D.E., 2-mercaptoethanol as a substrate for liver alcohol dehydrogenase, *Arch. Biochem. Biophys.* 179, 415–419, 1977; Opitz, H.G., Lemke, H, and Hewlett, G., Activation of T-cells by a macrophage or 2-mercaptoethanol-activated serum factor is essential for induction of a primary immune response to heterologous red cells *in vitro*, *Immunol. Rev.* 40, 53–77, 1978; Burger, M., An absolute requirement for 2-mercaptoethanol in the *in vitro* primary immune response in the absence of serum, *Immunology* 37, 669–671, 1979; Nealon, D.A., Pettit, S.M., and Henderson, A.R., Diluent pH and the stability of the thiol group in monothioglycerol, *N*-acetyl-L-cysteine, and 2-mercaptoethanol, *Clin. Chem.* 27, 505–506, 1981; Dahl, K.H. and McKinley-McKee, J.S., Enzymatic catalysis in the affinity labeling of liver alcohol dehydrogenase with haloacids, *Eur. J. Biochem.* 118, 507–513, 1981; Righetti, P.G., Tudor, G., and Glanazza, E., Effect of 2-mercaptoethanol on pH gradients in isoelectric focusing, *J. Biochem. Biophys. Methods* 6, 219–227, 1982; Soderberg, L.S. and Yeh, N.H., T-cells and the anti-trinitrophenyl antibody response to fetal calf serum and 2-mercaptoethanol, *Proc. Soc. Exp. Biol. Med.* 174, 107–113, 1983; Ochs, D., Protein contaminants of sodium dodecyl sulfate-polyacrylamide gels, *Anal. Biochem.* 135, 470–474, 1983; Schaefer, W.H., Harris, T.M., and Guengerich, F.P., Reaction of the model thiol 2-mercaptoethanol and glutathione with methylvinylmaleimide, a Michael acceptor with extended conjugation, *Arch. Biochem. Biophys.* 257, 186–193, 1987; Obiri, N. and Pruett, S.B., The role of thiols in lymphocyte responses: effect of 2-mercaptoethanol on interleukin 2 production, *Immunobiology* 176, 440–449, 1988; Gourgerot-Pocidalo, M.A., Fay, M., Roche, Y., and Chollet-Martin, S., Mechanisms by which oxidative injury inhibits the proliferative response of human lymphocytes to PHA. Effect of the thiol compound 2-mercaptoethanol, *Immunology* 64, 281–288, 1988; Fong, T.C. and Makinodan, T., Preferential enhancement by 2-mercaptoethanol of IL-2 responsiveness of T blast cells from old over young mice is associated with potentiated protein kinase C translocation, *Immunol. Lett.* 20, 149–154, 1989; De Graan, P.N., Moritz, A., de Wit, M., and Gispen, W.H., Purification of B-50 by 2-mercaptoethanol extraction from rat brain synaptosomal plasma membranes, *Neurochem. Res.* 18, 875–881, 1993; Carrithers, S.L. and Hoffman, J.L., Sequential methylation of 2-mercaptoethanol to the dimethyl sulfonium ion, 2-(dimethylthio)ethanol, *in vivo* and *in vitro*, *Biochem. Pharmacol.* 48, 1017–1024, 1994; Paul-Pretzer, K. and Parness, J., Elimination of keratin contaminant from 2-mercaptoethanol, *Anal. Biochem.* 289, 98–99, 2001; Adebiyi, A.P., Jin, D.H, Ogawa, T., and Muramoto,

K., Acid hydrolysis of protein in a microcapillary tube for the recovery of tryptophan, *Biosci. Biotechnol. Biochem.* 69, 255–257, 2005; Adams, B., Lowpetch, K., Throndycroft, F. et al., Stereochemistry of reactions of the inhibitor/substrates L- and D-β-chloroalanine with β-mercaptoethanol catalyzed by L-aspartate aminotransferase and D-amino acid amino-transferase, respectively, *Org. Biomol. Chem.* 3, 3357–3364, 2005; Layeyre, M., Leprince, J., Massonneau, M. et al., Aryldithioethyloxycarbonyl (Ardec): a new family of amine-protecting groups removable under mild reducing conditions and their applications to peptide synthesis, *Chemistry* 12, 3655–3671, 2006; Okun, I., Malarchuk, S., Dubrovskaya, E. et al., Screening for caspace-3 inhibitors: effect of a reducing agent on the identified hit chemotypes, *J. Biomol. Screen.* 11, 694–703, 2006; Aminian, M., Sivam, S., Lee, C.W. et al., Expression and purification of a trivalent pertussis toxin-diphtheria toxin-tetanus toxin fusion protein in *Escherichia coli*, *Protein Expr. Purif.* 51, 170–178, 2006.

(3-Mercaptopropyl)trimethoxysilane

3-(trimethoxysilyl)-1-propanethiol   196.34   Introduction of reactive sulfydryl onto glass (silane) surface.

(3-mercaptopropyl)-trimethoxysilane

Jung, S.K. and Wilson, G.S., Polymeric mercaptosilane-modified platinum electrodes for elimination of interferants in glucose biosensors, *Anal. Chem.* 68, 591–596, 1996; Mansur, H.S., Lobato, Z.P., Orefice, R.L. et al., Surface functionalization of porous glass networks: effects on bovine serum albumin and porcine insulin immobilization, *Biomacromolecules* 1, 479–497, 2000; Kumar, A., Larsson, O., Parodi, D., and Liang, Z., Silanized nucleic acids: a general platform for DNA immobilization, *Nucleic Acids Res.* 28, E71, 2000; Zhang, F., Kang, E.T., Neoh, K.G. et al., Surface modification of stainless steel by grafting of poly(ethylene glycol) for reduction in protein adsorption, *Biomaterials* 22, 1541–1548, 2001; Jia, J., Wang, B., Wu, A. et al., A method to construct a third-generation horseradish peroxidase biosensor: self-assembling gold nanoparticles to three-dimensional sol-gel network, *Anal. Chem.* 74, 2217–2223, 2002; Abdelghani-Jacquin, C., Abdelghani, A., Chmel, G. et al., Decorated surfaces by biofunctionalized gold beads: application to cell adhesion studies, *Eur. Biophys. J.* 31, 102–110, 2002; Ganesan, V. and Walcarius, A., Surfactant templated sulfonic acid functionalized silica microspheres as new efficient ion exchangers and electrode modifiers, *Langmuir* 20, 3632–3640, 2004; Crudden, C.M., Sateesh, M., and Lewis, R., Mercaptopropyl-modified mesoporous silica: a remarkable support for the preparation of a reusable, heterogeneous palladium catalyst for coupling to reactions, *J. Am. Chem. Soc.* 127, 10045–10050, 2005; Yang, L., Guihen, E., and Glennon, J.D., Alkylthiol gold nanoparticles in sol-gel-based open tabular capillary electrochromatography, *J. Sep. Sci.* 28, 757–766, 2005.

MES

1-morpholineethane-sulfonic Acid; 2-(4-morpholino) Ethane Sulfonate   198.2   A "Good" buffer.

4-Morpholineethanesulfonic acid, MES

Good, N.E., Winget, G.D., Winter, W. et al., Hydrogen ion buffers for biological research, *Biochemistry* 5, 467–477, 1966; Bugbee, B.G. and Salisbury, F.B., An evaluation of MES (2[*N*-morpholino]ethanesulfonic acid) and Amberlite 1RC-50 as pH buffers for nutrient growth studies, *J. Plant Nutr.* 8, 567–583, 1985; Kaushal, V. and Barnes, L.D., Effect of zwitterionic buffers on measurement of small masses of protein with bicinchoninic acid, *Anal. Biochem.* 157, 291–294, 1986; Grady, J.K., Chasteen, N.D., and Harris, D.C., Radicals from "Good's" buffers, *Anal. Biochem.* 173, 111–115, 1988; Le Hir, M., Impurity in buffer substances mimics the effect of ATP on soluble 5′-nucleotidase, *Enzyme* 45, 194–199, 1991; Pedrotti, B., Soffientini, A., and Islam, K., Sulphonate buffers affect the recovery of microtubule-associated proteins MAP1 and MAP2: evidence that MAP1A promotes microtubule assembly, *Cell Motil. Cytoskeleton* 25, 234–242, 1993; Vasseur, M.,

Frangne, R., and Alvarado, F., Buffer-dependent pH sensitivity of the fluorescent chloride-indicator dye SPQ, *Am. J. Physiol.* 264, C27–C31, 1993; Frick, J. and Mitchell, C.A., Stabilization of pH in solid-matrix hydroponic systems, *HortScience* 28, 981–984, 1993; Yu, Q., Kandegedara, A., Xu, Y., and Rorabacher, D.B., Avoiding interferences from Good's buffers: a contiguous series of noncomplexing tertiary amine buffers covering the entire range of pH 3–11, *Anal. Biochem.* 253, 50–56, 1997; Gelfi, C., Vigano, A., Curcio, M. et al., Single-strand conformation polymorphism analysis by capillary zone electrophoresis in neutral pH buffer, *Electrophoresis* 21, 785–791, 2000; Walsh, M.K., Wang, X., and Weimer, B.C., Optimizing the immobilization of single-stranded DNA onto glass beads, *J. Biochem. Biophys. Methods* 47, 221–231, 2001; Hosse, M. and Wilkinson, K.J., Determination of electrophoretic mobilities and hydrodynamic radii of three humic substances as a function of pH and ionic strength, *Environ. Sci. Technol.* 35, 4301–4306, 2001; Mash, H.E., Chin, Y.P., Sigg, L. et al., Complexation of copper by zwitterionic aminosulfonic (good) buffers, *Anal. Chem.* 75, 671–677, 2003; Ozkara, S., Akgol, S., Canak, Y., and Denizli, A., A novel magnetic adsorbent for immunoglobulin-g purification in a magnetically stabilized fluidized bed, *Biotechnol. Prog.* 20, 1169–1175, 2004; Hachmann, J.P. and Amshey, J.W., Models of protein modification in Tris-glycine and neutral pH Bis-Tris gels during electrophoresis: effect of pH, *Anal. Biochem.* 342, 237–345, 2005; Krajewska, B. and Ciurli, S., Jack bean (*Canavalia ensiformis*) urease. Probing acid-base groups of the active site by pH variation, *Plant Physiol. Biochem.* 43, 651–658, 2005; Zhao, G. and Chasteen, N.D., Oxidation of Good's buffers by hydrogen peroxide, *Anal. Biochem.* 349, 262–267, 2006.

Methanesulfonic Acid   96.11   Protein hydrolysis for amino acid analysis; deprotection during peptide synthesis; hydrolysis of protein substituents such as fatty acids.

Methylsulfonic acid   Tosylsulfonic acid
methanesulfonic acid

Simpson, R.J., Neuberger, M.R., and Liu, T.Y., Complete amino acid analysis of proteins from a single hydrolyzate, *J. Biol. Chem.* 251, 1936–1940, 1976; Kubota, M., Hirayama, T., Nagase, O., and Yajima, H., Synthesis of two peptides corresponding to an alpha-endophin and gamma-endorphin by the methanesulfonic acid deprotecting procedures, *Chem. Pharm. Bull.* 27, 1050–1054, 1979; Yajima, H., Akaji, K., Saito, H. et al., Studies on peptides. LXXXII. Synthesis of [4-Gln]-neurotensin by the methanesulfonic acid deprotecting procedure, *Chem. Pharm. Bull.* 27, 2238–2242, 1979; Sakuri, J. and Nagahama, M. Tryptophan content of *Clostridium perfringens* epsilon toxin, *Infect. Immun.* 47, 260–263, 1985; Malmer, M.F. and Schroeder, L.A., Amino acid analysis by high-performance liquid chromatography with methanesulfonic acid hydrolysis and 9-fluorenylmethyl-chloroformate derivatization, *J. Chromatog.* 514, 227–239, 1990; Weiss, M., Manneberg, M., Juranville, J.F. et al., Effect of the hydrolysis method on the determination of the amino acid composition of proteins, *J. Chromatog. A* 795, 263–275, 1998; Okimura, K., Ohki, K., Nagai, S., and Sakura, N., HPLC analysis of fatty acyl-glycine in the aqueous methanesulfonic acid hydrolysates of N-terminally fatty acylated peptides, *Biol. Pharm. Bull.* 26, 1166–1169, 2003; Wrobel, K., Kannamkumarath, S.S., Wrobel, K., and Caruso, J.A., Hydrolysis of proteins with methanesulfonic acid for improved HPLC-ICP-MS determination of seleno-methionine in yeast and nuts, *Anal. BioAnal. Chem.* 375, 133–138, 2003.

Methanol   Methyl Alcohol   32.04   Solvent.

Methylethyl Ketone (MEK)   2-butanal; 2-butanone   72.11   Solvent; with acid for cleavage of heme moiety of hemeproteins for preparation of apoproteins.

Methylethylketone

Teale, F.W., Cleavage of haem-protein link by acid methylethylketone, *Biochim. Biophys. Acta* 35, 543, 1959; Tran, C.D. and Darwent, J.R., Characterization of tetrapyridylporphyrinatozinc (II) apomyoglobin complexes as a potential photosynthetic model, *J. Chem. Soc. Faraday Trans. II*, 82, 2315–2322, 1986.

Methylglyoxal

Pyruvaldehyde; 2- 72.06
oxo-propanal

Derived from oxidative modification of triose phosphate during glucose metabolism; model for glycation of proteins; reacts with amino groups in proteins and nucleic acids; involved in advanced glycation endproducts.

Lysine

Methylglyoxal

N^6-carboxymethyllysine

Methylglyoxal-lysine dimer

Szabo, G., Kertesz, J.C., and Laki, K., Interaction of methylglyoxal with poly-L-lysine, *Biomaterials* 1, 27–29, 1980; McLaughlin, J.A., Pethig, R., and Szent-Gyorgyi, A., Spectroscopic studies of the protein-methylglyoxal adduct, *Proc. Natl. Acad. Sci. USA* 77, 949–951, 1980; Cooper, R.A., Metabolism of methylglyoxal in microorganisms, *Annu. Rev. Microbiol.* 38, 49–68, 1984; Richard, J.P., Mechanism for the formation of methylglyoxal from triosephosphates, *Biochem. Soc. Trans.* 21, 549–553, 1993; Riley, M.L. and Harding, J.J., The reaction of methylglyoxal with human and bovine lens proteins, *Biochim. Biophys. Acta* 1270, 36–43, 1995; Thornalley, P.J., Pharmacology of methylglyoxal: formation, modification of proteins and nucleic acids, and enzymatic detoxification — a role in pathogenesis and antiproliferative chemotherapy, *Gen. Pharmacol.* 27, 565–573, 1996; Nagaraj, R.H., Shipanova, I.N., and Faust, F.M., Protein crosslinking by the Maillard reaction. Isolation, characterization, and *in vivo* detection of a lysine–lysine crosslink derived from methylglyoxal, *J. Biol. Chem.* 271, 19338–19345, 1996; Shipanova, I.N., Glomb, M.A., and Nagaraj, R.H., Protein modification by methylglyoxal: chemical nature and synthetic mechanism of a major fluorescent adduct, *Arch. Biochem. Biophys.* 344, 29–34, 1997; Uchida, K., Khor, O.T., Oya, T. et al., Protein modification by a Maillard reaction intermediate methylglyoxal. Immunochemical detection of fluorescent 5-methylimidazolone derivatives *in vivo*, *FEBS Lett.* 410, 313–318, 1997; Degenhardt, T.P., Thorpe, S.R., and Baynes, J.W., Chemical modification of proteins by methylglyoxal, *Cell. Mol. Biol.* 44, 1139–1145, 1998; Izaguirre, G., Kikonyogo, A., and Pietruszko, R., Methylglyoxal as substrate and inhibitor of human aldehyde dehydrogenase: comparison of kinetic properties among the three isozymes, *Comp. Biochem. Physiol. B Biochem. Mol. Biol.* 119, 747–754, 1998; Lederer, M.O. and Klaiber, R.G., Crosslinking of proteins by Maillard processes: characterization and detection of lysine–arginine crosslinks derived from glyoxal and methylglyoxal, *Bioorg. Med. Chem.* 7, 2499–2507, 1999; Kalapos, M.P., Methylglyoxal in living organisms: chemistry, biochemistry, toxicology, and biological implications, *Toxicol. Lett.* 110, 145–175, 1999; Thornalley, P.J., Landborg, A., and Minhas, H.S., Formation of glyoxal, methylglyoxal, and 3-deoxyglucose in the glycation of proteins by glucose, *Biochem. J.* 344, 109–116, 1999; Nagai, R., Araki, T., Hayashi, C.M. et al., Identification of *N*-epsilon-(carboxyethyl)lysine, one of the methylglyoxal-derived AGE structures, in glucose-modified protein: mechanism for protein modification by reactive aldehydes, *J. Chromatog. B Analyt. Technol. Biomed. Life Sci.*788, 75–84, 2003.

Methyl Methane-thiosulfonate (MMTS)

S-methyl
Methanethiosul-
fonate

126.2

Modification of sulfhydryl groups.

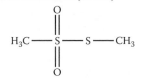

S-methyl methanethiosulfonate

Smith, D.J., Maggio, E.T., and Kenyon, G.L., Simple alkanethiol groups for temporary sulfhydryl groups of enzymes, *Biochemistry* 14, 766–771, 1975; Nishimura, J.S., Kenyon, G.L., and Smith, D.J., Reversible modification of the sulfhydryl groups of *Escherichia coli* succinic thiokinase with methanethiolating reagents, 5,5'-dithio-bis(2-nitrobenzoic acid), *p*-hydroxymercuribenzoate, and ethylmercurithiosalicylate, *Arch. Biochem. Biophys.* 170, 407–430, 1977; Bloxham, D.P., The chemical reactivity of the histidine-195 residue in lactate dehydrogenase thiomethylated at the cysteine-165 residue, *Biochem. J.* 193, 93–97, 1981; Gavilanes, F., Peterson, D., and Schirch, L., Methyl methanethiosulfate as an active site probe of serine hydroxymethyltransferase, *J. Biol. Chem.* 257, 11431–11436, 1982; Daly, T.J., Olson, J.S., and Matthews, K.S., Formation of mixed disulfide adducts as cysteine-281 of the lactose repressor protein affects operator- and inducer-binding parameters, *Biochemistry* 25, 5468–5474, 1986; Salam, W.H. and Bloxham, D.P., Identification of subsidiary catalytic groups at the active site of β-ketoacyl-CoA thiolase by covalent modification of the protein, *Biochim. Biophys. Acta* 873, 321–330, 1986; Stancato, L.F., Hutchison, K.A., Chakraborti, P.K. et al., Differential effects of the reversible thiol-reactive agents arsenite and methyl methanethiosulfonate on steroid binding by the glucocorticoid receptor, *Biochemistry* 32, 3739–3736, 1993; Hou, L.X. and Vollmer, S., The activity of *S*-thiolated modified creatine kinase is due to the regeneration of free thiol at the active site, *Biochim. Biophys. Acta* 1205, 83–88, 1994; Jensen, P.E., Shanbhag, V.P., and Stigbrand, T., Methanethiolation of the liberated cysteine residues of human α-2-macroglobulin treated with methylamine generates a derivative with similar functional characteristics as native β-2-macroglobulin, *Eur. J. Biochem.* 227, 612–616, 1995; Trimboli, A.J., Quinn, G.B., Smith, E.T., and Barber, M.J., Thiol modification and site-directed mutagenesis of the flavin domain of spinach NADH: nitrate reductase, *Arch. Biochem. Biophys.* 331, 117–126, 1996; Quinn, K.E. and Ehrlich, B.E., Methanethiosulfonate derivatives inhibits current through the rynodine receptor/channel, *J. Gen. Physiol.* 109, 225–264, 1997; Hashimoto, M., Majima, E., Hatanaka, T. et al., Irreversible extrusion of the first loop facing the matrix of the bovine heart mitochondrial ADP/ATP carrier by labeling the Cys(56) residue with the SH-reagent methyl methanethiosulfonate, *J. Biochem.* 127, 443–449, 2000; Spelta, V., Jiang, L.H., Bailey, R.J. et al., Interaction between cysteines introduced into each transmembrane domain of the rat P2X2 receptor, *Br. J. Pharmacol.* 138, 131–136, 2003; Britto, P.J., Knipling, L., McPhie, P., and Wolff, J., Thiol-disulphide interchange in tubulin: kinetics and the effect on polymerization, *Biochem. J.* 389, 549–558, 2005; Miller, C.M., Szegedi, S.S., and Garrow, T.A., Conformation-dependent inactivation of human betaine-homocysteine *S*-methyltransferase by hydrogen peroxide *in vitro*, *Biochem. J.* 392, 443–448, 2005.

N-Methylpyrrolidone

1-methyl-2-pyrrolidone   99.13   Polar solvent; transdermal transport of drugs.

N-Methylpyrrolidone

Barry, B.W. and Bennett, S.L., Effect of penetration enhancers on the permeation of mannitol, hydrocortisone, and progesterone through human skin, *J. Pharm. Pharmacol.* 39, 535–546, 1987; Forest, M. and Fournier, A., BOP reagent for the coupling of pGlu and Boc-His(Tos) in solid phase peptide synthesis, *Int. J. Pept. Protein Res.* 35, 89–94, 1990; Sasaki, H., Kojima, M., Nakamura, J., and Shibasaki, J., Enhancing effect of combining two pyrrolidone vehicles on transdermal drug delivery, *J. Pharm. Pharmacol.* 42, 196–199, 1990; Uch, A.S., Hesse, U., and Dressman, J.B., Use of 1-methyl-pyrrolidone as a solubilizing agent for determining the uptake of poorly soluble drugs, *Pharm. Res.* 16, 968–971, 1999; Zhao, F. Bhanage, B.M., Shirai, M., and Arai, M., Heck reactions of iodobenzene and methyl acrylate with conventional supported palladium catalysts in the presence of organic and/or inorganic bases without ligands, *Chemistry* 6, 843–848, 2000; Lee, P.J., Langer, R., and Shastri, V.P., Role of *n*-methyl pyrrolidone in the enhancement of aqueous phase transdermal transport, *J. Pharm. Sci.* 94, 912–917, 2005; Tae, G., Kornfield, J.A., and Hubbell, J.A., Sustained release of human growth hormone from *in situ* forming hydrogels using self-assembly of fluoroalkyl-ended poly(ethylene glycol), *Biomaterials* 26, 5259–5266, 2005; Babu, R.J. and Pandit, J.K., Effect of penetration enhancers on the transdermal delivery of bupranolol through rat skin, *Drug Deliv.* 12, 165–169, 2005; Luan, X. and Bodmeier, R., *In situ* forming microparticle system for controlled delivery of leupolide acetate: influence of the formulation and processing parameters, *Eur. J. Pharm. Sci.* 27, 143–149, 2006; Lee, P.J., Ahmad, N., Langer, R. et al., Evaluation of chemical enhancers in the transdermal delivery of lidocaine, *Int. J. Pharm.* 308, 33–39, 2006; Ruble, G.R., Giardino, O.X., Fossceco, S.L. et al., *J. Am. Assoc. Lab. Anim. Sci.* 45, 25–29, 2006.

MOPS

Betaine
1-Carboxy-*N, N, N*-trimethylamino inner salt

3-(*N*-morpholino)   209.3   A "Good" buffer.
Propanesulfonic
Acid;
4-morpholine-
propanesulfonic
Acid

3-(1-Pyridino)-1-[propanesulfonate

MOPS
3-(*N*-morpholino)propanesulfonate

Good, N.E., Winget, G.D., Winter, W. et al., Hydrogen ion buffers for biological research, *Biochemistry* 5, 467–477, 1966; Altura, B.M., Altura, B.M., Carella, A., and Altura, B.T., Adverse effects of Tris, HEPES, and MOPS buffers on contractile responses of arterial and venous smooth muscle induced by prostaglandins, *Prostaglandins Med.* 5, 123–130, 1980; Tadolini, B., Iron autoxidation in MOPS and HEPES buffers, *Free Radic. Res. Commun.* 4, 149–160, 1987; Tadolini, B. and Sechi, A.M., Iron oxidation in MOPS buffer. Effect of phosphorus-containing compounds, *Free Radic. Res. Commun.* 4, 161–172, 1987; Tadolini, B., Iron oxidation in MOPS buffer. Effect of EDTA, hydrogen peroxide, and FeCl$_3$, *Free Radic. Res. Commun.* 4, 172–182, 1987; Ishihara, H. and Welsh, M.J., Block by MOPS reveals a conformation change in the CFTR pore produced by ATP hydrolysis, *Am. J. Physiol.* 273, C1278–C1289, 1997; Schmidt, K., Pfeiffer, S., and Meyer, B., Reaction of peroxynitrite with HEPES or MOPS results in the formation of nitric oxide donors, *Free Radic. Biol. Med.* 24, 859–862, 1998; Hodges, G.R. and Ingold, K.U., Superoxide, amine buffers, and tetranitromethane: a novel free radical chain reaction, *Free Radic. Res.* 33, 547–550, 2000; Corona-Izquierdo, F.P. and Membrillo-Hernandez, J., Biofilm formation in *Escherichia coli* is affected by 3-(*N*-morpholino)propane sulfonate (MOPS), *Res. Microbiol.* 153, 181–185, 2002; Mash, H.E., Chin, Y.P., Sigg, L. et al., Complexation of copper by zwitterionic aminosulfonic (Good) buffers, *Anal. Chem.* 75, 671–677, 2003; Denizli, A., Alkan, M., Garipcan, B. et al., Novel metal-chelate affinity adsorbent for purification of immunoglobulin-G from human plasma, *J. Chromatog. B Analyt. Technol. Biomed. Life Sci.* 795, 93–103, 2003; Emir, S., Say, R., Yavuz, H., and Denizli, A., A new metal chelate affinity adsorbent for cytochrome C, *Biotechnol. Prog.* 20, 223–228, 2004; Cvetkovic, A., Zomerdijk, M., Straathof, A.J. et al., Adsorption of fluorescein by protein crystals, *Biotechnol. Bioeng.* 87, 658–668, 2004; Zhao, G. and Chasteen, J.D., Oxidation of Good's buffers by hydrogen peroxide, *Anal. Biochem.* 349, 262–267, 2006; Vrakas, D., Giaginis, C., and Tsantili-Kakoulidou, A., Different retention behavior of structurally diverse basic and neutral drugs in immobilized artificial membrane and reversed-phase high-performance liquid chromatography: comparison with octanol-water partitioning, *J. Chromatog. A* 1116, 158–164, 2006; de Carmen Candia-Plata, M., Garcia, J., Guzman, R. et al., Isolation of human serum immunoglobulins with a new salt-promoted adsorbent, *J. Chromatog. A* 1118, 211–217, 2006.

NBS

Tryptophan   *N*-bromosuccinimide ⟶   Oxindole derivative

N-   178   Protein modification
bromosuccinimide;   reagent; bromination of
1-bromo-2,5-   olefins; analysis of a
pyrrolidinedione   variety of other
compounds.

Sinn, H.J., Schrenk, H.H., Friedrich, E.A. et al., Radioiodination of proteins and lipoproteins using *N*-bromosuccinimide as oxidizing agent, *Anal. Biochem.* 170, 186–192, 1988; Tanemura, K., Suzuki, T., Nishida, Y. et al., A mild and efficient procedure for α-bromination of ketones using *N*-bromosuccinimide catalyzed by ammonium acetate, *Chem. Commun.* 3, 470–471, 2004; Lundblad, R.L., *Chemical Reagents for Protein Modification*, 3rd ed., CRC Press, Boca Raton, FL, 2004; Edens, G.J., Redox titration of antioxidant mixtures with *N*-bromosuccinimide as titrant: analysis by nonlinear least-squares with novel weighting function, *Anal. Sci.* 21, 1349–1354, 2005; Abdel-Wadood, H.M., Mohamed, H.A., and Mohamed, F.A., Spectrofluorometric determination of acetaminophen with *N*-bromosuccinimide, *J. AOAC Int.* 88, 1626–1630, 2005; Krebs, A., Starczewska, B., Purzanowska-Tarasiewicz, H., and Sledz, J., Spectrophotometric determination of olanzapine by its oxidation with *N*-bromosuccinimide and cerium(IV) sulfate, *Anal. Sci.* 22, 829–833, 2006; Braddock, D.C., Cansell, G., Hermitage, S.A., and White, A.J., Bromoiodinanes with a I(III)-Br bond: preparation, X-ray crystallography, and reactivity as electrophilic brominating agents, *Chem. Commun.* 13, 1442–1444, 2006; Chen, G., Sasaki, M., Li, X., and Yudin, A.K., Strained enamines as versatile intermediates for stereocontrolled construction of nitrogen heterocycles, *J. Org. Chem.* 71, 6067–6073, 2006; Braddock D.C., Cansell, G., and Hermitage, S.A., Ortho-substituted iodobenzenes as novel organocatalysts for the transfer of electrophilic bromine from *N*-bromosuccinimide to alkenes, *Chem. Commun.* 23, 2483–2485, 2006.

Neutral Red

Neutral red dye
N^8, N^8, -trimethyl-2, 8-phenazinediamine
monohydrochloride

N^8,N^8-3-trimethyl-2,8-phenazinediamine Monohydrochloride; CI 50040 — 288.78 — Cell viability assays (selective uptake into lysosomes); pH indicator; spectral probe.

Sawicki, W., Kieler, J., and Briand, P., Vital staining with neutral red and trypan blue of ³H-thymidine-labeled cells prior to autoradiography, *Stain Technol.* 42, 143–146, 1967; Barbosa, P. and Peters, T.M., The effects of vital dyes on living organisms with special reference to methylene blue and neutral red, *Histochem. J.* 3, 71–93, 1971; Modha, K., Whiteside, J.P., and Spier, R.E., The determination of cellular viability of hybridoma cells in microtitre plates: a colorimetric assay based on neutral red, *Cytotechnology* 13, 227–232, 1993; Lowik, C.W., Alblas, M.J., van de Ruit, M. et al., Quantification of adherent and nonadherent cell cultured I 96-well plates using the supravital stain neutral red, *Anal. Biochem.* 213, 426–433, 1993; Ciapetti, G., Granchi, D., Verri, E. et al., Application of a combination of neutral red and amido black staining for rapid, reliable cytotoxicity testing of biomaterials, *Biomaterials* 17, 1259–1264, 1996; Hall, J.O., Novakofski, J.E., and Beasley, V.R., Neutral red assay modification to prevent cytotoxicity and improve reproducibility using E-63 rat skeletal muscle cells, *Biotech. Histochem.* 73, 211–221, 1998; Valentin, I., Philippe, M., Lhuguenot, J., and Chagnon, M., Uridine uptake inhibition as a cytotoxicity test for a human hepatoma cell line (HepG2 cells): comparison with the neutral red assay, *Toxicology* 158, 127–139, 2001; Zuang, V., The neutral red release assay: a review, *Altern. Lab. Anim.* 29, 575–599, 2001; Choi, J.K. and Yoo, G.S., Fast protein staining in sodium dodecyl sulfate polyacrylamide gel using counter ion-dyes, Coomassie Brilliant Blue R-250 and neutral red, *Arch. Pharm. Res.* 25, 704–708, 2002; Wang, Z., Zhang, Z., Liu, D., and Dong, S., A temperature-dependent interaction of neutral red with calf thymus DNA, *Spectrochim. Acta A Mol. Biomol. Spectrosc.* 59, 949–956, 2003; Svendsen, C., Spurgeon, D.J., Hankard, P.K., and Weeks, J.M., A review of lysosomal membrane stability measured by neutral red retention: is it a workable earthworm biomarker? *Ecotoxicol. Environ. Saf.* 57, 20–29, 2004; Dubrovsky, J.G., Guttenberger, M., Saralegui, A. et al., Neutral red as a probe for confocal scanning microscopy studies of plant roots, *Ann. Bot.* 97, 1127–1138, 2006; Ni, Y., Lin, D., and Kokot, S., Synchronous fluorescence, UV-visible spectrophotometric, and voltammetric studies of the competitive interaction of bis(1,10-phenanthroline) copper(II) complex and netural red with DNA, *Anal. Biochem.* 352, 231–242, 2006.

NHS

N-hydroxysuccinimide

N-hydroxy-succinimide; 1-hydroxy-2,5-pyrrolidinedione — 111.1 — Use in preparation of active esters for modification of amino groups (with carbodiimide); structural basis for reagents for amino group modification.

Anderson, G.W., Callahan, F.M., and Zimmerman, J.E., Synthesis of *N*-hydroxysuccinimide esters of acyl peptides by the mixed anhydride method, *J. Am. Chem. Soc.* 89, 178, 1967; Lapidot, Y., Rappoport, S., and Wolman, Y., Use of esters of *N*-hydroxysuccinimide in the synthesis of *N*-acylamino acids, *J. Lipid Res.* 8, 142–145, 1967; Holmquist, B., Blumberg, S., and Vallee, B.L., Superactivation of neutral proteases: acylation with *N*-hydroxysuccinimide esters, *Biochemistry* 15, 4675–4680, 1976; 't Hoen, P.A., de Kort, F., van Ommen, G.J., and den Dunnen, J.T., Fluorescent labeling of cRNA for microarray applications, *Nucleic Acids Res.* 31, e20, 2003; Vogel, C.W., Preparation of immunoconjugates using antibody oligosaccharide moieties, *Methods Mol. Biol.* 283, 87–108, 2004; Cooper, M., Ebner, A., Briggs, M. et al., Cy3B: improving the performance of cyanine dyes, *J. Fluoresc.* 14, 145–150, 2004; Lundblad, R.L., *Chemical Reagents for Protein Modification*, 3rd ed., CRC Press, Boca Raton, FL, 2004; Zhang, R., Tang, M., Bowyer, A. et al., A novel pH- and ionic-strength-sensitive carboxy methyl dextran hydrogel, *Biomaterials* 26, 4677–4683, 2005; Tyan, Y.C., Jong, S.B., Liao, J.D. et al., Proteomic profiling of erythrocyte proteins by proteolytic digestion chip and identification using two-dimensional electrospray ionization tandem mass spectrometry, *J. Proteome Res.* 4, 748–757, 2005; Lovrinovic, M., Spengler, M., Deutsch, C., and Niemeyer, C.M., Synthesis of covalent DNA-protein conjugates by expressed protein ligation, *Mol. Biosyst.* 1, 64–69, 2005; Smith, G.P., Kinetics of amine modification of proteins, *Bioconjug. Chem.* 17, 501–506, 2006; Yang, W.C., Mirzael, H., Liu, X., and Regnier, F.E., Enhancement of amino acid detection and quantitation by electrospray ionization mass spectrometry, *Anal. Chem.* 78, 4702–4708, 2006; Yu, G., Liang, J., He, Z., and Sun, M., Quantum dot-mediated detection of gamma-aminobutyric acid binding sites on the surface of living pollen protoplasts in tobacco, *Chem. Biol.* 13, 723–731, 2006; Adden, N., Gamble, L.J., Castner, D.G. et al., Phosphonic acid monolayers for binding of bioactive molecules to titanium surfaces, *Langmuir* 22, 8197–8204, 2006.

Ninhydrin		1-*H*-indene-1,2,3-trione Monohydrate	178.14	Reagent for amino acid analysis; reagent for modification of arginine residues in proteins; reaction with amino groups and other nucleophiles such as sulfhydryl groups.

Ninhydrin

Duliere, W.L., The amino-groups of the proteins of human serum. Action of formaldehyde and ninhydrin, *Biochem. J.* 30, 770–772, 1936; Schwartz, T.B. and Engel, F.L., A photometric ninhydrin method for the measurement of proteolysis, *J. Biol. Chem.* 184, 197–202, 1950; Troll, W. and Cannan, R.K., A modified photometric ninhydrin method for the analysis of amino and imino acids, *J. Biol. Chem.* 200, 803–811, 1953; Moore, S. and Stein, W.H., A modified ninhydrin reagent for the photometric determination of amino acids and related compounds, *J. Biol. Chem.* 211, 907–913, 1954; Rosen, H., A modified ninhydrin colorimetric analysis for amino acids, *Arch. Biochem. Biophys.* 67, 10–15, 1957; Meyer, H., The ninhydrin reactions and its analytical applications, *Biochem. J.* 67, 333–340, 1957; Whitaker, J.R., Ninhydrin assay in the presence of thiol compounds, *Nature* 189, 662–663, 1961; Grant, D.R., Reagent stability in Rosen's ninhydrin method for analysis of amino acids, *Anal. Biochem.* 6, 109–110, 1963; Shapiro, R. and Agarwal, S.C., Reaction of ninhydrin with cytosine derivatives, *J. Am. Chem. Soc.* 90, 474–478, 1968; Moore, S., Amino acid analysis: aqueous dimethylsulfoxide as solvent for the ninhydrin reaction, *J. Biol. Chem.* 243, 6281–6283, 1968; McGrath, R., Protein measurement by ninhydrin determination of amino acids released by alkaline hydrolysis, *Anal. Biochem.* 49, 95–102, 1972; Lamothe, P.J. and McCormick, P.G., Role of hydrindantin in the determination of amino acids using ninhydrin, *Anal. Chem.* 45, 1906–1911, 1973; Quinn, J.R., Boisvert, J.G., and Wood, I., Semi-automated ninhydrin assay of Kjeldahl nitrogen, *Anal. Biochem.* 58, 609–614, 1974; Chaplin, M.R., The use of ninhydrin as a reagent for the reversible modification of arginine residues in proteins, *Biochem. J.* 155, 457–459, 1976; Takahashi, K., Specific modification of arginine residues in proteins with ninhydrin, *J. Biochem.* 80, 1173–1176, 1976; Yu, P.H. and Davis, B.A., Deuterium isotope effects in the ninhydrin reaction of primary amines, *Experientia* 38, 299–300, 1982; D'Aniello, A., D'Onofrio, G., Pischetola, M., and Strazzulo, L., Effect of various substances on the colorimetric amino acid–ninhydrin reaction, *Anal. Biochem.* 144, 610–611, 1985; Macchi, F.D., Shen, F.J., Keck, R.G., and Harris, R.J., Amino acid analysis, using postcolumn ninhydrin detection, in a biotechnology laboratory, *Methods Mol. Biol.* 159, 9–30, 2000; Moulin, M., Deleu, C., Larher, F.R., and Bouchereau, A., High-performance liquid chromatography determination of pipecolic acid after precolumn derivatization using domestic microwave, *Anal.*

Biochem. 308, 320–327, 2002; Pool, C.T., Boyd, J.G., and Tam, J.P., Ninhydrin as a reversible protecting group of amino-terminal cysteine, *J. Pept. Res.* 63, 223–234, 2004; Schulz, M.M., Wehner, H.D., Reichert, W., and Graw, M., Ninhydrin-dyed latent fingerprints as a DNA source in a murder case, *J. Clin. Forensic Med.* 11, 202–204, 2004; Buchberger, W. and Ferdig, M., Improved high-performance liquid chromatographic determination of guanidine compounds by precolumn derivatization with ninhydrin and fluorescence detection, *J. Sep. Sci.* 27, 1309–1312, 2004; Hansen, D.B., and Joullie, M.M., The development of novel ninhydrin analogues, *Chem. Soc. Rev.* 34, 408–417, 2005.

Nitric Acid	HNO$_3$	63.01	Strong acid.

p-Nitroaniline (PNA)   4-nitroaniline   138.13   Signal from cleavage of chromogenic substrate.

p-nitroaniline   *p*-nitrophenol

2-Nitrobenzylsulfenyl Chloride   *o*-nitrophenyl-   189.6   Modification of tryptophan sulfenyl Chloride   in proteins.

Tryptophan   2-Nitrobenzylsulfenyl chloride

RSH

2-Thiotryptophan

Fontana, A. and Scofone, E., Sulfenyl halides as modifying reagents for peptides and proteins, *Methods Enzymol.* 25B, 482–494, 1972; Sanda, A. and Irie, M., Chemical modification of tryptophan residues in ribonuclease form a *Rhizopus* sp., *J. Biochem.* 87, 1079–1087, 1980; De Wolf, M.J., Fridkin, M., Epstein, M., and Kohn, L.D., Structure-function studies of cholera toxin and its A and B protomers. Modification of tryptophan residues, *J. Biol. Chem.* 256, 5481–5488, 1981; Mollier, P., Chwetzoff, S., Bouet, F. et al., Tryptophan 110, a residue involved in the toxic activity but in the enzymatic activity of notexin, *Eur. J. Biochem.* 185, 263–270, 1989; Cymes, C.D., Iglesias, M.M., and Wolfenstein-Todel, C., Selective modification of tryptophan-150 in ovine placental lactogen, *Comp. Biochem. Physiol. B* 106, 743–746, 1993; Kuyama, H., Watanabe, M., Toda, C. et al., An approach to quantitate proteome analysis by labeling tryptophan residues, *Rapid Commun. Mass Spectrom.* 17, 1642–1650, 2003; Lundblad, R.L., *Chemical Reagents for Protein Modification*, 3rd ed., CRC Press, Boca Raton, FL, 2004; Matsuo, E., Toda, C., Watanabe, M., et al., Selective detection of 2-nitrobenzensulfenyl-labeled peptides by matrix-assisted laser desorption/ionization-time-of-flight mass spectrometry using a novel matrix, *Proteomics* 6, 2042–2049, 2006; Ou, K., Kesuma, D., Ganesan, K. et al., Quantitative labeling of drug-assisted proteomic alterations by combined 2-nitrobenzenesulfenyl chloride (NBS) isotope labeling and 2DE/MS identification, *J. Proteome Res.* 5, 2194–2206, 2006.

p-Nitrophenol   4-nitrophenol   139.11   Popular signal from indicator enzymes such as alkaline phosphatase.

Nitro Tetrazolium Blue   NBI, Nitro BT   817.7   Cytotoxicity determination based on intracellular reduction to formazan.

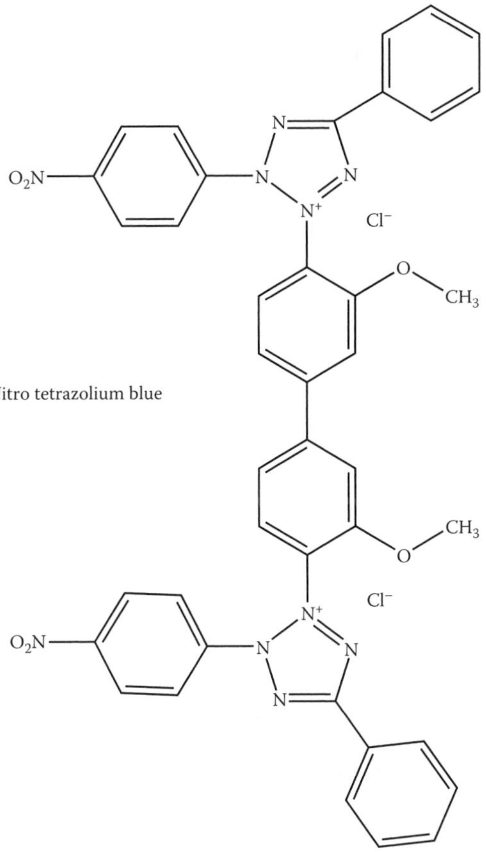

Nitro tetrazolium blue

Wieme, R.J., van Sande, M., Karcher, D. et al., A modified technique for direct staining with nitro-blue tetrazolium of lactate dehydrogenase iso-enzyme upon agar gel electrophoresis, *Clin. Chim. Acta* 7, 750–754, 1962; DeBari, V.A., Coste, J.F., and Needle, M.A., Direct spectrophotometric observation of intracellular nitro-blue tetrazolium and its formazan by multiple internal reflectance infrared spectroscopy, *Histochemistry* 45, 83–88, 1975; Fried, R., Enzymatic and nonenzymatic assay of superoxide dismutase, *Biochimie* 57, 657–660, 1975; DeBari, V.A. and Needle, M.A., Mechanism for transport of nitro-blue tetrazolium into viable and nonviable leukocytes, *Histochemistry* 56, 155–163, 1978; Ellsaesser, C., Miller, N., Lobb, C.J., and Clem, L.W., A new method for the cytochemical staining of cells immobilized in agarose, *Histochemistry* 80, 559–562, 1984; Walker, S.W., Howie, A.F., and Smith, A.F., The measurement of glycosylated albumin by reduction of alkaline nitro-blue tetrazolium, *Clin. Chim. Acta* 156, 197–206, 1986; Stegmaier, K., Corsello, S.M., Ross, K.N. et al., Gefitinib induces myeloid differentiation of acute myeloid leukemia, *Blood* 106, 2841–2848, 2005.

***n*-Octanol**   1-octanol;   130.23   Partitioning between octanol and water is used to determine lipophilicity; a factor in QSAR studies.
    Caprylic Alcohol

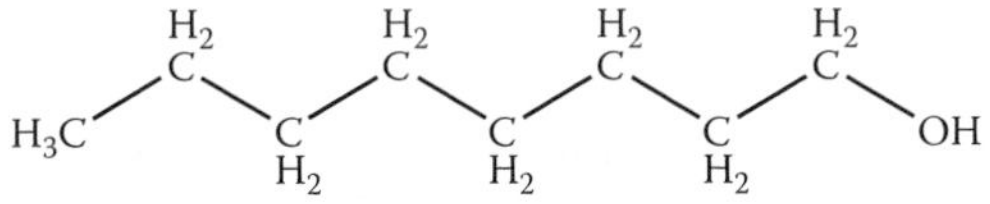

1-Octanol

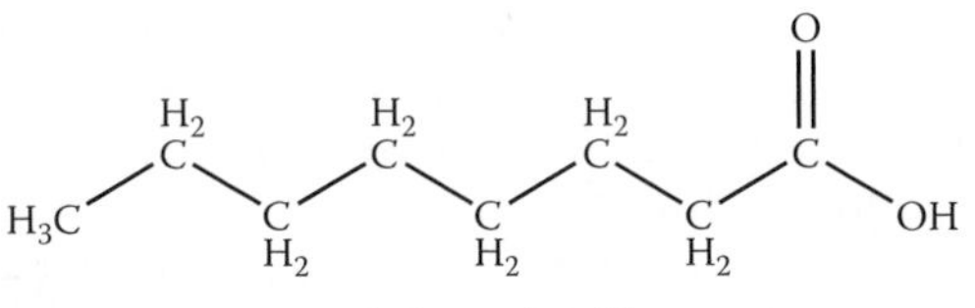

1-Octanoic acid

Marland, J.S. and Mulley, B.A., A phase-rule study of multiple-phase formation in a model emulsion system containing water, *n*-octanol, *n*-dodecane, and a non-ionic surface-active agent at 10 and 25 degrees, *J. Pharm. Pharmacol.* 23, 561–572, 1971; Dorsey, J.G. and Khaledi, M.G., Hydrophobicity estimations by reversed-phase liquid chromatography. Implications for biological partitioning processes, *J. Chromatog.* 656, 485–499, 1993; Vailaya, A. and Horvath, C., Retention in reversed-phase chromato-graphy: partition or adsorption? *J. Chromatog.* 829, 1–27, 1998; Kellogg, G.E. and Abraham, D.J., Hydrophobicity: is logP(o/w) more than the sum of its parts? *Eur. J. Med. Chem.* 35, 651–661, 2000; van de Waterbeemd,H., Smith, D.A., and Jones, B.C., Lipophilicity in PK design: methyl, ethyl, futile, *J. Comput. Aided Mol. Des.* 15, 273–286, 2001; Bethod, A. and Carda-Broch, S., Determination of liquid–liquid partition coefficients by separation methods, *J. Chromatog. A* 1037, 3–14, 2004.

Octoxynol   Triton X-100™;   Nonionic detergent;
                                                          Igepal CA-630™   surfactant.

Octoxynol, n = 5–15

Peroxynitrite

Petroleum Ether   Mixture of Pentanes   N/A
                                                     and Hexanes

Perchloric Acid   HClO$_4$ 100.5   Oxidizing agent.

1,10-Phenanthroline Monohydrate   *o*-phenanthroline   198.21   Chelating agent; inhibitor
                                                     Hydrate   for metalloproteinases; use
                                                                                       in design of synthetic
                                                                                       nucleases and proteases.

o-phenanthroline; 1,10-phenanthroline

Hoch, F.L., Willams, R.J., and Vallee, B.L., The role of zinc in alcohol dehydrogenases. II. The kinetics of the instantaneous reversible inactivation of yeast alcohol dehydrogenase by 1,10-phenanthroline, *J. Biol. Chem.* 232, 453–464, 1958; Sigman, D.S. and Chen, C.H., Chemical nucleases: new reagents in molecular biology, *Annu. Rev. Biochem.* 59, 207–236, 1990; Pan, C.Q., Landgraf, R., and Sigman, D.S., DNA-binding proteins as site-specific nucleases, *Mol. Microbiol.* 12, 335–342, 1994; Galis, Z.S., Sukhova, G.K., and Libby, P., Microscopic localization of active proteases by *in situ* zymography: detection of matrix metalloproteinase activity in vascular tissue, *FASEB J.* 9, 974–980, 1995; Papavassiliou, A.G., Chemical nucleases as probes for studying DNA–protein interactions, *Biochem. J.* 305, 345–357, 1995; Perrin, D.M., Mazumder, A., and Sigman, D.S., Oxidative chemical nucleases, *Prog. Nucleic Acid Res. Mol. Biol.* 52, 123–151, 1996; Sigman, D.S., Landgraf, R., Perrin, D.M., and Pearson, L., Nucleic acid chemistry of the cuprous complexes of 1,10-phenanthroline and derivatives, *Met. Ions Biol. Syst.* 33, 485–513, 1996; Cha, J., Pedersen, M.V., and Auld, D.S., Metal and pH dependence of heptapeptide catalysis by human matrilysin, *Biochemistry* 35, 15831–15838, 1996; Kidani, Y. and Hirose, J., Coordination chemical studies on metalloenzymes. II. Kinetic behavior of various types of chelating agents towards bovine carbonic anhydrase, *J. Biochem.* 81, 1383–1391, 1997; Marini, I., Bucchioni, L., Borella, P. et al., Sorbitol dehydrogenase from bovine lens: purification and properties, *Arch. Biochem. Biophys.* 340, 383–391, 1997; Dri, P., Gasparini, C., Menegazzi, R. et al., TNF-induced shedding of TNF receptors in human polymorphonuclear leukocytes: role of the 55-kDa TNF receptor and involvement of a membrane-bound and non-matrix metalloproteinase, *J. Immunol.* 165, 2165–2172, 2000; Kito, M. and Urade, R., Protease activity of 1,10-phenanthroline-copper systems, *Met. Ions Biol. Syst.* 38, 187–196, 2001; Winberg, J.O., Berg, E., Kolset, S.O. et al., Calcium-induced activation and truncation of promatrix metalloproteinase-9 linked to the core protein of chondroitin sulfate proteoglycans, *Eur. J. Biochem.* 270, 3996–4007, 2003; Butler, G.S., Tam, E.M., and Overall, C.M., The canonical methionine 392 of matrix metalloproteinase 2 (gelatinase A) is not required for catalytic efficiency or structural integrity: probing the role of the methionine-turn in the metzincin metalloprotease superfamily, *J. Biol. Chem.* 279, 15615–15620, 2004; Vauquelin, G. and Vanderheyden, P.M., Metal ion modulation of cystinyl aminopeptidase, *Biochem. J.* 390, 351–357, 2005; Schilling, S., Cynis, H., von Bohlen, A. et al., Isolation, catalytic properties, and competitive inhibitors of the zinc-dependent murine glutaminyl cyclase, *Biochemistry* 44, 13415–13424, 2005; Vik, S.B. and Ishmukhametov, R.R., Structure and function of subunit a of the ATP synthase of *Escherichia coli*, *J. Bioenerg. Biomembr.* 37, 445–449, 2005.

Phenol

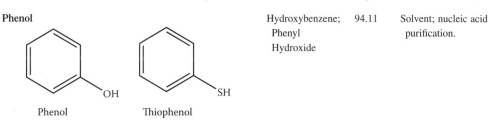

Phenol	Thiophenol

Hydroxybenzene; Phenyl Hydroxide   94.11   Solvent; nucleic acid purification.

Braun, W., Burrous, J.W., and Phillips, J.H., Jr., A phenol-extracted bacterial deoxyribonucleic acid, *Nature* 180, 1356–1357, 1957; Habermann, V., Evidence for peptides in RNA prepared by phenol extraction, *Biochim. Biophys. Acta* 32, 297–298, 1959; Colter, J.S., Brown, R.A., and Ellem, K.A., Observations on the use of phenol for the isolation of deoxyribonucleic acid, *Biochim. Biophys. Acta* 55, 31–39, 1962; Lust, J. and Richards, V., Influence of buffers on the phenol extraction of liver microsomal ribonucleic acids, *Anal. Biochem.* 20, 65–76, 1967; Yamaguchi, M., Dieffenbach, C.W., Connolly, R. et al., Effect of different laboratory techniques for guanidinium-phenol-chloroform RNA extraction on A260/A280 and on accuracy of mRNA quantitation by reverse transcriptase-PCR, *PCR Methods Appl.* 1, 286–290, 1992; Pitera, R., Pitera, J.E., Mufti, G.J., Salisbury, J.R., and Nickoloff, J.A., Sepharose spin column chromatography. A fast, nontoxic replacement for phenol: chloroform extraction/ethanol precipitation, *Mol. Biotechnol.* 1, 105–108, 1994; Finnegan, M.T., Herbert, K.E., Evans, M.D., and Lunec, J., Phenol isolation of DNA yields higher levels of 8-deoxodeoxyguanosine compared to pronase E isolation, *Biochem. Soc. Trans.* 23, 430S, 1995; Beaulieux, F., See, D.M., Leparc-Goffart, I. et al., Use of magnetic beads versus guanidium thiocyanate-phenol-chloroform RNA extraction followed by polymerase chain reaction for the rapid, sensitive detection of enterovirus RNA, *Res. Virol.* 148, 11–15, 1997; Fanson, B.G., Osmack, P., and Di Bisceglie, A.M., A comparison between the phenol-chloroform method of RNA extraction and the QIAamp viral RNA kit in the extraction of hepatitis C and GB virus-C/hepatitis G viral RNA from serum, *J. Virol. Methods* 89, 23–27, 2000; Kochl, S., Niederstratter, N., and Parson, W., DNA extraction and quantitation of forensic samples using the phenol-chloroform method and real-time PCR, *Methods Mol. Biol.* 297, 13–30, 2005; Izzo, V., Notomista, E., Picardi, A. et al., The thermophilic archaeon *Sulfolobus solfatarius* is able to grow on phenol, *Res. Microbiol.* 156, 677–689, 2005; Robertson, N. and Leek, R., Isolation of RNA from tumor samples: single-step guanidinium acid-phenol method, *Methods Mol. Biol.* 120, 55–59, 2006.

Phenoxyethanol

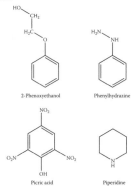

2-phenoxyethanol   138.16   Biochemical preservative; preservative in personal care products.

Nakahishi, M., Wilson, A.C., and Nolan, R.A., Phenoxyethanol: protein preservative for taxonomists, *Science* 163, 681–683, 1969; Frolich, K.W., Anderson, L.M., Knutsen, A., and Flood, P.R., Phenoxyethanol as a nontoxic substitute for formaldehyde in long-term preservation of human anatomical specimens for dissection and demonstration purposes, *Anat. Rec.* 208, 271–278, 1984.

Phenylglyoxal

Phenylglyoxal Hydrate	134.13	Modification of arginine residues.

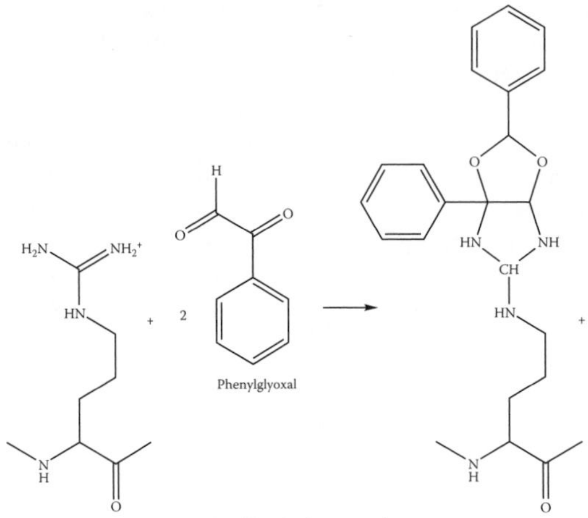

Arginine   Phenylglyoxal and reaction with arginine

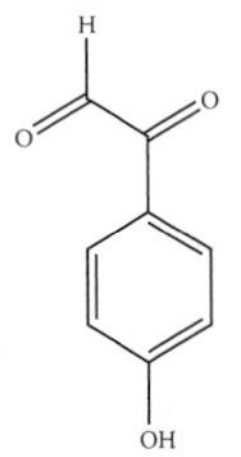

p-hydroxyphenylglyoxal 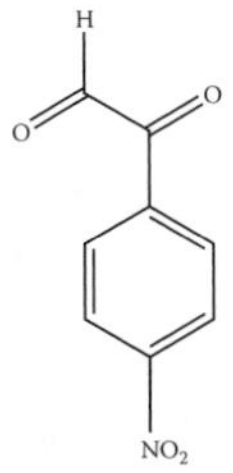   *p*-nitrophenylglyoxal

Takahashi, K., The reaction of phenylglyoxal with arginine residues in proteins, *J. Biol. Chem.* 243, 6171–6179, 1968; Bunzli, H.F. and Bosshard, H.R., Modification of the single arginine residue in insulin with phenylglyoxal, *Hoppe Seylers Z. Physiol. Chem.* 352, 1180–1182, 1971; Cheung, S.T. and Fonda, M.L., Reaction of phenylglyoxal with arginine. The effect of buffers and pH, *Biochem. Biophys. Res. Commun.* 90, 940–947, 1979; Srivastava, A. and Modak, M.J., Phenylglyoxal as a template site-specific reagent for DNA or RNA polymerases. Selective inhibition of initiation, *J. Biol. Chem.* 255, 917–921, 1980; Communi, D., Lecocq, R., Vanweyenberg, V., and Erneux, C., Active site labeling of inositol 1,4,5-triphosphate 3-kinase A by phenylglyoxal, *Biochem. J.* 310, 109–115, 1995; Eriksson, O., Fontaine, E., and Bernardi, P., Chemical modification of arginines by 2,3-butanedione and phenylglyoxal causes closure of the mitochondrial permeability transition pore, *J. Biol. Chem.* 273, 12669–12674, 1998; Redowicz, M.J., Phenylglyoxal reveals phosphorylation-dependent difference in the conformation of *Acanthamoeba* myosin II active site, *Arch. Biochem. Biophys.* 384, 413–417, 2000; Kucera, I., Inhibition by phenylglyoxal of nitrate transport in *Paracoccus denitrificans*; a comparison with the effect of a protonophorous uncoupler, *Arch. Biochem. Biophys.* 409, 327–334, 2003; Johans, M., Milanesi, E., Frank, M. et al., Modification of permeability transition pore arginine(s) by phenylglyoxal derivatives in isolated mitochondria and mammalian cells. Structure-function relationship of arginine ligands, *J. Biol. Chem.* 280, 12130–12136, 2005.

Phosgene

Carbonyl Chloride; Carbon Oxychloride	98.92	Reagent for organic synthesis; preparation of derivatives for analysis.

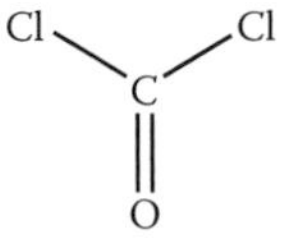

Phosgene

Wilchek, M., Ariely, S., and Patchornik, A., The reaction of asparagine, glutamine, and derivatives with phosgene, *J. Org. Chem.* 33, 1258–1259, 1968; Hamilton, R.D. and Lyman, D.J., Preparation of *N*-carboxy-α-amino acid anhydrides by the reaction of copper(II)-amino acid complexes with phosgene, *J. Org. Chem.* 34, 243–244, 1969; Pohl, L.R., Bhooshan, B.,

Whittaker, N.F., and Krishna, G., Phosgene: a metabolite of chloroform, *Biochem. Biophys. Res. Commun.* 79, 684–691, 1977; Gyllenhaal, O., Derivatization of 2-amino alcohols with phosgene in aqueous media: limitations of the reaction selectivity as found in the presence of *O*-glucuronides of alprenolol in urine, *J. Chromatog.* 413, 270–276, 1987; Gyllenhaal, O. and Vessman, J., Phosgene as a derivatizing reagent prior to gas and liquid chromatography, *J. Chromatog.* 435, 259–269, 1988; Noort, D., Hulst, A.G., Fidder, A., et al. *In vitro* adduct formation of phosgene with albumin and hemoglobin in human blood, *Chem. Res. Toxicol.* 13, 719–726, 2000; Lemoucheux, L. Rouden, J., Ibazizene, M. et al., Debenylation of tertiary amies using phosgene or triphosgen: an efficient and rapid procedure for the preparation of carbamoyl chlorides and unsymmetrical ureas. Application in carbon-11 chemistry, *J. Org. Chem.* 68, 7289–7297, 2003.

Picric Acid	2,4,6-trinitrophenol	229.1	Analytical reagent.

De Wesselow, O.L., The picric acid method for the estimation of sugar in blood and a comparison of this method with that of MacLean, *Biochem. J.* 13, 148–152, 1919; Newcomb, C., The error due to impure picric acid in creatinine estimations, *Biochem. J.* 18, 291–293, 1924; Davidsen, O., Fixation of proteins after agarose gel electrophoresis by means of picric acid, *Clin. Chim. Acta* 21, 205–209, 1968; Gisin, B.F., The monitoring of reactions in solid-phase peptide synthesis with picric acid, *Anal. Chim. Acta* 58, 248–249, 1972; Hancock, W.S., Battersby, J.E., and Harding, D.R., The use of picric acid as a simple monitoring procedure for automated peptide synthesis, *Anal. Biochem.* 69, 497–503, 1975; Vasiliades, J., Reaction of alkaline sodium picrate with creatinine: I. Kinetics and mechanism of formation of the mono-creatinine picric acid complex, *Clin. Chem.* 22, 1664–1671, 1976; Somogyi, P. and Takagi, H., A note on the use of picric acid-formaldehyde-glutaraldehyde fixative for correlated light and electron microscopic immunocytochemistry, *Neuroscience* 7, 1779–1783, 1982; Meyer, M.H., Meyer, R.A., Jr., Gray, R.W., and Irwin, R.L., Picric acid methods greatly overestimate serum creatinine in mice: more accurate results with high-performance liquid chromatography, *Anal. Biochem.* 144, 285–290, 1985; Knisley, K.A. and Rodkey, L.S., Direct detection of carrier ampholytes in immobilized pH gradients using picric acid precipitation, *Electrophoresis* 13, 220–224, 1992; Massoomi, F., Mathews, H.G., III, and Destache, C.J., Effect of seven fluoroquinolines on the determination of serum creatinine by the picric acid and enzymatic methods, *Ann. Pharmacother.* 27, 586–588, 1993.

Polysorbate

Polysorbates

	Tween 20		Nonionic detergent; surfactant.

Polyvinylpyrrolidone (PVP)

	Povidone	N/A	Pharmaceutical; excipient; phosphate analysis.

Morin, L.G. and Prox, J., New and rapid procedure for serum phosphorus using *o*-phenylenediamine as reductant, *Clin. Chim. Acta.* 46, 113–117, 1973; Ohnishi, S.T. and Gall, R.S., Characterization of the catalyzed phosphate assay, *Anal. Biochem.* 88, 347–356, 1978; Steige, H. and Jones, J.D., Determination of serum inorganic phosphorus using a discrete analyzer, *Clin. Chim. Acta.* 103, 123–127, 1980, Plaizier-Vercammen, J.A. and De Neve, R.E., Interaction of povidone with aromatic compounds. II: evaluation of ionic strength, buffer concentration, temperature, and pH by factorial analysis, *J. Pharm. Sci.* 70, 1252–1256, 1981; van Zanten, A.P. and Weber, J.A., Direct kinetic method for the determination of phosphate, *J. Clin. Chem. Clin. Biochem.* 25, 515–517, 1987; Barlow, I.M., Harrison, S.P., and Hogg, G.L., Evaluation of the Technicon Chem-1, *Clin. Chem.* 34, 2340–2344, 1988; Giulliano, K.A., Aqueous two-phase protein partitioning using textile dyes as affinity ligands, *Anal. Biochem.* 197, 333–339, 1991; Goldenheim, P.D., An appraisal of povidone-iodine and wound healing, *Postgrad. Med. J.*, 69 (Suppl. 3), S97–S105, 1993; Vemuri, S., Yu, C.D., and Roosdorp, N., Effect of cryoprotectants on freezing, lyophilization, and storage of lyophilized recombinant alpha 1-antitrypsin formulations, *PDA J. Pharm. Sci. Technol.* 48, 241–246, 1994; Anchordoquy, T.J. and Carpenter, J.F., Polymers protect lactate dehydrogenase during freeze-drying by inhibiting dissociation in the frozen state, *Arch. Biochem. Biophys.* 332, 231–238, 1996; Fleisher, W., and Reimer, K., Povidone-iodine in antisepsis — state of the art, *Dermatology* 195 (Suppl. 2), 3–9, 1997; Fernandes, S., Kim, H.S., and Hatti-Kaul, R., Affinity extraction of dye- and metal ion-binding proteins in polyvinalypyrrolidone-based aqueous two-phase system, *Protein Expr. Purif.* 24, 460–469, 2002; D'Souza, A.J., Schowen, R.L., Borchardt, R.T. et al., Reaction of a peptide with polyvinylpyrrolidone in the solid state, *J. Pharm. Sci.* 92, 585–593, 2003; Kaneda, Y., Tsutsumi, Y., Yoshioka, Y. et al., The use of PVP as a polymeric carrier to improve the plasma half-life of drugs, *Biomaterials* 25, 3259–3266, 2004; Art, G., Combination povidone-iodine and alcohol formulations more effective, more convenient versus formulations containing either iodine or alcohol alone: a review of the literature, *J. Infus. Nurs.* 28, 314–320, 2005; Yoshioka, S., Aso, Y., and Miyazaki, T., Negligible contribution of molecular mobility to the degradation of insulin lyophilized with poly(vinylpyrrolidone), *J. Pharm. Sci.* 95, 939–943, 2006.

Pyridine   Azine   79.10   Solvent.

Pyridine

Klingsberg, E. and Newkome, G.R., Eds., *Pyridine and Its Derivatives,* Interscience, New York, 1960; Schoefield, K., *Hetero-aromatic Nitrogen Compounds; Pyrroles and Pyridines,* Butterworths, London, 1967; Hurst, D.T., *An Introduction to the Chemistry and Biochemistry and Pyrimidines, Purines, and Ptreridines,* J. Wiley, Chichester, UK, 1980; Plunkett, A.O., Pyrrole, pyrrolidine, pyridine, piperidine, and azepine alkaloids, *Nat. Prod. Rep.* 11, 581–590, 1994; Kaiser, J.P., Feng, Y., and Bollag, J.M., Microbial metabolism of pyridine, quinoline, acridine, and their derivatives under aerobic and anaerobic conditions, *Microbiol. Rev.* 60, 483–498, 1996.

Pyridoxal-5-phosphate (PLP)   Pyridoxal-5-   247.14   Selective modification of
                                                         (dihydrogen   amino groups in proteins;
                                                         phosphate)   affinity label for certain
                                                                                      sites based on phosphate
                                                                                      group.

Lysine

Pyridoxal phosphate

NaBH₄ or NaBH₃CN

Hughes, R.C., Jenkins, W.T., and Fischer, E.H., The site of binding of pyridoxal-5′-phosphate to heart glutamic-aspartic transaminase, *Proc. Natl. Acad. Sci. USA* 48, 1615–1618, 1962; Finseth, R. and Sizer, I.W., Complexes of pyridoxal phosphate with amino acids, peptides, polylysine, and apotransaminase, *Biochem. Biophys. Res. Commun.* 26, 625–630, 1967; Pages, R.C., Benditt, E.P., and Kirkwood, C.R., Schiff base formation by the lysyl and hydroxylysyl side chains of collagen, *Biochem. Biophys. Res. Commun.* 33, 752–757, 1968; Whitman, W.B., Martin, M.N., and Tabita, F.R., Activation and regulation of ribulose bisphosphate carboxylase-oxygenase in the absence of small subunits, *J. Biol. Chem.* 254, 10184–10189, 1979; Howell, E.E. and Schray, K.J., Comparative inactivation and inhibition of the anomerase and isomerase activities of phosphoglucose isomerase, *Mol. Cell. Biochem.* 37, 101–107, 1981; Colanduoni, J. and Villafranca, J.J., Labeling of specific lysine residues at the active site of glutamine synthetase, *J. Biol. Chem.* 260, 15042–15050, 1985; Peterson, C.B., Noyes, C.M., Pecon, J.M. et al., Identification of a lysyl residue in antithrombin which is essential for heparin binding, *J. Biol. Chem.* 262, 8061–8065, 1987; Diffley, J.F., Affinity labeling the DNA polymerase alpha complex. Identification of subunits containing the DNA polymerase active site and an important regulatory nucleotide-binding site, *J. Biol. Chem.* 263, 19126–19131, 1988; Perez-Ramirez, B. and Martinez-Carrion, M., Pyridoxal phosphate as a probe of the cytoplasmic domains of transmembrane proteins: application to the nicotinic acetylcholine receptor, *Biochemistry* 28, 5034–5040, 1989; Valinger, Z., Engel, P.C., and Metzler, D.E., Is pyridoxal-5′-phosphate an affinity label for phosphate-binding sites in proteins? The case of bovine glutamate dehydrogenase, *Biochem. J.* 294, 835–839, 1993; Illy, C., Thielens, N.M., and Arlaud, G.J., Chemical characterization and location of ionic interactions involved in the assembly of the C1 complex of human complement, *J. Protein Chem.* 12, 771–781, 1993; Hountondji, C., Gillet, S., Schmitter, J.M. et al., Affinity labeling of *Escherichia coli* lysyl-tRNA synthetase with pyridoxal mono- and diphosphate, *J. Biochem.* 116, 502–507, 1994; Brody, S., Andersen, J.S., Kannangara, C.G. et al., Characterization of the different spectral forms of glutamate-1-semialdehyde aminotransferase by mass spectrometry, *Biochemistry* 34, 15918–15924, 1995; Kossekova, G., Miteva, M., and Atanasov, B., Characterization of pyridoxal phosphate as an optical label for measuring electrostatic potentials in proteins, *J. Photochem. Photobiol. B* 32, 71–79, 1996; Kim S.W., Lee, J., Song, M.S. et al., Essential active-site lysine of brain glutamate dehydrogenase isoproteins, *J. Neurochem.* 69, 418–422, 1997; Martin, D.L., Liu, H., Martin, S.B., and Wu, S.J., Structural features and regulatory properties of the brain glutamate decarboxylase, *Neurochem. Int.* 37, 111–119, 2000; Jaffe, M. and Bubis, J., Affinity labeling of the guanine nucleotide binding site of transducin by pyridoxal 5′-phosphate, *J. Protein Chem.* 21, 339–359, 2002.

| **Sodium Borohydride** | NaBH₄ | 37.83 | Reducing agent for Schiff bases; reduction of aldehydes; other chemical reductions. |

Chaykin, S., King, L., and Watson, J.G., The reduction of DPN+ and TPN+ with sodium borohydride, *Biochim. Biophys. Acta* 124, 13–25, 1966; Cerutti, P. and Miller, N., Selective reduction of yeast transfer ribonucleic acid with sodium borohydride, *J. Mol. Biol.* 26, 55–66, 1967; Tanzer, M.L., Collagen reduction by sodium borohydride: effects of reconstitution, maturation, and lathyrism, *Biochem. Biophys. Res. Commun.* 32, 885–892, 1968; Phillips, T.M., Kosicki, G.W., and Schmidt, D.E., Jr., Sodium borohydride reduction of pyruvate by sodium borohydride catalyzed by pyruvate kinase, *Biochim. Biophys. Acta* 293, 125–133, 1973; Craig, A.S., Sodium borohydride as an aldehyde-blocking reagent for electron microscope histochemistry, *Histochemistry* 42, 141–144, 1974; Miles, E.W., Houck, D.R., and Floss, H.G., Stereochemistry of sodium borohydride reduction of tryptophan synthase of *Escherichia coli* and its amino acid Schiff's bases, *J. Biol. Chem.* 257, 14203–14210, 1982; Kumar, A., Rao, P., and Pattabiraman, T.N., A colorimetric method for the estimation of serum glycated proteins based on differential reduction of free and bound glucose by sodium borohydride, *Biochem. Med. Metab. Biol.* 39, 296–304, 1988; Lenz, A.G., Costabel, U., Shaltiel, S., and Levine, R.L., Determination of carbonyl groups in oxidatively modified proteins by reduction with tritiated sodium borohydride, *Anal. Biochem.* 177, 419–425, 1989; Yan, L.J. and Sohal, R.S., Gel electrophoresis quantiation of protein carbonyls derivatized with tritiated sodium borohydride, *Anal. Biochem.* 265, 176–182, 1998; Azzam, T., Eliyahu, H., Shapira, L. et al., Polysaccharide-oligoamine-based conjugates for gene delivery, *J. Med. Chem.* 45, 1817–1824, 2002; Purich, D.L., Use of sodium borohydride to detect acyl-phosphate linkages in enzyme reactions, *Methods Enzymol.* 354, 168–177, 2002; Bald, E., Chwatko, S., Glowacki, R., and Kusmierek, K., Analysis of plasma thiols by high-performance liquid chromatography with ultraviolet detection, *J. Chromatog. A* 1032, 109–115, 2004; Eike, J.H. and Palmer, A.F., Effect of NABH₄ concentration and reaction time on physical properties of glutaraldehyde-polymerized hemoglobin, *Biotechnol. Prog.* 20, 946–952, 2004; Zhang, Z., Edwards, P.J., Roeske, R.W., and Guo, L., Synthesis and self-alkylation of isotope-coded affinity tag reagents, *Bioconjug. Chem.* 16, 458–464, 2005; Studelski, D.R., Giljum, K., McDowell, L.M., and Zhang, L., Quantitation of glycosaminoglycans by reversed-phase HPLC separation of fluorescent isoindole derivatives, *Glycobiology* 16, 65–72, 2006; Floor, E., Maples, A.M., Rankin, C.A. et al., A one-carbon modification of protein lysine associated with elevated oxidative stress in human substantia nigra, *J. Neurochem.* 97, 504–514, 2006; Kusmierek, K., Glowacki, R., and Bald, E., Analysis of urine for cysteine, cysteinylglycine, and homocysteine by high-performance liquid chromatography, *Anal. BioAnal. Chem.* 385, 855–860, 2006.

Sodium Chloride	Salt; NaCl	58.44	Ionic strength; physiological saline.

Sodium Cholate	430.55	Detergent.

Lindstrom, J., Anholt, R., Einarson, B. et al., Purification of acetylcholine receptors, reconstitution into lipid vesicles, and study of agonist-induced channel regulation, *J. Biol. Chem.* 255, 8340–8350, 1980; Gullick, W.J., Tzartos, S., and Lindstrom, J., Monoclonal antibodies as probes of acetylcholine receptor structure. 1. Peptide mapping, *Biochemistry* 20, 2173–2180, 1981; Henselman, R.A. and Cusanovich, M.A., The characterization of sodium cholate solubilized rhodopsin, *Biochemistry* 13, 5199–5203, 1974; Ninomiya, R., Masuoka, K., and Moroi, Y., Micelle formation of sodium chenodeoxycholate and solublization into the micelles: comparison with other unconjugated bile salts, *Biochim. Biophys. Acta* 1634, 116–125, 2003; Simoes, S.I., Marques, C.M., Cruz, M.E. et al., The effect of cholate on solubilization and permeability of simple and protein-loaded phosphatidylcholine/sodium cholate-mixed aggregates designed to mediate transdermal delivery of macromolecules, *Eur. J. Pharm. Biopharm.* 58, 509–519, 2004; Reis, S., Moutinho, C.G., Matos, C. et al., Noninvasive methods to determine the critical micelle concentration of some bile acid salts, *Anal. Biochem.* 334, 117–126, 2004; Nohara, D., Kajiura, T., and Takeda, K., Determination of micelle mass by electrospray ionization mass spectrometry, *J. Mass Spectrom.* 40, 489–493, 2005; Guo, J., Wu., T., Ping, Q. et al., Solublization and pharmacokinetic behaviors of sodium cholate/lecithin-mixed micelles containing cyclosporine A, *Drug Deliv.* 12, 35–39, 2005; Bottari, E., Buonfigli, A., and Festa, M.R., Composition of sodium cholate micellar solutions, *Ann. Chim.* 95, 479–490, 2005; Schweitzer, B., Felippe, A.C., Dal Bo, A. et al., Sodium dodecyl sulfate promoting a cooperative association process of sodium cholate with bovine serum albumin, *J. Colloid Interface Sci.* 298, 457–466, 2006; Burton, M.I., Herman, M.D., Alcain, F.J., and Villalba, J.M., Stimulation of polyprenyl 4-hydroxybenzoate transferase activity by sodium cholate and 3- [(cholamidopropyl)dimethylammonio]-1-propanesulfonate, *Anal. Biochem.* 353, 15–21, 2006; Ishibashi, A. and Nakashima, N., Individual dissolution of single-walled carbon nanotubes in aqueous solutions of steroid of sugar compounds and their Raman and near-IR spectral properties, *Chemistry,* 12, 7595–7602, 2006.

Sodium Cyanoborohydride	NaBH$_3$ (CN)	62.84	Reducing agent; considered more selective than NaBH$_4$.

Rosen, G.M., Use of sodium cyanoborohydride in the preparation of biologically active nitroxides, *J. Med. Chem.* 17, 358–360, 1974; Chauffe, L. and Friedman, M., Factors affecting cyanoborohydride reduction of aromatic Schiff's bases in proteins, *Adv. Exp. Med. Biol.* 86A, 415–424, 1977; Baues, R.J. and Gray, G.R., Lectin purification on affinity columns containing reductively aminated disaccharides, *J. Biol. Chem.* 252, 57–60, 1977; Jentoft, N. and Dearborn, D.G., Labeling of proteins by reductive methylation using sodium cyanoborohydride, *J. Biol. Chem.* 254, 4359–4365, 1979; Jentoft, N., and Dearborn, D.G., Protein labeling by reductive methylation with sodium cyanoborohydride: effect of cyanide and metal ions on the reaction, *Anal. Biochem.* 106, 186–190, 1980; Bunn, H.F. and Higgins, P.T., Reaction of monosaccharides with proteins: possible evolutionary significance, *Science* 213, 222–224, 1981; Geoghegan, K.F., Cabacungan, J.C., Dixon, H.B., and Feeney, R.E., Alternative reducing agents for reductive methylation of amino groups in proteins, *Int. J. Pept. Protein Res.* 17, 345–352, 1981; Habeeb, A.F., Comparative studies on radiolabeling of lysozyme by iodination and reductive methylation, *J. Immunol. Methods* 65, 27–39, 1983; Prakash, C. and Vijay, I.K., A new fluorescent tag for labeling of saccharides, *Anal. Biochem.* 128, 41–46, 1983; Acharya, A.S. and Sussman, L.G., The reversibility of the ketoamine linkages of aldoses with proteins, *J. Biol. Chem.* 259, 4372–4378, 1984; Climent, I., Tsai, L., and Levine, R.L., Derivatization of gamma-glutamyl semialdehyde residues in oxidized proteins by fluorescamine, *Anal. Biochem.* 182, 226–232, 1989; Hartmann, C. and Klinman, J.P., Reductive trapping of substrate to methylamine oxidase from *Arthrobacter* P1, *FEBS Lett.* 261, 441–444, 1990; Meunier, F. and Wilkinson, K.J., Nonperturbing fluorescent labeling of polysaccharides, *Biomacromolecules* 3, 858–864, 2002; Webb, M.E., Stephens, E., Smith, A.G., and Abell, C., Rapid screening by MALDI-TOF mass spectrometry to probe binding specificity at enzyme active sites, *Chem. Commun.* 19, 2416–2417, 2003; Sando, S., Matsui, K., Niinomi, Y. et al., Facile preparation of DNA-tagged carbohydrates, *Bioorg. Med. Chem. Lett.* 13, 2633–2636, 2003; Peelen, D. and Smith, L.M., Immobilization of anine-modified oligonucleotides on aldehyde-terminated alkanethiol monolayers on gold, *Langmuir* 21, 266–271, 2005; Mirzaei, H. and Regnier, F., Enrichment of carbonylated peptides using Girard P reagent and strong cation exchange chromatography, *Anal. Chem.* 78, 770–778, 2006.

Sodium Deoxycholate	Desoxycholic Acid, Sodium Salt	414.55	Detergent; potential therapeutic use with adipose tissue.

Bril, C., van der Horst, D.J., Poort, S.R., and Thomas, J.B., Fractionation of spinach chloroplasts with sodium deoxycholate, *Biochim. Biophys. Acta* 172, 345–348, 1969; Smart, J.E. and Bonner, J., Selective dissociation of histones from chromatin by sodium deoxycholate, *J. Mol. Biol.* 58, 651–659, 1971; Part, M., Tarone, G., and Comoglio, P.M., Antigenic and immunogenic properties of membrane proteins solubilized by sodium desoxycholate, papain digestion, or high ionic strength, *Immunochemistry* 12, 9–17, 1975; Johansson, K.E. and Wbolewski, H., Crossed immunoelectrophoresis, in the presence of tween 20 or sodium deoxycholate, or purified membrane proteins from *Acholeplasma laidlawii, J. Bacteriol.* 136, 324–330, 1978; Lehnert, T. and Berlet, H.H., Selective inactivation of lactate dehydrogenase of rat tissues by sodium deoxycholate, *Biochem. J.* 177, 813–818, 1979; Suzuki, N., Kawashima, S., Deguchi, K., and Ueta, N., Low-density lipoproteins form human ascites plasma. Characterization and degradation by sodium deoxycholate, *J. Biochem.* 87, 1253–1256, 1980; Robern, H., The application of sodium deoxycholate and Sephacryl S-200 for the delipidation and separation of high-density lipoprotein, *Experientia* 38, 437–439, 1982; Nedivi, E. and Schramm, M., The beta-adrenergic receptor survives solubilization in deoxycholate while forming a stable association with the agonist, *J. Biol. Chem.* 259, 5803–5808, 1984; McKernan, R.M., Castro, S., Poat, J.A., and Wong, E.H., Solubilization of the *N*-methyl-D-aspartate receptor channel complex from rat and porcine brain, *J. Neurochem.* 52, 777–785, 1989; Carter, H.R. Wallace, M.A., and Fain, J.N., Activation of phospholipase C in rabbit brain membranes by carbachol in the presence of GTP gamma S: effects of biological detergents, *Biochim. Biophys. Acta* 1054, 129–134, 1990; Shivanna, B.D. and Rowe, E.S., Preservation of the native structure and function of Ca2+-ATPase from sarcoplasmic reticulum: solubilization and reconstitution by new short-chain phospholipid detergent 1,2-diheptanoyl-*sn*-phosphatidylcholine, *Biochem. J.* 325, 533–542, 1997; Arnold, U. and Ulbrich-Hofmann, R., Quantitative protein precipitation from guandine hydrochloride-containing solutions by sodium deoxycholate/trichloroacetic acid, *Anal. Biochem.* 271, 197–199, 1999; Haque, M.E., Das, A.R., and Moulik, S.P., Mixed micelles for sodium deoxycholate and polyoxyethylene sobitan monooleate (Tween 80), *J. Colloid Interface Sci.* 217, 1–7, 1999; Srivastava, O.P. and Srivastava, K., Characterization of a sodium deoxycholate-activable proteinase activity associated with betaA3/A1-crystallin of human lenses, *Biochim. Biophys. Acta* 1434, 331–346, 1999; Rotunda, A.M., Suzuki, H., Moy, R.L., and Kolodney, M.S., Detergent effects of sodium deoxycholate are a major feature of an injectable phosphatidylcholine formulation used for localized fat dissolution, *Dermatol. Surg.* 30, 1001–1008, 2004; Asmann, Y.W., Dong, M., and Miller, L.J., Functional characterization and purification of the secretin receptor expressed in baculovirus-infected insect cells, *Regul. Pept.* 123, 217–223, 2004; Ranganathan, R., Tcacenco, C.M., Rosseto, R., and Hajdu, J., Characterization of the kinetics of phospholipase C activity toward mixed micelles of sodium deoxycholate and dimyristoyl-phophatidylcholine, *Biophys. Chem.* 122, 79–89, 2006.

Sodium Dodecylsulfate   Sodium Lauryl   288.38   Detergent.
                                Sulfate, SDS

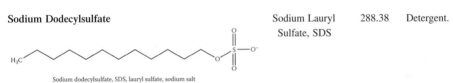

Sodium dodecylsulfate, SDS, lauryl sulfate, sodium salt

Shapiro, A.L., Vinuela, E., and Maizel, J.V., Jr., Molecular weight estimation of polypeptide chains by electrophoresis in SDS-polyacrylamide gels, *Biochem. Biophys. Res. Commun.* 28, 815–820, 1967; Shapiro, A.L., and Maizel, J.V., Jr., Molecular weight estimation of polypeptides by SDS-polyacrylamide gel electrophoresis: further data concerning resolving power and general considerations, *Anal. Biochem.* 29, 505–514, 1969; Weber, K. and Osborn, M., The reliability of molecular weight determinations of dodecyl sulfate-polyacryalmide gel electrophoresis, *J. Biol. Chem.* 244, 4406–4412, 1969; Weber, K. and Kuter, D.J., Reversible denaturation of enzymes by sodium dodecyl sulfate, *J. Biol. Chem.* 246, 4504–4509, 1971; de Haen, C., Molecular weight standards for calibration of gel filtration and sodium dodecyl sulfate-polyacrylamide gel electrophoresis: ferritin and apoferritin, *Anal. Biochem.* 166, 235–245, 1987; Smith, B.J., SDS polyacrylamide gel electrophoresis of proteins, *Methods Mol. Biol.* 32, 23–34, 1994; Guttman, A., Capillary sodium dodecyl sulfate-gel electrophoresis of proteins, *Electrophoresis* 17, 1333–1341, 1996; Bischoff, K.M., Shi, L., and Kennelly, P.J., The detection of enzyme activity following sodium dodecyl sulfate-polyacryalamide gel electrophoresis, *Anal. Biochem.* 260, 1–17, 1998; Maizel, J.V., SDS polyacrylamide gel electrophoresis, *Trends Biochem. Sci.* 35, 590–592, 2000; Robinson, J.M. and Vandre, D.D, Antigen retrieval in cells and tissues: enhancement with sodium dodecyl sulfate, *Histochem. Cell Biol.* 116, 119–130, 2001; Todorov, P.D., Kralchevsky, P.A., Denkov, N.D. et al., Kinetics of solublization of *n*-decane and benzene by micellar solutions of sodium dodecyl sulfate, *J. Colloid Interface Sci.* 245, 371–382, 2002; Zhdanov, S.A., Starov, V.M., Sobolev, V.D., and Velarde, M.G., Spreading of aqueous SDS solutions over nitrocellulose membranes, *J. Colloid Interface Sci.* 264, 481–489, 2003; Santos, S.F., Zanette, D., Fischer, H., and Itri, R., A systematic study of bovine serum albumin (BSA) and sodium dodecyl sulfate (SDS) interactions by surface tension and small angle X-ray scattering, *J. Colloid Interface Sci.* 262, 400–408, 2003; Biswas, A. and Das, K.P., SDS-induced structural changes in alpha-crystallin and its effect on refolding, *Protein J.* 23, 529–538, 2004; Jing, P., Kaneta, T., and Imasaka, T., On-line concentration of a protein using denaturation by sodium dodecyl sulfate, *Anal. Sci.* 21, 37–42, 2005; Choi, N.S., Hahm, J.H., Maeng, P.J., and Kim, S.H., Comparative study of enzyme activity and stability of bovine and human plasmins in electrophoretic reagents, β-mercaptoethanol, DTT, SDS, Triton X-100, and urea, *J. Biochem. Mol. Biol.* 38, 177–181, 2005; Miles, A.P.

and Saul, A., Quantifying recombinant proteins and their degradation products using SDS-PAGE and scanning laser densitometry, *Methods Mol. Biol.* 308, 349–356, 2005; Thongngam, M. and McClements, D.J., Influence of pH, ionic strength, and temperature on self-association and interactions of sodium dodecyl sulfate in the absence and presence of chitosan, *Langmuir* 21, 79–86, 2005; Romani, A.P., Gehlen, M.H., and Itri, R., Surfactant-polymer aggregates formed by sodium dodecyl sulfate, poly(*N*-vinyl-2-pyrrolidone), and poly(ethylene glycol), *Langmuir* 21, 1271–1233, 2005; Gudiksen, K.L., Gitlin, I., and Whitesides, G.M., Differentiation of proteins based on characteristic patterns of association and denaturation in solutions of SDS, *Proc. Natl. Acad. Sci. USA* 103, 7968–7972, 2006; Freitas, A.A., Paulo, L., Macanita, A.L, and Quina, F.H., Acid-base equilibria and dynamics in sodium dodecyl sulfate micelles: geminate recombination and effect of charge stabilization, *Langmuir* 22, 7986–7893, 2006.

Sodium Metabisulfite Sodium Bisulfite 190.1 Mild reducing agent; converts unmethylated cytosine residues to uracil residues (DNA methylation).

Miller, R.F., Small, G., and Norris, L.C., Studies on the effect of sodium bisulfite on the stability of vitamin E, *J. Nutr.* 55, 81–95, 1955; Hayatsu, H., Wataya, Y., Kai, K., and Iida, S., Reaction of sodium bisulfite with uracil, cytosine, and their derivatives, *Biochemistry* 9, 2858–2865, 1970; Seno, T., Conversion of *Escherichia coli* tRNATrp to glutamine-accepting tRNA by chemical modification with sodium bisulfite, *FEBS Lett.* 51, 325–329, 1975; Tasheva, B. and Dessev, G., Artifacts in sodium dodecyl sulfate-polyacrylamide gel electrophoresis due to 2-mercaptoethanol, *Anal. Biochem.* 129, 98–102, 1983; Draper, D.E., Attachment of reporter groups to specific, selected cytidine residues in RNA using a bisulfite-catalyzed transamination reaction, *Nucleic Acids Res.* 12, 989–1002, 1984; Oakeley, E.J., DNA methylation analysis: a review of current methodologies, *Pharmacol. Ther.* 84, 389–400, 1999; Geisler, J.P., Manahan, K.J., and Geisler, H.E., Evaluation of DNA methylation in the human genome: why examine it and what method to use, *Eur. J. Gynaecol. Oncol.* 25, 19–24, 2004; Thomassin, H., Kress, C., and Grange, T., MethylQuant: a sensitive method for quantifying methylation of specific cytosines within the genome, *Nucleic Acids Res.* 32, e168, 2004; Derks, S., Lentjes, M.H., Mellebrekers, D.M. et al., Methylation-specific PCR unraveled, *Cell. Oncol.* 26, 291–299, 2004; Galm, O. and Herman, J.G., Methylation-specific polymerase chain reaction, *Methods Mol. Biol.* 113, 279–291, 2005; Ogino, S., Kawasaki, T., Brahmandam, M. et al., Precision and performance characteristics of bisulfite conversion and real-time PCR (MethylLight) for quantitative DNA methylation analysis, *J. Mol. Diagn.* 8, 209–217, 2006; Yang, I., Park, I.Y., Jang, S.M. et al., Rapid quantitation of DNA methylation through dNMP analysis following bisulfite PCR, *Nucleic Acids Res.* 34, e61, 2006; Wischnewski, F., Pantel, K., and Schwazenbach, H., Promoter demethylation and histone acetylation mediate gene expression of MAGE-A1, -A2, -A3, and -A12 in human cancer cells, *Mol. Cancer Res.* 4, 339–349, 2006; Zhou, Y., Lum, J.M., Yeo, G.H. et al., Simplified molecular diagnosis of fragile X syndrome by fluorescent methylation-specific PCR and GeneScan analysis, *Clin. Chem.* 52, 1492–1500, 2006.

Succinic Anhydride Butanedioic Anhydride; 2,5-diketotetra-hydrofuran 100.1 Protein modification; dissociation of protein complexes.

Succinic anhydride + Lysine → N^6-succinyllysine

3, 4, 5, 6-Tetraphthalic anhydride + Lysine → N^6-3, 4, 5, 6-tetramethylphthaloyllysine

Habeeb, A.F., Cassidy, H.G., and Singer, S.J., Molecular structural effects produced in proteins by reaction with succinic anhydride, *Biochim. Biophys. Acta* 29, 587–593, 1958; Hass, L.F., Aldolase dissociation into subunits by reaction with succinic anhydride, *Biochemistry* 3, 535–541, 1964; Scanu, A., Pollard, H., and Reader, W., Properties of human serum low-density lipoproteins after modification by succinic anhydride, *J. Lipid Res.* 9, 342–349, 1968; Vasilets, I.M., Moshkov, K.A., and Kushner, V.P., Dissociation of human ceruloplasmin into subunits under the action of alkali and succinic anhydride, *Mol. Biol.* 6, 193–199, 1972; Tedeschi, H., Kinnally, K.W., and Mannella, C.A., Properties of channels in mitochondrial outer membrane, *J. Bioenerg. Biomembr.* 21, 451–459, 1989; Palacian, E., Gonzalez, P.J., Pineiro, M., and Hernandez, F., Dicarboxylic acid anhydrides as dissociating agents of protein-containing structures, *Mol. Cell. Biochem.* 97, 101–111, 1990; Pavliakova, D., Chu, C., Bystricky, S. et al., Treatment with succinic anhydride improves the immunogenicity of *Shigella flexneri* type 2a *O*-specific polysaccharide-protein conjugates in mice, *Infect. Immun.* 67, 5526–5529, 1999; Ferretti, V., Gilli, P., and Gavezzotti, A., X-ray diffraction and molecular simulation study of the crystalline and liquid states of succinic anhydride, *Chemistry* 8, 1710–1718, 2002.

Sucrose   342.30   Osmolyte; density gradient centrifugation.

Cann, J.R., Coombs, R.O., Howlett, G.J. et al., Effects of molecular crowding on protein self-association: a potential source of error in sedimentation coefficients obtained by zonal ultracentrifugation in a sucrose gradient, *Biochemistry* 33, 10185–10190, 1994; Camacho-Vanegas, O., Lorein, F., and Amaldi, F., Flat absorbance background for sucrose gradients, *Anal. Biochem.* 228, 172–173, 1995; Ben-Zeev, O. and Doolittle, M.H., Determining lipase subunit structure by sucrose gradient centrifugation, *Methods Mol. Biol.* 109, 257–266, 1999; Lustig, A., Engel, A., Tsiotis, G. et al., Molecular weight determination of membrane proteins by sedimentation equilibrium at the sucrose of nycodenz-adjusted density of the hydrated detergent micelle, *Biochim. Biophys. Acta* 1464, 199–206, 2000; Kim, Y.S., Jones, L.A., Dong, A. et al., Effects of sucrose on conformational equilibria and fluctuations within the native-state ensemble of proteins, *Protein Sci.* 12, 1252–1261, 2003; Srinivas, K.A., Chandresekar, G., Srivastava, R., and Puvanakrishna, R., A novel protocol for the subcellular fractionation of C3A hepatoma cells using sucrose-density gradient centrifugation, *J. Biochem. Biophys. Methods* 60, 23–27, 2004; Richter, W., Determining the subunit structure of phosphodiesterase using gel filtration and sucrose-density gradient centrifugation, *Methods Mol. Biol.* 307, 167–180, 2005; Cioni, P., Bramanti, E., and Strambini, G.B., Effects of sucrose on the internal dynamics of azurin, *Biophys. J.* 88, 4213–4222, 2005; Desplats, P., Folco, E. and Salerno, G.L., Sucrose may play an additional role to that of an osmolyte in *Synechocystis* sp. PCC 6803 salt-shocked cells, *Plant Physiol. Biochem.* 43, 133–138, 2005; Chen, L., Ferreira, J.A., Costa, S.M. et al., Compaction of ribosomal protein S6 by sucrose occurs only under native conditions, *Biochemistry* 21, 2189–2199, 2006.

Sulfuric Acid H_2SO_4 98.1   Strong acid; component of piranha solution with hydrogen peroxide.

TES   *N*-Tris(hydroxymethyl) Methyl-2-aminoethanesulfonic Acid   229.3   A "Good" buffer.

TES   TRIS

Good, N.E., Winget, G.D., Winter, W. et al., Hydrogen ion buffers for biological research, *Biochemistry* 5, 467–477, 1966; Itagaki, A. and Kimura, G., TES and HEPES buffers in mammalian cell cultures and viral studies: problem of carbon dioxide requirement, *Exp. Cell Res.* 83, 351–361, 1974; Bridges, S. and Ward, B., Effect of hydrogen ion buffers on photosynthetic oxygen evolution in the blue-green alga, *Agmenellum quadruplicatum*, *Microbios* 15, 49–56, 1976; Bailyes, E.M., Luzio, J.P., and Newby, A.C., The use of a zwitterionic detergent in the solubilization and purification of the intrinsic membrane protein 5′-nucleotidase, *Biochem. Soc. Trans.* 9, 140–141, 1981; Poole, C.A., Reilly, H.C., and Flint, M.H., The adverse effects of HEPES, TES, and BES zwitterionic buffers on the ultrastructure of cultured chick embryo epiphyseal chondrocytes, *In Vitro* 18, 755–765, 1982; Nakon, R. and Krishnamoorthy, C.R., Free-metal ion depletion by "Good's" buffers, *Science* 221, 749–750, 1983; del Castillo, J., Escalona de Motta, G., Eterovic, V.A., and Ferchmin, P.A., Succinyl derivatives of *N*-Tris (hydroxylmethyl) methyl-2-aminoethane sulphonic acid: their effects on the frog neuromuscular junction, *Br. J. Pharmacol.* 84, 275–288, 1985; Kaushal, V. and Varnes, L.D., Effect of zwitterionic buffers on measurement of small masses of protein with bicinchoninic acid, *Anal. Biochem.* 157, 291–294, 1986; Bhattacharyya, A. and Yanagimachi, R., Synthetic organic pH buffers can support fertilization of guinea pig eggs, but not as efficiently as bicarbonate buffer, *Gamete Res.* 19, 123–129, 1988; Veeck, L.L., TES and Tris (TEST)-yolk buffer systems, sperm function testing, and *in vitro* fertilization, *Fertil. Steril.* 58, 484–486, 1992; Kragh-Hansen, U. and Vorum, H., Quantitative analyses of the interaction between calcium ions and human serum albumin, *Clin. Chem.* 39, 202–208, 1993; Jacobs, B.R., Caulfield, J., and Boldt, J., Analysis of TEST (TES and Tris) yolk buffer effects of human sperm, *Fertil. Steril.* 63, 1064–1070, 1995; Stellwagne, N.C., Bossi, A., Gelfi, C., and Righetti, P.G., DNA and buffers: are there any noninteracting, neutral pH buffers? *Anal. Biochem.* 287, 167–175, 2000; Taylor, J., Hamilton, K.L., and Butt, A.G., HCO$_3^-$ potentiates the cAMP-dependent secretory response of the human distal colon through a DIDS-sensitive pathway, *Pflügers Arch.* 442, 256–262, 2001; Taha, M., Buffers for the physiological pH range: acidic dissociation constants of zwitterionic compounds in various hydroorganic media, *Ann. Chim.* 95, 105–109, 2005.

Tetrabutylammonium Chloride   277.9   Ion-pair reagent for extraction and HPLC.

Tetrabutylammonium chloride   *tert*-butrylhydroperoxide

Walseth, T.F., Graff, G., Moos, M.C., Jr., and Goldberg, N.D., Separation of 5′-ribonucleoside monophosphates by ion-pair reverse-phase high-performance liquid chromatography, *Anal. Biochem.* 107, 240–245, 1980; Ozkul, A. and Oztunc, A., Determination of naprotiline hydrochloride in tables by ion-pair extraction using bromthymol blue, *Pharmzie* 55, 321–322, 2000; Cecchi, T., Extended thermodynamic approach to ion interaction chromatography. Influence of the chain length of the solute ion; a chromatographic method for the determination of ion-pairing constants, *J. Sep. Sci* 28, 549–554, 2005; Pistos, C., Tsantili-Kakoulidou, A., and Koupparis, M., Investigation of the retention/pH profile of zwitterionic fluoroquinolones in reversed-phase and ion-interaction high-performance liquid chromatography, *J. Pharm. Biomed. Anal.* 39, 438–443, 2005; Choi, M.M., Douglas, A.D., and Murray, R.W., Ion-pair chromatographic separation of water-soluble gold monolayer-protected clusters, *Anal. Chem.* 78, 2779–2785, 2006; Saradhi, U.V., Prarbhakar, S., Reddy, T.J., and Vairamani, M., Ion-pair solid-phase extraction and gas chromatography mass spectrometric determination of acidic hydrolysis products of chemical warfare agents from aqueous samples, *J. Chromatog. A*, 1129, 9–13, 2006.

Tetrahydrofuran   Trimethylene   72.1   Solvent; template for Oxide   combinatorial chemistry.

Tetrahydrofuran

Leuty, S.J., Rapid dehydration of plant tissues for paraffin embedding; tetrahydrofuran vs. t-butanol, *Stain Technol.* 44, 103–104, 1969; Tandler, C.J. and Fiszer de Plazas, S., The use of tetrahydrofuran for delipidation and water solubilization of brain proteolipid proteins, *Life Sci.* 17, 1407–1410, 1975; Dressman, J.B., Himmelstein, K.J., and Higuchi, T., Diffusion of phenol in the presence of a complexing agent, tetrahydrofuran, *J. Pharm. Sci.* 72, 12–17, 1983; Diaz, R.S., Regueiro, P., Monreal, J., and Tandler, C.J., Selective extraction, solubilization, and reversed-phase high-performance liquid chromatography separation of the main proteins from myelin using tetrahydrofuran/water mixtures, *J. Neurosci. Res.* 29, 114–120, 1991; Santa, T., Koga, D., and Imai, K., Reversed-phase high-performance liquid chromatography of fullerenes with tetrahydrofuran-water as a mobile phase and sensitive ultraviolet or electrochemical detection, *Biomed. Chromatogr.* 9, 110–111, 1995; Lee, J., Kang, J.H., Lee, S.Y. et al., Protein kinase C ligands based on tetrahydrofuran templates containing a new set of phorbol ester pharmacophores, *J. Med. Chem.* 42, 4129–4139, 1999; Edwards, A.A., Ichihara, O., Murfin, S. et al., Tetrahydrofuran-based amino acids as library scaffolds, *J. Comb. Chem.* 6, 230–238, 2004; Baron, C.P., Refsgaard, H.H., Skibsted, L.H., and Andersen, M.L., Oxidation of bovine serum albumin initiated by the Fenton reaction — effect of EDTA, tert-butylhydroperoxide, and tetrahydrofuran, *Free Radic. Res.* 40, 409–417, 2006; Bowron, D.T., Finney, J.L., and Soper, A.K., The structure of liquid tetrahydrofuran, *J. Am. Chem. Soc.* 128, 5119–5126, 2006; Hermida, S.A., Possari, E.P., Souza, D.B. et al., 2′-deoxyguanosine, 2′-deoxycytidine, and 2′-deoxyadenosine adducts resulting from the reaction of tetrahydrofuran with DNA bases, *Chem. Res. Toxicol.* 19, 927–936, 2006; Li, A.C., Li, Y., Guirguis, M.S., Advantages of using tetrahydrofuran-water as mobile phases in the quantitation of cyclosporine A in monkey and rat plasma by liquid chromatography-tandem mass spectrometry, *J. Pharm. Biomed. Anal.* 43, 277–284, 2007.

Tetraphenylphosphonium Bromide   419.3   Membrane-permeable probe; determination of metal ions.

Tetraphenylphosphonium bromide

Boxman, A.W., Barts, P.W., and Borst-Pauwels, G.W., Some characteristics of tetraphenylphosphonium uptake into *Saccharomyces cerevisiae*, *Biochim. Biophys. Acta* 686, 13–18, 1982; Flewelling, R.F. and Hubbell, W.L., Hydrophobic ion interactions with membranes. Thermodynamic analysis of tetraphenylphosphonium binding to vesicles, *Biophys. J.* 49, 531–540, 1986; Prasad, R. and Hofer, M., Tetraphenylphosphonium is an indicator of negative membrane potential in *Candida albicans*, *Biochim. Biophys. Acta* 861, 377–380, 1986; Aiuchi, T., Matsunada, M., Nakaya, K., and Nakamura, Y., Calculation of membrane potential in synaptosomes with use of a lipophilic cation (tetraphenylphosphonium), *Chem. Pharm. Bull.* 37, 3333–3337, 1989; Nhujak T. and Goodall, D.M., Comparison of binding of tetraphenylborate and tetraphenylphosphonium ion to cyclodextrins studied by capillary electrophoresis, *Electrophoresis* 22, 117–122, 2001; Yasuda, K., Ohmizo, C., and Katsu, T., Potassium and tetraphenylphosphonium ion-selective electrodes for monitoring changes in the permeability of bacterial outer and cytoplasmic membranes, *J. Microbiol. Methods* 54, 111–115, 2003; Min, J.J., Biswal, S., Deroose, C., and Gambhir, S.S., Tetraphenylphosphonium as a novel molecular probe for imaging tumors, *J. Nucl. Med.* 45, 636–643, 2004.

Thioflavin T   Basic Yellow 1,   291   Dye for measurement of amyloid in tissue.
CI49005

Thioflavin T

Rogers, D.R., Screening for amyloid with the thioflavin T fluorescent method, *Am. J. Clin. Pathol.* 44, 59–61, 1965; Saeed, S.M. and Fine, G., Thioflavin T for amyloid, *Am. J. Clin. Pathol.* 47, 588–593, 1967; Levine, H., III, Stopped-flow kinetics reveal multiple phase of thioflavin T binding to Alzheimer beta (1–40) amyloid fibrils, *Arch. Biochem. Biophys.* 342,

306–316, 1997; De Ferrari, G.V., Mallender, W.D., Inestrosa, N.C., and Rosenberry, T.L., Thioflavin T is a fluorescent probe of the acetylcholinesterase peripheral site that reveals conformational interactions between the peripheral and acylation sites, *J. Biol. Chem.* 276, 23282–23287, 2001; Ban, T., Hamada, D., Hasegawa, K. et al., Direct observation of amyloid fibril growth monitored by thioflavin T fluorescence, *J. Biol. Chem.* 278, 16462–16465, 2003; Krebs, M.R., Bromley, E.H., and Donald, A.M., The binding of thioflavin T to amyloid fibrils: localization and implications, *J. Struct. Biol.* 149, 30–37, 2005; Khurana, R., Coleman, C., Ionescu-Zanetti, C. et al., Mechanisms of thioflavin T binding to amyloid fibrils, *J. Struct. Biol.* 151, 229–238, 2005; Darhal, N., Garnier-Suillerot, A., and Salerno, M., Mechanism of thioflavin T accumulation inside cells overexpressing P-glycoprotein or multidrug resistance-associated protein: role of lipophilicity and positive charge, *Biochem. Biophys. Res. Commun.* 343, 623–629, 2006; Eisert, R., Felau, L., and Brown, L.R., Methods for enhancing the accuracy and reproducibility of Congo red and thioflavin T assays, *Anal. Biochem.* 353, 144–146, 2006.

| **Thionyl Chloride** | Sulfurous Oxychloride | 118.97 | Preparation of acyl chlorides. |

Rodin, R.L. and Gershon, H., Photochemical alpha-chlorination of fatty acid chlorides by thionyl chloride, *J. Org. Chem.* 38, 3919–3921, 1973; DuVal, G., Swaisgood, H.E., and Horton, H.R, Preparation and characterization of thionyl chloride-activated succinamidopropyl-glass as a covalent immobilization matrix, *J. Appl. Biochem.* 6, 240–250, 1984; Molnar-Perl, I., Pinter-Szakacs, M., and Fabian-Vonsik, V., Esterification of amino acids with thionyl chloride acidified butanols for their gas chromatographic analysis, *J. Chromatog.* 390, 434–438, 1987; Stabel, T.J., Casele, E.S., Swaisgood, H.E., and Horton, H.R., Anti-IgG immobilized controlled pore glass. Thionyl chloride-activated succinamidopropyl-gas as a covalent immobization matrix, *Appl. Biochem. Biotechnol.* 36, 87–96, 1992; Chamoulaud, G. and Belanger, D., Chemical modification of the surface of a sulfonated membrane by formation of a sulfonamide bond, *Langmuir* 20, 4989–4895, 2004; Porjazoska, A.,Yilmaz, O.K., Baysal, K. et al., Synthesis and characterization of poly(ethylene glycol)-poly(D,L-lactide-co-glycolide) poly(ethylene glycol) tri-block co-polymers modified with collagen: a model surface suitable for cell interaction, *J. Biomater. Sci. Polym. Ed.* 17, 323–340, 2006; Gao, C., Jin, Z.Q., Kong, H. et al., Polyurea-functionalized multiwalled carbon nanotubes: synthesis, morphology, and Ramam spectroscopy, *J. Phys. Chem. B* 109, 11925–11932, 2005; Chen, G.X., Kim, H.S., Park, B.H., and Yoon, J.S., Controlled functionalization of multiwalled carbon nanotubes with various molecular-weight poly(L-lactic acid), *J. Phys. Chem. B* 109, 22237–22243, 2005.

| **Thiophosgene** | $CSCl_2$ | 115 |

| **Thiourea** | Thiocarbamide | 76.12 | Chaotropic agent; useful for membrane proteins; will react with haloacetyl derivatives such as iodoacetamide; protease inhibitor. |

Urea (carbamide)   Thiourea(thiocarbamide)

Maloof, F. and Soodak, M., Cleavage of disulfide bonds in thyroid tissue by thiourea, *J. Biol. Chem.* 236, 1689–1692, 1961; Gerfast, J.A., Automated analysis for thiourea and its derivatives in biological fluids, *Anal. Biochem.* 15, 358–360, 1966; Lippe, C., Urea and thiourea permeabilities of phospholipid and cholesterol bilayer membranes, *J. Mol. Biol.* 39, 588–590, 1966; Carlsson, J., Kierstan, M.P., and Brocklehurst, K., Reactions of L-ergothioneine and some other aminothiones with 2,2′- and 4,4′-dipyridyl disulphides and of L-ergothioneine with iodoacetamide, 2-mercaptoimidazoles, and 4-thioypyridones, thiourea, and thioacetamide as highly reactive neutral sulphur nucleophiles, *Biochem. J.* 139, 221–235, 1974; Filipski, J., Kohn K.W., Prather, R., and Bonner, W.M., Thiourea reverses crosslinks and restores biological activity in DNA treated with dichlorodiaminoplatinum (II), *Science* 204, 181–183, 1979; Wasil, M., Halliwell, B., Grootveld, M. et al., The specificity of thiourea, dimethylthiourea, and dimethyl sulphoxide as scavengers of hydroxyl radicals. Their protection of alpha-1-antiproteinase against inactivation by hypochlorous acid, *Biochem. J.* 243, 867–870, 1987; Doona, C.J. and Stanbury, D.M., Equilibrium and redox kinetics of copper(II)–thiourea complexes, *Inorg. Chem.* 35, 3210–3216, 1996; Rabilloud, T., Use of thiourea to increase the solubility of membrane proteins in two-dimensional electrophoresis, *Electrophoresis* 19, 758–760, 1998; Musante, L., Candiano, G., and Ghiggeri, G.M., Resolution of fibronectin and other uncharacterized proteins by two-dimensional polyacrylamide electrophoresis with thiourea, *J. Chromatog. B* 705, 351–356, 1998; Nagy, E., Mihalik, R., Hrabak, A. et al., Apoptosis inhibitory effect of the isothiourea compound, tri-(2-thioureido-S-ethyl)-amine, *Immunopharmacology* 47, 25–33, 2000; Galvani, M., Rovatti, L., Hamdan, M. et al., Protein alkylation in the presence/absence of thiourea in proteome analysis: a matrix-assisted laser desorption/ionization-time-of-flight-mass spectrometry investigation, *Electrophoresis* 22, 2066–2074, 2001; Castellanos-Serra, L. and Paz-Lago, D., Inhibition of unwanted proteolysis during sample preparation: evaluation of its efficiency in challenge experiments, *Electrophoresis* 23, 1745–1753, 2002; Tyagarajan, K., Pretzer, E., and Wiktorowicz, J.E., Thiol-reactive dyes for fluorescence labeling of proteomic samples, *Electrophoresis* 24, 2348–2358, 2003; Fuerst, D.E., and Jacosen, E.N., Thiourea-catalyzed enantioselective cyanosilylation of ketones, *J. Am. Chem. Soc.* 127, 8964–8965, 2005; Gomez, D.E., Fabbrizzi, L., Licchelli, M., and Monzani, E., Urea vs. thiourea in anion recognition, *Org. Biomol. Chem.* 3, 1495–1500, 2005; George, M., Tan, G., John, V.T., and Weiss, R.G., Urea and thiourea derivatives as low molecular-mass organochelators, *Chemistry* 11, 3243–3254, 2005; Limbut, W., Kanatharana, P., Mattiasson, B. et al., A comparative study of capacitive immunosensors based on self-assembled monolayers formed from thiourea, thioctic acid, and 3-mercaptopropionic acid, *Biosens. Bioelectron.* 22, 233–240, 2006.

TNBS

Trinitrobenzene Sulfonic Acid   293.2   Reagent for the determination of amino groups in proteins; also reacts with sulfydryl groups and hydrazides; used to induce animal model of colitis.

Lysine   +   2, 4, 6-trinitrobenzenesulfonic acid

Habeeb, A.F., Determination of free amino groups in proteins by trinitrobenzenesulfonic acid, *Anal. Biochem.* 14, 328–336, 1966; Goldfarb, A.R., A kinetic study of the reactions of amino acids and peptides with trinitrobenzenesulfonic acid, *Biochemistry* 5, 2570–2574, 1966; Scheele, R.B. and Lauffer, M.A., Restricted reactivity of the epsilon-amino groups of tobacco mosaic virus protein toward trinitrobenzenesulfonic acid, *Biochemistry* 8, 3597–3603, 1969; Godin, D.V. and Ng, T.W., Trinitrobenzenesulfonic acid: a possible chemical probe to investigate lipid–protein interactions in biological membranes, *Mol. Pharmacol.* 8, 426–437, 1972; Bubnis, W.A. and Ofner, C.M., III, The determination of epsilon-amino groups in soluble and poorly soluble proteinaceous materials by a spectrophotometric method using trinitrobenzenesulfonic acid, *Anal. Biochem.* 207, 129–133, 1992; Cayot, P. and Tainturier, G., The quantification of protein amino groups by the trinitrobenzenesulfonic acid method: a reexamination, *Anal. Biochem.* 249, 184–200, 1997; Neurath, M., Fuss, I., and Strober, W., TNBS-colitis, *Int. Rev. Immunol.* 19, 51–62, 2000; Lindsay, J., Van Montfrans, C., Brennen, F. et al., IL-10 gene therapy prevents TNBS-induced colitis, *Gene Ther.* 9, 1715–1721, 2002; Whittle, B.J., Cavicchi, M., and Lamarque, D., Assessment of anticolitic drugs in the trinitrobenzenesulfonic acid (TNBS) rat model of inflammatory bowel disease, *Methods Mol. Biol.* 225, 209–222, 2003; Necefli, A., Tulumoglu, B., Giris, M. et al., The effects of melatonin on TNBS-induced colitis, *Dig. Dis. Sci.* 51, 1538–1545, 2006.

TNM Tetranitromethane 196.03 Modification of tyrosine residues in proteins; crosslinking a side reaction as a reaction with cysteine; antibacterial and antiviral agent.

Sokolovsky, M., Riordan, J.F., and Vallee, B.L., Tetranitromethane. A reagent for the nitration of tyrosyl residues in proteins, *Biochemistry* 5, 3582–3589, 1966; Nishikimi, M. and Yagi, K., Reaction of reduced flavins with tetranitromethane, *Biochem. Biophys. Res. Commun.* 45, 1042–1048, 1971; Kunkel, G.R., Mehrabian, M., and Martinson, H.G., Contact-site crosslinking agents, *Mol. Cell. Biochem.* 34, 3–13, 1981; Rial, E. and Nicholls, D.G., Chemical modification of the brown-fat-mitochondrial uncoupling protein with tetranitromethane and *N*-ethylmaleimide. A cysteine residue is implicated in the nucleotide regulation of anion permeability, *Eur. J. Biochem.* 161, 689–694, 1986; Prozorovski, V., Krook, M., Atrian, S. et al., Identification of reactive tyrosine residues in cysteine-reactive dehydrogenases. Differences between liver sorbitol, liver alcohol, and *Drosophila* alcohol dehydrogenase, *FEBS Lett.* 304, 46–50, 1992; Gadda, G., Banerjee, A., and Fitzpatrick, P.F., Identification of an essential tyrosine residue in nitroalkane oxidase by modification with tetranitromethane,

Biochemistry 39, 1162–1168, 2000; Hodges, G.R. and Ingold, K.U., Superoxide, amine buffers, and tetranitro-methane: a novel free radical chain reaction, *Free Radic. Res.* 33, 547–550, 2000; Capeillere-Blandin, C., Gausson, V., Descamps-Latscha, B., and Witko-Sarsat, V., Biochemical and spectrophotometric significance of advanced oxidation protein products, *Biochim. Biophys. Acta* 1689, 91–102, 2004; Lundblad, R.L., *Chemical Reagents for Protein Modification*, CRC Press, Boca Raton, FL, 2004; Negrerie, M., Martin, J.L., and Nghiem, H.O., Functionality of nitrated acetylcholine receptor: the two-step formation of nitrotyrosines reveals their differential role in effectors binding, *FEBS Lett.* 579, 2643–2647, 2005; Carven, G.J. and Stern, L.J., Probing the ligand-induced conformational change in HLA-DR1 by selective chemical modification and mass spectrometry mapping, *Biochemistry* 44, 13625–13637, 2005.

Trehalose	α-D-glucopyrano-glucopyranosyl-1,1-α-D-glucopyranoside; Mycose	342.3	A nonreducing sugar that is found in a variety of organisms where it is thought to protect against stress such as dehydration; there is considerable interest in the use of trehalose as a stabilizer in biopharmaceutical proteins.

Elbein, A.D., The metabolism of alpha, alpha-trehalose, *Adv. Carbohydr. Chem. Biochem.* 30, 227–256, 1974; Wiemken, A., Trehalose in yeast, stress protectant rather than reserve carbohydrate, *Antonie Van Leeuwenhoek,* 58, 209–217, 1990; Newman, Y.M., Ring, S.G., and Colaco, C., The role of trehalose and other carbohydrates in biopreservation, *Biotechnol. Genet. Eng. Rev.* 11, 263–294, 1993; Panek, A.D., Trehalose metabolism — new horizons in technological applications, *Braz. J. Med. Biol. Res.* 28, 169–181, 1995; Schiraldi, C., Di Lernia, I., and De Rosa, M., Trehalose production: exploiting novel approaches, *Trends Biotechnol.* 20, 420–425, 2002; Elbein, A.D., Pan, Y.T., Pastuszak, I., and Carroll, D., New insights on trehalose: a multifunctional molecule, *Glycobiology* 13, 17R–27R, 2003; Gancedo, C. and Flores, C.L., The importance of a functional trehalose biosynthetic pathway for the life of yeasts and fungi, *FEMS Yeast Res.* 4, 351–359, 2004; Cordone, L., Cottone, G., Giuffrida, S. et al., Internal dynamics and protein-matrix coupling in trehalose-coated proteins, *Biochim. Biophys. Acta* 1749, 252–281, 2005.

Trichloroacetic Acid	163.4	Protein precipitant.

Trichloroacetic acid   Trifluoroacetic acid

Chang, Y.C., Efficient precipitation and accurate quantitation of detergent-solubilized membrane proteins, *Anal. Biochem.* 205, 22–26, 1992; Sivaraman, T., Kumar, T.K., Jayaraman, G., and Yu. C., The mechanism of 2,2,2-trichloroacetic acid-induced protein precipitation, *J. Protein Chem.* 16, 291–297, 1997; Arnold, U. and Ulbrich-Hoffman, R., Quantitative protein precipation form guandine hydrochloride-containing solutions by sodium deoxycholate/trichloroacetic acid, *Anal. Biochem.* 271, 197–199, 1999; Jacobs, D.I., van Rijssen, M.S., van der Heijden, R., and Verpoorte, R., Sequential solubilization of proteins precipitated with trichloroacetic acid in acetone from cultured *Catharanthus roseus* cells yields 52% more spots after two-dimensional electrophoresis, *Proteomics* 1, 1345–1350, 2001; Garcia-Rodriguez, S., Castilla, S.A., Machado, A., and Ayala, A., Comparison of methods for sample preparation of individual rat cerebrospinal fluid samples prior to two-dimensional polyacrylamide gel electrophoresis, *Biotechnol. Lett.* 25, 1899–1903, 2003; Chen, Y.Y., Lin, S.Y., Yeh, Y.Y. et al., A modified protein precipitation procedure for efficient removal of albumin from serum, *Electrophoresis* 26, 2117–2127, 2005; Zellner, M., Winkler, W., Hayden, H. et al., Quantitative validation of different protein precipitation methods in proteome analysis of blood platelets, *Electrophoresis* 26, 2481–2489, 2005; Carpentier, S.C., Witters, E., Laukens, K. et al., Preparation of protein extracts from recalcitrant plant tissues: an evaluation of different methods for two-dimensional gel electrophoresis analysis, *Proteomics* 5, 2497–2507, 2005; Manadas, B.J., Vougas, K., Fountoulakis, M., and Duarte, C.B., Sample sonication after trichloroacetic acid precipitation increases protein recovery from cultured hippocampal neurons, and improves resolution and reproducibility in two-dimensional gel electrophoresis, *Electrophoresis* 27, 1825–1831, 2006; Wang, A., Wu, C.J., and Chen, S.H., Gold nanoparticle-assisted protein enrichment and electroelution for biological samples containing low protein concentration — a prelude of gel electrophoresis, *J. Proteome Res.* 5, 1488–1492, 2006.

Triethanolamine

Triethanolamine   Triethanolamine hydrochloride

pKa approx. 9.5

Tris(2-hydroxyethyl) Amine   149.2   Buffer; transdermal transfer reagent.

Fitzgerald, J.W., The Tris-catalyzed isomerization of potassium D-glucose 6-*O*-sulfate, *Can. J. Biochem.* 53, 906–910, 1975; Buhl, S.N., Jackson, K.Y., and Graffunder, B., Optimal reaction conditions for assaying human lactate dehydrogenase pyruvate-to-lactate at 25, 30, and 37 degrees C, *Clin. Chem.* 24, 261–266, 1978; Myohanen, T.A., Bouriotas, V., and Dean, P.D., Affinity chromatography of yeast alpha-glucosidase using ligand-mediated chromatography on immobilized phenylboronic acids, *Biochem. J.* 197, 683–688, 1981; Shinomiya, Y., Kato, N., Imazawa, M., and Miyamoto, K., Enzyme immunoassay of the myelin basic protein, *J. Neurochem.* 39, 1291–1296, 1982; Arita, M., Iwamori, M., Higuchi, T., and Nagai, Y., 1,1,3,3-tetramethylurea and triethanolaminme as a new useful matrix for fast atom bombardment mass spectrometry of gangliosides and neutral glycosphingolipids, *J. Biochem.* 93, 319–322, 1983; Cao, H. and Preiss, J., Evidence for essential arginine residues at the active site of maize branching enzymes, *J. Protein Chem.* 15, 291–304, 1996; Knaak, J.B., Leung, H.W., Stott, W.T. et al., Toxicology of mono-, di-, and triethanolamine, *Rev. Environ. Contim. Toxicol.* 149, 1–86, 1997; Liu, Q., Li, X., and Sommer, S.S., pK-matched running buffers for gel electrophoresis, *Anal. Biochem.* 270, 112–122, 1999; Sanger-van de Griend, C.E., Enantiomeric separation of glycyl dipeptides by capillary electrophoresis with cyclodextrins as chiral selectors, *Electrophoresis* 20, 3417–3424, 1999; Fang, L., Kobayashi, Y., Numajiri, S. et al., The enhancing effect of a triethanolamine-ethanol-isopropyl myristate mixed system on the skin permeation of acidic drugs, *Biol. Pharm. Bull.* 25, 1339–1344, 2002; Musial, W. and Kubis, A., Effect of some anionic polymers of pH of triethanolamine aqueous solutions, *Polim. Med.* 34, 21–29, 2004.

Triethylamine

Triethylamine

N,N-diethylethanamine   101.2   Ion-pair reagent; buffer.

Brind, J.L., Kuo, S.W., Chervinsky, K., and Orentreich, N., A new reversed-phase, paired-ion thin-layer chromatographic method for steroid sulfate separations, *Steroids* 52, 561–570, 1988; Koves, E.M., Use of high-performance liquid chromatography-diode array detection in forensic toxicology, *J. Chromatog. A* 692, 103–119, 1995; Cole, S.R. and Dorsey, J.G., Cyclohexylamine additives for enhanced peptide separations in reversed-phase liquid chromatography, *Biomed. Chromatog.* 11, 167–171, 1997; Gilar, M. and Bouvier, E.S.P., Purification of crude DNA oligonucleotides by solid-phase extraction and reversed-phase high-performance liquid chromatography, *J. Chromatog. A* 890, 167–177, 2000; Loos, R. and Barcelo, D., Determination of haloacetic acids in aqueous environments by solid-phase extraction followed by ion-pair liquid chromatography-electrospray ionization mass spectrometric detection, *J. Chromatog. A* 938, 45–55, 2001; Gilar, M., Fountain, K.J., Budman, Y. et al., Ion-pair reversed-phase high-performance liquid chromatography analysis of oligonucleotides: retention prediction, *J. Chromatog. A* 958, 167–182, 2002; El-dawy, M.A., Mabrouk, M.M., and El-Barbary, F.A., Liquid chromatographic determination of fluoxetine, *J. Pharm. Biomed. Anal.* 30, 561–571, 2002; Yang, X., Zhang, X., Li, A. et al., Comprehensive two-dimensional separations based on capillary high-performance liquid chromatography and microchip electrophoresis, *Electrophoresis* 24, 1451–1457, 2003; Murphey, A.T., Brown-Augsburger, P., Yu, R.Z. et al., Development of an ion-pair reverse-phase liquid chromatographic/tandem mass spectrometry method for the determination of an 18-mer phosphorothioate oligonucleotide in mouse liver tissue, *Eur. J. Mass Spectrom.* 11, 209–215, 2005; Xie, G., Sueishi, Y., and Yamamoto, S., Analysis of the effects of protic, aprotic, and multi-component solvents on the fluorescence emission of naphthalene and its exciplex with triethylamine, *J. Fluoresc.* 15, 475–483, 2005.

Trifluoroacetic Acid   114.0   Ion-pair reagent; HLPC; peptide synthesis.

Rosbash, D.O. and Leavitt, D., Decalcification of bone with trifluoroacetic acid, *Am. J. Clin. Pathol.* 22, 914–915, 1952; Katz, J.J., Anhydrous trifluoroacetic acid as a solvent for proteins, *Nature* 174, 509, 1954; Uphaus, R.A., Grossweiner, L.I., Katz, J.J., and Kopple, K.D., Fluorescence of tryptophan derivatives in trifluoroacetic acid, *Science* 129, 641–643, 1959; Acharya, A.S., di Donato, A., Manjula, B.N. et al., Influence of trifluoroacetic acid on retention times of histidine-containing tryptic peptides in reverse phase HPLC, *Int. J. Pept. Protein Res.* 22, 78–82, 1983; Tsugita, A., Uchida, T., Mewes, H.W., and Ataka, T., A rapid vapor-phase acid (hydrochloric and trifluoroacetic acid) hydrolysis of peptide and protein, *J. Biochem.* 102, 1593–1597, 1987; Hulmes, J.D. and Pan, Y.C., Selective cleavage of polypeptides with trifluoroacetic acid: applications for microsequencing, *Anal. Biochem.* 197, 368–376, 1991; Eshragi, J. and Chowdhury, S.K., Factors affecting electrospray ionization of effluents containing trifluoroacetic acid for high-performance liquid chromatography/mass spectrometry, *Anal. Chem.* 65, 3528–3533, 1993; Apffel, A., Fischer, S., Goldberg, G. et al., Enhanced sensitivity for peptide mapping with electrospray liquid chromatography-mass spectrometry in the presence of signal suppression due to trifluoroacetic acid-containing mobiles phases, *J. Chromatog. A* 712, 177–190, 1995; Guy, C.A. and Fields, G.B., Trifluoroacetic acid cleavage and deprotection of resin-bound peptides following synthesis by Fmoc chemistry, *Methods Enzymol.* 289, 67–83, 1997; Morrison, I.M. and Stewart, D., Plant cell wall fragments released on solubilization in trifluoroacetic acid, *Phytochemistry* 49, 1555–1563, 1998; Yan, B., Nguyen, N., Liu, L. et al., Kinetic comparison of trifluoroacetic acid cleavage reactions of resin-bound carbamates, ureas, secondary amides, and sulfonamides from benzyl-, benzhydryl-, and indole-based linkers, *J. Comb. Chem.* 2, 66–74, 2000; Ahmad, A., Madhusudanan, K.P., and Bhakuni, V., Trichloroacetic acid- and trifluoroacetic acid-induced unfolding of cytochrome C: stabilization of a nativelike fold intermediate(1), *Biochim. Biophys. Acta* 1480, 201–210, 2000; Chen, Y., Mehok, A.R., Mant, C.T. et al., Optimum concentration of trifluoroacetic acid for reversed-phase liquid chromatography of peptide revisited, *J. Chromatog. A* 1043, 9–18, 2004.

Tris

Triethanolamine   pKa approx. 9.5   Triethanolamine hydrochloride

Triethylamine

Tris-(hydroxymethyl) Aminomethylmethane	121.14	Buffer.

Bernhard, S.A., Ionization constants and heats of Tris(hydroxymethyl)aminomethane and phosphate buffers, *J. Biol. Chem.* 218, 961–969, 1956; Rapp, R.D. and Memminger, M.M., Tris(hydroxymethyl)aminomethane as an electrophoresis buffer, *Am. J. Clin. Pathol.* 31, 400–403, 1959; Rodkey, F.L., Tris(hydroxymethyl)aminomethane as a standard for Kjeldahl nitrogen analysis, *Clin. Chem.* 10, 606–610, 1964; Oliver, R.W. and Viswanatha, T., Reaction of Tris(hydroxymethyl)aminomethane with cinnamoyl imidazole and cinnamoyltrypsin, *Biochim. Biophys. Acta* 156, 422–425, 1968; Douzou, P., Enzymology at subzero temperatures, *Mol. Cell. Biochem.* 1, 15–27, 1973; Fitzgerald, J.W., The Tris-catalyzed isomerization of potassium D-glucose 6-*O*-sulfate, *Can. J. Biochem.* 53, 906–910, 1975; Visconti, M.A. and Castrucci, A.M., Tris buffer effects on melanophore-aggregrating responses, *Comp. Biochem. Physiol. C* 82, 501–503, 1985; Stambler, B.S., Grant, A.O., Broughton, A., and Strauss, H.C., Influences of buffers on dV/dtmax recovery kinetics with lidocaine in myocardium, *Am. J. Physiol.* 249, H663–H671, 1985; Nakano, M. and Tauchi, H., Difference in activation by Tris(hydroxymethyl)aminomethane of Ca,Mg-ATPase activity between young and old rat skeletal muscles, *Mech. Aging Dev.* 36, 287–294, 1986; Oliveira, L., Araujo-Viel, M.S., Juliano, L., and Prado, E.S., Substrate activation of porcine kallikrein *N*- α derivatives of arginine 4-nitroanilides, *Biochemistry* 26, 5032–5035, 1987; Ashworth, C.D. and Nelson, D.R., Antimicrobial potentiation of irrigation solutions containing Tris-[hydroxymethyl] aminomethane-EDTA, *J. Am. Vet. Med. Assoc.* 197, 1513–1514, 1990; Schacker, M., Foth, H., Schluter, J., and Kahl, R., Oxidation of Tris to one-carbon compounds in a radical-producing model system, in microsomes, in hepatocytes, and in rats, *Free Radic. Res. Commun.* 11, 339–347, 1991; Weber, R.E., Use of ionic and zwitterionic (Tris/BisTris and HEPES) buffers in studies on hemoglobin function, *J. Appl. Physiol.* 72, 1611–1615, 1992; Veeck, L.L., TES and Tris (TEST)-yolk buffer systems, sperm function testing, and *in vitro* fertilization, *Fertil. Steril.* 58, 484–486, 1992; Shiraishi, H., Kataoka, M., Morita, Y., and Umemoto, J., Interaction of hydroxyl radicals with Tris (hydroxymethyl)aminomethane and Good's buffers containing hydroxymethyl or hydroxyethyl residues produce formaldehyde, *Free Radic. Res. Commun.* 19, 315–321, 1993; Vasseur, M., Frangne, R., and Alvarado, F., Buffer-dependent pH sensitivity of the fluorescent chloride-indicator dye SPQ, *Am. J. Physiol.* 264, C27–C31, 1993; Niedernhofer, L.J., Riley, M., Schnez-Boutand, N. et al., Temperature-dependent formation of a conjugate between Tris(hydroxymethyl) aminomethane buffer and the malondialdehyde-DNA adduct pyrimidopurinone, *Chem. Res. Toxicol.* 10, 556–561, 1997; Trivic, S., Leskovac, V., Zeremski, J. et al., Influence of Tris(hydroxymethyl)aminomethane on kinetic mechanism of yeast alcohol dehydrogenase, *J. Enzyme Inhib.* 13, 57–68, 1998; Afifi, N.N., Using difference spectrophotometry to study the influence of different ions and buffer systems on drug protein binding, *Drug Dev. Ind. Pharm.* 25, 735–743, 1999; AbouHaider, M.G. and Ivanov, I.G., Nonenzymatic RNA hydrolysis promoted by the combined catalytic activity of buffers and magnesium ions, *Z. Naturforsch.* 54, 542–548, 1999; Shihabi, Z.K., Stacking of discontinuous buffers in capillary zone electrophoresis, *Electrophoresis* 21, 2872–2878, 2000; Stellwagen, N.C, Bossi, A., Gelfi, C., and Righetti, P.G., DNA and buffers: are there any noninteracting, neutral pH buffers? *Anal. Biochem.* 287, 167–175, 2000; Burcham, P.C., Fontaine, F.R., Petersen, D.R., and Pyke, S.M., Reactivity of Tris(hydroxymethyl) aminomethane confounds immunodetection of acrolein-adducted proteins, *Chem. Res. Toxicol.* 16, 1196–1201, 2003; Koval, D., Kasicka, V., and Zuskova, I., Investigation of the effect of ionic strength of Tris-acetate background electrolyte on electrophoretic mobilities of mono-, di-, and trivalent organic anions by capillary electrophoresis, *Electrophoresis* 26, 3221–3231, 2005; Kinoshita, T., Yamaguchi, A., and Tada, T., Tris(hydroxymethyl)aminomethane-induced conformational change and crystal-packing contraction of porcine pancreatic elastase, *Acta Crystallograph. Sect. F Struct. Biol. Cryst. Commun.* 62, 623–626, 2006; Qi, Z., Li, X., Sun, D. et al., Effect of Tris on catalytic activity of MP-11, *Bioelectrochemistry* 68, 40–47, 2006.

| Tris-(2-carboxyethyl) phosphine | TCEP | 250.2 | Reducing agent. |

Tris(2-carboxyethyl)phosphine

Gray, W.R., Disulfide structures of highly bridged peptides: a new strategy for analysis, *Protein Sci.* 2, 1732–1748, 1993; Gray, W.R., Echistatin disulfide bridges: selective reduction and linkage assignment, *Protein Sci.* 2, 1749–1755, 1993; Han, J.C. and Han, G.Y., A procedure for quantitative determination of Tris(2-carboxyethyl)phosphine, an odorless reducing agent more stable and effective than dithiothreitol, *Anal. Biochem.* 220, 5–10, 1994; Wu, J., Gage, D.A., and Watson, J.T., A strategy to locate cysteine residues in proteins by specific chemical cleavage followed by matrix-assisted laser desorption/ionization-time-of-flight mass spectrometry, *Anal. Biochem.* 235, 161–174, 1996; Han, J., Yen. S., Han, G., and Han, F., Quantitation of hydrogen peroxide using Tris(2-carboxyethyl) phosphine, *Anal. Biochem.* 234, 107–109, 1996; Han, J., Clark, C., Han, G. et al., Preparation of 2-nitro-5-thiobenzoic acid using immobilized Tris(2-carboxyethyl) phosphine, *Anal. Biochem.* 268, 404–407, 1999; Anderson, M.T., Trudell, J.R., Voehringer, D.W. et al., An improved monobromobimane assay for glutathione utilizing Tris-(2-carboxyethyl)phosphine as the reductant, *Anal. Biochem.* 272, 107–109, 1999; Shafer, D.E., Inman, J.K. and Lees, A. Reaction of Tris(2-carboxyethyl)phosphine (TCEP) with maleimide and alpha-haloacyl groups: anomalous elution of TCEP by gel filtration, *Anal. Biochem.* 282, 161–164, 2000; Rhee, S.S. and Burke, D.H., Tris(2-carboxyethyl)phosphine stabilization of RNA: comparison with dithiothreitol for use with nucleic acid and thiophosphoryl chemistry, *Anal. Biochem.* 325, 137–143, 2004; Legros, C., Celerier, M.L., and Guette, C., An unusual cleavage reaction of a peptide observed during dithiothreitol and Tris(2-carboxyethyl)phosphine reduction: application to sequencing of HpTx2 spider toxin using nanospray tandem mass spectrometry, *Rapid Commun. Mass Spectrom.* 19, 1317–1323, 2004; Xu, G., Kiselar, J., He, Q., and Chance, M.R., Secondary reactions and strategies to improve quantitative protein footprinting, *Anal. Chem.* 77, 3029–3037, 2005; Valcu, C.M. and Schlink, K., Reduction of proteins during sample preparation and two-dimensional gel electrophoresis of woody plant samples, *Proteomics* 6, 1599–1605, 2006; Scales, C.W., Convertine, A.J., and McCormick, C.L., Fluorescent labeling of RAFT-generated poly(*N*-isopropylacrylamide) via a facile maleimide-thiol coupling reaction, *Biomacromolecules* 7, 1389–1392, 2006.

| **Urea** | Carbamide | 60.1 | Chaotropic agent. |

Edelhoch, H., The effect of urea analogues and metals on the rate of pepsin denaturation, *Biochim. Biophys. Acta* 22, 401–402, 1956; Steven, F.S. and Tristram, G.R., The denaturation of ovalbumin. Changes in optical rotation, extinction, and viscosity during serial denaturation in solution of urea, *Biochem. J.* 73, 86–90, 1959; Nelson, C.A. and Hummel, J.P., Reversible denaturation of pancreatic ribonuclease by urea, *J. Biol. Chem.* 237, 1567–1574, 1962; Herskovits, T.T., Nonaqueous solutions of DNA; denaturation by urea and its methyl derivatives, *Biochemistry* 2, 335–340, 1963; Subramanian, S., Sarma, T.S., Balasubramanian, D., and Ahluwalia, J.C., Effects of the urea–guanidinium class of protein denaturation on water structure: heats of solution and proton chemical shift studies, *J. Phys. Chem.* 75, 815–820, 1971; Strachan, A.F., Shephard, E.G., Bellstedt, D.U. et al., Human serum amyloid A protein. Behavior in aqueous and urea-containing solutions and antibody production, *Biochem. J.* 263, 365–370, 1989; Gervais, V., Guy, A., Teoule, R., and Fazakerley, G.V., Solution conformation of an oligonucleotide containing a urea deoxyribose residue in front of a thymine, *Nucleic Acids Res.* 20, 6455–6460, 1992; Smith, B.J., Acetic acid-urea polyacrylamide gel electrophoresis of proteins, *Methods Mol. Biol.* 32, 39–47, 1994; Buck, M., Radford, S.E., and Dobson, C.M., Amide hydrogen exchange in a highly denatured state. Hen egg-white lysozyme in urea, *J. Mol. Biol.* 237, 247–254, 1994; Shirley, B.A., Urea and guanidine hydrochloride denaturation curve, *Methods Mol. Biol.* 40, 177–190, 1995; Bennion, B.J. and Daggett, V., The molecular basis for the chemical denaturation of proteins by urea, *Proc. Natl. Acad. Sci. USA* 100, 5142–5147, 2003; Soper, A.K., Castner, E.W., and Luzar, A., Impact of urea on water structure: a clue to its properties as a denaturant? *Biophys.Chem.*105, 649–666, 2003; Smith, L.J., Jones, R.M., and van Gunsteren, W.F., Characterization of the denaturation of human alpha-1-lactalbumin in urea by molecule dynamics simulation, *Proteins* 58, 439–449, 2005; Idrissi, A., Molecular structure and dynamics of liquids: aqueous urea solutions, *Spectrochim. Acta A Mol. Biomol. Spectrosc.* 61, 1–17, 2005; Chow, C., Kurt, N., Murphey, R.M., and Cavagnero, S., Structural characterization of apomyoglobin self-associated species in aqueous buffer and urea solution, *Biophys. J.* 90, 298–309, 2006.

| **Vinyl Pyridine** | 4-vinylpyridine | 105.1 | Modification of cysteine residues in protein. |

4-vinylpyridine   2-vinylpyridine

| **Water** | Hydrogen Oxide | 18.0 | Solvent. |

Lumry, R. and Rajender, S., Enthalpy-entropy compensation phenomena in water solutions of proteins and small molecules: a ubiquitous property of water, *Biopolymers* 9, 1125–1227, 1970; Cooke, R. and Kuntz, I.D., The properties of water in biological systems, *Annu. Rev. Biophys. Bioeng.* 3, 95–126, 1974; Fettiplace, R. and Haydon, D.A., Water permeability of lipid membranes, *Physiol. Rev.* 60, 510–550, 1980; Lewis, C.A. and Wolfenden, R., Antiproteolytic aldehydes and ketones: substituent and secondary deuterium isotope effects on equilibrium addition of water and other nucleophiles, *Biochemistry* 16, 4886–4890, 1977; Wolfenden, R.V., Cullis, P.M., and Southgate, C.C., Water, protein folding, and the genetic code, *Science* 206, 575–577, 1979; Wolfenden, R., Andersson, L., Cullis, P.M., and Southgate, C.C., Affinities of amino acid side chains for solvent water, *Biochemistry* 20, 849–855, 1981; Cullis, P.M. and Wolfenden, R., Affinity of nucleic acid bases for solvent water, *Biochemistry* 20, 3024–3028, 1981; Radzicka, A., Pedersen, L., and Wolfenden, R., Influences of solvent water on protein folding: free energies of salvation of *cis* and *trans* peptides are nearly identical, *Biochemistry* 27, 4538–4541, 1988; Dzingeleski, G.D. and Wolfenden, R., Hypersensitivity of an enzyme reaction to solvent water,

Biochemistry 32, 9143–9147, 1993; Timasheff, S.N., The control of protein stability and association by weak interactions with water: how do solvents affect these processes? *Annu. Rev. Biophys. Biomol. Struct.* 22, 67–97, 1993; Wolfenden, R. and Radzcika, A., On the probability of finding a water molecule in a nonpolar cavity, *Science* 265, 936–937, 1994; Jayaram, B. and Jain, T., The role of water in protein–DNA recognition, *Annu. Rev. Biophys. Biomol. Struct.* 33, 343–361, 2004; Pace, C.N., Trevino, S., Prabhakaran, E., and Scholtz, J.M., Protein structure, stability, and solubility in water and other solvents, *Philos. Trans. R. Soc. Lond. B Biol. Sci.* 359, 1225–1234, 2004; Rand, R.P., Probing the role of water in protein conformation and function, *Philos. Trans. R. Soc. Lond. B Biol. Sci.* 359, 1277–1284, 2004; Bagchi, B., Water dynamics in the hydration layer around proteins and micelles, *Chem. Rev.* 105, 3179–3219, 2005; Raschke, T.M., Water structure and interactions with protein surfaces, *Curr. Opin. Struct. Biol.* 16, 152–159, 2006; Levy, Y. and Onuchic, J.N., Water mediation in protein folding and molecular recognition, *Annu. Rev. Biophys. Biomol. Struct.* 35, 389–415, 2006; Wolfenden, R., Degrees of difficulty of water-consuming reactions in the absence of enzymes, *Chem. Rev.* 106, 3379–3396, 2006.

4 A Listing of Log P Values, Water Solubility, and Molecular Weight for Some Selected Chemicals[a]

Compound	M.W.	Log P[b]	Water Solubility(gm/L)[c]
Acetamide	59.07	−1.26	2.25×10^3
Acetic Acid	60.05	−0.17	10×10^3
Acetic Anhydride	102.09	−0.58	1.2×10^2
Acetoacetic Acid	102.1	−0.98	1×10^3
Acetoin	88.11	−0.36	1×10^3
Acetone	58.08	−0.24	1×10^3
Acetophenone	120.15	1.58	6.13
N-Acetylcysteinamide	162.21	−0.29	5.8
N-Acetylcysteine		−0.64	
N-Aceylmethionine		−0.49	
Acetylsalicylic Acid	180.16	1.19	4.6
Acridine	179.22	3.40	0.03
Acrolein	56.06	−0.01	2.13×10^2
Acrylamide	71.08	−0.67	6.4×10^2
Adenine	135.13	−0.09	1.0
Adenosine	267.25	−1.05	8.2
Alanine	89.09	−2.96	1.7×10^2
Aldosterone		1.08	
9-Aminoacridine	194.23	2.74	0.02
4-Aminobenzoic Acid (p-aminobenzoic acid; PABA)	151.17	1.03	9.89
4-Aminobutyric Acid (γ-aminobutyric acid; GABA)	103.12	−3.17	1.3×10^3
6-Aminohexanoic Acid (ε-aminocaproic acid)	131.18	−2.95	5.05×10^2
Ammonium Picrate	246.14	−1.40	1.6×10^2
Aniline		0.9	
Anisole		2.11	
ANS (1-amino-2-naphthalene sulfonic acid)	222.25	−0.97	2.23
Anthracene		4.45	
Arabinose	150.13	−3.02	1×10^3
Arginine	174.20	−4.20	1.82×10^2
Ascorbic Acid	176.13	−1.64	1×10^3
Asparagine	132.12	−3.82	29.4
Aspartic Acid	133.10	−3.89	5.0
Barbital (5,5-diethylbarbituric acid)	184.20	0.65	7
Barbituric Acid	128.1	−1.47	
Benzamide	121.14	0.64	13.5
Benzamidine	120.16	0.65	27.9
Benzene	78.11	2.13	0.002
Benzoic Acid	122.12	1.87	3.4
Betaine	117.15	−4.93	6.11×10^2
Biuret (imidodicarbonic acid)	103.08		1.5
Bromoacetic Acid	138.95	0.41	93
2-Bromopropionic Acid	152.98	0.92	29.9
2,3-Butanediol	90.12	−0.36	7.6×10^2
2,3-Butanedione	86.09	−1.34	2×10^2
Butyl Urea	116.16	0.41	46.3
3-Butyl Hydroxy Urea	132.16	0.32	23.5
Cacodylic Acid	138.00	0.36	2×10^3
Carbon Tetrachloride	153.82	2.83	0.8
Chloroacetamide	93.51	−0.53	90
Chloroacetic Anhydride	170.98	−0.07	68

Chloroacetyl Chloride	112.94	−0.22	1.6×10^2
Chloroform	119.38	1.97	8
6-Chloroindole	151.60	3.25	0.1
p-Chloromercuribenzoic Acid	357.16	1.48	0.3
Chlorosuccinic Acid	152.54	−0.57	1.8×10^2
Cholesterol	386.67	8.74	0.9
Cholic Acid	405.58	2.02	0.2
Citric Acid	192.13	−1.72	5.92×10^2
Congo Red	696.68	2.63	1.2×10^2
Corticosterone		1.94	
Cortisone		2.88	
Creatine	132.14	−3.72	13.3
Creatinine	113.12	−1.76	80
Crotonaldehyde (2-butenal)	70.09	0.60	1.8×10^2
Cyanoacetic Acid	85.06	−0.76	7.7×10^2
Cyanogen	52.04	0.07	1.2×10^2
Cyanuric Acid	129.08	0.61	2
Cyclohexanone		0.81	
Cysteine	121.16	−2.49	1.1×10^2
Cystine	240.30	−5.08	0.2
Cytidine	243.22	−2.51	1.8×10^2
Cytosine	111.10	−1.73	8
Deoxycholic Acid	392.58	3.50	0.04
Deoxycorticosterone		2.88	
Dexamethasone		2.01	
Diazomethane	42.04	2.00	2
Dichloromethane		1.2	
Dicumarol	336.30	2.07	0.1
Diethyl Ether (ethyl ether; ether)	74.1	0.9	
Diethylsuberate	230.31	3.35	0.7
Diethylsulfone	122.19	−0.59	1.4×10^2
N,N-Diethyl Urea	116.2	0.1	4
Dihydroxyacetone	88.11	−0.49	16.2
Diketene	84.08	−0.39	5.3×10^2
Dimethylformamide		−1.04	
Dimethylguanidine	87.13	−0.95	1.6
Dimethylphthalate		1.56	
Dimethylsulfoxide	78.13	−1.35	1×10^3
1,4-Dinitrobenzene		1.47	
2,4-Dinitrophenol		1.55	
EDTA	292.25	−3.86	1
EDTA (sodium salt)	360.17	−13.17	1×10^3
Estradiol		2.69	
Ethanol (ethyl alcohol)	46.07	−0.31	$1 \times 10+3$
Ethylene Glycol	2.07	−1.36	1×10^3
Ethylene Oxide	44.05	−0.30	1×10^3
N-Ethylnicotinamide	150.18	0.31	41.2
N-Ethylthiourea	104.17	−0.21	24
Ethylurea	88.11	−0.74	26.4
Fluorescein	333.32	3.35	0.05
Fluoroacetone	76.07	−0.39	286
Folic Acid	441.41	−2.00	0.002
Formaldehyde	30.03	0.35	400
Formic Acid	48.03	−0.54	1×10^3
Galactose	180.16	−2.43	683
Glucose	180.16	−1.88	1.2×10^3
Glutamic Acid	147.10	−3.69	8.6
Glutamine	146.15	−3.64	41
Glycerol	92.10	−1.76	1×10^3
Glycine	75.10	−3.21	2.5×10^2
Glyoxal	58.04	−1.66	1×10^3
Glyoxylic Acid	74.04	−1.40	1×10^3
Guanidine	59.07	−1.63	1.8
Guanine	151.13	−0.91	2.1
Guanosine	283.25	−1.90	0.7
Hexanal	100.16	1.78	6
Hydroxyproline	131.13	−3.17	395
Hydroxyurea	76.06	−1.80	224
N-Hydroxyurea	104.11	−0.76	
N-Hydroxy-1-ethylurea	104.11	−0.10	7
Imidazole	68.08	−0.08	160
Indole	117.15	2.14	4
Inositol	180.16	−2.08	143
Iodoacetamide	184.96	−0.19	76
Isoleucine	131.18	−1.70	34

Isopropanol	60.10	0.05	1×10^3
Lactic Acid	90.08	−0.72	1×10^3
Lactose	342.30	−5.43	195
Leucine	131.18	−1.52	22
Linoleic Acid	280.45	7.05	0.00004
Lysine	146.19	−3.05	1×10^3
Maleic Anhydride	98.06	1.62	5
Maltose	342.30	−5.43	780
Mannitol	182.17	−3.10	216
Mercaptoacetic Acid	92.12	0.09	1×10^3
2-Mercaptobenzoic Acid	154.19	2.39	0.7
Methane	16.04	1.09	0.002
Methanol	32.04	−0.77	1×10^3
Methionine	149.21	−1.87	57
Methotrexate	454.45	−1.85	2.6
Methylene Blue	319.86	5.85	44
N-Methyl Glycine	89.09	−2.78	300
5-Methylindole	131.18	2.68	0.5
Methyl Isocyanate	57.05	0.79	29
Methylmalonic Acid	118.09	−0.83	680
Methyl Methacrylate	86.09	0.80	49
Methylmethane Sulfonate	110.13	−0.66	1×10^3
Methyl Thiocyanate	73.12	0.73	32
N-Methyl Thiourea	119.21	−0.69	240
Methyl Urea	74.08	−1.40	100
Naphthalene	128.17	3.29	220
Nicotinic Acid	123.11	0.36	18
Ornithine	132.16	−4.22	1×10^3
Orotic Acid	156.10	−0.83	2
Oxalic Acid	90.06	−2.22	
Oxindole	133.15	1.16	9
Palmitic Acid	256.43	7.17	0.0008
Paraldehyde	132.16	0.67	112
Pentobarbital	226.28	2.10	0.7
Phenol	94.11	1.46	83
Phenylalanine	165.19	−1.52	22
Phosgene	98.02	−0.71	475
Proline	115.13	−2.54	131
Propylamine	59.11	0.48	1×10^3
Propylene Oxide	58.08	0.03	595
Prostaglandin E2	352.48	2.82	0.006
Pyridine	79.10	0.65	1×10^3
Pyridoxal	203.63	−3.32	500
Pyridoxal-5-Phosphate	247.15	0.37	20
Pyridoxine	169.18	−0.77	282
Pyruvic Acid	88.06	−1.24	1×10^3
Ribose	150.13	−2.32	
Sarin	140.10	0.72	1×10^3
Serine	105.09	−3.07	425
Sorbic Acid	112.13	1.33	2
Sorbitol	182.17	−2.20	3×10^3
Stearic Acid	284.49	8.23	0.03
Succinic Anhydride	100.07	0.81	24
Succinimide	99.09	−0.85	196
Sucrose	342.30	−3.70	2.12×10^3
Testosterone	288.43	3.32	0.03
Tetrahydrofuran	72.11	0.46	1×10^3
Threonine	119.12	−2.94	97
Toluene	92.14	2.73	0.5
2,4,6-Trinitrobenzene	257.12	0.23	21
Tryptophan	204.23	−1.06	12
Urea	60.06	−2.11	545
Valine	117.15	−2.26	60

[a] Adapted from *Handbook of Physical Properties of Organic Chemicals*, Howard, P.H. and Meylan, W.M., Eds., CRC Press, Boca Raton, FL, 1997.

[b] Log P = log[concentration in 1-octanol]
       [concentration in water]

See Howard and Meylan and the following references for discussions of log P (log of partitioning coefficient for a substance between 1-octanol and water).

[c] Solubility values taken from various literature sources and in some cases are approximations.

REFERENCES

Abrahams, M.H., Du, C.M., and Platts, J.A., Lipophilicity of the nitrophenols, *J. Org. Chem.* 65, 7114–7718, 2000.

Avdeef, A., Physicochemical profiling (solubility, permeability, and charge state), *Curr. Top. Med. Chem.* 1, 277–351, 2001.

Chuman, H., Mori, A., and Tanaka, H., Prediction of the 1-octanol/H_2O partition coefficient, Log P, by *Ab Initio* calculations: hydrogen-bonding effect of organic solutes on Log P, *Analyt. Sci.* 18, 1015–1020, 2002.

Halling, P.J., Thermodynamic predictions for biocatalysis in nonconventional media: theory, tests, and recommendations for experimental design and analysis, *Enzyme Microb. Technol.* 16, 178–206, 1994.

Hansch, C. and Leo, A., *Exploring QSAR. Fundamentals and Applications in Chemistry and Biology*, American Chemical Society, Washington, DC, 1995.

Lipinski, C.A., Lombardo, F., Dominy, B.W., and Feeney, P.J., Experimental and computational approaches to estimate solubility and permeability in drug discovery and development settings, *Adv. Drug. Deliv. Rev.* 46, 3–26, 2001.

Uttamsingh, V., Keller, D.A., and Anders, M.W., Acylase I-catalyzed deacetylation of *N*-acetyl-L-cysteine and *S*-alkyl-*N*-acetyl-L-cysteines, *Chem. Res. Toxicol.* 11, 800–809, 1998.

Valko, K., Du, C.M., Bevan, C., Reynolds, D.P., and Abraham, M.H., Rapid method for the estimation of octanol/water partition coefficient (Log P_{oct}) from gradient RP-HPLC retention and a hydrogen bond acidity term ($\Sigma\alpha_2^H$), *Curr. Medicin. Chem.* 8, 1137–1146, 2001.

Yalkowsky, S.H. and He, Y., *Handbook of Aqueous Solubility Data*, CRC Press, Boca Raton, FL, 2003.

5 Protease Inhibitors and Protease Inhibitor Cocktails

While protease inhibitor cocktails have been in use for some time,[1] there are few rigorous studies examining their effect on proteolysis and very few concerned with proteolytic degradation during the processing of material for analysis or during purification.[2] It is usually assumed that proteolysis can be a problem and protease inhibitors or protease inhibitor cocktails are usually included as part of a protocol without the provision of justification. There are several excellent review articles in this area. Salveson and Nagase[3] discuss the inhibition of proteolytic enzymes in great detail including much practical information that should be considered in experimental design. The discussion of the relationship between inhibitor concentration, inhibitor/enzyme binding constants (association constants, binding constants, $t_{1/2}$, inhibition constants, etc.), and enzyme inhibition is of particular importance. For example, with a reversible enzyme inhibitor (such as benzamidine), if the K_i value is 100 nM, a 100 μM concentration of inhibitor would be required to decrease protease activity by 99.9%. Salveson and Nagase[3] also note the well-known differences in the reaction rates of inhibitors such as DFP and PMSF with the active site of serine proteases. DFP is much faster than PMSF with trypsin but equivalent rates are seen with chymotrypsin. PMSF is included in commercial protease inhibitor cocktails because of its lack of toxicity compared to DFP; 3,4-dichloroisocoumarin (3,4-DCI), as described by Powers and colleagues[4], is faster than either DFP or PMSF. Also enzyme inhibition occurs in the presence of substrate (proteins), which will influence the effectiveness of both irreversible and reversible enzyme inhibitors. In addition, some protease inhibitor cocktails include both PMSF and benzamidine. Benzamidine is a competitive inhibitor of trypticlike serine proteases and slows the rate of inactivation of such enzymes by reagents such as PMSF.[5] The investigator is also advised to consider the modification of proteins and other biological compounds by protease inhibitors in reactions not associated with proteases such as the modification of tyrosine by DFP or PMSF.[6] In addition, some of the protease inhibitors such as DFP and PMSF are subject to hydrolysis under conditions (pH $\geq$ 7.0) used for modification. For those unfamiliar with the history of DFP, DFP is a potent neurotoxin (inhibitor of acetyl cholinesterase) and should be treated with considerable care; a prudent investigator has a DFP repair kit in close proximity (weak base and pralidoxime-2-chloride [2-PAM]). Given these various issues, it is critical to validate that, in fact, the sample is being protected against proteolysis.

REFERENCES

1. Takei, Y., Marzi, I., Kauffman, F.C. et al., Increase in survival time of liver transplants by protease inhibitors and a calcium channel blocker, nisolidpine. *Transplantation* 50, 14–20, 1990.
2. Pyle, L.E., Barton, P., Fujiwara, Y., Mitchell, A., and Fidge, N., Secretion of biologically active human proapolipoprotein A-1 in a baculovirus-insect cell system: protection from degradation by protease inhibitors, *J. Lipid Res.* 36, 2355–2361, 1995.
3. Salveson, G. and Nagase, H., Inhibition of proteolytic enzymes, in *Proteolytic Enyzmes: Practical Approaches,* 2nd ed., R. Benyon and J.S. Bond, Eds., Oxford University Press, Oxford, UK, pp. 105–130, 2001.
4. Harper, J.W., Hemmi, K., and Powers, J.C., Reaction of serine proteases with substituted isocoumarins: discovery of 3,4-dichloroisocoumarin, a new general mechanism-based serine protease inhibitor, *Biochemistry* 24, 1831–1841, 1985.

5. Lundblad, R.L., A rapid method for the purification of bovine thrombin and the inhibition of the purified enzyme with phenylmethylsulfonyl fluoride, *Biochemistry* 10, 2501–2506, 1971.
6. Lundblad, R.L., *Chemical Reagents for Protein Modification*, CRC Press, Boca Raton, FL, 2004.

Characteristics of Selected Protease Inhibitors, Which Can Be Used in Protease Inhibitor Cocktails

Common Name	Other Nomenclature	M.W.	Primary Design
Amastatin	*N*-[(2*S*,3*R*)-3-amino-2-hydroxy-5-methyl hexanoyl]-L-valyl-L-valyl-L-aspartic Acid	529.0	Inhibitor of some aminopeptidases.

Amastatin

Amastatin is a complex peptidelike inhibitor of aminopeptidases obtained from *Actinoycetes* culture. Amastatin is a competitive inhibitor of aminopeptidase A, aminopeptidase M, and other aminopeptidases. Amastatin has been used for the affinity purification of aminopeptidases. Amastatin has been shown to inhibit amino acid iosomerases. Amastatin is structurally related to bestatin and has been described as an immunomodulatory factor. See Aoyagi, T., Tobe, H., Kojima, F. et al., Amastatin, an inhibitor of aminopeptidase A, produced by actinomycetes, *J. Antibiot.* 31, 636–638, 1978; Tobe, H., Kojima, F., Aoyagi, T., and Umezawa, H., Purification by affinity chromatography using amastatin and properties of he aminopeptidase A from pig kidney, *Biochim. Biophys. Acta* 613, 459–468, 1980; Rich, D.H., Moon, B.J., and Harbeson, S., Inhibition of aminopeptidases by amastatin and bestatin derivatives. Effect of inhibitor structure on slow-binding processes, *J. Med. Chem.* 27, 417–422 , 1984; Meisenberg, G. and Simmons, W.H., Amastatin potentiates the behavioral effects of vasopressin and oxytocin in mice, *Peptides* 5, 535–539, 1984; Wilkes, S.H. and Prescott, J.M., The slow, tight binding of bestatin and amastatin to aminopeptidases, *J. Biol. Chem.* 260, 13154–13162, 1985; Matsuda, N., Katsuragi, Y., Saiga, Y. et al., Effects of aminopeptidase inhibitors actinonin and amastatin on chemotactic and phagocytic responses of human neutrophils, *Biochem. Int.* 16, 383–390, 1988; Orawski, A.T. and Simmons, W.H., Dipeptidase activities in rat brain synaptosomes can be distinguished on the basis of inhibition by bestatin and amastatin: identification of a kyotrophin (Tyr-Arg)-degrading enzyme, *Neurochem. Res.* 17, 817–820, 1992; Kim, H. and Lipscomb, W.N., X-ray crystallographic determination of the structure of bovine lens leucine aminopeptidase complexed with amastatin: formation of a catalytic mechanism, featuring a gem-diolate transition state, *Biochemistry* 32, 8365–8378, 1993; Bernkop-Schnurch, A., The use of inhibitory agents to overcome the enzymatic barrier to perorally administered therapeutic peptides and proteins, *J. Control. Release* 52, 1–16, 1998; Fortin, J.P., Gera, L., Bouthillier, J. et al., Endogenous aminopeptidase N decreases the potency of peptide agonists and antagonists of the kinin B1 receptors in the rabbit aorta, *J. Pharmacol. Exp. Ther.* 312, 1169–1176, 2005; Olivo Rdo, A., Teixeira Cde, R., and Silveira, P.F., Representative aminopeptidases and prolyl endopeptidase from murin macrophages; comparative activity levels in resident and elicited cells, *Biochem. Pharmacol.* 69, 1441–1450, 2005; Gera. L., Fortin, J.P., Adam, A. et al., Discovery of a dual-function peptide that combines aminopeptidase N inhibition and kinin B1 receptor antagonism, *J. Pharmacol. Exp. Ther.* 317, 300–308, 2006; Krsyanovic, M., Brgles, M., Halassy, B. et al., Purification and characterization of the *l*,(*l*/*d*)-aminopeptidase from guinea pig serum, *Prep. Biochem. Biotechnol.* 36, 175–195, 2006; Torres, A.M., Tsampazi, M., Tsampazi, C. et al., Mammalian *l* to *d*-amino-acid-residue isomerase from platypus venom, *FEBS Lett.* 580, 1587–1591, 2006.

Aprotinin		6512	Protein protease inhibitor.

Basic pancreatic trypsin inhibitor; Kunitz pancreatic trypsin inhibitor; Trasylol®. This protein inhibits some but not all trypticlike serine proteinases and is included in some protease inhibitor cocktails. See Hulsemann, A.R., Jongejan, R.C., Rolien Raatgeep, H. et al., Epithelium removal and peptidase inhibition enhance relaxation of human airways to vasoactive intestinal peptide, *Am. Rev. Respir. Dis.* 147, 1483–1486, 1993; Cornelius, R.M. and Brash, J.L., Adsorption from plasma and buffer of single- and two-chain high molecular weight kininogen to glass and sulfonated polyurethane surfaces, *Biomaterials* 20, 341–350, 1999; Lafleur, M.A., Handsley, M.M., Knauper, V. et al., Endothelial tubulogenesis with fibrin gels specifically requires the activity of membrane-type-matrix metalloproteinases (MT-MMPs), *J. Cell Sci.* 115, 3427–3438, 2002; Shah, R.B., Palamakula, A., and Khan, M.A., Cytotoxicity evaluation of enzyme inhibitors and absorption enhancers in Caco-2 cells for oral delivery of salmon calcitonin, *J. Pharm. Sci.* 93, 1070–1982; Spens, E. and Häggerström, L., Protease activity in protein-free (NS) myeloma cell cultures, In Vitro *Cell Dev. Biol.* 41, 330–336, 2005. As it is a potent inhibitor of plasmin, aprotinin is frequently included in fibrin gel-based cultures to preserve the fibrin gel structure. See Ye, Q., Zund, G., Benedikt, P. et al., Fibrin gel as a three-dimensional matrix in cardiovascular tissue engineering, *Eur. J. Cardiothorac. Surg.* 17, 587–591, 2000; Krasna, M., Planinsek, F., Knezevic, M. et al., Evaluation of a fibrin-based skin substitute prepared in a defined keratinocyte medium, *Int. J. Pharm.* 291, 31–37, 2005; Sun, X.T., Ding, Y.T., Yan, X.G. et al., Antiangiogenic synergistic effect of basic fibroblast growth factor and vascular endothelial growth factor in an *in vitro* quantitative microcarrier-based three-dimensional fibrin angiogenesis system, *World J. Gastroenterol.* 10, 2524–2528, 2004; Gille, J., Meisner, U., Ehlers, E.M. et al., Migration pattern, morphology and viability of cells suspended in or sealed with fibrin glue: a histomorphology study, *Tissue Cell* 37, 339–348, 2005; Yao, L., Swartz, D.D., Gugino, S.F. et al., Fibrin-based tissue-engineered blood vessels: differential effects of biomaterial and culture parameters on mechanical strength and vascular reactivity, *Tissue Eng.* 11, 991–1003, 2005. Aprotinin is used therapeutically in the inhibition of plasmin activity both as a freestanding product and as a component of fibrin sealant products.

Benzamidine HC1

Benzamidine

156.61   Inhibitor of trypticlike serine proteases.

An aromatic amidine derivative (Markwardt, F., Landmann, H., and Walsmann, P., Comparative studies on the inhibition of trypsin, plasmin, and thrombin by derivatives of benzylamine and benzamidine, *Eur. J. Biochem.* 6, 502–506, 1968; Guvench, O., Price, D.J., and Brooks, C.L., III, Receptor rigidity and ligand mobility in trypsin-ligand complexes, *Proteins* 58, 407–417, 2005), which is used as a competitive inhibitor of trypticlike serine proteases. It is not a particularly tight-binding inhibitor and is usually used at millimolar concentrations. Ensinck, J.W., Shepard, C., Dudl, R.J., and Williams, R.H., Use of benzamidine as a proteolytic inhibitor in the radioimmunoassay of glucagon in plasma, *J. Clin. Endocrinol. Metab.* 35, 463–467, 1972; Bode, W. and Schwager, P., The refined crystal structure of bovine beta-trypsin at 1.8 Å resolution. II. Crystallographic refinement, calcium-binding site, benzamidine-binding site, and active site at pH 7.0., *J. Mol. Biol.* 98, 693–717, 1975; Nastruzzi, C., Feriotto, G., Barbieri, R. et al., Differential effects of benzamidine derivatives on the expression of *c-myc* and HLA-DR alpha genes in a human B-lymphoid tumor cell line, *Cancer Lett.* 38, 297–305, 1988; Clement, B., Schmitt, S., and Zimmerman, M., Enzymatic reduction of benzamidoxime to benzamidine, *Arch. Pharm.* 321, 955–956, 1988; Clement, B., Immel, M., Schmitt, S., and Steinman, U., Biotransformation of benzamidine and benzamidoxime *in vivo*, *Arch. Pharm.* 326, 807–812, 1993; Renatus, M., Bode, W., Huber, R. et al., Structural and functional analysis of benzamidine-based inhibitors in complex with trypsin: implications for the inhibition of factor Xa, tPA, and urokinase, *J. Med. Chem.* 41, 5445–5456, 1998; Henriques, R.S., Fonseca, N., and Ramos, M.J., On the modeling of snake venom serine proteinase interactions with benzamidine-based thrombin inhibitors, *Protein Sci.* 13, 2355–2369, 2004; Gustavsson, J., Farenmark, J., and Johansson, B.L., Quantitative determination of the ligand content in Benzamidine Sepharose® 4 Fast Flow media with ion-pair chromatography, *J. Chromatog. A* 1070, 103–109, 2005. Concentrated solutions of benzamidine will require pH adjustment prior to use.

Bestatin

Bestatin

N-[(2*S*,3*R*)-3-amino-2-hydroxy-1-oxo-4-phenylbutyl]-L-leucine

344.8   Aminopeptidase inhibitor; also described as a metalloproteinase inhibitor.

Bestatin is an inhibitor of some aminopeptidases and it was isolated from *Actinomycetes* culture. Bestatin was subsuently shown to have immunomodulatory activity and induces apoptosis in tumor cells. Bestatin is included in some proteaseinhibitor cocktails and has been demonstrated to inhibit intracellular protein degradation. See Umezawa, H., Aoyagi, T., Suda, H. et al., Bestatin, an inhibitor of aminopeptidase B, producted by actinomycetes, *J. Antibiot.* 29, 97–99, 1976; Suda, H., Takita, T., Aoyagi, T., and Umezawa, H., The structure of bestatin, *J. Antibiot.* 29, 100–101, 1976; Saito, M., Aoyagi, T., Umezawa, H., and Nagai, Y., Bestatin, a new specific inhibitor of aminopeptidases, enhances activation of small lymphocytes by concanavalin A, *Biochem. Biophys. Res. Commun.* 76, 526–533, 1976; Botbot, V. and Scornik, O.A., Degradation of abnormal proteins in intact mouse reticulocytes: accumulation of intermediates in the presence of bestatin, *Proc. Natl. Acad. Sci. USA* 76, 710–713, 1979; Botbol, V. and Scornik, O.A., Peptide intermediates in the degradation of cellular proteins. Bestatin permits their accumulation in mouse liver *in vivo*, *J. Biol. Chem.* 258, 1942–1949, 1983; Rich, D.H., Moon, B.J., and Harbeson, S., Inhibition of aminopeptidases by amastatin and bestatin derivatives. Effect of inhibitor structure on slow-binding processes, *J. Med. Chem.* 27, 417–422, 1984; Wilkes, S.H. and Prescott, J.M., The slow, tight binding of bestatin and amastatin to aminopeptidases, *J. Biol. Chem.* 260, 13154–13160, 1985; Patterson, E.K., Inhibition by bestatin of a mouse ascites tumor dipeptidase. Reversal by certain substrates, *J. Biol. Chem.* 264, 8004–8011, 1989; Botbol, V. and Scornik, O.A., Measurement of instant rates of protein degradation in the livers of intact mice by the accumulation of bestatin-induced peptides, *J. Biol. Chem.* 266, 2151–2157, 1991; Tieku, S. and Hooper, N.M., Inhibition of aminopeptidases N, A, and W. A re-evaluation of the actions of bestatin and inhibitors of angiotensin converting enzyme, *Biochem. Pharmacol.* 44, 1725–1730, 1992; Taylor, A., Peltier, C.Z., Torre, F.J., and Hakamian, N., Inhibition of bovine lens leucine aminopeptidase by bestatin: number of binding sites and slow binding of this inhibitor, *Biochemistry* 32, 784–790, 1993; Schaller, A., Bergey, D.R., and Ryan, C.A., Induction of wound response genes in tomato leaves by bestatin, an inhibitor of aminopeptidases, *Plant Cell* 7, 1893–1898, 1995; Nemoto, H., Ma, R., Suzuki, I.I., and Shibuya, M., A new one-pot method for the synthesis of alpha-siloxyamides from aldehydes or ketones and its application to the synthesis of (-)bestatin, *Org. Lett.* 2, 4245–4247, 2000; van Hensbergen, Y., Brfoxterman, H.J., Peters, E. et al., Aminopeptidase inhibitor bestatin stimulates microvascular endothelial cell invasion in a fibrin matrix, *Thromb. Haemost.* 90, 921–929, 2003; Stamper, C.C., Bienvenue, D.L., Bennett, B. et al., Spectroscopic and X-ray crystallographic characterization of bestatin bound to the aminopeptidase from *Aeromonas(Vibrio)proteolytica*, *Biochemistry* 43, 9620–9628, 2004; Zheng, W., Zhai, Q., Sun, J. et al., Bestatin, an inhibitor of aminopeptidases, provides a chemical genetics approach to dissect jasmonate signaling in *Aribidopsis*, *Plant Physiol.* 141, 1400–1413, 2006; Hui, M. and Hui, K.S., A novel aminopeptidase with highest preference for lysine, *Neurochem. Res.* 31, 95–102, 2006.

Cystatins

	Protein Inhibitors of Cysteine Proteases	Inhibitors of cysteine proteinases.

Cystatin refers to a diverse family of protein cysteine protease inhibitors. There are three general types of cystatins: Type 1 (stefens), which are primarily found in the cytoplasm but can appear in extracellular fluids; Type 2, which are secreted and found in most extracellular fluids; and Type 3, which are multidomain protease inhibitors containing carbohydrates and that include the kininogens. Cystatin 3 is used to measure renal function in clinical chemistry. See Barrett, A.J., The cystatins: a diverse superfamily of cysteine peptidase inhibitors, *Biomed. Biochim. Acta* 45, 1363–1374, 1986; Katunuma, N., Mechanisms and regulation of lysosomal proteolysis, *Revis. Biol. Cellular* 20, 35–61, 1989; Gauthier, F., Lalmanach, G., Moeau, T. et al., Cystatin mimicry by synthetic peptides, *Biol. Chem. Hoppe Seyler* 373, 465–470, 1992; Bobek, L.A. and Levine, M.J., Cystatins — inhibitors of cysteine proteineases, *Crit. Rev. Oral Biol. Med.* 3, 307–332, 1992; Calkins, C.C., and Sloane, B.F., Mammalian cysteine protease inhibitors: biochemical properties and possible roles in tumor progression, *Biol. Chem. Hoppe Seyler* 376, 71–80, 1995; Turk, B., Turk, V., and Turk, D., Structural and functional aspects of papainlike cysteine proteinases and their protein inhibitors, *Biol. Chem.* 378, 141–150, 1997; Kos, J., Stabuc, B., Cimerman, N., and Brunner, N., Serum cystatin C, a new marker of glomerular filtration rate, is increased during malignant progression, *Clin. Chem.* 44, 2556–2557, 1998; Vray, B., Hartman, S., and Hoebeke, J., Immunomodulatory properties of cystatins, *Cell. Mol. Life Sci.* 59, 1503–1512, 2002; Arai, S., Matsumoto, I., Emori, Y., and Abe, K., Plant seed cystatins and their target enzymes of endogenous and exogenous origin, *J. Agric. Food Chem.* 50, 6612–6617, 2002; Abrahamson, M., Alvarez-Fernandez, M., and Nathanson, C.M., Cystatins, *Biochem. Soc. Symp.* 70, 179–199, 2003; Dubin, G., Proteinaceous cysteine protease inhibitors, *Cell. Mol. Life Sci.* 62, 653–669, 2005; Righetti, P.G., Castagna, A., Antonucci, F. et al., Proteome analysis in the clinical chemistry laboratory: myth or reality? *Clin. Chim. Acta* 357, 123–139, 2005; Overall, C.M. and Dean, R.A., Degradomics: systems biology of the protease web. Pleiotropic roles of MMPs in cancer, *Cancer Metastasis Rev.* 25, 69–75, 2006; Kotsylfakis, M., Sá-Nunes, A., Francischetti, I.M.B. et al., Anti-inflammatory and immunosuppressive activity of sialostatin L, a salivary cystatin from Tick *Ixodes scapularis*, *J. Biol. Chem.* 281, 26298–26307, 2006.

DCI was developed by James C. Powers and coworkers at Georgia Institute of Technology (Harper, J.W., Hemmi, K., and Powers, J.C., Reaction of serine proteases with substituted isocoumarins: discovery of 3,4-dichloroisocoumarin, a new general mechanism-based serine protease inhibitor, *Biochemistry* 24, 1831–1841, 1985). This inhibitor is reasonably specific, although side reactions have been described. As with the sulfonyl fluorides and DFP, the modification is slowly reversible and enhanced by basic solvent conditions and/or nucleophiles. DCI has been used as a proteosome inhibitor. See Rusbridge, N.M. and

3,4-Dichloroisocoumarin DCI 215 Mechanism-based
inhibitor of serine
proteases.

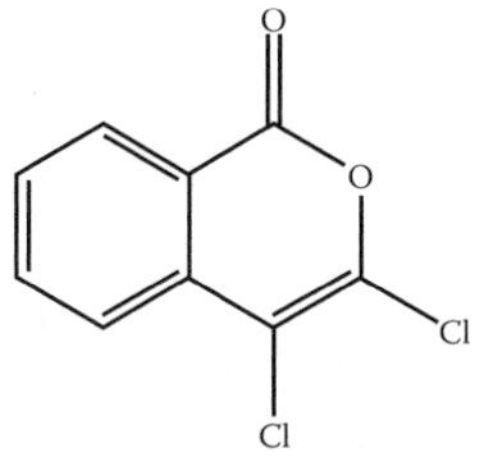

3,4-dichloroisocoumarin

Benyon, R.J., 3,4-dichloroisocoumarin, a serine protease inhibitor, inactivates glycogen phosphorylase b, *FEBS Lett.* 30, 133–136, 1990; Weaver, V.M., Lach, B., Walker, P.R., and Sikorska, M., Role of proteolysis in apoptosis: involvement of serine proteases in internucleosomal DNA fragmentation in immature thymocytes, *Biochem. Cell Biol.* 71, 488–500, 1993; Garder, A.M., Aviel, S., and Argon, Y., Rapid degradation of an unassembled immunoglobulin light chain is mediated by a serine protease and occurs in a pre-Golgi compartment, *J. Biol. Chem.* 268, 25940–25947, 1993; Lu, Q. and Mellgren, R.L., Calpain inhibitors and serine protease inhibitors can produce apoptosis in HL-60 cells, *Arch. Biochem. Biophys.* 334, 175–181, 1996; Adams, J. and Stein, R., Novel inhibitors of the proteosome and their therapeutic use in inflammation, *Annu. Rep. Med. Chem.* 31, 279–288, 1996; Olson, S.T., Swanson, R., Patston, P.A., and Bjork, I., Apparent formation of sodium dodecyl sulfate-stable complexes between serpins and 3,4-dichloroisocoumarin-inactivated proteinases is due to regeneration of active proteinase from the inactivated enzyme, *J. Biol. Chem.* 272, 13338–13342, 1997; Mesner, P.W., Bible, K.C., Martins, L.M. et al., Characterization of caspase processing and activation in HL-60 cell cytosol under cell-free conditions — nucleotide requirement and inhibitor profile, *J. Biol. Chem.* 274, 22635–22645, 1999; Kam, C.M., Hudig, D., and Powers, J.C., Granzymes (lymphocyte serine proteases): characterization with natural and synthetic substrates and inhibitors, *Biochem. Biophys. Acta* 1477, 307–323, 2000; Rivett, A.J. and Gardner, R.C., Proteosome inhibitors: from *in vitro* uses to clinical trials, *J. Pep. Sci.* 6, 478–488, 2000; Bogyo, M. and Wang, E.W., Proteosome inhibitors: complex tools for a complex enzyme, *Curr. Top. Microbiol. Immunol.* 268, 185–208, 2002; Powers, J.C., Asgian, J.L., Ekici, O.D., and James, K.E., Irreversible inhibitors of serine, cysteine, and threonine proteases, *Chem. Rev.* 102, 4639–4740, 2002; Pochet, L., Frederick, R., and Masereei, B., Coumarin and isocoumarin as serine protease inhibitors, *Curr. Pharm. Des.* 10, 3781–3796, 2004.

Diisopropyl Phosphosphorofluoridate DFP; Diisopropyl 184 Reaction at active site
Fluorophosphate serine.

Diisopropylphosphorofluoridate

Serine residue in protein

Disopropylphosphorylserine

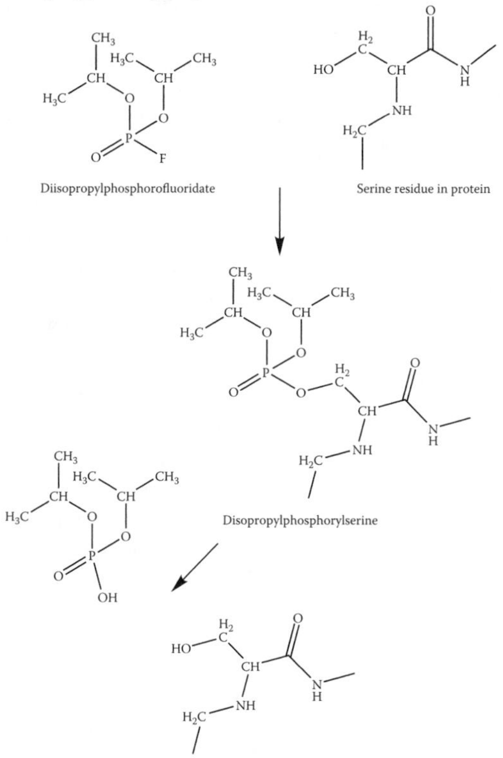

DFP was developed during World War II as a neurotoxin. DFP reacts with the active serine of serine proteases and was used to define the presence of this amino acid at the active sites of trypsin and chymotrypsin. DFP has been replaced by PMSF as a general reagent for inhibition of proteases although it is still used on occasion because of the ease of identification of the phosphoserine derivative. See Jansen, E.F., Jang, R., and Balls, A.K., The inhibition of purified, human plasma cholinesesterase with diisopropylfluorophosphate, *J. Biol. Chem.* 196, 247–253, 1952; Gladner, J.A. and Neurath, H.A., C-terminal groups in chymotrypsinogen and DFP-alpha-chymotrypsin in relation to the activation process, *Biochim. Biophys. Acta* 9, 335–336, 1952; Schaffer, N.K., May, S.C., Jr., and Summerson, W.H., Serine phosphoric acid from diisopropylphosphoryl chymotrypsin, *J. Biol. Chem.* 202, 67–76, 1953; Oosterbaan, R.A., Kunst, P., and Cohen, J.A., The nature of the reaction between diisopropylfluorophosphate and chymotrypsin, *Biochim. Biophys. Acta.* 16, 299–300, 1955; Wahlby, S., Studies on *Streptomyces griseus* protease. I. Separation of DFP-reacting enzymes and purification of one of the enzymes, *Biochim. Biophys. Acta* 151, 394–401, 1968; Hoskin, R.J. and Long, R.J., Purification of a DFP-hydrolyzing enzyme from squid head ganglion, *Arch. Biochem. Biophys.* 150, 548–555, 1972; Craik, C.S., Roczniak, S., Largman, C., and Rutter, W.J., The catalytic role of the active aspartic acid in serine proteases, *Science* 237, 909–913, 1987; D'Souza, C.A., Wood, D.D., She, Y.M., and Moscarello, M.A., Autocatalytic cleavage of myelin basic protein: an alternative to molecular mimicry, *Biochemistry* 44, 12905–12913, 2005. DFP is a potent neurotoxin and attention should be given to antidotes to organophosphates (Tuovinen, K., Kaliste-Korhonen, E., Raushel, F.M., and Hanninen, O., Phosphotriesterase, pralidoxime-2-chloride (2-PAM), and eptastigmine treatments and their combinations in DFP intoxication, *Toxicol. Appl. Pharmacol.* 141, 555–560, 1996; Auta, J., Costa, E., Davis, J., and Guidotti, A., Imidazenil: a potent and safe protective agent against diisopropyl fluorophosphate toxicity, *Neuropharmacology* 46, 397–403, 2004; Tuovinen, K., Organophosphate- induced convulsions and prevention of neuropathological damages, *Toxicology* 196, 31–39, 2004).

E-64

E-64 from *Aspergillus japonicus*

L-*trans*-epoxysuccinyl-leucylamide-(4-guanido)-butane or N-[N-(L-*trans*-carboxyoxiran-2-carbonyl)-L-leucyl]-agmatine

357.4

Inhibitor of sulfhydryl proteases.

E-64 is a reasonably specific inhibitor of sulfhydryl proteases and it functions by forming a thioether linkage with the active site cysteine. E-64 is frequently referred to as an inhibitor of lysosomal proteases and antigen processing. See Hashida, S., Towatari, T., Kominami, E., and Katunuma, N., Inhibition by E-64 derivatives of rat liver cathepsins B and cathepsin L *in vitro* and *in vivo*, *J. Biochem.* 88, 1805–1811, 1980; Grinde, B., Selective inhibition of lysosomal protein degradation by the thiol proteinase inhibitors E-64, Ep-459, and Ep-457 in isolated rat hepatocytes, *Biochim. Biophys. Acta* 701, 328–333, 1982; Barrett, A.J., Kembhavi, A.A., Brown, A.A. et al., L-*trans*-epoxysuccinyl-leucylamiodo (4-guanidino) butane (E-64) and its analogues as inhibitors of cysteine proteinases including cathepsins B, H, and L, *Biochem. J.* 201, 189–198, 1982; Ko, Y.M., Yamanaka, T., Umeda, M., and Suzuki, Y., Effects of thiol protease inhibitors on intracellular degradation of exogenous β-galactosidase in cultured human skin fibroblasts, *Exp. Cell Res.* 148, 525–529, 1983; Tamai, M., Matsumoto, K., Omura, S. et al., *In vitro* and *in vivo* inhibition of cysteine proteinases by EST, a new analog of E-64, *J. Pharmacobiodyn.* 9, 672–677, 1986; Shaw, E., Cysteinyl proteinases and their selective inactivation, *Adv. Enzymol. Relat. Areas Mol. Biol.* 63, 271–347, 1990; Mehdi, S., Cell-penetrating inhibitors of calpain, *Trends Biochem. Sci.* 16, 150–153, 1991; Min, K.S., Nakatsubo, T., Fujita, Y. et al., Degradation of cadmium metallothionein *in vitro* by lysosomal proteases, *Toxicol. Appl. Pharmacol.* 113 299–305, 1992; Schirmeister, T. and Klackow, A., Cysteine protease inhibitors containing small rings, *Mini Rev. Med. Chem.* 3, 585–596, 2003.

EACA

Epsilon-aminocaproic acid
6-aminohexanoic acid

Lysine

ε-aminocaproic Acid; 6-aminocaproic Acid; 6-aminohexanoic Acid; Amicar™

131.2

Analogue of lysine; inhibitor of trypsinlike enzymes such as plasmin.

EACA is an inhibitor of trypticlike serine proteases. It has been used as a hemostatic agent that functions by inhibiting fibrinolysis. It is included in some protease inhibitor cocktails. See Soter, N.A., Austen, K.F., and Gigli, I., Inhibition by epsilon-aminocaproic acid of the activation of the first component of the complement system, *J. Immunol.* 114, 928–932, 1975; Burden, A.C., Stacey, R., Wood, R.F., and Bell, P.R., Why do protease inhibitors enhance leukocyte migration inhibition to the antigen PPD? *Immunology* 35, 959–962, 1978; Nakagawa, H., Watanabe, K., and Sato, K., Inhibitory action of synthetic proteinase inhibitors and substrates on the chemotaxis of rat polymorphonuclear leukocytes *in vitro, J. Pharmacobiodyn.* 11, 674–678, 1988; Hill, G.E., Taylor, J.A., and Robbins, R.A., Differing effects of aprotinin and ε-aminocaproic acid on cytokine-induced inducible nitric oxide synthase expression, *Ann. Thorac. Surg.* 63, 74–77, 1997; Stonelake, P.S., Jones, C.E., Neoptolemos, J.P., and Baker, P.R., Proteinase inhibitors reduce basement membrane degradation by human breast cancer cell lines, *Br. J. Cancer* 75, 951–959, 1997; Sun, Z., Chen, Y.H., Wang, P. et al., The blockage of the high-affinity lysine-binding sites of plasminogen by EACA significantly inhibits prourokinase-induced plasminogen activation, *Biochim. Biophys. Acta* 1596, 182–192, 2002.

Ecotin   Broad-spectrum
protease inhibitor
derived from
Escherichia coli.

Ecotin is a broad-spectrum inhibitor of serine proteases that can be engineered to enhance inhibition of specific enzymes. See McGrath, M.E., Hines, W.M., Sakanari, J.A. et al., The sequence and reactive site of ecotin. A general inhibitor of pancreatic serine proteases from *Escherichia coli, J. Biol. Chem.* 266, 6620–6625, 1991; Erpel, T., Hwang, P., Craik, C.S. et al., Physical map location of the new *Escherichia coli* gene eco, encoding the serin protease inhibitor ecotin, *J. Bacteriol.* 174, 1704, 1992; Wang, C.I., Yang, Q., and Craik, C.S., Isolation of a high affinity inhibitor of urokinase-type plasminogen activator by phage display of ecotin, *J. Biol. Chem.* 270, 12250–12256, 1995; Yang, S.Q., Wang, C.T., Gilmor, S.A. et al., Ecotin: a serine protease inhibitor with two distinct and interacting binding sites, *J. Mol. Biol.* 279, 945–957, 1998; Gilmor, S.A., Takeuchi, T., Yang, S.Q. et al., Compromise and accommodation in ecotin, a dimeric macromolecular inhibitor of serine proteases, *J. Mol. Biol.* 299, 993–1003, 2000; Eggers, C.T., Wang, S.X., Fletterick, R.J., and Craik, C.S., The role of ecotin dimerization in protease inhibition, *J. Mol. Biol.* 308, 975–991, 2001; Wang, B., Brown, K.C., Lodder, M. et al., Chemical-mediated site-specific proteolysis. Alteration of protein–protein interaction, *Biochemistry* 41, 2805–2813, 2002; Stoop, A.A. and Craik, C.S., Engineering of a macromolecular scaffold to develop specific protease inhibitors, *Nat. Biotechnol.* 21, 1063–1068, 2003; Eggers, C.T., Murray, I.A., Delmar, V.A. et al., The periplasmic serine protease inhibitor ecotin protects bacteria against neutrophil elastase, *Biochem. J.* 379, 107–118, 2004.

Ethylenediamine Tetraacetic Acid   EDTA   292.2   Metal ion chelator;
inhibitor of
metalloenzymes.

Edetic acid; EDTA; ethylenediaminetetraacetic acid;
N, N′-1, 2-ethanediaminediylbis-[*N*-(carboxymethylglycine)]

(Ethylenedinitrilo)tetraacetic acid (ethylenediamine tetraacetic acid) chelates metal ions with a preference for divalent cations. EDTA functions as an inhibitor of metalloproteinases. See Manna, S.K., Bhattacharya, C., Gupta, S.K., and Samanta, A.K., Regulation of interleukin-8 receptor expression in human polymorphonuclear neutrophils, *Mol. Immunol.* 32, 883–893, 1995; Martin-Valmaseda, E.M., Sanchez-Yague, Y., Marcos, R., and Lianillo, M., Decrease in platelet, erythrocyte, and lymphocyte acetylcholinesterase activities due to the presence of protease inhibitors in the storage buffers, *Biochem. Mol. Biol. Int.* 41, 83–91, 1997; Oh-Ishi, M., Satoh, M., and Maeda, T., Preparative two-dimensional gel electrophoresis with agarose gels in the first dimension for high molecular mass proteins, *Electrophoresis* 21, 1653–1669, 2000; Shah, R.B., Palamakula, A., and Khan, M.A., Cytotoxicity evaluation of enzyme inhibitors and absorption enhancers in caco-2 cells for oral delivery of salmon calcitonin, *J. Pharm. Sci.* 93, 1070–1082, 2004; Pagano, M.R., Paredi, M.E., and Crupkin, M., Cytoskeletal ultrastructure and lipid composition of I-Z-I fraction in muscle from pre- and post-spawned female hake (*Meriluccius hubbsi*), *Comp. Biochem. Physiol. B Biochem. Mol. Biol.* 141, 13–21, 2005; Wei, G.X. and Bobek, L.A., Human salivary mucin MUC7 12-mer-L and 12-mer-D peptides: antifungal activity in saliva, enhancement of activity with protease inhibitor cocktail or EDTA, and cytotoxicity to human cells, *Antimicrob. Agents Chemother.* 49, 2336–2342, 2005.

Iodoacetamide

Iodoacetamide   Iodoacetic acid

185   Primary reaction with
       sulfhydryl groups
       and slower reaction
       with other protein
       nucleophiles.

Iodoacetic acid and iodoacetamide can both be used to modify nucleophiles in proteins. The chloro- and bromo-derivatives can be used as well but the rate of modification is slower. The haloacetyl function can also be used as the reactive function for more complex derivatives. Iodoacetamide is neutral compared to iodoacetic acid and is less influenced by the local environment of the reactive nucleophile. See Janatova, J., Lorenz, P.E., and Schechter, A.N., Third component of human complement: appearance of a sulfhydryl group following chemical or enzymatic inactivation, *Biochemistry* 19, 4471–4478, 1980; Haas, A.L., Murphey, K.E., and Bright, P.M., The inactivation of ubiquitin accounts for the inability to demonstrate ATP, ubiquitin-dependent proteolysis in liver extracts, *J. Biol. Chem.* 260, 4694–4703, 1985; Molla, A., Yamamoto, T., and Maeda, H., Characterization of 73 kDa thiol protease from *Serratia marcescens* and its effect on plasma proteins, *J. Biochem.* 104, 616–621, 1988; Wingfield, P., Graber, P., Turcatti, G. et al., Purification and characterization of a methionine-specific aminopeptidase from *Salmonella tyrphimurium*, *Eur. J. Biochem.* 180, 23–32, 1989; Kembhavi, A.A., Buttle, D.J., Rauber, P., and Barrett, A.J., Clostripain: characterization of the active site, *FEBS Lett.* 283, 277–280, 1991; Jagels, M.A., Travis, J., Potempa, J. et al., Proteolytic inactivation of the leukocyte C5a receptor by proteinases derived from *Porphyromas gingivalis*, *Infect. Immun.* 64, 1984–1991, 1996; Tanksale, A.M., Vernekar, J.V., Ghatge, M.S., and Deshpande, V.V., Evidence for tryptophan in proximity to histidine and cysteine as essential to the active site of an alkaline protease, *Biochem. Biophys. Res. Commun.* 270, 910–917, 2000; Karki, P., Lee, J., Shin, S.Y. et al., Kinetic comparison of procapase-3 and caspases-3, *Arch. Biochem. Biophys.* 442, 125–132, 2005. The haloalkyl derivatives do react with thiourea and are perhaps less reliable than maleimides.

LBTI

Lima Bean Trypsin   6500   Protein protease
Inhibitor   inhibitor.

Lima bean trypsin inhibitor is a protein/peptide with unusual stability. It is stable to heat (90°C for 15 minutes at pH 7 with no loss of activity) and acid (the original purification uses extraction with ethanol and dilute sulfuric acid). This is a reflection of the high content of cystine resulting in a "tight" structure. As a Bowman–Birk inhibitor, LBTI has seven disulfide bonds (Weder, J.K.P. and Hinkers, S.C., Complete amino acid sequence of the Lentil trypsin-chymotrypin inhibitor LCI-1.7 and a discussion of atypical binding sites of Bowman–Birk inhibitors, *J. Agric. Food Chem.* 52, 4219–4226, 2004). LBTI also inhibits both trypsin and chymotrypsin (Krahn, J. and Stevens, F.C., Lima bean trypsin inhibitor. Limited proteolysis by trypsin and chymotrypsin, *Biochemistry* 27, 1330–1335, 1970) as well as various other serine proteases. For additional information, see Fraenkel-Conrat, H., Bean, R.C., Ducay, E.D., and Olcott, H.S., Isolation and characterization of a trypsin inhibitor from lima beans, *Arch. Biochem. Biophys.* 37, 393–407, 1952; Stevens, F.C. and Doskoch, E., Lima bean protease inhibitor: reduction and reoxidation of the disulfide bonds and their reactivity in the trypsin-inhibitor complex, *Can. J. Biochem.* 51, 1021–1028, 1973; Nordlund, T.M., Liu, X.Y., and Sommer, J.H., Fluorescence polarization decay of tyrosine in lima bean trypsin inhibitor, *Proc. Natl. Acad. Sci. USA* 83, 8977–8981, 1986; Hanlon, M.H. and Liener, I.E., A kinetic analysis of the inhibition of rat and bovine trypsins by naturally occurring protease inhibitors, *Comp. Biochem. Physiol. B* 84, 53–57, 1986; Xiong, W., Chen, L.M., Woodley-Miller, C. et al., Identification, purification, and localization of tissue kallikrein in rat heart, *Biochem. J.* 267, 639–646, 1990; Briseid, K., Hoem, N.O., and Johannesen, S., Part of prekallikrein removed from human plasma together with IgG-immunoblot and functional tests, *Scand. J. Clin. Lab. Invest.* 59, 55–63, 1999; Yamasaki, Y., Satomi, S., Murai, N. et al., Inhibition of membrane-type serine protease 1/matriptase by natural and synthetic protease inhibitors, *J. Nutr. Sci. Vitaminol.* 49, 27–32, 2003.

Leupeptin

(ac/pr-LeuLeuArginal)

Transition-state inhibitor of proteinase.

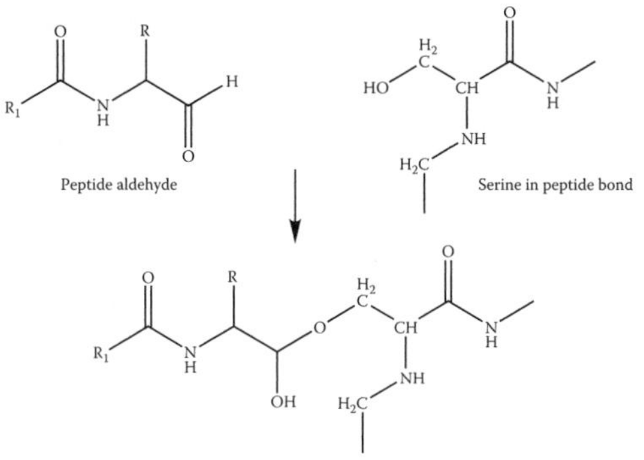

Peptide aldehyde

Serine in peptide bond

Stabilized tetrahedral aldol

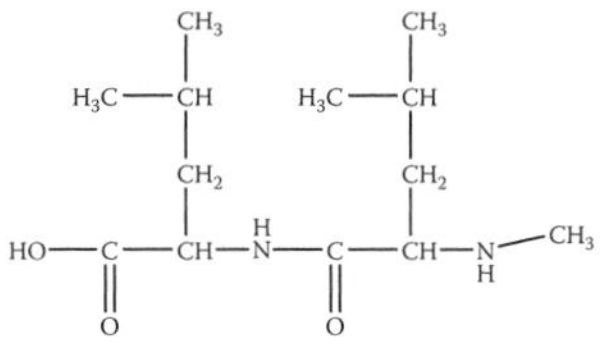

Leupeptide A

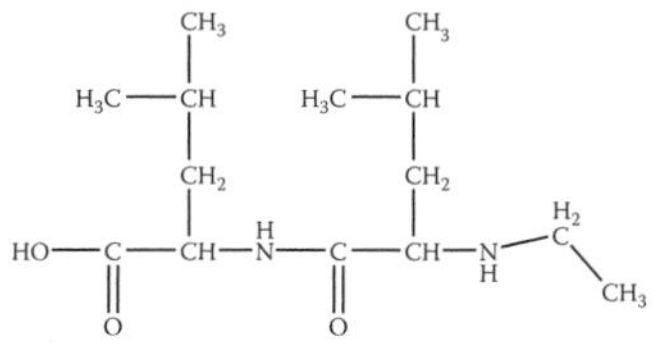

Leupeptide B

A tripeptide aldehyde (ac/pr-LeuLeuArginal) proteinase inhibitor isolated from *Actinomycetes*. It is a relatively common component of protease inhibitor cocktails used to preserve proteins during storage and purification. See Alpi, A. and Beevers, H., Proteinases and enzyme stability in crude extracts of castor bean endosperm, *Plant Physiol.* 67, 499–502, 1981; Ratajzak, T., Luc, T., Samec, A.M., and Hahnel, R., The influence of leupeptin, molybdate, and calcium ions on estrogen receptor stability, *FEBS Lett.* 136, 115–118, 1981; Takei, Y., Marzi, I., Kauffman, F.C. et al., Increase in survival time of liver transplants by protease inhibitors and a calcium channel blocker, nisoldipine, *Transplantation* 50, 14–20, 1990; Satoh, M., Hosoi, S., Miyaji, M. et al., Stable production of recombinant pro-urokinase by human lymphoblastoid Namalwa KJM-1 cells: host-cell dependency of the expressed-protein stability, *Cytotechnology* 13, 79–88, 1993; Hutchesson, A.C., Hughes, C.V., Bowden, S.J., and Ratcliffe, W.A., *In vitro* stability of endogenous parathyroid hormone-related protein in blood and plasma, *Ann. Clin. Biochem.* 31, 35–39, 1994; Agarwal, S. and Sohal, R.S., Aging and proteolysis of oxidized proteins, *Arch. Biochem. Biophys.* 309, 24–28, 1994; Yamada, T., Shinnoh, N., and Kobayashi, T., Proteinase inhibitors suppress the degradation of mutant adrenoleukodytrophy proteins but do not correct impairment of very long chain fatty acid metabolism in adrenoleukodystrophy fibroblasts, *Neurochem. Res.* 22, 233–237, 1997; Bi, M. and Singh, J., Effect of buffer pH, buffer concentration, and skin with or without enzyme inhibitors on the stability of [Arg(9)]-vasopressin, *Int. J. Pharm.* 197, 87–93, 2000; Bi, M. and Singh, J., Stability of luteinizing hormone-releasing hormone: effects of pH, temperature, pig skin, and enzyme inhibitors, *Pharm. Dev. Technol.* 5, 417–422, 2000; Ratnala, V.R., Swarts, H.G., VanOostrum, J. et al., Large-scale overproduction, functional purification, and ligand affinities of the His-tagged human histamine H1 receptor, *Eur. J. Biochem.* 271, 2636–2646, 2004.

(*p*-Amidinophenyl) Methanesulfonyl Fluoride

aPMSF

163

Reaction at active site serine.

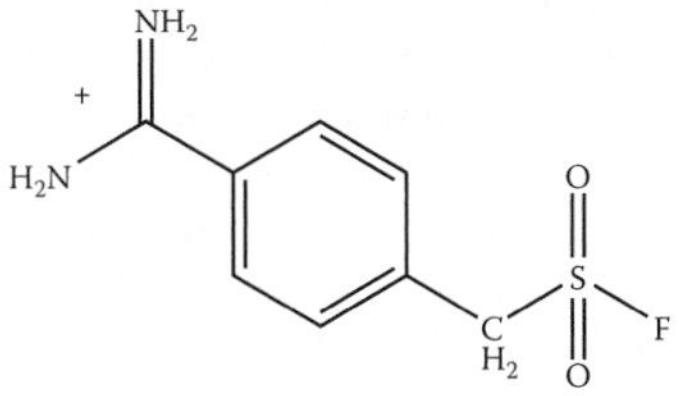

(*p*-amidinophenyl) methanesulfonyl fluoride

(*p*-amidinophenyl) methanesulfonyl fluoride was developed by Bing and coworkers (Laura, R., Robison, D.J., and Bing, D.H., [*p*-Amidinophenyl] methanesulfonyl fluoride, an irreversible inhibitor of serine proteases, *Biochemistry* 19, 4859–4864, 1980) to improve the specificity of PMSF for trypticlike enzymes. aPMSF readily reacts with trypsin but is only poorly reactive with chymotrypsin. See Katz, I.R., Thorbecke, G.J., Bell, M.K. et al., Protease-induced immunoregulatory activity of platelet factor 4, *Proc. Natl. Acad. Sci. USA* 83, 3491–3495, 1986; Unson, C.G. and Merrifield, R.B., Identification of an essential serine residue in glucagon: implications for an active site triad, *Proc. Natl. Acad. Sci. USA* 91, 454–458, 1994; Nikai, T., Komori, Y., Kato, S., and Sugihara, H., Bioloical properties of kinin-releasing enzyme from *Trimeresurus okinavensis(himehabu)* venom, *J. Nat. Toxins* 7, 23–35, 1998; Ishidoh, K., Takeda-Ezaki, M., Watanabe, S. et al., Analysis of where and which types of proteinases participate in lysosomal proteinase processing using balifomycin A1 and *Helicobacter pylori* Vac A toxin, *J. Biochem.* 125, 770–779, 1999; Komori, Y., Tatematsu, R., Tanida, S., and Nikai, T., Thrombin-like enzyme, flavovilase, with kinin-releasing activity from *Trimesurus flavoviridis(habu)* venom, *J. Nat. Toxins* 10, 239–248, 2001; Luo, L.Y., Shan, S.J., Elliott, M.B. et al., Purification and characterization of human kallikrein 11, a candidate prostate and ovarian cancer biomarker, from seminal plasma, *Clin. Cancer Res.* 12, 742–750, 2006. Reaction at a residue other than a serine has not been demonstrated although it is not unlikely that, as with DFP and PMSF, reaction could occur at a serine residue.

p-(Aminoethyl) Benzene Sulfonyl Fluoride	AEBSF; 4-(2-aminoethyl) Benzene Sulfonyl Fluoride (Pefabloc™ SC)	165 Reaction at active site serine.

4(2-aminoethyl) benzenesulfonyl fluoride

This reagent was developed to improve the reactivity of PMSF. It was originally considered to be somewhat more effective than PMSF; however, AEBSF has been shown to be somewhat promiscuous in its reaction pattern and care is suggested in its use during sample preparation. See Su, B., Bochan, M.R., Hanna, W.L. et al., Human granzyme B is essential for DNA fragmentation of susceptible target cells, *Eur. J. Immunol.* 24, 2073–2080, 1994; Helser, A., Ulrichs, K., and Muller-Ruchholtz, W., Isolation of porcine pancreatic islets: low trypsin activity during the isolation procedure guarantees reproducible high islet yields, *J. Clin. Lab. Anal.* 8, 407–411, 1994; Dentan, C., Tselepis, A.D., Chapman, M.J., and Ninio, E., Pefabloc, 4-[2-aminoethyl'benzenesulfonyl fluoride, is a new potent nontoxic and irreversible inhibitor of PAF-degrading acetylhydrolase, *Biochim. Biophys. Acta* 1299, 353–357, 1996; Sweeney, B., Proudfoot, K., Parton, A.H. et al., Purification of the T-cell receptor zeta-chain: covalent modification by 4-(2-aminoethyl)-benzenesulfonyl fluoride, *Anal. Biochem.* 245, 107–109, 1997; Diatchuk, V., Lotan, O., Koshkin, V. et al., Inhibition of NADPH oxidase activation by 4-(2-aminoethyl)benzenesulfonyl fluoride and related compounds, *J. Biol. Chem.* 272, 13292–13301, 1997; Chu, T.M. and Kawinski, E., Plasmin, subtilisin-like endoproteases, tissue plasminogen activator, and urokinase plasminogen activator are involved in activation of latent TGF-beta 1 in human seminal plasma, *Biochem. Biophys. Res. Commun.* 253, 128–134, 1998; Guo, Z.J., Lamb, C., and Dixon, R.A., A serine protease from suspension-cultured soybean cells, *Phytochemistry* 47, 547–553, 1998; Wechuck, J.B., Goins, W.F., Glorioso, J.C., and Ataai, M.M., Effect of protease inhibitors on yield of HSV-1-based viral vectors, *Biotechnol. Prog.* 16, 493–496, 2000; Baszk, S., Stewart, N.A., Chrétien, M., and Basak, A., Aminoethyl benzenesulfonyl fluoride and its hexapeptide (AC-VFRSLK) conjugate are both *in vitro* inhibitors of subtilisin kexin isozyme-1, *FEBS Lett.* 573, 186–194, 2004; King, M.A., Halicka, H.D., and Dzrzynkiewicz, Z., Pro- and anti-apoptotic effects of an inhibitor of chymotrypsin-like serine proteases, *Cell Cycle* 3, 1566–1571, 2004; Odintsova, E.S., Buneva, V.N, and Nevinsky, G.A., Casein-hydrolyzing activity of sIGA antibodies from human milk, *J. Mol. Recog.* 18, 413–421, 2005; Solovyan, V.T. and Keski-Oja, J., Proteolytic activation of latent TGF-beta precedes caspase-3 activation and enhances apoptotic death of lung epithelial cells, *J. Cell Physiol.* 207, 445–453, 2006.

Pepstatin

685.9   Acid protease
inhibitor.

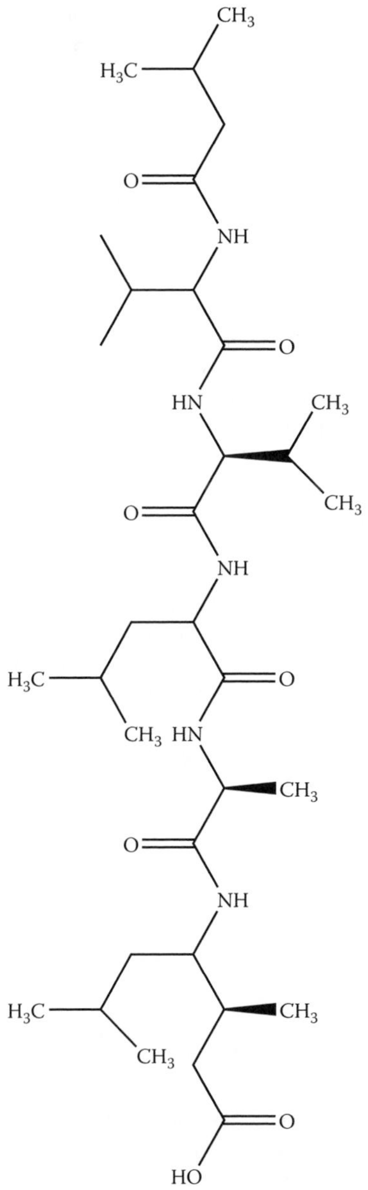

Pepstatin

A group of pentapeptide acid protease inhibitors isolated from *Streptomeyces* (Umezawa, H., Aoyagi, T., Morishima, H. et al., Pepstatin, a new pepsin inhibitor produced by *Actinomycetes*, *J. Antibiot.* 23, 259–262, 1970; Aoyagi, T., Kunimoto, S., Morichima, H. et al., Effect of pepstatin on acid proteases, *J. Antibiot.* 24, 687–694, 1971). Pepstatins are frequently included in protease inhibitor cocktails and used for the stabilization of proteins during extraction, storage, and purification. See Takei, Y., Marzi, I., Kaufmann, F.C. et al., Increase in survival time of liver transplants by protease inhibitors and a calcium channel blocker, nisoldipine, *Transplantation* 50, 14–20, 1990; Liang, M.N., Witt, S.N., and McConnell, H.M., Inhibition of class II MHC-peptide complex formation by protease inhibitors, *J. Immunol. Methods* 173, 127–131, 1994; Deng, J., Rudick, V., and Dory, L., Lysosomal degradation and sorting of apolipoprotein E in macrophages, *J. Lipid Res.* 36, 2129–2140, 1995; Wang, Y.K., Lin, H.H., and Tang, M.J., Collagen gel overlay induces two phases of apoptosis in MDCK cells, *Am. J. Physiol. Cell Physiol.* 280, C1440–C1448, 2001; Lafleur, M.A., Handsley, M.M., Knaupper, V. et al., Endothelial tubulogenesis within fibrin gels specifically requires the activity of membrane-type-matrix-metalloproteinases (MT-MMPs), *J. Cell Sci.* 155, 3427–3438, 2002.

Phenanthroline Monohydrate   1,10-phenanthroline   198.2   Metal ion chelator;
                                                                                                                                   inhibitor of
                                                                                                                                   metalloenzymes;
                                                                                                                                   specificity for zinc-
                                                                                                                                   metalloenzymes.

o-Phenanthroline
1, 10-Phenanthroline

1,10-phenanthroline, *o*-phenanthroline: an inhibitor of metalloproteinases and a reagent for the detection of ferrous ions. See Felber, J.P., Cooobes, T.L., and Vallee, B.L., The mechanism of inhibition of carboxypeptidase A by 1,10-phenanthroline, *Biochemistry* 1, 231–238, 1962; Hakala, M.T. and Suolinna, E.M., Specific protection of folate reductase against chemical and proteolytic inactivation, *Mol. Pharmacol.* 2, 465–480, 1966; Latt, S.A., Holmquist, B., and Vallee, B.L., Thermolysin: a zinc metalloenzyme, *Biochem. Biophys. Res. Commun.* 37, 333–339, 1969; Berman, M.B. and Manabe, R., Corneal collagenases: evidence for zinc metalloenzymes, *Ann. Ophthalmol.* 5, 1993–1995, 1973; Seltzer, J.L., Jeffrey, J.J., and Eisen, A.Z., Evidence for mammalian collagenases as zinc ion metalloenzymes, *Biochim. Biophys. Acta* 485, 179–187, 1977; Krogdahl, A. and Holm, H., Inhibition of human and rat pancreatic proteinases by crude and purified soybean trypsin inhibitor, *J. Nutr.* 109, 551–558, 1979; St. John, A.C., Schroer, D.W., and Cannavacciuolo, L., Relative stability of intracellular proteins in bacterial cells, *Acta. Biol. Med. Ger.* 40, 1375–1384, 1981; Kitjaroentham, A., Suthiphongchai, T., and Wilairat, P., Effect of metalloprotease inhibitors on invasion of red blood cells by *Plasmodium falciparum*, *Acta Trop.* 97, 5–9, 2006; Thwaite, J.E., Hibbs, S., Tritall, R.W., and Atkins, T.P., Proteolytic degradation of human antimicrobioal peptide LL-37 by *Bacillus anthracis* may contribute to virulence, *Antimicrob. Agents Chemother.* 50, 2316–2322, 2006.

Phenylmethylsulfonyl Fluoride   PMSF   174   Reaction at active site
                                                                                                                    serine.

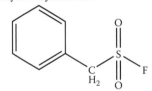

Phenylmethylsulfonyl fluoride (PMSF)

Phenylmethylsulfonyl fluoride was developed by David Fahrney and Allen Gold and inhibits serine proteases such as trypsin and chymotrypsin in a manner similar to DFP. The rate of modification of trypsin and chymotrypsin with PMSF is similar to that observed with DFP; however, the reaction with acetylcholinesterase with PMSF is much less than that of DFP ($>6.1 \times 10^{-2}$ M^{-1}min^{-1} vs. 1.3×10^4 M^{-1}min^{-1-})(Fahrney, D.E. and Gold, A.M., Sulfonyl fluorides as inhibitors of esterases. I. Rates of reaction with acetylcholinesterase, α-chymotrypsin, and trypsin, *J. Amer. Chem. Soc.* 85, 997–1000, 1963). For other applications see Lundblad, R.L., A rapid method for the purification of bovine thrombin and the inhibition of the purified enzyme with phenylmethylsulfonyl fluoride, *Biochemistry* 10, 2501–2506, 1971; Pringle, J.R., Methods for avoiding proteolytic artefacts in studies of enzymes and other proteins from yeasts, *Methods Cell Biol.* 12, 149–184, 1975; Bendtzen, K., Human leukocyte migration inhibitory factor (LIF). I. Effect of synthetic and naturally occurring esterase and protease inhibitors, *Scand. J. Immunol.* 6, 125–131, 1977; Carter, D.B., Efird, P.H., and Chae, C.B., Chromatin-bound proteases and their inhibitors, *Methods Cell Biol.* 19, 175–190, 1978; Hubbard, J.R. and Kalimi, M., Influence of proteinase inhibitors on glucocorticoid receptor properties: recent progress and future perspectives, *Mol. Cell. Biochem.* 66, 101–109, 1985; Kato, T., Sakamoto, E., Kutsana, H. et al., Proteolytic conversion of STAT3alpha to STAT3gamma in human neutrophils: role of granule-derived serine proteases, *J. Biol. Chem.* 279, 31076–31080, 2004; Cho, I.H., Choi, E.S., Lim, H.G., and Lee, H.H., Purification and characterization of six fibrinolytic serine proteases from earthworm *Lumbricus rubellus*, *J. Biochem. Mol. Biol.* 37, 199–205, 2004; Khosravi, J., Diamandi, A., Bodani, U. et al., Pitfalls of immunoassay and sample for IGF-1: comparison of different assay methodologies using fresh and stored serum samples, *Clin. Biochem.* 38, 659–666, 2005; Shao, B., Belaaouaj, A., Velinde, C.L. et al., Methionine sulfoxide and proteolytic cleavage contribute to the inactivation of cathepsin G by hypochlorous acid: an oxidative mechanism for regulation of serine proteinases by myeloperoxidase, *J. Biol. Chem.* 260, 29311–29321, 2005; Pagano, M.R., Paredi,

M.E., and Crupkin, M., Cytoskeletal ultrastructural and lipid composition of 1-Z-1 fraction in muscle from pre- and post-spawned female hake (*Merluccius hubbsi*), *Comp. Biochem. Physiol. B Biochem. Mol. Biol.*141, 13–21, 2005. Although PMSF is reasonably specific for reaction with the serine residue at the active site of serine proteinases, as with DFP, reaction at tyrosine has been reported (De Vendittis, E., Ursby, T., Rullo, R. et al., Phenylmethanesulfonyl fluoride inactivates an archeael superoxide dismutase by chemical modification of a specific tyrosine residue. Cloning, sequencing, and expression of the gene coding for *Sulfolobus solfataricus* dismutase, *Eur. J. Biochem.* 268, 1794–1801, 2001). PMSF does have solubility issues and usually ethanol or another suitable water-miscible organic solvent is used to introduce this reagent. On occasion, the volume of ethanol required influences the reaction (see Bramley, T.A., Menzies, G.S., and McPhie, C.A., Effects of alcohol on the human placental GnRH receptor system, *Mol. Hum. Reprod.* 5, 777–783, 1999).

SBTI　　　　　　　　　　　　　　　　　　　　Soybean Trypsin　　　21,500　　Protein protease
　　　　　　　　　　　　　　　　　　　　　　　　Inhibitor　　　　　　　　　　　inhibitor.

Soybean trypsin inhibitor (SBTI, STI) usually refers to the inhibitor first isolated by Kunitz (Kunitz, M., Crystalline soybean trypsin inhibitor, *J. Gen. Physiol.* 29, 149–154, 1946; Kunitz, M., Crystalline soybean trypsin inhibitor. II. General properties, *J. Gen. Physiol.* 30, 291–310, 1947). This material is described as the Kunitz inhibitor and is reasonably specific for trypticlike enzymes. There are other protease inhibitors derived from soybeans; the Bowman–Birk inhibitor (Birk, Y., The Bowman–Birk inhibitor. Trypsin and chymotrypsin-inhibitor from soybeans, *Int. J. Pept. Protein Res.* 25, 113–131, 1985; Birk, Y., Protein proteinase inhibitors in legume seeds — overview, *Arch. Latinoam. Nutr.* 44 (4 Suppl. 1), 26S–30S, 1996) is the best known and, unlike the Kunitz inhibitor, inhibits both trypsin and chymotrypsin; the Bowman–Birk inhibitor is also a double-headed inhibitor having two reactive sites (see Frattali, V. and Steiner, R.F., Soybean inhibitors. I. Separation and some properties of three inhibitors from commercial crude soybean trypsin inhibitor, *Biochemistry* 7, 521–530, 1968; Frattali, V. and Steiner, R.F., Interaction of trypsin and chymotrypsin with a soybean proteinase inhibitor, *Biochem. Biophys. Res. Commun.* 34, 480–487, 1969; Krogdahl, A. and Holm, H., Inhibition of human and rat pancreatic proteinases by crude and purified soybean trypsin inhibitor, *J. Nutr.* 109, 551–558, 1979). Soybean trypsin inhibitor (Kunitz) is used as a model protein (Liu, C.L., Kamei, D.T., King, J.A. et al., Separation of proteins and viruses using two-phase aqueous micellar systems, *J. Chromatog. B* 711, 127–138, 1998; Higgs, R.E., Knierman, M.D., Gelfanova, Y. et al., Comprehensive label-free method for the relative quantification of proteins from biological samples, *J. Proteome Res.* 4, 1442–1450, 2005). The broad specificity of the Kunitz inhibitor for trypticlike serine proteases provides the basis for its use in the demonstration of protease processing steps (Hansen, K.K., Sherman, P.M., Cellars, L. et al., A major role for proteolytic and proteinase-activated receptor-3 in the pathogenesis of infectious colitis, *Proc. Natl. Acad. Sci. USA* 102, 8363–8368, 2005).

Tosyl-lysine Chloromethyl Ketone　　　　　　TLCK; 1-chloro-3-　　369.2　　Reaction at active site
　　　　　　　　　　　　　　　　　　　　　　　　tosylamido-7-amino-　(HCl)　　histidine residues of
　　　　　　　　　　　　　　　　　　　　　　　　2-heptanone　　　　　　　　　trypsinlike serine
　　　　　　　　　　　　　　　　　　　　　　　　　　　　　　　　　　　　　　proteases.

Tosyl-lysine chloromethyl ketone

Tosyl-lysine chloromethyl ketone (TLCK) was developed by Elliott Shaw and colleagues (Shaw, E., Mares-Guia, M., and Cohen, W., Evidence of an active center histidine in trypsin through use of a specific reagent, 1-chloro-3-tosylamido-7-amido-2-heptanone, the chloromethyl ketone derived from N-αtosyl-L-lysine, *Biochemistry* 4, 2219–2224, 1965). As with TPCK, reaction is not absolutely specific for trypticlike serine proteases (Earp, H.S., Austin, K.S., Gillespie, G.Y. et al., Characterization of distinct tyrosine-specific protein kinases in B and T lymphocytes, *J. Biol. Chem.* 260,

4351–4356, 1985; Needham, L. and Houslay, M.D., Tosyl-lysyl chloromethylketone detects conformational changes in the catalytic unit of adenylate cyclase induced by receptor and G-protein stimulation, *Biochem. Biophys. Res. Commun.* 156, 855–859, 1988). Reaction of this chloroalkyl compound with sulfydryl groups would be expected and it is possible that other protein nucleophilic centers would react, although this has not been unequivocally demonstrated. Attempts to synthesize the direct arginine analogue were unsuccessful; it was possible to make more complex arginine derivatives such as Ala-Phe-Arg-CMK, which was more effective with human plasma Kallikrein than the corresponding lysine derivatives (Ki = 0.078 μM vs. M vs. 4.9 μM) (Kettner, C. and Shaw, E., Synthesis of peptides of arginine chloromethyl ketone. Selective inactivation of human plasma kallikrein, *Biochemistry* 17, 4778–4784, 1978).

Tosyl-phenylalanine Chloromethyl Ketone

Tosyl phenylalanine chloromethylketone

TPCK; L-1-tosylamido-2-phenylethyl Chloromethyl Ketone)

351.9

Reaction at active site histidine residues of chymotrypsinlike serine proteases.

Tosyl-phenylalanine chloromethyl ketone (TPCK) was developed by Guenther Schoellmann and Elliott Shaw (Schoellmann, G. and Shaw, E., Direct evidence for the presence of histidine in the active center of chymotrypsin, *Biochemistry* 2, 252–255, 1963). TPCK was developed as an affinity label (Plapp, B.V., Application of affinity labeling for studying structure and function of enzymes, *Methods Enzymol.* 87, 469–499, 1982) where binding to chymotrypsin is driven by the phenyl function with subsequent alkylation of the active site histidine. The chloroalkyl function was selected to reduce reactivity with other protein nucleophiles such as cysteine. TPCK does undergo a slow rate of hydrolysis to form the corresponding alcohol. TPCK inactivates proteases with chymotrypsinlike specificity. The rate of inactivation is relatively slow but is irreversible; reaction rates can be enhanced by a more elaborate peptide chloromethyl ketone structure. In the case of cucumisin, a plant serine proteinase, TPCK did not result in inactivation while inactivation was achieved with Z-Ala-Ala-Pro-Phe-chloromethyl ketone (Yonezawa, H., Uchikoba, T., and Kaneda, M., Identification of the reactive histidine of cucumisin, a plant serine protease: modification with peptidyl chloromethyl ketone derivative of peptide substrate, *J. Biochem.* 118, 917–920, 1995). There is, however, significant reaction of TPCK with other proteins at residues other than histidine (see Rychlik, I., Jonak, J., and Sdelacek, J., Inhibition of the EF-Tu factor by L-1-tosylamido-2-phenylethyl chloromethyl ketone, *Acta Biol. Med. Ger.* 33, 867–876, 1974); TPCK has been described as an inhibitor of cysteine proteinases (Bennett, M.J., Van Leeuwen, E.M., and Kearse, K.P., Calnexin association is not sufficient to protect T cell receptor proteins from rapid degradation in CD4+CD8+ thymocytes, *J. Biol. Chem.* 273, 23674–23680, 1998). TPCK has been suggested to react with a lysine residue in aminoacylase (Frey, J., Kordel, W., and Schneider, F., The reaction of aminoacylase with chloromethylketone analogs of amino acids, *Z. Naturforsch.* 32, 769–776, 1966). Other reactions continue to be described (McCray, J.W. and Weil, R., Inactivation of interferons: halomethyl ketone derivatives of phenylalanine as affinity labels, *Proc. Natl. Acad. Sci. USA* 79, 4829–4833, 1982; Conseiller, E.C. and Lederer, F., Inhibition of NADPH oxidase by aminoacyl chloromethane protease inhibitors in phorbol-ester-stimulated human-neutrophils-A reinvestigation — are proteases really involved in the activation process? *Eur. J. Biochem.* 183, 107–114, 1989; Borukhov, S.I. and Strongin, A.Y., Chemical modification of the recombinant human α-interferons and β-interferons, *Biochem. Biophys. Res. Commun.* 167, 74–80, 1990; Gillibert, M., Dehry, Z., Terrier, M. et al., Another biological effect of tosylphenylalanylchloromethane (TPCK): it prevents p47(phox) phosphorylation and translocation upon neutrophil stimulation, *Biochem. J.* 386, 549–556, 2005).

Peptide Halomethyl Ketones: While TPCK and TLCK represented a major advance in modifying active site residues in serine proteases, slow and relatively nonspecific reaction was a problem. The development of tripeptide halomethyl ketones provided a major advance in the value of such derivatives as presented in some specific examples below. However, even with these derivatives, reactions occur with "unexpected" enzymes. More general information can be obtained from the following references: Poulos, T.L., Alden, R.A., Freer, S.T. et al., Polypeptide halomethyl ketones bind to serine proteases as analogs of the tetrahedral intermediate. X-ray crystallographic comparison of lysine- and phenylalanine-polypeptide chloromethyl ketone-inhibited subtilisin, *J. Biol. Chem.* 251, 1097–1103, 1976; Powers, J.C., Reaction of serine proteases with halomethyl ketones, *Methods Enzymol.* 46, 197–208, 1977; Navarro, J., Abdel Ghany, M., and Racker, E., Inhibition of tyrosine protein kinases by halomethyl ketones, *Biochemistry* 21, 6138–6144, 1982; Conde, S., Perez, D.I., Martinez, A. et al., Thienyl and phenyl α-halomethyl ketones: new inhibitors of glycogen synthase kinase (GSK-3β) from a library of compound searching, *J. Med. Chem.* 46, 4631–4633, 2003.

Peptide Fluoromethyl Ketones: Fluoroalkyl derivatives of the peptide chloromethyl ketones have been prepared in an attempt to improve specificity by reducing nonspecific alkylation at cysteine residues (Rasnick, D., Synthesis of peptide fluoromethyl ketones and the inhibition of human cathepsin B, *Anal. Biochem.* 149, 461–465, 1985). Nonspecific reaction with sulfydryl groups such as those in glutathione was reduced; there was still reaction with active site cysteine although at a slower rate than with the chloroalkyl derivative (16,200 $M^{-1}s^{-1}$ vs. 45,300 $M^{-1}s^{-1}$; $T_{1/2}$ 21.9 min. vs. 5.1 min.). Reaction also occurred with serine proteases (Shaw, E., Angliker, H., Rauber, P. et al., Peptidyl fluoromethyl ketones as thiol protease inhibitors, *Biomed. Biochim. Acta* 45, 1397–1403, 1986) where the modification occurred at a histidine residue (Imperiali, B. and Abeles, R.H., Inhibition of serine proteases by peptide fluoromethyl ketones, *Biochemistry* 25, 3760–3767, 1986). The trifluoromethyl derivative was also an inhibitor but formed a hemiacetal derivative. The peptide fluoromethyl ketone, z-VAD-FMK, has proved to be a useful inhibitor of caspases

| **D-Phe-Pro-Arg-chloromethyl Ketone** | PPACK | Reaction at active site histidine residues of trypsinlike serine proteases. |

D-Phe-Pro-Arg-chloromethyl ketone was one of the first complex peptide halomethyl ketones synthesized. These derivatives have the advantage of increased reaction rate and specificity (see Williams, E.B. and Mann, K.G., Peptide chloromethyl ketones as labeling reagents, *Methods Enzymol.* 222, 503–513, 1993; Odake, S., Kam, C.M., and Powers, J.C., Inhibition of thrombin by arginine-containing peptide chloromethyl ketones and bis chloromethyl ketone-albumin conjugates, *J. Enzyme Inhib.* 9, 17–27, 1995; Lundblad, R.L., Bergstrom, J., De Vreker, R. et al., Measurement of active coagulation factors in Autoplex®-T with colorimetric active site-specific assay technology, *Thromb. Haemostas.* 80, 811–815, 1998). With chymotrypsin, CHO-PheCH$_2$Cl, $k_{obsv.}/[I] = 0.55$ $M^{-1}s^{-1}$ and Boc-Ala-Gly-Phe-CH$_2$Cl, $k_{obsv.}/[I] = 3.34$ $M^{-1}s^{-1}$ (Kurachi, K., Powers, J.C., and Wilcox, P.E., Kinetics of the reaction of chymotrypsin A α with peptide chloromethyl ketones in relation to subsite specificity, *Biochemistry* 12, 771–777, 1973. See also Ketter, C. and Shaw, E., The selective affinity labeling of factor Xa by peptides of arginine chloromethyl ketone, *Thromb. Res.* 22, 645–652, 1981; Shaw, E., Synthetic inactivators of kallikrein, *Adv. Exp. Med. Biol.* 156, 339–345, 1983; McMurray, J.S. and Dyckes, D.F., Evidence for hemiketals as intermediates in the inactivation of serine proteinases with halomethyl ketones, *Biochemistry* 25, 2298–2301, 1986). There is a similar peptide chloromethyl ketone, PPACK II (D-Phe-Phe-Arg-CMK), which has been used to stabilize B-type natriuretic peptide (BNP) in plasma samples (Belenky, A., Smith, A., Zhang, B. et al., The effect of class-specific protease inhibitors on the stabilization of B-type natriuretic peptide in human plasma, *Clin. Chim. Acta* 340, 163–172, 2004).

z-VAD-FMK   Benzyloxycarbonyl-   Inhibitor of caspases.
                                  Val-Ala-Asp(OMe)
                                  Fluoromethyl Ketone

z-VADFMK

Benzyloxycarbonyl-Val-Ala-Asp(OMe) fluoromethyl ketone (z-VAD-FMK) is a peptide halomethyl ketone used for the inhibition of caspases and related enzymes. Because z-VAD-FMK is neutral, it passes the cell membrane and can inhibit intracellular proteolysis and is useful in understanding the role of caspases and related enzymes in cellular function. See Zhu, H., Fearnhead, H.O., and Cohen, G.M., An ICE-like protease is a common mediator of apoptosis induced by diverse stimuli in human monocytes THP.1 cells, *FEBS Lett.* 374, 303–308, 1995; Mirzoeva, O.K., Yaqoob, P., Knox,

K.A., and Calder, P.C., Inhibition of ICE-family cysteine proteases rescues murine lymphocytes from lipoxygenase inhibitor-induced apoptosis, *FEBS Lett.* 396, 266–270, 1996; Slee, E.A., Zhu, H., Chow, S.C. et al., Benzyloxycarbonyl-Val-Ala-Asp(OMe) fluoromethylketone (z-VAD.FMK) inhibits apoptosis by blocking the processing of CPP32, *Biochem. J.* 315, 21–24, 1996; Gottron, F.J., Ying, H.S., and Choi, D.W., Caspase inhibition selectively reduces the apoptotic component of oxygen-glucose deprivation-induced cortical neuronal cell death, *Mol. Cell. Neurosci.* 9, 159–169, 1997; Longthorne, V.L. and Williams, G.T., Caspase activity is required for commitment to Fas-mediated apoptosis, *EMBO J.* 16, 3805–3812, 1997; Hallan, E., Blomhoff, H.K., Smeland, E.B., and Long, J., Involvement of ICE (Caspase) family in gamma-radiation-induced apoptosis of normal B lymphocytes, *Scand. J. Immunol.* 46, 601–608, 1997; Polverino, A.J. and Patterson, S.D., Selective activation of caspases during apoptotic induction in HL-60 cells. Effects of a tetrapeptide inhibitor, *J. Biol. Chem.* 272, 7013–7021, 1997; Cohen, G.M., Caspases: the executioners of apoptosis, *Biochem. J.* 328, 1–16, 1997; Sarin, A., Haddad, E.K., and Henkart, P.A., Caspase dependence of target cell damage induced by cytotoxic lymphocytes, *J. Immunol.* 161, 2810–2816, 1998; Nicotera, P., Leist, M., Single, B., and Volbracht, C., Execution of apoptosis: converging or diverging pathway? *Biol. Chem.* 380, 1035–1040, 1999; Grfaczyk, P.P., Caspase inhibitors as anti-inflammatory and antiapoptotic agents, *Prog. Med. Chem.* 39, 1–72, 2002; Blankenberg, F., Mari, C., and Strauss, H.W., Imaging cell death *in vivo*, *Q. J. Nucl. Med.* 47, 337–348, 2003; Srivastava, A., Henneke, P., Visintin, A. et al., The apoptotic response to pneumolysin in Toll-like receptor 4 dependent and protects against pneumococcal disease, *Infect. Immun.* 73, 6479–6489, 2005; Clements, K.M., Burton-Wurster, N., Nuttall, M.E., and Lust, G., Caspase-3/7 inhibition alters cell morphology in mitomycin-C treated chondrocytes, *J. Cell Physiol.* 205, 133–140, 2005; Coward, W.R., Marie, A., Yang, A. et al., Statin-induced proinflammatory response in mitrogen-activated peripheral blood mononuclear cells through the activation of caspases-1 and IL-18 secretion in monocytes, *J. Immunol.* 176, 5284–5292, 2006.

[a] The protease inhibitor cocktails referred to herein are not to be confused with the protease inhibitor cocktails that are used for therapy for patients who have Acquired Immune Deficiency Syndrome (AIDS).

GENERAL REFERENCES FOR INHIBITORS OF PROTEOLYTIC ENZYMES

Albeck, A. and Kliper, S., Mechanism of cysteine protease inactivation by peptidyl epoxides, *Biochem. J.* 322, 879–884, 1997.

Banner, C.D. and Nixon, R.A., Eds., *Proteases and Protease Inhibitors in Alzheimer's Disease Pathogenesis,* New York Academy of Sciences, New York, 1992.

Barrett, A.J. and Salvesen, G., Eds., *Protease Inhibitors,* Elsevier, Amsterdam, NL, 1986.

Bernstein, N.K. and James, M.N., Novel ways to prevent proteolysis — prophytepsin and proplasmepsin II, *Curr. Opin. Struct. Biol.* 9, 684–689, 1999.

Birk, Y., Ed., *Plant Protease Inhibitors: Significance in Nutrition, Plant Protection, Cancer Prevention, and Genetic Engineering,* Springer, Berlin, 2003.

Cheronis, J.C.D. and Repine, J.E., *Proteases, Protease Inhibitors, and Protease-Derived Peptides: Importance in Human Pathophysiology and Therapeutics,* Birkhäuser Verlag, Basel, Switzerland, 1993.

Church, F.C., Ed., *Chemistry and Biology of Serpins,* Plenum Press, New York, 1997.

Frlan, R. and Gobec, S., Inhibitors of cathepsin B, *Curr. Med. Chem.* 13, 2309–2327, 2006.

Giglione, C., Boularot, A., and Meinnel, T., Protein *N*-terminal excision, *Cell. Mol. Life Sci.* 61, 1455–1474, 2004.

Johnson, S.L. and Pellechhia, M., Structure- and fragment-based approaches to protease inhibition, *Curr. Top. Med. Chem.* 6, 317–329, 2006.

Kim, D.H., Chemistry-based design of inhibitors for carboxypeptidase A, *Curr. Top. Med. Chem.* 4, 1217–1226, 2004.

Lowther, W.T., and Matthews, B.W., Structure and function of the methionine aminopeptidases, *Biochim. Biophys. Acta* 1477, 157–167, 2000.

Magnusson, S., Ed., *Regulatory Proteolytic Enzymes and Their Inhibitors,* Pergamon Press, Oxford, UK, 1986.

Powers, J.C. and Harper, J.W., Inhibition of serine proteinases, in *Proteinase Inhibitors,* Barrett, A.J. and Salvesen, G., Elsevier, Amsterdam, chapter 3, pp. 55–152.

Saklatvala, J., and Nagase, H., Eds., *Proteases and the Regulation of Biological Processes*, Portland Press, London, UK, 2003.

Shaw, E., Cysteinyl proteinases and their selective inactivation, *Adv. Enzymol. Relat. Areas Mol. Biol.* 63, 271–347, 1990.

Stennicke, H.R. and Salvesen, G.S, Chemical ligation — an unusual paradigm in protease inhibition, *Mol. Cell.* 21, 727–728, 2006.

Tam, T.F., Leung-Toung, R., Li, W. et al., Medicinal chemistry and properties of 1,2,4-thiadiazoles, *Mini Rev. Med. Chem.* 5, 367–379, 2005.

Vogel, R., Trautschold, I., and Werle, E., *Natural Proteinase Inhibitors*, Academic Press, New York, 1968.

6 List of Buffers

Common Name	Chemical Name	M.W.	Properties and Comment
ACES	2-[2-amino-2-oxyethyl)-amino]ethanesulfonic Acid	182.20	One of the several "Good" buffers.

Tunnicliff, G. and Smith, J.A., Competitive inhibition of gamma-aminobutyric acid receptor binding by *N*-hydroxy-ethylpiperazine-*N'*-2-ethanesulfonic acid and related buffers, *J. Neurochem.* 36, 1122–1126, 1981; Chappel, D.J., *N*-[(carbamoylmethyl)amino] ethanesulfonic acid improves phenotyping of α-1-antitrypsin by isoelectric focusing on agarose gel, *Clin. Chem.* 31, 1384–1386, 1985; Liu, Q., Li, X., and Sommer, S.S., pk-matched running buffers for gel electrophoresis, *Anal. Biochem.* 270, 112–122, 1999; Taha, M., Buffers for the physiological pH range: acidic dissociation constants of zwitterionic compounds in various hydroorganic media, *Ann. Chim.* 95, 105–109, 2005.

Cacodylic Acid	Dimethylarsinic Acid	138.10	Buffer salt in neutral pH range; largely replaced because of toxicity.

McAlpine, J.C., Histochemical demonstration of the activation of rat acetylcholinesterase by sodium cacodylate and cacodylic acid using the thioacetic acid method, *J. R. Microsc. Soc.* 82, 95–106, 1963; Jacobson, K.B., Murphy, J.B., and Das Sarma, B., Reaction of cacodylic acid with organic thiols, *FEBS Lett.* 22, 80–82, 1972; Travers, F., Douzou, P., Pederson, T., and Gunsalus. I.C., Ternary solvents to investigate proteins at subzero temperature, *Biochimie* 57, 43–48, 1975; Young, C.W., Dessources, C., Hodas, S., and Bittar, E.S., Use of cationic disc electrophoresis near neutral pH in the evaluation of trace proteins in human plasma, *Cancer Res.* 35, 1991–1995, 1975; Chirpich, T.P., The effect of different buffers on terminal deoxynucleotidyl transferase activity, *Biochim. Biophys. Acta* 518, 535–538, 1978; Nunes, J.F., Aguas, A.P., and Soares, J.O., Growth of fungi in cacodylate buffer, *Stain Technol.* 55, 191–192, 1980; Caswell, A.H. and Bruschwig, J.P., Identification and extraction of proteins that compose the triad junction of skeletal muscle, *J. Cell Biol.* 99, 929–939, 1984; Parks, J.C. and Cohen, G.M., Glutaraldehyde fixatives for preserving the chick's inner ear, *Acta Otolaryngol.* 98, 72–80, 1984; Song, A.H. and Asher, S.A., Internal intensity standards for heme protein UV resonance Raman studies: excitation profiles of cacodylic acid and sodium selenate, *Biochemistry* 30, 1199–1205, 1991; Henney, P.J., Johnson, E.L., and Cothran, E.G., A new buffer system for acid PAGE typing of equine protease inhibitor, *Anim. Genet.* 25, 363–364, 1994; Jezewska, M.J., Rajendran, S., and Bujalowski, W., Interactions of the 8-kDa domain of rat DNA polymerase beta with DNA, *Biochemistry* 40, 3295–3307, 2001; Kenyon, E.M. and Hughes, M.F., A concise review of the toxicity and carcinogenicity of dimethylarsinic acid, *Toxicology* 160, 227–236, 2001; Cohen, S.M., Arnold, L.L., Eldan, M. et al., Methylated arsenicals: the implications of metabolism and carcinogenicity studies in rodents to human risk management, *Crit. Rev. Toxicol.* 99–133, 2006.

HEPES	4-(2-hydroxyethyl)-1-piperizineethanesulfonic Acid		A "Good" buffer; reagent purity has been an issue; metal ion binding must be considered; there are buffer-specific effects that are poorly understood; component of tissue-fixing technique.

Good, N.E., Winget, G.D., Winter, W. et al., Hydrogen ion buffers for biological research, *Biochemistry* 5, 467–477, 1966; Turner, L.V. and Manchester, K.L., Interference of HEPES with the Lowry method, *Science* 170, 649, 1970; Chirpich, T.P., The effect of different buffers on terminal deoxynucleotidyl transferase activity, *Biochim. Biophys. Acta* 518, 535–538, 1978; Tadolini, B., Iron autoxidation in MOPS and HEPES buffers, *Free Radic. Res. Commun.* 4, 149–160, 1987; Simpson, J.A., Cheeseman, K.H., Smith, S.E., and Dean, R.T., Free-radical generation by copper ions and hydrogen peroxide. Stimulation by HEPES buffer, *Biochem. J.* 254, 519–523, 1988; Abas, L. and Guppy M., Acetate: a contaminant in HEPES buffer, *Anal. Biochem.* 229, 131–140, 1995; Schmidt, K., Pfeiffer, S., and Mayer, B., Reaction of peroxynitrite with HEPES or MOPS results in the formation of nitric oxide donors, *Free Radic. Biol. Med.* 24, 859–862, 1998; Wiedorn, K.H., Olert, J., Stacy, R.A. et al., HOPE — a new fixing technique enables preservation and extraction of high molecular weight DNA and RNA of >20 kb from paraffin-embedded tissues. HEPES-glutamic acid buffer mediated organic solvent protection effect, *Pathol. Res. Pract.* 198, 735–740, 2002; Fulop, L., Szigeti, G., Magyar, J. et al.,

Differences in electrophysiological and contractile properties of mammalian cardiac tissues bathed in bicarbonate- and HEPES-buffered solutions, *Acta Physiol. Scand.*178, 11–18, 2003; Mash, H.E., Chin, Y.P., Sigg, L. et al., Complexation of copper by zwitterionic aminosulfonic (good) buffers, *Anal. Chem.* 75, 671–677, 2003; Sokolowska, M. and Bal, W., Cu(II) complexation by "non-coordinating" *N*-2-hydroxyethylpiperazine-*N*'-ethanesulfonic acid (HEPES buffer), *J. Inorg. Biochem.* 99, 1653–1660, 2005; Zhao, G. and Chasteen, N.D., Oxidation of Good's buffers by hydrogen peroxide, *Anal. Biochem.* 349, 262–267, 2006; Hartman, R.F. and Rose, S.D., Kinetics and mechanism of the addition of nucleophiles to alpha,beta-unsaturated thiol esters, *J. Org. Chem.* 71, 6342–6350, 2006.

| MES | 1-morpholineethane-sulfonic Acid; 2- | 198.2 | A "Good" buffer. |
| | (4-morpholino) Ethane Sulfonate | | |

Good, N.E., Winget, G.D., Winter, W. et al., Hydrogen ion buffers for biological research, *Biochemistry* 5, 467–477, 1966; Bugbee, B.G. and Salisbury, F.B., An evaluation of MES (2(*N*-morpholino ethanesulfonic acid) and Amberlite 1RC-50 as pH buffers for nutrient growth studies, *J. Plant Nutr.* 8, 567–583, 1985; Kaushal, V. and Barnes, L.D., Effect of zwitterionic buffers on measurement of small masses of protein with bicinchoninic acid, *Anal. Biochem.* 157, 291–294, 1986; Grady, J.K., Chasteen, N.D., and Harris, D.C., Radicals from "Good's" buffers, *Anal. Biochem.* 173, 111–115, 1988; Le Hir, M., Impurity in buffer substances mimics the effect of ATP on soluble 5'-nucleotidase, *Enzyme* 45, 194–199, 1991; Pedrotti, B., Soffientini, A., and Islam, K., Sulphonate buffers affect the recovery of microtubule-associated proteins MAP1 and MAP2: evidence that MAP1A promotes microtubule assembly, *Cell Motil. Cytoskeleton* 25, 234–242, 1993; Vasseur, M., Frangne, R., and Alvarado, F., Buffer-dependent pH sensitivity of the fluorescent chloride-indicator dye SPQ, *Am. J. Physiol.* 264, C27–C31, 1993; Frick, J. and Mitchell, C.A., Stabilization of pH in solid-matrix hydroponic systems, *HortScience* 28, 981–984, 1993; Yu, Q., Kandegedara, A., Xu, Y., and Rorabacher, D.B., Avoiding interferences from Good's buffers: a contiguous series of noncomplexing tertiary amine buffers covering the entire range of pH 3–11, *Anal. Biochem.* 253, 50–56, 1997; Gelfi, C., Vigano, A., Curcio, M. et al., Single-strand conformation polymorphism analysis by capillary zone electrophoresis in neutral pH buffer, *Electrophoresis* 21, 785–791, 2000; Walsh, M.K., Wang, X., and Weimer, B.C., Optimizing the immobilization of single-stranded DNA onto glass beads, *J. Biochem. Biophys. Methods* 47, 221–231, 2001; Hosse, M. and Wilkinson, K.J., Determination of electrophoretic mobilities and hydrodynamic radii of three humic substances as a function of pH and ionic strength, *Environ. Sci. Technol.* 35, 4301–4306, 2001; Mash, H.E., Chin, Y.P., Sigg, L. et al., Complexation of copper by zwitterionic aminosulfonic (good) buffers, *Anal. Chem.* 75, 671–677, 2003; Ozkara, S., Akgol, S., Canak, Y., and Denizli, A., A novel magnetic adsorbent for immunoglobulin-g purification in a magnetically stabilized fluidized bed, *Biotechnol. Prog.* 20, 1169–1175, 2004; Hachmann, J.P. and Amshey, J.W., Models of protein modification in Tris-glycine and neutral pH Bis-Tris gels during electrophoresis: effect of pH, *Anal. Biochem.* 342, 237–345, 2005; Krajewska, B. and Ciurli, S., Jack bean (*Canavalia ensiformis*) urease. Probing acid-base groups of the active site by pH variation, *Plant Physiol. Biochem.* 43, 651–658, 2005; Zhao, G. and Chasteen, N.D., Oxidation of Good's buffers by hydrogen peroxide, *Anal. Biochem.* 349, 262–267, 2006.

MOPS	3-(*N*-morpholino) Propanesulfonic	209.3	A "Good" buffer.
	Acid;		
	4-morpholine-propanesulfonic Acid		

Good, N.E., Winget, G.D., Winter, W. et al., Hydrogen ion buffers for biological research, *Biochemistry* 5, 467–477, 1966; Altura, B.M., Altura, B.M., Carella, A., and Altura, B.T., Adverse effects of Tris, HEPES, and MOPS buffers on contractile responses of arterial and venous smooth muscle induced by prostaglandins, *Prostaglandins Med.* 5, 123–130, 1980; Tadolini, B., Iron autoxidation in MOPS and HEPES buffers, *Free Radic. Res. Commun.* 4, 149–160, 1987; Tadolini, B. and Sechi, A.M., Iron oxidation in MOPS buffer. Effect of phosphorus containing compounds, *Free Radic. Res. Commun.* 4, 161–172, 1987; Tadolini, B., Iron oxidation in MOPS buffer. Effect of EDTA, hydrogen peroxide, and FeCl$_3$, *Free Radic. Res. Commun.* 4, 172–182, 1987; Ishihara, H. and Welsh, M.J., Block by MOPS reveals a conformation change in the CFTR pore produced by ATP hydrolysis, *Am. J. Physiol.* 273, C1278–C1289, 1997; Schmidt, K., Pfeiffer, S., and Meyer, B., Reaction of peroxynitrite with HEPES or MOPS results in the formation of nitric oxide donors, *Free Radic. Biol. Med.* 24, 859–862, 1998; Hodges, G.R. and Ingold, K.U., Superoxide, amine buffers, and tetranitromethane: a novel free radical chain reaction, *Free Radic. Res.* 33, 547–550, 2000; Corona-Izquierdo, F.P. and Membrillo-Hernandez, J., Biofilm formation in *Escherichia coli* is affected by 3-(*N*-morpholino) propane sulfonate (MOPS), *Res. Microbiol.* 153, 181–185, 2002; Mash, H.E., Chin, Y.P., Sigg, L. et al., Complexation of copper by zwitterionic aminosulfonic (Good) buffers, *Anal. Chem.* 75, 671–677, 2003; Denizli, A., Alkan, M., Garipcan, B. et al., Novel metal-chelate affinity adsorbent for purification of immunoglobulin-G from human plasma, *J. Chromatog. B Analyt. Technol. Biomed. Life Sci.* 795, 93–103, 2003; Emir, S., Say, R., Yavuz, H., and Denizli, A., A new metal chelate affinity adsorbent for cytochrome C, *Biotechnol. Prog.* 20, 223–228, 2004; Cvetkovic, A., Zomerdijk, M., Straathof, A.J. et al., Adsorption of fluorescein by protein crystals, *Biotechnol. Bioeng.* 87, 658–668, 2004; Zhao, G. and Chasteen,

J.D., Oxidation of Good's buffers by hydrogen peroxide, *Anal. Biochem.* 349, 262–267, 2006; Vrakas, D., Giaginis, C., and Tsantili-Kakoulidou, A., Different retention behavior of structurally diverse basic and neutral drugs in immobilized artificial membrane and reversed-phase high-performance liquid chromatography: comparison with octanol-water partitioning, *J. Chromatog. A* 1116, 158–164, 2006; de Carmen Candia-Plata, M., Garcia, J., Guzman, R. et al., Isolation of human serum immunoglobulins with a new salt-promoted adsorbent, *J. Chromatog. A* 1118, 211–217, 2006.

Phosphate   Buffers, physiological solution.

Phosphate buffers are among the most common buffers used for biological studies. The use of phosphate solutions in early transfusion medicine led to the discovery of the importance of calcium ions in blood coagulation (Hutchin, P., History of blood transfusion: a tercentennial look, *Surgery* 64, 685–700, 1968). Phosphate-buffer saline (PBS; generally 0.01 M sodium phosphate — 0.14 M NaCl, pH 7.2. An incredible variation in PBS exists so it is necessary to verify composition — the only common factor that this writer finds is 0.01 M [10 mM] phosphate) is extensively used. Sodium phosphate buffers are the most common, but there is extensive use of potassium phosphate buffers and mixtures of sodium and potassium. Unfortunately, many investigators simply refer to phosphate buffers without respect to counter ion. Also, investigators will prepare a stock solution of sodium phosphate (usually sodium dihydrogen phosphate [sodium phosphate, monobasic] or disodium hydrogen phosphate [sodium phosphate, dibasic]) and adjust pH as required with (usually) hydrochloric acid or sodium hydrogen. This is not preferable and, if used, must be described in the text to permit other investigators to repeat the experiment. pH changes in phosphate buffers during freezing can be dramatic due to precipitation of phosphate buffer salts (van den Berg, L. and Rose, D., Effect of freezing on the pH and composition of sodium and potassium phosphate solutions: the reciprocal system KH_2PO_4-Na_2PO_4-H_2O, *Arch. Biochem. Biophys.* 81, 319–329, 1959; Murase, N. and Franks, F., Salt precipitation during the freeze-concentration of phosphate buffer solutions, *Biophys. Chem.* 34, 393–300, 1989; Pikal-Cleland, K.A. and Carpenter, J.F., Lyophilization-induced protein denaturation in phosphate buffer systems: monomeric and tetrameric beta-galactosidase, *J. Pharm. Sci.* 90, 1255–1268, 2001; Gomez, G., Pikal, M., and Rodriguez-Hornedo, N., Effect of initial buffer composition on pH changes during far-from-equilibrium freezing of sodium phosphate buffer solutions, *Pharm. Res.* 18, 90–97, 2001; Pikal-Cleland, K.A., Cleland, J.L., Anchorodoquy, T.J., and Carpenter, J.F., Effect of glycine on pH changes and protein stability during freeze-thawing in phosphate buffer systems, *J. Pharm. Sci.* 91, 1969–1979, 2002). Phosphate binds divalent cations in solutions and can form insoluble salts. Phosphate influences biological reactions by binding cations such as calcium, platinum, and iron (Staum, M.M., Incompatibility of phosphate buffer in 99^m Tc-sulfur colloid containing aluminum ion, *J. Nucl. Med.* 13, 386–387, 1972; Frank, G.B., Antagonism by phosphate buffer of the twitch ions in isolated muscle fibers produced by calcium-free solutions, *Can. J. Physiol. Pharmacol.* 56, 523–526, 1978; Hasegawa, K., Hashi, K., and Okada, R., Physicochemical stability of pharmaceutical phosphate buffer solutions. I. Complexation behavior of Ca(II) with additives in phosphate buffer solutions, *J. Parenter. Sci. Technol.* 36, 128–133, 1982; Abe, K., Kogure, K., Arai, H., and Nakano, M., Ascorbate-induced lipid peroxidation results in loss of receptor binding in Tris, but not in phosphate, buffer. Implications for the involvement of metal ions, *Biochem. Int.* 11, 341–348, 1985; Pedersen, H.B., Josephsen, J., and Keerszan, G., Phosphate buffer and salt medium concentrations affect the inactivation of T4 phage by platinum(II) complexes, *Chem. Biol. Interact.* 54, 1–8, 1985; Kuzuya, M., Yamada, K., Hayashi, T. et al., Oxidation of low-density lipoprotein by copper and iron in phosphate buffer, *Biochim. Biophys. Acta* 1084, 198–201, 1991). Also see Wolf, W.J., and Sly, D.A., Effects of buffer cations on chromatography of proteins on hydroxylapatite, *J. Chromatog.* 15, 247–250, 1964; Taborsky, G., Oxidative modification of proteins in the presence of ferrous ion and air. Effect of ionic constituents of the reaction medium on the nature of the oxidation products, *Biochemistry* 12, 1341–1348, 1973; Millsap, K.W., Reid, G., van der Mei, H.C., and Busscher, H.J., Adhesion of *Lactobacillus* species in urine and phosphate buffer to silicone rubber and glass under flow, *Biomaterials* 18, 87–91, 1997; Gebauer, P. and Bocek, P., New aspects of buffering with multivalent weak acids in capillary zone electrophoresis: pros and cons of the phosphate buffer, *Electrophoresis* 21, 2809–2813, 2000; Gebauer, P., Pantuikova, P., and Bocek, P., Capillary zone electrophoresis in phosphate buffer — known or unknown? *J. Chromatog. A* 894, 89–93, 2000; Buchanan, D.D., Jameson, E.E., Perlette, J. et al., Effect of buffer, electric field, and separation time on detection of aptamers-ligand complexes for affinity probe capillary electrophoresis, *Electrophoresis* 24, 1375–1382, 2003; Ahmad, I., Fasihullah, Z., and Vaid, F.H., Effect of phosphate buffer on photodegradation reactions of riboflavin in aqueous solution, *J. Photochem. Photobiol. B* 78, 229–234, 2005.

TES   *N*-Tris(hydroxymethyl)methyl-2-   229.3   A "Good" buffer.
                         aminoethane-sulfonic Acid

Good, N.E., Winget, G.D., Winter, W. et al., Hydrogen ion buffers for biological research, *Biochemistry* 5, 467–477, 1966; Itagaki, A. and Kimura, G., TES and HEPES buffers in mammalian cell cultures and viral studies: problem of carbon dioxide requirement, *Exp. Cell Res.* 83, 351–361, 1974; Bridges, S. and Ward, B., Effect of hydrogen ion buffers on photosynthetic oxygen evolution in the blue-green alga, *Agmenellum quadruplicatum*, *Microbios* 15, 49–56, 1976; Bailyes, E.M., Luzio, J.P., and Newby, A.C., The use of a zwitterionic detergent in the solubilization and purification of the intrinsic membrane protein 5′-nucleotidase, *Biochem. Soc. Trans.* 9, 140–141, 1981; Poole, C.A., Reilly, H.C., and Flint, M.H., The adverse effects of HEPES, TES, and BES zwitterionic buffers on the ultrastructure of cultured chick embryo epiphyseal chondrocytes, In Vitro 18, 755–765, 1982; Nakon, R. and Krishnamoorthy, C.R., Free-metal ion depletion by "Good's" buffers, *Science* 221, 749–750, 1983; del Castillo, J., Escalona de Motta, G., Eterovic, V.A., and Ferchmin, P.A., Succinyl derivatives of *N*-Tris (hydroxylmethyl) methyl-2-aminoethane sulphonic acid: their effects on the frog neuromuscular junction, *Br. J. Pharmacol.* 84, 275–288, 1985; Kaushal, V. and Varnes, L.D., Effect of zwitterionic buffers on measurement of small masses of protein with bicinchoninic acid, *Anal. Biochem.* 157, 291–294, 1986; Bhattacharyya, A. and Yanagimachi, R., Synthetic organic pH buffers can support fertilization of guinea pig eggs, but not as efficiently as bicarbonate buffer, *Gamete Res.* 19, 123–129, 1988; Veeck, L.L., TES and Tris (TEST)-yolk buffer systems, sperm function testing, and *in vitro* fertilization, *Fertil. Steril.* 58, 484–486, 1992; Kragh-Hansen, U. and Vorum, H., Quantitative analyses of the interaction between calcium ions and human serum albumin, *Clin. Chem.* 39, 202–208, 1993; Jacobs, B.R., Caulfield, J., and Boldt, J., Analysis of TEST (TES and Tris) yolk buffer effects of human sperm, *Fertil. Steril.* 63, 1064–1070, 1995; Stellwagne, N.C., Bossi, A., Gelfi, C., and Righetti, P.G., DNA and buffers: are there any noninteracting, neutral pH buffers? *Anal. Biochem.* 287, 167–175, 2000; Taylor, J., Hamilton, K.L., and Butt, A.G., HCO₃⁻ potentiates the cAMP-dependent secretory response of the human distal colon through a DIDS-sensitive pathway, *Pflügers Arch.* 442, 256–262, 2001; Taha, M., Buffers for the physiological pH range: acidic dissociation constants of zwitterionic compounds in various hydroorganic media, *Ann. Chim.* 95, 105–109, 2005.

Triethanolamine   Tris(2-hydroxyethyl)amine   149.2   Buffer; transdermal transfer reagent.

Fitzgerald, J.W., The Tris-catalyzed isomerization of potassium D-glucose 6-*O*-sulfate, *Can. J. Biochem.* 53, 906–910, 1975; Buhl, S.N., Jackson, K.Y., and Graffunder, B., Optimal reaction conditions for assaying human lactate dehydrogenase pyruvate-to-lactate at 25, 30, and 37 degrees C, *Clin. Chem.* 24, 261–266, 1978; Myohanen, T.A., Bouriotas, V., and Dean, P.D., Affinity chromatography of yeast alpha-glucosidase using ligand-mediated chromatography on immobilized phenylboronic acids, *Biochem. J.* 197, 683–688, 1981; Shinomiya, Y., Kato, N., Imazawa, M., and Miyamoto, K., Enzyme immunoassay of the myelin basic protein, *J. Neurochem.* 39, 1291–1296, 1982; Arita, M., Iwamori, M., Higuchi, T., and Nagai, Y., 1,1,3,3-tetramethylurea and triethanolaminne as a new useful matrix for fast atom bombardment mass spectrometry of gangliosides and neutral glycosphingolipids, *J. Biochem.* 93, 319–322, 1983; Cao, H. and Preiss, J., Evidence for essential arginine residues at the active site of maize branching enzymes, *J. Protein Chem.* 15, 291–304, 1996; Knaak, J.B., Leung, H.W., Stott, W.T. et al., Toxicology of mono-, di-, and triethanolamine, *Rev. Environ. Contim. Toxicol.* 149, 1–86, 1997; Liu, Q., Li, X., and Sommer, S.S., pK-matched running buffers for gel electrophoresis, *Anal. Biochem.* 270, 112–122, 1999; Sanger-van de Griend, C.E., Enantiomeric separation of glycyl dipeptides by capillary electrophoresis with cyclodextrins as chiral selectors, *Electrophoresis* 20, 3417–3424, 1999; Fang, L., Kobayashi, Y., Numajiri, S. et al., The enhancing effect of a triethanolamine-ethanol-isopropyl myristate mixed system on the skin permeation of acidic drugs, *Biol. Pharm. Bull.* 25, 1339–1344, 2002; Musial, W. and Kubis, A., Effect of some anionic polymers of pH of triethanolamine aqueous solutions, *Polim. Med.* 34, 21–29, 2004.

Triethylamine   *N,N*-diethylethanamine   101.2   Ion-pair reagent; buffer.

Brind, J.L., Kuo, S.W., Chervinsky, K., and Orentreich, N., A new reversed-phase, paired-ion, thin-layer chromatographic method for steroid sulfate separations, *Steroids* 52, 561–570, 1988; Koves, E.M., Use of high-performance liquid chromatography-diode array detection in forensic toxicology, *J. Chromatog. A* 692, 103–119, 1995; Cole, S.R. and Dorsey, J.G., Cyclohexylamine additives for enhanced peptide separations in reversed-phase liquid chromatography, *Biomed. Chromatog.* 11, 167–171, 1997; Gilar, M., and Bouvier, E.S.P., Purification of crude DNA oligonucleotides by solid-phase extraction and reversed-phase high-performance liquid chromatography, *J. Chromatog. A* 890, 167–177, 2000; Loos, R. and Barcelo, D., Determination of haloacetic acids in aqueous environments by solid-phase extraction followed by ion-pair liquid chromatography-electrospray ionization mass spectrometric detection, *J. Chromatog. A* 938, 45–55, 2001; Gilar, M., Fountain, K.J., Budman, Y. et al., Ion-pair, reversed-phase, high-performance liquid chromatography analysis of oligonucleotides: retention prediction, *J. Chromatog. A.* 958, 167–182, 2002; El-dawy, M.A., Mabrouk, M.M., and El-Barbary, F.A., Liquid chromatographic determination of fluoxetine, *J. Pharm. Biomed. Anal.* 30, 561–571, 2002; Yang, X., Zhang, X., Li, A. et al., Comprehensive two-dimensional separations based on capillary high-performance liquid chromatography and microchip electrophoresis, *Electrophoresis* 24, 1451–1457, 2003; Murphey, A.T., Brown-Augsburger, P., Yu, R.Z. et al., Development of an ion-pair reverse-phase liquid chromatographic/tandem mass spectrometry method for the determination of an 18-mer phosphorothioate oligonucleotide in mouse liver tissue, *Eur. J. Mass Spectrom.* 11, 209–215, 2005; Xie, G., Sueishi, Y., and Yamamoto, S., Analysis of the effects of protic, aprotic, and multicomponent solvents on the fluorescence emission of naphthalene and its exciplex with triethylamine, *J. Fluoresc.* 15, 475–483, 2005.

Tris   Tris(hydroxymethyl)   121.14   Buffer.
                           aminomethylmethane

Bernhard, S.A., Ionization constants and heats of Tris(hydroxymethyl)aminomethane and phosphate buffers, *J. Biol. Chem.* 218, 961–969, 1956; Rapp, R.D. and Memminger, M.M., Tris(hydroxymethyl)aminomethane as an electrophoresis buffer, *Am. J. Clin. Pathol.* 31, 400–403, 1959; Rodkey, F.L., Tris(hydroxymethyl)aminomethane as a standard for Kjeldahl nitrogen analysis, *Clin. Chem.* 10, 606–610, 1964; Oliver, R.W. and Viswanatha, T., Reaction of Tris(hydroxymethyl) aminomethane with cinnamoyl imidazole and cinnamoyltrypsin, *Biochim. Biophys. Acta* 156, 422–425, 1968; Douzou, P., Enzymology at subzero temperatures, *Mol. Cell. Biochem.* 1, 15–27, 1973; Fitzgerald, J.W., The Tris-catalyzed isomerization of potassium D-glucose 6-*O*-sulfate, *Can. J. Biochem.* 53, 906–910, 1975; Visconti, M.A. and Castrucci, A.M., Tris buffer effects on melanophore aggregating responses, *Comp. Biochem. Physiol. C* 82, 501–503, 1985; Stambler, B.S., Grant, A.O., Broughton, A., and Strauss, H.C., Influences of buffers on dV/dtmax recovery kinetics with lidocaine in myocardium, *Am. J. Physiol.* 249, H663–H671, 1985; Nakano, M. and Tauchi, H., Difference in activation by Tris(hydroxymethyl)aminomethane of Ca,Mg-ATPase activity between young and old rat skeletal muscles, *Mech. Aging. Dev.* 36, 287–294, 1986; Oliveira, L., Araujo-Viel, M.S., Juliano, L., and Prado, E.S., Substrate activation of porcine kallikrein *N*-α derivatives of arginine 4-nitroanilides, *Biochemistry* 26, 5032–5035, 1987; Ashworth, C.D. and Nelson, D.R., Antimicrobial potentiation of irrigation solutions containing Tris(hydroxymethyl)aminomethane-EDTA, *J. Am. Vet. Med. Assoc.* 197, 1513–1514, 1990; Schacker, M., Foth, H., Schluter, J., and Kahl, R., Oxidation of Tris to one-carbon compounds in a radical-producing model system, in microsomes, in hepatocytes, and in rats, *Free Radic. Res. Commun.* 11, 339–347, 1991; Weber, R.E., Use of ionic and zwitterionic (Tris/BisTris and HEPES) buffers in studies on hemoglobin function, *J. Appl. Physiol.* 72, 1611–1615, 1992; Veeck, L.L., TES and Tris (TEST)-yolk buffer systems, sperm function testing, and *in vitro* fertilization, *Fertil. Steril.* 58, 484–486, 1992; Shiraishi, H., Kataoka, M., Morita, Y., and Umemoto, J., Interaction of hydroxyl radicals with Tris(hydroxymethyl)aminomethane and Good's buffers containing hydroxymethyl or hydroxyethyl residues produce formaldehyde, *Free Radic. Res. Commun.* 19, 315–321, 1993; Vasseur, M., Frangne, R., and Alvarado, F., Buffer-dependent pH sensitivity of the fluorescent chloride-indicator dye SPQ, *Am. J. Physiol.* 264, C27–C31, 1993; Niedernhofer, L.J., Riley, M., Schnez-Boutand, N. et al., Temperature-dependent formation of a conjugate between Tris(hydroxymethyl)aminomethane buffer and the malondialdehyde-DNA adduct pyrimidopurinone, *Chem. Res. Toxicol.* 10, 556–561, 1997; Trivic, S., Leskovac, V., Zeremski, J. et al., Influence of Tris(hydroxymethyl)aminomethane on kinetic mechanism of yeast alcohol dehydrogenase, *J. Enzyme Inhib.* 13, 57–68, 1998; Afifi, N.N., Using difference spectrophotometry to study the influence of different ions and buffer systems on drug protein binding, *Drug Dev. Ind. Pharm.* 25, 735–743, 1999; AbouHaider, M.G. and Ivanov, I.G., Nonenzymatic RNA hydrolysis promoted by the combined catalytic activity of buffers and magnesium ions, *Z. Naturforsch.* 54, 542–548, 1999; Shihabi, Z.K., Stacking of discontinuous buffers in capillary zone electrophoresis, *Electrophoresis* 21, 2872–2878, 2000; Stellwagen, N.C, Bossi, A., Gelfi, C., and Righetti, P.G., DNA and buffers: are there any noninteracting, neutral pH buffers? *Anal. Biochem.* 287, 167–175, 2000; Burcham, P.C., Fontaine, F.R., Petersen, D.R., and Pyke, S.M., Reactivity of Tris(hydroxymethyl)aminomethane confounds immunodetection of acrolein-adducted proteins, *Chem. Res. Toxicol.* 16, 1196–1201, 2003; Koval, D., Kasicka, V., and Zuskova, I., Investigation of the effect of ionic strength of Tris-acetate background electrolyte on electrophoretic mobilities of mono-, di-, and trivalent organic anions by capillary electrophoresis, *Electrophoresis* 26, 3221–3231, 2005; Kinoshita, T., Yamaguchi, A., and Tada, T., Tris(hydroxymethyl)aminomethane-induced conformational change and crystal-packing contraction of porcine pancreatic elastase, *Acta Crystallograph. Sect. F Struct. Biol. Cryst. Commun.* 62, 623–626, 2006; Qi, Z., Li, X., Sun, D. et al., Effect of Tris on catalytic activity of MP-11, *Bioelectrochemistry* 68, 40–47, 2006.

7 Organic Name Reactions Useful in Biochemistry and Molecular Biology

AKABORI AMINO ACID REACTION

Reaction in the presence of hydrazine yields hydrazides which can be coupled to aromatic aldehydes

Bose, A.K., *et al.*, Microwave enhanced Akabori reaction for peptide analysis, *J.Am.Soc.Mass Spectrom.* **13**, 839-850, 2002

Originally devised as a method for the conversion of amino acids or amino acid esters to aldehydes. The Akabori reaction has been modified for use in the determination of C-terminal amino acids by performing the reaction in the presence of hydrazine and for the production of derivatives useful for mass spectrometric identification. See Ambach, E. and Beck, W., Metal-complexes with biologically important ligands. 35. Nickel, cobalt, palladium, and platinum complexes with Schiff-bases of α-amino acids — a contribution to the mechanism of the Akabori reaction, *Chemische Berichte-Recueil* 118, 2722–2737, 1985; Bose, A.K., Ing, Y.H., Pramanik, B.N. et al., Microwave-enhanced Akabori reaction for peptide analysis, *J. Am. Soc. Mass Spectrom.* 13, 839–850, 2002; Pramanik, B.N., Ing, Y.H., Bose, A.K. et al., Rapid cyclopeptide anaylsis by microwave-enhanced Akabori reaction, *Tetrahedron Lett.* 44, 2565–2568, 2003; Puar, M.S., Chan, T.M., Delgarno, D. et al., Sch 486058: a novel cyclic peptide of actinomycete origin, *J. Antibiot.* 58, 151–154, 2005.

ALDOL CONDENSATION

Aldol condensation

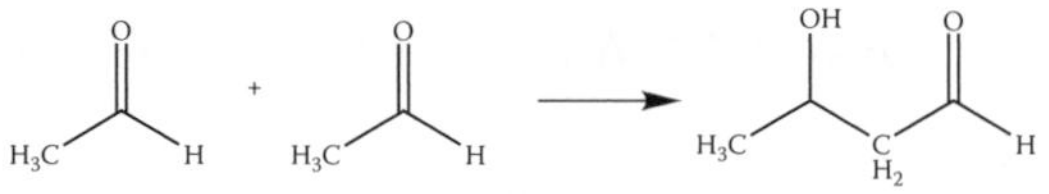

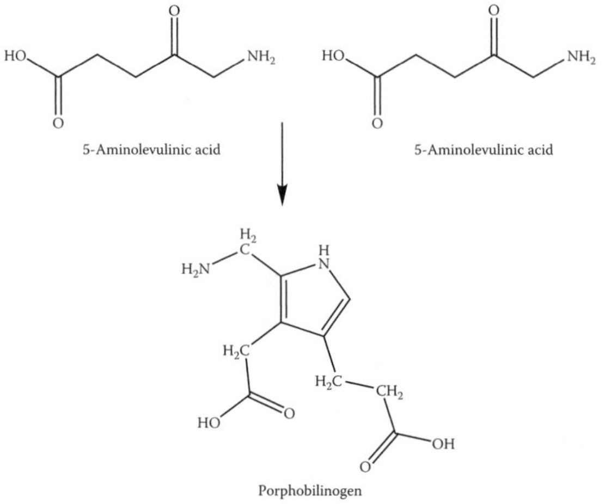

5-Aminolevulinic acid

5-Aminolevulinic acid

Porphobilinogen

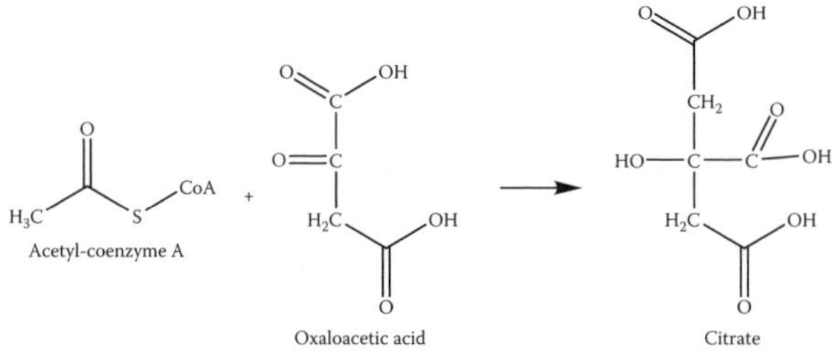

Acetyl-coenzyme A

Oxaloacetic acid

Citrate

Citrate synthase
an aldol-like condensation

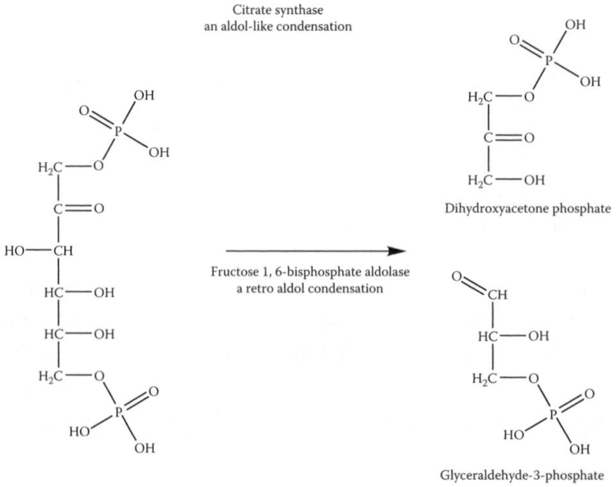

Dihydroxyacetone phosphate

Fructose 1, 6-bisphosphate aldolase
a retro aldol condensation

Glyceraldehyde-3-phosphate

Condensation of one carbonyl compound with the enol/enolate form of another to form an α-hydroxyaldehyde; the base-catalyzed reaction proceeds via the enolate form while the acid-catalyzed reaction proceeds via the enol form. The basic chemistry of the aldol condensation is observed in several enzymatic reactions including citrate synthase, fructose-1,6-bisphosphate aldolase, and 2-keto-4-hydroxyglutarate aldolase. See Lane, R.S., Hansen, B.A., and Dekker, E.E., Sulfhydryl groups in relation to the structure and catalytic activity of 2-oxo-4-hydroxyglutarate aldolase from bovine liver, *Biochim. Biophys. Acta* 481, 212–221, 1977; Evans, D.A. and McGee, L.R., Aldol diastereoselection. Zirconium enolates. Product selective, enolate structure independent condensations, *Tetrahedron Lett.* 21, 3975–3978, 1980; Grady, S.R., Wang, J.K., and Dekker, E.E., Steady-state kinetics and inhibition studies of the aldol condensation reaction catalyzed by bovine liver and *Escherichia coli* 2-keto-4-hydroxyglutarate aldolase, *Biochemistry* 20, 2497–2502, 1981; Rokita, S.E., Srere, P.A., and Walsh, C.T., 3-fluoro-3-deoxycitrate: a probe for mechanistic study of citrate-utilizing enzymes, *Biochemistry* 21, 3765–3774, 1982; Frere, R., Nentwich, M., Gacond, S. et al., Probing the active site of *Pseudomonas aeruginosa* porphobilinogen synthase using newly developed inhibitors, *Biochemistry* 45, 8243–8253, 2006; Dalsgaard, T.K., Nielsen, J.H., and Larsen, L.B., Characterization of reaction products formed in a model reaction between pentanal and lysine-containing oligopeptides, *J. Agric. Food Chem.* 54, 6367–6373, 2006. A crossed aldol refers to a condensation reaction with two different aldehydes/ketones; the second aldehyde frequently is formaldehyde as it cannot react with itself although this is not a requirement (Kiehlman, E. and Loo, P.W., Orientation in crossed aldol condensation of chloral with unsymmetrical aliphatic ketones, *Canad. J. Chem.* 49, 1588, 1971; Findlay, J.A., Desai, D.N., and McCaulay, J.B., Thermally induced crossed aldol condensations, *Canad. J. Chem.* 59, 3303–3304, 1981; Esmaelli, A.A., Tabas, M.S., Nasseri, M.A., and Kazemi, F., Solvent-free crossed aldol condensation of cyclic ketones with aromatic aldehydes assisted by microwave irradiation, *Monatshefte fur Chemie* 136, 571–576, 2005).

AMADORI REARRANGEMENT

Amadori rearrangement

A reaction following the formation of the unstable reaction product between an aldehyde (reducing sugar) and an amino group (formation of a Schiff base, an aldimine), which results in a more stable ketoamine. The Amadori rearrangement is part of the Malliard reaction, which is also called the Browning reaction, and can result in the formation of advanced glycation endproducts. See Amadori, M., Products of the condensation between glucose and p-phenetidine, *Atti. Accad. Nazl. Lincei* 2, 337, 1925; Hodge, J.E., The Amadori rearrangement, *Adv. Carbohydrate Chem.* 10, 169–205, 1955; Acharya, A.S. and Manning, J.M., Amadori rearrangement of glyceraldehyde-hemoglobin Schiff based adducts. A new procedure for the determination of ketoamine adducts in proteins, *J. Biol. Chem.* 255, 7218–7224, 1980; Acharya, A.S. and Manning, J.M., Reaction of glycoaldehyde with proteins: latent crosslinking potential of α-hydroxyaldehydes, *Proc. Natl. Acad. Sci. USA* 80, 3590–3594, 1983; Roper, H., Roper, S., and Meyer, B., Amadori- and N-nitroso-Amadori compounds and their pyrolysis products. Chemical, analytical, and biological aspects, *IARC Sci. Publ.* 57, 101–111, 1984; Baynes, J.W., Watkins, N.G., Fisher, C.I. et al., The Amadori product on protein: structure and reactions, *Prog. Clin. Biol. Res.* 304, 43–67, 1989; Nacharaju, P. and Acharya, A.S., Amadori rearrangement potential of hemoglobin at its glycation sites is dependent on the three-dimensional structure of protein, *Biochemistry* 31, 12673–12679, 1992; Zyzak, D.V., Richardson, J.M., Thorpe, S.R., and Baynes, J.W., Formation of reactive intermediates from Amadori compounds under physiological conditions, *Arch. Biochem. Biophys.* 316, 547–554, 1995; Khalifah, R.G., Baynes, J.W., and Hudson, B.G., Amadorins: novel post-Amadori inhibitors of advanced glycation reactions, *Biochem. Biophys. Res. Commun.* 257, 251–258, 1999; Davidek, T., Clety, N., Aubin, S., and Blank, I., Degradation of the Amadori compound *N*-(1-deoxy-D-fructose-1-yl)glycine in aqueous model system, *J. Agric. Food Chem.* 50, 5472–5479, 2002.

BAEYER–VILLIGER REACTION

Baeyer–Villiger reaction

The oxidation of a ketone by a peroxy acid to yield an ester. This reaction is catalyzed by bacterial monooxygenases and has proved useful in preparing optically pure esters and lactones. See Ryerson, C.C., Ballou, D.P., and Walsh, C., Mechanistic studies on cyclohexanone oxygenase, *Biochemistry* 21, 2644–2655, 1982; Bolm, C., Metal-catalyzed asymmetric oxidations, *Med. Res. Rev.* 19, 348–356, 1999; Zambianchi, F., Pasta, P., Carrea, G. et al., Use of isolated cyclohexanone monoox-ygenase from recombinant *Escherichia coli* as a biocatalyst for Baeyer–Villiger and sulfide oxida-tions, *Biotechnol. Bioeng.* 78, 489–496, 2002; Alphand, V., Carrea, G., Wohlgemuth, R. et al., Towards large-scale synthetic application of Baeyer–Villiger monooxygenase, *Trends Biotechnol.* 21, 318–323, 2003; Walton, A.Z. and Stewart, J.D., Understanding and improving NADPH-depen-dent reactions by nongrowing *Escherichia coli* cells, *Biotechnol. Prog.* 20, 403–411, 2004; Malito, E., Alfieri, A., Fraaije, M.W., and Mattevi, A., Crystal structure of a Baeyer–Villiger monooxyge-nase, *Proc. Natl. Acad. Sci. USA* 101, 13157–13162, 2004; ten Brink, G.J., Arends, I.W., and Sheldon, R.A., The Baeyer–Villiger reaction: new developments toward greener procedures, *Chem. Rev.* 104, 4105–4124, 2004; Boronat, M., Corma. A., Renz, M. et al., A multisite molecular mechanism for Baeyer–Villiger oxidations on solid catalysts using environmentally friendly H_2O_2 as oxidant, *Chemistry* 11, 6905–6915, 2005; Mihovilovic, M.D., Rudroff, E., Winninger, A. et al., Microbial Baeyer–Villiger oxidation: stereopreference and substrate acceptance of cyclohexanone monooxygenase mutants prepared by directed evolution, *Org. Lett.* 8, 1221–1224, 2006; Baldwin, C.V. and Woodley, J.M., On oxygen limitation in a whole cell biocatalytic Baeyer–Villiger oxidation process, *Biotechnol. Bioeng.* 95, 362–369, 2006.

BECKMANN REARRANGEMENT

Oxime   Amide

Beckmann rearrangement

An acid (protic or Lewis) catalyzed conversion of an oxime to a substituted carboxylic amide. See Darling, C.M. and Chen, C.P., Rearrangement of *N*-benzyl-2-cyano-(hydroxyimino)acetamide,

J. Pharm. Sci. 67, 860–861, 1978; Gayen, A.K. and Knowles, C.O., Penetration and fate of methomyl and its oxime metabolite in insects and two spotted spider mites, *Arch. Environ. Contam. Toxicol.* 10, 55–67, 1981; Mangold, J.B., Mangold, B.L., and Spina, A., Rat liver aryl sulfotrans-ferase-catalyzed sulfation and rearrangement of 9-fluorenone oxime, *Biochim. Biophys. Acta* 874, 37–43, 1986; De Luca, L., Giacomelli, G., and Procheddu, A., Beckmann rearrangement of oximes under very mild conditions, *J. Org. Chem.* 67, 6272–6274, 2002; Torisawa, Y., Nishi, T., and Minamikawa, J., A study on the conversion of indanones into carbostyrils, *Bioorg. Med. Chem.* 11, 2205–2209, 2003; Furuya, Y., Ishihara, K., and Yamamoto, H., Cyanuric chloride as a mild and active Beckmann rearrangement catalyst, *J. Am. Chem. Soc.* 127, 11240–11241, 2005; Yamabe, S., Tsuchida, N., and Yamazaki, S., Is the Beckmann rearrangement a concerted or stepwise reaction? A computational study, *J. Org. Chem.* 70, 10638–10644, 2005; Ichino, T., Arimoto, H., and Uemura, D., Possibility of a non–amino acid pathway in the biosynthesis of marine-derived oxazoles, *Chem. Commun.* 16, 1742–1744, 2006.

BENZOIN CONDENSATION

The conversion of benzaldehyde to benzoin (aromatic α-hydroxyketones) via cyanide-mediated condensation; other aromatic aldehydes can participate in this reaction. See Iding, H., Dunnwald, T., Greiner, L. et al., Benzoylformate decarboxylase from *Pseudomonas putida* as stable catalyst for the synthesis of chiral 2-hydroxy ketones, *Chemistry* 6, 1483–1495, 2000; White, M.J. and Leeper, F.J., Kinetics of the thiazolium ion-catalyzed benzoin condensation, *J. Org. Chem.* 66, 5124–5131, 2001; Dunkelmann, P., Kolter-Jung, D., Nitsche, A. et al., Development of a donor-acceptor concept for enzymatic cross-coupling reactions of aldehydes: the first asymmetric cross-benzoin condensation, *J. Am. Chem. Soc.* 124, 12084–12085, 2002; Pohl, M., Lingen, B., and Muller, M., Thiamin-diphosphate-dependent enzymes: new aspects of asymmetric C–C bond for-mation, *Chemistry* 8, 5288–5295, 2002; Wildemann, H., Dunkelmann, P., Muller, M., and Schmidt, B., A short olefin metathesis-based route to enantiomerically pure arylated dihydropyrans and α,β-unsaturated δ-valero lactones, *J. Org. Chem.* 68, 799–804, 2003; Murry, J.A., Synthetic meth-odology utilized to prepare substituted imidazole p38 MAP kinase inhibitors, *Curr. Opin. Drug Discov. Devel.* 6, 945–965, 2003; Reich, B.J., Justice, A.K., Beckstead, B.T. et al., Cyanide-catalyzed cyclizations via aldamine coupling, *J. Org. Chem.* 69, 1357–1359, 2004; Sklute, G., Oizerowich, R., Shulman, H., and Keinan, E., Antibody-catalyzed benzoin oxidation as a mecha-nistic probe for nucleophilic catalysis by an active site lysine, *Chemistry* 10, 2159–2165, 2004; Breslow, R., Determining the geometries of transition states by use of antihydrophobic additives in water, *Acc. Chem. Res.* 37, 471–478, 2004.

CANNIZZARO REACTION

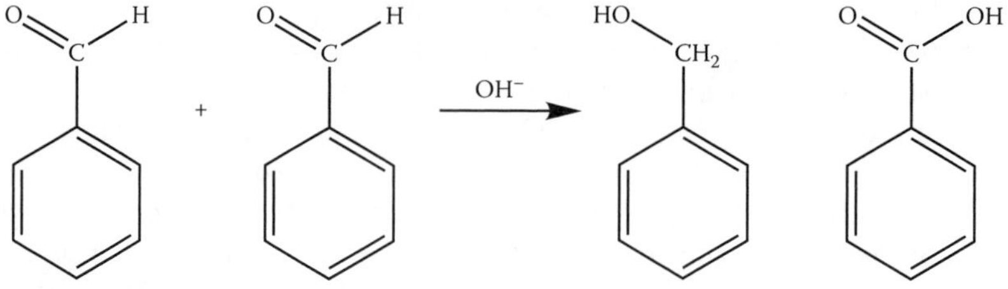

Cannizzaro reaction

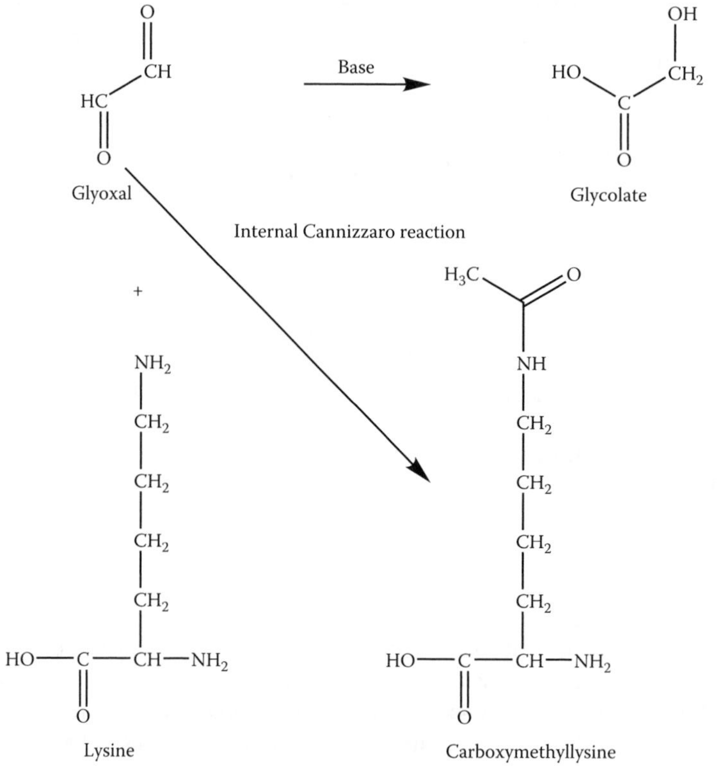

Glyoxal — Base → Glycolate

Internal Cannizzaro reaction

Lysine   Carboxymethyllysine

Base-catalyzed disproportionation of an aldehyde to yield a carboxylic acid and the corresponding alcohol; if an α-hydrogen is present, an aldol condensation is a competing reaction. See Hazlet, S.E. and Stauffer, D.A., Crossed Cannizzaro reactions, *J. Org. Chem.* 27, 2021–2024, 1962; Entezari, M.H. and Shameli, A.A., Phase-transfer catalysis and ultrasonic waves. I. Cannizzaro reaction, *Ultrason. Sonochem.* 7, 169–172, 2000; Matin, M.M., Sharma, T., Sabharwal, S.G., and Dhavale, D.D., Synthesis and evaluation of the glycosidase inhibitory activity of 5-hydroxy substituted isofaomine analogues, *Org. Biomol. Chem.* 3, 1702–1707, 2005; Zhang, L., Wang, S., Zhou, S. et al., Cannizzaro-type disproportionation of aromatic aldehydes to amides and alcohols by using either a stoichiometric amount or a catalytic amount of lanthanide compounds, *J. Org. Chem.* 71, 3149–3153, 2006. Intramolecular Cannizzaro reactions have been described (Glomb, M.A. and Monnier, V.M., Mechanism of protein modification by glyoxal and glycoaldehyde, reactive intermediates of the Maillard reaction, *J. Biol. Chem.* 270, 10017–10026, 1995; Russell, A.E., Miller, S.P., and Morken, J.P., Efficient Lewis acid catalyzed intramolecular Cannizzaro reaction, *J. Org. Chem.* 65, 8381–8383, 2000; Schramm, C. and Rinderer, B., Determination of cotton-bound glyoxal via an internal Cannizzaro reaction by means of high-performance liquid chromatography, *Anal. Chem.* 72, 5829–5833, 2000).

CLAISEN CONDENSATION

Claisen condensation

The base-catalyzed condensation of two moles of an ester to give a β-keto ester. Claisen condensations are more favorable with thioesters. This reaction is of great importance in the biosynthesis of fatty acids and polyketides. See Haapalainen, A.M., Meriläinen, G., and Wierenga, R.K., The thiolase superfamily: condensing enzymes with diverse reaction specificities, *Trends Biochem. Sci.* 31, 64–71, 2006. For general issues, see Dewar, M.J. and Dieter, K.M., Mechanism of the chain extension step in the biosynthesis of fatty acids, *Biochemistry* 27, 3302–3308, 1988; Clark, J.D., O'Keefe, S.J., and Knowles, J.R., Malate synthase: proof of a stepwise Claisen condensation using the double-isotope fractionation test, *Biochemistry* 27, 5961–5971, 1988; Nicholson, J.M., Edafiogho, I.O., Moore, J.A. et al., Cyclization reactions leading to β-hydroxyketo esters, *J. Pharm. Sci.* 83, 76–78, 1994; Lee, R.E., Armour, J.W., Takayama, K. et al., Mycolic acid biosynthesis: definition and targeting of the Claisen condensation step, *Biochim. Biophys. Acta* 1346, 275–284, 1997; Shimakata, T. and Minatogawa, Y., Essential role of trehalose in the synthesis and subsequent metabolism of corynomycolic acid in *Corynebacterium matruchotil, Arch. Biochem. Biophys.* 380, 331–338, 2000; Olsen, J.G., Madziola, A., von Wettstein-Knowles, P. et al., Structures of β-ketoacyl-acyl carrier protein synthase I complexed with fatty acids elucidate its catalytic machinery, *Structure* 9, 233–243, 2001; Klavins, M., Dipane, J., and Babre, K., Humic substances as catalysts in condensation reactions, *Chemosphere* 44, 737–742, 2001; Heath, R.J. and Rock, C.O., The Claisen condensation in biology, *Nat. Prod. Rep.* 19, 581–596, 2002; Takayama, K., Wang, C., and Besra, G.S., Pathway to synthesis and processing of mycolic acids in *Mycobacterium tuberculosis, Clin. Microbiol. Rev.* 18, 81–101, 2005; Ryu, Y., Kim, K.J., Roessner, C.A., and Scott, A.I., Decarboxylative Claisen condensation catalyzed by *in vitro* selected ribozymes, *Chem. Commun.* 13, 1439–1441, 2006; Kamijo, S. and Dudley, G.B., Claisen-type condensation of vinylogous acyl triflates, *Org. Lett.* 8, 175–177, 2006.

CLAISEN REARRANGEMENT

Chorismic Acid   Prephenic Acid

Zhang Z. and Bruice T.C., Temperature dependence of the structure of the substrate and active site of the *Thermus thermophilus* chorismate mutase E-S complex, *Biochemistry* **45**, 8562-8567, 2006

The rearrangement of an allyl vinyl ether, the nitrogen or sulfur analogue, or an allyl aryl ether to yield a γ,δ-unsaturated ketone or an *o*-allyl substituted phenol. See Hilvert, D., Carpenter, S.H., Nared, K.D., and Auditor, M.T., Catalysis of concerted reactions by antibodies: the Claisen rearrangement, *Proc. Natl. Acad. Sci. USA* 85, 4953–4955, 1988; Campbell, A.P., Tarasow, T.M., Massefski, W. et al., *Proc. Natl. Acad. Sci. USA* 90, 8663–8667, 1993; Swiss, K.A. and Firestone, R.A., Catalysis of Claisen rearrangement by low molecular weight polyethylene(1), *J. Org. Chem.* 64, 2158–2159, 1999; Berkowitz, D.B., Choi, S., and Maeng, J.H., Enzyme-assisted asymmetric total synthesis of (-)-podopyllotoxin and (-)-picropodophyllin, *J. Org. Chem.* 65, 847–860, 2000; Itami, K. and Yoshida, J., The use of hydrophilic groups in aqueous organic reactions, *Chem. Rec.* 2, 213–224, 2002; Martin Castro, A.M., Claisen rearrangement over the past nine decades, *Chem. Rev.* 104, 2939–3002, 2004; Sparano, B.A., Shahi, S.P., and Koide, K., Effect of binding and conformation on fluorescence quenching in new 2′,7′-dichlorofluorescein derivatives, *Org. Lett.* 6, 1947–1949, 2004; Davis, C.J., Hurst, T.E., Jacob, A.M., and Moody, C.J., Microwave-mediated Claisen rearrangement followed by phenol oxidation: a simple route to naturally occurring 1,4-benzoquinones. The first synthesis of verapliquinones A and B and panicein A., *J. Org. Chem.* 70, 4414–4422, 2005; Wright, S.K., DeClue, M.S., Mandal, A. et al., Isotope effects on the enzymatic and nonenzymatic reactions of chorismate, *J. Am. Chem. Soc.* 127, 12957–12964, 2005; Declue, M.S., Baldridge, K.K., Kast, P., and Hilvert, D., Experimental and computational investigation of the uncatalyzed rearrangement and elimination reactions of isochorismate, *J. Am. Chem. Soc.* 128, 2043–2051, 2006; Zhang, X. and Bruice, T.C., Temperature dependence of the structure of the substrate and active site of the *Thermus thermophilus* chorismate mutase E-S complex, *Biochemistry* 45, 8562–8567, 2006.

CRIEGEE REACTION

Mostly the reaction of a peroxyacid with a tertiary alcohol to form a ketone and an alcohol. The intermediate peroxyester is an intermediate (Criegee adduct or Criegee intermediate) in the Baeyer–Villiger reaction. The Criegee intermediate is important in the ozonolysis of alkenes including fatty acids. See Leffler, J.E. and Scrivener, F.E., Jr., The decomposition of cumyl peracetate in nonpolar solvents, *J. Org. Chem.* 37, 1794–1796, 1978; Srisankar, E.V. and Patterson, L.K., Reactions of ozone with fatty acid monolayer: a model system for disruption of lipid molecular assemblies by ozone, *Arch. Environ. Health* 34, 346–349, 1979; Grammer, J.C., Loo, J.A., Edmonds, C.G. et al., Chemistry and mechanism of vanadate-promoted photooxidative cleavage of myosin, *Biochemistry* 35, 15582–15592, 1996; Krasutsky, P.A., Kolomitsyn, I.V., Kiprof P. et al., Observation of a stable carbocation in a consecutive Criegee rearrangement with trifluoroperacetic acid, *J. Org. Chem.* 65, 3926–3992, 1996; Carlqvist, P., Eklund, P., Hult, K., and Brinck, T., Rational design of a lipase to accommodate catalysis of Baeyer–Villiger oxidation with hydrogen peroxide, *J. Mol. Model.* 9, 164–171, 2003; Deeth, R.J. and Bugg, T.D., A density functional investigation of the extradiol cleavage mechanism in non-heme iron catechol dioxygenase, *J. Biol. Inorg. Chem.* 8, 409–418,

2003; Krasutsky, P.A., Kolomitsyn, I.V., Krasutsky, S.G., and Kiprof, P., Double- and triple-consecutive O-insertion into *tert*-butyl and triarylmethyl structures, *Org. Lett.* 6, 2539–2542, 2004.

CURTIUS REARRANGEMENT

The conversion of a carboxylic acid to an amine via an acid–acid intermediate. See Inouye, K., Watanabe, K., and Shin, M., Formation and degradation of urea derivatives in the azide method of peptide synthesis. Part 1. The Curtius rearrangement and urea formation, *J. Chem. Soc.* 17, 1905–1911, 1977; Chorev, M., and Goodman, M., Partially modified retro-inverso peptides. Comparative Curtius rearrangements to prepare 1,1-diaminoalkane derivatives, *Int. J. Pept. Protein Res.* 21, 258–268, 1983; Sasmal, S., Geyer, A., and Maier, M.E., Synthesis of cyclic peptidomimetics from aldol building blocks, *J. Org. Chem.* 67, 6260–6263, 2002; Kedrowski, B.L., Synthesis of orthogonally protected (R)- and (S)-2-methylcysteine via an enzymatic desymmerization and Curtius rearrangement, *J. Org. Chem.* 68, 5403–5406, 2003; Englund, E.A., Gopi, H.N., and Appella, D.H., An efficient synthesis of a probe for protein function: 2,3-diaminopropionic acid with orthogonal protecting groups, *Org. Lett.* 6, 213–215, 2004; Spino, C., Tremblay, M.C., and Gobout, C., A stereodivergent approach to amino acids, amino alcohols, or oxazolidonones of high enantiomeric purity, *Org. Lett.* 6, 2801–2804, 2004; Brase, S., Gil, C., Knepper, K., and Zimmerman, V., Organic azides: an exploding diversity of a unique class of compounds, *Angew. Chem. Int. Ed. Engl.* 44, 5188–5240, 2005; Lebel, H. and Leogane, O., Boc-protected amines via a mild and efficient one-pot Curtius rearrangement, *Org. Lett.* 7, 4107–4110, 2005.

DAKIN REACTION

Conversion of an aromatic ketone or aldehyde to a phenolic derivative with alkaline hydrogen peroxide. The mechanism is thought to be similar to the Baeyer–Villiger reaction, possibly proceeding through a peroxyacid intermediate. The presence of an amino group or a hydroxyl group in the position *para* to the carbonyl function is required. See Corforth, J.W. and Elliott, D.F., Mechanism of the Dakin and West reaction, *Science* 112, 534–535, 1950.

DAKIN–WEST REACTION

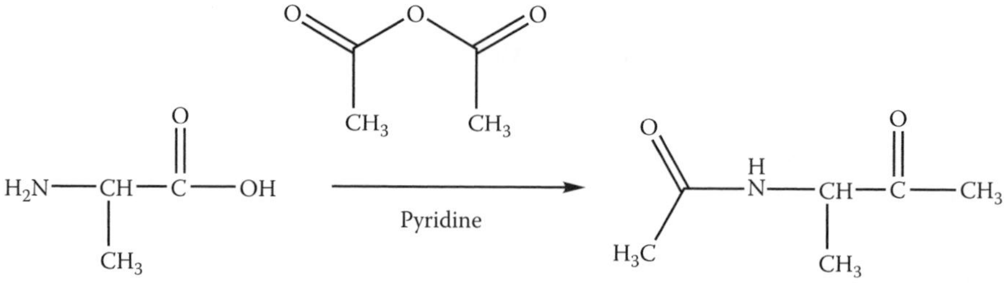

Dakin–West reaction

Conversion of amino acids to acetamidoketones via the action of acetic anhydride in a base where a carboxyl group is replaced by an acyl group in a reaction proceeding through an oxazolone intermediate. This reaction has been used for the synthesis of enzyme inhibitors and receptor antagonists. See Angliker, H., Wikstrom, P., Rauber, P. et al., Synthesis and properties of peptidyl derivatives of arginylfluoromethanes, *Biochem. J.* 256, 481–486, 1988; Cheng, L., Goodwin, C.A., Schully, M.F. et al., Synthesis and biological activity of ketomethylene pseudopeptide analogues as thrombin inhibitors, *J. Med. Chem.* 35, 3364–3369, 1992; Godfrey, A.B., Brooks, D.A., Hay, L.A. et al., Application of the Dakin–West reaction for the synthesis of oxazole-containing dual PPARα/γ agonists, *J. Org. Chem.* 68, 2623–2632, 2003; Loksha, Y.M., el-Barbary, A.A., el-Barbary, M.A. et al., Synthesis of 2-(aminocarbonylmethylthio)-1*H*-imidazoles as novel Capravirine analogues, *Bioorg. Med. Chem.* 13, 4209–4220, 2005.

DIELS–ALDER CONDENSATION

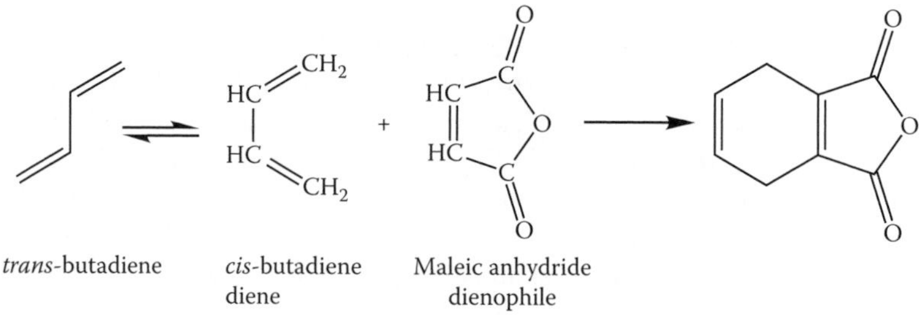

trans-butadiene	*cis*-butadiene	Maleic anhydride
	diene	dienophile

Diels–Alder condensation

A cycloaddition reaction between a conjugated diene and an alkene resulting in the formation of an alkene ring; construction of a six-membered ring with multiple stereogenic centers resulting in a chiral molecule. See Wasserman, A., *Diels-Alder Reactions: Organic Background and Physico-Chemical Aspects*, Elsevier, Amsterdam, Netherlands, 1965; Fringuelli, F. and Taticchi, A., *The Diels-Alder Reaction: Selected Practical Methods*, John Wiley & Sons, Chichester, UK, 2002; Stocking, E.M. and Williams, R.M., Chemistry and biology of biosynthetic Diels–Alder reactions, *Angew. Chem. Int. Ed.* 42, 3078–3115, 2003. See also Waller, R.L. and Recknagel, R.O., Determination of lipid conjugated dienes with tetracyanoethylene-[14]C: significance for study of the pathology of lipid peroxidation, *Lipids* 12, 914–921, 1977; Melucci, M., Barbarella, G., and Sotgiu, G., Solvent-free, microwave-assisted synthesis of thiophene oligomers via Suzuki coupling, *J. Org. Chem.* 67, 8877–8884, 2002; Breslow, R., Determining the geometries of transition states by use of antihydrophobic additives in water, *Acc. Chem. Res.* 37, 471–478, 2004; Conley, N.R., Hung, R.J., and Willison, C.G., A new synthetic route to authentic *N*-substituted aminomaleimides, *J. Org. Chem.* 70, 4553–4555, 2005; Boul, P.J., Reutenauer, P., and Lehn, J.M., Reversible Diels–Alder reactions for the generation of dynamic combinatorial libraries, *Org. Lett.* 7, 15–18, 2005. Catalytic antibodies have been used for Diels–Alder reactions (Suckling, C.J., Tedford, C.M., Proctor, G.R. et al., Catalytic antibodies: a new window on protein chemistry, *Ciba Found. Symp.* 159, 201–208, 1991; Meekel, A.A., Resmini, M., and Pandit, U.K., Regioselectivity and enantioselectivity in an antibody-catalyzed hetero Diels–Alder reaction, *Bioorg. Med. Chem.* 4, 1051–1057, 1996; Romesberg, F.E., Spiller, B., Schultz, P.G., and Stevens, R.C., Immunological origins of binding and catalysis in a Diels-Alderase antibody, *Science* 279, 1934–1940, 1998; Romesberg, F.E. and Schultz, P.G., A mutational study of a Diels-Alderase catalytic antibody, *Bioorg. Med. Chem. Lett.* 9, 1741–1744, 1999; Chen, J., Deng, Q., Wang, R. et al., Shape complementarity binding-site dynamics and transition state stabilization: a theoretical study of Diels–Alder catalysis by antibody IE9, *Chem. Bio. Chem.* 1, 255–261, 2000; Kim, S.P., Leach, A.G., and Houk, K.N., The origins of noncovalent catalysis of intermolecular Diels–Alder reactions by cyclodextrins, self-assembling capsules, antibodies, and RNAses, *J. Org. Chem.* 67, 4250–4260, 2002; Cannizzaro, C.E., Ashley, J.A., Janda, K.D., and Houk, K.N., Experimental determination of the absolute enantioselectivity of an antibody-catalyzed Diels–Alder reaction and theoretical explorations of the origins of stereoselectivity, *J. Am. Chem. Soc.* 125, 2489–2506, 2003).

EDMAN DEGRADATION

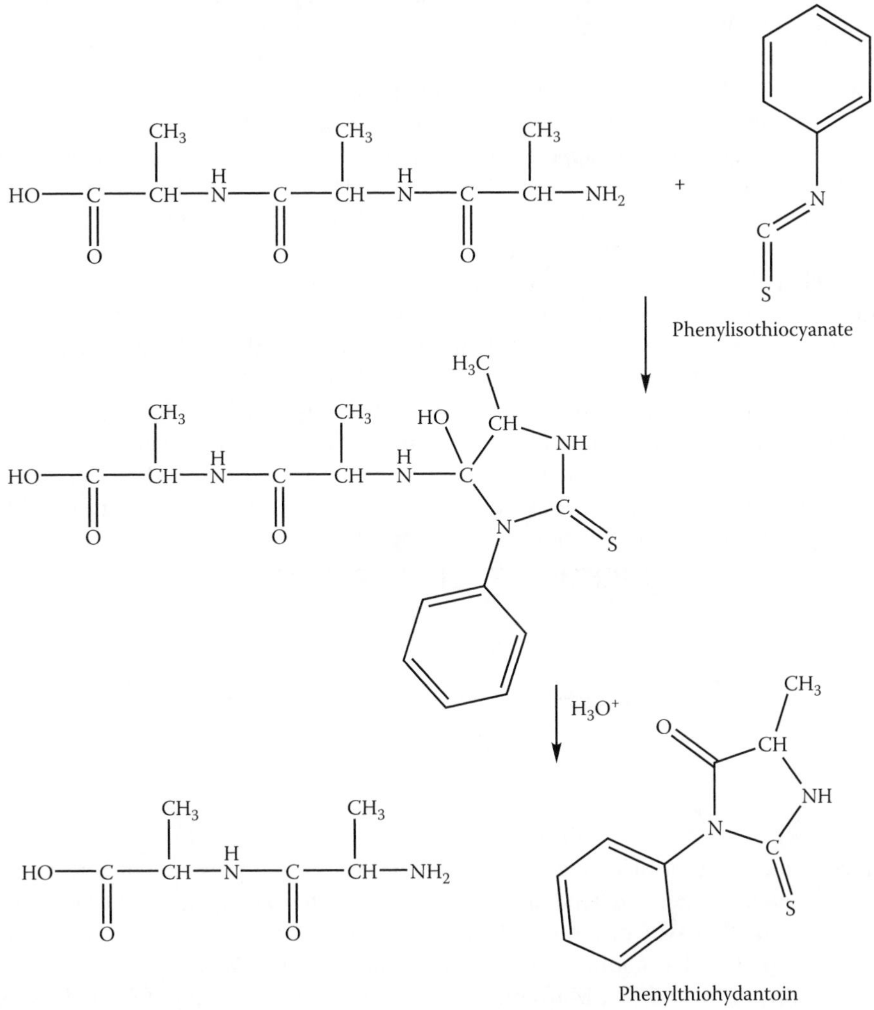

Phenylisothiocyanate

Phenylthiohydantoin

The stepwise degradation of a peptide chain from the amino terminal via reaction with phenylisothiocyanate. This process is used for the chemical determination of the amino acid sequence of a peptide or protein. See Edman, P., Sequence determination, *Mol. Biol. Biochem. Biophys.* 9, 211–255, 1970; Heinrikson, R.L., Application of automated sequence analysis to the understanding of protein structure and function, *Ann. Clin. Lab. Sci.* 8, 295–301, 1978; Tsugita, A., Developments in protein microsequencing, *Adv. Biophys.* 23, 91–113, 1987; Han, K.K. and Martinage, A., Post-translational chemical modifications of proteins — III. Current developments in analytical procedures of identification and quantation of post-translational chemically modified amino acid(s) and its derivatives, *Int. J. Biochem.* 25, 957–970, 1993; Masiarz, F.R. and Malcolm, B.A., Rapid determination of endoprotease specificity using peptide mixtures and Edman degradation analysis, *Methods Enzymol.* 241, 302–310, 1994; Gooley, A.A., Ou, K., Russell, J. et al., A role for Edman degradation in proteome studies, *Electrophoresis* 18, 1068–1072, 1997; Wurzel, C. and Wittmann-Liebold, B., A wafer-based micro reaction system for the Edman degradation of proteins and peptides, *J. Protein Chem.* 17, 561–564, 1998; Walk, T.B., Sussmuth, R., Kempter, C. et al., Identification of unusual amino acids in peptides using automated sequential Edman degradation coupled to direct detection by electrospray-ionization mass spectrometry, *Biopolymers* 49, 329–340, 1999; Lauer-Fields, J.L., Nagase, H., and Fields, G.B., Use of Edman degradation sequence analysis and matrix-assisted laser desorption/ionization mass spectrometry in designing substrates for matrix metalloproteinases, *J. Chromatog. A* 890, 117–125, 2000; Hajdu, J., Neutze, R., Sjogren, T. et al., Analyzing protein functions in four dimensions, *Nat. Struct. Biol.* 7, 1006–1012, 2000; Shively, J.E., The chemistry of protein sequence analysis, *EXS* 88, 99–117, 2000; Wang, P., Arabaci, G., and Pei, D., Rapid sequencing of library-derived peptides by partial Edman degradation and mass spectrometry, *J. Comb. Chem.* 3, 251–254, 2001; Brewer, M., Oost, T., Sukonpan, C. et al., Sequencing hydroxylethyleneamine-containing peptides via Edman degradation, *Org. Lett.* 4, 3469–3472, 2002; Sweeney, M.C. and Pei, D., An improved method for rapid sequencing of support-bound peptides by partial Edman degradation and mass spectrometry, *J. Comb. Chem.* 5, 218–222, 2003; Buda, F., Ensing, B., Gribnau, M.C., and Baerends, E.J., O_2 evolution in the Fenton reaction, *Chemistry* 9, 3436–3444, 2003; Liu, Q., Berchner-Pfannschmidt, U., Moller, U. et al., A Fenton reaction at the endoplasmic reticulum is involved in the redox control of hypoxia-inducible gene expression, *Proc. Natl. Acad. Sci. USA* 101, 4302–4307, 2004; Maksimovic, V., Mojovic, M., Neumann, G., and Vucinic, Z., Nonenzymatic reaction of dihydroxyacetone with hydrogen peroxide enhanced via a Fenton reaction, *Ann. N.Y. Acad. Sci.* 1048, 461–465, 2005; Lu, C. and Koppenol, W.H., Inhibition of the Fenton reaction by nitrogen monoxide, *J. Biol. Inorg. Chem.* 10, 732–738, 2005; Baron, C.P., Refsgaard, H.H., Skibsted, H., and Andersen, M.L., Oxidation of bovine serum albumin initiated by the Fenton reaction — effect of EDTA, *tert*-butylhydroperoxide and tetrahydrofuran, *Free Radic. Res.* 40, 409–417, 2006; Thakkar, A., Wavreille, A.S., and Pei, D., Traceless capping agent for peptide sequencing by partial Edman degradation and mass spectrometry, *Anal. Chem.* 78, 5935–5939, 2006.

ESCHWEILER–CLARK REACTION

Eschweiler–Clarke reaction

The reductive methylation of amines with formaldehyde in the presence of formic acid. See Lindeke, B., Anderson, B., and Jenden, D.J., Specific deuteromethylation by the Eschweiler–Clark reaction. Synthesis of differently labelled variants of trimethylamine and their use of the preparation of labelled choline and acetylcholine, *Biomed. Mass Spectrom.* 3, 257–259, 1976; Boldavalli, F., Bruno, O., Mariani, E. et al., Esters of *N*-methyl-*N*-(2-hydroxyethyl or

3-hydroxypropyl)-1,3,3-trimethylbicyclo[2.2.1] heptan-2-endo-amine with hypotensive activity, *Farmaco* 42, 175–183, 1987; Lee, S.S., Wu, W.N., Wilton, J.H. et al., Longiberine and *O*-methyllogiberine, dimeric protoberberine-benzyl tetrahydroisoqunioline alkaloids from *Thalictrum longistrylum*, *J. Nat. Prod.* 62, 1410–1414, 1999; Suma, R. and Sai Prakash, P.K., Conversion of sertraline to *N*-methyl sertraline in embalming fluid: a forensic implication, *J. Anal. Toxicol.* 30, 395–399, 2006. The reaction can be accomplished with sodium borohydride or sodium cyanoborohydride and is related to the reductive methylation/alkylation of lysine residues in proteins (Lundblad, R.L., *Chemical Reagents for the Modification of Proteins*, 3rd ed., CRC Press, Boca Raton, FL, 2004).

FAVORSKII REARRANGEMENT

Favorskii rearrangement

The rearrangement of an α-ketone in the presence of an alkoxide to form a carboxylic ester; cyclic α-ketones undergo ring contraction. See March, J., *Advanced Organic Chemistry: Reactions, Mechanisms, and Structures,* 3rd ed., John Wiley & Sons, New York, 1985; Gardner, H.W., Simpson, T.D., and Hamberg, M., Mechanism of linoleic acid hydroperoxide reaction with alkali, *Lipids* 31, 1023–1028, 1996; Xiang, L., Kalaitzis, J.A., Nilsen, G. et al., Mutational analysis of the enterocin Favorskii biosynthetic rearrangement, *Org. Lett.* 4, 957–960, 2002; Zhang, L. and Koreeda, M., Stereocontrolled synthesis of kelsoene by the homo-Favorskii rearrangement, *Org. Lett.* 4, 3755–3788, 2002; Grainger, R.S., Owoare, R.B., Tisselli, P., and Steed, J.W., A synthetic alternative to the type-II intramolecular 4 + 3 cycloaddition, *J. Org. Chem.* 68, 7899–7902, 2003.

FENTON REAGENT/REACTION

$$H_2O_2 + Fe^{2+} \longrightarrow OH\cdot + {}^-OH$$

The reaction of ferrous ions and hydrogen peroxide to yield a hydroxyl radical. See Aust, S.D., Morehouse, L.A., and Thomas, C.E., Role of metals in oxygen radical reactions, *J. Free Radic. Biol. Med.* 1, 3–25, 1985; Goldstein, S., Meyerstein, D., and Czapski, G., The Fenton reagents, *Free Radic. Biol. Med.* 15, 435–445, 1993; Wardman, P. and Candeias, L.P., Fenton chemistry: an introduction, *Radiat. Res.* 145, 523–531, 1996; Held, K.D., Sylvester, F.C., Hopcia, K.L., and Biaglow, J.E., Role of Fenton chemistry in the thiol-induced toxicity and apopotosis, *Radiat. Res.* 145, 542–553, 1996; Merli, C., Petrucci, E., Da Pozzo, A., and Pernetti, M., Fenton-type treatment: state of the art, *Ann. Chim.* 93, 761–770, 2003; Groves, J.T., High-valent iron in chemical and biological oxidations, *J. Inorg. Biochem.* 100, 434–447, 2006.

FISCHER CARBENE COMPLEXES

A Fischer carbene complex consists of a transition metal with a formal carbon–metal bond containing a carbene in the singlet state; stabilization of the carbene is provided by the metal interaction. The Fischer carbene complex is electrophilic at the carbene carbon as opposed to the Schrock complex which is in the triplet state and nucleophilic at the carbene carbon. The Fischer carbene complex is highly reactive and is used in many synthetic procedures. An example is provided by the α,β-unsaturated carbenepentacarbonylchromium complex (de Meijere, A., Schirmer, H., and Duetsch, M., Fischer carbene complexes as chemical multitalents: the incredible range of products from carbene-pentacarbonylmetal α,β-unsaturated complexes, *Angew. Chem. Int. Ed.* 39. 3964–4002, 2000). See also Salmain, M., Blais, J.C., Tran-Huy, H. et al., Reaction of hen egg-white lysozyme with Fischer-type metallocarbene complexes. Characterization of the conjugates and determination of the metal complex binding sites, *Eur. J. Biochem.* 268, 5479–5487, 2001; Merlic, C.A. and Doroh, B.C., Amine-catalyzed coupling of aldehydes and ketenes derived from Fischer carbene complexes: formation of beta-lactones and enol ethers, *J. Org. Chem.* 68, 6056–6069, 2003; Barluenga, J., Santamaria, J., and Tomas, M., Synthesis of heterocycles via group VI Fischer carbene complexes, *Chem. Rev.* 104, 2259–2283, 2004; Barluenga, J., Fananas-Mastral, M., and Aznar, F., A new synthesis of allyl sulfoxides via nucleophilic addition of sulfinyl carbanions to group 6 Fischer carbene complexes, *Org. Lett.* 7, 1235–1237, 2005; Lian, Y. and Wulff, W.D., Iron in the service of chromium: the *o*-benzannulation of *trans,trans*-dienyl Fischer carbene complexes, *J. Am. Chem. Soc.* 127, 17162–17163, 2005; Barluenga, J., Mendoza, A., Dieguez, A. et al., Umpolung reactivity of alkenyl Fischer carbene complexes, copper enolates, and electrophiles, *Angew. Chem. Int. Ed. Engl.* 45, 4848–4850, 2006; Samanta, D., Sawoo, S., and Sarkar, A., *In situ* generation of gold nanoparticles on a protein surface: Fischer carbene complex as reducing agent, *Chem. Commun.* 32, 3438–3440, 2006; Rawat, M., Prutyanov, V., and Wulff, W.D., Chromene chromium carbene complexes in the syntheses of naphthoyran and naphthopyrandione units present in photochromic materials and biologically active natural products, *J. Am. Chem. Soc.* 128, 11044–11053, 2006. For general information on carbenes including Fischer carbene complexes and Schrock carbene complexes, see Bertrand, G., Ed., *Carbene Chemistry: From Fleeting Intermediates to Powerful Reagents*, Fontis Media/Marcel Dekker, New York, 2002.

FISCHER INDOLE SYNTHESIS

Fischer indole synthesis

The thermal conversion of arylhydrazones in the presence of a protic acid or a Lewis acid to form an indole ring. See Owellen, R.J., Fitzgerald, J.A., Fitzgerald, B.M. et al., The cyclization phase of the Fischer indole synthesis. The structure and significance of Pleininger's intermediate, *Tetrahedron Lett.* 18, 1741–1746, 1967; Kim, R.M., Manna, M., Hutchins, S.M. et al., Dendrimer-supported combinatorial chemistry, *Proc. Natl. Acad. Sci. USA* 93, 10012–10017, 1996; Brase, S., Gil, C., and Knepper, K., The recent impact of solid-phase synthesis on medicinally relevant benzoannelated nitrogen heterocycles, *Bioorg. Med. Chem.* 10, 2415–2437, 2002; Rosenbaum, C., Katzka, C., Marzinzik, A., and Waldmann, H., Traceless Fischer indole synthesis on the solid phase, *Chem. Commun.* 15, 1822–1823, 2003; Mun, H.S., Ham, W.H., and Jeong, J.H., Synthesis of 2,3-disubstituted indole on solid phase by the Fischer indole synthesis, *J. Comb. Chem.* 7, 130–135, 2005; Narayana, B., Ashalatha, B.V., Vijaya Raj, K.K. et al., Synthesis of some new biologically active 1,3,4-oxadiazolyl nitroindole and a modified Fischer indole synthesis of ethyl nitro indole-2-carboxylates, *Bioorg. Med. Chem.* 13, 4638–4644, 2005; Schmidt, A.M. and Eilbracht, P., Tandem hydroformylation-hydrazone formation-Fischer indole synthesis: a novel approach to tryptamides, *Org. Biomol. Chem.* 3, 2333–2343, 2005; Linnepe Nee Kohling, P., Schmidt, A.M., and Eilbracht, P., 2,3-disubstituted indoles from olefins and hydrazines via tandem hydroformylation-Fischer indole synthesis and skeletal rearrangement, *Org. Biomol. Chem.* 4, 302–313, 2006; Landwehr, J., George, S., Karg, E.M. et al., Design and synthesis of novel 2-amino-5-hydroxyindole derivatives that inhibit human 5-lipooxygenase, *J. Med. Chem.* 49, 4327–4332, 2006.

FRIEDEL-CRAFTS REACTION

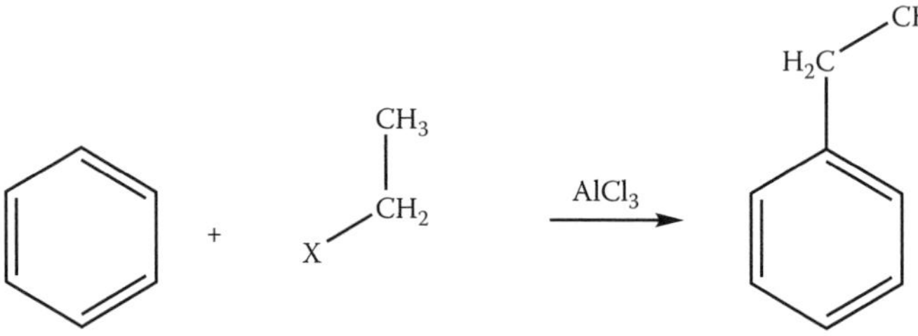

Friedel-Crafts alkylation

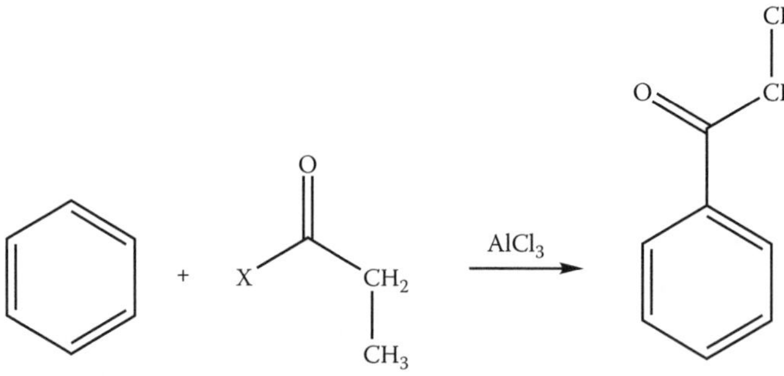

Friedel-Crafts acylation

The alkylation of an aromatic ring by an alkyl halide (order of reactivity F>Cl>Br>I) in the presence of a strong Lewis acid such as aluminum chloride; the acylation of an aromatic ring by an acyl halide (order of reactivity usually is I>Br>Cl>F) in the presence of a strong Lewis acid. Acids and acid anhydrides can replace the acyl halides. A related reaction is the Derzen–Nenitzescu ketone synthesis. See Olah, G.A., *Friedel-Crafts Chemistry*, John Wiley & Sons, New York, 1973; Roberts, R.M. and Khalaf, A.A., *Friedel-Crafts Alkylyation Chemistry: A Century of Discovery*, Marcel Dekker, New York, 1989. See also Retey, J., Enzymatic catalysis by Friedel–Crafts-type reactions, *Naturwissenschaften* 83, 439–447, 1996; White, E.H., Darbeau, R.W., Chen, Y. et al., A new look at the Friedel–Crafts alkylation reaction(1), *J. Org. Chem.* 61, 7986–7987, 1996; Studer, J., Purdie, N., and Krouse, J.A., Friedel–Crafts acylation as a quality control assay for steroids, *Appl. Spectrosc.* 57, 791–796, 2003; Retey, J., Discovery and role of methylidene imidazolone, a highly reactive electrophilic prosthetic group, *Biochim. Biophys. Acta* 1647, 179–184, 2003; Bandini, M., Melloni, A., and Umani-Ronchi, A., New catalytic approaches in the stereoselective Friedel–Crafts alkylation reaction, *Angew. Chem. Int. Ed. Engl.* 43, 550–556, 2004; Poppe, L. and Retey, J., Friedel–Crafts-type mechanism for the enzymatic elimination of ammonia from histidine and phenylalanine,

Angew. Chem. Int. Ed. Engl. 44, 3668–3688, 2005; Keni, M. and Tepe, J.J., One-pot Friedel–Crafts/ Robinson–Gabriel synthesis of oxazoles using oxazolone templates, *J. Org. Chem.* 70, 4211–4213, 2005; Movassaghi, M. and Ondrus, A.E., Enantioselective total synthesis of tricyclic myrmicarin alkaloids, *Org. Lett.* 7, 4423–4426, 2005; Paizs, C., Katona, A., and Retey, J., The interaction of heteroaryl-acrylates and alanines with phenylalanine ammonia-lyase form parsley, *Chemistry* 12, 2739–2744, 2006. Cuprous ions have been observed to promote a Friedel–Crafts acylation reaction (Kozikowski, A.P. and Ames, A., Copper(I) promoted acylation reactions. A transition metal-mediated version of the Friedel–Crafts reaction, *J. Am. Chem. Soc.* 102, 860–862, 1980).

FRIEDLÄNDER SYNTHESIS

Friedlander synthesis

The base-catalyzed formation of quinoline derivatives by condensation of an *o*-aminobenzaldehyde with a ketone; also referred to as the Friedländer quinoline synthesis. The general utility of the reaction is somewhat limited by the availability of *o*-aminobenzaldehyde derivatives. See Maguire, M.P., Sheets, K.R., McVety, K. et al., A new series of PDGF receptor tyrosine kinase inhibitors: 3-substituted quinoline derivatives, *J. Med. Chem.* 37, 2129–2137, 1994; Lindstrom, S., Friedländer synthesis of the food carcinogen 2-amino-1-methyl-6-phenylimidazo[4,5-b]pyridine, *Acta Chem. Scand.* 49, 361–363, 1995; Gladiali, S., Chelucci, G., Mudadu, M.S. et al., Friedländer synthesis of chiral alkyl-substituted 1,10-phenanthrolines, *J. Org. Chem.* 66, 400–405, 2001; Patteux, C., Levacher, V., and Dupas, G., A novel traceless solid-phase Friedländer synthesis, *Org. Lett.* 5, 3061–3063, 2003; McNaughton, B.R. and Miller, B.L., A mild and efficient one-step synthesis of quinolines, *Org. Lett.* 5, 4257–4259, 2003; Yasuda, N., Hsiao, Y., Jensen, M.S. et al., An efficient synthesis of an $\alpha_v\beta_3$ antagonist, *J. Org. Chem.* 69, 1959–1966, 2004.

FRIES REARRANGEMENT

Rearrangement of a phenolic ester to yield *o*- and *p*-acylphenols. The distribution of products between the *ortho* and *para* acyl derivates depends on reaction conditions. With the presence of a solvent and a Lewis acid, the *para* product is preferred; with the photolytic process or at high temperature in the absence of solvent, the *ortho* derivative is preferred. See Sen, A.B. and Bhattacharji, S., Fries' rearrangement of aliphatic esters of β-naphthol, *Curr. Sci.* 20, 132–133, 1951; Iwasaki, S., Photochemistry of imidazolides. I. The photo-Fries-type rearrangement of *N*-substituted imidazoles, *Helv. Chim. Acta* 59, 2738–2752, 1976; Castell, J.V., Gomez, M.J., Mirabet, V. et al., Photolytic degradation of benorylate: effects of the photoproducts on cultured hepatocytes, *J. Pharm. Sci.* 76, 374–378, 1987; Climent, M.J. and Miranda, M.A., Gas chromatographic-mass spectrometric study of photodegradation of carbamate pesticides, *J. Chromatog. A* 738, 225–231, 1996; Kozhevnikova, E.F., Derouane, E.G., and Kozhevnikov, I.V., Heteropoly acid as a novel efficient catalyst for Fries rearrangement, *Chem. Commun.* 11, 1178–1179, 2002; Dickerson, T.J., Tremblay, M.R., Hoffman, T.Z. et al., Catalysis of the photo-Fries reaction: antibody-mediated stabilization of high-energy states, *J. Am. Chem. Soc.* 125, 15395–15401, 2003; Seijas, J.A., Vazquez-Tato, M.P., and Carballido-Reboredo, R., Solvent-free synthesis of functionalized flavones under microwave irradiation, *J. Org. Chem.* 70, 2855–2858, 2005; Canle Lopez, M., Fernandez, M.I., Rodriguez, S. et al., Mechanisms of direct and TiO$_2$-photocatalyzed degradation of phenylurea herbicides, *Chemphyschem* 6, 2064–2074, 2005; Slana, G.B., de Azevedo, M.S., Lopes, R.S. et al., Total syntheses of oxygenated brazanquinones via regioselective homologous anionic Fries rearrangement of benzylic *O*-carbamates, *Beilstein J. Org. Chem.* 2, 1, 2006.

GABRIEL SYNTHESIS

Gabriel synthesis

The conversion of an alkyl halide to alkyl amine mediated by potassium phthalimide. The intermediate product of the reaction of the alkyl halide and phthalimide is hydrolyzed to the product amine by acid or by reflux in ethanolic hydrazine. See Mikola, H. and Hanninen, E., Introduction of aliphatic amino and hydroxy groups to keto steroids using *O*-substituted hydroxylamines, *Bioconjugate Chem.* 3, 182–186, 1992; Groutas, W.C., Chong, L.S., Venkataraman, R. et al., Mechanism-based inhibitors of serine proteinases based on the Gabriel–Colman rearrangement, *Biochem. Biophys. Res. Commun.* 194, 1491–1499, 1993; Konig, S., Ugi, I., and Schramm, H.J., Facile syntheses of C$_2$-symmetrical HIV-1 protease inhibitor, *Arch. Pharm.* 328, 699–704, 1995; Zhang, X.X. and Lippard, S.J., Synthesis of PDK, a novel porphyrin-linked dicarboxyate ligand, *J. Org. Chem.* 65, 5298–5305, 2000; Scozzafava, A., Saramet, I., Banciu, M.D., and Supuran, C.T., Carbonic anhydrase activity modulators: synthesis of inhibitors and activators incorporating 2-substituted-thiazol-4-yl-methyl scaffolds, *J. Enzyme Inhib.* 16, 351–358, 2001; Nicolaou, K.C., Hao, J., Reddy, M.V. et al., Chemistry and biology of diazonamide A: second total synthesis and biological investigations, *J. Am. Chem. Soc.* 126, 12897–12906, 2004; Remond, C., Plantier-Royon, R., Aubry, N., and O'Donohue, M.J., An original chemoenzymatic route for the synthesis of β-D-galactofuranosides using an α-L-arabinofuranosidase, *Carbohydr. Res.* 340, 637–644, 2005; Pulici, M., Quartieri, F., and Felder, E.R., Trifluoroacetic acid anhydride-mediated solid-phase version of the Robison–Gabriel synthesis of oxazoles, *J. Comb. Chem.* 7, 463–473, 2005.

GREISS REACTION

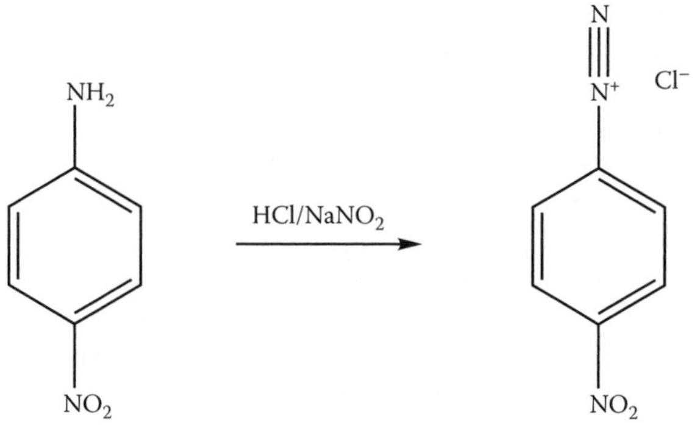

Greiss reaction

Greiss reaction as used for the measurement of nitrite

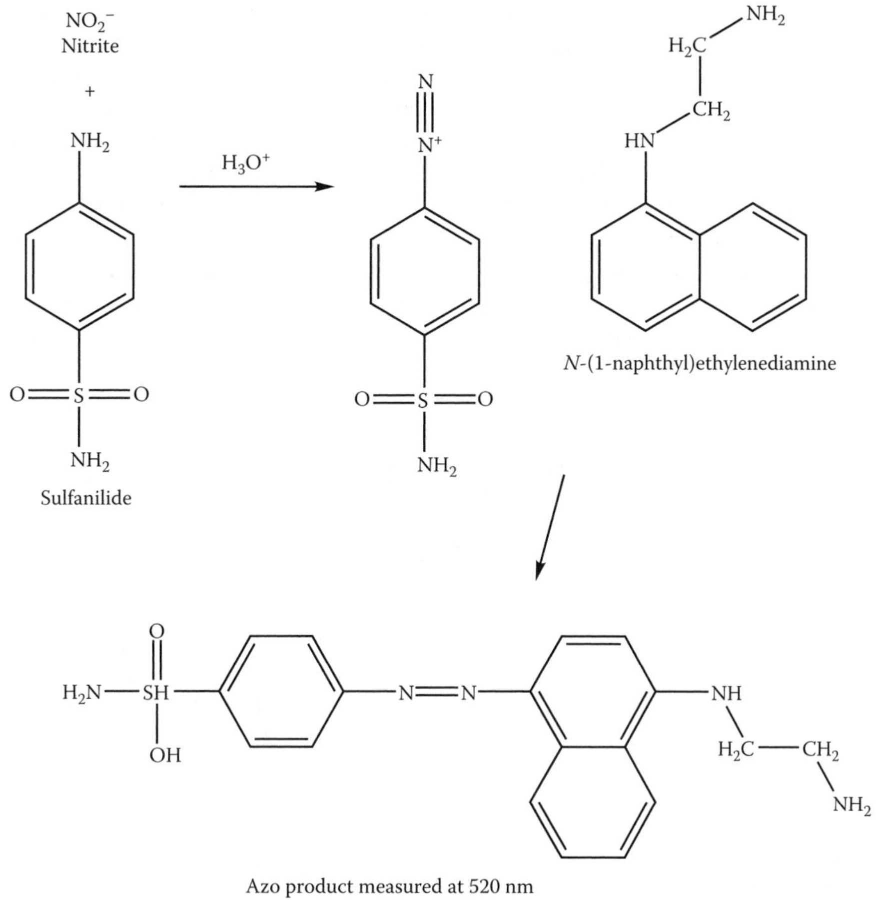

N-(1-naphthyl)ethylenediamine

Sulfanilide

Azo product measured at 520 nm

Diazotization of aromatic amines; used for the assay of nitrites in nitric oxide research. The assay for nitrates uses diazotization of sulfanilamide with subsequent coupling to an aromatic amine (*N*-1-naphthylethylenediamine) to form a chromophoric azo derivative. See Greenberg, S.S., Xie, J., Spitzer, J.J. et al., Nitro containing L-arginine analogs interfere with assays for nitrate and nitrite, *Life Sci.* 57, 1949–1961, 1995; Pratt, P.F., Nithipatikom, K., and Campbell, W.B., Simultaneous determination of nitrate and nitrite in biological samples by multichannel flow injection analysis, *Anal. Biochem.* 231, 383–386, 1995; Tang, Y., Han, C., and Wang, X., Role of nitric oxide and prostaglandins in the potentiating effects of calcitonin gene-related peptide on lipopolysaccharide-induced interleukin-6 release from mouse peritoneal macrophages, *Immunology* 96, 171–175, 1999; Baines, P.B., Stanford, S., Bishop-Bailey, D. et al., Nitric oxide production in meningococcal disease is directly related to disease severity, *Crit. Care. Med.* 27, 1163–1165, 1999; Rabbani, G.H., Islam, S., Chowdhury, A.K. et al., Increased nitrite and nitrate concentrations in sera and urine of patients with cholera or shigellosis, *Am. J. Gastroenterol.* 96, 467–472, 2001; Lee, R.H., Efron, D., Tantry, U., and Barbul, A., Nitric oxide in the healing wound: a time-course study, *J. Surg. Res.* 101, 104–108, 2001; Stark, J.M., Khan, A.M., Chiappetta, C.L. et al., Immune and functional role of nitric oxide in a mouse model of respiratory syncytial virus infection, *J. Infect. Dis.* 191, 387–395, 2005; Bellows, C.F., Alder, A., Wludyka, P., and Jaffe, B.M., Modulation of macrophage nitric oxide production by prostaglandin D2, *J. Surg. Res.* 132, 92–97, 2006. Diazotization of aromatic amines is also used for the the modification of proteins (Lundblad, R.L., *Chemical Reagents for Protein Modification*, CRC Press, Boca Raton, FL, 2004; Kennedy, J.H., Kricka, L.J., and Wilding, P., Protein–protein coupling reactions and the application of protein conjugates, *Clin. Chim. Acta* 70, 1–31, 1976; Sinnott, M.L., Affinity labeling via deamination reactions, *CRC Crit. Rev. Biochem.* 12, 327–372, 1982; Blair, A.H. and Ghose, T.I., Linkage of cytotoxic agents to immunoglobulins, *J. Immunol. Methods* 59, 129–143, 1983). While alkyl azides are unstable, carbonyl azides such as diazoacetyl derivatives have been used in the modification of proteins (Lundblad, R.L. and Stein, W.H., On the reaction of diazoacetyl compounds with pepsin, *J. Biol. Chem.* 244, 154–160, 1969; Keilova, H. and Lapresle, C., Inhibition of cathepsin E by diazoacetyl-norleucine methyl ester, *FEBS Lett.* 9, 348–350, 1970; Giraldi, T. and Nisi, C., Effects of cupric ions on the antitumour activity of diazoacetyl-glycine derivatives, *Chem. Biol. Interact.* 11, 59–61, 1975; Kaehn, K., Morr, M., and Kula, M.R., Inhibition of the acid proteinase from *Neurospora crassa* by diazoacetyl-DL-norleucine methyl ester, 1,2-epoxy-3-[4-nitrophenoxy]propane and pepstatin, *Hoppe Seylers Z. Physiol. Chem.* 360, 791–794, 1979; Ouihia, A., René, L., Guilhem, J. et al., A new diazoacylating reagent: preparation, structure, and use of succinimidyl diazoacetate, *J. Org. Chem.* 58, 1641–1642, 1993).

GRIGNARD REAGENT OR GRIGNARD REACTION

The reaction of alkyl or aryl halides with magnesium in dry ether to yield derivatives, which can be used in a variety of organic synthetic reactions. See Nagano, T. and Hayashi, T., Iron-catalyzed Grignard cross-coupling with alkyl halides possessing beta-hydrogens, *Org. Lett.* 6, 1297–1299, 2004; Querner, C., Reiss, P., Bleuse, J., and Pron, A., Chelating ligands for nanocrystals' functionalization, *J. Am. Chem. Soc.* 126, 11574–11582, 2004; Agarwal, S. and Knolker, H.J., A novel pyrrole synthesis, *Org. Biomol. Chem.* 2, 3060–3062, 2004; Hatano, M., Matsumara, T., and Ishihara, K., Highly alkyl-selective addition to ketones with magnesiumate complexes derived from Grignard reagents, *Org. Lett.* 7, 573–576, 2005; Itami, K., Higashi, S., Mineno, M., and Yoshida, J., Iron-catalyzed cross-coupling of alkenyl sulfides with Grignard reagents, *Org. Lett.* 7, 1219–1222, 2005; Wang, X.J., Zhang, L., Sun, X. et al., Addition of Grignard reagents to aryl chlorides: an efficient synthesis of aryl ketones, *Org. Lett.* 7, 5593–5595, 2005; Hoffman-Emery, F., Hilpert, H., Scalone, M., and Waldmeier, F., Efficient synthesis of novel NK1 receptor antagonists: selective 1,4-additional of Grignard reagents to 6-chloronicotinic acid derivatives, *J. Org. Chem.* 71, 2000–2008, 2006; Werner, T. and Barrett, A.G., Simple method for the preparation of esters from Grignard reagents and alkyl 1-imidazolecarbox-ylates, *J. Org. Chem.* 71, 4302–4304, 2006; Demel, P., Keller, M., and Breit, B., *o*-DPPB-directed copper-mediated and -catalyzed allylic substitution with Grignard reagents, *Chemistry* 12, 6669–6683, 2006.

KNOEVENAGEL REACTION OR KNOEVENAGEL CONDENSATION

Knoevenagel condensation

EWG = electron-withdrawing group such as CHO, COOH, COOR, CN, NO$_2$

An amine-catalyzed reaction between active hydrogen compounds of the type Z-CH$_2$-Z, where Z can be a CHO, COOH, COOR, NO$_2$, SOR, or related electron withdrawing groups and an aldehyde or ketone. For example, the reaction of malonic acid or malonic acid esters and an aldehyde or ketone to yield an α,β-unsaturated derivative. With malonic acid (Z is carboxyl group), decarbox-ylation occurs *in situ*. See March, J., *Advanced Organic Chemistry: Reactions, Mechanisms, and Structure*, 3rd ed., John Wiley & Sons, New York, 1985; Klavins, M., Dipane, J., and Babre, K., Humic substances as catalysts in condensation reactions, *Chemosphere* 44, 737–742, 2001; Lai. S.M., Martin-Aranda, R., and Yeung, K.L., Knoevenagel condensation reaction in a membrane bioreactor, *Chem. Commun.* 2, 218–219, 2003; Pivonka, D.E. and Empfield, J.R., Real-time *in situ* Ramen analysis of microwave-assisted organic reactions, *Appl. Spectrosc.* 58, 41–46, 2004; Stro-hmeier, G.A., Haas, W., and Kappe, C.O., Synthesis of functionalized 1,3-thiazine libraries com-bining solid-phase synthesis and post-cleavage modification reactions, *Chemistry* 10, 2919–2926, 2004; Wirz, R., Ferri, D., and Baiker, A., ATR-IR spectroscopy of pendant NH$_2$ groups on silica involved in the Knoevenagel condensation, *Langmuir* 22, 3698–3706, 2006.

LEUCKART REACTION

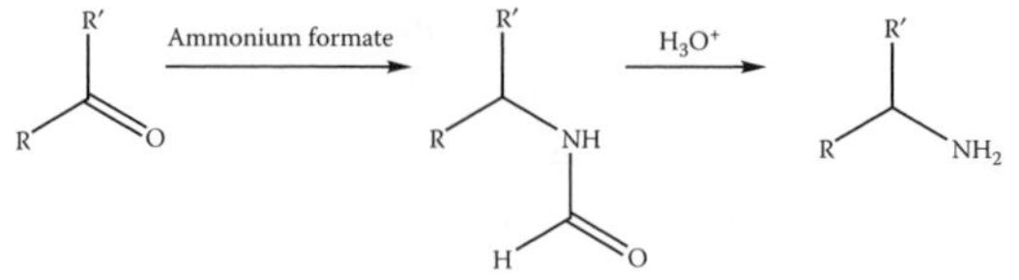

Leuckart reaction

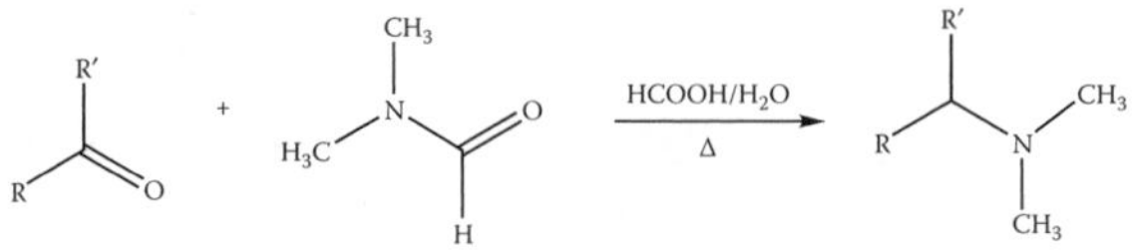

The reductive amination of carbonyl groups by ammonium formate or amine salts of formic acid; formamides may also be used in the reaction. See Matsueda, G.R. and Stewart, J.M., A *p*-methyl-benzhydrylamine resin for improved solid-phase synthesis of peptide amides, *Peptides* 2, 45–50, 1981; Agwada, V.C. and Awachie, P.I., Intermediates in the Leuckart reaction of benzophenone with formamide, *Tetrahedron Lett.* 23, 779–780, 1982; Loupy, A., Monteux, D., Petit, A. et al., Toward the rehabilitation of the Leuckart reductive amination reaction using microwave technology, *Tetrahedron Lett.* 37, 8177–8180, 1996; Adger, B.M., Dyer, U.C., Lennon, I.C. et al., A novel synthesis of *tert*-leucine via a Leuckart-type reaction, *Tetrahedron Lett.* 38, 2153–2154, 1997; Lejon, T. and Helland, I., Effect of formamide in the Leuckart reaction, *Acta Chem. Scand.* 53, 76–78, 1999; Kitamura, M., Lee, D., Hayashi, S. et al., Catalytic Leuckart–Wallach type reductive amination of ketones, *J. Org. Chem.* 67, 8685–8687, 2002; Swist, M., Wilamowski, J., and Parc-zewski, A., Basic and neutral route specific impurities in MDMA prepared by different synthesis methods. Comparison of impurity profiles, *Forensic Sci. Int.* 155, 100–111, 2005; Tournier, L. and Zard, S.Z., A practical variation on the Leuckart reaction, *Tetrahedron Lett.* 46, 971–973, 2005.

LOSSEN REARRANGEMENT

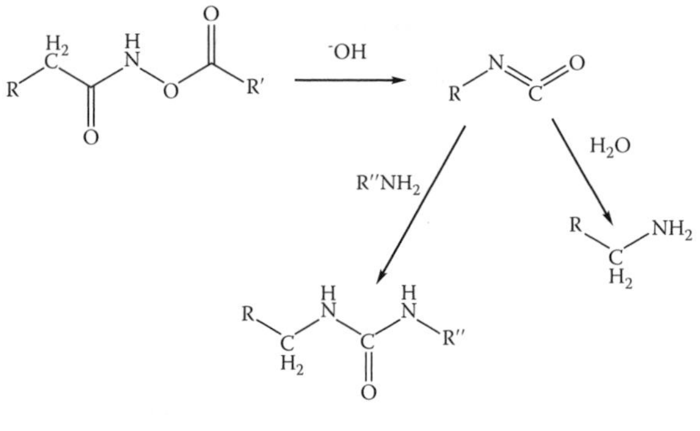

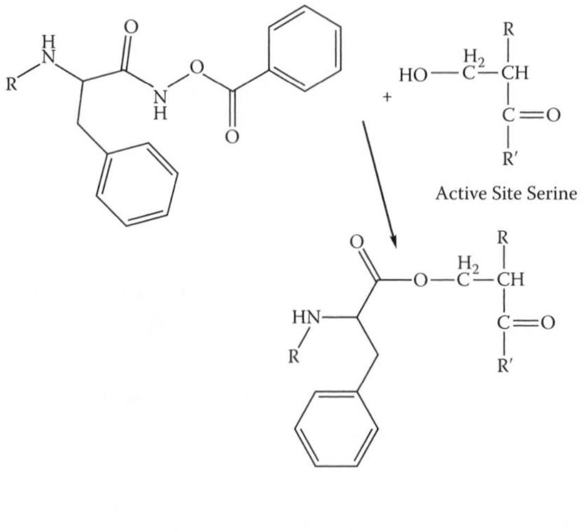

Active Site Serine

The formation of isocyanates on heating of *O*-acyl derivatives of hydroxamic acids or treatment by base. The isocyanate frequently adds water *in situ* to form an amine one carbon shorter than the parent compound; in the presence of amines, there is the formation of ureas. See Andersen, W., The synthesis of phenylcarbamoyl derivatives by Lossen rearrangement of dibenzohydroxamic acid, *C. R. Trav. Lab. Carlsberg.* 30, 79–103, 1956; Gallop, P.M., Seifter, S., Lukin, M., and Meilman, E., Application of the Lossen rearrangement of dintirophenylhydroxamates to analysis of carboxyl groups in model compounds and gelatin, *J. Biol. Chem.* 235, 2619–2627, 1960; Hoare, D.G., Olson, A., and Koshland, D.E., Jr., The reaction of hydroxamic acids with water-soluble carbodiimides. A Lossen rearrangement, *J. Am. Chem. Soc.* 90, 1638–1643, 1968; Dell, D., Boreham, D.R., and Martin, B.K., Estimation of 4-butoyphenylacetohydroxamic acid utilizing the Lossen rearrangement, *J. Pharm. Sci.* 60, 1368–1370, 1971; Harris, R.B. and Wilson, I.B., Glutamic acid is an active site residue of angiotensin I-converting enzyme. Use of the Lossen rearrangement for identification of dicarboxylic acid residues, *J. Biol. Chem.* 258, 1357–1362, 1983; Libert, R., Draye, J.P., Van Hoof, F. et al., Study of reactions induced by hydroxylamine treatment of esters for organic acids and of 3-ketoacids: application to the study of urines from patients under valproate therapy, *Biol. Mass. Spectrom.* 20, 75–86, 1991; Neumann, U. and Gutschow, M., *N*-(sulfonyloxy)phthal-imides and analogues are potent inactivators of serine proteases, *J. Biol. Chem.* 269, 21561–21567, 1994; Steinmetz, A.C., Demuth, H.U., and Ringe, D., Inactivation of subtilisin Carlsberg by *N*-[(*t*-butoxycarbonyl) alanylprolyl-phenylalanyl]-*O*-benzoylhydroxyl-amine: formation of a covalent enzyme-inhibitor linkage in the form of a carbamate derivative, *Biochemistry* 33, 10535–10544, 1994; Needs, P.W., Rigby, N.M., Ring, S.G., and MacDougall, A.J., Specific degradation of pectins via a carbodiimide-mediated Lossen rearrangement of methyl esterified galacturonic acid residues, *Carbohydr. Res.* 333, 47–58, 2001.

MAILLARD REACTION

N-substituted glycosamine

Amadori product
N-substituted 1-amino-2-deoxy-2-ketose

The reaction of amino groups with carbonyl groups resulting in the formation of complex products. This process is involved in the tanning of leather and the Browning reaction, which is considered unique to the reaction of carbohydrates with proteins and is a critical aspect of food preparation. The Maillard reaction involves the nonenzymatic reaction of sugars with proteins and the formation of advanced glycation endproducts (AGE products). The Maillard reaction results in the formation of a number of reaction products. See Dills, W.J., Jr., Protein fructosylation: fructose and the Maillard reaction, *Am. J. Clin. Nutr.* 58 (Suppl. 5), 779S–787S, 1993; Chuyen, N.V., Maillard reaction and food processing. Application aspects, *Adv. Exp. Med. Biol.* 434, 213–235, 1998; van Boekel, M.A., Kinetic aspects of the Maillard reaction: a critical review, *Nahrung* 45, 150–159, 2001; Horvat, S. and Jakas, A., Peptide and amino acid glycation: new insights into the Maillard reaction, *J. Pept. Sci.* 10, 119–137, 2004; Fay, L.B. and Brevard, H., Contribution of mass spectrometry to the study of the Maillard reaction in food, *Mass Spectrom. Rev.* 24, 487–507, 2005; Yaylayan, V.A., Haffenden, L., Chu, F.L., and Wnorowski, A., Oxidative pyrolysis and post-pyrolytic derivatization techniques for the total analysis of Maillard model systems: investigations of control parameters of Maillard reaction pathways, *Ann. N.Y. Acad. Sci.* 1043, 41–54, 2005; Monnier, V.M., Mustata, G.T., Biemel, K.L. et al., Crosslinked of the extracellular matrix by the Maillard reaction in aging and diabetes: an update on "a puzzle nearing resolution," *Ann. N.Y. Acad. Sci.* 1043, 533–544, 2005; Matiacevich, S.B., Santagapita, P.R., and Buera, M.P., Fluorescence from the Maillard reaction and its potential applications in food science, *Crit. Rev. Food Sci. Nutr.* 45, 483–495, 2005; van Boekel, M.A., Formation of flavour compounds in the Maillard reaction, *Biotechnol. Adv.* 24, 230–233, 2006.

MALAPRADE REACTION

Malaprade reaction

Periodic cleavage of a diol; although this term is seldomly used for this extremely common reaction, it would appear to be the correct term. Periodic acid is used for the diol cleavage in aqueous solvent while lead tetraacetate can be used in organic solvents. The reaction also occurs in an amine group vicinal to a hydroxyl function. The term *Malaprade reaction* has been used more in description of analytical techniques for organic diols such as gluconic acid or in the assay of periodate. See Belcher, R., Dryhurst, G., and MacDonal, A.M., Submicro-methods for analysis of organic compounds. 22. Malaprade reaction, *Journal of the Chemical Society* (July), 3964, 1965; Chen, K.,P., Determination of calcium gluconate by selective oxidation with periodate, *J. Pharm. Sci.* 73, 681–683, 1984; Verma, K.K., Gupta, D., Sanghi, S.K., and Jain, A., Spectrophotometric determination of periodate with amodiaquine dihydrochloride and its application to the indirect determination of some organic compounds via the Malaprade reaction, *Analyst* 112, 1519–1522, 1987; Nevado, J.J.B. and Gonzalez, P.V., Spectrophotometric determination of periodate with salicyaldehyde guanylhydrazone — indirect determination of some organic compounds using the Malaprade reaction, *Analyst* 114, 243–244, 1989; Jie, N.,Q., Yang, D.L., Zhang, Q.N. et al., Fluorometric determination of periodate with thiamine and its application to the determination of ethylene glycol and glycerol, *Anal. Chim. Acta* 359, 87–92, 1998; Guillan-Sans, R. and Guzman-Chozas, M.,

The thiobarbituric acid (TBA) reaction in foods, a review, *Crit. Rev. Food Sci. Nutrition* 38, 315–330, 1998; Pumera, M., Jelinek, I., Jindrich, J. et al., Determination of cyclodextrin content using periodate oxidation by capillary electrophoresis, *J. Chromatog. A* 891, 201–206, 2000; Afkhami, A. and Mosaed, F., Kinetic determination of periodate based on its reaction with ferroin and its application to the indirect determination of ethylene glycol and glycerol, *Microchemical J.* 68, 35–40, 2001; Afkhami, A. and Mosaed, F., Sensitive kinetic-spectrophotometric determination of trace amounts of periodate ion, *J. Anal. Chem.* 58, 588–593, 2003; Mihovilovic, M.D., Spina, M., Muller, B., and Stanetty, P., Synthesis of carbo- and heterocyclic aldehydes bearing an adjacent donor group — ozonolysis versus OsO_4/KIO_4-oxidation, *Monatshefte für Chemie* 135, 899–909, 2004.

MALONIC ESTER SYNTHESIS

Malonic ester synthesis

The synthesis of a variety of derivatives taking advantage of the reactivity (acidity) of the methylene carbon in malonic esters. The malonic ester synthesis is related to the acetoacetic ester synthesis and the Knoevenagel reaction. See Mizuno, Y., Adachi, K., and Ikeda, K., Studies on condensed systems of aromatic nitrogenous series. XIII. Extension of malonic ester synthesis to the heterocyclic series, *Pharm. Bull.* 2, 225–234, 1954; Beres, J.A., Varner, M.G., and Bria, C., Synthesis and cyclization of dialkylmalonuric esters, *J. Pharm. Sci.* 69, 451–454, 1980; Kinder, D.H., Frank, S.K., and Ames, M.M., Analogues of carbamyl asparate as inhibitors of dihydroorotase: preparation of boronic acid transition-state analogues and a zinc chelator carbamylhomocysteine, *J. Med. Chem.* 33, 819–823, 1990; Groth, T. and Meldal, M., Synthesis of aldehyde building blocks protected as acid labile *N*-boc-*N.O*-acetals: toward combinatorial solid phase synthesis of novel peptide isosteres, *J. Comb. Chem.* 3, 34–44, 2001; Hachiya, I., Ogura, K., and Shimizu, M., Novel 2-pyridine synthesis via nucleophilic addition of malonic esters to alkynyl imines, *Org. Lett.* 4, 2755–2757, 2002; Strohmeier, G.A., Haas, W., and Kappe, C.O., Synthesis of functionalized 1,3-thiazine libraries combining solid-phase synthesis and post-cleavage modification methods, *Chemistry* 10, 2919–2926, 2004.

MANNICH REACTION

Eschenmoser's Salt

Condensation of an amine with a carbonyl compound that can exist in an enol form and a carbonyl compound that cannot exist as an enol. The reaction frequently uses formaldehyde as the carbonyl compound not existing as an enol for condensing with a secondary amine in the first phase of the reaction. See Britton, S.B., Caldwell, H.C., and Nobles, W.L., The use of 2-pipecoline in the Mannich reaction, *J. Am. Pharm. Assoc. Am. Pharm. Assoc.* 43, 641–643, 1954; Nobles, W.L. and Thompson, B.B., Application of the Mannich reaction to sulfones. I. Reactive methylene moiety of sulfones, *J. Pharm. Sci.* 54, 576–580, 1965; Thompson, B.B., The Mannich reaction. Mechanistic and technological considrations, *J. Pharm. Sci.* 57, 715–733, 1968; Nobles, W.L. and Potti, N.D., Studies on the mechanism of the Mannich reaction, *J. Pharm. Sci.* 57, 1097–1103, 1968; Delia, T.J., Scovill, J.P., Munslow, W.D., and Burckhalter, J.H., Synthesis of 5-substituted aminomethyluracils via the Mannich reaction, *J. Med. Chem.* 19, 344–346, 1976; List, B., Pojarliev, P., Biller, W.T., and Martin, H.J., The proline-catalyzed direct asymmetric three-component Mannich reaction: scope, optimization, and application to the highly enantioselective synthesis of 1,2-amino alcohols, *J. Am. Chem. Soc.* 124, 827–833, 2002; Palomo, C., Oiarbide, M., Landa, A. et al., Design and synthesis of a novel class of sugar-peptide hybrids: *C*-linked glyco β-amino acids through a stereoselective "acetate" Mannich reaction as the key strategic element, *J. Am. Chem. Soc.* 124, 8637–8643, 2002; Cordova, A., The direct catalytic asymmetric Mannich reaction, *Acc. Chem. Res.* 37, 102–112, 2004; Azizi, N., Torkiyan, L., and Saidi, M.R., Highly efficient one-pot three-component Mannich reaction in water catalyzed by heteropoly acids, *Org. Lett.* 8, 2079–2082, 2006; Matsuo, J., Tanaki, Y., and Ishibashi, H., Oxidative Mannich reaction of *N*-carbobenzoxy amines 1,3-dicarbonyl compounds, *Org. Lett.* 8, 4371–4374, 2006.

MEERWEIN REACTION

The reaction of an aryl diazonium halide with an aliphatic unsaturated compound to yield an α-halo-β-phenyl alkene and alkanes. The reaction is performed in the presence of cupric ions. The presence of an electron-withdrawing group is useful in promoting the reactivity of the alkene. See Kochi, J.K., The Meerwein reaction. Catalysis by cuprous chloride, *J. Am. Chem. Soc.* 77, 5090, 1955; Morales, L.A. and Eberlin, M.N., The gas-phase Meerwein reaction, *Chemistry* 6, 897–905, 2000; Riter, L.S., Meurer, E.C., Handberg, E.S. et al., Ion/molecule reactions performed in a miniature cylindrical ion trap mass spectrometer, *Analyst* 128, 1112–1118, 2003; Meurer, E.C., Chen, H., Riter, L.S. et al., Meerwein reaction of phosphonium ions with epoxides and thioepoxides in the gas phase, *J. Am. Soc. Mass Spectrom.* 15, 398–405, 2004; Meurer, E.C. and Eberlin, M.N., The atmospheric pressure Meerwein reaction, *J. Mass Spectrom.* 41, 470–476, 2006.

MICHAEL ADDITION (MICHAEL CONDENSATION)

Michael addition/Michael condensation

Reaction of cysteine with *N*-ethylmaleimide as a Michael addition reaction

4-HNE

Formally a 1,4 addition/conjugate addition of a resonance-stabilized carbanion (the reaction of an active methylene compound such as a malonate and an α,β-unsaturated carbonyl compound or the reaction of a nucleophile with an activated unsaturated system; a carbanion defined as an anion with an even number of electrons). The addition of a nucleophile to a conjugated double bond. See Flavin, M. and Slaughter, C., Enzymatic elimination from a substituted four-carbon amino acid coupled to Michael addition of a β-carbon to an electrophilic double bond. Structure of the reaction product, *Biochemistry* 5, 1340–1350, 1966; Fitt, J.J. and Gschwend, H.W., α-alkylation and Michael addition of amino acid — a practical method, *J. Org. Chem.* 42, 2639–2641, 1977; Powell, G.K., Winter, H.C., and Dekker, E.E., Michael addition of thiols with 4-methyleneglutamic acid: preparation of adducts, their properties, and presence in peanuts, *Biochem. Biophys. Res. Commun.* 105, 1361–1367, 1982; Wang, M., Nishikawa, A., and Chung, F.L., Differential effects of thiols on DNA modifications via alkylation and Michael addition by α-acetoxy-*N*-nitrosopyrrolidine, *Chem. Res. Toxicol.* 5, 528–531, 1992; Jang, D.P., Chang, C.W., and Uang, B.J., Highly diastereoselective Michael addition of α-hydroxy acid derivatives and enantioselective synthesis of (+)-crobarbatic acid, *Org. Lett.* 3, 983–985, 2001; Naidu, B.N., Sorenson, M.E., Connolly, T.P., and Ueda, Y., Michael addition of amines and thiols to dehydroalanine amides: a remarkable rate acceleration in water, *J. Org. Chem.* 68, 10098–10102, 2003; Ooi, T., Doda, K., and Maruoka, K., Highly enantioselective Michael addition of silyl nitronates to α,β-unsaturated aldehydes catalyzed by designer chiral ammonium bifluorides: efficient access to optically active γ-nitro aldehydes and their enol silyl ethers, *J. Am. Chem. Soc.* 125, 9022–9023, 2003; Weinstein, R., Lerner, R.A., Barbas, C.F., III, and Shabat, D., Antibody-catalyzed asymmetric intramolecular Michael addition of aldehydes and ketones to yield the disfavored *cis*-product, *J. Am. Chem. Soc.* 127, 13104–13105, 2005; Ding, R., Katebzadeh, K., Roman, L. et al., Expanding the scope of Lewis acid catalysis in water: remarkable ligand acceleration of aqueous ytterbium triflate catalyzed Michael addition reactions, *J. Org. Chem.* 71, 352–355, 2006; Pansare, S.V. and Pandya, K., Simple diamine- and triamine-protonic acid catalysts for the enantioselective Michael addition of cyclic ketones to nitroalkenes, *J. Am. Chem. Soc.* 128, 9624–9625, 2006; Dai, H.X., Yao, S.P., and Wang, J., Michael addition of pyrimidine with disaccharide acrylates catalyzed in organic medium with lipase M from *Mucor javanicus*, *Biotechnol. Lett.* 28, 1503–1507, 2006. One of the best examples in biochemistry is the modification of cysteine residues with *N*-alkylmaleimide derivatives (Lundblad, R.L., *Chemical Reagents for Protein Modification*, 3rd ed., CRC Press, Boca Raton, FL, 2004; Heitz, J.R., Anderson, C.D., and Anderson, B.M., Inactivation of yeast alcohol dehydrogenase by *N*-alkylmaleimides, *Arch. Biochem. Biophys.* 127, 627–636, 1968; Smyth, D.B. and Tuppy, H., Acylation reactions with cyclic imides, *Biochim. Biophys. Acta* 168, 173–180, 1968; Lusty, C.J. and Fasold, H., Characterization of sulfhydryl groups of actin, *Biochemistry* 8, 2933–2939, 1969; Bowes, T.J. and Gupta, R.S., Induction of mitochondrial fusion of cysteine-alklyators ethacrynic acid and *N*-ethylmaleimide, *J. Cell Physiol.* 202, 796–804, 2005).

Another important example of the Michael addition in biochemistry and molecular biology is the reaction of 4-hydroxynon-2-enal with amines and sulfydryl groups (Winter, C.K., Segall, H.J., and Haddon, W.F., Formation of cyclic adducts of deoxyguanosine with the aldehyde *trans*-4-hydroxy-2-hexenal and *trans*-4-hydroxy-2-nonenal *in vitro*, *Cancer Res.* 46, 5682–5686, 1986; Sayre, L.M., Arora, P.K., Iyer, R.S., and Salomon, R.G., Pyrrole formation from 4-hydroxynonenal and primary amines, *Chem. Res. Toxicol.* 6, 19–22, 1993; Hartley, D.P., Ruth, J.A., and Petersen, D.R., The hepatocellular metabolism of 4-hydroxynonenal by alcohol dehydrogenase, aldehyde dehydrogenase, and glutathione-*S*-transferase, *Arch. Biochem. Biophys.* 316, 197–205, 1995; Engle, M.R., Singh, S.P., Czernik, P.J. et al., Physiological role of mGSTA4-4, a glutathione-*S*-transferase metabolizing 4-hydroxynonenal: generation and analysis of mGst4 null mouse, *Toxicol. Appl. Pharmacol.* 194, 296–308, 2004).

REFORMATSKY REACTION

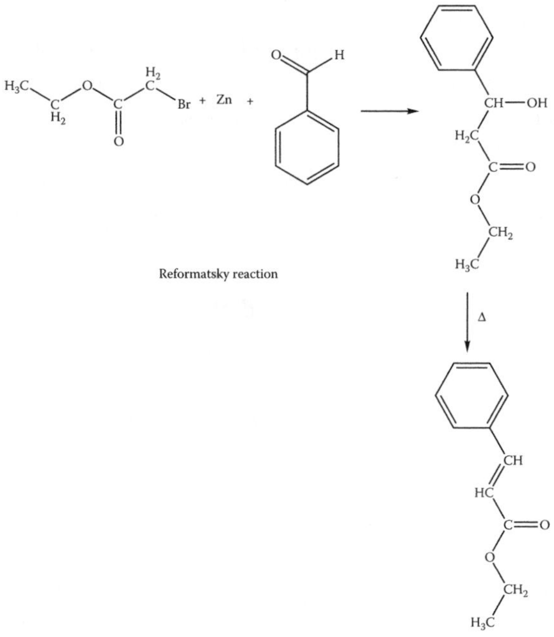

Reformatsky reaction

Formation of a complex between zinc and an α-bromoester, followed by condensation with an aldehyde yielding a β-hydroxyester; an α,β-unsaturated ester via dehydration follows the condensation reaction. See Tanabe, K., Studies on vitamin A and its related compounds. II. Reformatsky reaction of β-cyclocitral with methyl γ-bromosenecioate, *Pharm. Bull.* 3, 25–31, 1955; Ross, N.A. and Bartsch, R.A., High-intensity ultrasound-promoted Reformatsky reactions, *J. Org. Chem.* 68, 360–366, 2003; Jung, J.C., Lee, J.H., and Oh., S., Synthesis and antitumor activity of 4-hydroxy-coumarin derivatives, *Bioorg. Med. Chem. Lett.* 14, 5527–5531, 2004; Kloetzing, R.J., Thaler, T., and Knochel, P., An improved asymmetric Reformatsky reaction mediated by (−)-*N,N*-dimethy-laminoisoborneol, *Org. Lett.* 8, 1125–1128, 2006; Moume, R., Laavielle, S., and Karoyan, P., Efficient synthesis of β$_2$-amino acid by homologation of α-amino acids involving the Reformatsky reaction and Mannich-type imminium electrophile, *J. Org. Chem.* 71, 3332–3334, 2006.

RITTER REACTION

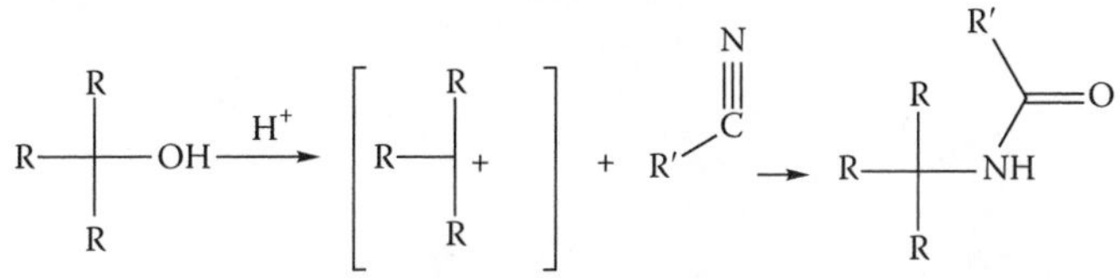

carbonium ion

Acid-catalyzed nucleophilic addition of a nitrile to a carbenium ion generated from alcohol (usually tertiary; primary alcohols other than benzyl alcohol will not react), yielding an amide. Sanguigni, J.A. and Levine, R., Amides from nitriles and alcohols by the Ritter reaction, *J. Med. Chem.* 53, 573–574, 1964; Radzicka, A. and Konieczny, M., Studies on the Ritter reaction. I. Synthesis of 3-/5-bartbituryl/-1propanesulfonic acids with anti-inflammatory activity, *Arch. Immunol. Ther. Exp.* 30, 421–432, 1982; Van Emelen, K., De Wit, T., Hoornaert, G.J., and Compernolle, F., Diastereoselective intramolecular

Ritter reaction: generation of a *cis*-fused hexahydro-4a*H*-indeno[1,2-*b*] pyridine ring system with 4a,9b-diangular substituents, *Org. Lett.* 2, 3083–3086, 2000; Concellon, J.M., Reigo, E., Suarez, J.R. et al., Synthesis of enantiopure imidazolines through a Ritter reaction of 2-(1-aminoalkyl)azirdines with nitriles, *Org. Lett.* 6, 4499–4501, 2004; Feske, B.D., Kaluzna, I.A., and Stewart, J.D., Enantiodivergent, biocatalytic routes to both taxol side chain antipodes, *J. Org. Chem.* 70, 9654–9657, 2005; Crich, D. and Patel, M., On the nitrile effect in L-rhamnopyranosylation, *Carbohydr. Res.* 341, 1467–1475, 2006; Fu, Q. and Li, L., Neutral loss of water from the b ions with histidine at the *C*-terminus and formation of the c ions involving lysine side chains, *J. Mass. Spectrom.* 41, 1600–1607, 2006.

SCHIFF BASE

The formation of an unstable derivative generally between a carbonyl (usually an aldehyde) and an amino group. The Schiff base can be converted to a stable derivative by reduction with sodium borohydride or sodium cyanoborohydride; Schiff bases appear to be resistant to reduction with sulfhydryl-base reducing agents such as 2-mercaptoethanol or dithiothreitol and phosphines. Schiff bases are involved in a diverse group of biochemical events including the interaction of pyridoxal phosphate with proteins, the interaction of reducing carbohydrates with proteins in reaction leading to AGE products, and reductive alkylation of amino groups in proteins. See Feeney, R.E., Blankenhorn, G., and Dixon, H.B., Carbonyl-amine reactions in protein chemistry, *Adv. Protein. Chem.* 29, 135–203, 1975; Metzler, D.E., Tautomerism in pyridoxal phosphate and in enzymatic catalysis, *Adv. Enzymol. Relat. Areas Mol. Biol.* 50, 1–40, 1979; Puchtler, H. and Meloan, S.N., On Schiff's bases and aldehyde-fuchsin: a review from H. Schiff to R.D. Lillie, *Histochemistry* 72, 321–332, 1981; O'Donnell, J.P., The reaction of amines with carbonyls: its significance in the nonezymatic metabolism of xenobiotics, *Drug. Metab. Rev.* 13, 123–159, 1982; Stadtman, E.R., Covalent modification reactions are marking steps in protein turnover, *Biochemistry* 29, 6232–6331, 1990; Tuma, D.J., Hoffman, T., and Sorrell, M.F., The chemistry of aldehyde-protein adducts, *Alcohol Alcohol Suppl.* 1, 271–276, 1991; Hargrave, P.A., Hamm, H.E., and Hofmann, K.P., Interaction of rhodopsin with the G-protein, transducin, *Bioessays* 15, 43–50, 1993; Chen, H. and Rhodes, J., Schiff base–forming drugs: mechanisms of immune potentiation and therapeutic potential, *J. Mol. Med.* 74, 497–504, 1996; Yim, M.B., Yim, H.S., Lee, C. et al., Protein glycation: creation of catalytic sites for free radication generation, *Ann. N.Y. Acad. Sci.* 928, 48–53, 2001; Gramatikova, S., Mouratou, B., Stetefeld, J. et al., Pyridoxal-5'-phosphate-dependent catatlytic antibodies, *J. Immunol. Methods* 269, 99–110, 2002; Schaur, R.J., Basic aspects of the biochemical reactivity of 4-hydroxynonenal, *Mol. Aspects Med.* 24, 149–159, 2003; Kurtz, A.J. and Lloyd, R.S., 1, N^2-deoxyguanosine adducts of acrolein, crotonaldehyde, and *trans*-4-hydroxynonenal crosslink to peptides via Schiff base linkage, *J. Biol. Chem.* 278, 5970–5975, 2003; Kandori, H., Hydration switch model for the proton transfer in the Schiff base region of bacteriorhodopsin, *Biochim. Biophys. Acta* 1658, 72–79, 2004; Hadjoudis, E. and Mavridis, I.M., Photochromism and thermochromism of Schiff bases in the solid state: structural aspects, *Chem. Soc. Rev.* 33, 579–588, 2004; Stadler, R.H., Acrylamide formation in different foods and potential strategies for reduction, *Adv. Expt. Med. Biol.* 561, 157–169, 2005. There is interesting material on Schiff bases in inorganic chemistry (Nakoji, M., Kanayama, T., Okino, T., and Takemoto, Y., Chiral phosphine-free Pd-mediated asymmetric allylation of prochiral enolate with a chiral phase-transfer catalyst, *Org. Lett.* 2, 3329–3331, 2001; Walther, D., Fugger, C., Schreer, H. et al., Reversible fixation of carbon dioxide at nickel[0] centers: a route for large organometallic rings, dimers, and tetramers, *Chemistry* 7, 5214–5221, 2001; Benny, P.D., Green, J.L., Engelbrecht, H.P., Reactivity and rhenium[V] oxo Schiff base complexes with phosphine ligands: rearrangement and reduction reactions, *Inorg. Chem.* 44, 2381–2390, 2005).

SCHMIDT REACTION/SCHMIDT REARRANGMENT

Used to describe the reaction of carboxylic acids, aldehyde and ketones (carbonyl compounds), and alcohols/alkenes with hydrazoic acid. Reaction with carboxylic acids yields amines, carbonyl compounds yield amides in a reaction involving a rearrangement, and alcohols/azides yield alkyl azides. See Rabinowitz, J.L., Chase, G.D., and Kaliner, L.F., Isotope effects in the decarboxylation of 1-^{14}C-dicarboxylic acids studied by means of the Schmidt reaction, *Anal. Biochem.* 19, 578–583, 1967; Iyengar, R., Schildknegt, K., and Aube, J., Regiocontrol in an intramolecular Schmidt reaction: total synthesis of (+)-aspidospermidine, *Org. Lett.* 2, 1625–1627, 2000; Sahasrabudhe, K., Gracias, V., Furness, K. et al., Asymmetric Schmidt reaction of hydroxyalkyl azides with ketones, *J. Am. Chem. Soc.* 125, 7914–7922, 2003; Wang, W., Mei, Y., Li, H., and Wang, J., A novel pyrrolidine imide-catalyzed direct formation of α,β-unsaturated ketones from unmodified ketones and aldehydes, *Org. Lett.* 7, 601–604, 2005; Brase, S., Gil., C., Knepper, K., and Zimmerman, V., Organic azides: an exploding diversity of a unique class of compounds, *Angew. Chem. Int. Ed. Engl.* 44, 5188–5240, 2005; Lang, S. and Murphy, J.A., Azide rearrangements in electron-deficient systems, *Chem. Soc. Rev.* 35, 146–156, 2006; Zarghi, A., Zebardast, T., Hakimion, F. et al., Synthesis and biological evaluation of 1,3-diphenylprop-2-en-1-ones possessing a methanesulfonamido or an azido pharmacophore as cyclooxygenase-1/-2 inhibitors, *Bioorg. Med. Chem.* 14, 7044–7050, 2006.

UGI CONDENSATION

A four-component (aldehyde, amine, isocyanide, and a carboxyl group) condensation resulting in an α-aminoacyl amide. See Liu, X.C., Clark, D.S., and Dordick, J.S., Chemoenzymatic construction of a four-component Ugi combinatorial library, *Biotechnol. Bioeng.* 69, 457–460, 2000; Bayer, T.,

Riemer, C., and Kessler, H., A new strategy for the synthesis of cyclopeptides containing diami-noglutaric acid, *J. Pept. Sci.* 7, 250–261, 2001; Crescenzi, V., Francescangeli, A., Renier, D., and Bellini, D., New crosslinked and sulfated derivatives of partially deacylated hyaluronan: synthesis and preliminary characterization, *Biopolymers* 64, 86–94, 2002; Liu, L., Ping Li, C., Cochran, S., and Ferro, V., Application of the four-component Ugi condensation for the preparation of glyco-conjugate libraries, *Bioorg. Med. Chem. Lett.* 14, 2221–2226, 2004; Bu, H., Kjoniksen, A.L., Knudsen, K.D., and Nystrom, B., Rheological and structural properties of aqueous alginate during gelation via the Ugi multicomponent condensation reaction, *Biomacromolecules* 5, 1470–1479, 2004; Tempest, P.A., Recent advances in heterocycle generation using the efficient Ugi multiple-component condensation reaction, *Curr. Opin. Drug Discov. Devel.* 8, 776–788, 2005.

WITTIG OLEFINATION

Wittig reaction/Wittig olefination   Ylide

Synthesis of an alkene from the reaction of an aldehyde or ketone with an ylide (internal salt) generated from a phosphophonium salt. See Jorgensen, M., Iversen, E.H., and Madsen, R., A convenient route to higher sugars by two-carbon chain elongation using Wittig/dihydroxylation reactions, *J. Org. Chem.* 66, 4625–4629, 2001; Magrioti, V. and Constantinou-Kokotou, V., Syn-thesis of (S)-α-amino oleic acid, *Lipids* 37, 223–228, 2002; van Staden, L.F., Gravestock, D., and Ager, D.J., New developments in the Peterson olefination reaction, *Chem. Soc. Rev.* 31, 195–200, 2002; Han, H., Sinha, M.K., D'Sousa, L.J. et al., Total synthesis of 34-hydroxyasimicin and its photoactive derivative for affinity labeling of the mitochondrial complex I, *Chemistry* 10, 2149–2158, 2004; Rhee, J.U. and Krische, M.J., Alkynes as synthetic equivalents to stabilized Wittig reagents: intra- and intermolecular carbonyl olefinations catalyzed by Ag(1), BF_3, and HBF_4, *Org. Lett.* 7, 2493–2495, 2005; Ermolenko, L. and Sasaki, N.A., Diastereoselective synthesis of all eight *l*-hexoses from L-ascorbic acid, *J. Org. Chem.* 71, 693–703, 2006; Halim, R., Brimble, M.A., and Merten, J., Synthesis of the ABC tricyclic fragment of the pectenotoxins via stereocon-trolled cyclization of a γ-hydroxyepoxide appended to the AB spiroacetal unit, *Org. Biomol. Chem.* 4, 1387–1399, 2006; Phillips, D.J., Pillinger, K.S., Li, W. et al., Desymmetrization of diols by a tandem oxidation/Wittig olefination reaction, *Chem. Commun.* 21, 2280–2282, 2006; Modica, E., Compostella, F., Colombo, D. et al., Stereoselective synthesis and immunogenic activity of the C-analogue of sulfatide, *Org. Lett.* 8, 3255–3258, 2006.

Index

A

A, 1
A 23187, 1
AAA, 1
AAAA, 1
AAG box, 1
AAS, 1
AAT, 1
AAV, 1
ABA, 1
Abbreviated New Drug Application (ANDA), 27
Abbreviations, 1–26
ABC, 1
ABC transporter, 27
ABC-transporter proteins, 1
ABE, 1
Abl, 1
Ablation, 27–28
ABRC, 1
ABRE, 1
Abscisic acid, 28
Absolute oils. *See* Essential oils
Absorption, 28
Abzymes. *See* Catalytic antibodies
7-ACA, 1
AcCho, 1
AcChoR, 1
Accuracy, 28
Accurate Mass Tag (AMT), 28
ACES, 1, 249, 349
Acetamide, 327
Acetic acid, 249–250, 327
Acetic anhydride, 250, 327
Acetoacetic acid, 327
Acetoin, 327
Acetone, 250–251, 327
Acetonitrile, 251
Acetophenone, 327
2-(acetoxy)benzoic acid, 252–253
Acetyl chloride, 251
1-acetyl-1*H*-imidazole, 252
N-acetyl-L-cysteine, 251–252, 327
N-Acetylcysteinamide, 327
N-acetylcysteine, 251–252, 327
N-Acetylimidazole, 252
N-Acetylmethionine, 327
Acetylsalicylic acid, 252–253, 327
Ach, 1
AchR, 1
Acid ammonium carbonate, 254–255
ACME, 1
Acridine, 327

Acrolein, 327
Acrylamide, 253, 327
Acrylodan, 1
ACS, 1
ACSF, 1
ACTH, 1
Active ingredient, 28
Active sequence collection (ACS), 28
Activity-based proteomics, 29
Acute phase proteins, 29–30
ADA, 1
ADAM, 1
ADAM-TS, 1, 30
ADCC, 1
Adenine, 327
Adenosine, 327
ADH, 1
Adjuvant, 30
ADME, 1
ADME-Tox, 1
AdoMet, 1
Adrenomedullin, 30–31
Adsorption, 31
Advanced glycation endproducts (AGE), 31
AEBSF, 340
AEC, 1
Aeration, 31
Aerosol, 32
Affibody, 32
Affinity proteomics, 32–33
AFLP, 1
AFM, 1
Agar/Agarose, 33
AGE, 1
Aggregation, 33–34
AGO, 1
Agonist, 34
AGP, 1
AID, 1
Akabori amino acid reaction, 355–356
AKAP, 2
Akt, 2
Akt, 2
Alanine, 327
Albumin, 34–35
Aldol condensation, 356–357
Aldosterone, 327
Algorithm, 36
ALL, 2
Alloantibody, 36
Alloantigen, 36
Allosteric, 37
ALP, 2

L

Labile zinc, 143–144
Lactarcins, 144
Lactic acid, 329
Lactoferrin, 144
Lactose, 329
LAK, 13
LATE-PCR, 13
LB, 13
LBTI, 338
LC_{50}, 13
LC-MS, 13
Lck, 13
LCR, 13
LCST, 13
LD, 13
LD_{50}, 13
LDL, 13
LECE, 13
Lectin, 144
LED, 13
Lek, 13
Leucine, 329
Leuckart reaction, 376
Leupeptin, 339
LFA, 13
LGIC, 13
LH, 13
LIF, 13
LIM, 13
Lima bean trypsin inhibitor, 338
LINE, 13
Linkage group, 145
Linoleic acid, 329
Lipofection, 145
Lipophilic, 145
Liposomes, 145–146
LLE, 13
LLOD, 13
LLOQ, 13
lnRNP, 13
Localized surface plasmon resonance, 146
Locus control region, 146
LOD, 13
Log P values, 327–329
LOLA, 14
Longin domain, 146–147
LOQ, 14
Lossen rearrangement, 376–377
LP, 14
LPA, 14
LPH, 14
LPS, 14
LRP, 14
LSPR, 14
LTB_4, 14
LTH, 14
Ltk, 14
LTR, 14
LUCA, 14

Luminescence, 147
Lutheran glycoprotein, 147
Lysine, 329
Lysosomes, 147–148

M

M, 14
M13, 14
M-CSF, 14
Mab, 14
MAB_q, 14
MAC, 14
Macrolide, 148
Macrophage, 148
Macropinocytosis, 148–149
MAD, 14
Madin-Darby Canine Kidney, 149
Maf, 14
MAGE, 14
Maillard reaction, 149–150, 377–378
Major groove, 150
Major histocompatibility complex (MHC), 150–151
Malaprade reaction, 378–379
MALDI-TOF, 14
Maleic anhydride, 296, 329
Malonic ester synthesis, 379
Maltose, 329
Mannich reaction, 380
Mannitol, 329
MAP, 14
MAPK, 14
MAPKK, 14
MAPKKK, 14
MAPPER, 151
MAR, 14
Mass spectrometer, 151
Mast cell, 151
MB, 14
Mb, 14
mb, 14
MBL, 14
MBP, 14
MCA, 14
MCAT, 14
MCD, 14
MCM, 14
MCS, 14
MDA, 14
MDCK, 14
MDMA, 14
Meerwein reaction, 380–381
MEF, 14
MEF-2, 14
MEGA-8, 14
MEGA-10, 14
MEK, 14
MELC, 14
MELK, 14
MEM, 14
Mer, 14

Q

Q-TOF, 19
QA, 19
QC, 19
QSAR, 19
QTL, 19
Quadruple mass spectrometry, 189–190
Quantum dots, 190
Quantum yield, 190
Quelling, 190–191
8-quinolinol, 291

R

RA, 19
RAB-GAP, 19
RACE, 19
Radical, 107
Raf, 19
RAGE, 19
Raman scattering, 191
Raman spectroscopy, 191
RAMP, 19
Randomization, 191
RANK, 19
RANK-L, 19
Rap, 19
Rap1, 19
RAPD, 19
RARE, 19
RAS, 19
RC, 19
RCA, 19
RCCX, 19
RCFP, 19
RCP, 19
RCR, 19
rDNA, 19
REA, 20
Real-Time PCR, 192
Receptor activator of NF-κB (RANK), 193
Receptor activity modifying proteins (RAMP), 193
Receptor tyrosine kinase, 194
Receptorome, 193
Receptors for AGE (RAGE), 193–194
Receptosome, 194–195
Reformatsky reaction, 383
Refractive index, 195
Regulators of G-protein signaling, 195–196
Regulatory transcription factors, 196
Rel, 20
REMI, 20
Resolution, chromatographic, 196–197
Resurrection plants, 197
RET, 20
Retention time, 197–198
Retention volume, 198
Retromer, 198
Retropseudogenes, 188
Retrotranslocation, 198

Reverse immunology, 198–199
Reverse micelle, 199
Reverse proteomic, 199
Reverse transcriptase, 199–200
Reverse transcriptase-polymerase chain reaction (RT-PCR), 200
RF, 20
R_f, 19
Rfactor, 20
RFID, 20
RFLP, 20
RGD, 20
RGS, 20
RHD, 20
Rheb, 20
Rho factor, 200–201
RhoA, 20
Rhomboid, 201–202
RI, 20
Ribonuclease III, 205
Ribose, 329
Riboswitch, 202
Ring-finger domains, 202
Ring-finger proteins, 202
RIP, 20
RIS, 20
RISC, 20
RIT, 20
Ritter reaction, 383–384
RM, 20
dsRNA, 20
rRNA, 20
shRNA, 20
siRNA, 20
snRNA, 20
snoRNA, 20
stRNA, 20
RNA-induced silencing complex (RISC), 202–203
RNA interference, 203
RNA isolation, 203–204
RNA polymerase, 204–205
RNA splicing, 205
RNAase, 20
RNAi, 20
hpRNAi, 20
RNAse, 20
RNAse III, 20, 205
RNC, 20
snRNP, 20
RNS, 20
RO, 20
ROCK, 20
ROESY, 20
ROK, 20
Rolling, 205
Ron, 20
ROS, 20
Ros, 20
RP, 20
RP-CEC, 20
RP-HPLC, 20